Recent Advances in
Life-Testing
and Reliability

A Volume in Honor of
Alonzo Clifford Cohen, Jr.

Edited by
N. Balakrishnan
McMaster University
Hamilton, Ontario, Canada

CRC Press
Boca Raton   London   Tokyo

Library of Congress Cataloging-in-Publication Data

Recent advances in life-testing and reliability : a volume in honor of
  Alonzo Clifford Cohen, Jr. / edited by N. Balakrishnan.
      p.   cm.
  Includes bibliographical references.
  ISBN 0-8493-8972-0 (acid-free paper)
  1. Accelerated life testing--Statistical methods.   2. Reliability
(Engineering)--Statistical methods. I. Balakrishnan, N., 1956–
II. Cohen, A. Clifford, 1911–
TA169.3.R43 1995
620′.004--dc20
DNLM/DLC
for Library of Congress   95-4006
                                                                                                                          CIP

ALONZO CLIFFORD COHEN, JR.

Preface

Alonzo Cifford Cohen, Jr., has made pioneering contributions to the areas of life-testing and reliability, and especially so to the problem of inference under censored and truncated samples. This is clearly evident from his lifetime publications list and the numerous citations his publications have received over the past four decades.

My association with Clifford began in 1987 when I invited him to contribute an article for the Special Issue of *Communication in Statistics - Theory and Methods* (Vol. 17, No. 7) that I edited on *Order Statistics and Applications*. We have kept in close touch with one another since then. In 1989, we began collaborating on a manuscript entitled *Order Statistics and Inference: Estimation Methods* with the primary aim of highlighting several methods of estimation based on order statistics. Upon successful completion, this was published by Academic Press in 1991.

We have continued to have close ties personally as well as academically. I have benefitted greatly from him on both these grounds. While I admire him for his sincerity and dedication in his work, I appreciate very much his humility and kindness.

The idea for this volume came in 1991 while I was attending Clifford's 80th birthday celebration held at the Department of Statistics of the University of Georgia that Clifford founded. When I sent the invitation letters to numerous leading researchers, I received overwhelming support and cooperation from all of them. I attribute this to the stature and respect that Clifford enjoys from his peers, fellow researchers, and the statistical community in general.

My sincere thanks go to all the authors who have contributed to this volume, and provided great support and encouragement through out the course of this project. Special thanks go to Mrs. Debbie Iscoe for the excellent typesetting of the entire volume, and to Mrs. Faneeta M. Richardson (Customer Service Supervisor, Olan Mills Studio, Chattanooga, Tennessee) for providing permission to reproduce the photograph of Clifford. Thanks are offered to John Wiley & Sons, Ltd., the American Statistical Association and the American Society for Quality Control for providing permission to reproduce some previously published tables and figures. Table 9.8 on p. 169 has been reprinted from Hamada, M. (1993), *Quality and Reliability Engineering International*, 9, 7-13, with permission from John Wiley & Sons, Ltd. Tables 20.3-20.4 on pp. 368-369 and Figures 20.1-20.4 on pp. 374-376 have been reproduced from Doganoksoy, N. and Schmee, J. (1993), *Technometrics*, 35, 175-184, with permission from *Technometrics* under the copyright (1993) by the American Statistical Association and the American Society for

Quality Control. Final thanks are due to Mr. Wayne Yuhasz (Editor, CRC Press) for the invitation to undertake this project, and to Ms. Nora Konopka for her assistance in the production of the volume.

It has been a nice experience corresponding with all the authors involved, and it is with great pleasure that I dedicate this volume to **Alonzo Clifford Cohen, Jr.**. I sincerely hope that this work will be of interest to researchers, applied statisticians and graduate students working in the areas of life-testing and reliability.

N. Balakrishnan
Hamilton, Ontario, Canada

August 1994

Contents

List of Contributors

Jon E. Anderson, Department of Statistics, University of Georgia, 240 Statistics Building, Athens, GA 30602.
E-mail: *jon@marie.stat.uga.edu*

Barry C. Arnold, Department of Statistics, University of California, Riverside, CA 92521.
Ph: (909)787-5939, Fax: (909)787-3286, E-mail: *barnold@ucrstat2.ucr.edu*

N. Balakrishnan, Department of Mathematics and Statistics, McMaster University, Hamilton, Ontario, Canada L8S 4K1.
Ph: (905)525-9140 ext. 23420, Fax: (905)522-1676, E-mail: *bala@mcmail.cis.mcmaster.ca*

D. Bandyopadhyay, AT&T, Room No. 4A 210Q, 900 Route 202/206 North, Bedminster, NJ 07921.

Richard E. Barlow, IEOR Department, University of California, 4177 Etcheverry Hall, Berkeley, CA 94720.
Ph: (415)642-3833, Fax: (415)643-8982, E-mail: *barlow@bayes.Berkeley.edu*

Asit P. Basu, Department of Statistics, University of Missouri-Columbia, 222 Math Sciences Bldg., Columbia, MO 65211-0001.
Ph: (314)882-8283, Fax: (314)882-9676, E-mail: *Basu@stat.missouri.edu*

Tim Bedford, Department of Mathematics and Informatics, Delft University of Technology, P.O. Box 5031, 2600 GA Delft, Mekelweg 4, 2628 CD Delft, The Netherlands.
Ph: +31 15 78 16 35, Fax: +31 15 78 72 55, E-mail: *T.bedford@twi.tudelft.nl*

Claudio Benski, Merlin Gerin DRD/A2, 38050 Grenoble Cedex, France.
Ph: (33)7657-9383, Fax: (33)7657-9860, E-mail: *benski@merlin-gerin.fr*

Gouri K. Bhattacharyya, Department of Statistics, University of Wisconsin-Madison, 1210 West Dayton Street, Madison, WI 53706-1685.
Ph: (608)262-2598.

xvi

Emmanuel Cabau, Merlin Gerin DRD/A2, 38050 Grenoble Cedex, France.
Ph: (33)7657-6060, Fax: (33)7657-9860, E-mail: *cabau@merlin-gerin.fr*

Ping-Shing Chan, Department of Statistics, The Chinese University of Hong Kong,
Shatin, New Territories, Hong Kong.
E-mail: *chan@sparc2.sta.cuhk.hk*

Gemai Chen, Department of Mathematics and Statistics, University of Regina, Regina,
Canada S4S 0A2.
Ph: (306)585-4353, Fax: (306)585-5205

Roger Cooke, Department of Mathematics and Informatics, Delft University of Technology, P.O. Box 5031, 2600 GA Delft, Mekelweg 4, 2628 CD Delft, The Netherlands.
Ph: +31 15 78 16 35, Fax: +31 15 78 72 55, E-mail: *R.Cooke@twi.tudelft.nl*

Necip Doganoksoy, Statistician, Management Science and Statistics Program, Information Technology Laboratory, Research and Development Center, General Electric Company, P.O. Box 8, K14C43, Schenectady, NY 12301.
Ph: (518)387-5319, Fax: (518)387-5714, E-mail: *DOGANOKSOY@CRD.GE.COM*

Nader Ebrahimi, Department of Mathematical Sciences, Division of Statistics, Northern
Illinois University, DeKalb, IL 60115-2854.
Ph: (815)753-6884

Max Engelhardt, Scientific Specialist, Statistics & Reliability Engineer, EG&G Idaho,
Inc., P.O. Box 1625, Idaho Falls, ID 83415.
Ph: (208)526-2100, Fax: (208)526-5647, E-mail: *MEE@inel.gov*

James Evans, U.S. Forest Products Lab., Madison, WI.

Alec A. Feinberg, The Analytic Sciences Corporation, 55 Walkers Brook Drive, Reading, MA 01867-3297.
Ph: (617)942-2000, Fax: (617)942-7100

Christian G. Garrigoux, Department of Mathematics, ITESM, Sucursal de Correos J,
Monterrey, N.L. 64849, Mexico.
E-mail: *cgarrigo@mtecv2.mty.itesm.mx*

Diane I. Gibbons, NAO Research & Development Center, Mathematics Department,
30500 Mound Road, Box 9055, Warren, Michigan 48090-9055.
Ph: (313)986-1102, Fax: (313)986-0124, E-mail: *diane@diane.ma.gmr.com*

Gregory J. Gibson, The Analytic Sciences Corporation, 55 Walkers Brook Drive, Reading, MA 01867-3297.
Ph: (617)942-2000, Fax: (617)942-7100

David Green, U.S. Forest Products Lab., Madison, WI.

Michael Hamada, Department of Statistics and Actuarial Science and The Institute
for Improvement in Quality and Productivity, University of Waterloo, Waterloo, Ontario,

Canada K2L 3G1.
Ph: (519)888-4740, Fax: (519)746-5524, E-mail: *mshamada@math.uwaterloo.ca*

Jonathan R.M. Hosking, IBM, Mathematical Sciences Department, Thomas J. Watson Research Center, P.O. Box 218, Yorktown Heights, New York 10598.
Ph: (914)945-1031, Fax: (914)945-3434, E-mail: *hosking@watson.ibm.com*

Richard A. Johnson, Department of Statistics, University of Wisconsin-Madison, 1210 West Dayton Street, Madison, WI 53706-1685.
Ph: (608)262-2598, Fax: (608)262-0032

Markos V. Koutras, Department of Statistics, University of Athens, Mitsaki 35, 111 41 Athens, Greece.
Ph: 301-7284-213, Fax: 301-7243-502, E-mail: *mkoutras@atlas.uoa.ariadne-t.gr*

K.B. Kulasekera, Department of Mathematical Sciences, Clemson University, Martin Hall, Box 34107, Clemson, South Carolina 29634-1907.
Ph: (803)656-3434

Harry F. Martz, Staff Member, Group A-1 MS F600, Los Alamos National Laboratory, Los Alamos, NM 87545.
Ph: (505)667-2687, Fax: (505)665-5204

Gary C. McDonald, NAO Research & Development Center, Head, Operating Sciences Department, 30500 Mound Road, Box 9055, Warren, Michigan 48090-9055.
E-mail: *gmcdonald@cmsa.gmr.com*

William Q. Meeker, Center for Nondestructive Evaluation, Department of Statistics, Iowa State University, 326 Snedecor Hall, Ames, IA 50011.
Ph: (515)294-5336, Fax: (515)294-2456, E-mail: *wqmeeker@iastate.edu*

Peter R. Nelson, Department of Mathematical Sciences, Clemson University, Martin Hall, Box 34107, Clemson, South Carolina 29634-1907.
Ph: (803)656-2882, Fax: (803)656-5230, E-mail: *NPETER@math.CLEMSON.edu*

Wayne Nelson, Statistical Consultant, 739 Huntingdon Drive, Schenectady, NY 12309.
Ph: (518)346-5138, E-mail: *nelsonw@crd.ge.com*

Stavros G. Papastavridis, Department of Statistics, University of Athens, Mitsaki 35, 111 41 Athens, Greece.
Ph: 301-7284-282, Fax: 301-7243-502, E-mail: *spapast@atlas.uoa.ariadne-t.gr*

Robert L. Parker, Department of Mathematics and Statistics, The University of New Mexico, Humanities Building 419, Albuquerque, NM 87131-1141.

Edsel A. Peña, Department of Mathematics and Statistics, Bowling Green State University, Bowling Green, OH 43403.
Ph: (419)372-7461, E-mail: *pena@andy.bgsu.edu*

Kyriakos I. Petakos, University of Athens, Mitsaki 35, 111 41 Athens, Greece.
Ph: 301-7284-213, Fax: 301-7243-502

Richard R. Prairie, Mechanical Engineering Department, University of New Mexico, Albuquerque, NM 87131-1141.

Philip Prescott, Faculty of Mathematical Studies, University of Southampton, Highfield, Southampton S09 5NH, England, U.K.
Ph: (0703)59500, Fax: (0703)593939, E-mail: *postmaster@uk.ac.soton.maths*

Jeffrey A. Robinson, Research & Development Center, General Motors Corporation, 30500 Mound Road, Box 9055, Warren, Michigan 48090-9055.
Ph: (810)986-1101, Fax: (810)986-0124, E-mail: *jeff@math.ma.gmr.com*

Ananda Sen, Department of Mathematical Sciences, Oakland University, Rochester, MI 48309.
E-mail: *sen@vela.acs.oakland.edu*

Richard L. Smith, Department of Statistics, The University of North Carolina, Chapel Hill, NC 27599-3260.
E-mail: *rs@stat.unc.edu*

Kazuyuki Suzuki, Communication & Systems Engineering Department, The University of Electro-Communications, 1-5-1 Chofugaoka, Chofu-shi, Tokyo 182, Japan.
Ph: 81-44-945-6531, Fax: 81-44-945-4305, E-mail: *suzuki@cocktail.cas.uec.ac.jp*

Chris P. Tsokos, Department of Mathematics and Statistics, University of South Florida, Tampa, FL 33620.
Ph: (813)961-1992

Lonnie C. Vance, NAO Research & Development Center, Cadillac Car Division, 30500 Mound Road, Box 9055, Warren, Michigan 48090-9055.

Alan Widom, Department of Physics, Northeastern University, Boston, MA.

Nelson O. Wood, Member of the Technical Staff, The Analytic Sciences Corporation, 1342 South Douglas Boulevard, Midwest City, OK 73130.
Ph: (405)737-3300, Fax: (405)732-9600

Shelemyahu Zacks, Department of Mathematical Sciences, Binghamton University, P.O. Box 6000, Binghamton, NY 13902-6000.
Ph: (607)777-2147, E-mail: *BG0525@BINGVMB.bitnet*

William J. Zimmer, Department of Mathematics and Statistics, University of New Mexico, Albuquerque, NM 87131.
Ph: (505)277-4803

List of Tables

List of Figures

0

Alonzo Clifford Cohen, Jr.

For many years, A(lonzo) Clifford Cohen, Jr., an Emeritus Professor of Statistics at the University of Georgia in Athens, Georgia, has been recognized as a leader and a pioneer in the development of statistical methodology for the analysis of life span data, where samples are often truncated and/or censored. He is also recognized for his work in the field of Quality Control. He is the author or the coauthor of 74 research publications including three books that are listed here. His research interests originated as a graduate student at the University of Michigan prior to WWII. After a six-year interruption as an Ordnance Officer during WWII in the grades of Captain, Major, and Lt. Colonel, he came to the University of Georgia in June 1947 as an Associate Professor of Mathematics and Statistics.

Military service included three years as a Director of Statistical Quality Control at Picatinny Arsenal, a manufacturing arsenal for the manufacture of artillery ammunition. In this capacity, he came under the influence and guidance of Dr. Walter A. Shewhart of the Bell Telephone Laboratories and of Leslie Simon (Captain, Major, Lieutenant Colonel, Colonel, General) of the Ballistic Research Laboratories, Aberdeen Proving Ground, Maryland. Other military service included eighteen months as a Statistical Analyst with the Maintenance Division of Headquarters Army Supply Forces at the Pentagon in Washington, and eighteen months in Europe. European duty consisted of assignments as Instructor of Mathematics and Statistics at Biarritz American University in Biarritz, France; Assistant to the Ordnance Officer of Western Base Section in Paris, France; Assistant to the Ordnance Officer of Headquarters US Forces, European Theater in Frankfort, Germany; and as a Battalion Commander for the operation of ammunition depots in Bamberg, Germany. Following release from Active Duty, he remained in the Army Reserve forces, and was subsequently promoted to the rank of Colonel.

Soon after arriving at the University of Georgia, with encouragement from Professor Tomlinson Fort, Head, Department of Mathematics, he embarked on a research program that has received financial support from the Army Research Office, NASA, the NSF, and the University of Georgia.

He has served as a consultant to the Operations Analysis Office, Headquarters US Air Force, the Ballistic Research Laboratories at Aberdeen Proving Ground, and to private industry. During a six-month leave-of-absence from University of Georgia in 1951, he served as an Operations Analyst with the Fifth Air Force in Japan and Korea during the Korean War.

He is a member of numerous professional organizations. These include ISI, IMS, ORSA, Sigma Xi, Phi Kappa Phi, Eta Kappa Nu, and Tau Beta Pi. He is a fellow and founding member of ASQC, a fellow of ASA, and AAAS. He was a member of the SREB Committee on Statistics for twenty-three years.

He was instrumental in founding the University of Georgia Institute of Statistics in

1958, and the University of Georgia Department of Statistics in 1964.

He was born near the Mississippi Gulf Coast in Ten Mile (Stone County), Mississippi, to parents Bessie Davis and Alonzo Clifford Cohen, Sr. on September 4, 1911. At the time of his birth, Ten Mile was an unincorporated sawmill (company owned) town in Harrison County, where his father was manager of the company commissary. In later years, the northern half of Harrison County, which included Ten Mile, was split off and became Stone County.

Clifford was the oldest of four children. He had a brother, Raymond, and two sisters, Virginia and Bessie Mae. At age five, the family moved to Norfield after a brief sojourn in Perkinston. He entered the public school system in Norfield. In 1920, the family moved to Brookhaven, the county seat of Lincoln County, Mississippi, with a population of approximately 6,000, where his father acquired an interest in a veneer mill. That is where Clifford and his siblings grew up and finished High School.

At an early age, he demonstrated an aptitude for mathematics, and for things mechanical and electrical. Mathematics and Physics were his best subjects in high school. His outside interests included Boy Scouts, radio, automobiles and music. He assembled the third radio receiver in Brookhaven and became skilled in assembling, repairing and installing radios. He earned spending money during high school days from these activities. His first car was a Model T Ford stripdown. He learned to play the clarinet and became a member of a small municipal band.

He graduated from Brookhaven High School in 1928 and entered the Alabama Polytechnic Institute (now Auburn University) in Auburn, Alabama, that fall as a student in Electrical Engineering. Outside activities included playing clarinet in the Auburn Band, working as a reporter on the college newspaper, the Plainsman, and membership in the Auburn Amateur Radio Club. He graduated with a BS degree in Electrical Engineering in May 1932, during the darkest days of the Great Depression. In addition to his academic schedule as a student, he was also enrolled in the Reserve Officer Training Corps program, and upon graduation he received a certificate, which on his 21st birthday was exchanged for a reserve commission as a Second Lieutenant in the Army Corps of Engineers. While a student at Auburn, he obtained both an amateur and a commercial radio operator's license.

During the summer following graduation, even though jobs were extremely scarce as a consequence of the depression, he found employment as an Assistant Engineer with Radio Station WSFA in Montgomery, Alabama. At the end of the summer, he returned to Auburn to accept a Teaching Assistantship in mathematics which enabled him to study toward a Masters degree, a degree that was awarded in 1933.

In September 1933, he entered the Westinghouse Student Engineer Training Program in Pittsburgh. In March of 1934, the program was drastically curtailed when the economy failed to improve as expected, and he was placed on furlough. Soon thereafter, he found employment with the Mine Safety Appliance Company in Pittsburgh.

On June 16, 1934, he was married in Catlettsburg, Kentucky, to Dorothy Armour of Columbus, Georgia. The couple lived in Wilkinsburg, Pennsylvania, until August at which time they drove back to Columbus in response to orders for two weeks of active duty reserve officer training at Fort Benning, Georgia.

While at Benning, an offer to teach mathematics at Auburn was accepted, and the couple moved to Auburn without returning to Pittsburgh. Unfortunately, the depression impacted on the State of Alabama to the extent that Auburn was unable to meet 1934-35 payrolls as contracted. Checks for the months of September-November were for only half the contracted amount. The only full check for the year came in December in time for Christmas.

A daughter, Judith, was born on March 22, 1934, and soon thereafter, orders were received for a six-month tour of active Army Reserve duty with the Civilian Conservation Corps (CCC). After reporting to Ft. McClellan, Alabama, Second Lietenant Cohen was assigned to a CCC Camp in Ecru, Mississippi, where he served as a Junior Officer. After promotion from Second to First Lieutenant, subsequent tours of CCC duty, while on leave from Auburn, were spent in camps at Hollandale, Mississippi, Haughton, Louisiana, and at Thibodaux, Louisiana where he was camp commander. He was assigned during the summer of 1938 as a Mess Inspector for a group of camps located in South Carolina, North Carolina, and in North Georgia.

In June of 1939, he enrolled as a graduate student at the University of Michigan in Ann Arbor, Michigan in pursuit of a Ph.D. degree in mathematics (statistics). He became a student of Cecil Craig, Harry Carver, Paul Dwyer, and Arthur Copeland. He had previously attended the University of Michigan summer session of 1937. He, his wife and daughter remained in Ann Arbor until September 1940, at which time he accepted an appointment as Assistant Professor of Mathematics at Michigan State University in East Lansing, Michigan. This appointment enabled him to complete a dissertation under the direction of Professor Cecil Craig at Ann Arbor, and he received his Ph.D. on June 21, 1941. While in East Lansing, he received a promotion in the Army Reserve to the rank of Captain.

He was ordered to active duty in the Ordnance Corps of the Army effective June 22, 1941, the day following receipt of his Ph.D. Degree and the day that Hitler's armies invaded Russia. After reporting to Maxwell Field in Montgomery, Alabama, he was sent to Camp Polk, Louisiana, for training with the Third Armored Division. At the request of Major Leslie Simon (later, to become Lieutenant General), he was transferred in August 1941 to Picatinny Arsenal, New Jersey.

A second daughter, Susan, was born at Picatinny in 1942. He remained at Picatinny for three years, and in the Spring of 1944, after a promotion to the rank of Major, he was transferred to the Maintenance Division of Headquarters Army Supply Forces in the Pentagon as a statistical analyst. After a promotion to the rank of the Lieutenant Colonel, and after the surrender of both Germany and Japan, he was sent to Biarritz, France in August 1945 to teach statistics at Biarritz American University. This was one of the universities established by the Army to facilitate the orderly transfer of troops to the Pacific, when fighting ceased in Europe.

He was released from active WWII duty in March 1947, and he returned to Michigan State University. After teaching at Michigan State University for the spring quarter of 1947, he came to the University of Georgia in Athens in June 1947 as an Associate Professor of Mathematics and Statistics in the Mathematics Department.

He was promoted to Full Professor in 1952. A third daughter, Deborah, (Debbie) was born in 1952. He retired from active teaching in 1978, but he and his wife, Dorothy continue to live in Athens, Georgia, where he is still active in research.

Lifetime Publications of *A. CLIFFORD COHEN, JR.*

1. The Numerical Computation of the Product of Conjugate Imaginary Gamma Functions, *Annals of Mathematical Statistics*, **11**, 1940, 213-218.

2. Quality Control Through Sampling Inspections, A Manual for Ordnance Inspectors, 1942, Picatinny Arsenal.

3. On a Quality Report to Management, *Industrial Quality Control*, **V**, 1948, 9-10.

4. A Note on Truncated Distributions, *Industrial Quality Control*, **VI**, 1949, 22.

5. On Estimating the Mean and Standard Deviation of Truncated Normal Distributions, *Journal of the American Statistical Association*, **44**, 1949, 518-525.

6. Estimating Parameters of Pearson Type III Populations from Truncated Samples, *Journal of the American Statististical Association*, **45**, 1950, 411-423.

7. Estimating the Mean and Variance of Normal Populations from Singly Truncated and Doubly Truncated Samples, *Annals of Mathematical Statistics*, **21**, 1950, 557-560.

8. Estimating Parameters of Logarithmic-Normal Distributions by the Method of Maximum Likelihood, *Journal of the American Statistical Association*, **46**, 1951, 206-212.

9. Estimation of Parameters in Truncated Pearson Frequency Distributions, *Annals of Mathematical Statistics*, **22**, 1951, 256-265.

10. The Frequency of Selecting Samples for Attribute Testing, *Industrial Quality Control*, **VIII**, 1951, 2-44.

11. On Estimating the Mean and Variance of Singly Truncated Normal Frequency Distributions from the First Three Sample Moments, *Annals of the Institute of Statistical Mathematics*, **III**, 1951, 32-44.

12. Estimating Parameters in Truncated Pearson Frequency Distributions Without Resort to Higher Moments, *Biometrika*, **40**, 1953, 50-57.

13. Tables of Pearson-Lee-Fisher Functions of Singly Truncated Normal Distributions (with John Woodward), *Biometrics*, **9**, 1953, 489-497.

14. Estimation of the Poisson Parameter from Truncated Samples and from Censored Samples, *Journal of the American Statistical Association*, **49**, 1954, 158-168.

15. On Some Functions Involving Mill's Ratio (with D.F. Barrow), *Annals of Mathematical Statistics*, **25**, 1954, 405-408.

16. An Advanced Problem Proposed for A Compound Normal Distribution, Problem 4616, *American Mathematical Monthly*, **61**(10), 1954, 718. Solution by P.J. Burke, *American Mathematical Monthly*, **63**(2), 1956, 129.

17. Truncated and Censored Samples from Normal Distributions, *Convention Transactions of the American Society for Quality Control*, 1955, 27-36.

18. Restriction and Selection in Bivariate Normal Distributions, *Journal of the American Statistical Association*, **50**, 1955, 884-893.

19. Maximum Likelihood Estimation of the Dispersion Parameter of a Chi-Distributed Radial Error from Truncated and Censored Samples with Applications to Target Analysis, *Journal of the American Statistical Association*, **50**, 1955, 1122-1135.

20. Censored Samples from Truncated Normal Distributions, *Biometrika*, **42**, 1955, 516-519.

21. On the Solution of Estimation Equations for Truncated and Censored Samples from Normal Populations, *Biometrika*, **44**, 1957, 226-236.

22. Restriction and Selection in Multinormal Distributions, *Annals of Mathematical Statistics*, **28**, 1957, 731-741.

23. Simplified Estimators for the Normal Distribution when Samples are Singly Censored or Truncated, *Technometrics*, **1**, 1959, pp. 217-237.

24. Simplified Estimators for the Normal Distribution when Samples are Singly Cen-

sored or Truncated, *Bulletin of the International Statististical Institute*, **XXXVII**(3), 1960, 251-269.

25. Misclassified Data from a Binomial Population, *Technometrics*, **2**, 1960, 109-113.

26. Estimating the Parameters of a Modified Poisson Distribution, *Journal of the American Statistical Association*, **55**, 1960, 139-143.

27. Estimation in the Truncated Poisson Distribution when Zeros and Some Ones are Missing, *Journal of the American Statistical Association*, **55**, 1960, 342-348.

28. Estimating the Parameter in a Conditional Poisson Distribution, *Biometrics*, **16**, 1960, 203-211.

29. Estimation in the Poisson Distribution when Samples Values of $c+1$ are Sometimes Erroneously Reported as c, *Annals of the Institute of Statistical Mathematics*, **XI**, 1960, 189-193.

30. An Extension of a Truncated Poisson Distribution, *Biometrics*, **16**, 1960, 446-450.

31. Estimating the Poisson Parameter from Samples that are Truncated on the Right, *Technometrics*, **3**, 1961, 433-438.

32. Tables for Maximum Likelihood Estimates; Singly Truncated and Singly Censored Samples, *Technometrics*, **3**, 1961, 535-541.

33. Estimation in Mixtures of Discrete Distributions, *Proceedings of McGill University Symposium on Discrete Distributions*, Aug. 1963, 373-378.

34. Progressively Censored Samples in Life Testing, *Technometrics*, 5, 1963, 327-339.

35. Maximum Likelihood Estimation in the Weibull Distribution Based on Complete and on Censored Samples, *Technometrics*, **7**, 1965, 579-588.

36. Reliability in Complex Systems, *Transactions of the American Society for Quality Control Technical Conference*, 1966, 667-675.

37. Discussion of "Estimation of Parameters for a Mixture of Normal Distribution" by Victor Hasselblad, *Technometrics*, **8**, 1966, 445-446.

38. Life Testing and Early Failure, Query 18, *Technometrics*, **8**, 1966, pp. 539-545.

39. A Note on Certain Discrete Mixed Distributions, *Biometrics*, **22**, 1966, 566-571.

40. Estimation in Mixtures of Two Normal Distributions, *Technometrics*, **9**, 1967, 15-28.

41. Curtailed Attribute Sampling, *Technometrics*, **12**, 1970, 295-298.

42. Estimating the Bernoulli Parameter and the Average Sample Size from Truncated Samples, *Random Counts in Physical Science 1*, Penn State University Press, 1970, 127-134.

43. Quality Control at Picatinny Arsenal, 1934-1945, (with L.E. Simon), *Transactions of the American Society for Quality Control 25th Annual Technical Conference*, Chicago, 1971, 273-279.

44. Picatinny Arsenal 1934-1945; Quality Control Movement Parallels War Effort (with L.E. Simon), *Quality Progress*, 4(5,6), 1971, 70-74.

45. Estimation in a Poisson Process Based on Combinations of Complete and Truncated Samples, *Contributed Papers, 38th Session ISI*, Washington, 1971, 85-89.

46. Estimation in a Poisson Process Based on Combinations of Complete and Truncated Samples, *Technometrics*, **14**, 1972, 841-846.

47. Estimation in the Exponential Distribution (with F. Russel Helm), *Technometrics*, **15**, 1973, 415-418.

48. The Reflected Weibull Distribution, *Technometrics*, **15**, 1973, 867-873.

49. Multi-Censored Samples in the Three-parameter Weibull Distribution, *Proceedings of the 39th Session ISI*, Vienna, **1**, 1973, 277-282.

50. Estimation in the Three-Parameter Gamma Distribution, *Contributed Papers, 40th Session ISI*, Warsaw, 1975, 170-173.

51. Multi-Censored Sampling in the Three-Parameter Weibull Distribution, *Technometrics*, **17**, 1975, 347-351.

52. Progressively Censored Samples in the Three Parameter Log-Normal Distribution, *Technometrics*, **18**, 1976, 99-103.

53. Progressively Censored Sampling in the Three-Parameter Gamma Distribution (with Nicholas J. Norgaard), *Technometrics*, **19**, 1977, 333-340.

54. Estimating Mean Life Using Progressively Censored Samples, *Proceedings of the AAMI 12th Annual Meeting*, 1977, 200.

55. Estimation in the Three-Parameter Lognormal Distribution (with Betty J. Whitten), *Journal of the American Statistical Association*, **75**, 1980, 399-404.

56. Percentiles and other Characteristics of the Four-Parameter Generalized Gamma Distribution (with Betty J. Whitten), *Communications in Statistics - Simulation and Computation* **10**, 1981, 175-218.

57. Estimation of Lognormal Distributions (with Betty J. Whitten), *American Journal of Mathematical and Management Sciences*, **1**, 1981, 139-153.

58. Modified Moment and Maximum Likelihood Estimators for Parameters of the Three-Parameter Gamma Distribution (with Betty J. Whitten), *Communications in Statistics - Simulation and Computation* **11**(2), 1982, 197-214.

59. Modified Maximum Likelihood and Modified Moment Estimators for the Three-parameter Weibull Distribution (with Betty J. Whitten), *Communications in Statistics - Theory and Methods* **11**(23), 1982, 2631-2656.

60. The Standardized Inverse Gaussian Distribution; Tables of the Cumulative Probability Function (with Micah Y. Chan and Betty J. Whitten), *Communications in Statistics - Simulation and Computation*, **12**(4), 1983, 423-442.

61. Modified Maximum likelihood and Modified Moment Estimators for the Three-Parameter Inverse Gausian Distribution (with Micah Y. Chan and Betty J. Whitten), *Communications in Statistics - Simulation and Computation*, **13**(1), 1984, 47-68.

62. Estimation in the Singly Truncated Weibull Distribution with a Unknown Truncation Point (with Dusit Charenkavanich), *Communications in Statistics - Theory and Methods*, **13**(7), 1984, 843-857.

63. Modified Moment Estimation for the Three-Parameter Weibull Distribution (with Betty J. Whitten and Y. Ding), *Journal of Quality Technology*, **16**, 1984, 159-167.

64. Modified Moment Estimation for the Three-Parameter Lognormal Distribution (with Betty J. Whitten and Y. Ding), *Journal of Quality Technology*, **17**, 1985, 92-99.

65. Estimation in the Three-Parameter Inverse Gaussian Distribution (with Betty J. Whitten), *Contributed Papers 45th Session ISI*, Amsterdam, **1**, 1985, 257-258.

66. Modified Moment Estimation for the Three-Parameter Inverse Gaussian Distribution (with Betty J. Whitten), *Journal of Quality Technology*, **17**, 1985, 147-154.

67. Modified Moment Estimation for the Three-Parameter Gamma Distribution (with

Betty J. Whitten), *Journal of Quality Technology*, **18**, 1986, 53-62.

68. A Pseudo Complete Sample Technique for Estimation from Censored Samples (with Betty J. Whitten and Vani Sundaraiyer), *Communications in Statistics - Theory and Methods*, **17**(7), 1988, 2239-2258.

69. Three Parameter Estimation, Chapter 4 of *Lognormal Distributions – Theory and Applications*, Edited by E.L. Crow and Kunio Shimizu, Marcel Dekker, Inc., New York, 1988, 113-137.

70. Censored, Truncated and Grouped Estimation, Chapter 5 of *Lognormal Distributions – Theory and Applications*, Edited by E.L. Crow and Kunio Shimizu, Marcel Dekker, Inc., New York, 1988, 139-172.

71. *Parameter Estimation in Reliability and Life Span Models* (with Betty J. Whitten), Marcel Dekker, Inc., New York, 1988.

72. *Order Statistics and Inference: Estimation Methods* (with N. Balakrishnan), Academic Press, San Diego, 1991.

73. *Truncated and Censored Samples – Theory and Applications*, Marcel Dekker, Inc., New York, 1991.

74. MLEs Under Censoring and Truncation and Inference, Chapter 4 of *The Exponential Distribution: Theory, Methods and Applications*, Edited by N. Balakrishnan and A.P. Basu, Gordon & Breach, Langhorne, Pennsylvania, 1994 (In Press).

Part I

Reliability

1

Analysis of Reliability Data Using Subsurvival Functions and Censoring Models

Roger Cooke and Tim Bedford

Delft University of Technology

ABSTRACT The problem of failure rate estimation is considered in the light of data acquired from the ATV reporting system. A number of different models for preventive maintenance are discussed which make the failure rate identifiable, as are more general bounding methods which do not require identifiability. These techniques are illustrated on ATV data which is shown in graphs of subsurvival functions. These graphs show, amongst other things, the effect of different maintenance policies on identical systems.

CONTENTS

1.1 Introduction

Component reliability data bases give failure rates corresponding to competing failure modes. It is not unusual to distinguish up to 10 failure modes, typically grouped into "critical", "degraded" and/or "incipient"[1] Degraded or incipient failure events are typically events discovered during tests and maintenance. Though not failed, a component is found in a condition which, in the judgment of the maintenance crew, might cause failure before the next regular inspection. Such components are taken out of service and repaired to as good as new. They may be replaced by a spare and re-enter service in a different socket if the system uses several identical components. Good maintenance strives to intercept critical failures while losing as little effective life as possible. Log books yield "component-socket histories". A stylized example of component socket history for socket # 1 is shown in Table 1.1. For the 311 pilot valves there are 8 sockets in each plant whose function is identical. The data from all 8 sockets are pooled. Time out of service for each socket is negligible.

If we let $x_1, \ldots, x_m$ denote the sojourn times ending in corrective maintenance, and $z_1, \ldots, z_n$ the sojourn times ending in preventive maintenance, then from the above data

---

[1]For reliability data base design see [8], [3], and [1] . The latter source describes the ATV reporting system from which data analyzed here is drawn.

TABLE 1.1
Typical component history record

component socket	date	activity	cause
pilot valve 311-1	1-1-75	enters service	
pilot valve 311-1	1-1-80	critical failure	description
pilot valve 311-1	1-6-87	prev. maint.	description
pilot valve 311-1	15-8-89	critical failure	description
pilot valve 311-1	15-11-93	prev. maint.	description
.	.	.	.
.	.	.	.
.	.	.	.

we would compute (in months) $x_1 = 60$, $x_2 = 26.5$, $z_1 = 90$, $z_2 = 51$.

From such data we would like to compute the "naked failure rate"[2] for critical failure, that is, the failure rate which would be observed if no preventive maintenance were undertaken. In addition we would like to have ways for assessing the effectiveness of maintenance policy. Current practice assumes that competing failure modes operate independently. When the competing modes are associated with degraded or incipient failures discovered during maintenance, this is equivalent to assuming that the maintenance crew performs repairs on components randomly, seldom a reasonable supposition.

We introduce concepts analyzing censored failure data when it cannot be assumed that competing failure modes operate independently. The ideas presented here were first worked out in research under contract with the European Space Agency [5] from which sections of this are taken, and in research supported by the Swedish Nuclear directorate [11].

Section 1.2 discusses models for censored reliability data with identifiable critical failure variable. Section 1.3 addresses the problem of assessing uncertainty bounds for failure rate estimates from censored data. Section 1.4 applies the concepts developed here to component socket data from the Swedish ATV component reporting system. A final section draws conclusions.

1.2 Identifiable Life Distributions Under Right Censoring

1.2.1 Definitions and notation

In this section we are concerned with two competing risks or failure modes. X is called the "critical failure" or "corrective maintenance" mode, and Z is called the "preventive maintenance" variable. For simplicity we assume that $P(X = Z) = 0$. We anticipate a possible dependence between Z and X. Independent copies of $Y = (\min\{X, Z\}, 1_{\{X < Z\}})$ are observed, that is, we observe the least of X and Z, and observe which it is. A third mode is "removal from service due to termination of observation period"; which is assumed independent of the others. This mode will also be called random right censoring. Random right censoring is addressed briefly in Section 1.2.4 and for the most part we concentrate on X and Z.

If F_X is the cumulative distribution function of X, then $S_X = 1 - F_X$ is called the *survival function* of X. If F_X has a density (unless stated otherwise, we assume

---

[2] We assume that the failure rate observed if no maintenance were performed is the marginal critical failure rate. As "marginal" means "unimportant" to the uninitiated, engineers show distinctly greater interest in the *naked* failure rates.

throughout that all distributions have densities) then the failure rate $r_X(t)$ of X is $r_X(t) = f_X(t)/S_X(t) = -(dS_X/dt)/S_X(t)$. Since $d\ln(S_X) = dS_X/S_X$, we have

$$S_X(t) = \exp\left\{-\int_0^t r_X(t)dt\right\} = \exp\left\{\int_0^t (dS_X/dt)/S_X dt\right\}. \tag{1.2.1}$$

The function $S_X^*(t) = P\{X > t \text{ and } X < Z\}$ is called the *subsurvival function* of X (see [13]). Note that S_X^* depends on Z, though this fact is suppressed in the notation. Note also that if S_X^* is continuous at 0 then $S_X^*(0) = P(X < Z)$. If X and Z are independent then one has

$$S_X^*(t) + S_Z^*(t) = P\{X > t \text{ and } Z > t\} \tag{1.2.2}$$
$$= P\{X > t\}P\{Z > t\} = S_X(t)S_Z(t)$$

DEFINITION 1.2.1 *Real functions S_1^* and S_2^* on $[0, \infty)$, form a continuous subsurvival pair if*

1. *S_1^* and S_2^* are non-negative non-increasing continuous with $S_1^*(0) < 1$, $S_2^*(0) < 1$,*
2. *$\lim_{t\to\infty} S_1^*(t) = 0, \lim_{t\to\infty} S_2^*(t) = 0$,*
3. *$S_1^*(0) + S_2^*(0) = 1$.*

For two random variables X and Z as above, the functions S_X^* and S_Z^* clearly form a subsurvival pair. Assuming, as above that a sequence of i.i.d. copies of Y are observed, the empirical subsurvival functions contain all the information in the data; that is any function of the data can be written as a function of the empirical subsurvival functions. The *conditional subsurvival function* is the subsurvival function, conditioned on the event that the failure mode in question is manifested. With continuity of S_X^* and S_Y^* at zero,

$$P(X > t \text{ and } X < Z \mid X < Z) = S_X^*(t)/S_X^*(0) ,$$

and a similar equation holds for $S_Z^*(t)/S_Z^*(0)$.

Closely related to the notion of subsurvival functions is the probability of censoring beyond time t,

$$\Phi(t) := P(Z < X \mid Z \wedge X > t) = \frac{S_Z^*(t)}{S_X^*(t) + S_Z^*(t)} .$$

This function seems to have some diagnostic value, enabling one to distinguish between certain kinds of censoring models. Note that $\Phi(0) = P(Z < X)$ for continuous subsurvival functions.

The *subdistribution functions* for X and Z are

$$F_X^*(t) = P(X \le t \text{ and } X < Z) = S_X^*(0) - S_X^*(t) \tag{1.2.3}$$
$$F_Z^*(t) = P(Z \le t \text{ and } Z < X) = S_Z^*(0) - S_Z^*(t) . \tag{1.2.4}$$

From data one can only estimate either the subsurvival functions (or, equivalently, the subdistribution functions). However, one would often like to have an estimate for the marginal distribution function F_X. Without additional assumptions on the joint distribution of X and Z, ie. assumptions about the interrelation of corrective and preventative maintenance, it is impossible to estimate this marginal ([14]). By making extra assumptions, it is possible to restrict oneself to a subclass of models in which this marginal *is* identifiable. In Section 1.3 ways are considered of finding bounds for F_X without making such extra assumptions.

1.2.2 Independent competing failure modes

The main theorem for independent competing risks has been in the folklore in various forms for some time. A more general version of the theorem stated below is found in [14].

THEOREM 1.2.2

(1) Let X and Z be independent life variables, with F_X and F_Z continuous. Let X' and Z' be independent life variables such that $S_X^* = S_{X'}^*$ and $S_Z^* = S_{Z'}^*$; then $F_X = F_{X'}$ and $F_Z = F_{Z'}$.

(2) If S_1^* and S_2^* are a continuous subsurvival pair then there exist independent life variables X and Z such that $S_X^* = S_1^*$ and $S_Z^* = S_2^*$.

By observing independent copies of $Y = (\min\{X, Z\}, 1_{\{X<Z\}})$ we can estimate the subsurvival functions. *Assuming* independence of X and Z we can then use the above theorem to identify the survival functions of X and Z. Of course, the assumption that X and Z are independent may be not be true, and in this case the survival functions gotten in this way would NOT be correct. Moreover, no statistical test based only on the censored observations could possibly test the independence assumption since, according to Theorem 1.2.2(2) *any* censored observations can be explained by an independent model. References for independent competing risks are [6], [7] and [9].

1.2.3 Random sign censoring

Perhaps the simplest dependent competing failure mode model which leads to identifiable life distributions without restricting their form is *random sign censoring* (called *age-dependent* censoring in [4], where the theorems reported here are proved). Consider a component subject to right censoring, where X denotes the time at which a component would expire if not censored. Suppose that the event that the component's life be censored is independent of the age X at which the component would expire, but *given* that the component is censored, the time at which it is censored may depend on X. The following intuitive picture suggests how such models might arise in practice. Suppose a component emits a warning before it fails. Suppose that if the warning is seen, the component will be pulled off line and repaired (corresponding to incipient or degraded failures) but if the warning is unheeded, the component goes on to fail. In random signs censoring, the event that the warning is heeded is independent of the component's age. This situation is captured in the following definition.

DEFINITION 1.2.3 *Let X and Z be life variables with $Z = X - W\delta$, where $0 < W < X$ is a random variable and δ is a random variable taking values $\{1, -1\}$, with X and δ independent. The variable $Y = (\min\{X, Z\}, 1_{\{X<Z\}})$ is called a random sign censoring of X by Z.*

Note that

$$S_X^*(t) = P(X > t \text{ and } \delta = -1) = P(X > t)P(\delta = -1) \qquad (1.2.5)$$
$$= S_X(t)P(Z > X) = S_X(t)S_X^*(0)$$

and

$$S_Z^*(t) = P(X - W > t \text{ and } \delta = 1).$$

Note also that $P(Z > X)$ and S_X^* can be estimated from observing independent copies of Y. Eq. (1.2.5) says that under random signs censoring S_X is equal to the conditional subsurvival function of X.

The following result tells us when given subsurvival functions are consistent with a random sign censoring model.

THEOREM 1.2.4

Let (S_1^*, S_2^*) be a continuous subsurvival pair with S_1^*, S_2^* strictly monotonic; then the following are equivalent:

1. *There exist random variables W, δ and X with X and δ independent such that*

$$S_1^*(t) = P(X > t \ and \ \delta = -1)$$
$$S_2^*(t) = P(X - W > t \ and \ \delta = 1).$$

2. *For all $t > 0$,*

$$\frac{S_1^*(t)}{S_1^*(0)} > \frac{S_2^*(t)}{S_2^*(0)}.$$

3. *For all $t > 0$, $\Phi(0) > \Phi(t)$.*

If extra information about the relationship between X and W is known, then more may be said about the form of $\Phi(t)$. For example,

1. Suppose that for sufficiently large X, $W = a$, where a is a positive constant. Then writing $\alpha = S_X^*(0)/S_Z^*(0)$, we have

$$\Phi(t) = 1/(1 + \alpha S_X(t)/S_X(t + a)),$$

for t sufficiently large. If additionally we assume that X is exponential with parameter λ, then

$$\Phi(t) = 1/(1 + \alpha \exp(\lambda a))$$

for t sufficiently large.

2. If $W = aX$ for a positive constant $a < 1$, then

$$\Phi(t) = 1/(1 + \alpha S_X(t)/S_X(t/(1 - a))).$$

For a large class of distributions, including the exponential, $S_X(t)/S_X(t/(1-a)) \to \infty$ as $t \to \infty$, so that in this case $\Phi(t) \to 0$ as $t \to \infty$.

1.2.4 Exponential models

In this section we assume that the life variable X follows an exponential distribution, and we are interested in models for (X, Z) under which the variable X is identifiable. A continuous subsurvival pair is always consistent with an independent model, but it is not always consistent with an independent exponential model. We can derive a fairly sharp criterion for exponentiality in terms of the subsurvival functions using the following simple result [4]:

THEOREM 1.2.5

Let X and Z be independent life variables, then any two of the following imply the others:

(i) $S_X(t) = \exp\{-\lambda t\}$

(ii) $S_X^*(t) = \lambda \exp\{-(\lambda + \gamma)t)\}/(\lambda + \gamma)$

(iii) $S_Z^*(t) = \gamma \exp\{-(\lambda + \gamma)t)\}/(\lambda + \gamma)$

(iv) $S_Z(t) = \exp\{-\gamma t\}$

Hence if X and Z are independent exponential life variables with failure rates λ and γ, then the two normalized subsurvival functions decrease exponentially and are identical, $S_X^*(t)/S_X^*(0) = S_Z^*(t)/S_Z^*(0) = e^{-(\lambda+\gamma)t}$, and the function Φ is constant, $\Phi(t) = \mu/(\lambda+\mu)$. Assuming continuity and strict monotonicity, Theorem 1.2.4 says that a random signs censoring model exists if and only if for all t, $\Phi(0) > \Phi(t)$. This suggests that if the independent exponential model if true, then the random signs model will be difficult to reject from data. Painfully, the consequences of these two models for the inferences regarding the variable X can be quite different.

Consider a set of independent copies of $Y = (\min\{X, Z\}, 1_{\{X<Z\}})$. Let $x_1, \ldots x_m$ be the uncensored observations, and let $z_1, \ldots z_n$ be the censored observations. Assume that X is exponential with parameter λ. Under random sign censoring the population of observed failures are statistically equivalent to the original population, hence a consistent estimator for λ is

$$\frac{m}{\sum_{i=1}^m x_i}. \tag{1.2.6}$$

Recall now the maximum likelihood estimator of the failure rate in the case of independent censoring,

$$\frac{m}{\sum_{i=1}^m x_i + \sum_{j=1}^n z_j}. \tag{1.2.7}$$

Using the independent model when random sign censoring applies can lead to a drastically wrong estimate of the failure rate, for then the estimate of λ converges with probability 1 to

$$\frac{1}{\frac{1}{\lambda} + \frac{P(Z<X)}{P(X<Z)}E(Z|Z<X)}.$$

Since typically $P(X < Z) << P(Z < X)$, we see that this estimate will typically substantially differ from the true value. In particular, if $E(Z|Z < X) \approx E(X)$ then this estimate is roughly $\lambda P(X < Z)$.

Random Clipping (RC)

Perhaps the simplest model of interaction between an exponential life process X and another process Z is gotten by assuming that X is *always* censored by a random variable $Z = X - W$. More specifically, we assume that X is exponential and for some positive random variable W independent of X, we observe $X - W$. W may be thought of as a warning which a component emits prior to expiring at time X. Of course W may be greater than X, which we interpret as censoring at birth.

Let us suppose that censors at birth are simply not recorded. Suppose in other words, that components emitting warnings at birth are simply repaired until the warning disappears, and that the false start is not recorded as an incipient failure at time 0. Indeed, this is what usually happens. We call the variable $X - W$ *given* $X - W > 0$ a random clipping (RC) of X.

The next theorem says that, regardless of the distribution of W, the distribution of $X - W$ *given* that $X - W > 0$, is identical to that of X; in particular, $\lambda \sim n/\sum z_i$ (no failures are observed and $N = n$).

THEOREM 1.2.6
Let X be exponential with parameter λ, let $W > 0$ be a random variable independent of X, and $U = X - W$. Then, conditional on $U > 0$, U has the same distribution as X.

PROOF For $u \in (-\infty, \infty)$ and $v = \max\{0, -u\}$ we have

$$dF_U(u) = \int_v^\infty \lambda e^{-\lambda(u+w)} dF_W(w).$$

If $u \geq 0$, then $v = 0$ and

$$f_U(u) = \lambda e^{-\lambda u} \mathcal{L}_W(\lambda),$$

where $\mathcal{L}_W$ denotes the Laplace transform of W. Normalizing this over $u \in [0, \infty)$ yields the desired result. ∎

In the coming sections we introduce a number of different censoring models in which the underlying lifetime distribution is exponential, and in which the exponential parameter λ is identifiable. In each case a consistent estimator of λ is determined. It is not possible to give an MLE in these cases as a more explicit specification of the distribution of the warning variable is required for determination of the likelihood function.

Bounded random warning - random inspection (BRWRI)

We consider an exponential variable X with failure rate λ subject to random inspections. We assume that the component emits a warning at time $X - W$, before expiring at time X. If an inspection occurs in the interval $(X - W, X)$, then the component is pulled off for preventive maintenance, and we say that X is censored by $Z = $ "time of first inspection in $(X - W, X)$". If no inspection occurs in $(X - W, X)$ then the component is observed to fail at time X, and we stipulate that $Z = \infty$. If $W > X$ then the component is censored at birth. We are interested in simple models for inspection and warning processes which have physical plausibility and which enable estimation of the failure rate of X from the observations.

We consider a random inspection policy whereby the expected number of inspections in any time interval of length I is proportional to I, and such that the number of inspections in disjoint time intervals are independent. In other words, inspections are described by a Poisson distribution with intensity γ:

$$P(k \text{ inspections in } (t, t + I)) = e^{\gamma I}(\gamma I)^k/k!.$$

The probability of no inspections in $(t, t + I)$ is $e^{-\gamma I}$, and the probability of at least one inspection in this interval is $1 - e^{-\gamma I}$. We assume that the life variable X is independent of the inspection process.

We consider the warning variable W independent of X. The possibility of censoring at birth complicates the computations considerably, and this is of dubious value since censors at birth are seldom recorded as such. In cases where W is typically much smaller than X, the phenomena near the component's birth are not important. We therefore seek a model which avoids these complications near birth, yet which captures the overall behavior of independent warning times W with $W << X$. We assume therefore that W is bounded by some number d, and that no warning is emitted by the component if it expires before d. More specifically, we assume $W < d$, W and X independent given $X \geq d$, and $W = 0$ if $X < d$. The variable $Y = (\min\{X, Z\}, 1_{\{X<Z\}})$ is called $BRWRI(d, \gamma)$. For $X \geq d$, the event of censoring is independent of X, as in random signs censoring; for $X < d$, however the probability of censoring is zero.

The analysis of S_Z^* is rather complicated and explicitly involves the distribution of W. However, we can extract useful information from S_X^*.

$$P(X < Z) = S_X^*(0) = P(X < Z \text{ and } X \leq d) + P(X < Z \text{ and } X > d)$$
$$= 1 - e^{-\lambda d} + e^{-\lambda d}\mathcal{L}_W(\gamma),$$

where, as above, $\mathcal{L}_W$ denotes the Laplace transform of W. The subsurvival function for X is given by:

$$S_X^*(t) = \begin{cases} \int \int_t^\infty \lambda e^{-\lambda u} e^{-\gamma w} \, du \, dF_W(w) = e^{-\lambda t}\mathcal{L}_W(\gamma) & \text{for } t > d; \\ e^{-\lambda t} - e^{-\lambda d} + e^{-\lambda d}\mathcal{L}_W(\gamma) & \text{for } t \leq d. \end{cases}$$

The conditional expectation $E(X \mid X < Z)$ is just the integral of the conditional subsurvival function:

$$E(X \mid X < Z) = (1/S_X^*(0)) \int S_X^*(t) \, dt$$

$$= (1/S_X^*(0)) \left(\frac{1}{\lambda}(1 - e^{-\lambda d} + e^{-\lambda d}\mathcal{L}_W(\gamma)) - de^{-\lambda d}(1 - \mathcal{L}_W(\gamma)) \right)$$

$$= (1/\lambda) - dP(Z < X)/P(X < Z). \tag{1.2.8}$$

Letting $\sim$ denote asymptotic equivalence as the total number N of observations goes to infinity, using well known convergence properties, and omitting summation indices when these are clear from the context we collect these remarks in the following theorem.

THEOREM 1.2.7
Let $z_1, \ldots z_n$ be censoring times and $x_1, \ldots x_m$ be failure times of $N = n + m$ independent realizations of the $BRWRI(d, \gamma)$ variable $Y = (\min\{X, Z\}, 1_{\{X<Z\}})$ where X is exponentially distributed with parameter λ. Then

$$\left(\frac{\sum x_i}{m} + d\frac{n}{m} \right)^{-1} \sim \lambda,$$

and

$$d = \sup\{c : P(Z < X | X < c) = 0\}.$$

PROOF   We have

$$P(X < Z) \sim m/N, \text{ and } E(X \mid X < Z) \sim \sum x_i/m.$$

Using $e^{-\lambda d} = P(X > d)$, the result follows by substituting in (1.2.8).   ∎

The statistic estimating the failure rate is less than that in the case of random signs censoring, but the difference goes to zero as d goes to zero.

Random warning - constant inspection (RWCI)

We suppose that X is exponentially distributed with failure rate λ, and is inspected at regular intervals I. If a warning is seen at the i-th inspection, then X is censored at time $Z = Ii$. We suppose that there is some positive random variable W independent of X, such that X's warning is emitted at time $X - W$. In other words, X is censored by Z, with $Z = iI$ if $I(i-1) < X - W \le Ii < X$; and X is observed at time X if no inspection occurs in the interval $[X - W, X]$. In this case we stipulate that $Z = \infty$. If W exceeds I with positive probability, then it becomes impossible to estimate the failure rate of X from censored data. We therefore consider only the case that $P(W \le I) = 1$. Censors at birth are now possible, but censoring more than one inspection interval of the component's life is excluded. The subsurvival function S_Z^* is not continuous and has an atom at zero.

It is convenient to measure time in units of I, so that $I = 1$. The variable $Y = (\min\{X, Z\}, 1_{\{X<Z\}})$ is $RWCI(W)$. Recall for $0 < p < 1$,

$$\sum_{j>i} p^j = p^{i+1}/(1-p); \quad \sum_{i\ge 0}\sum_{j>i} p^j = p/(1-p)^2.$$

THEOREM 1.2.8
Let $x_1, \ldots x_m$ be the observed failures and $z_1, \ldots z_n$ the observed censors from a population of $N = m+n$ independent realizations of $RWCI(W)$ variable $Y = (\min\{X, Z\}, 1_{\{X<Z\}})$; then

(i) $P(Z \le X) = (1 - \mathcal{L}_W(\lambda))/(1 - e^{-\lambda}) \sim n/N$
(ii) $\lambda \sim \ln(1 + n/\sum z_i)$
(iii) $\Phi(i) = \frac{(1-\mathcal{L}_W(\lambda))}{\mathcal{L}_W(\lambda)(e^\lambda - 1)} = e^{-\lambda}P(Z \le X)/\mathcal{L}_W(\lambda)$
(iv) $S_Z^*(i)/P(Z \le X) = e^{-\lambda(i+1)}; S_X^*(i)/P(X < Z) = e^{-\lambda i}$.

PROOF   (i): Counting censors at birth, we have:

$$n/N \sim P(Z \le X) = \sum_{i\ge 0} e^{-\lambda i} - \int e^{-\lambda(i+w)}dF_W = (1 - \mathcal{L}_W(\lambda))/(1 - e^{-\lambda}).$$

(ii)

$$S_Z^*(i) = P(Z > i \text{ and } Z < X)$$

$$= \sum_{j>i} e^{-\lambda j} - \int e^{-\lambda(j+w)}dF_W = e^{-\lambda(i+1)}(1 - \mathcal{L}_W(\lambda))/(1 - e^{-\lambda}).$$

and

$$\sum z_i/N \sim \sum_{i\ge 0} S_Z^*(i) = (1 - \mathcal{L}_W(\lambda))e^{-\lambda}/(1 - e^{-\lambda})^2$$

$$= P(Z \le X)e^\lambda/(1 - e^{-\lambda}).$$

Hence, $\sum z_i/n \sim e^{-\lambda}/(1 - e^{-\lambda})$ from which (ii) follows.
Using

$$S_X^*(i) = P(X > i \text{ and } X < Z) = \sum_{j\ge i}\int e^{-\lambda(j+w)}dF_W - e^{-\lambda(j+1)}$$

$$= (\mathcal{L}_W(\lambda) - e^{-\lambda})e^{-\lambda i}/(1 - e^{-\lambda});$$

$$\text{and} \quad S_Z^*(i) + S_X^*(i) = \mathcal{L}_W(\lambda)e^{-\lambda i},$$

(iii) and (iv) are straightforward calculations. ∎

The failure rate of X can be estimated independently of the distribution of W, in fact we can estimate $\mathcal{L}_W(\lambda)$ also, as (iii) shows. Note that this model is not consistent with random signs censoring as the probability of censoring is not independent of a component's age, for suppose that $W < d < 1$ with probability one, then the probability of X being censored when $j + d < X < (j + 1)$ is zero.

Constant warning - constant inspection (CWCI)

A constant is statistically independent of every random variable. As a special case of Theorem 1.2.8, we may therefore consider the case that W is a constant, $W = d < 1$. In such a model, a component issues a warning at time $X - d$ before expiring at time X. Such a model is termed $CWCI(d)$.

The Laplace transform of a constant d is simply $e^{-\lambda d}$. We have

$$P(Z \leq X) = (1 - e^{-\lambda d})/(1 - e^{-\lambda}) ,$$

$$\lambda \sim \ln(1 + n/\sum z_i) ,$$

$$\Phi(i) = (1 - e^{-\lambda d})/(e^{-\lambda d}(e^{\lambda} - 1)).$$

Proportional warning - constant inspection (PWCI)

This model is similar to the previous model except that the warning is emitted at time X/δ if the component expires at X; with δ a constant, $1 < \delta$. A process of censored observations is called $PWCI(\delta)$. Exact solutions for this model are difficult. However if $i \leq 1/(\delta - 1)$ then $\delta i \leq i + 1$, and if δ is close to one, then most contributions will come from inspection periods i such that $\delta i \leq i + 1$. Let "$\approx$" denote "equal up to effects for $i > 1/(\delta - 1)$". As before, we set $I = 1$. Finally let $\sim$ denote asymptotic equivalence as the total number of observations goes to infinity. Note that there is no censoring at birth.

THEOREM 1.2.9
Let $x_1, \ldots x_m$ be the observed failures and $z_1, \ldots z_n$ the observed censors from a population of $N = m + n$ independent realizations of PWCI variable $Y = (\min\{X, Z\}, 1_{\{X < Z\}})$, and let X be exponentially distributed with parameter λ, then:

(i) $\quad \lambda \approx -\ln\left(1 - \dfrac{2nN}{\sum z_j + nN + n^2}\right)$

(ii) $\quad \delta \approx -(1/\lambda)\ln\left(1 - \dfrac{2nN}{\sum z_i + nN - n^2}\right)$

(iii) $\quad \Phi(i) = e^{i\lambda(\delta - 1)}/(e^{\lambda} - 1) - 1/(e^{\lambda\delta} - 1).$

PROOF (i) The component is censored in the i-th observation period if and only if $i \leq X < i\delta$, thus

$$n/N \sim P(Z \leq X) = \left(\sum_{1 \leq 1/(\delta-1)} e^{-\lambda i} - e^{-\lambda \delta i} \right) + e^{-\lambda/(\delta-1)}$$

$$\approx \sum_{i \geq 0} e^{-\lambda i} - e^{-\lambda \delta i}$$

$$= \frac{1}{1 - e^{-\lambda}} - \frac{1}{1 - e^{-\lambda \delta}} = Q_1 - Q_2.$$

Further,

$$\sum_{i>0} z_j/N \sim \sum_{i>0} S_Z^*(i) \approx \sum_{i \geq 0} \sum_{j>i} e^{-\lambda j} - e^{-\lambda \delta j} = e^{-\lambda} Q_1^2 - e^{-\lambda} Q_2^2.$$

Note that the inner summation runs over $j > i$, as $S_Z^*(i) = P(Z > i$ and $Z < X)$. Adding the "$j = i$" term to the summation results in:

$$n/N + \sum z_j/N \sim P(Z < X) + \sum_{i \geq 0} S_Z^*(i) \approx \sum_{i \geq 0} \sum_{j \geq i} e^{-\lambda j} - e^{-\lambda \delta j}$$

$$= Q_1^2 - Q_2^2 = (Q_1 - Q_2)(Q_1 + Q_2).$$

Hence dividing by the first equation and solving for Q :

$$Q_1 + Q_2 \sim 1 + \sum z_i/n;$$

$$2Q_1 = n/N + 1 + \sum z_i/n;$$

$$\lambda \approx -\ln\left(1 - \frac{2nN}{\sum z_j + nN + n^2}\right).$$

(ii) The expression for δ results from solving for Q :

$$2Q_2 \sim 1 + \sum z_i - n/N;$$

$$\delta \approx -(1/\lambda)\ln\left(1 - \frac{2nN}{\sum z_j + nN - n^2}\right).$$

(iii) Using

$$S_Z^*(i) = \sum_{j>i} e^{-\lambda i} - e^{-\lambda \delta i} \text{ and } S_X^*(i) = \sum_{j \geq i} e^{-\lambda \delta i} - e^{-\lambda(i+1)},$$

one computes:

$$S_Z^*(i) + S_X^*(i) = e^{-\lambda \delta i};$$

$$\Phi(i) = S_Z^*(i)/(S_Z^*(i) + S_X^*(i))$$

$$= e^{i\lambda(\delta-1)}/(e^\lambda - 1) - 1/(e^{\lambda \delta} - 1).$$

∎

Another estimate of λ can be obtained using the subsurvival function for X, if we make a certain simplification. If X is observed to fail, this can only occur between δi and $(i+1)$

for some inspection time i. Suppose the failure time is recorded as $i+1$; suppose in other words, that failures could only be discovered at the next inspection. We should then have:

$$\sum x_i/N \sim \sum_{i\geq 0}\sum_{j\geq 0} e^{-\lambda\delta i} - e^{-\lambda(i+1)} = Q_2^2 - e^{-\lambda}Q_1^2;$$

$$\sum x_i/N + \sum z_i/N \approx (1 - e^{-\lambda\delta})Q_2^2 = Q_2;$$

$$Q_1 = n/N + (\sum x_i + \sum z_i)/N;$$

$$\lambda \approx -\ln\left(1 - \frac{N}{n + \sum x_i + \sum z_i}\right).$$

This latter estimate involves upwards rounding off the failure times, but on the other hand it uses all of the observations, and is therefore less sensitive for sampling fluctuations.

The behavior of the PWCI model can be intuitively understood as follows; for larger values of X, the time during which the warning is on, $X(\delta-1)$ gets longer and it becomes more probable that the component will be censored. The observed failures times tend to be drawn from the lower end of the distribution; whereas the observed maintenances tend to censor the older components. This suggests intuitively that $S_Z^*(t)/S_Z^*(0) > S_X^*(t)/S_X^*(0)$. For PWCI variables the function $\Phi(i)$ is increasing. Indeed, since $\delta > 1$, statement (iii) shows that this is the case.

We must note, however, that no failures would be observed for times X such that $X - X/\delta > 1$, since the warning period is then so large that inspection cannot miss it. In other words, we should see no failures beyond $x = \delta/(\delta - 1)$. Substituting $j = 0$ in Theorem 1.2.9(iii), we find to the first order:

$$\Phi(0) \simeq (\delta - 1)/(\lambda\delta) \simeq P(Z \leq X).$$

Note also that the variable X/δ follows a Weibull distribution if X is exponentially distributed. The errors which enter with "$\approx$" in the estimation of λ are on the order of $e^{-\lambda/(\delta-1)}/(\delta - 1)$.

Proportional warning - random inspection (PWRI)

We consider an exponential variable with parameter λ which issues a warning at time X/δ if the component expires at time X. Inspection follows a Poisson distribution with intensity γ. If an inspection occurs while the warning is issued, the component is preventively maintained at time $Z = $ "first inspection beyond X/δ"; if no inspection occurs in the interval $[X/\delta, X]$ then $Z = \infty$ and the component fails at X. The variable $Y = (\min\{X, Z\}, 1_{\{X<Z\}})$ is called PWRI(γ). The analysis of S_Z^* is rather complicated, but simple results can be obtained only from S_X^*.

THEOREM 1.2.10
If X is exponential with parameter λ and $x_1, \ldots x_m; z_1, \ldots z_n, N = n+m$ are independent observations of $PWRI(\gamma)$ variable $Y = (\min\{X, Z\}, 1_{\{X<Z\}})$, then

$$\lambda \sim m^2/(N\sum x_i).$$

PROOF   Put $\alpha = (\delta - 1)/\delta$. Then

$$P(X < Z) = P\{\text{no inspection in } (X/\delta, X)\}$$

$$= \int \lambda e^{-(\lambda+\gamma\alpha)t} dt = \lambda/(\lambda + \alpha) \sim m/N.$$

Further,

$$\sum x_i/N \sim \int S_X^*(t) dt = \int \int \int_t^\infty \lambda e^{-(\lambda+\gamma\alpha)u} du\ dt = \lambda/(\lambda+\alpha)^2;$$

$$\lambda \sim m^2/(N \sum x_i).$$

∎

Expressions can be obtained for $S_Z^*(t)$ and $\Phi(t)$, but no simple pattern emerges. It is interesting to compare the two proportional warning models PWCI and PWRI for some representative values. We measure time in units of the inspection interval I, and assume $\gamma = 1$ so that the expected number of inspections per unit time is one. Putting $\alpha = (\delta - 1)/\delta$, we have to the first order:

$$P_{PWCI}(X < Z) = (\lambda - \alpha)/\lambda;$$

$$P_{PWRI}(X < Z) = \lambda/(\lambda + \alpha).$$

Filling in some representative numbers, $\delta = 1.02, \lambda = 0.03$, we find

$$P_{PWCI}(X < Z) = 0.57 P_{PWRI}(X < Z),$$

so that the probability of failure under regular inspection is about half the probability under Poisson inspection. On the other hand if the probability of failure is observed and one estimates λ, then assuming a PWRI model tends to give a lower failure rate than assuming a PWCI model. Of course the latter constrains the length of the observed failure times, whereas PWRI does not.

Random right censoring

Methods for incorporating random right censoring have not yet been studied in detail. To analyze the data in Section 3 a rough procedure was introduced to take account of the effects of random right censoring. Consider an exponential variable subject to independent random right censoring. Referring to (1.2.7), we can write the maximal likelihood estimate of the failure rate under independent censoring λ_c as a function of the estimate λ_f which we would have made if we had interpreted the observed failures as if they were uncensored realizations of X, plus a correction term depending on the total time of censored observations $\sum_{i=1}^n z_i$ and the number of failures $m = N - \#$ censored observations:

$$\frac{1}{\lambda_c} = \frac{1}{\lambda_f} + \frac{\text{(total time of censored observations)}}{\#\ \text{observations - }\#\ \text{censored observations}}.$$

In dealing with critical failure, X, preventive maintenance, Z and random right censoring, say C, we consider observations of $X \wedge Z$ as uncensored observations leading to a failure rate estimate λ_f as given by any of the preceding models. The total time of censored observations is $\sum_{i=1}^q c_i$, where q is the number of censored observations, that is the number of service sojourns ended by terminating the observation period.

1.3   Uncertainty Bounds

It is essential to distinguish two types of uncertainty, namely uncertainty due to non-identifiability and and uncertainty due to sampling fluctuations.

1.3.1   Uncertainty due to non-identifiability: bounds in the absence of sampling fluctuations

In [12] Peterson derived lower and upper bounds on the survival function S_X by observable quantities by noting that

$$P(X \leq t \text{ and } X \leq Z) \leq P(X \leq t) \leq P(X \leq t \text{ or } Z \leq t),$$

that is:

$$1 - F_X^*(t) \geq S_X(t) \geq S_X^*(t) + S_Z^*(t) = F_{\min}(t) = P(X \wedge Z \leq t).$$

He showed that these bounds are sharp[3] in the following sense. For all t and all $\epsilon > 0$ there exist joint distributions on pairs of life variables (X_1, Z_1) and (X_2, Z_2) such that

$$S_{X_1}^* = S_{X_2}^* = S_X^*;$$

$$1 - F^*(t) - \epsilon \leq S_1(t) \text{ and } 1 - F_{\min}(t) + \epsilon \geq S_2(t).$$

If F is the cumulative distribution function for X, these bounds can also be written as

$$F^*(t) \leq F(t) \leq F_{\min}(t).$$

If X follows an exponential distribution with failure rate λ, we therefore have

$$\lambda = 1/E(X) \leq 1/E(X \wedge Z).$$

Hence, under the assumption of exponentiality, observed data may be used to give an upper bound on λ.

The above cited result of Peterson does *not* say that any distribution function between F^* and $F_{\min}$ is a possible distribution of X. A more general version of the following result is proved in [2]:

THEOREM 1.3.1
If F is a cumulative distribution function satisfying

$$F^*(t) < F(t) < F_{\min}(t),$$

then there is a joint distribution for (X, Z) with F as marginal distribution for X if and only if for all t_1 and t_2, with $t_1 < t_2$,

$$F(t_1) - F^*(t_1) \leq F(t_2) - F^*(t_2).$$

In other words, the distance between $F(t) - F^*(t)$ must be an increasing function of t. Bedford and Meilijson call the representation of F as the sum of two monotone functions $F^* + (F - F^*)$ a *comonotone representation*, and the functional bounds for F which one obtains from the above theorem the *comonotone bounds*.

---

[3] It is noted in [2] that Peterson's statement is true only under certain continuity conditions.

As an illustration of the potentially significant difference between the pointwise Peterson and the comonotone bounds consider the following examples, taken from [2].

Censored below the mean, uncensored above

Let X be exponentially distributed with failure rate $\lambda = 1$. If X is less than one then it is censored with probability one, and the censoring variable Z is equal to 0. If $X > 1$ then it is not censored, that is $X < Z$.

Now, hypothesising exponentiality of X, comonotonicity *identifies* the failure rate because the inequality between the densities ($\lambda e^{-\lambda x} \geq e^{-x}$ for all $x > 1$) only admits $\lambda = 1$. The upper Peterson bound is sharp for this case and claims that $\lambda \leq 1$, but the lower bound is not, since

$$1 - e^{-\lambda x} \geq e^{-1} - e^{-x}$$

only implies that $\lambda \geq 0.1355$. In other words, the Peterson bound claims $E(X)$ to be between 1 and 7.38 while comonotonicity identifies this mean as 1.

As an illustration of a case where comonotonicity contributes next to nothing beyond the pointwise Peterson bounds, consider the following example.

Independent censoring

Let X and Z be independent, exponentially distributed with parameters λ_1 and λ_2 respectively.

Hypothesising exponentiality of X, the Peterson bounds claim that its failure rate λ satisfies $\lambda \in [\lambda_1, \lambda_1 + \lambda_2]$. Comonotonicity improves the result by merely ruling out $\lambda = \lambda_1 + \lambda_2$.

A better appreciation of the data is afforded by considering the time average failure rates. Recall that the failure rate $r_X(t)$ for X is given by:

$$r_X(t) = -dS_X(t)/S_X(t) = d(\ln(S_X(t))/dt$$

so that the time average failure rate is

$$-\ln(S_X(t))/t = (1/t) \int_0^t r_X(u)du.$$

Of course, if $S_X(t) = \exp(-\lambda t)$, then the time average failure rate is just the (constant) failure rate λ. Applying this transformation the Peterson bounds become:

$$\mathrm{lmin}(t) = -\ln(1 - F^*(t))/t \leq (1/t) \int_0^t r_X(u)du \leq -\ln(1 - F_{\min}(t))/t = \mathrm{lmax}(t).$$

As $t \to \infty$, $F^*(t) \to P(X \leq Z) < 1$ so that $\mathrm{lmin}(t) \to 0$. Hence the lower bound on the admissible values of the time average failure rate decreases as time becomes large. The Peterson bounds are the endpoints of the intersection for all t of the admissible time average failure rates:

$$\text{Peterson bounds } = \cap_{t=1}^{\infty} [\mathrm{lmax}(t), \mathrm{lmin}(t)] = [\max_t (\mathrm{lmin}(t)), \min_t (\mathrm{lmax}(t))].$$

1.3.2   Accounting for sampling fluctuations

To apply the Peterson or comonotone bounds to data, we must take proper account of the fact that only empirical versions of the subsurvival functions are available. We discuss a number of approaches.

Estimate based on E_{min}

The quantity $E_{min} = E(X \wedge Z)$ can be estimated from the data and its sample estimate under reasonable assumptions is asymptotically normal. If X is exponential with failure rate λ we have $E(X) = 1/\lambda \geq E_{min}$ This leads to an upper bound on λ.

Sampling fluctuations of Peterson Bounds

We write the Peterson bounds as:

$$\text{lmin}(t) = -\ln(1 - F^*(t))/t \leq (1/t) \int_0^t r_X(u)du \leq -\ln(1 - F_{min}(t))/t = \text{lmax}(t).$$

For each t, the probabilities $F^*(t)$ and $F_{min}(t)$ can be estimated from the data. Classical confidence bounds on these estimates can be substituted into the above expression to yield classical confidence bounds for the time average failure rate of X, *for each time t*, denoted $\text{lbmin}(t)$ and $\text{ubmax}(t)$. If X follows an exponential distribution with failure rate λ, then for all t,

$$\text{lmin}(t) \leq \lambda \leq \text{lmax}(t);$$

For fixed t, $[\text{lbmin}(t), \text{ubmax}(t)]$ represent a 90% classical confidence bound for λ. In this sense, plotting $\text{lbmin}(t)$ and $\text{ubmax}(t)$ give a picture of the sample fluctuations of the Peterson bounds.

The problem with t-wise confidence bounds is that we do not know how to combine them for different values of t. There are two procedures for obtaining confidence bounds for λ independent of t; one based on compound hypothesis testing, the other on the Kolmogorov-Smirnov statistic.

Compound hypothesis tests

We consider only the lower Peterson bound. Let $p_i = P(i - 1 < X \leq i, X < Z)$ for $i = 1, \ldots m$. The lower Peterson bound may be written as a set of inequality constraints on $p_1, \ldots p_m$:

constraint set C :

$$p_1 \leq 1 - e^{-\lambda},$$

$$p_1 + p_2 \leq 1 - e^{-2\lambda},$$

$$p_1 + p_2 + p_3 \leq 1 - e^{-3\lambda},$$

$$\vdots$$

$$p_1 + p_2 + \ldots + p_m \leq 1 - e^{-m\lambda},$$

$$p_1 + \ldots p_{m+1} = 1,$$

$$p_i \geq 0 \ (1 = 1, \ldots m + 1).$$

Put $\underline{p} = (p_1, \ldots p_{m+1})$, where $p_{m+1} = 1 - p_1 - \ldots p_m$. Suppose we have N independent observations of the variable $Y = (X \wedge Z, 1_{X<Z})$. Let $\underline{s} = (s_1, \ldots s_{m+1})$ where $s_i =$

$\#\{i | i - 1 < X \le 1, X < Z\}/N$, $i = 1, \ldots, m$ and $s_{m+1} = 1 - s_1 - \ldots - s_m$. Fix a value λ and ask which probability vector $\underline{p}$ satisfying constraint set C has the greatest likelihood given the data $(\underline{s}, N)$? This can be solved by finding a probability vector $\underline{p}$ which solves a constrained optimization problem:

$$minimize \; Q(\underline{p}) = 2N \sum s_i \ln(s_i/p_i) \; subject \; to \; constraint \; set \; C$$

The quantity $Q(\underline{p})$ is proportional to the log likelihood ratio for the hypothesis that the data has been generated by independent samples from $\underline{p}$. The asymptotic sampling distribution of $Q(\underline{p})$ is Chi square with $N - 1$ degrees of freedom.

Kolmogoroff-Smirnov Bounds

Unlike the bounds in the previous section, the bounds considered below make no assumption regarding the form of the unknown distribution of X, but rather makes use of the Kolmogoroff-Smirnov statistic. This procedure is described in detail in [2]; a very brief description will suffice here.

Recall that the Kolmogoroff-Smirnov test of a hypothesized distribution function F is based on calculating the maximum absolute deviation between the transformed empirical distribution function $\hat{F}(F^{-1}(t))$ based on N observations and the uniform distribution function. For large N, this deviation exceeds $1.35/\sqrt{N}$ with probability approximately 5%.

Suppose we observe N independent realizations of $\{X \wedge Z, 1_{X<Z}\}$, and let F^* be the empirical subcumulative distribution of X defined above. Consider the hypothesis that the (unobserved) marginal cumulative distribution of X is G. Considering all possible joint distributions of X and Z, we can find the subcumulative G^* compatible with G which is closest to the observed distribution $\mathcal{F}^*$. In [2] it is proved that the maximum over t of the deviation $| G^*(t) - \mathcal{F}^*(t) |$ is smaller than the maximum over t of $| G(t) - \mathcal{F}(t) |$, where $\mathcal{F}$ is the (unobserved) empirical distribution of X.

This enables us to use the Kolmogoroff-Smirnov statistic to reject certain hypothesized marginals for X without making assumptions on the dependence structure between X and Z.

Choosing G from the class of exponential distributions, the test allows us to reject certain failure rates λ. In [2] it is shown that the remaining failure rates that have not been rejected form an interval $[\lambda_{kl}, \lambda_{ku}]$. The Kolmogoroff-Smirnov test tells us that with 95% confidence, the interval of possible failure rates that would be obtained with a given number of observations is contained in $[\lambda_{kl}, \lambda_{ku}]$.

The Kolmogorov-Smirnov procedure and the E_{min} upper bound have been implemented and applied to data from the ATV system, given results shown in Section 1.3.3. The compound hypothesis test has not yet been fully implemented, but initial results suggest that it yields reasonable results.

1.3.3 Comparison of bounds for ATV data

Using data from the ATV system, we can compare the E_{min} bounds and the Kolmogoroff-Smirnov bounds. Table 1.2 presents the results, for a description of the systems see Section 1.3.3. The variable of interest is corrective maintenance, which is roughly comparable to critical failure aggregated over all failure modes (e.g. fail to open, fail to close, spurious operation, etc.). The censoring Z is preventive maintenance and random right censoring is due to termination of observation time.

The 312 feedwater pump data aggregates many dissimilar systems, and it is anticipated that this data is not consistent with an exponential distribution for critical failure. With the Kolmogorov-Smirnov bounds the 312 data for station 2 is just barely consistent with exponentiality, whereas the station 1 data is not. For the other systems exponentiality seems to be quite consistent with the data, with the possible exception of the station 2 311 pilot valves, where the feasibility bounds are rather narrow.

Note finally that these bounds agree quite well with regard to the upper bound for λ. In light of the differences in derivation this agreement is significant. It demonstrates the possibility of extracting meaningful bounds from censored life data. $1/E_{\min}$ yields only upper bounds on λ. The upper and lower values reflect sampling fluctuations and make no assumptions on the form of the distribution of X. Lower bounds on λ are harder to come by and require assumptions on the distribution of X.

1.4   Application of Subsurvival Concepts to ATV Data

The concepts presented in the foregoing have been applied to component socket data from the ATV reporting system made available by the Swedish nuclear directorate. The data are taken from two nuclear power stations; station 1 has one reactor and station 2 has two reactors. The reactors are identical in design and fabrication, but have small differences with regard to maintenance policies. At each station, data from the following systems were analyzed:

> 311 inner isolation valves
> 311 outer isolation valves
> 311 pilot valves
> 314 relief valves to containment
> 314 relief valves to sump
> 314 pilot valves
> 312 feedwater pumps
> 323 core spray cooling
> 221 control rod drives

(see [5] for descriptions of these systems). The 312 feedwater pump system was included to create a non-homogeneous data set, the sockets correspond to pumps, valves, batteries, heat exchangers, etc. From discussions with maintenance engineers at the plants, it became clear that random inspection plays no significant role in these plants. The data were analyzed relative to calendar time by occularization to verify that no effect of calendar time could be observed. The following models were used to compute critical failure rates:

> Independent exponential (InEx)
> Random Signs (RnSg)
> Constant Warning-Constant Inspection (CWCI)
> Proportional Warning-Constant Inspection (PWCI)
> Proportional Warning-Random Inspection (PWRI)

Data for the 311, 312, and 314 systems at station 2 are given in Figures 1.1, 1.3, 1.4, 1.5, 1.6, 1.8, and 1.9, each of which contains three plots. These show (i) the subsurvival functions in months for corrective and preventive maintenance, and for random censors; (ii) conditional subsurvival functions in months for corrective and preventive maintenance with the conditional probability for preventive maintenance beyond time t, $(P/(P+C))$; and (iii) the time average failure rate bounds. These plots are also given for the 311 outer isolation valves (Figure 1.2) and for the 314 pilot valves (Figure 1.7) from station 1.

TABLE 1.2
Bounds for failure rates λ for critical failure $1/E_{min}$ are 90% confidence bounds for the upper bound for λ; $\lambda_{kl}, \lambda_{ku}$ are one-sided lower and upper Kolmogoroff-Smirnov 95% bounds. "*" means the bounds are not consistent with an exponential distribution, ie. the lower bound is higher than the upper bound.

	$1/E_{min}$ 90% confidence on upper bound		λ_{kl}, Kolmog. lower	λ_{ku} Smirnov upper
Station 1 311 inner isolation valves	.0376	.0639	.0064	.0636
Station 2 311 inner isolation valves	.0259	.0388	.0031	.0388
Station 1 311 outer isolation valves	.0544	.0825	.0200	.0799
Station 2 311 outer isolation valves	.0499	.0684	.011	.0618
Station 1 311 pilot valves	.015	.0174	.0045	.0167
Station 2 311 pilot valves	.0222	.0249	.0100	.0232
Station 1 314 relief valves to containment	.0285	.0419	.0120	.0392
Station 2 314 relief valves to containment	.0335	.0415	.0072	.0405
Station 1 314 relief valves to sump	.0113	.0193	.0040	.0156
Station 2 314 relief valves to sump	.0309	.0406	.0098	.0348
Station 1 314 pilot valves	.0161	.0196	.0097	.0173
Station 2 314 pilot valves	.0166	.0188	.0052	.0165
Station 1 312 feedwater pumps	.0173	.0201	-*	-*
Station 2 312 feedwater pumps	.018	.0201	-*	-*
Station 1 323 core spray cooling	.0104	.0128	.00088	.01127
Station 2 323 core spray cooling	.01	.012	.00020	.01000
Station 1 rod drives/control rods	.01	.0119	.00005	.0082
Station 2 rod drives/control rods	.0089	.0107	.00035	.0067

The plots (ii) also show estimates of the critical failure rate corresponding to the above five models. Also shown are the number of critical failures ($\#C$), the number of preventive maintenances ($\#P$), the number of random censors ($\#$ Cnsr), and the total number of observations ($\#$). An upper bound for the critical failure rate is the inverse of the expectation of the minimum of critical failure preventive maintenance or random censoring. This upper bound is given as "$1/E_{\min}$" with 90% classical confidence bounds derived from the central limit theorem.

The graphs "lmin" and "lmax" in (iii) are bounds for the time average failure rates obtained by transforming the Peterson bounds as discussed in Section 1.3. The graphs "lbmin" and "ubmax" are the 5% lower and 95% upper t-wise confidence bounds associated with $\mathrm{lmin}(t)$ and $\mathrm{lmax}(t)$ respectively.

The data afford many useful illustrations of the concepts introduced in the foregoing sections. A few are selected and discussed below.

Censoring models

The most significant observation from the data is this: for systems involving a moderate to large amount of data, the conditional subsurvival function for critical failure is above that for preventive maintenance. The pilot valves and feed water pumps all involve upwards of 300 observations, and these, with the exception[4] of station 2 314 pilot valves, all exhibit this behavior. This pattern was predicted on the random signs model for competing failure modes, and we find here a strong initial indication in favor of the random signs model.

Comparison of Station 1 and Station 2

Inspecting the subsurvival functions for *all* systems the following consistent picture emerged. The subsurvival function for critical failure is *above* that for preventive maintenance at station 1, but at station 2 critical failure is *below* preventive maintenance. At the same time the conditional subsurvival function for preventive maintenance is further away from that for critical failure at station 1 than at station 2. This indicates that the maintenance at station 2 is more effective in catching incipient failures, and at the same time is better able to couple the preventive maintenance times to the failure times.

Exponentiality and heterogeneous data

Most of the data seems broadly consistent with the assumption that critical failure has a constant failure rate. Exceptions are the 312 system and the 314 pilot valves for station 1.

The data from the 312 feed water pump system is made to be very heterogeneous. The feed water pump system is very large and the socket histories constituting this data come from pumps, valves, heat exchangers, etc., which are expected to have very different failure characteristics. The pronounced upswing for lower time values suggests a mixture of a few components with very high failure rates and other components with low failure rates.

The 314 pilot valves appear to be heterogeneous for station 1, but not for station 2. Focusing on station 1, the non-homogeneity is concentrated very near the origin, say before 10 months, (the curving behavior in the 312 pictures extend out to 100 months). The picture suggests that there may be spurious reporting of very early failures. We noted

---

[4]We do not know whether this exception is statistically significant, as tests for the random signs hypothesis have not yet been developed (such tests can be easily developed, but time is required to select the most appropriate test).

that unsuccessfully completed maintenance appears in the ATV data as a separate event. Note that the 314 pilot valves at station 2 do not look inhomogeneous. This supports the conjecture that spurious reporting of early failures might explain the picture for station 1.

1.5   Conclusions

The data which has been graphically presented here suggests a number of conclusions (which at this moment have not been checked by statistical tests).

1. The independent exponential model is a poor model for analyzing competing failure mode data from component socket history data, since the conditional subsurvival functions are rarely identical and exponential (as required by Theorem 1.2.5).

2. The random sign model accords well with the data analyzed here, and seems physically more plausible than a independent censoring model.

3. In cases where the data is not known to be strongly heterogeneous, the exponential model for the naked failure processes is broadly consistent with the censored life data.

4. Graphs are a good tool for quickly reading the drift of data. The efficacy of maintenance (alternatively the predominance of degraded/incipient failure over critical failure) can be gleaned from the subsurvival graphs (i). The interaction of competing failure modes can be inferred from the conditional subsurvival graphs (ii). Possible non-homogeneity of the data, indicating different failure rates for different component sockets can be read from the time average failure rates (iii).

The research team originally planned to compare the numbers given in the above table with failure rate estimates in the T-Book, which are based on the same data. This turned out to be impossible for a number of reasons, including

- The T-Book estimates are based on data up to 1987, where as this data goes up to 1993.
- The T-Book distinguishes more failure modes (e.g. failure to open, failure to close, spurious opening, spurious closure, etc.).
- The T-Book contains very many zero failure rate estimates.

In spite of these difficulties we may report one overall observation: the confidence bounds reported here are generally narrower than bounds in the T-Book. The ratio between the 50% and 95% values in the T-Book is often more than 100. We hasten to add, however, that no simple interpretation can be placed on this fact. The explanation may lie in the choice of statistical methods for generating bounds, and it may also lie in sparsity of data caused by disaggregation.

Acknowledgments

The authors would like to thank I. Meilijson and L. Meester for their contributions to the ESA project reported in [5], R.E. Barlow for his helpful and constructive criticism of the prior version of this paper, J. Paulsen of the Risø National Laboratory for help in analyzing the data and the Swedish Nuclear Authority SKI for allowing use of their data.

References

1. *T-Book Reliability Data of Components in Nordic Nuclear Power Plants*, ATV office, Vattenfall AB, S-162 87 Vallingby Sweden, 1992.

2. Bedford, T. and Meilijson, I. (1993), A characterisation of marginal distributions of (possibly dependent) lifetime variables which right censor each other, *Preprint of the Faculty of Technical Mathematics and Informatics*, Delft University of Technology, 93-116.

3. *Guidelines for Process Equipment Reliability Data*, Center for Chemical Process Safety of the American Institute of Chemical Engineers, New York, 1989.

4. Cooke, R.M. (1993), The total time on test statistic and age-dependent censoring, *Statistics & Probability Letters*, **18**, 307-312.

5. Cooke, R., Bedford, T., Meilijson, I., and Meester, L. (1993), Design of reliability data bases for aerospace applications, *Report to the European Space Agency*, Department of Mathematics, Report 93-110, Delft University of Technology.

6. Cox, D.R. (1959), The analysis of exponentially distributed life-times with two types of failure, *Journal of the Royal Statistical Society, Series B*, **21**, 414-421.

7. David, H.A. and Moeschberger, M.L. (1978), *The Theory of Competing Risks*, Griffin, London.

8. Fragola, J.R. (1987), Reliability data bases: The current picture, *Hazard Prevention*, Jan/Feb 1987, 24-29.

9. Gail, M. (1975), A review and critique of some models used in competing risk analysis, *Biometrics*, **31**, 209-222.

10. Johanson, G. and Fragola, J. (1982), Synthesis of the data base for the Ringhals 2 PRA using the Swedish ATV data system, *Presented at ANS/ENS International Meeting on Thermal Reactor Safety*, Chicago.

11. Paulsen, J. and Cooke, R. (1994), Concepts for measuring maintenance performance and methods for analyzing competing failure modes, *Proceedings of ESREL-94*, La Baule, France.

12. Peterson, A.V. (1976), Bounds for a joint distribution function with fixed subdistribution functions: Application to competing risks, *Proceedings of the National Academy of Sciences - USA*, **73**, 11-13.

13. Peterson, A.V. (1977), Expressing the Kaplan-Meier estimator as a function of empirical subsurvival functions, *Journal of the American Statistical Association*, **72**, 854-858.

14. Tsiatis, A. (1975), A nonidentifiablility aspect in the problem of competing risks, *Proceedings of the National Academy of Sciences - USA*, **72**, 20-22.

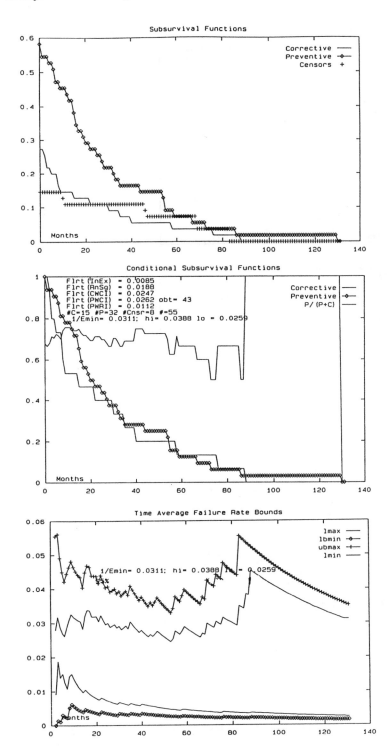

FIGURE 1.1
Failure data for the 311 inner isolation valves, station 2

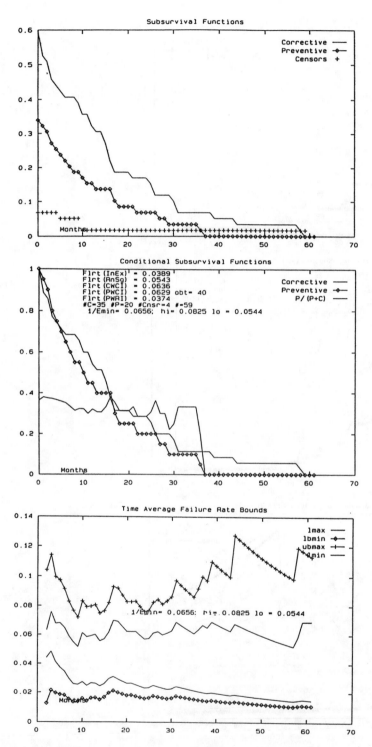

FIGURE 1.2
Failure data for the 311 outer isolation valves, station 1

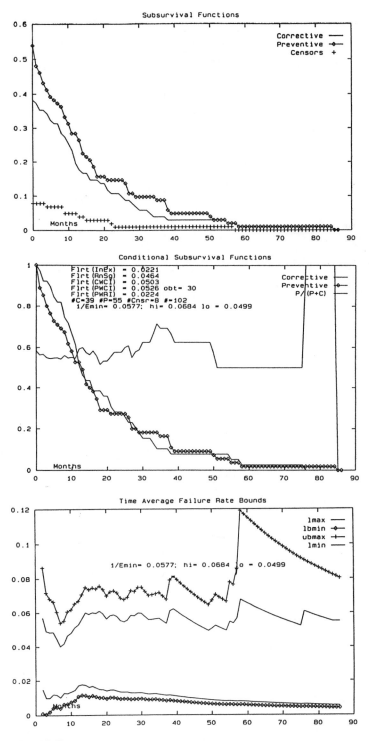

FIGURE 1.3
Failure data for the 311 outer isolation valves, station 2

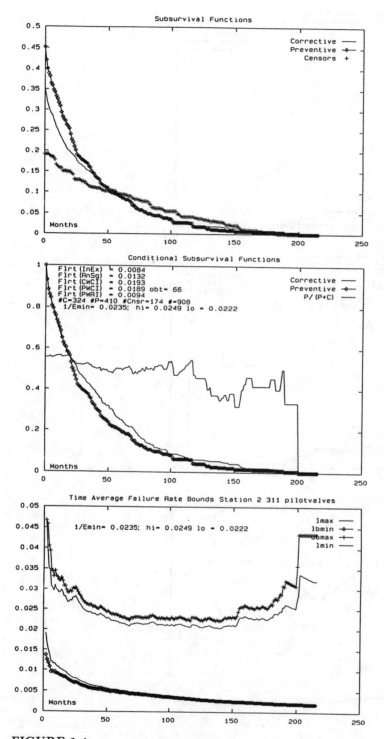

FIGURE 1.4
Failure data for the 311 pilot valves, station 2

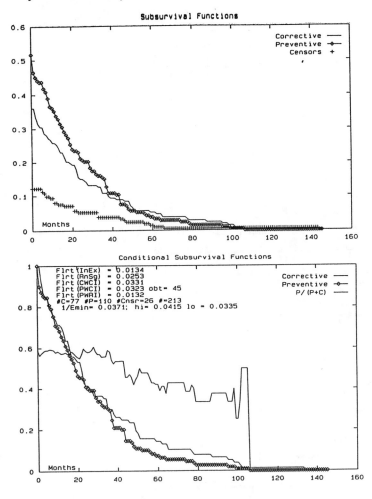

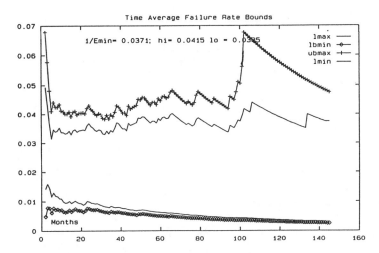

FIGURE 1.5
Failure data for the 314 containment relief valves, station 2

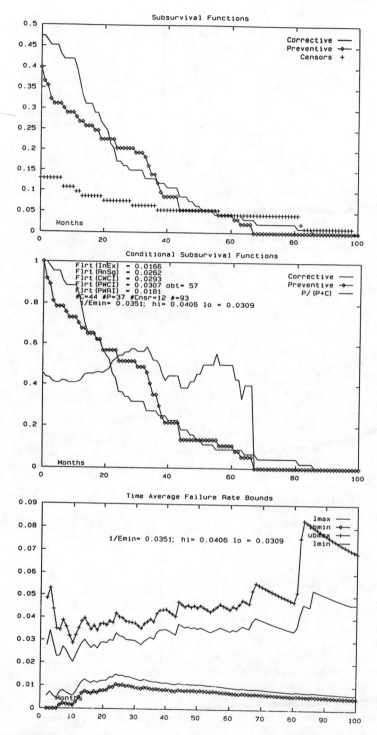

FIGURE 1.6
Failure data for the 314 sump relief valves, station 2

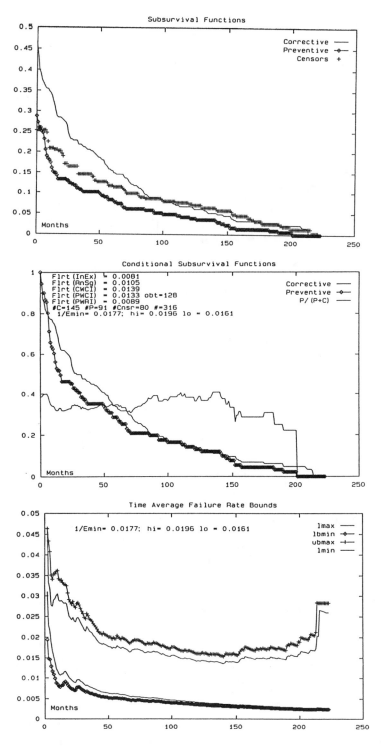

FIGURE 1.7
Failure data for the 314 pilot valves, station 1

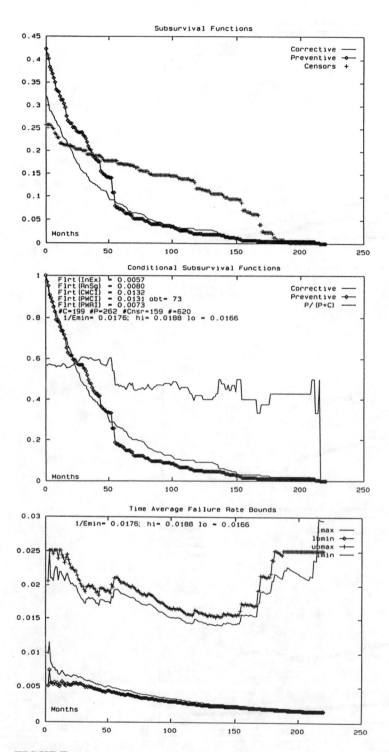

FIGURE 1.8
Failure data for the 314 pilot valves, station 2

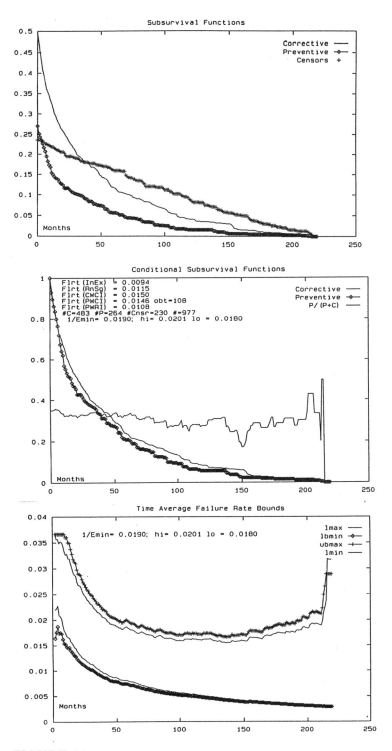

FIGURE 1.9
Failure data for the 312 feedwater pumps, station 2

2

A Bayesian Approach to the Analysis of Reliability Data Bases

Richard E. Barlow

University of California, Berkeley

ABSTRACT The report on "Design of reliability data bases for aerospace applications" by [1] is a well documented discussion of existing failure data analysis techniques. It also provides new methodology, using empirical subsurvival functions in a novel way to analyze Swedish nuclear power plant data. An "operational bayesian" approach [4] would require a somewhat different methodology. We develop an operational methodology for this problem.

CONTENTS

2.1 Introduction

The objective of data analysis is the acquisition of information. Information has been defined as "anything which changes belief". We first summarize the points of general agreement.

- Data needs to be summarized and interpreted. It is necessary to have a framework for this summarization and interpretation.

- A priori judgments are required before beginning a data analysis. In the report [1], for example, outer isolation valve experiences in stations 1 and 2 were not considered a priori exchangeable although valve experiences for a specified station were considered a priori exchangeable.

- Probability distributions are the right way to measure uncertainty and should be used to analyze data in general.

- Methods for data analysis are needed to provide information which can be used by a variety of decision makers and for a variety of purposes.

2.2 The Operational Bayesian Approach

Consider the problem of failure data involving just two types of component removal:

1. removal due to critical failure
2. removal for incipient or degraded failure.

Let *potentially observable* components be labeled $\{1, 2, ..., N\}$. Let $I_i = 1$ if component i were to be removed due to critical failure and $I_i = 0$ if component i were to be removed due to incipient or degraded failure. Let Z_i be the time of removal of component i. We can potentially observe (Z_i, I_i) for $i = 1, 2, ..., N$.

Suppose we judge components a priori exchangeable with respect to (Z, I), that is, $(Z_1, I_1), (Z_2, I_2), ..., (Z_N, I_N)$ are a priori exchangeable random vectors. At this point we must identify operational parameters relevant to our problem. Parameters are operational in the sense that they must be functions of a finite number of observable or potentially observable quantities. Also we don't want to create parameters needlessly. Non-parametric methods employ uncountably many parameters! Our approach to probability is that it is just another analytical tool and is based on judgment (see [2]).

We should ask the question: How would a knowledgeable engineer summarize the data? For example, the engineer might consider one or more of the following data summaries sufficient:

i) counts of the numbers of different types of removals;

ii) averages of removal intervals of each type;

iii) maxima and minima of removal intervals;

iv) counts of the numbers of different types of removals in specified intervals.

Counts of the numbers of different types of removals would correspond to attribute data while averages of removal intervals of each type would correspond to measurement data.

Dichotomous Data

We begin by considering a finite "population" of N observables. Later we will use n for the number of data vectors *actually observed*. In the case when i) and ii) are deemed sufficient data summaries by the engineer and only two types of removals are recorded, let

$$\sum_{i=1}^{N} Z_i I_i = \theta_1(N) \sum_{i=1}^{N} I_i \tag{2.2.1}$$

$$\sum_{i=1}^{N} Z_i (1 - I_i) = \theta_0(N) \sum_{i=1}^{N} (1 - I_i) \tag{2.2.2}$$

$$\sum_{i=1}^{N} I_i = N \rho(N). \tag{2.2.3}$$

Then $\theta_1(N)$ is a "parameter" corresponding to the average time to a critical failure removal, $\theta_0(N)$ is the average time to a preventive removal and $\rho(N)$ is the proportion of critical failure removals among all removals. Let

$$(\mathbf{z}_N, \mathbf{I}_N) = \big((z_1, I_1), (z_2, I_2), \ldots, (z_N, I_N)\big).$$

The next step is to make use of the sufficiency judgment. Namely that, a priori, any predictive density we might consider has the property that

$$p(\mathbf{z}_N, \mathbf{I}_N) = p(\mathbf{z}_N', \mathbf{I}_N')$$

when

$$\sum_{i=1}^{N} z_i I_i = \sum_{i=1}^{N} z_i' I_i'$$

$$\sum_{i=1}^{N} z_i(1 - I_i) = \sum_{i=1}^{N} z_i'(1 - I_i')$$

$$\sum_{i=1}^{N} I_i = \sum_{i=1}^{N} I_i'.$$

Note that no independence judgment is made regarding the two types of failures. The probability invariance judgment is made on observables, namely $(\mathbf{z}_N, \mathbf{I}_N)$.

For a sample of size n with data vectors $(\mathbf{z}_n, \mathbf{I}_n)$, our problem is to calculate

$$p(\mathbf{z}_n, \mathbf{I}_n | \theta_1, \theta_0, \rho)$$

in the limiting case, when the "population size" $N \to \infty$. By the de Finetti representation theorem, the marginal random vector pairs will be independent.

From sufficiency considerations, the de Finetti exchangeability representation theorem, and Cauchy's equation we can show that

$$p(z_i, I_i | \theta_1, \theta_0, \rho) = \left[\frac{\rho}{\theta_1}\right]^{I_i} \exp\left\{-\frac{z_i I_i}{\theta_1}\right\} \left[\frac{1-\rho}{\theta_0}\right]^{1-I_i} \exp\left\{-\frac{z_i(1 - I_i)}{\theta_0}\right\}.$$

In the case of a censored observation, y_i; i.e. the current component in a specified socket has operated y_i time units without being removed, then the likelihood will also contain the term

$$p(y_i | \theta_1, \theta_0, \rho) = \rho \exp\left\{-\frac{y_i}{\theta_1}\right\} + (1 - \rho) \exp\left\{-\frac{y_i}{\theta_0}\right\}.$$

The fundamental difference between this approach and the usual statistical analysis is that here sufficient statistics are first identified relative to the problem of interest and then an appropriate likelihood model for data analysis is derived. In the usual analysis, a class of distributions is first chosen and then the sufficient statistics are identified!

Multi-attribute Data

This case is a straightforward generalization of the dichotomous case. In this case we use the notation $(z_i, I_{i1}, I_{i2}, \ldots, I_{im})$ if z_i is observed together with observable attributes $j = 1, 2, \ldots, m$. Let

$$I_{ij} = \begin{cases} 1 & \text{if } z_i \text{ is observed together with attribute } j \\ 0 & \text{otherwise.} \end{cases}$$

where

$$\sum_{j=1}^{m} I_{ij} = 1.$$

We also need to introduce parameters $\theta_1, \theta_2, \ldots, \theta_m$ and $\rho_1, \rho_2, \ldots, \rho_m$ where $\theta_j > 0$, $0 \leq \rho_j \leq 1$ for $j = 1, 2, \ldots, m$ and

$$\sum_{j=1}^{m} \rho_j = 1.$$

Bayes' Formula

To complete an operational bayesian analysis we need to assess a distribution for $(\theta_1, \theta_0, \rho)$, say $\pi(\theta_1, \theta_0, \rho)$. With this we can calculate $p(\theta_1, \theta_0, \rho | \mathbf{z}_N, \mathbf{I}_N)$ as well as the predictive distribution $p(z_{n+1}, I_{n+1} | \mathbf{z}_N, \mathbf{I}_N)$ and many other probabilities of interest.

Since $(\theta_1(N), \theta_0(N), \rho(N))$ is operationally defined, we start by considering $\pi_N(\theta_1(N), \theta_0(N), \rho(N))$. This probability function should encapsulate any opinion or information we may have concerning maintenance policies and censorship.

Random Preventive Maintenance

In this case we could have

$$\rho(N) = \frac{\theta_1(N)}{\theta_1(N) + \theta_0(N)}$$

for example, while $\theta_1(N) \amalg \theta_0(N)$. Only $\pi_N(\theta_1(N))$ and $\pi_N(\theta_0(N))$ are left to be assessed.

Deterministic Preventive Maintenance

In this case, z_i given $I_i = 0$ is known and the distribution of $\rho(N)$ is determined by the distribution of z_i given $I_i = 1$ and the distribution of $\theta_1(N)$

2.3   Points of Disagreement

The main differences between the usual approach to the analysis of reliability data bases and the operational bayesian approach are

1. the meaning of probability, and
2. the uses of probability.

The meaning of probability

For simplicity let us first ignore censoring. The underlying assumption in the usual approach is that there is a probability distribution corresponding to observed failure times. This probability distribution **exists** independent of the analyst. It is not a probability distribution fashioned in the brain of the analyst. Furthermore, were this distribution known, random failure times could be simulated using a computer. Because of this assumption, classical technical procedures for analyzing data can be evaluated using Monte Carlo techniques.

The above assumption contradicts the basic philosophy and arguments of Bruno de Finetti for whom probability does not exist, at least objectively, independent of the analyst. His argument is given in two volumes and is specifically directed toward the above assumption. We argue that probability is just another analytical tool and is based on judgment. It is not some physical mechanism operating in the real world. This changes dramatically the way in which data is analyzed.

The use of probability in inductive inference

The report on "Design of reliability data bases for aerospace applications" by [1] uses (defective) conditional probability functions called subsurvival functions. These are calculated using the "laws" of probability and there is no quarrel with the purely mathematical results which are documented. The empirical subsurvival functions, on the other hand, are based on data and in particular, the order statistics (which are sufficient under the exchangeability judgment). By construction, these functions have the mathematical properties of (defective) survival distributions. They are, in a sample theoretic sense, **maximum likelihood estimates** for distributions in the class of continuous survival distributions. The only sample theoretic justification that I know of is **consistency**. That is, convergence to the "true" distribution as the sample size approaches infinity. As noted in the previous discussion of the meaning of probability, the relevance of the "true" distribution is disputed. A bayesian could of course "adopt" this estimate as his/her predictive probability distribution for future failure times.

We have given a bayesian procedure valid under the specific conditions given. However many additional procedures for conditions likely to be encountered would have to be developed to provide a really useful methodology for reliability data bases.

Acknowledgements

This research was partially supported by the Department of Mathematics and Informatics, Delft University of Technology and the U.S. Air Force Office of Scientific Research (F49620-93-1-001) grant to the University of California at Berkeley.

This is to acknowledge my indebtedness to Roger Cooke, Tim Bedford and Jan Noortwijk at the Delft University of Technology for their help and encouragement.

Appendix

We wish to show that when $(\theta_1(N), \theta_0(N), \rho(N))$ converges in measure as $N \to \infty$, then we have the following limiting marginal distribution for (z_i, I_i):

$$p(z_i, I_i | \theta_1, \theta_0, \rho) = \left[\frac{\rho}{\theta_1}\right]^{I_i} \exp\left\{-\frac{z_i I_i}{\theta_1}\right\} \left[\frac{1-\rho}{\theta_0}\right]^{1-I_i} \exp\left\{-\frac{z_i(1-I_i)}{\theta_0}\right\}.$$

Sufficiency

Let $x_i = (z_i I_i, z_i(1 - I_i), I_i)$ and $\mathbf{T}(\mathbf{x}_N) = \sum_{i=1}^{N} x_i$. The argument proceeds by first showing that,

$$\left((z_1, I_1), (z_2, I_2), \cdots, (z_n, I_n)\right) \amalg \left(\theta_1(N), \theta_0(N), \rho(N)\right) \Big| \mathbf{T}(\mathbf{x}_n).$$

The following lemma is a very general result.

LEMMA 2.3.1
We assume for all $n \leq N$ that

$$p_n(\mathbf{x}_n) = p_n(\mathbf{x}_n')$$

when

$$\mathbf{T}(\mathbf{x}_n) = \mathbf{T}(\mathbf{x}_n')$$

where $\mathbf{x}_n = (x_1, \cdots, x_n)$ is an n-vector whose coordinates may also be vectors. $\mathbf{T}(\mathbf{x}_n)$ can be thought of as a "sufficient statistic" for $p_n(\cdot)$. Then

$$\{\mathbf{x}_n\} \amalg \mathbf{T}(\mathbf{x}_N) \big| \mathbf{T}(\mathbf{x}_n)$$

with respect to $p_N(\cdot)$. That is, $\mathbf{T}(\mathbf{x}_n)$ is sufficient for $\mathbf{T}(\mathbf{x}_N)$ with respect to $p_N(\cdot)$.

PROOF   Let $p_N(\mathbf{x}_N | \mathbf{T}(\mathbf{x}_N) = \mathbf{t}_N) = 1/V(\mathbf{t}_N)$; i.e. a constant for fixed $\mathbf{t}_N$. Then

$$p_N(\mathbf{t}_n, x_{n+1}, \cdots, x_N | \mathbf{T}(\mathbf{x}_N) = \mathbf{t}_N)$$

$$= \int_{\{x_1,\ldots,x_n | \mathbf{T}(\mathbf{x}_n)=\mathbf{t}_n\}} p_N(x_1, \cdots, x_n, \cdots, x_N | \mathbf{T}(\mathbf{x}_N) = \mathbf{t}_N) \, dx_1 \, dx_2 \, \ldots \, dx_n$$

$$= \frac{1}{V(\mathbf{t}_N)} \int_{\{x_1,\ldots,x_n | \mathbf{T}(\mathbf{x}_n)=\mathbf{t}_n\}} dx_1 \, dx_2 \, \ldots \, dx_n = \frac{V(\mathbf{t}_n)}{V(\mathbf{t}_N)}.$$

Hence

$$p_N(x_1, \cdots, x_n | \mathbf{T}(\mathbf{x}_n) = \mathbf{t}_n, x_{n+1}, \cdots, x_N, \mathbf{T}(\mathbf{x}_N) = \mathbf{t}_N)$$

$$= \frac{p_N(x_1, \cdots, x_n, \mathbf{t}_n, x_{n+1}, \cdots, x_N | \mathbf{T}(\mathbf{x}_N) = \mathbf{t}_N)}{p_N(\mathbf{t}_n, x_{n+1}, \cdots, x_N | \mathbf{T}(\mathbf{x}_N) = \mathbf{t}_N)}$$

$$= \frac{1}{V(\mathbf{t}_n)} = p_n(x_1, \cdots, x_n | \mathbf{T}(\mathbf{x}_n) = \mathbf{t}_n).$$

This in turn implies

$$\{\mathbf{x}_n\} \amalg \mathbf{T}(\mathbf{x}_N) | \mathbf{T}(\mathbf{x}_n).$$

i.e. $\mathbf{T}(\mathbf{x}_n)$ is sufficient for $\mathbf{T}(\mathbf{x}_N)$. ∎

The de Finetti Representation

The sufficiency conditions imply that

$$\Big((z_1, I_1), (z_2, I_2), \cdots, (z_N, I_N) \Big)$$

are exchangeable. Suppose now that the sufficiency conditions apply for all N. We now invoke the de Finetti exchangeability representation theorem [2] to assert that

$$\lim_{N \to \infty} p_N(\mathbf{z}_n, \mathbf{I}_n | \theta_1(N), \theta_0(N), \rho(N)) = \prod_{i=1}^n p(z_i, I_i | \theta_1, \theta_0, \rho).$$

The Likelihood Characterization

Since $z_i = z_i I_i + z_i (1 - I_i)$ we can let $x_i = (z_i I_i, z_i(1 - I_i), I_i)$, $\mathbf{x}_n = (x_1, \cdots, x_n)$ and

$$\mathbf{T}(\mathbf{x}_n) = \sum_{i=1}^n x_i.$$

Now fix n. From sufficiency and the de Finetti theorem we have

$$p_n \left(\sum_{i=1}^n x_i \middle| \theta_1, \theta_0, \rho \right) = \prod_{i=1}^n p(x_i | \theta_1, \theta_0, \rho).$$

since $\mathbf{T}(\mathbf{x}_n) = \sum_{i=1}^n x_i$ is sufficient for $(\theta_1, \theta_0, \rho)$ in the limiting case.
    Fixing $(\theta_1, \theta_0, \rho)$, let

$$\phi_n \left(\sum_{i=1}^n x_i \right) = \log \left\{ p_n \left(\sum_{i=1}^n x_i \middle| \theta_1, \theta_0, \rho \right) \right\}$$

and

$$\lambda(x_i) = \log\{p(x_i | \theta_1, \theta_0, \rho)\}.$$

Then, in this notation,

$$\phi_n \left(\sum_{i=1}^n x_i \right) = \sum_{i=1}^n \lambda(x_i).$$

Invoking the generalized Cauchy equality theorem of [3] [see Appendix], it follows that

$$\lambda(x_i) = \alpha + \beta \cdot x_i,$$

where α is a real number, β and x_i are vectors with 3 real coordinates.
    In the limit using (2.2.1), (2.2.2) and (2.2.3) we have

$$E(z_i | I_i = 1) = \theta_1$$

$$E(z_i | I_i = 0) = \theta_0$$

$$E(I_i) = \rho.$$

Hence

$$p(z_i, I_i | \theta_1, \theta_0, \rho) = \left[\frac{\rho}{\theta_1}\right]^{I_i} \exp\left\{-\frac{z_i I_i}{\theta_1}\right\} \left[\frac{1-\rho}{\theta_0}\right]^{1-I_i} \exp\left\{-\frac{z_i(1-I_i)}{\theta_0}\right\}.$$

∎

The following generalization of Cauchy's equation is in [3, page 239] and is repeated here for reference convenience.

THEOREM 2.3.2
Let G be a Borel subset of $\mathbb{R}^d$, with positive Lebesgue measure. Let λ be a Borel function defined almost everywhere on G. Let $G^{(1)} = G$ and $G^{(n+1)} = G^{(n)} + G$. Let ϕ_n be a Borel function defined almost everywhere on $G^{(n)}$. Suppose that for each n, for almost all x_i in G:

$$\phi_n\left(\sum_{i=1}^{n} x_i\right) = \sum_{i=1}^{n} \lambda(x_i).$$

Then there is a real number α, and a vector $\beta \in \mathbb{R}^d$ such that $\lambda(x) = \alpha + \beta \cdot x$ for Lebesgue-almost all $x \in G$.

References

1. Cooke, R., Bedford, T., Meilijson, I., Meester, L. (1993), Design of reliability data bases for aerospace applications, *Technical Report*, Technical University of Delft, Holland, 93-110.

2. De Finetti, B. (1937), Foresight: Its logical laws, its subjective sources, *Annales de l'Institute Henri Poincaré*, **7**, 1-68, English translation In *Studies in Subjective Probability*, (Eds., H.E. Kyburg, Jr. and H.E.Smokler), 1980, Second edition, Krieger, Huntington, New York. 53-118.

3. Diaconis, P. and Freedman, D.A. (1990), Cauchy's equation and de Finetti's theorem, *Scandinavian Journal of Statistics*, **17**, 235-250.

4. Mendel, M.B. (1989), Development of Bayesian parametric theory with applications to control, *Ph.D. Thesis*, Department of Mechanical Engineering, Massachusetts Institute of Technology.

3

Analysis of Repair Data With Two Measures of Usage

Wayne Nelson

Consultant, Schenectady, NY

ABSTRACT Some repairable products have more than one measure of usage. The example here concerns the reliability of blood analyzers, whose repairable failures are affected both by the hours they are on and the number of specimens they process. This paper presents a simple new model and data analysis methods for estimating the effect on reliability of each measure of usage. Suppose that repairable analyzer i processes s_i specimens in t_i hours. Then its number Y_i of repairs is assumed to have a Poisson distribution with mean $M_i = \nu s_i + \lambda t_i$. Using data from a number of analyzers, this paper presents estimates and confidence limits for the model parameters ν and λ and other analyses.

Keywords: Poisson, Recurrence data, Reliability

CONTENTS

3.1 Introduction

Background. Various products have more than one measure of usage. Gas turbine usage consists of running hours and number of start-ups. Lightbulb usage consists of hours on and the number of on-off cycles. Previous modeling and analysis of life data on such products with two usages involved fitting a regression model where one usage is the dependent variable and the other is an independent variable. This unequal treatment of the two usages is unappealing. The model for repair data presented here treats both usages equally.

  Example. The model and data analyses here were developed for the following application to blood analyzers, which measure blood chemistries. Field data and bench testing of such analyzers revealed that their repair rates depend on both the hours they are on and the number of specimens they process. Nelson's [4] plot of test data indicated that sample functions for the mean cumulative number of repairs are linear. This suggested that repairs might be described with a Poisson process. To investigate the effect of each usage, engineers tested two groups of analysers. Each Group 1 analyzer processed

10 specimens per hour, and each Group 2 analyzer processed 30 specimens per hour. Group 1 analyzers accumulated $Y_1 = 27$ repairs while processing a total of $s_1 = 12.4$ thousand specimens in a total of $t_1 = 1.24$ thousand hours. Similarly, Group 2 analyzers accumulated $Y_2 = 42$ repairs while processing a total of $s_2 = 32.4$ thousand specimens in a total of $t_2 = 1.08$ thousand hours. This paper presents a model for such data, maximum likelihood (ML) estimates and confidence limits for model parameters, and other analyses. The fitted model is used to predict repair rates of analyzers under different operating conditions.

Overview. Section 3.2 presents a simple new model for such data. Section 3.3 derives the ML estimates and confidence limits for model parameters and applies them to the blood analyzer data. Section 3.4 discusses the results and their extensions.

3.2   The Model

Suppose that Y_i repairs occur on a blood analyzer (or a homogeneous group of analyzers) that processed s_i specimens and ran t_i hours. The number Y_i is assumed to have a Poisson distribution with an expected number of repairs

$$M_i = \nu s_i + \lambda t_i \; .$$

Here ν and λ are model parameters to be estimated from data (Y_i, s_i, t_i) on I systems.

Assumptions. In addition to the usual assumptions for a Poisson process, the model above involves further assumptions. First, for a particular analyzer i, the number s_i of specimens and the hours t_i may be accumulated in any way and M_i is the same; that is, the analyzer may pass through any increasing (s, t) values going from its starting value $(0,0)$ to (s_i, t_i). Briefly stated, M_i does not depend on the path to (s_i, t_i). Second, the analyzers are all assumed to have the same ν and λ values. Third, $Y_1, Y_2, \ldots, Y_I$ are assumed to be statistically idependent.

3.3   Maximum Likelihood Theory

This section presents ML theory for estimates and confidence limits for model parameters. As is well known, ML estimates and confidence limits have good statistical properties. Topics include the sample (log) likelihood, the normal equations and ML estimates, the Fisher and covariance matrices, and approximate confidence limits. Nelson [3] provides background on theory for ML estimation and LR tests.

Likelihood. Suppose system i has Y_i occurrences in usage (s_i, t_i), $i = 1, 2, \ldots, I$. Then the likelihood for system i is the Poisson probability

$$L_i = f(Y_i) = (1/Y_i!) M_i^{Y_i} \exp(-M_i) \; .$$

The corresponding log likelihood is

$$\mathcal{L}_i = \ln(L_i) = \ln(1/Y_i!) + Y_i \ln(\nu s_i + \lambda t_i) - (\nu s_i + \lambda t_i) \; .$$

The sample log likelihood is

$$\mathcal{L} = \mathcal{L}_1 + \mathcal{L}_2 + \cdots + \mathcal{L}_I$$

$$= -\sum_i \ln(Y_i!) - (\nu S + \lambda T) + \sum_i Y_i \ln(\nu s_i + \lambda t_i)$$

where $S = s_1 + s_2 + \cdots + s_I$ and $T = t_1 + t_2 + \cdots + t_I$ are the total usages.

ML Estimates. The ML estimates $\widehat{\nu}$ and $\widehat{\lambda}$ for ν and λ are the parameter values that maximize the sample log likelihood. They are found by setting its partial derivatives with respect to ν and λ equal to zero and solving for $\widehat{\nu}$ and $\widehat{\lambda}$. These so-called normal equations are

$$0 = \partial \mathcal{L}/\partial \nu = -S + \sum_i Y_i s_i / (\nu s_i + \lambda t_i) \; ,$$

$$0 = \partial \mathcal{L}/\partial \lambda = -T + \sum_i Y_i t_i / (\nu s_i + \lambda t_i) \; .$$

For $I > 2$ systems, these equations must be solved iteratively for $\widehat{\nu}$ and $\widehat{\lambda}$. Also, $\widehat{\nu}$ and $\widehat{\lambda}$ can be obtained by numerically maximizing the sample log likelihood by iteration.

Two Systems. For $I = 2$ systems, as for the two groups of blood analyzers, the likelihood equations have explicit solutions

$$\widehat{\nu} = [(Y_1/t_1) - (Y_2/t_2)] / [(s_1/t_1) - (s_2/t_2)] \; ,$$

$$\widehat{\lambda} = [(Y_1/s_1) - (Y_2/s_2)] / [(t_1/s_1) - (t_2/s_2)] \; .$$

For the blood analyzers,

$$\widehat{\nu} = [(27/1.24) - (42/1.08)]/[(12.4/1.24) - (32.4/1.08)]$$

$$= 0.8557 \text{ repairs per thousand-specimens} \; ,$$

$$\widehat{\lambda} = [(27/12.4) - (42/32.4)]/[(1.24/12.4) - (1.08/32.4)]$$

$$= 13.22 \text{ repairs per thousand-hours} \; .$$

Thus the fitted model is $M(s,t) = 0.8557\,s + 13.22\,t$. Also, for $I = 2$, $\widehat{\nu}$ and $\widehat{\lambda}$ are moment estimators obtained by setting each observed number Y_i of occurences equal to its expected number M_i and solving

$$Y_1 = \nu\,s_1 + \lambda\,t_1 \; , \qquad Y_2 = \nu\,s_2 + \lambda\,t_2 \; .$$

For $I = 2$, it is easy to show that $\widehat{\nu}$ and $\widehat{\lambda}$ are unbiased estimators and have exact theoretical variances and covariance

$$V(\widehat{\nu}) = [M_1\,t_2^2 + M_2\,t_1^2] / [t_1\,s_2 - t_2\,s_1]^2,$$

$$V(\widehat{\lambda}) = [M_1\,s_2^2 + M_2\,s_1^2] / [t_1\,s_2 - t_2\,s_1]^2 \; ,$$

$$V(\widehat{\nu}, \widehat{\lambda}) = -[M_1\,s_2 t_2 + M_2\,s_1 t_1] / [t_1 s_2 - t_2 s_1]^2 \; ,$$

where $M_i = \nu\,s_i + \lambda\,t_i$. Note that each (co)variance is infinite if $s_1/t_1 = s_2/t_2$ (specimens per hour); thus these ratios must differ. For the blood analyzers, the corresponding estimates of these variances are

$$v(\widehat{\nu}) = [27(1.08)^2 + 42(1.24)^2] / [1.24 \times 32.4 - 1.08 \times 12.4]^2 = 0.1339 \; ,$$

$$v(\widehat{\lambda}) = [27(32.4)^2 + 42(12.4)^2] / [1.24 \times 32.4 - 1.08 \times 12.4]^2 = 48.51 \; ,$$

$$v(\widehat{\nu}, \widehat{\lambda}) = -[27(32.4)1.08 + 42(12.4)1.24] / [1.24 \times 32.4 - 1.08 \times 12.4]^2 = -2.217 \; ;$$

here Y_i is substituted for its expectation M_i in the previous equations. The estimate of the correlation coefficient of $\hat{\nu}$ and $\hat{\lambda}$ is $r = -2.217/(0.1339 \times 48.51)^2 = -0.870$.

Fisher Matrix.   For any I, the negative partial derivatives for the Fisher information matrix are

$$-\partial^2 \mathcal{L}/\partial \nu^2 = \sum_i Y_i s_i^2/M_i^2 \;,$$

$$-\partial^2 \mathcal{L}/\partial \lambda^2 = \sum_i Y_i t_i^2/M_i^2 \;,$$

$$-\partial^2 \mathcal{L}/\partial \nu \partial \lambda = \sum_i Y_i s_i t_i/M_i^2 \;.$$

Their expectations are

$$F_{\nu\nu} = E\{-\partial^2 \mathcal{L}/\partial \nu^2\} = \sum_i s_i^2/M_i \;,$$

$$F_{\lambda\lambda} = E\{-\partial^2 \mathcal{L}/\partial \lambda^2\} = \sum_i t_i^2/M_i \;,$$

$$F_{\nu\lambda} = E\{-\partial^2 \mathcal{L}/\partial \nu \partial \lambda\} = \sum_i s_i t_i/M_i.$$

The Fisher matrix is

$$\boldsymbol{F} = \begin{bmatrix} F_{\nu\nu} & F_{\nu\lambda} \\ F_{\nu\lambda} & F_{\lambda\lambda} \end{bmatrix} \;.$$

Covariance Matrix.   For any I, the theoretical asymptotic covariance matrix of $\hat{\nu}$ and $\hat{\lambda}$ is the inverse of the Fisher matrix, namely,

$$\begin{bmatrix} V(\hat{\nu}) & V(\hat{\nu},\hat{\lambda}) \\ V(\hat{\nu},\hat{\lambda}) & V(\hat{\lambda}) \end{bmatrix} = \boldsymbol{F}^{-1} \;.$$

This matrix is estimated by replacing ν and λ in the Fisher matrix by their estimates $\hat{\nu}$ and $\hat{\lambda}$.

Confidence Limits.   For any I, based on the asymptotic normal distributions of $\hat{\nu}$ and $\hat{\lambda}$, two-sided approximate $100P\%$ confidence limits for ν and λ are

$$\nu_L = \hat{\nu} - z_{P'}[v(\hat{\nu})]^{1/2} \;, \quad \nu_U = \hat{\nu} + z_{P'}[v(\hat{\nu})]^{1/2} \;,$$

$$\lambda_L = \hat{\lambda} - z_{P'}[v(\hat{\lambda})]^{1/2} \;, \quad \lambda_U = \hat{\lambda} + z_{P'}[v(\hat{\lambda})]^{1/2} \;.$$

where $z_{P'}$ is the standard normal $P' = (1 + P)/2$ fractile.   For the blood analyzers, approximate 95% confidence limits are

$$\nu_L = 0.8557 - 1.960[0.1339]^{1/2} = 0.138 \text{ repairs per thou-specimens} \;,$$

$$\nu_U = 0.8557 + 1.960[0.1339]^{1/2} = 1.573 \text{ repairs per thou-specimens} \;,$$

$$\lambda_L = 13.22 - 1.960[48.51]^{1/2} = -0.43 \text{ repairs per thou-hours} \;,$$

$$\lambda_U = 13.22 + 1.960[48.51]^{1/2} = 26.87 \text{ repairs per thou-hours} \;.$$

λ_L is negative, since the model mathematically allows negative λ (and ν) values, although they may not be physically plausible.

3.4 Extensions

Purpose. This section presents extensions of the results.

Literature. The model here is a Poisson regression model. It is possible that this model has appeared in statistical literature on epidemiological or other applications. Its use in this reliability application appears new.

More Usages. The theory above for two measures of usage extends easily to any number.

LR Limits. As described by Doganaksoy [1], confidence limits based on the normal approximation (like those above) generally have drawbacks:

1. They may have values outside the assumed parameter range, for example, a negative lower limit for ν or λ, which would generally be positive in most applications.

2. They are too short, and their actual confidence level is below $100P\%$.

3. The actual error probabilities for the interval being above or below the true parameter value can be quite unequal and far from $(1-P)/2$. Thus such a limit may not be suitable as a one-sided limit with confidence $100(1+P)/2\%$.

Likelihood Ratio (LR) limits for parameters generally are better. 1) Such limits cannot have values outside the parameter range. 2) Their confidence level is closer to the intended $100P\%$. 3) Their error probabilites above and below are much closer to equal and to $(1-P)/2$. The LR limits are relatively easy to calculate for the model here.

Prediction. In some applications, one wants a prediction and prediction limits for a future number Y' of occurrences during an amount of useage (s', t') where Y' is statistically independent of previous data (Y_i, s_i, t_i), $i = 1, 2, \ldots, I$. Y' may be the future number for a single system or a fleet. The obvious predictor is $Y'' = \hat{\nu}\, s' + \hat{\lambda}\, t'$. $100P\%$ prediction limits based on a normal approximation to the distribution of the prediction error $(Y'' - Y')$ are

$$Y'_L = Y'' - z_{P'}[v(Y'' - Y')]^{1/2}\, , \qquad Y'_U = Y'' + z_{P'}[v(Y'' - Y')]^{1/2}$$

where $z_{P'}$ is the standard normal $P' = (1+P)/2$ fractile and $v(Y'' - Y')$ is the obvious estimate of the true prediction error variance

$$V(Y'' - Y') = V(Y'') + V(Y')$$
$$= V(\hat{\nu})s'^2 + V(\hat{\lambda})t'^2 + 2V(\hat{\nu}, \hat{\lambda})s'\,t' + [\nu\, s' + \lambda\, t']\, ,$$

since $V(Y') = \nu\, s' + \lambda\, t'$. It may be possible to use a LR test or interval to obtain a more accurate approximate interval. Hahn and Meeker [2] discuss prediction intervals in general.

No Effect. It is possible that a supposed usage (say, t) has no effect on the repair rate; that is, $\lambda = 0$ and Y_i is Poisson with mean $M_i = \nu\, s_i$. The hypothesis $\lambda = 0$ can be tested with the following LR test. Under $\lambda = 0$, the sample log likelihood for $Y_1, Y_2, \ldots, Y_I$ is

$$\mathcal{L}(\nu, 0) = -\sum_i \ln(Y_i!) + \sum_i Y_i \ln(s_i) + Y \ln(\nu) - \nu S$$

where $Y = Y_1 + Y_2 + \cdots + Y_I$ and $S = s_1 + s_2 + \cdots + s_I$. The ML estimate for ν is $\hat{\nu}' = Y/S$, the usual estimate for a Poisson rate. The corresponding maximum log likelihood is

$$\mathcal{L}(\hat{\nu}', 0) = -\sum_i \ln(Y_i!) + \sum_i Y_i \ln(s_i) + Y \ln(Y/S) - Y\, .$$

For the blood analyzers, $\widehat{\nu}' = 1.540$ and $\mathcal{L}(\widehat{\nu}', 0) = 174.861$, omitting terms with $Y_i!$, which drop out of Q_λ below. The maximum log likelihood for the model with both ν and λ is $\mathcal{L}(\widehat{\mu}, \widehat{\lambda})$, which can be evaluated from the previous formula for $\mathcal{L}(\nu, \lambda)$. For the analyzers, $\mathcal{L}(\widehat{\nu}, \widehat{\lambda}) = 176.970$, omitting terms with $Y_i!$. The LR test statistic is

$$Q_\lambda = 2[\mathcal{L}(\widehat{\nu}, \widehat{\lambda}) - \mathcal{L}(\widehat{\nu}', 0)] .$$

Under the hypothesis $\lambda = 0$, the distribution of Q_λ is approximately chi-square with 1 degree of freedom. The hypothesis test is

- o if $Q_\lambda > \chi^2(C; 1)$, there is statistically significant (convincing) evidence that $\lambda = 0$ is false; otherwise,
- o if $Q_\lambda \leq \chi^2(C; 1)$, the data are consistent with $\lambda = 0$;

here $\chi^2(C; 1)$ is the C fractile of the chi-square distribution with 1 degree of freedom. For the analyzers, $Q_\lambda = 2[176.970 - 174.861] = 4.218 > 3.841 = \chi^2(0.95; 1)$. Hence λ differs from 0 statistically significantly (convincingly) at the 5% level. With obvious modifications, this test can be used to test the hypothesis $\nu = 0$. Then $Q_\nu = 5.688 > 3.841 = \chi^2(0.95; 1)$. Hence ν differs from 0 statistically significantly (convincingly) at the 5% level.

Test of Fit. The basic model $M_i = \nu\, s_i + \lambda\, t_i$ can be regarded as a regression equation fitted to the data (Y_i, s_i, t_i), $i = 1, 2, \ldots, I$. For $I > 2$, the following LR test can be used to assess whether the model adequately fits the data. First assume that the (s_i, t_i) are all distinct; if not, the LR test must be modified in the obvious way. Then the alternative to the model is that $M_1, M_2, \ldots, M_I$ have distinct values that do not all fall on the equation. Under this alternative, the sample log likelihood is

$$\mathcal{L}(M_1, M_2, \ldots, M_I) = -\sum_i \ln(Y_i!) + \sum_i Y_i \ln(M_i) - (M_1 + \cdots + M_I) .$$

The ML estimates under the alternative are $m_i = Y_i$, and the corresponding maximum log likelihood is

$$\mathcal{L}(m_1, m_2, \ldots, m_I) = -\sum_i \ln(Y_i!) + \sum_i Y_i \ln(Y_i) - (Y_1 + \ldots + Y_I) .$$

Under the basic model, the maximum log likelihood is $\mathcal{L}(\widehat{\nu}, \widehat{\lambda})$ given above. Then the LR test statistic is

$$Q_F = 2[\mathcal{L}(m_1, m_2, \ldots, m_I) - \mathcal{L}(\widehat{\nu}, \widehat{\lambda})] .$$

Under the model, the distribution of Q_F is approximately chi-square with $I - 2$ degrees of freedom. The hypothesis test is

- o if $Q_F > \chi^2(C; I - 2)$, there is statistically significant (convincing) evidence that the model does not adequately fit the data; otherwise,
- o if $Q_F \leq \chi^2(C; I - 2)$, the model is consistent with the data;

here $\chi^2(C; I - 2)$ is the C fractile of the chi-square distribution with $I - 2$ degrees of freedom.

Homogeneous Parameters. An obvious generalization of the model is that $M_i = \nu_i\, s_i + \lambda_i\, t_i$, $i = 1, 2, \ldots, I$; that is, each system has its own values of the parameters, rather than common values ν and λ. A hypothesis test for this would be useful. Unfortunately this model has $2 \times I$ parameters, and there are only I data points (Y_i, s_i, t_i) to estimate them. This could be overcome by observing each system for two usages (s_i, t_i) and

(s_i', t_i'), where s_i/t_i is not equal to s_i'/t_i', and using the obvious LR test. Also, this could be overcome by using a components-of-variance model where each ν_i and λ_i comes from a separate, say, Gamma distribution; then under the hypothesis of single common ν and λ values, the variances of the Gamma distributions are zero.

Test Plans. A test plan consists of the number I of test systems and their test usages (s_i, t_i), which may be subject to constraints; for example, each $s_i < s_0$ or $t_1 + t_2 + \cdots + t_I < T_0$. A test plan could be chosen to minimize a variance, for example, $V(\widehat{\nu})$, $V(\widehat{\lambda})$, $V(\widehat{\nu} s + \widehat{\lambda} t)$, or a prediction variance $V(Y'' - Y')$. Also, a plan could be chosen to obtain the locally most powerful LR test among the tests above. Such plans can be obtained in a straight-forward manner. Also, plans can be extended to situations where each system is observed for two (or more) usages.

Failure Modes. The blood analyzers had many failure modes. The engineers proposed that some modes may depend more on the number of specimens and others more on the hours on. This might also be modeled.

References

1. Doganaksoy, Necip (1994), Likelihood ratio confidence intervals in life-data analysis, In *Recent Advances in Life-Testing and Reliability* (Ed., N. Balakrishnan), CRC Press, Boca Raton, Florida.

2. Hahn, G.J. and Meeker, W.Q. (1991), *Statistical Intervals - A Guide for Practitioners*, John Wiley & Sons, New York, (800)879-4539.

3. Nelson, Wayne (1982), *Applied Life Data Analysis*, John Wiley & Sons, New York, (800)879-4539.

4. Nelson, Wayne (1988), Graphical analysis of system repair data, *Journal of Quality Technology*, **20**, 24-35.

4

Assessing the Effect of In-Service Inspections on the Reliability of Degrading Components

Christian G. Garrigoux and William Q. Meeker

ITESM, México
Iowa State University

ABSTRACT In this paper we combine a reliability model and an inspection model to provide a means of planning in-service inspections. Inspections are planned so as to keep the hazard function for the component below a specified threshold. Although our degradation model is more general, the example involves fatigue failure with a random effects model to describe unit-to-unit differences in degradation and reliability. The random effects in this example are initial crack size and two Paris law crack-growth parameters. Although the inspection model is again more general, our example uses, as is common in the airline industry, given periodic inspection opportunities and a probability of detection (POD) function. The POD is often given as a simple function of crack size.

Keywords: Fatigue failure, Hazard function, Nondestructive inspections

CONTENTS

4.1   Introduction

Machines or systems that should not be allowed to fail may have components subject to a degradation that could lead to failure. For such machines or systems there are two main strategies to handle the problem of controlling reliability.

1. One may decide on theoretical or practical grounds that the critical component will be retired from service at a fixed predetermined time. Of course, this horizon time, also called design life, is chosen so that the probability of an in-service failure is acceptably low. This policy, standard for critical components that cannot be inspected while in service, is still often in use. The method, however, may be inefficient in the sense that many of the stronger components in the product population are retired when they could stay safely in service for a much longer time.

2. Recent developments in methods of Nondestructive Evaluation (NDE) give the opportunity to inspect parts while in service to determine if they should be retired from service or if they could safely remain until the next inspection. This decision, of course, depends on some criteria as to the meaning of "safe."

It is therefore of interest to study the possible properties and advantages of such nondestructive inspections, e.g., how NDE can help to maintain an adequate level of reliability. Treatments of the component replacement problem, with different approaches, may be found in a number of books , e.g., [6] and [1], etc.

We consider in this paper the problem of assessing the useful life of components that may fail due to degradation. Inspections are to be performed on the components during service life. At each inspection, the component may be removed or remain in service, depending on the observed level of degradation. We provide formulas for reliability features like the time-to-failure probability distribution, the probability to be removed at the inspections, and the hazard function. Failure time density and probabilities of removal are necessary inputs in the further computation of the hazard function. We describe a computer program that evaluates these formulas and draws useful graphs for the reliability analysis of such degrading components. The computer code is written in the S language [3], with calls to Fortran subroutines for the heavy computations. Similar formulas have been developed in more specific contexts and with different approaches in a number of papers (e.g., [7], [5], [11], [9]).

A numerical example shows how inspections may be optimized so that the hazard function does not exceed a critical level with a minimum number of inspections.

4.2   Model

The key features of the model considered in this paper are described next.

1. The amount of degradation is measured by means of one level-of-degradation parameter.

2. This level of degradation parameter evolves through time by obeying a known degradation law.

3. The level of degradation eventually crosses a specified failure-defining value. Generally this level of degradation is not equivalent with a catastrophic failure but is, instead, a level of degradation at which the system should not operate either because of poor performance or reduced safety.

4. While in service, inspections are performed on the component in order to assess the current level of degradation.

5. A prespecified set of points in time is given where inspections may take place. These points are called "inspection opportunities."

6. If the observed level of degradation is considered as unacceptable at the inspection time, the component is removed from service.

7. If the component is not removed at an inspection it will eventually either cross the critical level or reach its design life or horizon time, where it is "retired" independent of its level of degradation.

8. Three parts of the degradation process are stochastic:

- The level-of-degradation parameter for a component entering in service (initial condition) is random from component to component with known probability distribution.

- The level-of-degradation parameter evolves over time following a known degradation law. This degradation law is described by a vector of degradation-law parameters considered to be random from unit-to-unit (but fixed over time) with a known joint probability distribution.

- Measurement errors occur so that the observed level-of-degradation parameter is a random variable with known probability distribution conditional on a given actual (unobservable) level of degradation.

4.3   Reliability Figures

4.3.1   Probability Distribution of the Service Life T

A unit's service life may end upon removal at the discrete points in time corresponding to the inspections/horizon or, continuously upon failure at any point between two successive inspections. We state here the expressions of the discrete and continuous part of the service life length (T) probability distribution $g_T(t)$ that are associated with these events, and we interpret them intuitively.

The continuous part of the service life length distribution

The formula for the failure density corresponding to the continuous part of the service life length distribution is

$$g_T(t) = \int_{D\underline{w}} g_{\underline{w}}[\phi_{\underline{w}}(t)] \left| \frac{d\phi_{\underline{w}}(t)}{dt} \right| \prod_{j=1}^{i} P_{NR}[\underline{w}, \phi_{\underline{w}}(t), t_j] \, dG(\underline{w})$$

$$\text{for } t_i < t < t_{i+1} \text{ and } i = 1, \ldots, N, \tag{4.3.1}$$

$$g_T(t) = \int_{D\underline{w}} g_{\underline{w}}[\phi_{\underline{w}}(t)] \left| \frac{d\phi_{\underline{w}}(t)}{dt} \right| dG(\underline{w}) \qquad \text{for } t_0 < t < t_1 \, ,$$

where $\phi_{\underline{w}}(t)$ stands for the initial level of degradation parameter (ϕ) that grows to failure size at time t when the vector of degradation law parameters has the values $(\underline{w})$, and N is the total number of inspections. We denote the starting time of the degradation process by t_0 and the horizon time by t_{N+1}. Here, $g_{\underline{w}}$ is the initial level of degradation probability density function conditional on $\underline{w}$, $D(\underline{w})$ denotes the domain of all values of $(\underline{w})$, and $P_{NR}[\underline{w}, \phi_{\underline{w}}(t), t_j]$ represents the probability not to remove at inspection time t_j a component with parameters $(\underline{w}, \phi)$ heading to fail at time t. Finally, G is the joint cumulative probability distribution of the degradation law parameters.

We provide detailed justification for equation (4.3.1) in the appendix. Intuitively, however, the equation can be explained in the following manner. For given values of the vector of degradation law parameters $\underline{w}$, consider the interval $d\phi_{\underline{w}}$ of initial level of degradation values such that failure will occur in the time interval dt, i.e., for the given $\underline{w}$, dt is the mapping of $d\phi_{\underline{w}}$ on the failure size line by means of the degradation law. Therefore, in the absence of inspections, and with $g[\phi_{\underline{w}}(t)]$ representing the probability density of $\phi_{\underline{w}}(t)$, the probability to fail in dt is $g[\phi_{\underline{w}}(t)] \left| \frac{d\phi_{\underline{w}}(t)}{dt} \right|$, by a probability transformation argument. Furthermore, $\phi_{\underline{w}}$ and $\underline{w}$ could be stochastically dependent so that this probability density is denoted $g_{\underline{w}}[\phi_{\underline{w}}(t)]$.

Next, in presence of inspections, we extend the previous intuitive consideration by saying that in order to fail in dt, for given $\underline{w}$, we need to start in the right $d\phi_{\underline{w}}$ and pass all inspections successfully. Now, the events of passing inspections, conditional on $(\underline{w}, \phi)$, are independent. Hence, the probability of passing all inspections on the way to failure in dt is simply the product of the probabilities of passing each of these inspections for given $\underline{w}$ and $\phi_{\underline{w}}$. The product in the integrand of (4.3.1) represents this conditional probability, so that the whole integrand stands for the probability of failing in a given dt for certain values of the $\underline{w}$. Finally, the total probability of failing in dt is obtained by integrating over the domain of the $\underline{w}$.

Discrete part of the service life length distribution

The discrete part of the service life length distribution gives $\Pr(T = t_i)$, the probability that removal occurs at the ith inspection (which takes place at time t_i):

$$\Pr\{T = t_i\} = \begin{cases} \displaystyle\int_{D(\underline{w})} \int_0^{\phi_{\underline{w}}(t_i)} \prod_{j=1}^{i-1} P_{NR}[\underline{w}, \phi, t_j] \, P_R[\underline{w}, \phi, t_i] \, g_{\underline{w}}[\phi] \, d\phi \, dG(\underline{w}) \\ \qquad \text{for } i = 2, \dots, N, \\[2ex] \displaystyle\int_{D(\underline{w})} \int_0^{\phi_{\underline{w}}(t_1)} P_R[\underline{w}, \phi, t_1] \, g_{\underline{w}}[\phi] d\phi \, dG(\underline{w}) \qquad \text{for } i = 1, \end{cases} \tag{4.3.2}$$

where $P_R[\underline{w}, \phi, t_i]$ is the complementary probability of $P_{NR}[\underline{w}, \phi, t_j]$ defined above. This last formula is also used for the probability of reaching the horizon, by setting an inspection at $t_{N+1} = $ Horizon time, with $P_R \equiv 1$.

We provide detailed justification for equation (4.3.2) in the appendix. Intuitively, however, the equation can be explained in the following manner. In order for the component to be removed at t_i, we consider all the combinations of the parameter values such that the item is heading to fail after t_i. This is done by integrating over the whole $\underline{w}$ domain and then, by integrating on values of ϕ such that $t > t_i$, conditional on $\underline{w}$. Furthermore, conditional on $(\underline{w}, \phi)$, the item reaches t_i and is removed there if it is not removed at any of the inspections prior to t_i and if it is finally observed in the removal region at t_i. Thus, the probability of this conditional event is the product of the corresponding non-removal probabilities P_{NR} with a final removal probability P_R. When removed at the first inspection ($i = 1$), the formula simplifies as indicated because there are no inspections prior to the removal.

4.3.2   Hazard function

The hazard function $\lambda(t)$ is calculated from

$$\lambda(t) = \frac{g_T(t)}{1 - F_T(t)},$$

where $F_T(t)$, the cumulative probability of having the component ending its service life before time t, may be obtained from

$$F_T(t) = \sum_{j=0}^{i} \Pr\{T = t_j\} + \int_0^t g_T(t)dt , \qquad (4.3.3)$$

with $t \in (t_i, t_{i+1})$, $i = 0, \ldots, N$, and where we define $Pr\{T = t_0\} = 0$.

Note: Given the partly discrete and partly continuous character of the service life length probability distribution, this hazard function has the special feature that the integral of the numerator g_T over all values of T up to t does not add up to the denominator evaluated at t. The probabilistic meaning of this hazard function remains however the standard one, i.e., $\lambda(t)dt$ may be interpreted as the probability of failing in the next infinitesimal interval dt, conditional on having survived up to t.

4.4   Data Transformations

Let $\theta = (w, \phi)$ represent the vector of all degradation parameters. Then, as mentioned earlier, θ is random from item to item. Now, the multivariate probability distribution of these parameters will not generally follow a standard multivariate distribution. We, therefore, need a mathematical transformation of the sample values of the parameters so that a standard model can be fitted conveniently to the transformed parameters. Thus, we will be able to describe the underlying multivariate distribution of the degradation parameters population by means of a few probability distribution parameters. We give next the steps in these data transformations that we have found to be adequate to describe the random effect distributions in our applications.

Transformation 1 (T_1): Let θ_i, $i = 1, \ldots, n$, denote the length-q vectors of degradation parameter estimates for a sample of n items used to estimate materials properties. These vectors are transformed by T_1 so that the transformed parameters vectors $\theta_i^{[1]}$, $i = 1, \ldots, n$, can be adequately modeled by a sample from a known multivariate distribution. Typically, T_1 will be a multivariate Box-Cox transformation, and the transformed parameter vectors are modeled as a sample from a multivariate normal distribution.

Model fitting: Once the n random-effect parameter vectors are transformed, a model is fitted to them. We use a parametric multivariate distribution to describe the distribution of the random-effect parameter vectors with a relatively small number of parameters. For example, if the multivariate Box-Cox transformation is used, the resulting multivariate distribution is normal and defined by its vector of means $\mu_{\theta^{[1]}}$ and covariance matrix $\Sigma_{\theta^{[1]}}$.

Transformation 2 (T_2): A discrete representation of the multivariate probability distribution for degradation parameters is necessary for the numerical integration needed to compute (4.3.1) and (4.3.2). For that purpose, it is convenient to use a multilinear transformation of the random parameters $\theta^{[1]}$ such that the distribution of the transformed parameters $\theta^{[2]}$ is still multivariate normal, but now uncorrelated. This allows the multivariate probability distribution to be covered by a rectangular grid of points with as few as possible of the grid cells covering zones where the parameter values have small probabilities.

Probability distribution discretization: Based on the multivariate normal distribution, probabilities, say $p_{j_1,\ldots,j_q}$ with $j_i = 1,\ldots,r$, are associated to a rectangular grid of values of the transformed parameters $(\theta^{[2]}_{1,j_1},\ldots,\theta^{[2]}_{q,j_q})$ with the range of each parameter, except ϕ, discretized, for convenience, with the same number r of intervals.

Original parameter probabilities: Finally, we return to the original parameter space $\underline{\theta}$ by applying the inverse transformations of T_2 and T_1 to the grid of points $(\theta^{[2]}_{1,j_1},\ldots,\theta^{[2]}_{q,j_q})$. We now have a discretized representation of the multivariate probability distribution of the non-transformed degradation parameters with the probabilities $p_{j_1,\ldots,j_q}$ associated to the vectors $(\theta_{1,j_1},\ldots,\theta_{q,j_q})$. We use this representation to compute the integrals in (4.3.1) and (4.3.2).

4.5   Example

In this section, we describe a specific example where we have applied our model to the assessment of a set of in-service inspection plans. We present the inputs and outputs to the model and the corresponding computer program. We also comment on some of the interesting aspects of the graphical outputs from our model.

4.5.1   Description of the problem

Our example describes the fatigue crack growth that is expected to occur at a high-stress point of a rod-spring. Although the system in which the rod-spring was to be installed is proprietary, we can describe the application of this component in some detail. The level-of-degradation parameter is the size of a crack located within the rod-spring to be manufactured out of powdered metal 304 stainless steel. The powdered metal 304 stainless steel from which the rod-springs are machined will, at the time of manufacture, have, at a microscopic level, a characteristic grain structure. Under cyclic loading, a detectable crack will initiate out of this grain structure at a point of high-stress [a stress map for a part can be computed by using either finite element methods (FEM) or boundary element methods (BEM)]. We follow conventional material-science notation and let a denote crack size (and thus $a = \phi$). This crack will grow, with each use-cycle, until cycling stops or until the component fractures. A classical model in materials science, used to describe fatigue crack growth, is the "Paris law," a differential equation relating crack length (a) and number of loading cycles (t) as follows

$$\frac{da}{dt} = C\left[\Delta K(a)\right]^m . \tag{4.5.1}$$

The degradation law parameters $\underline{w} = (C, m)$ are materials properties and could be estimated, independently of geometry and applied stress levels, from any number of standard fatigue crack-growth tests (e.g., see [10]). The function $K(a)$ is known as the stress intensity function and is used to describe the other non-materials factors that would affect the crack-growth rate. These include part geometry and applied stress. For a buried, isolated crack with circular shape, subjected to a pure tensile stress σ, this relation is given by

$$K(a) = \frac{2}{\pi}\sigma\sqrt{a\pi}.$$

In other configurations, the $2/\pi$ factor in this expression may be replaced by an appropriate configuration factor; see [2]. More specifically, the function $\Delta K(a)$ in (4.5.1) is the difference between maximum and minimum of the stress intensity factor during a cycle. We take the minimum stress as being 0 so that $\Delta\sigma = \sigma$ and the Paris law may be written

$$\frac{da}{dt} = C\left(\frac{2}{\pi}\sigma\sqrt{a\,\pi}\right)^m. \tag{4.5.2}$$

For more details see, for example, [2].

4.5.2 Inputs

Necessary inputs for the calculations are:

1. The initial crack size distribution $f_{A_0}(a_0)$,
2. The crack growth law (Paris law) relating the crack size a to the number of cycles t, a measure of the elapsed time, and to the initial crack size a_0,
3. The probability of detection for a crack of size a,
4. A vector of potential inspection times,
5. The failure crack size a_f.

Note that, for generality, the model in this paper considers a probability of removal model based on the theoretical possibility to actually measure the crack (using a measurement error model) at inspection and to compare this measurement with a critical remove/not-remove decision threshold. In our example, however, we follow the most common current practice in nondestructive evaluation of critical components: when a crack is detected, the unit is removed from service. In this case, the input to our model is simply a function that gives the probability of detection (POD) as a function of crack size a.

4.5.3 Choices for Inputs

1. The crack growth law was modeled with the unidimensional Paris law in (4.5.1). This particular form is appropriate for the axially-symmetric geometry of the rod-spring.
2. The initial conditions of the Paris law are described by the initial crack size probability density f_{A_0}, modeled in this example as a discretized normal with its .001 quantile at .01 mm and .999 quantile at .05 mm. This distribution was determined on the basis of experimental micro-crack data on powdered-metal 304 stainless steel conducted by E. Chen of Northwestern University and expert judgment by materials scientists. Because of the data available to us we had to assume that the initial crack size is independent of the Paris law parameters. If information on a more general multivariate distribution were available, it could be used with our model. For purposes of the numerical analysis, the initial crack size distribution was discretized into 149 intervals of equal sizes.
3. The data shown in Figures 4.1 and 4.2 give observed values of m and C obtained from fatigue tests on powdered-metal 304 stainless steel. These data were provided to us by S. M. McGuire of Northwestern University. The need for the log transformation was suggested by Figures 4.1 and 4.2 and maximum likelihood estimation of the Box-Cox transformation parameters, as described in [8]. Because of the large number of measurements taken on each of the 16 cracks and to keep the computations simple,

we used the sample mean and sample covariance matrix of the transformed data to estimate the parameters of the bivariate normal distribution for m and $\log(C)$ giving a mean vector $\underline{\mu}_{\underline{\theta}[1]} = (1.68, 0.189)$ and a variance-covariance matrix

$$\underline{\Sigma}_{\underline{\theta}[1]} = \begin{bmatrix} 0.775 & -2.53 \\ -2.53 & 8.55 \end{bmatrix}$$

In [8], the authors suggest a more elaborate method of estimating $\underline{\Sigma}_{\underline{\theta}[1]}$ that could be used to correct for measurement error.

In the numerical analysis, the bivariate normal distribution of $\underline{\theta}^{[2]}$ (uncorrelated due to the T_2 transformation described in Section 4.4) was modeled with a 5×5 grid of points. Numerical experiments with a finer grid of points showed that this level of discretization was adequate to give approximately 2 significant figures of accuracy in the quadrature approximation. More accuracy could be obtained at the cost of more computer time.

4. Typically, in practice, the value of applied stress [σ in (4.5.1)] on a component is chosen to maximize the performance of the system, subject to constraints on length of life. In other situations σ is set at the minimum value giving the desired performance for the system and then components are inspected or replaced frequently enough to maintain the needed system reliability. In the current application it was determined that $\sigma = 2.2 \times 10^{-8}$ would provide the needed level of performance for the application involving the rod-spring.

5. The probability of detection for a particular ultrasonic inspection scheme was modeled as a truncated type I smallest extreme-value distribution:

$$P_R(\underline{w}, \phi, t) = \text{POD}(a) = 1 - \exp\{-\exp[(a - \mu)/\sigma]\}, \, a > 0$$
$$= .01 \qquad\qquad\qquad\qquad\qquad\qquad a = 0$$

where $\mu = 0.11261$ and $\sigma = 0.02448$. This POD function specifies that the probability of a false alarm (detection of a crack of zero length) is .01 and the probability of detecting a crack of size .15 mm is .99. This POD function was based on modeling work being done at the Iowa State University Center for Nondestructive Evaluation (e.g. [4]).

6. A time range of .5 million of cycles was chosen with 20 equally-spaced inspection opportunities, the first one being set at time zero. At "inspection opportunity 20," an inspection is set with a probability of detecting a crack equal to 1 so that it acts as the horizon time (i.e., the unit will be retired even if a crack has not been detected).

7. The "failure size" of the crack was fixed at $a_f = .20$ mm. This is a crack size that was determined to be safely below that which would cause a catastrophic failure of the system in which the rod-spring was installed. Although the system containing the rod-spring would be likely to continue to operate with a crack of this size, we assume that the operators of the system would rather replace the rod-spring if it has a crack that is larger than $a_f = .20$ mm.

4.5.4   Presentation of Outputs

Given the previous inputs, $g_T(t)$ and $\lambda(t)$ were calculated and plotted for several schemes of inspections. Note that the discretization of $f_{A_0}(a_0)$ implies a discretization of $g_T(t)$ and $\lambda(t)$. In fact the plots of $\lambda(t)$ are in units of the probability of failing in the next cycle. From the figures, however, the intervals are small enough to make these distributions look smooth.

4.5.5 Comments on Graphical Output

- In Figure 4.3, we have, for the sake of comparison, computed the basic model output, assuming no inspections, with the variability in the Paris law parameters assumed to be equal to 0. The "initial crack size distribution" $f_{A_0}(a_0)$ is plotted along the vertical axis at $t = 0$. The time-to-crossing density $g_T(t)$ is the curve above the horizontal line drawn at $a_f = .20$ mm. The discretization of $f_{A_0}(a_0)$ shows up as the cracks grow and their different slopes separate them. The first A_0 value generated from the discretized $f_{A_0}(a_0)$ distribution leads to a failure time shortly before the seventeenth inspection opportunity (i.e., IO_{17}).
  In this no-inspection case, the "probability density" $g_T(t)$ corresponds to the nonlinear transformation of the initial crack size distribution random variable A_0 through the solution of the Paris Law differential equation (except for the discretization and the truncation at the horizon).

- Figure 4.4 shows the additional spread induced into the time-of-crossing distribution when the variability of C and m is taken into account. In the figure, we show the cracks growing only for a central combination of C and m (otherwise the curves would cross and the plot would be less informative). We will use this more complete model for the rest of our discussion. Again, with no inspection, this is equivalent to using the solution of (4.5.1) to map the joint distribution of the three random variables (A_0, C and m) into the time-to-cross random variable T.

- Figure 4.5 shows the hazard function $\lambda(t)$ with no inspections. We see that $\lambda(t)$ crosses the critical hazard value (1.2×10^{-7}) just after IO_7.

- Based on a criterion like, say, minimizing the number of inspections, and in order to keep the hazard function below 1.2×10^{-7}, we add one inspection at IO_7. The results are shown in Figures 4.6 and 4.7. The effect is to remove from service at this point any rod springs for which a crack is detected. The derivative (or density) of the POD function is shown on the same scale as the initial crack size distribution. As seen in Figures 4.6 and 4.7, the time-to-cross density becomes truncated at IO_7 and consequently $\lambda(t)$ becomes much smaller just after IO_7. We see, however, that $\lambda(t)$ grows again rapidly and will cross the critical hazard value just after IO_9.

- Trying to keep the hazard function from going too far above 1.2×10^{-7}, Figures 4.8 and 4.9 show the effect of having inspections at IO_7 and IO_9.

- Finally, Figure 4.10 shows the effect of having inspections at IO_7 and IO_9, IO_{11}, and IO_{13}. We see that it is possible to keep $\lambda(t)$ below 1.2×10^{-7}, but that after each succeeding inspection, $\lambda(t)$ grows quickly, in effect, catching up to the point where it would have been if there had been no earlier inspections.

It is clear, that for a given set of parameters, $\lambda(t)$ could eventually cross its critical value before the next inspection opportunity. Theoretically, this could be avoided by changing the parameters of the initial crack size distribution, or the crack growth law and its parameters (e.g., by changing materials). Other alternatives include changing probability of detection function (by changing the inspection method) or, by decreasing the spacing between inspection opportunities.

4.6   Summary and Discussion of Possible Extensions

In this paper we have described and illustrated a reliability/inspection model for critical components in systems that need to be maintained such that the probability of a failure is acceptably small.

The work that we have presented has a number of interesting and important extensions. For example,

- We have shown how to evaluate the effect of different inspections schemes on a component's hazard function. Operationally we might want to choose inspections according to some specific criterion such as minimizing the maximum of the hazard function or (given costs of repair, replacement, and inspection) minimizing total cost of component operation.

- In our model we have allowed for the use of inspection data at an inspection opportunity to make the decision about when to replace the component or not. This is the way in which such decisions are usually made in real applications of nondestructive evaluation. An important extension of this method would be to combine all of the previous inspections to make the decision at the current inspection time.

- We have assumed that a component would be subject to a prespecified inspection scheme that would be determined by analytical and economic analyses at the time that the component is put into service. If continuous inspection information is available, it might be useful to allow a "dynamic" form of inspection policy in which information available from current and past inspections would be used not only to decide whether to continue with the component in operation, but also to decide on the time of the next inspection.

- Answers from models such as ours will, of course, depend on the inputs. In practice one should conduct sensitivity analyses to assess the effect that perturbations to the various inputs will have on the outputs. For example, we have used parameters estimated from data as inputs to our model. As such, the outputs represent only point-predictions of actual system output. In [8], the authors use bootstrap methods to estimate the effect of sampling error and compute confidence intervals for points on a time-to-failure distributions, based on degradation data. These same ideas could be applied to construct confidence intervals for the outputs from our model.

Acknowledgements

We would like to thank Otto Buck, Brian Moran, and Les Schmerr for describing to us some of the ideas behind fatigue failure and practical nondestructive inspection processes and other aspects of the example in Section 4.5. Pradipta Sarkar made some helpful suggestions on an earlier version of this paper.

Appendix: Derivation of the Probability Distribution of the Service Life T

A.1 Derivation of the continuous part of the service life length distribution

In this section we derive the expression in (4.3.1). The probability of failing in (t', t) may be expressed as

$$\Pr\{T \in (t', t)\} = \int_{t'}^{t} g_T(u)\, du.$$

To derive the expression for $g_T(u)$, we analyze the event $\{T \in (t', t)\}$, with $t_i < t' < t < t_{i+1}$, where t_i and t_{i+1} are any two successive inspection times. The event

$$\{T \in (t', t)\}$$

is equivalent to the intersection of the following two events:

Event E_1: The component starts its service life with a level of degradation and degradation law parameters leading to failure in the interval (t', t). That is, the elements of the vector of initial level of degradation and degradation growth parameters $\underline{\theta}$ have values such that the failure time $t(\underline{\theta}) \in (t', t)$. We denote this event by

$$E_1 = \{\underline{\Theta} = \underline{\theta} \ni t(\underline{\theta}) \in (t', t)\}. \tag{A.1.1}$$

Event E_2: The component successfully passes all inspections up to the ith inspection time inclusive. This may be expressed as

$$E_2 = \bigcap_{j=1}^{i} \{\underline{X}_j = \underline{x}_j \in \text{NRR}_j\}, \tag{A.1.2}$$

where NRR stands for the non-removal region so that $\{\underline{X}_j = \underline{x}_j \in \text{NRR}_j\}$ corresponds to these values of the random variables $\underline{X}_j$, (the vector of all data from inspections $1, \ldots, j$), such that inspection j is successfully passed. This is so because the non removal at inspection j could depend on the whole set of current and past observations $\underline{x}_j = (\underline{x}_1 \cdots \underline{x}_j)$.

Note that for the trivial case where failure occurs before the first inspection, we simply have

$$\{T \in (t', t)\} = E_1.$$

Now, using (A.1.1) and (A.1.2), $\Pr\{T \in (t', t)\}$ can be expressed analytically in terms of the CDF $G(\underline{\theta}, \underline{x}_i)$, when $i = 1, \ldots, N$, and $G(\underline{\theta})$, when $i = 0$, where t_0 denotes the starting time of the degradation process. Thus, we write

$$\Pr\{T \in (t', t)\} = \int_{\underline{\theta} \ni t(\underline{\theta}) \in (t', t)} \int_{\underline{x}_1 \in \text{NRR}_1} \cdots \int_{\underline{x}_i \in \text{NRR}_i} dG(\underline{\theta}, \underline{x}_i)$$

for $i = 1, \ldots, N$, and

$$\Pr\{T \in (t', t)\} = \int_{\underline{\theta} \ni t(\underline{\theta}) \in (t', t)} dG(\underline{\theta}) \qquad \text{for } i = 0,$$

where the domains of integrations come directly from the definitions of the events E_1 and E_2. Next, one of the parameters in $\underline{\theta}$, say ϕ, which describes the initial level of degradation, is isolated. Also, observations $\underline{X}_j$ are independent from each other, conditional on $\underline{\theta}$. Hence,

$$\Pr\{T \in (t', t)\} = \int_{D\underline{w}} \int_{\phi\underline{w} \ni t(\phi) \in (t', t)} \left\{ \prod_{j=1}^{i} \int_{\underline{x}_j \ni \mathrm{NRR}_j} dG(\underline{x}_j | \underline{\theta}) \right\} dG_{\underline{w}}(\phi) \, dG(\underline{w})$$

for $i = 1, \ldots, N$, and

$$\Pr\{T \in (t', t)\} = \int_{D\underline{w}} \int_{\phi\underline{w} \ni t(\phi) \in (t', t)} dG_{\underline{w}}(\phi) \, dG(\underline{w}) \qquad \text{for } i = 0,$$

where the G functions represent the cumulative distribution functions of their arguments. In particular, $dG_{\underline{w}}(\phi)$ is subscripted because the distribution of ϕ may be stochastically conditional on $\underline{w}$. In turn, the subscript $\underline{w}$ for ϕ indicates that the initial level of degradation values that lead to failure at a certain time t depend deterministically on the degradation trajectories. Now, since the integral in the product represents the probability of not being removed at inspection j conditional on $(\underline{w}, \phi)$ values, we may use for this integral the more meaningful notation $P_{NR}[\underline{w}, \phi, t_j]$, where NR stands for non-removal, and where the arguments in brackets indicate when (inspection time t_j), and on which actual level of degradation $(\underline{w}, \phi)$, measurements are taken. Thus, we have

$$\Pr\{T \in (t', t)\} = \int_{D\underline{w}} \int_{\phi\underline{w} \ni t(\phi) \in (t', t)} \left\{ \prod_{j=1}^{i} P_{NR}[\underline{w}, \phi, t_j] \right\} g_{\underline{w}}(\phi) \, d\phi \, dG(\underline{w})$$

for $i = 1, \ldots, N$, and

$$\Pr\{T \in (t', t)\} = \int_{D\underline{w}} \int_{\phi\underline{w} \ni t(\phi) \in (t', t)} g_{\underline{w}}(\phi) d\phi \, dG(\underline{w}) \qquad \text{for } i = 0.$$

Besides, if ϕ is assumed to have a (most physically plausible) monotonic decreasing dependence on time conditional on $\underline{w}$, i.e., failure time increases as the initial level of degradation parameter decreases, we can express ϕ as a function of t and parameterize in t instead of ϕ. This gives

$$\Pr\{T \in (t', t)\} = \int_{D\underline{w}} \int_{t'}^{t} \left\{ \prod_{j=1}^{i} P_{NR}[\underline{w}, \phi_{\underline{w}}(u), t_j] \right\} g_{\underline{w}}[\phi_{\underline{w}}(u)] \left| \frac{d\phi_{\underline{w}}(u)}{du} \right| du \, dG(\underline{w})$$

for $i = 1, \ldots, N$, and

$$\Pr\{T \in (t', \ t)\} = \int\limits_{D\underline{w}} \int\limits_{t'}^{t} g_{\underline{w}}[\phi_{\underline{w}}(u)] \left| \frac{d\phi_{\underline{w}}(u)}{du} \right| du \, dG(\underline{w}) \qquad \text{for } i = 0.$$

This reparameterization enables differentiation of $\Pr\{T \in (t', \ t)\}$ with respect to t in order to obtain $g_T(t)$. We get

$$g_T(t) = \int\limits_{D\underline{w}} g_{\underline{w}}[\phi_{\underline{w}}(t)] \left| \frac{d\phi_{\underline{w}}(t)}{dt} \right| \prod_{j=1}^{i} P_{NR}[\underline{w}, \ \phi_{\underline{w}}(t), \ t_j] \, dG(\underline{w})$$

$$\text{for } t_i < t < t_{i+1} \ \text{ and } \ i = 1, \ \dots, N,$$

$$g_T(t) = \int\limits_{D\underline{w}} g_{\underline{w}}[\phi_{\underline{w}}(t)] \left| \frac{d\phi_{\underline{w}}(t)}{dt} \right| dG(\underline{w}) \qquad \text{for } t_0 < t < t_1.$$

A.2 Derivation of the discrete part of the service life length distribution

We now obtain an expression for $\Pr\{T = t_i\}$, the probability that a removal occurs at inspection opportunity t_i, as given in (4.3.2). As in the previous section, the event $\{T = t_i\}$ may be viewed as the intersection of three simpler events that we proceed to describe.

Event E_1': the component starts its service life with degradation parameters $\underline{\theta}$ such that failure would occur after t_i. We have

$$E_1' = \{\underline{\Theta} = \underline{\theta} \ni t(\underline{\theta}) > t_i\}, \tag{A.2.1}$$

with notations similar to these in the previous section.

Event E_2': the component passes successfully all inspections prior to t_i. We have

$$E_2' = \bigcap_{j=1}^{i-1} \{\underline{X_j} = \underline{x_j} \in \mathrm{NRR}_j\}. \tag{A.2.2}$$

where NRR_j is the non-removal region at inspection j.

Event E_3': the component is removed at t_i.

$$E_3' = \{\underline{X_i} = \underline{x_i} \in \mathrm{RR}_i\}, \tag{A.2.3}$$

where RR_i is the removal region at inspection i.

Now, using (A.2.1), (A.2.2), (A.2.3), $\Pr\{T = t_i\}$ can be expressed analytically in terms of the CDF $G(\underline{\theta}, \underline{x_i})$. We write

$$\Pr\{T = t_i\} =$$

$$\begin{cases} \displaystyle\int\limits_{\underline{\theta} \ni t(\underline{\theta}) > t_i} \int\limits_{\underline{x_1} \in \mathrm{NRR}_1} \cdots \int\limits_{\underline{x_{i-1}} \in \mathrm{NRR}_{i-1}} \int\limits_{\underline{x_i} \in \mathrm{RR}_i} dG(\underline{\theta}, \underline{x_i}), & \text{for } i = 2, \dots, N, \\[4ex] \displaystyle\int\limits_{\underline{\theta} \ni t(\underline{\theta}) > t_i} \int\limits_{\underline{x_1} \in \mathrm{RR}_1} dG(\underline{\theta}, \underline{x_1}), & \text{for } i = 1, \end{cases}$$

where the domains of integrations come directly from the definitions of the events E_1', E_2' and E_3'. Next, and as in the previous section, we isolate the initial level of degradation parameter ϕ in $\underline{\theta} = (\underline{w}, \phi)$, and take advantage of the independence of the observations $\underline{x}_j$ conditional on $(\underline{w}, \phi)$. Thus,

$$\Pr\{T = t_i\} =$$

$$
\begin{cases}
\displaystyle\int_{D\underline{w}} \int_{\phi_{\underline{w}} \ni t(\phi) > t_i} \prod_{j=1}^{i-1} \int_{\underline{x}_j \in \mathrm{NRR}_j} dG[\underline{x}_j | (\underline{w}, \phi)] \int_{\underline{x}_i \in \mathrm{RR}_i} dG[\underline{x}_i | (\underline{w}, \phi)] \, dG_{\underline{w}}(\phi) \, dG(\underline{w}), \\
\qquad\qquad \text{for } i = 2, \dots, N, \\[2ex]
\displaystyle\int_{D\underline{w}} \int_{\phi_{\underline{w}} \ni t(\phi) > t_i} \int_{\underline{x}_i \in \mathrm{RR}_i} dG[\underline{x}_i | (\underline{w}, \phi)] \, dG_{\underline{w}}(\phi) \, dG(\underline{w}), \qquad \text{for } i = 1.
\end{cases}
$$

Finally, we note again that the integrals over non-removal or removal regions calculate the probabilities of falling in these regions and we change the notations accordingly. Also, we recall the assumption of monotonic decreasing dependence of the initial level of degradation ϕ on the failure time t and put the minimum possible value of the initial level of degradation at 0. Thus, with the same notation as in the first part of this appendix, we get

$$
\Pr\{T = t_i\} =
\begin{cases}
\displaystyle\int_{D(\underline{w})} \int_0^{\phi_{\underline{w}}(t_i)} \prod_{j=1}^{i-1} P_{NR}[\underline{w}, \phi, t_j] \, P_R[\underline{w}, \phi, t_i] \, g_{\underline{w}}[\phi] \, d\phi \, dG(\underline{w}), \\
\qquad\qquad \text{for } i = 2, \dots, N, \\[2ex]
\displaystyle\int_{D(\underline{w})} \int_0^{\phi_{\underline{w}}(t_1)} P_R[\underline{w}, \phi, t_1] \, g_{\underline{w}}[\phi] \, d\phi \, dG(\underline{w}) \qquad \text{for } i = 1.
\end{cases}
$$

References

1. Anders, G.J. (1990), *Probability Concepts in Electric Power Systems*, John Wiley & Sons, New York.

2. Anderson, T.L. (1991), *Fracture Mechanics*, CRC Press, Boston, MA.

3. Becker, R.A., Chambers, J.M., and Wilks, A.R. (1988), *The New S Language*, Wadsworth & Brooks/Cole, Pacific Grove, CA.

4. Gray, T.A. and Thompson, R.B. (1986), Use of models to predict ultrasonic reliability, In *Review of Progress in Quantitative Nondestructive Evaluation*, **5**, (Eds., D.O. Thompson and D.E. Chimenti), Plenum Press, New York.

5. Harris, D.O. and Lim, E.Y. (1983), Applications of a Probabilistic Fracture Mechanics Model to the Influence of In-Service Inspection on Structural Reliability, In *Probabilistic Fracture Mechanics and Fatigue Methods: Applications for Structural Design and Maintenance*, ASTM STP 798, (Eds., J.M. Bloom and J.C. Ekvall), 19-41, American Society for Testing and Materials, Philadelphia.

6. Jorgenson, D.W., McCall, J.J., and Radner, R. (1967), *Optimal Replacement Policy*, Rand McNally & Company, Chicago.

7. Kitagawa, H. and Hisada, T. (1977), Reliability analysis of structures under periodic non-destructive inspection, *Pressure Vessel Technology*, Pt 1, ASME, New York.

8. Lu, C.J. and Meeker, W.Q. (1993), Using degradation measures to estimate a time-to-failure distribution, *Technometrics*, **35**, 161-174.

9. Provan, J.W. (Ed.) (1987), *Probabilistic Fracture Mechanics and Reliability*, Martinus Nijhoff Publishers, Hingham, MA.

10. Trantina, G.G. and Johnson, C.A. (1983), Probabilistic Defect Size Analysis Using Fatigue and Cyclic Crack Growth Rate Data, In *Probabilistic Fracture Mechanics and Fatigue Methods: Applications for Structural Design and Maintenance*, ASTM STP 798, (Eds., J.M. Bloom and J.C. Ekvall), 67-78, American Society for Testing and Materials, Philadelphia.

11. Yang, J.N. and Chen, S. (1985), Fatigue reliability of gas turbine engine components under scheduled inspection maintenance, *Journal of Aircraft*, **22**, 415-422.

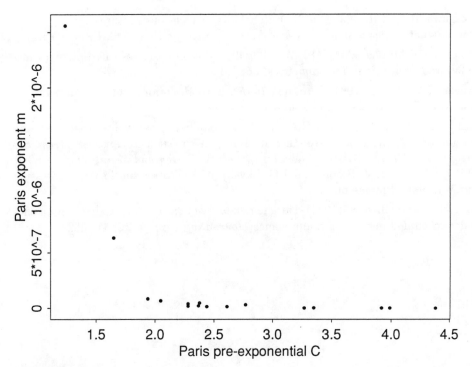

FIGURE 4.1
Scatter plot of Paris law parameters m and C

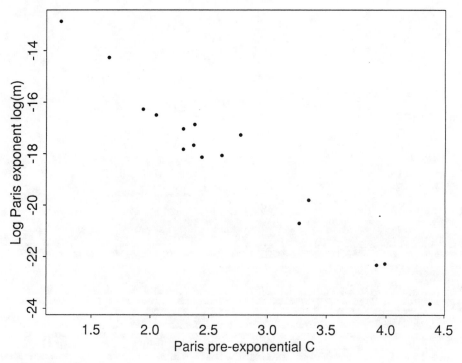

FIGURE 4.2
Scatter plot of Paris law parameters m and $\log(C)$

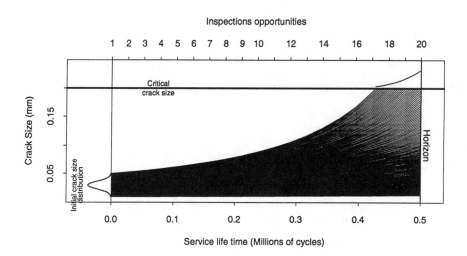

FIGURE 4.3
Reliability model output with no variability in Paris law parameters and no inspections

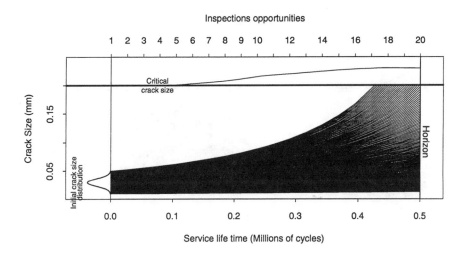

FIGURE 4.4
Reliability model output with variability in Paris law parameters and no inspections

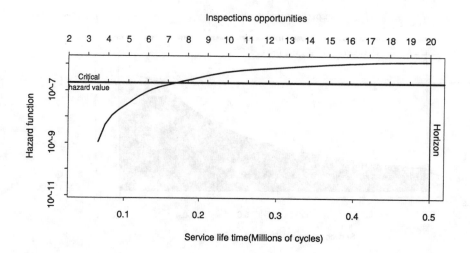

FIGURE 4.5
Hazard function with variability in Paris law parameters and no inspections

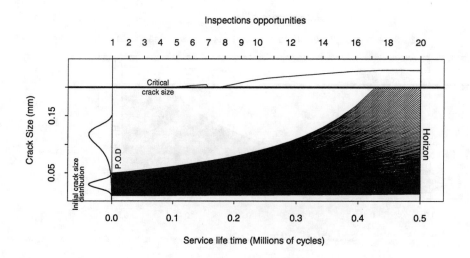

FIGURE 4.6
Reliability model output with variability in Paris law parameters and an inspection at IO_7

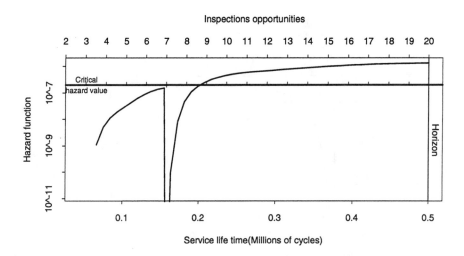

FIGURE 4.7
Hazard function with variability in Paris law parameters and an inspection at IO_7

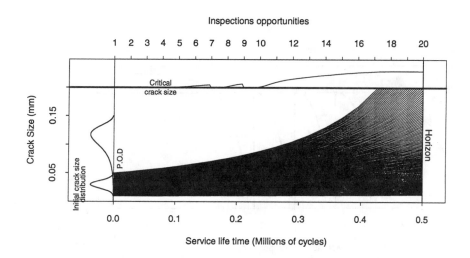

FIGURE 4.8
Reliability model output with variability in Paris law parameters and inspections at IO_7 and IO_9

C.G. Garrigoux and W.Q. Meeker

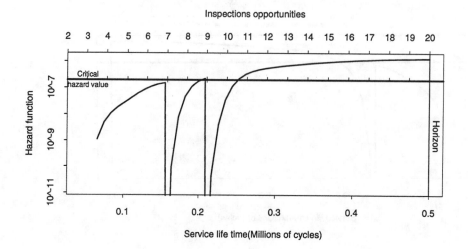

FIGURE 4.9
Hazard function with variability in Paris law parameters and inspections at IO_7 and IO_9

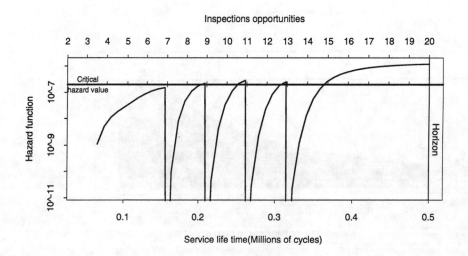

FIGURE 4.10
Hazard function with variability in Paris law parameters and inspections at IO_7, IO_9, IO_{11}, and IO_{13}

5

Models and Analyses for the Reliability of A Single Repairable System

Max Engelhardt

Idaho National Engineering Laboratory and EG&G Idaho, Inc.

ABSTRACT This article reviews methods of estimation and inference which are available for analyzing data from a single repairable system. Most of the results are based on the assumption that the cumulative number of failures is distributed according to a special type of Poisson process called a power law process. Some graphical and nonparametric methods are also discussed as well as several models which are related to power law processes. Strengths and weaknesses of the various models are discussed.

Keywords: Estimation, Inference, Left censoring, Left truncation, Power law process

CONTENTS

5.1 Introduction

Much of the theory of reliability deals with nonrepairable systems or devices, and it emphasizes the study of lifetime models. A *nonrepairable system* can fail only once, and a lifetime model such as the Weibull distribution models the distribution of the times at which individuals from such a population of systems fail. On the other hand, a *repairable system* can be repaired and placed back in service. Thus, a model for repairable systems must allow for a whole sequence of repeated failures, and it should be capable of reflecting changes in the reliability of the system as it ages.

A repairable system is often modeled by means of a counting process. Let $N(t)$ represent the number of failures of a repairable system in the time interval $[0, t]$. It follows that $N(t)$ is nonnegative and integer-valued, and if $t > s$, the difference $N(t) - N(s)$ is the number of failures in the interval $(s, t]$. Other familiar characterizations can be given in terms of successive failure times $T_1, T_2, \ldots, T_n, \ldots$, and also in terms of interfailure times, $X_n = T_n - T_{n-1}$.

An often used approach to the analysis of data from repairable systems involves parametric assumptions which reflect important aspects of the system being modeled. For example, if a system is repaired to "like new" condition following each failure, then it might be reasonable to assume that the times between failures are independent and identically distributed. This would correspond to assuming that the system is modeled by a renewal process.

A different kind of situation, commonly encountered with repairable systems, involves changes in the reliability of the system as it ages. For example, when a complex system is in the development stage, early prototypes will often contain design flaws. During the early testing phase, design changes are made to correct such problems. If the development program is succeeding, one would expect a tendency toward longer times between failures. When this occurs, such systems are said to be undergoing *reliability growth*. On the other hand, if a system is deteriorating and it is given only minimal repairs when it fails, one would expect a tendency toward shorter times between failures as the system ages.

Poisson processes are often used in the modeling of repairable systems. A common approach in characterizing such a process is to specify a set of axioms or properties which describe the probabilistic behavior of $N(t)$, such as the property of *independent increments*; that is, the number of failures in disjoint time intervals are stochastically independent. The independent increments assumption greatly simplifies the derivation of results required for statistical analyses, but it is not always reasonable in practice. For a discussion of a typical set of axioms which define a Poisson process, see e.g. [60]. A simple and rather intuitive characterization, discussed by [23], involves the mean function $M(t) = E[N(t)]$. A counting process is called *regular* if $M(t)$ is continuous. It can be shown that a regular counting process is a Poisson process if and only if it has independent increments and no simultaneous failures. Of course, for a Poisson process with mean function $M(t)$ the number of occurrences in an interval $(s, t]$ is Poisson distributed, with mean $M(t) - M(s)$.

If $M(t)$ is also differentiable, then

$$\nu(t) = \frac{d}{dt} M(t)$$

is called the *rate of occurrence of failures* (ROCOF), also sometimes called the *recurrence rate*. For regular processes, there is another interpretation of $\nu(t)$, namely

$$\nu(t) = \lim_{\Delta t \to 0} \frac{P[N(t + \Delta t) - N(t) \geq 1]}{\Delta t} .$$

This limit is called the *intensity function*, and for more general counting processes the intensity function and the derivative of the mean function are not necessarily the same. However, for regular Poisson processes they are the same.

Some asymptotic properties of $\nu(t)$ are derived by [13] and [36].

The best known case of a Poisson process is a *homogeneous Poisson process* (HPP), in which case the intensity function is constant, say $\nu(t) = \lambda$. However, much of the recent work on repairable systems involves Poisson processes with nonconstant intensity functions. Such a process is usually called a *nonhomogeneous Poisson process* (NHPP). Of course, an NHPP would be capable of modeling systems which are undergoing either reliability growth or deterioration. In particular, if the intensity function $\nu(t)$ is decreasing, the successive times between failures tend to get longer, and if it is increasing they tend to get shorter.

Much of the recent work on modeling and analysis of repairable systems is based on the assumption of a special type of NHPP known as a *power law process*. The mean value function of a power law process has the form

$$M(t) = \lambda t^{\beta} .$$

The parameterization $M(t) = (t/\theta)^\beta$, with θ a scale parameter, is also common. The parameter β is often called a *shape* parameter, or *power* parameter. If $\beta = 1$ the process is an HPP, which is often used to model renewal. Otherwise, a power law process provides a model for a system whose reliability changes as it ages. If $\beta > 1$, $M(t)$, is increasing, and the model is appropriate for deteriorating systems since times between failures tend to get shorter with time. On the other hand, when $\beta < 1$, $M(t)$ is decreasing, and the times between failures tend to get longer with time, which might be expected with reliability growth. In other words, the system is under development, and improvements are made when the system fails in order to improve reliability. Although, this model is sometimes applied to study reliability trends in a single system, it can also be used to infer properties of a population of identical systems. Even when only one such system exists, it is still possible to consider it as representative of a conceptual population of identical systems. This model is also known in the literature as a *Weibull process*, although this terminology has led to misconceptions about the model. In [8], the authors state that ". . . the prevalent terminology could scarcely be more misleading if it had been designed to mislead – specifically, it has engendered such deep-seated misconceptions that it is extraordinarily difficult to supplant it with improved nomenclature." The name Weibull process derives primarily from the resemblance of the intensity function of the process to the hazard function of a Weibull distribution. In particular the intensity function has the form

$$\nu(t) = \lambda \beta t^{\beta - 1} .$$

Other names for this model which have appeared in the literature of repairable systems are the Weibull restoration process by [19], the nonhomogeneous Poisson process with Weibull intensity function by [28], the Rasch-Weibull process by [48] and the power-intensity process by [36]. Other details on the history of this model are given in a review paper by [56].

It is necessary to cease taking further observations at some point. In general, the process is said to be *right time truncated* if it is observed for a fixed length of time, and it is said to be *right failure truncated* if it is observed until a fixed number of failures have occurred. With right failure truncation, the data consists simply of the set of observed failure times, whereas with right time truncation the number of occurrences in the interval of observation is also a random part of the data set. That is, with right failure truncation the system is observed until a fixed number, say n, failures have occurred, and if all the failure times are observed, then the data consists simply of the ordered times. On the other hand, with right time truncation the process is observed for a fixed length of time, say τ, in which case the number of failures, $N(\tau) = n$. If no failures are observed in the interval $[0, \tau]$, then the data set includes only the value $n = 0$. However, if some failures are observed, then the data set consists of the ordered times, as well as n.

Until fairly recently, most of the work on the power law process has been based on the assumption that all of the occurrence times are available from the start of the experiment (time zero) until the experiment is terminated by right truncation. However, several recent papers have dealt with analyses of power law processes with less than complete data. These data can be categorized as follows: (a) *Left truncation* in which failure-times are missing from the left side of the observation interval and the number of missing observations is not known, (b) *Left censoring* in which failure-times are missing from the left side of the observation interval and the number of missing observations is known, (c) *Gapped data* in which failure-times from intervals within the observation interval are unavailable or not used, and (d) *Grouped data* in which no exact failure-times are available, but the numbers of failures in disjoint subintervals are known.

Much of the emphasis in this article will be to review these results. Some of the methodology will be included and possible extensions to new situations will be discussed. Other topics also discussed include graphical and nonparametric approaches with repairable system data, and models related to the power law process.

5.2   Point Estimation

The maximum likelihood estimators (MLE's) of the power law process in the cases of right time and failure truncation, have been available in the literature for several years. Articles dealing with the MLE's with both types of right truncated data include [28], [29], [40], [48], [45], [11], and [33]. Another article [47] suggests some advantages for a least squares-type estimator of β. In [28], the author provides methods for handling data from either a single system or a sample of systems. Much of the literature of the power law process deals with data obtained from observing a single system, and the primary focus in the rest of this article will be on the case of a single system. For example, a power law process for a single system with right time truncation at time τ, and $N = n$ observed failure times, $t_1 < \cdots < t_n$, in the interval $[0, \tau]$, has the likelihood function

$$L = (\lambda \beta)^n \left[\prod_{i=1}^{n} t_i \right]^{\beta - 1} \exp(-\lambda \tau^\beta) \tag{5.2.1}$$

if $n \geq 1$ and $0 < t_1 < \cdots < t_n < \tau$. The MLE's of β and λ for right time truncated data at time τ are

$$\hat{\beta} = n \bigg/ \sum_{i=1}^{n} \ln(\tau/t_i) \ , \quad \hat{\lambda} = n / \tau^{\hat{\beta}} \ .$$

Similar expressions are obtained for the MLE's in the case of right failure truncated data by replacing τ with the last observed failure time t_n,

$$\hat{\beta} = n \bigg/ \sum_{i=1}^{n} \ln(t_n/t_i) \ , \quad \hat{\lambda} = n / t_n^{\hat{\beta}} \ .$$

The following subsections discussion extensions of the MLE's to cases with missing data of various types.

5.2.1   Estimation with Left Truncated Data

The problem of estimating parameters for a power law process in the case of left truncation was considered by [28], but properties of the estimators were not investigated, and inference procedures such as tests and confidence intervals were not considered for this case. A recent article by [39] dealt with maximum likelihood fitting of the power law model to left truncated data.

Let $0 < \tau_1 < \tau_2 < \infty$, and suppose r failures are observed in the interval $(\tau_1, \tau_2]$, but nothing is known about possible failures in the interval $[0, \tau_1]$. Let $R = N(\tau_2) - N(\tau_1)$, and denote by r the observed value of R. We say that a process is *left truncated* at time τ_1 and *right truncated* at time τ_2 if the only available data are the occurrence times in the interval $[\tau_1, \tau_2]$, say $t_1 < \cdots < t_r$. In other words, the right truncation time, τ_2, is selected for the purpose of stopping observation and analyzing the data, while the left truncation time, τ_1, is a point below which failure times were not observed. In this section we consider the situation in which the number of failures below τ_1 is unknown, which is typical in the case of truncation. However, in Section 5.3.2, we will also consider what happens if the number of failures less than τ_1 is known.

The likelihood function corresponds to the joint density of the number of failures $R = r$ and the failure times, $T_1, \ldots, T_r$, for a truncated power law process. For brevity, a vector notation for the r values of the failure times, $\boldsymbol{t} = (t_1, \ldots, t_r)$ will be used.

The truncated counting process $N_1(t)$ can be expressed in terms of the original counting process as $N_1(t) = 0$ if $0 \leq t < \tau_1$ and $N_1 = N(t) - N(\tau_1)$ if $\tau_1 \leq t$. In other words, $N_1(t)$ is a counting process which ignores failures times which occur in the interval $[0, \tau_1)$, but counts all failures in the interval $[\tau_1, \infty)$. The process $N_1(t)$ is not a power law process as defined previously, but it is a Poisson process with mean function $M_1(t) = 0$ if $0 \leq t < \tau_1$ and $M_1(t) = \lambda(\beta - \tau_1^\beta)$ if $\tau_1 \leq t$.

By differentiation we obtain the intensity function which has the form $\nu_1(t) = 0$ if $0 \leq t < \tau_1$ and $\nu_1(t) = \lambda \beta t^{\beta-1}$ if $\tau_1 \leq t$. Now $R = N_1(\tau_2)$ is the number of failure times counted in the interval $[0, \tau_2]$ by the left truncated process, and it follows from well known results about Poisson processes that conditional on $R = r$, the failure times, $T_1 < \cdots < T_r$, are distributed as order statistics of a random sample of size r from a distribution with density function of the form $f_1(t) = \nu_1(t)/M_1(\tau_2)$. This yields

$$f_1(t) = \frac{\beta t^{\beta-1}}{(\tau_2^\beta - \tau_1^\beta)} \tag{5.2.2}$$

if $\tau_1 \leq t \leq \tau_2$, and zero otherwise. It follows that the conditional density function of the observed failure times $\boldsymbol{T} = (T_1, \ldots, T_r)$ given $R = r$ is

$$f(\boldsymbol{t}|r) = r! \, \frac{\beta^r \left[\prod_{i=1}^r t_i\right]^{\beta-1}}{(\tau_2^\beta - \tau_1^\beta)^r}$$

if $\tau_1 \leq t_1 < \cdots < t_r \leq \tau_2$, and zero otherwise. Since R is Poisson distributed with mean $\lambda(\tau_2^\beta - \tau_1^\beta)$, the joint density of $T_1, \ldots, T_r$ and R and hence the likelihood function is obtained, after some simplification, from the relationship $L = f(\boldsymbol{t}|r)P[R = r]$.

Thus, if a power law process has truncation from the left at time τ_1 and time truncation from the right at time τ_2, and with $R = r$ observed failures at times $t_1 < \cdots < t_r$, in the interval $[\tau_1, \tau_2]$, then the likelihood function is given by

$$L = (\lambda \beta)^r \, \tilde{t}^{r(\beta-1)} \exp[-\lambda(\tau_2^\beta - \tau_1^\beta)] \tag{5.2.3}$$

if $r \geq 1$, $\tau_1 < t_1 < \cdots < t_r < \tau_2$ and where the *geometric mean* of the observed failure times is

$$\tilde{t} = \left[\prod_{i=1}^r t_i\right]^{1/r} .$$

Similarly, we will denote the geometric mean of the truncation times as $\tilde{\tau} \equiv \sqrt{\tau_1 \tau_2}$. It is possible from this to compute the MLE's under certain conditions. For $r \geq 1$, the MLE's $\hat{\lambda}$ and $\hat{\beta}$ are obtained by the standard approach of equating the partial derivatives of the log-likelihood to zero and solving the resulting pair of equations.

As a result, if a power law process is truncated from the left at τ_1 and truncated from the right at τ_2, the MLE's are solutions of:

$$\hat{\lambda} = r / (\tau_2^{\hat{\beta}} - \tau_1^{\hat{\beta}}) ,$$

$$\frac{\tau_2^{\hat\beta}\ln\tau_2 - \tau_1^{\hat\beta}\ln\tau_1}{\tau_2^{\hat\beta} - \tau_1^{\hat\beta}} - \frac{1}{\hat\beta} = \ln\tilde{t}\ . \tag{5.2.4}$$

The MLE's exist if and only if $\tilde{t} > \tilde{\tau}$. Conditions under which the solutions exist follow from the that the MLE of β is the same as the conditional MLE of β given $R = r$. Thus, the failure times in the interval $[\tau_1, \tau_2]$ can be transformed into variables which are distributed conditionally as an ordered random sample from a truncated exponential distribution. In particular, we know that $T_1 < \cdots < T_r$ are distributed, conditional on $R = r$, as order statistics for a random sample of size r from a distribution with density function $f_1(t)$ as given by (5.2.2). We now consider the transformation

$$x_i = \ln(\tau_2/t_{r-i+1}) \tag{5.2.5}$$

for $i = 1, \ldots, r$, and note that this reverses the order of the variables and that the range of the transformed variables $\boldsymbol{X} = (X_1, \ldots, X_r)$ is from $0 = \ln(\tau_2/\tau_2)$ to $\psi \equiv \ln(\tau_2/\tau_1)$. Thus, the conditional density of the X_i's is

$$g(\boldsymbol{x}|r) = r!\ \frac{\beta^r \exp\left(-\beta \sum_{i=1}^r x_i\right)}{[1 - \exp(-\beta\psi)]^r}$$

if $0 \le x_1 < \cdots < x_r \le \psi$, and zero otherwise. This is the same as the joint density of the order statistics for a sample of size r from a *truncated exponential* distribution with parameter $\xi = 1/\beta$ and right truncation point ψ. The MLE of ξ for this distribution was derived by [31]. In particular, they found the MLE of ξ to be the implicit solution $\hat\xi$ of an equation involving the sample mean $\bar{x}$ and the truncation point ψ. A solution to the equation was shown to exist if $0 < \bar{x}/\psi < 1/2$, with no solution if $1/2 \le \bar{x}/\psi$. If we relate this to the original problem of estimating β for a power law process, the result is equivalent to (5.2.4) when $\tilde{\tau} = \sqrt{\tau_1\tau_2} < \tilde{t}$.

5.2.2   Estimation with Left Censored Data

A recent article [37] dealt with ML fitting a power law process to left censored data from a single system.

Left Time Censored Data

Consider the case in which it is known how many failures have occurred before some fixed time τ_1, but the exact values of the failure times are only known after τ_1. Suppose now there are k failures in the interval $[0, \tau_1]$, and r failures in the interval $(\tau_1, \tau_2]$. The numbers k and r are observed values of random variables $K = N(\tau_1)$ and $R = N(\tau_2) - N(\tau_1)$. For convenience we will define $N = N(\tau_2)$, and note that $N = K + R$. We say that a process is *left time censored* at time τ_1 and *right time truncated* at time τ_2 if the only available data are k the number of failures in $[0, \tau_1)$, and the exact occurrence times in the interval $[\tau_1, \tau_2]$, $t_1 < \cdots < t_r$. In other words, τ_1 is a point below which it is not possible to observe the exact failure times, but the number of unobserved failures k is known. It is possible to make use of some of the results about left truncation in this case.

Recall that $f(\boldsymbol{t}, r)$, which is also the likelihood function for $\boldsymbol{T} = (T_1, \ldots, T_r)$ and R, is given by (5.2.3). Furthermore, the K failures in the interval $[0, \tau_1)$ occur independently of failures in the interval $[\tau_1, \tau_2]$, and $K \sim \text{POI}(\lambda\tau_1^\beta)$. Thus, the joint density function of

T, R and K is $f(t, r, k) = f(t, r)P[K = k]$. It follows that the likelihood function in this case can be expressed as

$$L = (\beta/\tau_2)^r \left[(\tau_1/\tau_2)^k \prod_{i=1}^{r}(t_i/\tau_2) \right]^{\beta-1} \times (\lambda \tau_2^\beta)^{r+k}(\tau_1/\tau_2)^k \exp(-\lambda \tau_2^\beta)/k! \quad (5.2.6)$$

if $k \geq 0$, $r \geq 1$; $\tau_1 < t_1 < \cdots < t_r < \tau_2$. The joint MLE's, which can be derived in the standard way are

$$\hat\lambda = (r + k)/\tau_2^{\hat\beta},$$

$$\hat\beta = r \Big/ \left[\sum_{i=1}^{r} \ln(\tau_2/t_i) + k \ln(\tau_2/\tau_1) \right]. \quad (5.2.7)$$

As in the case of left truncation, when $r = 0$, there are no observed failure times and the MLE's of λ and β do not exist.

Left Failure Censored Data

In the case of left failure censored data from a single system, the left censoring time is the smallest observed failure time. Thus, in this case there is no fixed value τ_1, and the formulas can be simplified using the notation $\tau_2 = \tau$. Let $N = N(\tau)$, and suppose there are $N = n$ failures in the interval $[0, \tau]$, and it is known that k failures have occurred before the first exact failure time is observed. Thus, although only the last r failures, say $t_1 < \cdots < t_r$ are observed, n, the total number of failures by time τ, is known. We say in this case that the data is *left failure censored*.

The likelihood function is obtained by a different approach than with left time censoring. First consider the complete set of failure times, including the ones which are not observed, denoted by $T_1^* < \cdots < T_n^*$. It follows from results cited above about Poisson processes applied to a power law process that conditional on $N = n$, the failure times $T_1^*, \ldots, T_n^*$ are distributed as order statistics of a random sample of size n from a distribution with density function $f(t) = (\beta/\tau)(t/\tau)^{\beta-1}$ if $0 \leq t \leq \tau$, and zero otherwise. Thus, the conditional density function of the failure times $\boldsymbol{T}^* = (T_1^*, \ldots, T_n^*)$ given $N = n$ is

$$f(t^*|n) = n!\,(\beta/\tau)^n \left[\prod_{i=1}^{n}(t_i^*/\tau) \right]^{\beta-1}$$

if $0 \leq t_1^* < \cdots < t_n^* \leq \tau$, and zero otherwise. Integrating successively with respect to t_1^* through t_k^* yields the joint density function of), $\boldsymbol{T} = (T_1, \ldots, T_r) = (T_{k+1}^*, \ldots, T_n^*)$,

$$f(t|n) = \frac{n!}{k!}\,(t_1/\tau)^k(\beta/\tau)^r \times \left[(t_1/\tau)^k \prod_{i=1}^{r}(t_i/\tau) \right]^{\beta-1}$$

if $0 \leq t_1 < \cdots < t_r \leq \tau$ and $r \geq 1$. Since $N \sim \mathrm{POI}(\lambda \tau^\beta)$, it follows that the joint density of $\boldsymbol{T} = (T_1, \ldots, T_r)$ and N is obtained as $f(t, n) = f(t|n)f_N(n)$, yielding the likelihood

$$L = (\beta/\tau)^r \left[(t_1/\tau)^k \prod_{i=1}^{r}(t_i/\tau) \right]^{\beta-1} \times (\lambda \tau^\beta)^{r+k}(t_1/\tau)^k \exp(-\lambda \tau^\beta)/k! \quad (5.2.8)$$

if $k \geq 0$, $r \geq 1$ and $0 < t_1 < \cdots < t_r < \tau$. Thus, if a power law process is left failure censored and right time truncated, the MLE's of λ and β are

$$\hat{\lambda} = (r + k) / \tau^{\hat{\beta}} ,$$

$$\hat{\beta} = r / \left[\sum_{i=1}^{r} \ln(\tau/t_i) + k \ln(\tau/t_1) \right] . \tag{5.2.9}$$

Gap Analysis

Another variation on fitting a power law process to truncated data, considered by [20], is called "gap analysis". Instead of the truncation occurring on the left of the time interval, it occurs in the middle. One possible motivation for this situation would be if some temporary perturbation of a system caused an unusually high number of failures during an observation period. Even though the failure times during this period might be known, it might be prudent to ignore them since they are not typical of the system under standard operating conditions.

Suppose s_1 and s_2 are two fixed times in the time interval $(0, \tau]$ such that $0 < s_1 < s_2 \leq \tau$, and let $0 < x_1 < x_2 < \cdots < x_{n_1} \leq s_1$ be the failure times over the interval $(0, s_1]$ and let $s_2 < y_1 < y_2 < \cdots < y_{n_2} \leq \tau$ be the failure times over $(s_2, \tau]$. The approaches discussed above can be modified to obtain the likelihood function for such data. In particular, over the interval $(0, s_1]$, the likelihood L_1 is of the form given by (5.2.1) with right truncation at s_1, n replaced by n_1, and the t_i's replaced by the x_i's. Furthermore, over the interval $(s_2, \tau]$, the likelihood L_2 is of the form given by (5.2.3) with left truncation at s_2, right truncation at τ, r replaced by n_2, and the t_i's replaced by the y_i's. By the property of independent increments, the likelihood of the gapped data is of the form $L = L_1 \times L_2$. The MLE's of λ and β, obtained by the standard procedure are solutions to the equations

$$\hat{\lambda} = \frac{n_1 + n_2}{s_1^{\hat{\beta}} + (\tau^{\hat{\beta}} - s_2^{\hat{\beta}})} , \quad \hat{\lambda} = \frac{n_1 + n_2}{\{\hat{\lambda} s_1^{\hat{\beta}} \ln(s_1) + \tau^{\hat{\beta}} \ln(\tau) - s_2^{\hat{\beta}} \ln(s_2) - S\}}$$

where $S = \sum_{i=1}^{n_1} \ln(x_i) + \sum_{j=1}^{n_2} \ln(y_j)$.

It is obvious that the methodology of constructing likelihoods discussed in this section could be extended, if necessary to multiply gapped data. That is, if more than one subinterval is missing from the basic observation interval $(0, \tau]$, the likelihoods L_i defined by (5.2.1) and (5.2.3) could be used, with suitable modification of the notation, to write down the likelihoods for the disjoint intervals in which data is available, and the overall likelihood would be the product $L = \prod_i L_i$. Of course, the complexity of the equations which need to be solved will increase with the number of gaps in the observation interval.

Grouped Data

Situations sometimes occur in which no exact failure times are available, but rather the number of failures occurring in disjoint time intervals are the only available data. In [30], the authors considered this problem for a power law process. Suppose the time interval

$(0, \tau]$ is partitioned into k subintervals which have endpoints defined by $0 = \tau_0 < \tau_1 < \tau_2 < \cdots < \tau_k = \tau$, and denote by N_i the number of failures in the ith subintervals $(\tau_{i-1}, \tau_i]$ for $i = 1, \ldots, k$. Using the independent increment property, it follows that $N_1, N_2, \ldots, N_k$ are independent Poisson random variables with respective means $\lambda(\tau_i^\beta - \tau_{i-1}^\beta)$, and conditional on $N = n$, the joint distribution of $N_1, N_2, \ldots, N_{k-1}$ is multinomial having parameters n and $p_i = (\tau_i^\beta - \tau_{i-1}^\beta)/\tau_k^\beta$ if $i = 1, \ldots, k - 1$. Since $N \sim \text{POI}(\lambda \tau^\beta)$, the joint pdf of the first $k - 1$ counts $\boldsymbol{N} = (N_1, N_2, \ldots, N_{k-1})$ and the total count N is obtained from the fact that $f(\boldsymbol{n}, n) = f(\boldsymbol{n}|n) f_N(n)$, and the likelihood function is

$$L = \prod_{i=1}^{k} \frac{(\tau_i^\beta - \tau_{i-1}^\beta)^{n_i}}{n_i!} \times \lambda^n \exp(-\lambda \tau^\beta) . \qquad (5.2.10)$$

By equating the partials of the log-likelihood to zero, and solving for $\hat{\lambda}$ and $\hat{\beta}$, the MLE's are solutions to the equations

$$\hat{\lambda} = n / \tau^{\hat{\beta}} , \quad \sum_{i=1}^{k} n_i \frac{\tau_i^{\hat{\beta}} \ln(\tau_i) - \tau_{i-1}^{\hat{\beta}} \ln(\tau_{i-1})}{\tau_i^{\hat{\beta}} - \tau_{i-1}^{\hat{\beta}}} - n \ln(\tau) = 0$$

with the convention that $0 \times \ln(0) \equiv 0$.

The problem of performing a gap analysis with grouped data was also considered by [19]. Suppose, for example, it is decided that the interval $(\tau_{i-1}, \tau_j]$ should be gapped. The failures counts N_i in the intervals with $i \neq j$, conditional on $N^* = N - n_j = n^*$, have the multinomial distribution with parameters n^* and $p_i^* = p_i/(1 - p_j)$ for $i \neq j$. Furthermore, N^* has a Poisson distribution with expectation $E(N^*) = \lambda[\tau^\beta - (\tau_j^\beta - \tau_{j-1}^\beta)]$. Based on the resulting likelihood function, equations defining the MLE's can be obtained. We will not display them here.

Estimates of the Intensity Function

By the invariance property of MLE's, the MLE of a function $g(\beta, \lambda)$ can be obtained by substitution of the MLE's into the function. For example, the MLE of the intensity function $\nu(t)$ is

$$\hat{\nu}(t) = \hat{\lambda} \hat{\beta} t^{\hat{\beta}-1} .$$

Several recent papers have dealt with various alternative estimates of the intensity function evaluated at the right truncation point ($t = t_n$ for failure truncation and $t = \tau$ for time truncation), but without any left truncation or censoring.

Right Failure Truncated Data. In [55], the authors compared efficiencies relative to the MLE of several estimators of $\nu(T_n)$. Conditional on $T_n = t_n$, the MLE of $\nu \equiv \nu(t_n)$ is $\hat{\nu} = n\hat{\beta}/t_n$. One class of estimators considered consisted of estimators of the form

$$\hat{\nu}_{c_n} \equiv c_n \hat{\beta} / T_n .$$

Of course, when $c_n = n$, this gives the MLE. Other choices also included in the study were $c_n = (n-1)(n-2)/n$ and $c_n = (n-2)(n-3)/n$. It was shown by [48] that these constants give, respectively, unbiased and *minimum mean square error* (MMSE) estimators within this class when $\beta = 1$. Another estimator from this class, with $c_n = n - 1$, due to [41], called a *quasi-Bayes* estimator. A final class of estimators considered in that study and denoted by $\tilde{\nu}_{c_n, \alpha}$ is based on first performing a size α test of the null hypothesis $H_0 : \beta = 1$.

If the test does not reject H_0, then use the MLE of ν with known $\beta = 1$, $\tilde{\nu}_{c_n,\alpha} \equiv n/t_n$, and when H_0 is rejected, let $\tilde{\nu}_{c_n,\alpha} \equiv c_n\hat{\beta}/T_n$. The main conclusion from this study was that the MMSE estimator performed well in the sense of having high efficiency in a wide variety of situations, even when $\beta \neq 1$.

Right Time Truncated Data. In [58], the authors compared efficiencies relative to the conditional MLE (conditional on N) of several estimators of $\nu(\tau)$. Conditional on N, the MLE of $\nu \equiv \nu(\tau)$ is $\hat{\nu} = N\hat{\beta}/\tau$. One class of estimators considered was of the form

$$\hat{\nu}_{c_N} \equiv c_N\hat{\beta}/\tau .$$

Of course, when $c_N = N$, this corresponds to the conditional MLE. Other choices considered, analogous with [55], were $c_N = n - 1$, $c_N = (N - 1)(N - 2)/N$, and also $c_N = (N - 2)(N - 3)/N$. Also, by analogy, a class of estimators based on an initial test of $H_0 : \beta = 1$ was considered. Since some of the estimators in these classes do not exist unless $N > 3$, the properties of all estimators in the study were evaluated conditional on $N \geq 3$. The main conclusion from this study was that the estimator which performed best in a wide variety of situations was the one with the sequence $c_N = (N-1)(N-2)/N$ and without the initial test of H_0. Some estimators in the class with an initial test of H_0 were found to have high efficiency in some cases, but low efficiency in other cases.

Another estimator included in this study was $\bar{\nu} = N/\tau$, which is the MLE in the HPP case, $\beta = 1$. Although one would not expect this estimator to perform well overall, it performed fairly well when the expected number of events was small, say $E(N) = \lambda\tau^\beta \leq 10$.

Two other papers by the authors in [57] and [59] deal in greater detail with the general question "What is the effect on point and interval estimation of the intensity function at the time of truncation if an HPP is assumed when the true process is a non-HPP power law process?". The additional case of failure truncated data with $\bar{\nu} = n/t_n$ was also considered. As one might expect, the effect of assuming the wrong model was found to be more pronounced when β is not close to 1. Also, as discovered in an earlier study, $\bar{\nu}$ performs well provided the data set is not too large.

5.3   Inference Procedures

In [11] and also in [28], [29], *uniformly most powerful unbiased* (UMPU) tests of β and $\nu(\tau_2)$ were derived for right time truncation data. The variables $V = \hat{\beta}/\beta$ and $W = (\hat{\theta}/\theta)$ are pivotal quantities where $\hat{\beta}$ and $\hat{\theta}$ are the MLE's with right failure truncation. In [28], the author shows that $2\,r\,\hat{\beta}/\beta \sim \chi^2[2(r - 1)]$, and the tabled percentiles $w_{r,\gamma}$ for the pivotal quantity $W = (\hat{\theta}/\theta)^{\hat{\beta}}$ are provided by [40]. Further distributional properties are derived by [45].

Tests for differences of parameters based on two or more independent estimates are derived by [44] and [14]. These tests will not be presented here.

It was also shown by [28] that for the case of right time truncation, with $\hat{\beta}$ and $\hat{\theta}$ the MLE's for that case, that $2\,r\,\hat{\beta}/\beta \sim \chi^2(2r)$. However, distributional properties for $(\hat{\theta}/\theta)^{\hat{\beta}}$ are somewhat more difficult to obtain, since the pivotal property no longer holds with time truncation. In [11], the authors derived an approximation which makes use of Finkelstein's tables. For example, an approximate $100 \times (1 - \alpha)\%$ lower confidence limit for θ is

$$\theta_L \simeq \hat{\theta}\left\{r[(r + 1)w_{r+1,1-\alpha}]^{-r/(r+1)}\right\}^{1/\hat{\beta}} .$$

Prediction intervals for future observations of a power law process were derived by [45]. The results of Lee and Lee rely on numerical integration to solve for percentiles of the distribution of the appropriate pivotal quantity. In [34], the authors derived an explicit expression and a convenient F-approximation for the distribution of the appropriate pivotal quantity. These results will not be presented here.

Sequential Probability Likelihood Tests (SPRTs) for the power parameter β with the scale parameter θ an unknown nuisance parameter were derived by [12]. Such tests can be specified in terms of a sequence of statistics $\{\hat{\beta}_n\}$, which has the same form as the usual (fixed n) MLE of β for the right failure truncated case. However, in this setting n is a realization of a random variable N which is the number of observations required to make a decision using a SPRT. In [12] it was proved that this sequential procedure will terminate in a finite number of steps. The Wald approximation for the OC function and an approximation for the average *sample number* (ASN) are also derived. A sequential K-sample test and 2-sample test for difference of power parameters of two independent Weibull processes are also provided in that article but they will not be given here.

A number of goodness-of-fit (GOF) tests have been suggested, mostly based on adaptations of GOF tests for uniform and exponential distributions in the standard random sampling framework. In [28], the author provides tables for the Cramer-von Mises test which substitutes for the unknown parameter β an unbiased estimate.

Another approach to testing GOF is based on a result of [48] that for data from a right failure truncated power law process, the transformation $x_i = -\ln(t_i/t_n)$ for $1 \leq i \leq n-1$ yields data distributed as the order statistics in a random sample of size $n-1$ from the exponential distribution with mean $\theta = 1/\beta$. Thus, any GOF test for the exponential distribution with unknown mean (but known origin) can be applied to the transformed event times. An analogous transformation for right time truncated data, which is a special case of (5.2.5), is $x_i = -\ln(t_i/\tau)$ for $1 \leq i \leq n$. In this case, conditional on n, the x_i's are distributed as the order statistics in a random sample of size n from the exponential distribution. Both failure and time truncated cases, and tables required to run the GOF tests are discussed by [14]. Other related results on GOF tests are discussed by [54].

5.3.1 Inferences with Left Truncated Data

In [39], the authors dealt with interval estimation and tests of hypotheses with left truncated data from a power law process. The results from Section 5.2.1 can be used with the reparametrization based on the substitution

$$\lambda(\tau_2^\beta - \tau_1^\beta) = (\tau_2/\beta)\nu[1 - (\tau_1/\tau_2)^\beta] \ .$$

It follows from (5.2.3) that the density of $\boldsymbol{T} = (T_1, \ldots, T_r)$ and R given $R \geq 1$ is

$$f_c(\boldsymbol{t}, r) = c(\nu, \beta)\nu^r \left[\prod_{i=1}^r (t_i/\tau_2)\right]^{\beta-1} \tag{5.3.1}$$

if $1 \leq r$ and $\tau_1 \leq t_1 < \cdots < t_r \leq \tau_2$, and zero otherwise, where $c(\nu, \beta)$ is a constant. Notice that (5.3.1) can be expressed in the regular exponential form of [46]; namely

$$f_c(\boldsymbol{t}, r) = c(\nu, \beta)h(\boldsymbol{t}, r)e^{r \ln(\nu) + w \beta} \tag{5.3.2}$$

where $W = \sum_{i=1}^r \ln(T_i/\tau_2)$. Thus, the pair (R, W) is complete and sufficient for (ν, β). Furthermore, it is possible to derive UMPU tests for each parameter with the other parameter an unknown nuisance parameter. Each test is expressed as a conditional test of one of the statistics given the other.

Inferences for β

The procedure for β, which is based on the conditional distribution of W given $R = r$ can be expressed in terms of the related statistic $S = -W = \sum_{i=1}^{r} X_i$, where for each i, $X_i = \ln(\tau_2/T_i)$. The advantage is that due to (5.2.5) S is distributed as the sum of order statistics for a random sample of size r from a truncated exponential distribution with $\xi = 1/\beta$. Of course, this is also the distribution of the sum of the random sample, since it involves all of the variables. The distribution of this sum was derived by [17], and in the present notation we obtain the conditional density of S given $R = r$,

$$f_{S|r}(s; \beta) = \frac{e^{-\beta s}}{[(1 - e^{-\beta \psi})/\beta]^r (r-1)!} \times \sum_{i=0}^{m} \binom{r}{i} (-1)^i (s - i\psi)^{r-1} \qquad (5.3.3)$$

if $0 < s \le r\psi$ with $m = m(s) = [s/\psi]$ where $[\cdot]$ is the greatest integer function.

Any test for the parameter of the truncated exponential distribution can be easily adapted, using the fact that small values of $\xi = 1/\beta$ correspond to large values of β, and visa versa. This suggests how UMPU tests can be constructed for β. For example, if a power law process has truncation from the left at time τ_1 and time truncation from the right at time τ_2, with $R = r$ observed failure times, $t_1 < \cdots < t_r$, in the interval $[\tau_1, \tau_2]$, then a UMPU test of size α of $H_0 : \beta \le \beta_0$ versus $H_a : \beta > \beta_0$ is to reject H_0 if $s \le s_\alpha(r; \beta_0)$ where $s_\alpha(r; \beta)$ is the $\alpha \times$ 100th percentile of (5.3.3). Similarly, a size α UMPU test of $H_0 : \beta \le \beta_0$ versus $H_a : \beta \le \beta_0$ is to reject the null hypothesis H_0 if $s \ge s_{1-\alpha}(r; \beta_0)$. A size α test of $H_0 : \beta = \beta_0$ versus $H_a : \beta \ne \beta_0$ is to reject H_0 if either $s \le s_{\alpha/2}(r; \beta_0)$ or else $s \ge s_{1-\alpha/2}(r; \beta_0)$. Notice that it is not claimed that the two-tailed test is UMPU. In order to achieve unbiasedness for a two-tailed test, an additional constraint is needed. This is usually rather difficult to achieve, and in practice equal-tail tests are often used instead. We will not pursue this point further. Also, confidence limits can also be obtained by inverting tests of hypotheses, but this is not a trivial matter since there is no pivotal quantity for β, and (5.3.3) is a rather complicated function of .

A Beta approximation was derived by [15] which is useful in constructing approximate tests for β. Specifically, the distribution of the variable $S/(r\psi) = \bar{X}/\psi$ can be approximated by a Beta distribution with parameters

$$a = \frac{r\mu[\mu(1 - \mu) - \sigma^2/r]}{\sigma^2} \quad , \quad b = \frac{r(1 - \mu)[\mu(1 - \mu) - \sigma^2/r]}{\sigma^2}$$

where μ and σ are the mean and variance of the distribution of X/ψ, namely

$$\mu = \eta \left\{ 1 - \frac{\exp(-1/\eta)}{\eta[1 - \exp(-1/\eta)]} \right\} \quad , \quad \sigma^2 = \eta^2 \left\{ 1 - \frac{\exp(-1/\eta)}{\eta^2[1 - \exp(-1/\eta)]^2} \right\}$$

with $\eta = 1/(\beta\psi)$. A convenient variation on this is to base a test on an approximate F-statistic,

$$Y = \left(\frac{a}{b}\right) \left(\frac{r\psi}{S} - 1\right) \sim F(2b, 2a) \qquad (5.3.4)$$

Tests for the parameter β can be expressed in terms of (5.3.4). For example, an approximate size α test of $H_0 : \beta \le \beta_0$ versus $H_a : \beta > \beta_0$ is obtained by rejecting H_0 if $y \ge f_\alpha(2b, 2a)$.

Inferences for $\nu = \nu(\tau_2)$

It is also possible to derive UMPU tests of the parameter $\nu \equiv \nu(\tau_2)$ with β an unknown nuisance parameter. According to the theory of Lehmann, such tests will be conditional tests, based on the conditional distribution of R given $S = s$. It is also necessary to condition on $R \geq 1$. If a power law process is left truncated at time τ_1 and right truncated at time τ_2, with $R = r$ failure times $T_1, \ldots, T_r$ in the interval $[\tau_1, \tau_2]$, and if $S = \sum_{i=1}^{r} \ln(\tau_2/T_i)$, then the conditional probability mass function of R given $S = s$ and $R \geq 1$ is

$$f_{R|s}(r; \nu) = \frac{(\tau_2 \nu)^r}{c(\nu)(r-1)!} \times \sum_{i=0}^{m} \frac{(-1)^i}{i!(r-i)!} (s - i\psi)^{r-1} \tag{5.3.5}$$

if $r > m = [s/\psi]$ and $s > 0$ where $c(\nu)$ is the constant

$$\sum_{j=m+1}^{\infty} \frac{(\tau_2 \nu)^j}{(j-1)!} \sum_{i=0}^{m} \frac{(-1)^i}{i!(j-i)!} (s - i\psi)^{j-1} .$$

For a power law process with left time truncation at time τ_1 and right time truncation at time τ_2, a size α UMPU test of $H_0 : \nu \leq \nu_0$ versus $H_0 : \nu > \nu_0$ is to reject H_0 if $\sum_{j=r}^{\infty} f_{R|s}(j; \nu_0) \leq \alpha$. As is often the case with a discrete distribution, it is more convenient to express the critical region in terms of a cumulative probability rather than trying to express it in terms of percentiles of the null distribution. Similarly, a size α UMPU test of $H_0 : \nu \geq \nu_0$ versus the alternative $H_a : \nu < \nu_0$ is to reject H_0 if $\sum_{j=1}^{\infty} f_{R|s}(j; \nu_0) \leq \alpha$. It is also possible to construct a two-tailed test for ν. A size α test of the null hypothesis $H_0 : \nu = \nu_0$ versus $H_a : \nu \neq \nu_0$ is to reject H_0 if either $\alpha/2 \geq \sum_{j=1}^{r} f_{R|s}(j; \nu_0)$ or else $\alpha/2 \geq \sum_{j=r}^{\infty} f_{R|s}(j; \nu_0)$. Strictly speaking, this two-tailed test is not unbiased unless an additional constraint is also satisfied, but a test with is free of the nuisance parameter β is obtained regardless of this constraint. Since R is discrete, it will usually not be possible to construct an exact size α test without using randomization. In general, rather than randomization, it is recommended to use the above test. Although this test is not exact, it is conservative. Although (5.3.5) is rather complicated, it is not too difficult to evaluate with the aid of a computer.

5.3.2 Inferences with Left Censored Data

In [37], the authors dealt with interval estimation and tests of hypotheses with left censored data from a power law process. There are two situations, left time censoring and left failure censoring. We first consider the case of left failure censoring, since some of these results provide approximations for left time truncation.

Left Failure Censoring

As in Section 5.2.2, the left censoring time is the smallest observed failure time. Thus, in this case there is no fixed value τ_1, and the formulas can be simplified using the notation $\tau_2 = \tau$. Let $N = N(\tau)$, and suppose there are $N = n$ failures in the interval $[0, \tau]$, and it is known that k failures have occurred before the first exact failure time is observed. Thus, although only the last r failures, say $t_1 < \cdots < t_r$ are observed, n, the total number of failures by time τ, is known. In order to have at least one failure observed in the interval

$[0, \tau]$, necessarily $R \geq 1$. Thus, it is appropriate to condition on $R \geq 1$. The joint density of $T_1, \ldots, T_r$ and R given $R \geq 1$ differs from (5.2.8) only by a multiplicative constant. Based on this, it is possible to construct optimal confidence limits and tests of hypotheses. It is desirable to reparametrize first in terms of β and the intensity at the right truncation point, say $\nu = \nu(\tau)$. Since the intensity function is $\nu(t) = \lambda \beta t^{\beta-1} = (\beta/t)\lambda t^{\beta-1}$, for the value $t = \tau$ we can substitute $\lambda \tau^\beta = (\tau/\beta)\nu$. The joint density of $\boldsymbol{T} = (T_1, \ldots, T_r)$ and R given $R \geq 1$ can be written in the regular exponential form of [46],

$$f_c(\boldsymbol{t}, r) = c(\nu, \beta)h(\boldsymbol{t}, r)e^{r \ln(\nu) + w\beta} \tag{5.3.6}$$

where the variables w and r are values of statistics $W = \sum_{i=1}^{r} \ln(T_i/\tau) + k \ln(T_1/\tau)$ and $R = N(\tau) - k$. As a consequence, it follows that the pair (R, W) is complete and sufficient for the pair (ν, β). It is also possible to derive UMPU tests for each parameter with the other parameter an unknown nuisance parameter.

Inferences for β. A size α UMPU test of the null hypothesis $H_0 : \beta \leq \beta_0$ versus $H_a : \beta > \beta_0$ is to reject H_0 if $2r \beta_0 / \hat{\beta} \leq \chi_\alpha^2(2r)$. This seen by defining a statistic related to the MLE $\hat{\beta}$ as $S = \sum_{i=1}^{r} \ln(\tau/T_i) + k \ln(\tau/T_1) = r/\hat{\beta}$, and noting that $S = -W$. Conditional on $N = n$, $T_1, \ldots, T_r$, are distributed as the largest r out of n order statistics from a distribution with cumulative distribution function $F(t) = (t/\tau)^\beta$ for $0 < t < \tau$. Defining $U_i = F(T_i) = (T_i/\tau)^\beta$, the random variables $U_1, \ldots, U_r$ are distributed as the largest r out of n order statistics from a uniform distribution on the interval $(0, 1)$. The transformation $x_i = -\ln(u_{r-i+1})/\beta$ reverses the order of the T_i's, and conditional on $N = n$ the X_i's are distributed as the smallest r out of n ordered observations in a random sample from an exponential distribution with mean $\xi = 1/\beta$. Thus, $X_1 < \cdots < X_r$ are distributed as an ordered random sample with type II censoring on the right from an exponential distribution with mean $\xi = 1$. Consequently, $2\beta S = 2r \hat{\xi}/\xi$, and it follows from well known results about type II censored samples from an exponential distribution that given $N = n$, $2r \beta/\hat{\beta} = 2\beta S \sim \chi^2(2r)$. Lehmann's test for this alternative is also conditional on $N = n$, with rejection for large W, which corresponds to small S. Similarly, a size α UMPU test of the null hypothesis $H_0 : \beta \geq \beta_0$ versus $H_a : \beta < \beta_0$ is to reject H_0 if $2r \beta_0/\hat{\beta} \geq \chi_{1-\alpha}^2(2r)$. Although two-tailed UMPU tests can also be constructed they are more complicated since they involve an additional constraint. However, the simpler equal-tailed tests are easily constructed from ordinary chi square tables, and they are often preferred for this reason. Of course, confidence intervals can also be derived. For example, a $(1 - \alpha) \times 100$ percent confidence interval for β is

$$\hat{\beta} \chi_{1-\alpha/2}^2(2r) / 2r < \beta < \hat{\beta} \chi_{\alpha/2}^2(2r) / 2r .$$

Inferences for $\nu = \nu(\tau)$. It is also possible to derive UMPU tests for the parameter $\nu = \nu(\tau)$, with β an unknown nuisance parameter. Using Lehmann's approach, such tests are conditional tests, based on R given $S = s$. Since conditional on $R = r$, $2\beta S \sim \chi^2(2r)$, the density of R conditional on $S = s$ and $R \geq 1$ is

$$f_{R|s}(r; \nu) = \frac{[(s\tau)\nu]^r / [(r - 1)!(r + k)!]}{\sum_{j=1}^{\infty}[(s\tau)\nu]^j / [(j - 1)!(j + k)!]} \tag{5.3.7}$$

for $s > 0$ and $r \geq 1$. From (5.3.7) it is possible to construct tests for $\nu = \nu(\tau)$. For example, for a power law process with left failure censoring and right time truncation at time τ, a size α UMPU test of $H_0 : \nu \geq \nu_0$ versus the alternative $H_a : \nu < \nu_0$ is to reject the null hypothesis H_0 if $\sum_{j=1}^{r} f_{R|s}(j; \nu_0) \leq \alpha$. Upper one-tailed and two-tailed tests can be constructed in a similar manner. Of course, it is also possible to construct confidence limits by inverting such tests. For example, limits for a conservative $(1 - \alpha) \times 100$ percent confidence interval are obtained as solutions $\nu = \nu_1$ and $\nu = \nu_2$ to $\alpha_1 = \sum_{j=r}^{\infty} f_{R|s}(j; \nu)$

and $\alpha_2 = \sum_{j=1}^{r} f_{R|s}(j; \nu)$ with $\alpha = \alpha_1 + \alpha_2$, $\alpha_1 > 0$ and $\alpha_2 > 0$. Solutions to these equations cannot be obtained explicitly, but they can be found numerically with the aid of a computer program.

Another possible approach involves an asymptotic normal approximation. The details of the approximation are given in the Appendix and a summary is given here. In particular, it is possible to derive approximations for the conditional mean and variance which apply for large $h = s\tau\nu$, namely

$$E(R|s) \simeq \sqrt{h} + \frac{1 - 2k}{4}, \quad Var(R|s) \simeq \frac{\sqrt{h}}{2}.$$

If we define $\mu_0 = \sqrt{h} + (1 - 2k)/4$, $\sigma_0^2 = \sqrt{h}/2$ and $Z = (R - \mu_0)/\sigma_0$, then, as shown in the Appendix, the limiting distribution of Z is standard normal as $h \to \infty$. Consequently, limits for an approximate $(1 - \alpha) \times 100$ percent confidence interval for ν are obtained as solutions to the equations

$$\pm z_{\alpha/2} = \frac{r + \frac{1}{2} - [\sqrt{s\tau\nu} + (1 - 2k)/4]}{\sqrt{\sqrt{s\tau\nu}/2}}. \tag{5.3.8}$$

The solutions to (5.3.8), say ν_L and ν_U, have the form

$$\frac{1}{s\tau} \times \left[A + \frac{z_{\alpha/2}^2}{4} \pm \frac{z_{\alpha/2}}{2} \sqrt{2A + \frac{z_{\alpha/2}^2}{4}} \right]^2 \tag{5.3.9}$$

where $A \equiv (r + 1/4 + k/2)$. A study of this approximation for several values of h and k, and using nominal confidence levels of 90 and 95 percent, showed that the approximation is good for $h \geq 10$ and the best results occur when $k \leq 5$.

Left Time Censoring

Following the notation of Section 5.3.1, it is possible to derive the joint density of $T_1, \ldots, T_r$, R and $K = N - R$ given $R \geq 1$. The joint density of $T = (T_1, \ldots, T_r)$, R and K given $R \geq 1$ is given by (5.2.6). This density can be written in an exponential form, but not a regular form. Specifically, (5.2.6) can be written in a form similar to (5.3.2), namely

$$f_c(t, r, k) = c(\nu, \beta) h(t, r, k) \times e^{r \ln(\nu) + w\beta + k \ln(\nu/\beta)} \tag{5.3.10}$$

where the variables w, k and r are values of $W = \sum_{i=1}^{r} \ln(T_i/\tau_2) + k \ln(\tau_1/\tau_2)$, $K = N(\tau_1)$ and $R = N - K$. As a consequence, it follows that the triple (R, K, W), is sufficient for the pair (ν, β). This is not an ideal situation, since the dimension of the sufficient statistic is higher than the dimension of the parameter space, and thus, the statistic is not complete. As a consequence, Lehmann's theory does not apply, and whatever inference procedures are obtained may not be optimal.

Some algebraic manipulation with this function yields an alternate form in which the exponent is $n \ln(\nu) + w\beta + k \ln(1/\beta)$. This still leaves three statistics to construct inference procedures for two parameters, but it suggests a more natural grouping where N is associated with ν, and the pair (K, W) is associated with β. We will consider some inference procedures based on this grouping.

Inferences for β. Equation (5.2.5) provides a way of transforming the failure times above τ_1 into the first r ordered observations from an exponential distribution. This approach is easily adapted for left censored data. Thus, conditional on $N = n$, the transformed data $x_1 < \cdots < x_r$, obtained from (5.2.5), are distributed as a Type I censored random sample with from an exponential distribution with mean $\xi = 1/\beta$, with $k = n-r$ a known number of censored values. This suggests the possibility of adapting methods of inference for the mean ξ of an exponential distribution with Type I censoring such as those discussed by [14]. For example, since $\beta = 1/\xi$, the reciprocal of a lower confidence limit for ξ would correspond to an upper confidence limit for β, and tests of hypotheses can be similarly modified.

There are two possible approaches, one based on the MLE $\hat{\beta}$, and the other based on the number of observed failures, $R = r$, but not on the exact failure times. Although R is Poisson distributed, its conditional distribution given $N = n$ is binomial with parameters n and $p = 1 - (\tau_1/\tau_2)^\beta$. Consequently, a number of common inference procedures for the binomial parameter p are available for analyzing β. Note that a small value of β corresponds to a small value of p, in which case r, the observed value of R, will tend to be small. Thus, for example, with time censored data, a size α test of $H_0 : \beta \leq \beta_0$ versus $H_a : \beta > \beta_0$ rejects the null hypothesis H_0 if $B(r, n, p_0) \leq \alpha$ with $p_0 = 1 - (\tau_1/\tau_2)^{\beta_0}$. Of course, for large n asymptotic normal results for p (or $q = 1 - p$) can be used. For example, approximate $(1 - \alpha) \times 100$ percent confidence limits for β are of the form

$$\ln \left(\hat{q} \pm z_{\alpha/2} \sqrt{\hat{p}\hat{q}/n} \right) / \psi$$

for $\hat{q} = k/n$, $\hat{p} = 1 - \hat{q}$, and $\psi = \ln(\tau_2/\tau_1)$. Although this is relatively simple and convenient to apply, an obvious matter of concern is the amount of efficiency lost by not using the actual failure times. This question, for the corresponding exponential problem was studied by [18]. By comparing asymptotic variances, an asymptotic efficiency of about 96 percent was found when $p \leq 0.5$, which in the present problem corresponds to $\beta \leq \ln 2/\psi$. For large β, there is an approximate normal test, also found in [18], for the exponential problem which uses all of the data. When adapted to the problem of testing β for a power law process, the test statistic is

$$Z = \frac{V\sqrt{np}}{\sqrt{1 - 2(q \ln q)V/p + qV^2}} \tag{5.3.11}$$

with $V = \beta/\hat{\beta} - 1$ and $p = 1 - e^{-\beta\psi}$. Conditional on both $N = n$ and $R > 0$, this statistic has an approximate standard normal distribution, with the best approximation occurring when $p \geq 0.5$ (or when $\beta \leq \ln 2/\psi$), which is the range where the test based only on r is least efficient. Thus, it seems reasonable to use (5.3.11) for testing large values of β, and R for small values of β. For example, if $\beta_0 \geq (\ln 2)/\psi$, and z_0 is an observed value of (5.3.11), an approximate size α test of $H_0 : \beta \leq \beta_0$ versus $H_a : \beta > \beta_0$ is to reject H_0 if $z_0 \leq z_\alpha$.

Inferences for $\nu = \nu(\tau_2)$. The best way to proceed in constructing a test for $\nu = \nu(\tau_2)$ is not as clear as it was for β. The remarks following (5.3.10) suggest using a conditional test given the pair (K, W) since the conditional distribution of any other statistic will be free of the nuisance parameter β. Note also that

$$S \equiv \sum_{i=1}^{r} \ln(\tau_2/T_i) + K \ln(\tau_2/\tau_1) = -W .$$

Thus, conditioning on K and W is equivalent to conditioning on K and S. Now recall that in Section 5.3.2 convenient approximate confidence limits for ν were provided for use with left failure censored data. Although a procedure derived for use with failure censoring is not theoretically correct when applied with time censored data since k is not fixed but an observed value of a random variable K, if the amount of censoring is not too large, a reasonable approximation might result. The proposed method of inference for ν for the case of left time censoring is based on using (5.3.9) for computing confidence bounds, except that now we condition on the statistic $S = R/\hat{\beta}$ where $\hat{\beta}$ is the MLE under left time censoring given by (5.2.7) rather than the left failure censored version. A study of this procedure was conducted in [37] based on computer simulation with left time censored data. The relative frequency of coverage with intervals based on (5.3.9) was performed with the modifications mentioned above. From this study, it appears that the approximation works fairly well provided that τ_1 is fairly small relative to τ_2, say $\tau_1/\tau_2 \leq 1/3$.

5.4 Related Models

It is the case in many applications that the power law process is not appropriate. A number of alternative models which are related in some way to power law processes have appeared in the literature of repairable systems.

5.4.1 Renewal Processes

As noted in Section 5.1, a power law process with $\beta = 1$ is a HPP with intensity λ. It is well known that the times between failures for a HPP are stochastically independent exponentially distributed random variables with mean $1/\lambda$. In general, if times between failures for a nonterminating counting process are stochastically independent and identically distributed, the process is called a *renewal process*.

In order to discuss renewal processes it is helpful to make a distinction between failure times of a system measured from the initial start-up time of the system (*global time*), and time measured since the last system failure (*local time*). If a repairable system is modeled by a renewal process, the system is in "like-new" condition immediately after failure and repair. Thus, for a system modeled by a renewal process, system reliability depends in a simple way on local time, but in a complex way on global time. On the other hand, for a system modeled by a power law process with $\beta \neq 1$ (or any other NHPP), system reliability depends, in a simple way, on global time. We will denote cumulative failure times (global time) by T_i, and times between successive failure times (local time) by $X_i = T_i - T_{i-1}$ with $T_0 \equiv 0$.

A general development of renewal processes is given by [25], and applications involving reliability of repairable systems are discussed in [8]. Some recent articles also discuss various aspects of renewal processes as models for repairable systems.

In [3], the author argues that a renewal process with Weibull distributed times between failures is an implausible repair model. The primary concern is the behavior of the hazard function

$$h_X(x) = \frac{f_X(x)}{1 - F_X(x)}$$

where $f_X(x)$ and $F_X(x)$ are the probability density function (pdf) and cumulative distribution function (CDF), respectively, of X. The CDF of a Weibull variable X is

$$F_X(x) = 1 - \exp[-(x/\theta)^\kappa]$$

if $x > 0$, and zero otherwise. The parameters κ and θ are shape and scale parameters, respectively. Of course, if $\kappa = 1$, the distribution of X is exponential, in which case $h_X(x)$ is constant. Furthermore, a renewal process has exponential times between failures if and only if it is a HPP. However, $\kappa < 1$ implies $h_X(0) = \infty$, which would model a repairable system with very low reliability immediately following each repair. On the other hand $\kappa > 1$ implies $h_X(0) = 0$, and the system would have a fairly high reliability immediately after each repair. It is suggested by Ascher that a more plausible generalization of the HPP for repair to like-new condition is a renewal process with distribution of local times such that $0 < h_X(0) < \infty$ and $0 < h_X(x) < \infty$.

A more general argument is made in another article by [6] where it is stated that "A renewal process, in general, is not a good model for a repairable system. The fact that most repairs involve the replacement of only a small portion of a system's constituent parts, makes it implausible that such repairs renew the system to its original condition." In this same article Ascher also discusses advantages and disadvantages of both the HPP and NHPP as a models for repairable systems. He states, "Some authors have gone to great lengths to emphasize that the NHPP will seldom, if ever, be an exact model for a repairable system. They are correct: there will be some effect due to previously performed repairs, so the independent increments property will not hold exactly. (Since the HPP also has the independent increments property, it never will be an exact model either.) Nevertheless, the NHPP is a good first-order model for an improving/deteriorating repairable system... Overall, then, the NHPP is an exceptionally useful model for repairable systems analysis,..."

Other concerns resulting from misconceptions in the reliability literature are discussed in an article by [9]. They point out that problems in the interpretation and analysis of reliability data often stem from confusing the hazard function $h_X(x)$ of a nonrepairable system with the intensity function $\nu(t)$ of a repairable system. As noted by [62], both of these functions are often erroneously called "failure rate". For further discussion on related matters see [2], [4], and [5].

A different point is made by [7], about the potential for ambiguities in the interpretation of failure data. The basic point was that greater care should be taken in distinguishing between data generated by repeated failures by a single repairable system, as opposed to the order statistics derived from the times to failure of n separate nonrepairable items. The following quote of [53] was given in support of this point: "There is a basic ambiguity in the theory of spacings caused by radically different assumptions which can be placed on the . . . process [of times to or between events]".

5.4.2   Compound Power Law Processes

In some applications, a *compound Poisson distribution* is used as an alternative to the ordinary Poisson distribution for count data for a single system. Such a compound distribution, which has a negative binomial form, occurs when the population consists of components with Poisson distributed failure counts, but with intensities that vary from component to component according to a gamma distribution. For example, in a population of repairable components the intensities can differ from one component to the next because they operate in diverse environments, differences due to manufacturing variation, or any other factor which might cause differences in reliability among population members.

A *compound power law process* was derived by [35] with intensity function parametrized in terms of the parameters λ and β. When data are obtained from a single system and inferences are made only for that system, then a power law process with fixed values of the parameters is an appropriate model. If the parameter λ varies from system to system, then a compound model is more suitable.

We consider the following set of assumptions: (a) The population consists of a heterogeneous collection of systems such that the failures of each system are distributed according to a power law process, but with values of λ which differ from system to system. That is the counting process for the number of failures in time t for each individual system has density of the form

$$f(n|\lambda) = \frac{(\lambda t^\beta)^n \exp(-\lambda t^\beta)}{n!} \qquad (5.4.1)$$

if $n = 0, 1, \ldots$, and zero otherwise. (b) Each system in the population has the same power parameter β, but the parameter λ varies according to a gamma density with shape parameter k and scale parameter γ

$$g(\lambda) = \frac{\lambda^{k-1} \exp(-\lambda/\gamma)}{\gamma^k \Gamma(k)} \qquad (5.4.2)$$

if $\lambda > 0$, and zero otherwise. If assumptions (a) and (b) hold for the population, the discrete density of $N(t)$ is

$$f(n) = \binom{n+k-1}{n} \frac{(\gamma t^\beta)^n}{(1 + \gamma t^\beta)^{n+k}} \qquad (5.4.3)$$

if $n = 0, 1, \ldots$, and zero otherwise. The density (5.4.3) is obtained by integrating the "joint density" of $N(t)$ and λ, obtained as the product of (5.4.1) and (5.4.2).

The distribution given by (5.4.3) corresponds to a special form of the negative binomial distribution with parameters: k and $p = 1/(1 + \gamma t^\beta)$. It follows from standard results for the negative binomial distribution that the mean and variance of $N(t)$ are

$$E[N(t)] = \gamma t^\beta \ ,$$
$$\text{var}[N(t)] = k\gamma t^\beta (1 + \gamma t^\beta) \ . \qquad (5.4.4)$$

Obviously, $\text{var}[N(t)] > E[N(t)]$. In the limit, however, as γ approaches 0 with $k\gamma$ fixed, $\text{var}[N(t)] = E[N(t)]$, which is a well-known property of Poisson distributed random variables. The negative binomial density, given by (5.4.3), approaches the Poisson density, given by (5.4.1). In the limit as γ approaches 0 with $k\gamma$ fixed variance $k\gamma^2$ of the distribution of intensities (5.4.2) approaches zero while the mean, $k\gamma$, remains constant. This limiting behavior suggests that when k is large and $k\gamma$ is small, the intensities of the systems don't differ much, and data from the compound distribution (5.4.3) are hard to distinguish from Poisson data. This can cause some difficulties in estimating the parameters of the model. We will not attempt to discuss estimation and inferences here, but the MLE's and some inference procedures are given in [35] for the special case in which all processes in the sample are right time truncated at the same point τ.

5.4.3 Modulated Power Law Processes

Another counting process related to the power law process, called a *modulated power law process* is defined in [42]. This is a special case of a type of stochastic process defined [21] called the *inhomogeneous gamma process*.

Suppose now that shocks to the system occur according to a power law process with mean function $M(t) = (t/\theta)^\beta$, and suppose that rather than a failure occurring at every shock, one occurs at every kth shock. If $0 < t_1 < \cdots < t_n$ are the first n failure times;

$t = (t_1, \ldots, t_n)$ then their joint density function is

$$f(t) = \frac{\prod_{i=1}^{n} \nu(t_i)\{M(t_i) - M(t_{i-1})\}^{k-1} e^{-M(t_i)}}{[\Gamma(k)]^n}.$$

As noted in [42] for this process, there is no reason to restrict the parameter k to be integer-valued, since $f(t)$ is well defined mathematically for any real $k > 0$. They also note that when $k = 1$, the model reduces to the power law process, and when $\beta = 1$ the model reduces to a renewal process in which the times between failures are gamma distributed which shape parameter k.

In [42] it is argued that a renewal process seems overly optimistic since it models repair to like-new condition, while a power law process with increasing intensity seems overly pessimistic. The modulated power law model is a generalization which includes both power law processes and a special class of renewal processes, and can be viewed as a compromise between the two models. It might also be argued that a power law process with decreasing intensity is optimistic about the effect of repairs, relative to renewal.

The MLE's of β, k and θ are derived [42], but we will not attempt to present them here. Confidence intervals and tests of hypotheses for the parameters are not presently available.

5.4.4   Other Special NHPP's

A generalization of the power law model was proposed by [43] involving NHPP's with intensity function of the form

$$\nu(t) = \lambda \beta t^{\beta-1} \exp(\delta t) \tag{5.4.5}$$

This model also includes another class of NHPP's with intensity function of the form

$$\nu(t) = \exp(\gamma + \delta t)$$

which is discussed by [26]. The term *log linear* is used by [8], because the logarithm of $\nu(t)$ is a linear function of t.

The general model (5.4.5) reduces to the power law case if $\delta = 0$, and the log linear case if $\beta = 1$. Tests were derived by [43], in this more general framework, for testing the adequacy of the simpler power law and log linear models. A test for increasing intensity, usually attributed to Laplace, along with several others, including a test based on the MLE of β for the power law model, were studied for a variety of intensities, including the power law and log linear intensities in articles by [16], [38], and [24]. Applications of the power law, log linear and also of a linear intensity in probabilistic risk assessment are discussed by [10].

As noted by [27] if λ, β and δ are all positive, and $\beta < 1$, a *bathtub* or U-shaped curve is obtained.

5.5   Nonparametric and Graphical Methods

It was demonstrated by [32], using plots of data from real repairable systems, that a model that is related to the power law process fits many systems adequately. It was shown that on log-log paper a plot of cumulative operating time t versus $t/N(t)$, where $N(t)$ is the number of failures before time t, was nearly linear for many of these systems. Such behavior would be expected for data from a power law process. In [32] it was assumed that the average number of failures before time t is a particular increasing function of t, and thus his model is deterministic. Of course, the power law model can be viewed as a stochastic analog of the model used in [32].

In [22], the author derived estimates for the intensity function as well as tests for the null hypothesis of HPP versus the alternative of an NHPP with increasing intensity function within the class of nondecreasing intensity functions.

The results discussed in the previous sections provide methods for fitting this model, but sometimes there are situations in which one or more of the basic assumptions do not seem reasonable. For example, the property of independent increments would be a dubious assumption in many applications.

5.5.1   Nonparametric Estimation

A method for plotting data on the cumulative numbers of recurrent failures for *any number* of systems versus system age is given by [50]. The method yields an estimate $M^*(t)$ which is a nonparametric estimate of the mean $M(t)$ of the cumulative number of repairs at age t. The method has the advantage that it can be applied to data from systems with different censoring ages. Another advantage of this approach is that the mathematical assumptions required are not as strong as those associated with commonly used stochastic models such as an NHPP.

There are three basic assumptions: (a) The population mean cumulative function is finite over the range of interest. (b) The sample systems are a random sample from the population of systems. (c) Censoring of system histories is random. In other words, the repair histories of all systems in the sample are stochastically independent of their censoring ages. It is noted by [49] that assumption (b) fails to hold if prototype systems are tested in the laboratory to predict performance of production systems in actual use. It is also noted that assumption (c) is satisfied if the censoring ages in the sample are randomly assigned to the sample repair histories, and it is not satisfied if some systems are retired early because they had undergone severe use. Under assumptions (a) through (c), in [50] it is shown that $M^*(t)$ is an unbiased estimate of $M(t)$.

A generalization in [50] extends the method to apply with a function $C(t)$ which is the mean of cumulative total repair cost by age t. An unbiased estimate $C^*(t)$ of $C(t)$ is obtained under the same assumptions required of $M(t)$. The function $C(t)$ is a generalization of $M(t)$ since they agree when the repair cost is a constant standard unit (e.g. $1/repair).

A report by [51] provides a computer program for constructing plots, computing estimates and confidence limits for $M(t)$ or $C(t)$, and several interesting examples. One of the examples involves a mixture of two Poisson processes, providing an example in which the independent increments assumption, which is a basic property of Poisson processes, does not hold.

Another report by [52] provides a method and a computer program to compare two samples of recurrent failure data.

5.5.2   Nonparametric Testing

A nonparametric test, known as the *Mann test*, for testing the null hypothesis of a renewal process versus a process with increasing rate of occurrence, is discussed by [8] and [61]. The Mann test is based on the tendency for interarrival times to become shorter if the rate of occurrence is increasing. This test is performed by counting *reverse arrangements* among the chronologically ordered interfailure times X_i of the process. Specifically, a reverse arrangement occurs whenever $X_i < X_j$ for $i < j$. The Mann test is based on the statistic R which is defined as the total number of reverse arrangements. In other words, R is computed by comparing every interarrival time with every later interarrival time. Fewer reverse arrangements, would be expected if interarrival times are tending to get shorter, and thus a small value of R would indicate an increasing rate of occurrence.

It should be noted that when the data are obtained by observing failures during a fixed period of time, the exact value of the interarrival time from the th to the $(n+1)$st failure will not be known. It might be possible to increase the power of the test by using the fact that no events occurred between the th occurrence and the end of the observation period, but for simplicity one might simply disregard this additional information. It can be shown that R is approximately normal if $n > 10$, and tables exist for smaller n.

5.6 Conclusions

A number of models for repair data from a single repairable system have been discussed along with their associated analyses.

Poisson processes, both homogeneous and nonhomogeneous cases, have received much attention and will continue to receive attention in repairable systems analysis due to the availability and simplicity of statistical techniques. Due to the nature of the data which is often available, one area of analysis which would benefit from further work is grouped data, in which only the counts of failures in fixed time intervals can be observed. Also, additional work on related models would be useful. In particular, more work on models which don't assume independent increments would be useful. A particular problem of interest is to extend the analysis of a compound power law process to the case of time intervals which are not all of equal length.

Another area of future work should be the development of new methods, or the enhancement of existing methods of nonparametric and graphical analyses.

Acknowledgements

This work was supported by the U.S. Department of Energy Contract DE-AC07-76ID0157. In particular, the author is grateful for partial support by the Center for Reliability and Risk Assessment of EG&G Idaho, Inc.

Appendix

The conditional density function (5.3.7) can be expressed in terms of modified Bessel functions, whose properties are discussed in [1],

$$f_{R|s}(r;\nu) = \frac{(s\tau\nu)^{r+\frac{k-1}{2}}}{(r-1)!(r+k)!I_{k+1}(2\sqrt{s\tau\nu})} \tag{A.1}$$

where $I_m(z)$ represents a modified Bessel function of order m. This is based on the series

$$I_m(z) = (z/2)^m \sum_{j=1}^{\infty} \frac{(z^2/4)^{j-1}}{(j-1)!(j+m-1)!} \tag{A.2}$$

Using the substitution $z = 2\sqrt{y}$ and simplification, it follows that

$$\sum_{j=1}^{\infty} \frac{y^j}{(j-1)(j+m-1)!} = y^{-\frac{m-2}{2}} I_m(2\sqrt{y}) . \tag{A.3}$$

Using (A.3), it is possible to derive the following expression for the conditional mean:

$$E(R|s) = 1 + \sqrt{s\tau\nu}\, \frac{I_{k+2}(2\sqrt{s\tau\nu})}{I_{k+1}(2\sqrt{s\tau\nu})} \tag{A.4}$$

To simplify the notation, we let $h = s\tau\nu$ and consider the first factorial moment

$$E(R-1|s) = \sum_{r=1}^{\infty} \frac{(r-1)h^{r+(k-1)/2}}{(r-1)!(r+k)!I_{k+1}(2\sqrt{h})}$$

$$= \frac{\sum_{r=2}^{\infty}[h^{r+(k-1)/2}]/(r-2)!(r+k)!}{I_{k+1}(2\sqrt{h})}$$

$$= \frac{h^{(k+1)/2}\sum_{j=1}^{\infty} h^j/(j-1)!(j+k+1)!}{I_{k+1}(2\sqrt{h})}$$

$$= \sqrt{h}\, \frac{I_{k+2}(2\sqrt{h})}{I_{k+1}(2\sqrt{h})} \tag{A.5}$$

from which (A.4) follows. Based on (A.5) it is also possible to obtain an approximation for the conditional moment, which holds for large h, namely

$$E(R|s) \simeq \sqrt{h} + \frac{1-2k}{4} \tag{A.6}$$

This is based on the following asymptotic approximation for modified Bessel functions, which holds for large z,

$$I_m(z) \sim \frac{e^z}{\sqrt{2\pi z}}\left[1 - \frac{4m^2-1}{8z} + \dots\right] . \tag{A.7}$$

Thus,

$$E(R-1|s) \simeq \sqrt{h} \; \frac{1 - \frac{4(k+2)^2-1}{16\sqrt{h}} + \dots}{1 - \frac{4(k+1)^2-1}{16\sqrt{h}} + \dots}$$

$$= \sqrt{h} \left[1 - \frac{4(k+2)^2-1}{16\sqrt{h}} + \frac{4(k+1)^2-1}{16\sqrt{h}} + \dots \right]$$

$$= \sqrt{h} + \frac{1-2k}{4} - 1 + \dots$$

from which (A.6) follows. Similarly, it can be shown that

$$E[(R-1)(R-2)|s] = \sqrt{h} \; \frac{I_{k+3}(2\sqrt{h})}{I_{k+1}(2\sqrt{h})} \tag{A.8}$$

and, consequently that

$$E(R^2|s) = E[(R-1)(R-2)|s] + 3E(R|s) - 2$$

$$= \sqrt{h} \; \frac{I_{k+3}(2\sqrt{h}) + 3\,I_{k+2}(2\sqrt{h})}{I_{k+1}(2\sqrt{h})} + 1 \tag{A.9}$$

Based on (A.9), we have the following approximation for the conditional variance:

$$Var(R|s) \simeq \frac{\sqrt{h}}{2} \tag{A.10}$$

Thus, if we define $\mu_0 = \sqrt{h} + (1-2k)/4$, $\sigma_0^2 = \sqrt{h}/2$ and $Z = (R-\mu_0)/\sigma_0$, then we can derive the conditional moment generating function of Z as

$$M_{Z|s}(t) = \exp\left[-(t/\sigma_0)(\mu_0 + \frac{k-1}{2} \right]$$

$$\times \; \frac{I_{k+1}\left(2\sqrt{h}\exp(t/\sigma_0) \right)}{I_{k+1}(2\sqrt{h})} \tag{A.11}$$

Using the approximation (A.7) with (A.11), it can be shown that

$$M_{Z|s}(t) \rightarrow \exp\left(\frac{t^2}{2} \right)$$

as $h \rightarrow \infty$, which implies that Z has a standard normal limiting distribution. Consequently, limits for an approximate $(1-\alpha) \times 100$ percent confidence interval for ν are obtained as solutions to the equations

$$\pm z_{\alpha/2} = \frac{r + \frac{1}{2} - \mu_0}{\sigma_0}$$

$$= \frac{r + \frac{1}{2} - [\sqrt{s\tau\nu} + (1-2k)/4]}{\sqrt{\sqrt{s\tau\nu}/2}} \tag{A.12}$$

The solutions to (A.12), say ν_L and ν_U, have the form given by (5.3.9).

References

1. Abramowitz, M. and Stegun, I.E. (1970), *Handbook of Mathematical Functions*, National Bureau of Standards, Washington, D.C.

2. Ascher, H. (1987), MIL-STD-781C: A vicious circle, *IEEE Transactions on Reliability*, R-36, **4**, 397-402.

3. Ascher, H. (1987), Comments on: Sequential tests of hypotheses for system- reliability modeled by a 2-parameter Weibull distribution, *IEEE Transactions on Reliability*, R-36, **4**, 390-391.

4. Ascher, H. (1990), Comment on 'The nonhomogeneous Poisson process – A model for the reliability of complex repairable systems' by G. Härtler, *Microelectronics & Reliability*, Letter to the Editor.

5. Ascher, H. (1991), Comments on 'Constant failure rate – A paradigm in transition' by J. A. McLinn, *Quality and Reliability Engineering International*, Short Communication.

6. Ascher, H. (1992), Basic probabilistic and statistical concepts for maintenance of parts and systems, *IMA Journal of Mathematics Applied in Business & Industry*, **3**, 153-167.

7. Ascher, H. (1992), Comparison of nonparametric estimators for the renewal function, *Applied Statistics*, **41**, Letter to the Editor, 598-600.

8. Ascher, H. and Feingold, H. (1984), *Repairable Systems Reliability*, Marcel Dekker, New York.

9. Ascher, H., Lin, T.Y., and Siewiorek, D.P., (1992), Modification of: Error log analysis: Statistical modeling and heuristic trend analysis, *IEEE Transactions on Reliability*, **41**, 599-607.

10. Atwood, C.L. (1992), Parametric estimation of time-dependent failure rates for probabilistic risk assessment, *Reliability Engineering and System Safety*, **37**, 181-194.

11. Bain, L.J. and Engelhardt, M. (1980), Inferences on the parameters and current system reliability for a time truncated Weibull process, *Technometrics*, **22**, 421-426.

12. Bain, L.J. and Engelhardt, M. (1982), Sequential probability ratio tests for the shape parameter of a nonhomogeneous Poisson process, *IEEE Transactions on Reliability*, **31**, 79-83.

13. Bain, L.J. and Engelhardt, M. (1986), On the asymptotic behavior of the mean time between failures for repairable systems, In *Reliability and Quality Control* (Ed., A.P. Basu), 1-7, North-Holland, Amsterdam.

14. Bain, L.J. and Engelhardt, M. (1991), *Statistical Analysis of Reliability and Life-Testing Models*, Second edition, Marcel Dekker, New York.

15. Bain, L.J., Engelhardt, M., and Wright, F.T. (1977), Inferential procedures for the truncated exponential distribution, *Communications in Statistics - Theory and Methods*, **6**, 103-111.

16. Bain, L.J., Engelhardt, M., and Wright, F.T. (1985), Tests for an increasing trend in the intensity of a Poisson process: A power study, *Journal of the American Statistical Association*, **80**, 419-422.

17. Bain L.J. and Weeks, D.L. (1964), A note on the truncated exponential distribution, *Annals of Mathematical Statistics*, **35**, 1366-1367.

18. Bartholomew, D J. (1963), The sampling distribution of an estimate arising in life-testing, *Technometrics*, **5**, 361-374.

19. Bassin, W.M. (1973), A Bayesian optimal overhaul interval model for the Weibull restoration process, *Journal of the American Statistical Association*, **68**, 575-578.

20. Basu, A.P. and Crow, L. (1991), Reliability Growth Estimation with Missing Data – I, Invited talk presented at the Joint Statistical Meetings of the American Statistical Association, Biometric Society and Institute of Mathematical Statistics, Atlanta, GA, August 1991.

21. Berman, M. (1981), Inhomogeneous and modulated gamma processes, *Biometrika*, **68**, 143-152.

22. Boswell, M.T. (1966), Estimating and testing trend in a stochastic process of Poisson type, *Annals of Mathematical Statistics*, **37**, 1564-1573.

23. Çinlar, E. (1975), *Introduction to Stochastic Processes*, Prentice-Hall, Englewood Cliffs, NJ.

24. Cohen, A. and Sackrowitz, H.B. (1993), Evaluating tests for increasing intensity of a Poisson process, *Technometrics*, **35**, 446-448.

25. Cox, D.R. (1962), *Renewal Theory*, Methuen, New York.

26. Cox, D.R. and Lewis, P.A.W. (1978), *The Statistical Analysis of Series of Events*, Chapman and Hall, London.

27. Crevecoeur, G. U. (1993), A model for the integrity assessment of aging repairable systems, *IEEE Transactions on Reliability*, **42**, 148-155.

28. Crow, L. (1974), Reliability analysis of complex, repairable systems, In *Reliability and Biometry*, (Eds., F. Proschan and R. J. Serfling), 379-410, SIAM, Philadelphia.

29. Crow, L. (1982), Confidence interval procedures for the Weibull process with applications to reliability growth, *Technometrics*, **24**, 67-72.

30. Crow, L. and Basu, A.P. (1988), Reliability growth estimation with missing data - II, *Proceedings of the Annual Reliability and Maintainability Symposium*.

31. Deemer, W.L. and Votaw, D.F. (1955), Estimation of parameters of truncated or censored exponential distributions, *Annals of Mathemathical Statistics*, **26**, 498-504.

32. Duane, J.T. (1964), Learning curve approach to reliability, *IEEE Transactions on Aerospace*, **2**, 563-566.

33. Engelhardt, M. (1988), Weibull Processes, In *Encyclopedia of Statistical Sciences*, (Eds., N. L. Johnson and S. Kotz), Vol. 9, 557-561, John Wiley & Sons, New York.

34. Engelhardt, M. and Bain, L.J. (1978), Prediction intervals for the Weibull process, *Technometrics*, **20**, 167-169.

35. Engelhardt, M. and Bain, L.J. (1987), Statistical analysis of a compound power-law model for repairable systems, *IEEE Transactions on Reliability*, **36**, 392-396.

36. Engelhardt, M. and Bain, L.J. (1986), On the mean time between failures for re-pairable systems, *IEEE Transactions on Reliability*, **35**, 419-422.

37. Engelhardt, M. and Bain, L.J. (1992), Statistical analysis of a Weibull process with left-censored data, *Proceedings of the NATO Advanced Studies Workshop on Survival Analysis and Related Topics*. (Eds., J.P. Klein and P. Goel), NATO ASI, Series E: Applied Sciences - Vol. 211, Kluwer Academic Publishers, Dordrecht, The Netherlands.

38. Engelhardt, M., Guffey, J.M., and Wright, F.T. (1990), Tests for positive jumps in the intensity of a Poisson process: A power study, *IEEE Transactions on Reliability*, **39**, 356-360.

39. Engelhardt, M., Williams, D.H., and Bain, L.J. (1993), Statistical analysis of a power-law process with left-truncated data, In *Advances in Reliability: A Selection of Papers from the International Conference on Reliability*, (Ed., A. P. Basu), Elsevier Science Publishers, Dordrecht, The Netherlands.

40. Finkelstein, J.M. (1976), Confidence bounds on the parameters of the Weibull process, *Technometrics*, **18**, 115-117.

41. Higgins, J.J. and Tsokos, C.P. (1981), A quasi-Bayes estimate of the failure intensity of a reliability growth model, *IEEE Transactions on Reliability*, **R-30**, 471-475.

42. Lakey, M.J. and Rigdon, S.E. (1992), *The Modulated Power Law Process*, 1992–ASQC Quality Congress Transactions–Nashville, 559-563.

43. Lee, L. (1980), Testing adequacy of the Weibull and log-linear rate models for a Poisson process, *Technometrics*, **22**, 195-199.

44. Lee, L. (1980), Comparing rates of several independent Weibull processes, *Technometrics*, **22**, 427-430.

45. Lee, L. and Lee, S.K. (1978), Some results on inference for the Weibull process, *Technometrics*, **20**, 41-45.

46. Lehmann, E.L. (1959), *Testing of Statistical Hypotheses*, John Wiley & Sons, New York.

47. Molitor, J.T. and Rigdon, S.E. (1993), *Proceedings of the American Statistical Association*, SPES, **1**, 1-4.

48. Møller, S.K. (1976), The Rasch-Weibull process, *Scandinavian Journal of Statistics*, **3**, 107-115.

49. Nelson, W. (1987), *Simple Plots of Repair Data*, Paper presented at the International Conference on Quality Control 1987, October 20-23, Tokyo.

50. Nelson, W. (1988), Graphical analysis of system repair data, *Journal of Quality Technology*, **20**, 24-35.

51. Nelson, W. and Doganaksoy, N. (1989), A computer program for an estimate and confidence limits for the mean cumulative function for cost or number of repairs of repairable products, *General Electric Research and Development TIS Report 89CRD239*.

52. Nelson, W. and Doganaksoy, N. (1991), A method and computer program MCFDIFF to compare two samples of repair data, *General Electric Research and Development TIS Report 91CRD172*.

53. Pyke, R. (1972), Spacings revisited, *Proceedings of 6th Berkeley Symposium on Mathematical Statistics and Probability* (Eds., L. LeCam, J. Neyman and E. Scott), Vol. I, 417-427, Berkeley, University of California Press.

54. Rigdon, S.E. (1989), Testing goodness- of-fit for the power law process, *Communications in Statistics - Theory and Methods*, **18**, 4665-4676.

55. Rigdon, S.E. and Basu, A.P. (1988), Estimating the intensity function of the Weibull process at the current time: Failure-truncation case, *Journal of Statistical Computation and Simulation*, **30**, 17-38.

56. Rigdon, S.E. and Basu, A.P. (1989), The power law process: A model for the reliability of repairable systems, *Journal of Quality Technology*, **21**, 251-260.

57. Rigdon, S.E. and Basu, A.P. (1989), Mean square errors of estimators of the intensity function of a nonhomogeneous Poisson process, *Statistics & Probability Letters*, **8**, 445-449.

58. Rigdon, S.E. and Basu, A.P. (1990), Estimating the intensity function of the Weibull process at the current time: Time truncated case, *Communications in Statistics - Theory and Methods*, **9**, 1079-1104.

59. Rigdon, S.E. and Basu, A.P. (1990), The effect of assuming a homogeneous Poisson process when the true process is a power law process, *Journal of Quality Technology*, **22**, 111-117.

60. Ross, S.M. (1983), *Stochastic Processes*, John Wiley & Sons, New York.

61. Tibor, C. (1993), Some parameter-free tests for trend and their applications to reliability analysis, *Reliability Engineering and System Safety*, **41**, 225-230.

62. Thompson, W.A. (1981), On the foundations of reliability, *Technometrics*, **23**, 1-13.

6

Sequential Procedures in Software Reliability Testing

Shelemyahu Zacks

Binghamton University

ABSTRACT Sequential procedures in software reliability testing are special types of sequential stopping rules, due to the special kind of "experiment", associated with fault detection and rectification during software testing. The present article discusses this sequential testing procedure in the context of time-domain and data-domain reliability models.

Keywords: Faults detection, Sequential stopping rules, Software reliability

CONTENTS

6.1   Introduction and Summary

Software is a complex system of programs, routines and symbolic languages that control the functioning of the hardware of a computer, and direct its operation.

Software design begins with functional specifications and continues into translated sequences of instructions. The development process consists of at least nine steps, from problem statement to run-time check of executable codes, and final documentation. Much has been written on this subject, and the reader is referred to books on software engineering. More complicated is the notion of software reliability (see [36]). A specific software is considered reliable, if it performs its specified functions or tasks according to defined standards or expectations. Deviations from the specified standards in execution of functional requirements are caused by **faults** in parts of the software which relate to these functions. Such deviations degrade the reliability of the software. The question is how to quantify the notion, and arrive at an operational index of software reliability. In the present article we present statistical formulations, which are connected with modeling the process of fault detection and correction during final testing (in the certification phase).

Testing of a particular software is designed to activate certain of its branches (chains of modules) in order to detect, via the system failures, the faults which cause it. This can be done by randomly choosing samples of functional requirements for which the software is designed (see [10]). The certification (testing) process records the execution times (CPU) for test case samples, which are run against successive increments. The interfailure times of the system are recorded. The most important questions are, which modules should be tested and how long one should continue this testing process? The answer depends on the objectives, cost of testing, cost of failure in the field (after release of the software) etc. Moreover, assessment of the reliability of the software at any time during the test process depends on the model assumed for the activation of faulty modules. Indeed, if a certain module of a given software contains faults but is never activated (applied) then these faults will never be revealed. Most of the models in the literature pertaining to this matter are the so called "**time domain**" models, which consider the time till a fault is activated as a random variable realized according to some stochastic process. The other types of models are called "**data domain**" models. These models consider the software system as a finite population of functional units. The realization of faults is due to selection of faulty units according to some sampling scheme. In Section 6.2 we discuss these models and set up the framework for further developments.

The focus of attention in the present article is the sequential stopping problem, to answer the question: when should one stop testing and release the software. Not much is available in the literature on sequential testing for software reliability. In [12] and a recent study of [13], the authors provide optimal stopping rule for some time domain model. In Section 6.3 we study and demonstrate the performance of sequential stopping rules for time domain models which are discussed in Section 6.2. In Section 6.4 we study such sequential stopping rules for data domain models. We consider quantal prediction models, which have not yet been discussed in the literature on software reliability, but as demonstrated have high potential for yielding interesting results. We modify also the capture-recapture sampling scheme to suit the present testing problem. Although the purely sequential strategy may not always be readily applicable, it can be modified to group sequential sampling, or multi-stage sampling, as appropriate to the special circumstances.

The sequential stopping for software reliability problems, which are discussed in the present paper are different from the classical sequential procedures. In the classical procedures, one can generate new independent realizations of a random variable X, as long as one wishes to continue testing. In the software reliability context, if faults are detected and corrected during testing, the random variables that can be observed are **not** independent. The time required until additional faults are observed increases stochastically. Eventually it becomes very difficult or impossible to detect new faults. This fact requires the development of new statistical procedures.

6.2   Models for Software Reliability

6.2.1   Time Domain Models

Consider a list (ordered set) of a large number N^* of functional requirements. Each element in this list activates a certain branch of the software. We say that the execution of a functional requirement is successful if the system does not fail. If the system fails an immediate diagnosis and modification (correction) takes place. If at a given time N functional requirements in the list could activate branches with faults, we say that the software includes N faults. Time domain models assume that the time (measured on a cumulative CPU clock) until a functional requirement having a fault is activated follows a random stochastic process, which is generally taken as a non-homogeneous Poisson process. Different faults are assumed to be connected with **independent** processes having different failure rate functions. Thus, system failure times follow a random **competing risk** model. In [20], the authors assumed that all the elements leading to faults are associated with the same homogeneous Poisson process, with failure rate (intensity) Λ. According to their model, the time T_i $(i = 1, 2, \ldots, N)$, between the $(i-1)$st and the i-th failure, is a random variable having an exponential distribution with mean

$$\mu_i = 1/(\Lambda(N - i + 1)), \quad i = 1, 2, \ldots, N \ . \tag{6.2.1}$$

In the time interval $[0, t]$ one observes $J(t)$ failures if $T_1 + \ldots + T_{J(t)} \leq t < T_1 + \ldots + T_{J(t)+1}$, where $J(t) \equiv 0$ if $T_1 > t$. the **software reliability** is defined as the probability that the system will operate without failure for a specified length of time τ. Thus, according to the Jelinski-Moranda model, the software reliability at time t is

$$R(\tau; t) = \exp\{-\tau\Lambda(N - J(t) + 1)\} \ . \tag{6.2.2}$$

The problem in determining $R(\tau; t)$ is that the parameters Λ and N are unknown. Jelinski and Moranda suggested to estimate Λ and N by the maximum likelihood estimators (MLE). In [25], the authors criticized the above model, which assigns every faulty functional requirement the same failure rate Λ, and which is stated in terms of the parameter N; arguing that it is not interesting to estimate N, but the attention should be focused on reliability estimation. Thus, the Littlewood-Verall model [25] assumes that the inter-failure times $T_1, T_2, \ldots$ are conditionally independent and exponentially distributed with means $M_i = 1/\Lambda_i$, $i = 1, 2, \ldots$ given $\Lambda_1, \Lambda_2, \ldots$. Furthermore, $\Lambda_1, \Lambda_2, \ldots$ are independent random variables having Gamma distributions with densities

$$g_i(\lambda; \psi(i), \nu) = \frac{(\psi(i))^\nu}{\Gamma(\alpha)} \lambda^{\nu-1} \exp\{-\psi(i)\lambda\}, \quad 0 < \lambda < \infty \ . \tag{6.2.3}$$

and where $\psi(i) = \beta_0 + \beta_1 i$, $i = 1, 2, \ldots$. Notice that, according to this model, the predictive distribution of T_i is Pareto with p.d.f.

$$f_i(t; \alpha, \psi(i)) = \frac{\nu}{\psi(i)} \left(1 + \frac{t}{\psi(i)}\right)^{-(\nu+1)}, \quad 0 < t < \infty \ . \tag{6.2.4}$$

Thus, if $\nu > 1$ and $\beta_0, \beta_1 > 0$, the MTBF's of the i-th interfailure time is $E\{T_i\} = \frac{\beta_0 + \beta_1 i}{\nu - 1}$, $i = 1, 2, \ldots$. This forms an increasing sequence (reliability growth). Moreover, the software reliability function at time t, given $J(t)$, is

$$R(\tau; t, J(t)) = \left[\frac{\beta_0 + \beta_1 J(t)}{\beta_0 + \beta_1 J(t) + \tau}\right]^\nu, \quad \tau \geq 0 \ . \tag{6.2.5}$$

This model requires the estimation of the parameters β_0, β_1 and α from the failure data.

In [10], the authors considered geometrically increasing MTBF's. They assumed that the interfailure times $T_1, T_2, \ldots$ are independent, having exponential distributions with MTBF's

$$\mu_i = \mu/\alpha^{i-1} , \quad i = 1, 2, \ldots \tag{6.2.6}$$

where μ is the initial MTBF of the software, and $1/\alpha$, $0 < \alpha < 1$, is the growth rate at each engineering modification. The authors called this model the "Certification Model". The reliability function at time t, given $J(t)$, according to this model is

$$R(\tau; t, J(t)) = \exp\{-\tau\alpha^{J(t)}/\mu\} \ . \tag{6.2.7}$$

This model requires the estimation of two parameters μ and α from the failure data. Several variations of the above time domain models have been studied in the literature. In [11], the author provided general survey of dozens of models with references and main characteristics. In [24], the authors considered the time domain model of Jelinski and Moranda within a Bayesian framework, and showed that the Littlewood-Verall model and that in [17] can be obtained as special cases.

It should be remarked, before concluding the present section, that all the above models, which assume independent interfailure times are simplified models. Actually, the statistical estimation and stopping problems, connected with software reliability models, belong to the class of estimating the parameters of some life distributions when the information is based on **time censored order statistics**, and the **sample size is unknown**. In the Jelinski Moranda model, for example, the $J(t)$ failure times $0 < t_1 < \ldots < t_{J(t)} \leq t$, are the first $J(t)$ order statistics in a sample of N i.i.d. random variables, having a common exponential distribution with parameter Λ. The problem is to estimate the unknown sample size, N, and the parameter Λ from the observed data. From this point of view the papers by [2], [3], and [4] are all related to this field of research.

6.2.2   Data Domain Models

Data domain models are not based on the assumption of the existence of random time processes for the activation of faulty functional requirements, but develop the theory on the model of sampling surveys. Accordingly, the model is that the population consists of N^* elements (the functional requirements). N unknown elements of the list are "faulty" and the parameter N is unknown. Elements are drawn from the list (at random or not). Faulty elements are modified and corrected. Let J_n be the number of faulty elements among the first n elements drawn, then the reliability after testing n elements can be defined as

$$R_n(J_n) = \left(1 - \frac{N - J_n}{N^*}\right)^{n_0} , \qquad (6.2.8)$$

where n_0 is a specified integer. If $P_n = (N - J_n)/N^*$ is the proportion of the remaining faulty elements, after sampling n elements then, when n_0 is large,

$$R_n(P_n) \approx \exp\{-n_0 P_n\} . \qquad (6.2.9)$$

There are data-domain models in the literature (see [39]) which follow the models of estimating the number of different species in a population by the capture-recapture methods (see [42]). According to the capture-recapture method, any fault detected in the software is "tagged" (documented but not corrected) and replaced into the population (list). Thus, the system may fail again due to faults which have already been observed. After each test trial the number of unobserved faults in the software is estimated.

We introduce a new class of data domain models. These are **models with covariates**. Let $\mathcal{P} = \{u_1, u_2, \ldots, u_{N^*}\}$ be the population of functional requirements. Each such functional requirement defines a module of software elements. We specify a certain number, p, of covariates $x_1, \ldots, x_p$, which are associated with the modules. These are quantified variables like size of module, length of preparation time, complexity index, etc. We assume that the values of the covariates are **known** for all the elements of $\mathcal{P}$. Let J_i denote the number of faults in the i-th element of $\mathcal{P}$, and let $y_i = 1$ if $J_i \geq 1$ and $y_i = 0$, otherwise. The values of J_i and y_i $(i = 1, \ldots, N^*)$ are unknown prior to testing of a module. We introduce, however, probabilistic models, which related the covariates to the distributions of J_i and y_i $(i = 1, \ldots, N^*)$. With the models we can predict the risk associated with leaving modules untested. These predictions lead to sampling rules (generally not random), which are designed to minimize the terminal risk of the system. See the formulation in Section 6.4 below.

6.3 Sequential Stopping for Time Domain Models

In the present section we discuss the problem of sequential estimation and testing, focussing on the problem of stopping times only. For a given (parametric) model we have to decide when to stop testing the software, and, at termination of testing, to state at an appropriate confidence level what is the software reliability.

The decision concerning the optimal stopping time involves balancing the cost of testing versus the cost (penalty) for failures in the field. If the cost of testing is negligible compared to the cost of failure in the field, we may wish to stop as soon as there is sufficiently high confidence level that the reliability $R(\tau; t, J(t))$ is greater than a specified value ρ_0. We illustrate this first with regard to the Certification Model.

6.3.1 Sequential Stopping and Estimation For the Certification Model

The Certification Model assumes that the interfailure times are independent exponential, with $E\{T_i\} = \mu/\alpha^{i-1}$, $i = 1, 2, \ldots$. Thus, when α and μ are known, the reliability exceeds ρ_0 right after $k_0(\mu, \alpha)$ modifications, where,

$$k_0(\mu, \alpha) = 1 + \text{int}\left[\frac{\log(-\log \rho_0) + \log \mu - \log \tau}{\log \alpha}\right], \qquad (6.3.1)$$

int[·] denotes the integer part of the term within [·]. For example, if $\mu = 100$, $\tau = 200$ [time units] and $\rho_0 = 0.99$, we obtain the following values of $k_0(\mu, \alpha)$, as α varies:

α	0.70	0.75	0.80	0.85	0.90	0.95
$k_0(\mu, \alpha)$	15	19	24	33	51	104

When μ and α are unknown, we could estimate them and base the stopping rule on these estimates. For example, given k interfailure times, $T_1, \ldots, T_k$, the likelihood function is

$$L_k(\mu, \alpha) = \frac{\alpha^{k(k-1)/2}}{\mu^k} \exp\left\{-\frac{1}{\mu}\sum_{i=1}^{k} \alpha^{i-1}T_i\right\}. \qquad (6.3.2)$$

Let $\hat{\mu}_k$ and $\hat{\alpha}_k$ be the MLE's after k failures. We can show that

$$\hat{\mu}_k = \frac{1}{k}\sum_{i=1}^{k}(\hat{\alpha}_k)^{i-1}T_i, \qquad (6.3.3)$$

and $\hat{\alpha}_k$ is a root of the polynomial equation

$$\sum_{i=0}^{k-1} b_{i,k}\alpha^i = 0, \qquad (6.3.4)$$

where

$$b_{i,k} = \frac{k-1-2i}{k-1}T_{i+1}, \quad i = 0, \ldots, k-1. \qquad (6.3.5)$$

TABLE 6.1
Means and standard-errors of 500 MLE from simulated samples of size $k = 20$,
$\alpha = .7(.05).95, .975$ and $\mu = 100^*$

α	MLE	
	$\hat{\alpha}$	$\hat{\mu}$
.7	.701	105.21
	.030	46.95
.75	.751	105.08
	.030	44.33
.80	.801	105.73
	.033	47.80
.85	.850	104.14
	.033	46.76
.90	.900	103.23
	.036	47.21
.95	.951	105.08
	.038	44.33
.975	.976	105.73
	.040	47.80

The sequence $\{b_{i,k}; i = 0, \ldots, k - 1\}$ has one change of sign. Hence Eq. (6.3.4) has a **unique** positive real root $\hat{\alpha}_k$ (see [19, p. 442]). If this root is in $(0, 1)$ then it is the unique MLE. The value of $\hat{\alpha}_k$ can be found by Newton-Raphson iterative equation, starting with $\alpha_0 > 1$.

The inverse of the Fisher information matrix on (μ, α) after k observations is

$$\Psi_k = \frac{12}{k(k^2 - 1)} \begin{bmatrix} \frac{(k-1)(2k-1)}{6}\mu^2 & \frac{k-1}{2}\mu\alpha \\ \frac{k-1}{2}\mu\alpha & \alpha^2 \end{bmatrix} . \tag{6.3.6}$$

The MLE of $R_k(\tau; \mu, \alpha)$ after k observations has the asymptotic variance

$$AV\{\hat{R}_k(\tau; \hat{\mu}_k, \hat{\alpha}_k)\} = \frac{4}{k} R_k^2(\tau; \mu, \alpha) \frac{\tau^2 \alpha^{2k}}{\mu^2} , \tag{6.3.7}$$

where $R_k(\tau; \mu, \alpha) = \exp\{-\tau\alpha^k/\mu\}$.

In Table 6.1 we present the means and standard errors of 500 MLE's of α and μ obtained from simulated samples of size $k = 20$.

The asymptotic standard error for $\hat{\mu}$, for the parameters of Table 6.1, as obtained from Eq. (6.3.6), is A.S.E. $\{\hat{\mu}\} = 43.095$. This value is slightly below the simulated estimates of Table 6.1. On the other hand, the asymptotic standard errors of $\hat{\alpha}$ are very close to those of Table 6.1. Indeed from Eq. (6.3.6) we obtain

α	0.70	0.75	0.80	0.85	0.90	0.95
A.S.E.	0.027	0.029	0.031	0.033	0.035	0.037

We see that the asymptotic theory of maximum likelihood estimation provides good estimates even for samples of size $k = 20$.

We consider now the effect of the sequential stopping on the MLE's.

Consider the stopping variable

$$K = \text{least } k \geq 1 \text{ such that } R_k(\tau; \hat{\mu}_k, \hat{\alpha}_k) \geq \rho_0 . \tag{6.3.8}$$

At stopping, we estimate the reliability by $R_K(\tau; \hat{\mu}_k, \hat{\alpha}_K)$. Simulations could provide information on the sampling distributions of K, $\hat{\mu}_k$ and $\hat{\alpha}_K$.

TABLE 6.2
Summary statistics for 100 simulations of K, $\hat{\alpha}_K$, $\hat{\mu}_K$, when $\alpha = 0.925$, $\mu = 100$ and $\rho_0 = 0.99$

Statistic	K	$\hat{\alpha}_K$	$\hat{\mu}_K$
Mean	68.45	0.924	100.96
Median	69.00	0.924	99.60
Std. Dev.	3.34	0.006	23.14

In Table 6.2 we provide some summary statistics for 100 simulation runs of K, $\hat{\alpha}_K$ and $\hat{\mu}_K$, for the case of $\alpha = .925$, $\mu = 100$ and $\rho_0 = 0.99$. The empirical distributions of $\hat{\alpha}_K$ and $\hat{\mu}_K$ were found to be close to normal. Kolmogorov-Smirnov test of normality resulted with P-values greater than 0.15. It is interesting also to note that the simulation estimates of the S.E. $\{\hat{\alpha}_K\}$ and S.E. $\{\hat{\mu}_K\}$ are in excellent agreement with those obtained from Eq. (6.3.6), by substituting $k = E\{K\}$.

6.3.2 Bayes Sequential Stopping For the Littlewood-Verall Model

In the present section we formulate the sequential decision problem for a variation of the Littlewood-Verall model with $\psi(i) = \psi$ (constant) for all $i = 1, 2, \ldots, N$. Accordingly, let $T_1, \ldots, T_N$ be independent random variables having a common Pareto distribution with p.d.f.

$$f(t \mid \nu, \psi) = \frac{\nu}{\psi}\left(1 + \frac{t}{\psi}\right)^{-(\nu+1)}, \qquad 0 \le t < \infty . \tag{6.3.9}$$

These random variables designate the time till failure of the N fault requirements.

Let $T_{(1)} \le T_{(2)} \le \ldots \le T_{(N)}$ be the corresponding order statistic. $\tau_i = T_{(i)}$ is the i-th observed failure time of the system. After n failures we have observed the realization of $\tau_1 < \tau_2 < \ldots < \tau_n$. If the $(n+1)$st failure has not yet occurred we can potentially observe $(N - n)$ future failures at future epochs $\tau_{n+1} < \ldots < \tau_N$. The value of N is, however, unknown. Obviously $N \ge n$. If $N = n$ we will not observe additional failures.

Let $u = \log\left(1 + \frac{t}{\psi}\right)$ be a time transformation, and let $U_i = \log\left(1 + \frac{\tau_i}{\psi}\right)$, $i = 1, \ldots, N$. Obviously $U_1 < U_2 < \ldots < U_N$. Notice that $U = \log\left(1 + \frac{T}{\psi}\right)$ has an exponential distribution, with mean $1/\nu$. Thus, $U_{(1)} < U_{(2)} < \ldots < U_{(N)}$ are the N order statistics of a random sample of size N from this exponential distribution. Moreover, in order to apply this transformation the parameter ψ should be known. Let $J(u)$ denote the number of U_i values smaller than u, i.e.,

$$J(u) = \sum_{i=1}^{N} I\{U_i \le u\}, \qquad 0 \le u < \infty , \tag{6.3.10}$$

where $I\{A\}$ is the indicator variable, assuming the value 1 on A, and the value 0 otherwise.

Let $W_n = \sum_{i=0}^{n} U_i$, with $U_0 \equiv 0$. Define the stochastic process $\{W(u); 0 \le u < \infty\}$, as

$$W(u) = \sum_{n=0}^{N-1} I\{U_n \le u < U_{n+1}\}W_n + W_N I\{U_N \le u\}.$$

The likelihood function of (N, ν) at (transformed) time u is

$$L(N, \nu; u, J(u), W(u)) = \frac{N!}{(N - J(u))!}\nu^{J(u)} \exp\{-\nu W(u) - \nu(N - J(u))u\} . \tag{6.3.11}$$

From Eq. (6.3.11) we imply that $\{(J(u), W(u)); 0 \leq u\}$ is a minimal sufficient process.

In a Bayesian framework we assume a prior joint distribution for the unknown parameters (N, ν). We assume here for example that N and ν are priorly independent; that N has a prior Poisson distribution with mean η, and that ν has a prior gamma distribution $G\left(\frac{1}{\xi}, \alpha\right)$; i.e., $E\{\nu\} = \alpha\xi$, $0 < \alpha$, $\xi < \infty$.

The prior p.d.f. of (N, ν) is

$$h(N, \nu)d\nu = e^{-\eta} \frac{\eta^N}{N!} \frac{1}{\Gamma(\alpha)\xi^\alpha} \nu^{k-1} e^{-\nu/\xi} d\nu, \quad 0 < \nu < \infty, \quad N = 0, 1, 2, \dots \quad (6.3.12)$$

The posterior p.d.f. of (N, ν) at the transformed time u is

$$
\begin{aligned}
&h(N, \nu \mid u, J(u), W(u)) \\
&= \frac{\eta^{N-J(u)} \nu^{\alpha+J(u)-1} \exp\left\{-\nu\left[W(u) + \frac{1}{\xi} + (N - J(u))u\right]\right\}}{\Gamma(\alpha + J(u))(N - J(u))! \sum_{m=0}^{\infty} \frac{\eta^m}{m!}\left[W(u) + \frac{1}{\xi} + mu\right]^{-(\alpha+J(u))}}, \quad (6.3.13)
\end{aligned}
$$

for $N \geq J(u)$, $0 < \nu < \infty$.

Let $M(u) = N - J(u)$, $0 < u < \infty$, be the number of remaining faults at time u. Obviously $M(u_1) \geq M(u_2)$ for all $u_1 < u_2$ and $\lim_{u \to \infty} M(u) = 0$ a.s. Let $\eta(u, J(u), W(u))$ denote the posterior expectation of $M(u)$, at time u, given $(J(u), W(u))$.

Define, for $\eta, \lambda, x, u > 0$,

$$\Psi(\eta, \lambda, x, u) = \sum_{m=0}^{\infty} \frac{\eta^m}{m!}(x + um)^{-\lambda}. \quad (6.3.14)$$

Obviously, $0 < \Psi(\eta, \lambda, x, u) < \exp\{\eta\} < \infty$. In terms of this function we can write the posterior expectation of $M(u)$ as

$$\eta(u, J(u), W(u)) = \eta \frac{\Psi\left(\eta, \alpha + J(u), W(u) + \frac{1}{\xi} + u, u\right)}{\Psi\left(\eta, \alpha + J(u), W(u) + \frac{1}{\xi}, u\right)}. \quad (6.3.15)$$

Recall that $J(u)$ and $W(u)$ are random step functions with jumps at the (transformed) failure points $U_1, U_2, \dots$.

One may encounter computational difficulties if η is large. We suggest to use the approximation

$$\eta(u, i, W) \cong \eta \frac{\int_{\eta-4\sqrt{\eta}}^{\eta+4\sqrt{\eta}} \exp\left\{-\frac{1}{2\eta}(x - \eta)^2\right\}\left(W + \frac{1}{\xi} + (x + 1)u\right)^{-(\alpha+i)} dx}{\int_{\eta-4\sqrt{\eta}}^{\eta+4\sqrt{\eta}} \exp\left\{-\frac{1}{2\eta}(x - \eta)^2\right\}\left(W + \frac{1}{\xi} + xu\right)^{-(\alpha)} dx}$$

when $\eta > 50$. The integrals in this formula can be evaluated numerically.

LEMMA 6.3.1

The posterior expectation $\eta(u, J(u), W(u))$ is a decreasing function of u in intervals between failure points, i.e., if $U_i < u_1 < u_2 < U_{i+1}$ then $\eta(u_1, i, W_i) \geq \eta(u_2, i, W_i)$ for all $i = 1, 2, \dots$ where $W_i = \sum_{j=1}^{i} U_j$.

PROOF   The posterior p.d.f. of M, given $J(u) = i$, $W(u) = W_i$ is

$$\pi(m \mid u, i, W_i) = \frac{\eta^m}{m!}\left(W_i + \frac{1}{\xi} + um\right)^{-(i+\alpha)}$$

TABLE 6.3
Values of the posterior expectation

i	U_i	W_i	$\eta(U_i, i, W_i)$	$\eta(U_{i+1}, i, W_i)$
1	0.0349	0.0349	9.50532	8.99670
2	0.1047	0.1396	8.57648	8.15402
3	0.1729	0.3125	7.70003	6.80413
4	0.3500	0.6624	6.35925	5.49504
5	0.5495	1.2120	5.20678	4.98426
6	0.6006	1.8126	4.79722	1.22756
7	1.6674	3.4800	2.03731	1.83647
8	1.7586	5.2386	2.32969	2.11596
9	1.8612	7.0998	2.45796	1.83194
10	2.2050	9.3049	2.17824	0.19968*

* Computed at $U = 5$.

$$\div \Psi \left(\eta, \alpha + i, W_i + \frac{1}{\xi}, u \right) . \tag{6.3.16}$$

The family $\mathcal{P}_{i,W_i} = \{ \pi(\cdot \mid u, i, W_i), U_i < u < U_{i+1} \}$ has the monotone likelihood ratio property. Indeed, for $U_i < u' < u'' < U_{i+1}$,

$$\frac{\pi(m \mid u'', i, W_i)}{\pi(m \mid u', i, W_i)} = \frac{\Psi \left(\eta, \alpha + i, W_i + \frac{1}{\xi}, u' \right)}{\Psi \left(\eta, \alpha + i, W_i + \frac{1}{\xi}, u'' \right)} \cdot \left(\frac{W_i + \frac{1}{\xi} + u'm}{W_i + \frac{1}{\xi} + u''m} \right)^{i+\alpha} .$$

This ratio is a decreasing function of m. Hence, by Karlin's Lemma (see [49, p. 124]), the expected value of M is a decreasing function of u. ∎

In Table 6.3 we present the function $\eta(u, J(u), W(u))$ for 10 values of U_i, simulated from the exponential distribution with mean 1. The Bayesian parameters are $\eta = 10$, $\alpha = 1$ and $\xi = 1$.

We see in Table 6.3 the discontinuities of $\eta(u; J(u), W(u))$ at the failure points. The values between $\eta(U_i, i, W_i)$ and $\eta(U_{i+1}, i, W_i)$ can be approximated by linear interpolation.

In order to determine optimal stopping times, let $K[\$]$ be the cost (penalty) of finding a fault in the field (after release of the software), and let $c[\$]$ be the cost of testing per unit of transformed time. Thus, the posterior risk if testing is stopped at time u is

$$R(u; J(u), W(u)) = cu + K\eta(u, J(u), W(u)) . \tag{6.3.17}$$

The following is a reasonable stopping rule:
  SR1: **At time u compute**

$$\tilde{U}_1(u) = \arg \min_{t \geq u} R(t, J(u), W(u)) . \tag{6.3.18}$$

  Stop as soon as $\tilde{U}_1(u) = u$.
We denote this stopping time by u_1^0.

In Table 6.4 we illustrate SR1 on the data of Table 6.2, with $c = 0.2K$.

Since there are initially $N = 10$ faults, there are no more observed failures after $U_{10} = 2.205$ ($U_{11} = \infty$). Hence, the testing stops at $u_1^0 = 4.905$. At stopping the Bayesian estimator of $M(4.905)$ is $\eta(4.905, 10, 9.305) = 0.2149$ with posterior risk of $R(4.91, 10, 9.305) = 1.196$.

TABLE 6.4
Values of $\check{U}_i$, $N = 10$, $\alpha = 1$, $\xi = 1$, $\eta = 10$

i	U_i	W_i	$\check{U}_i$
1	0.035	0.035	2.335
2	0.105	0.140	8.605
3	0.173	0.312	4.773
4	0.350	0.662	3.650
5	0.550	1.212	3.250
6	0.601	1.813	3.101
7	1.667	3.480	3.767
8	1.759	5.239	4.159
9	1.861	7.100	4.561
10	2.205	9.305	4.905

TABLE 6.5
Results of 10 simulation runs, for SR1 with $N = 20$, $\nu = 1$, $\eta = 20$, $\xi = 1$, $\alpha = 1$, $K = \$100$.

i	$c = 0.1 \cdot K$		$c = K$	
	u_1^0	AC^*	u_1^0	AC
1	5.981	59.81	3.781	378.1
2	4.913	49.13	3.213	321.3
3	7.677	76.77	4.777	477.7
4	5.920	59.20	3.397	439.7
5	6.315	63.15	3.175	517.5
6	6.391	63.91	3.144	514.4
7	8.566	85.66	5.266	526.6
8	7.326	73.26	3.658	565.8
9	6.333	63.33	3.567	456.7
10	5.426	54.26	3.526	352.6

*AC denotes actual cost $= cu^0 + K \cdot M$

In Table 6.5 we present the results of 10 simulation runs of the above stopping rule, when the system contains initially $N = 20$ faults and $\nu = 1$. The Bayesian parameters are $\eta = 20$, $\xi = 1$ and $\alpha = 1$, and $c = 0.1 \cdot K$.

The average actual cost when $c = 0.1K$ is $\bar{AC} = 64.85$ while, when $C = K$ we obtained $\bar{AC} = 455.04$. This is intuitively clear, since, when the cost of testing is small relative to the cost of fault in the field, one can prolong the testing period and correct most of the faults.

Another possible stopping rule is

SR2: **Let**

$$\tilde{U}_2(u) = \inf\{t : t \geq u, R(t, J(u), W(u)) < 1\} \qquad (6.3.19)$$

Stop as soon as $u = \tilde{U}_2(u)$.

That is, after the i-th failure ($i = 1, 2, \ldots$) we find the smallest value of $u \geq U_i$ for which $\eta(u, i, W_i) < 1$. If there is no additional failure by this time we stop testing.

Let u_2^0 denote the stopping time according to SR2. In Table 6.6 we present the results of 10 simulation runs of SR2, with parameters equal to those of Table 6.5. The actual cost is computed for $K = 100[\$]$ and $c = 0.1 \cdot K$.

The average actual cost in Table 6.6 is $\bar{AC} = 123.29$.

We see that under similar conditions SR1 yields an average actual cost of 64.58 while SR2 yields an average actual cost of 123.29. Stopping rule SR1 is a better stopping rule for

TABLE 6.6
Results of 10 simulation runs for SR2, with $\nu = 1$, $\alpha = 1$, $\xi = 1$, $N = 20$, $\eta = 20$, $K = 100$, $c = 0.1K$.

i	u_2^0	AC
1	3.271	132.71
2	2.913	29.13
3	4.077	40.77
4	3.097	130.97
5	2.875	228.75
6	2.844	228.44
7	4.066	40.66
8	3.458	234.58
9	3.367	133.67
10	3.326	33.26

this type of penalty structure (actual cost). It is of interest to investigate the relationship between SR1 and a Bayes sequential optimal stopping rule. This is generally a difficult problem. In [12] and [13], the authors studied the structure of optimal stopping rules, but their model is different.

If one decides to stop at time u, the posterior risk at time u is $R(u, J(u), W(u))$. The alternative is to continue testing, following the optimal strategy.

Let $R^0(u, J(u), W(u)) = R(\tilde{U}_1(u), J(u), W(u))$, where

$$R(\tilde{U}_1(u), J(u), W(u)) = \min_{t \geq u} R(t, J(u), W(u)) \ . \tag{6.3.20}$$

Let $\rho(u, J(u), W(u))$ denote the minimal Bayes risk, following the optimal strategy. $\rho(u, J(u), W(u))$ satisfies the Dynamic Programming Equation (DPE)

$$\rho(u, J(u), W(u)) = \min\{R^0(u, J(u), W(u)),$$
$$\inf_{t \geq u}[ct + E\{\rho(t, J(t), W(t)) \mid u, J(u), W(u))]\} \tag{6.3.21}$$

where the expectation on the r.h.s. of (6.3.21) is with respect to the **predictive** distribution of $(J(t), W(t))$, given $(u, J(u), W(u))$. Notice that $\rho(u, J(u), W(u)) \leq R^0(u, J(u), W(u))$ for all $0 < u < \infty$.

Consider two cases for $i = 1, \ldots, N - 1$,

    (i) $U_i < u < U_{i+1} < \tilde{U}_1(u)$;

    (ii) $U_i < u < \tilde{U}_1(u) \leq U_{i+1}$.

In case (i), if one stops before U_{i+1}, at U^*, $U_i < u \leq U^* < U_{i+1}$, then, since $J(U^*) = J(U_i) = i$ and $W(U^*) = W_i$, $R(U^*, J(U^*), W(U^*)) > R^0(u, J(u), W(u)) \geq \rho(u, J(u), W(u))$. Thus, the optimal stopping rule cannot stop before U_{i+1}. In Case (ii) SR1 stops at $\tilde{U}_1(u) < U_{i+1}$. As in case (i), the optimal strategy cannot stop before time $\tilde{U}_1(u)$. Thus, we have shown that, as long as SR1 rules to continue testing, so does the optimal stopping rule. The optimal stopping rule may, however, continue testing beyond the time point u_1^0. We can show, however, that SR1 is **asymptotically optimal**, i.e.,

$$\lim_{u \to \infty} R^0(u, J(u), W(u)) = \lim_{u \to \infty} \rho(u, N, W_N),$$

a.s. [N]. It is interesting to characterize the optimal stopping rule for small values of u, but this will not be done here.

6.3.3   Empirical Bayesian Analysis For The Littlewood-Verall Model

Consider again the Littlewood-Verall model according to which the time till failure of the
N faults are, conditional on N, i.i.d. random variables with a Pareto p.d.f. (6.3.9). As
before, we assume that ψ is known, and make the time transformation $u = \log\left(1 + \frac{t}{\psi}\right)$.
Given n failures, at transformed times $U_1 < U_2 < \ldots < U_n$ the marginal likelihood of
the total number of initial faults, N, is given by (6.3.11). If we consider ν known and
use a Poisson prior distribution for N, with mean η, then the posterior distribution of
$M(u)$ is also Poisson, with mean $\eta(u) = \eta e^{-\nu u}$. It is interesting that, in contrast to the
results of the previous section, if ν is known, then the Bayesian inference on the number
of remaining faults $M(u)$, for any prior distribution of N, is independent of the number
of failures up to time u, $J(u)$, and on the times of failure $U_1, \ldots, U_{J(u)}$. Thus, when ν is
known, the optimal stopping time for the posterior risk

$$\rho(u; \eta, \nu) = cu + K\eta e^{-\nu u} \tag{6.3.22}$$

is

$$u_3^0 = -\frac{1}{\nu} \log\left(\frac{c}{K\eta\nu}\right) . \tag{6.3.23}$$

In the empirical Bayesian analysis we estimate the prior parameters η, ν from the data.
We provide here the main results, the interested reader should see [48] for further details.

The **anticipated likelihood** function for (η, ν), under the Poisson prior, is the prior
expected value of (6.3.11), namely

$$L^*(\eta, \nu; u, J(u), W(u)) = (\eta\nu)^{J(u)} \exp\{-\eta(1 - e^{-\nu u}) - \nu W(u)\} . \tag{6.3.24}$$

We first notice that, as long as $J(u) = 0$

$$L^*(\eta, \nu, 0, 0) = \exp\{-\eta(1 - e^{-\nu u})\} . \tag{6.3.25}$$

In this case the MLE of η and ν are $\hat{\eta}(u) = 0$ and $\hat{\nu}(u) = 0$. These values are, however,
inadmissible. For values of u for which $J(u) \geq 1$, the MLE of η and ν are

$$\hat{\eta}(u) = \frac{J(u)}{1 - e^{-u\hat{\nu}(u)}} , \tag{6.3.26}$$

and $\hat{\nu}(u)$ is the root ν of the equation

$$\nu = \frac{1}{E(u) + \frac{ue^{-u\nu}}{1 - e^{-u\nu}}} , \tag{6.3.27}$$

where $E(u) = \frac{W(u)}{J(u)}$. Notice that $E(u)$ is the average of the U_i values at time u. Also,

$$\lim_{u \to \infty} \hat{\nu}(u) = \left(\lim_{u \to \infty} E(u)\right)^{-1} = \left(\frac{1}{N} \sum_{i=1}^{N} U_i\right)^{-1} , \quad \text{a.s. } [N] . \tag{6.3.28}$$

Thus, $\hat{\nu}(u)$ is asymptotically the classical MLE of the parameter ν of the exponential
distribution, which is a function of the complete sample of N failure times.

We illustrate these MLE with the simulated data of Table 6.2.

At time $u = 0.5$ we have $J(u) = 4$ and $W(u) = 0.662$. Thus, $E(0.5) = 0.1655$. The
l.h.s. of (6.3.27) is monotonically decreasing in ν, and is equal to $E(0.5)$ at $\nu \doteq 4.35$.
Thus, $\hat{\nu}(0.5) = 4.35$ and $\hat{\eta}(0.5) = \frac{4}{1 - e^{-0.5 \times 4.35}} = 4.5127$. At $u = 2.0$, $J(u) = 9$ and
$W(u) = 7.100$. Thus $E(2.0) = 0.7889$, $\hat{\nu}(2.0) = 0.667$ and $\hat{\eta}(2.0) = 11.3411$. In this

TABLE 6.7
Sample statistics for 100 simulation runs of the stopping rule 6.3.25

	Mean	Median	StDev.	Min	Max	Q_1	Q_3
$\hat{\nu}$	0.840	0.804	0.261	0.377	1.647	0.686	0.976
$\hat{\eta}$	22.66	21.387	3.312	19.186	36.546	20.398	24.098
n_s	19.89	20.000	0.345	18.000	20.000	20.000	20.000
u_3^0	6.733	6.419	1.978	3.524	12.485	5.456	7.485
AC	78.33	67.06	33.55	35.26	253.80	57.78	90.72

example $N = 10$, and $\nu = 1$. The MLE of $\hat{\nu}(u)$ is consistent estimator of ν as u and $N \to \infty$. However, for small values of N it converges in distribution to a random variable as $u \to \infty$. For the data of Table 6.2, the MLE of $\hat{\nu}(u) \to 1.07$ as $u \to \infty$.

A reasonable stopping rule, based on the MLE is to estimate η and ν after each failure, and substitute the MLE estimates in Eq. (6.3.23) to obtain the MLE of u_3^0. Thus, after the i-th failure, $i = 1, 2, \ldots$ let $\hat{\eta}_i = \hat{\eta}(U_i)$ and $\hat{\nu}_i = \hat{\nu}(U_i)$. Similarly, let

$$u_{3,i}^0 = -\frac{1}{\hat{\nu}_i} \log \left(\frac{c}{K \hat{\eta}_i \hat{\nu}_i} \right), \quad i = 1, 2, \ldots \tag{6.3.29}$$

Stop at the smallest value of $u_{3,i}^0 < U_{i+1}$ for some i. Since $i \leq N$, and $U_N < \infty$. This stopping rule yields finite stopping times a.s. [N]. To illustrate the performance of this stopping rule, we present the results of 100 simulation runs, with the parameters $N = 20$, $c = 0.1K$, $K = 100[\$]$, $\nu = 1$. Notice that the MLE of η, $\hat{\eta}$ is an empirical Bayes estimator of N. We denote by n_s the number of failures observed at stopping. u_3^0 is the stopping time and $AC = 20u_3^0 + 100(20 - n_s)$ is the actual cost. In Table 6.7 we present sample statistics for these quantities, obtained from the simulations.

Q_1 and Q_3 are, respectively, the first and third quartiles.

Comparing the mean actual cost, AC, to that of Table 6.6, we see that the empirical Bayes approach has the potential for considerable savings.

6.4 Sequential Stopping for Data Domain Models

As described in Section 6.2.2, data domain models are completely different from frequency domain models. In the present section we show two different approaches. The first is based on sequential prediction theory for finite population quantities (see [5]). The second approach is based on sequential estimation of the number of defective items in a finite population, adapted to the software testing problem. The notation in the following sections might be conflicting with that of the previous ones. We will try to minimize this conflict.

6.4.1 Sequential Testing Using Quantal Prediction Models

In [7], the authors studied the problem of optimal stopping times in a rectifying sampling inspection. The classical formulation of the problem is the following one. Let $\mathcal{P}$ be a finite population of N^* elements, having proportion θ, $0 < \theta < 1$, of defective ones. The loss due to releasing $\mathcal{P}$ with $M = N^*\theta$ defectives is $KM[\$]$. The cost of inspecting (and rectifying) n elements is $cn[\$]$. All defective elements found during an inspection are replaced by good ones (rectification). The risk after inspecting n elements is

$$R(\theta; n) = (N^* - n)\theta K + cn = MK - n(\theta K - c) . \qquad (6.4.1)$$

If $\theta K > c$ then it is optimal to inspect all N^* elements. Otherwise, it is optimal to inspect none. The problem is that the exact value of θ is unknown. Chernoff and Ray provided approximations to the optimal Bayes sequential stopping rule, assuming that the unknown value of θ has a prior beta distribution. In the present section we show the sequential stopping rule problem, when different elements of $\mathcal{P}$ have different penalty values for being released while having a fault, and having different inspection cost. In addition, we assume that a vector $\boldsymbol{x} = (x_1, \ldots, x_p)'$ of known covariates is available for each element of the population, $\mathcal{P}$.

Let $u_1, u_2, \ldots, u_{N^*}$ be the elements of $\mathcal{P}$. With the i-th element $(i = 1, \ldots, N^*)$ we associate the values y_i, $\boldsymbol{x}_i = (x_{i1}, \ldots, x_{ip})'$, K_i, c_i, where $y_i = 1$ if the element is **with** faults, and $y_i = 0$ otherwise; $\boldsymbol{x}_i$ is a vector of p known covariates; K_i is the penalty (in the field) due to faults in the i-th element and c_i is the cost of inspecting and rectifying the i-th elements. The values of y_i $(i = 1, \ldots, N^*)$ are unknown. If an element is inspected, the value of y_i is recorded, and the element is rectified. Let $\theta(\boldsymbol{x}) = P[y = 1 \mid \boldsymbol{x}]$. The values of the covariates are related to y through a model called a **quantal prediction model**. Examples are the **logistic regression model**

$$\theta(\boldsymbol{x}) = \exp(\boldsymbol{\gamma}'\boldsymbol{x})/(1 + \exp(\boldsymbol{\gamma}'\boldsymbol{x})) , \qquad (6.4.2)$$

or the **normit (probit) regression model**

$$\theta(\boldsymbol{x}) = \Phi(\boldsymbol{\gamma}'\boldsymbol{x}) , \qquad (6.4.3)$$

where $\boldsymbol{\gamma} = (\gamma_1, \ldots, \gamma_p)'$ are regression coefficients. Both in (6.4.2) and (6.4.3) we assume that $\boldsymbol{\gamma}'\boldsymbol{x}$ can take values between $-\infty$ to ∞. If the regression coefficients, $\boldsymbol{\gamma}$, are known we can compute the risk of not inspecting the lot, namely,

$$R_0 = \sum_{i=1}^{N^*} K_i \theta(\boldsymbol{x}_i'\boldsymbol{\gamma}) , \qquad (6.4.4)$$

versus the cost of inspecting n elements, $1 \le n \le N^*$, plus the remaining risk. Thus, if $\boldsymbol{s} = \{i_1, \ldots, i_n\}$ denotes a sample of n elements, set

$$R_1(\boldsymbol{s}) = \sum_{i \in \boldsymbol{s}} c_i + \sum_{j \in \boldsymbol{r}} K_j \theta(\boldsymbol{x}_j'\boldsymbol{\gamma}) , \qquad (6.4.5)$$

where $\boldsymbol{r} = \mathcal{P} - \boldsymbol{s}$. Equivalently,

$$R_1(\boldsymbol{s}) = R_0 - \sum_{i \in \boldsymbol{s}} (K_i \theta(\boldsymbol{x}_i'\boldsymbol{\gamma}) - c_i) . \qquad (6.4.6)$$

Obviously, only elements for which $\theta(\boldsymbol{x}'\boldsymbol{\gamma}) > c/K$ should be inspected. If not all such elements can be inspected, then choose a sample $\boldsymbol{s}^0$ for which $\sum_{i \in \boldsymbol{s}^0}(K_i \theta(\boldsymbol{x}_i'\boldsymbol{\gamma}) - c_i)$ is maximal (with respect to all possible samples).

The problem is to devise adaptive selection and stopping rules, when $\boldsymbol{\gamma}$ is unknown, and to test the sensitivity of the procedures to the choice of model.

We consider now a concrete example. The units u_i are modules. Each unit has one covariate x_i, which is its size (number of code lines) or degree of complexity. $y_i = 0$ if the unit is faultless and $y_i = 1$ otherwise. According to the prediction models, $y_1, \ldots, y_{N^*}$ are mutually independent with

$$\theta(x; \gamma) = e^{-\gamma/x}, \quad 0 < \gamma < \infty . \qquad (6.4.7)$$

Notice that $\theta(x; \gamma)$ is decreasing in γ and increasing in x.

We assume that the cost structure is

$$c_i = cx_i, \quad i = 1, 2, \ldots$$

and $\qquad\qquad\qquad\qquad\qquad\qquad\qquad\qquad\qquad\qquad\qquad$ (6.4.8)

$$K_i = Kx_i, \quad i = 1, 2, \ldots .$$

$\qquad\qquad\qquad\qquad\qquad\qquad\qquad\qquad\qquad\qquad\qquad\qquad\qquad$ (6.4.9)

Thus,

$$\sum_{i \in s}(K_i \theta(x_i; \gamma) - c_i) = \sum_{i \in s} x_i(K\theta(x_i; \gamma) - c) . \qquad (6.4.10)$$

This quantity is maximized by choosing for the sample n units having the **largest** x_i values, assuming that $c = 0$.

Given a sample of size n, the likelihood function of γ is

$$L(\gamma; \boldsymbol{x}_n, \boldsymbol{y}_n) = \prod_{i=1}^{n} e^{-\gamma y_i/x_i}(1 - e^{-\gamma/x_i})^{1-y_i} . \qquad (6.4.11)$$

Accordingly, the score function for γ is

$$\begin{aligned} S(\gamma) &= \frac{\partial}{\partial \gamma} \log L(\gamma; \boldsymbol{x}_n, \boldsymbol{y}_n) \\ &= \sum_{i=1}^{n} \frac{e^{-\gamma/x_i}}{x_i(1 - e^{-\gamma/x_i})} - \sum_{i=1}^{n} \frac{y_i}{x_i(1 - e^{-\gamma/x_i})} . \end{aligned} \qquad (6.4.12)$$

The MLE of γ is the root of the equation

$$\sum_{i=1}^{n} \frac{e^{-\gamma/x_i}}{x_i(1 - e^{-\gamma/x_i})} = \sum_{i=1}^{n} \frac{y_i}{x_i(1 - e^{-\gamma/x_i})} . \qquad (6.4.13)$$

The MLE does not exist if $\sum_{i=1}^{n} y_i = 0$, since $e^{-\gamma/x}/(1 - e^{-\gamma/x})$ is a decreasing function of γ with $\lim_{\gamma \to \infty} e^{-\gamma/x}/(1 - e^{-\gamma/x}) = 0$.

When $\sum_{i=1}^{n} y_i > 0$, the MLE of γ is unique and can be obtained numerically. The asymptotic variance of the MLE is the inverse of the Fisher information function which is

$$I_n(\gamma) = \sum_{i=1}^{n} \frac{e^{-\gamma/x_i}}{x_i^2(1 - e^{-\gamma/x_i})} . \qquad (6.4.14)$$

A simpler, m.s. consistent estimator of γ, can be obtained by the method of moment equation.

Let $p_n = \frac{1}{n} \sum_{i=1}^{n} y_i$, and consider the equation

$$\frac{1}{n} \sum_{i=1}^{n} e^{-\gamma/x_i} = p_n . \qquad (6.4.15)$$

Let $\hat{\gamma}_n$ be the unique root of Eq. (6.4.15). This is the moment equation estimator of γ. Thus,

$$p_n = \frac{1}{n} \sum_{i=1}^{n} e^{-\hat{\gamma}_n/x_i}$$

$$= \frac{1}{n} \sum_{i=1}^{n} e^{-\gamma/x_i} - (\hat{\hat{\gamma}}_n - \gamma) \cdot \frac{1}{n} \sum_{i=1}^{n} \frac{1}{x_i} e^{-\gamma/x_i} + o_p(\hat{\hat{\gamma}}_n - \gamma) \ . \qquad (6.4.16)$$

Hence,

$$V\{p_n\} = \frac{1}{n^2} \sum_{i=1}^{n} e^{-\gamma/x_i}(1 - e^{-\gamma/x_i})$$

$$= E\{(\hat{\hat{\gamma}}_n - \gamma)^2\} \cdot \frac{1}{n^2} \left(\sum_{i=1}^{n} \frac{1}{x_i} e^{-\gamma/x_i} \right)^2 + o\left(\frac{1}{n}\right)$$

or, the mean-squared error (MSE) of $\hat{\hat{\gamma}}_n$ is

$$E\{(\hat{\hat{\gamma}}_n - \gamma)^2\} = \frac{\sum_{i=1}^{n} e^{-\gamma/x_i}(1 - e^{-\gamma/x_i})}{\left(\sum_{i=1}^{n} \frac{1}{x_i} e^{-\gamma/x_i} \right)^2} + o\left(\frac{1}{n}\right) \ . \qquad (6.4.17)$$

We assume that all values of x are in a closed interval $[x', x'']$, hence $E(\hat{\hat{\gamma}}_n - \gamma)^2 = O\left(\frac{1}{n}\right)$. This implies that $\hat{\hat{\gamma}}_n$ converges to γ in mean square, as $n \to \infty$, which is the m.s. consistency. The value of $\hat{\hat{\gamma}}_n$ can be used as an initial solution for the MLE, $\hat{\gamma}_n$. The MLE, $\hat{\gamma}_n$, is a BAN estimator (see [46, p. 244]), and a lower confidence limit, for large sample is

$$\hat{\gamma}_n^{(L)} = \hat{\gamma}_n - 2\mathrm{ASE}\{\hat{\gamma}_n\},$$

where

$$\mathrm{ASE}\{\hat{\gamma}_n\} = \left(\frac{1}{I_n(\hat{\gamma}_n)} \right)^{1/2} \ . \qquad (6.4.18)$$

The following is a reasonable sequential stopping rule. We first order the units according to their x-values, from largest to smallest. Let $x_{(1)} \geq x_{(2)} \geq \ldots \geq x_{(N^*)}$ be the ordered x-values. We take initially a sample of n_1 units, having largest x-values. Let $\hat{\gamma}_{n_1}$ be the MLE of γ based on this sample. The following stopping rule is suggested.

SRP: **After observing n^* units, compute**

$$R_{n^*} = K \exp\{-\hat{\gamma}_{n^*}^{(L)}/x_{(n^*+1)}\} - c \ . \qquad (6.4.19)$$

Stop at the smallest n^*, such that $R_{n^*} < 0$, or at N^*.

6.4.2   Sequential Capture-Recapture

As explained in Section 6.2.2, the capture-recapture method is a sequence of sampling of functional units (modules) from the population list. All units in which faults are found are "tagged" and replaced in the population. The objective is to estimate the size, N, of the "faulty sub-population". There is a vast literature on capture-recapture sampling for estimating population size. The studies in the literature cannot be applied without modification, since in the original problem every unit in the sample which is not tagged belongs to the population of interest, while in the software testing, only faulty units are of interest. The number of tagged or untagged faulty units in a random sample, is a random variable. We will modify here the formulation and results of [45] to the software testing.

Consider a population $\mathcal{P}$ of size N^*, with N faulty units. The number, N, as well as the faulty units, are unknown. Let n_i^* ($i = 1, 2, \ldots$) be the size of the i-th sample, which is assumed to be a random sample **without** replacement from $\mathcal{P}$, and let J_i be the number of faulty elements in the i-th sample. All the faulty elements are "tagged" (but not corrected). Let U_i be the number of untagged faulty units in the i-th sample, $U_i \leq J_i$, and let $T_k = \sum_{j=k}^k U_j$, $S_k = \sum_{j=1}^k J_j$ and $\tilde{N}_k = \sum_{j=1}^i n_j^*$. After observing a sample, it is returned to the population $\mathcal{P}$, and another sample is drawn from $\mathcal{P}$. We assume that, given $n_1^*, n_2^*, \ldots, n_k^*$, the k samples are conditionally independent. Notice that T_k is the number of distinct faulty units observed in the first k samples. $\tilde{N}_k - T_k$ is the number of units without faults, observed in the first k samples. Let $M_k = N - T_k$ denote the number of untagged faulty units in $\mathcal{P}$ after the first k samples, $k = 1, 2 \ldots$. Let $\boldsymbol{u}^{(k)} = (u_1, \ldots, u_k)'$, $\boldsymbol{J}^{(k)} = (J_1, \ldots, J_k)$, $\boldsymbol{n}^{(k)} = (n_1^*, \ldots, n_k^*)$ and $\mathcal{D}_k = \{\boldsymbol{u}^{(k)}, \boldsymbol{J}^{(k)}, \boldsymbol{n}^{(k)}\}$ denote the sample statistics. The likelihood of N, given $\mathcal{D}_k$ is

$$L(N; \mathcal{D}_k) = \prod_{j=1}^k \frac{\binom{N-T_{j-1}}{U_j}\binom{T_{j-1}}{J_j - U_j}\binom{N^* - N}{n_j^* - J_j}}{\binom{N^*}{n_j^*}}, \qquad (6.4.20)$$

where $T_0 \equiv 0$. Let $\theta = \frac{N}{N^*}$. θ is an unknown parameter. If the maximal sample size is considerably smaller than N^* we can approximate the likelihood of θ, given the data by,

$$L^*(\theta; \mathcal{D}_k) = \begin{cases} \prod_{j=1}^k \frac{n_j^*}{U_j!(J_j - U_j)!(n_j^* - J_j)!} \cdot \left(\theta - \frac{T_{j-1}}{N^*}\right)^{U_j} \left(\frac{T_{j-1}}{N^*}\right)^{J_j - U_j} (1 - \theta)^{n_j^* - J_j}, \\ \qquad \theta \geq \frac{T_{k-1}}{N^*} \\ 0, \qquad \theta < T_{k-1}/N^* \end{cases} \qquad (6.4.21)$$

This is the likelihood function of θ, when sampling is **random with replacement**. The likelihood function (6.4.15) can be factored to

$$L^*(\theta; \mathcal{D}_k) = G(\theta; \mathcal{D}_k) \cdot D_k, \qquad (6.4.22)$$

where

$$D_k = \prod_{j=1}^k \frac{n_j^*}{U_j!(J_j - U_j)!(n_j^* - J_j)!} \left(\frac{T_{j-1}}{N^*}\right)^{J_j - U_j} \qquad (6.4.23)$$

and

$$G(\theta, \mathcal{D}_k) = \begin{cases} \theta^{T_k} \cdot (1 - \theta)^{(\tilde{N}_k - S_k)} \prod_{j=1}^k \left(1 - \frac{T_{j-1}}{N^*\theta}\right)^{U_j}, & \text{if } \theta < \frac{T_{k-1}}{N^*} \\ 0, & \text{if } \theta \leq \frac{T_{k-1}}{N^*}. \end{cases} \qquad (6.4.24)$$

Notice that D_k does not depend on θ and that $G(\theta; \mathcal{D}_k)$ is the likelihood kernel. If we assign θ a prior p.d.f. on $(0, 1)$, say $h(\theta)$, then the **posterior** p.d.f. of θ, given $\mathcal{D}_k$, is

$$h(\theta \mid \mathcal{D}_k) = \frac{h(\theta)G(\theta; \mathcal{D}_k)}{\int_0^1 h(\theta)G(\theta; \mathcal{D}_k)d\theta}. \qquad (6.4.25)$$

A prior beta p.d.f. seems to be appropriate for this case.

The MLE of θ, after k samples, can be obtained from $G(\theta; \mathcal{D}_k)$. Let

$$l_k(\theta) = \log G(\theta; \mathcal{D}_k)$$

$$= T_k \log \theta + (\tilde{N}_k - S_k) \log(1 - \theta)$$

$$+ \sum_{j=1}^k u_j \log\left(1 - \frac{T_{j-1}}{N^*\theta}\right), \quad \text{if } \theta < \frac{T_{k-1}}{N^*}. \qquad (6.4.26)$$

Thus, the score function, for $\theta > \frac{T_{k-1}}{N^*}$, is

$$S_k(\theta) = \frac{\partial}{\partial \theta} l_k(\theta) = \frac{T_k}{\theta} - \frac{\tilde{N}_k - S_k}{1 - \theta} + \frac{1}{\theta} \sum_{j=1}^{k} \frac{u_j T_{j-1}}{N^* \theta - T_{j-1}} . \qquad (6.4.27)$$

Let θ^* be the root of the equation $S(\theta) \equiv 0$, or the root of

$$\theta^* = \frac{T_k + \sum_{j=1}^{k} \frac{U_j T_{j-1}}{N^* \theta^* - T_{j-1}}}{\tilde{N}_k - S_k + T_k + \sum_{j=1}^{k} \frac{U_j T_{j-1}}{N^* \theta^* - T_{j-1}}} . \qquad (6.4.28)$$

The MLE of θ is

$$\hat{\theta}_k = \max \left\{ \frac{T_{k-1}}{N^*}, \theta^* \right\} . \qquad (6.4.29)$$

Notice that $\{T_j; j = 1, 2, \ldots\}$ is non-decreasing converging to $N^* \theta$ a.s. $[\theta]$. θ^* converges to 0 a.s. Hence, $\lim_{k \to \infty} \hat{\theta}_k = \theta$ a.s. This is the strong consistency of the MLE. Finally, the MLE of $M_k(\theta)$ is

$$\hat{M}_k = N^* \hat{\theta}_k - T_k . \qquad (6.4.30)$$

A stopping rule similar to (6.4.19) can be based on this MLE estimator.

References

1. Arsenault, J.E. and Roberts, J.A. (Eds.) (1980), *Reliability And Maintainability of Electronic Systems*, Computer Science Press, Rockville, MD.

2. Blumenthal, S. (1977), Estimating population size with truncated sampling, *Communications in Statistics*, **A6**, 197-308.

3. Blumenthal, S., Dahiya, R.C., and Gross, A.J. (1978), Estimating the complete sample size from an incomplete Poisson sample, *Journal of the American Statistical Association*, **73** 182-187.

4. Blumenthal, S. and Marcus, R. (1975), Estimating population size with exponential failures, *Journal of the American Statistical Association*, **70**, 913-922.

5. Bolfarine, H. and Zacks, S. (1992), *Prediction Theory For Finite Populations*, Springer-Verlag, New York.

6. Boyd, S.C. and Ural, H. (1991), On the complexity of generating optimal test sequences, *IEEE Transactions on Software Engineering*, **17**, 976-978.

7. Chernoff, H. and Ray, S.N. (1965), A Bayes sequential sampling inspection plan, *Annals of Mathematical Statistics*, **36**, 1387-1407.

8. Christensen, D.A. and Huang, S.T. (1987), Code inspection management using statistical control limits, *Proceedings of the National Communications Forum*, **41**, 1095-1100.

9. Christenson, D.A., Huang, S.T., Lamperez, A.J., and Smith, D.P. (1992), Statistical methods applied to software, In *Total Quality Management for Software*, (Eds., G. Schulmeyer and J. McManus), Van Nostrand and Reinhold Co., New York.

10. Currit, P.A., Dyer, M., and Mills, H.D. (1983), *Certifying The Reliability of Software*, IBM, Federal Systems Division, TR 86.0005.

11. Dale, C.J. (1982), *Software Reliability Evaluation Methods*, British Aerospace Dynamics Group. Stevenage, Hertfordshire, England.

12. Dalal, S.R. and Mallows, C.L. (1988), When should one stop testing software?, *Journal of the American Statistical Association*, **83**, 872-879.

13. Fakhre-Zakeri, I. and Slud, E. (1994), Optimal stopping of sequential size dependent search, *Technical Report*, Department of Mathematics, University of Maryland, College Park.

14. Freedman, R.S. (1991), Testability of software components, *IEEE Transactions on Software Engineering*, **17**, 553-564.

15. Forman, E.H. and Singpurwalla, N.D. (1977), An empirical stopping rule for debugging and testing computer software, *Journal of the American Statistical Association*, **72**, 750-757.

16. Forman, E.H. and Singpurwalla, N.D. (1979), Optimal time intervals for testing hypotheses on computer software errors, *IEEE Transactions on Reliability*, **R-28**, 250-253.

17. Goel, A.L. and Okumoto, K. (1979), Time dependent error detection rate model for software, *IEEE Transactions on Reliability*, **R-28**, 206-211.

18. Goyden, M. (1989), The software lifecycle with Ada: A command and control application, *Proceedings of TRI-Ada 89*, Pittsburgh, PA.

19. Henrici, P. (1974), *Applied and Computational Complex Analysis*, Vol. I, John Wiley & Sons, New York.

20. Jelinski, Z. and Moranda, P.B. (1972), Software reliability research, In *Statistical Computer Performance Evaluation*, (Ed., W. Freiberger), Academic Press, San Diego.

21. Kenett, R.S. and Pollak, M. (1986), A semi-parametric approach to testing for reliability growth with an application to software systems, *IEEE Transactions on Reliability*, **R-35**, 304-311.

22. Kenett, R.S. (1992), Understanding the software process, In *Total Quality Management for Software*, (Eds., G. Schulmeyer and J. McManus), Van Nostrand and Reinhold Co., New York.

23. Kubat, P. and Koch, H.S. (1983), Managing test-procedures to achieve reliable software, *IEEE Transactions on Reliability*, **R-32**, 293-303.

24. Langberg, N. and Singpurwalla, N.D. (1982), Unification of some software reliability models via the Bayesian approach, *SIAM Journal of Scientific and Statistical Computing*, **6**, 781-790.

25. Littlewood, B. and Verall, J.L. (1973), A Bayesian reliability growth model for computer software, *Journal of the Royal Statistical Society, Series B*, **22**, 332-346.

26. Littlewood, B. and Verall, J.L. (1974), A Bayesian reliability model with a stochastically monotone failure rate, *IEEE Transactions on Reliability*, **R-23**, 108-114.

27. Littlewood, B. (1980), Theories of software reliability: How good are they and how can they be improved?, *IEEE Transactions on Software Engineering*, **SE6(5)**, 489-500.

28. Littlewood, B. (1981), Stochastic reliability growth: A model for fault removal in computer programs and hardware designs, *IEEE Transactions on Reliability*, **R-30**, 313-320.

29. Littlewood, B. and Verall, J.L. (1981), On the likelihood function of a debugging

model for computer software reliability, *IEEE Transactions on Reliability*, **R-30**, 145-148.

30. McGarvey, R. (1990), ABET - A standard for Ada in a test environment, *Proceedings of TRI-Ada 90*, Baltimore, MD.

31. Meinhold, R.J. and Singpurwalla, N.D. (1983), Bayesian analysis of a commonly used model for detecting software failures, *The Statistician*, **32**, 168-173.

32. Moranda, P.B. (1975), A comparison of software error-rate models, *Proceedings of 4th Annual Texas Conference on Computing Systems*, University of Texas, Austin, Texas.

33. Moranda, P.B. (1975), Prediction of software reliability during debugging, *Proceedings of 1975 Annual Reliability and Maintainability Symposium*, Washington, D.C.

34. Musa, J.D. (1975), A theory of software reliability and its application, *IEEE Transactions on Software Engineering*, **SE1 (3)**, 312-327.

35. Musa, J.D. (1979), Validity of execution time theory of software reliability, *IEEE Transactions on Reliability*, **R-28**, 181-191.

36. Musa, J.D., Iannino, A., and Okumoto, K. (1987), *Software Reliability Measurement, Prediction, Applications*, McGraw-Hill, New York.

37. Ramamoorthy, C.V. and Bastani, F. (1982), Software reliability - status and perspectives, *IEEE Transactions on Software Engineering*, **SE8**, 334-371.

38. Randall, B. (1978), Reliability issues in computer system design, *ACM Computing Surveys*, **10**, 123-165.

39. Schick, G.J. and Wolverton, R.W. (1974), Achieving reliability in large scale software systems, *Proceedings of 1974 Reliability and Maintainability Symposium*, Los Angeles, California, 302-319.

40. Schick, G.J. and Wolverton, R.W. (1978), An analysis of competing software reliability models, *IEEE Transactions on Software Engineering*, **SE4**, 1104-120.

41. Scholz, F.W. (1986), Software reliability modelling and analysis, *IEEE Transactions on Software Engineering*, **12**, 25-31.

42. Seber, G.A.F. (1985), *The Estimation of Animal Abundance*, second edition, Griffin, London.

43. Stefanski, L.A. (1982), An approach of renewal theory to software reliability, *Proceedings of the 27th Conference on the Design of Experiments in Army Research*, ARO Report **82-2**, 101-118.

44. Sukert, A.N. (1977), An Investigation of software reliability models, *Proceedings of 1977 Annual Reliability and Maintainability Symposium*, 473-484.

45. Zacks, S., Pereira, C.A. de B., and Leite, J.G. (1990), Bayes sequential estimation of the size of a finite population, *Journal of Statistical Planning and Inference*, **25**, 363-380.

46. Zacks, S. (1971), *The Theory of Statistical Inference*, John Wiley & Sons, New York.

47. Zacks, S. (1981), *Parametric Statistical Inference: Basic Theory and Modern Approaches*, Pergamon Press, Oxford, England.

48. Zacks, S. (1994), Sequential testing of reliability system with change points, *Journal of Applied Statistical Science*, **1**, 13-18.

7

Prediction of Fatigue Life that Would Result If Defects Are Eliminated

Wayne Nelson

Statistical Consultant, Schenectady, NY

CONTENTS

7.1   Introduction

Fatigue specimens of some materials can be inspected after failure to determine whether the defect that caused the failure was on the surface of the specimen or in the interior. For a particular alloy, failures caused by surface defects tended to occur earlier than those caused by interior defects. In an effort to improve the fatigue life of the alloy, metallurgists proposed process changes to reduce surface defects. This expository paper presents analyses of past data to predict the maximum improvement in fatigue life that would result if surface defects were entirely eliminated. Thus the potential value of process or product changes to eliminate any kind of defect in a material can be evaluated *before* changes are made. The analysis methods are applied to data on a nickel base super alloy. However, the methods apply to other defects and materials, including dielectrics and semiconductors. This paper is intended to introduce metallurgists and other materials engineers to these methods.

The following sections of this paper present

7.2 The Data

7.3 The Competing Modes Model

7.4 Graphical Analyses

7.5 Concluding Remarks

7.2   The Data

The methods are illustrated with fatigue data from a single vendor and a 0.4 inch diameter cylindrical specimens. Figure 7.1 shows the (base 10) log of fatigue life at strain ranges of 0.66% and 0.78%.

The location of the defect that caused failure was identified as an interior defect or as a surface defect. Both at 0.66 and 0.78 strain ranges, failures due to surface defects tend to occur earlier than those due to interior defects. At a strain range of 0.66, most failures are due to interior defects. At a strain range of 0.78, the overwhelming majority of the failures are due to surface defects.

The main question is how much would fatigue life be improved if surface defects were eliminated by process improvements, assuming interior defects are unaffected?

7.3   The Competing Modes Model

In the analyses in the following section, each specimen is regarded as having a number of cycles to failure due to an interior defect and a number of cycles to failure due to a surface defect. In addition, it is assumed that the two failure times for a specimen are statically *independent* random observations from the corresponding distributions of cycles to failure for surface defects and interior defects. Of course, one observes only the earlier of the two failure times that a particular specimen has. The mathematical statement of the model follows. Suppose that the fraction of specimens surviving an age t if only failure cause k operates is $R_k(t)$, $k = 1, 2$. Then the fraction $R(t)$ of specimens surviving age t where both causes of failure operate is

$$R(t) = R_1(t) \times R_2(t) .$$

This is called the product rule for reliability (survival probability). This model for the failure times of the specimen is called the *series-system model* for independent competing failure modes. It is described in more detail in [3] and [5].

If the two different times to failure due to the two types of defects are not statistically independent, they are likely to be positively correlated. If such a positive correlation exists, the estimate given here for the distribution of cycles to failure when there are no surface defects tends to be optimistic. Thus, the distribution given here is the best possible distribution that could result from total elimination of failures due to surface defects.

7.4   Graphical Analyses

A distribution of cycles to failure due to a particular cause can be graphically estimated by means of probability plotting as described in [2], [3, Chap. 4], and [5]. The plots can be made by hand or by a computer program for life data analyses, such as STATPAC [4].

It is not valid to use the cycles to failure for just interior defects to estimate the distribution of cycles to failure from interior defects. The reason is that many specimens did not run long enough to fail from an interior defect but failed earlier from a surface defect. The estimate of the distribution of cycles to interior failure must take into account the running times on specimens without an interior defect failure. Then the life data on interior defects consist of the cycles on specimens that failed from interior defects and the cycles on specimens that were not run long enough to fail from an interior defect. Such life data with intermixed running and failure times are called *multiply censored* and require special data analysis methods. It is assumed that, if surface defects are eliminated, the distribution of cycles to failure from an interior defect is unaffected.

Similarly, the estimate of the distribution of cycles to failure from a surface defect must take into account specimens that failed from interior defects.

Distribution of Fatigue Life When Both Causes Act

Figures 7.2 and 7.3 show lognormal probability plots of the cycles to failure when both causes operate. The probability scale shows the cumulative percentage failed as function of cycles on the log scale. Both plots are curved. This indicates that the lognormal distribution does not describe cycles to failure at either strain range when both causes of failure operate. These plots yield graphical estimates of the 50% and 1% points of the life distributions. The estimates of these percentiles are valid regardless whether a sample is plotted on lognormal or other probability paper. For the simple situation with two causes of failure, the sample cumulative distribution function when both failure mode distributions are lognormal will not be straight on such a plot. Instead, the plot will be curved and concave downward. On Figure 7.2 for strain range 0.66, the plot has a double curvature. On Figure 7.3 for strain range 0.78, the plot has a curvature concave upward. Such curvature suggests either that one of the distributions of the two causes is not lognormal or else some specimens are not prone to failures from both causes; this is consistent with the possibility that some specimens do not contain surface defects that would produce failures.

A straight line on such a lognormal plot represents a lognormal distribution. Such a straight line was fitted to each curved plot. The corresponding standard deviations of log cycles to failure indicate the scatter in the fatigue data even though the lognormal distribution does not fit. At 0.66 strain range, the standard deviation of log life is 0.32, and the standard deviation of log life at a strain range of 0.78 is 0.24. Thus, there is considerably less scatter in the data at the strain range of 0.78. As shown later, this difference in the scatter is due entirely to surface defects, as the scatter in the cycles to failure for interior defects is the same at both strain ranges.

Distribution of Fatigue Life with Only Interior Defects (Surface Defects Eliminated)

Figures 7.4 and 7.5 show lognormal probability plots of the distributions of cycles to failure due to an interior defect for strain ranges of 0.66 and 0.78. These are the life distributions that would result if all surface failures were eliminated through improved manufacture.

Table 7.1 shows estimates of the 1% point (early failure) and 50% point (typical life) of the distribution at strain ranges of 0.66 and 0.78 for both the previous distribution of fatigue life and for the distribution of fatigue life that would result if surface defects were entirely eliminated. The cycles to failure in this table are rough graphical estimates taken from the plots and are based on a limited number of specimens. Metallurgists had had the opinion that elimination of surface defects would increase life by a factor of three to ten.

Both Figures 7.4 and 7.5 are relatively straight plots, suggesting the traditional log-

TABLE 7.1
Estimate of cycles to failure

	0.66 Strain Range		0.78 Strain Range	
	Surface and Interior Present	Surface Defects Eliminated	Surface and Interior Present	Surface Defects Eliminated
1% point:	5,500	12,000	4,000	5,500
50% point:	40,000	50,000	7,000	22,000

normal distribution adequently fits cycles to failure due to interior defects. At a strain range of 0.66, the plot is not quite straight but shows a long tail towards low cycles to failure. Also, the standard deviation of log cycles to failure is approximately the same at both strain ranges. For 0.66, the standard deviation is 0.20, and for 0.78, the standard deviation is 0.22. The respective median log cycles to failure are 4.68 and 4.38. The estimates of the medians and standard deviations are from a least squares fit of a straight line to the plots.

It is significant that the two standard deviations are the same for this one cause of failure, whereas they are not the same for the combined effects of surface and interior defects. The usual simplest model for fatigue life of a material assumes that the standard deviation of log cycles to failure is the same at any stress or, equivalently, at any strain range. Thus, the cycles to failure due to interior defects is better described by such a standard model than is cycles to failure for surface defects. The calculations and competing modes model that yield these plots are based on the assumption that specimens which failed due to a surface defect were not run long enough to fail from an interior defect. That is, such specimens are regarded as removed from test (or censored in statistical terminology) before they failed due to an interior defect. The long lower tail at 0.66 strain range may possibly be due to a few surface defects misclassified as interior defects. It is important to reexamine the specimens with these early failures to determine if they could be surface defect failures.

Distribution of Fatigue Life with Only Surface Defects (Interior Defects Not Acting)

Figures 7.6 and 7.7 show lognormal probability plots of the cycles to failure if only surface defects were present for strain ranges of 0.66 and 0.78. These plots show that the distributions of cycles to failure at the two strain ranges for surface defects are complex. Note that the two log scales for cycles to failure differ and there is greater scatter in the data at a strain range of 0.66, although the graph makes it appear that the scatter is less. The difference in the scatter can be gauged from the estimate of the standard deviation of a lognormal distribution fitted to the data. For a strain range of 0.66, the standard deviation of log cycles is 0.53, and for a strain range of 0.78, the standard deviation of log cycles is 0.21. Thus, the scatter in log cycles to surface failure is much greater at the low strain range. Note also that the standard deviation 0.21 at 0.78 strain range is comparable to the standard deviation for interior failures.

Figure 7.6 for a strain range of 0.66 shows a plot with double curvature. The reason for this is difficult to determine. A simple explanation, perhaps, is that the lognormal distribution does not adequately describe such data. The lower tail of the data is much shorter than that for a lognormal distribution. That is, failures occur later in reality than a lognormal distribution would predict. Also note the flat spot in the distribution around the 50% point; this is an odd appearance of the data and may be worth investigating further to determine what those specimens have in common. As before, 50% and 1% points can be estimated graphically from these two plots.

If a fraction of the specimens is not subject to failure due to a surface defect, the plots would have a certain appearance. In particular, the plots would be curved upwards and would have an asymptote that is a vertical line corresponding to some percentage less than 100%. That percentage is the percentage of specimens that have surface defects that cause failure. The rest of the specimens do not have a surface defect that causes failure. For a strain range of 0.66, this asymptote is difficult to determine and could be anywhere from 50% on up. For a strain range of 0.78, the asymptote is somewhere between 90 and 95%.

The double curvature of the two plots indicates that the Weibull or another parametric distribution will not fit the data on cycles to surface failure any better than the lognormal distribution.

7.5 Concluding Remarks

The preceding results may be summarized as follows. The distributions of cycles to failure from an interior defect are adequently described by a standard model for fatigue data, namely, a lognormal life distribution with a constant standard deviation of log life (that is, one that does not depend on strain range). The distributions of cycles to failure due to a surface defect are much more complex and require further understanding. For this cause of failure, the scatter in life is much greater at low strain range than at high strain range.

A more elaborate analysis would involve fitting a model with fatigue curves for the life distribution for each failure mode as a function of strain range and any other variables. In [1], [2], and [5], the author described graphical and numerical methods (maximum likelihood) for such fitting. Such a fitting is difficult for surface defects, as the distribution of cycles to failure due to a surface defect cannot be readily represented by any standard mathematical distribution.

Acknowledgements

The author thanks Mr. Irwin Mortman, Manager of the Data Book and Specifications Unit of the General Electric Company Aircraft Engine Business, for originally sponsoring the data analyses. Publication of this paper was sponsored by General Electric Company Corporate Research & Development.

References

1. Nelson, Wayne (1974), Analysis of accelerated life test data with a mix of failure modes by maximum likelihood, *General Electric Company Corporate Research & Development TIS Report 74CRD160.*

2. Nelson, Wayne (1975), Graphical analysis of accelerated life test data with a mix of failure modes, *IEEE Transactions on Reliability*, **R-24**, 230-237.

3. Nelson, Wayne (1982), *Applied Life Data Analysis*, John Wiley & Sons, New York.

4. Nelson, Wayne, Morgan, C.B., and Caporal, P. (1978), 1979 STATPAC simplified – a short introduction to how to run STATPAC, a general statistical package for data analysis, *General Electric Company Corporate Research & Development TIS Report 78CRD276*.

5. Nelson, Wayne (1990), *Accelerated Testing: Statistical Models, Test Plans, and Data Analyses*, John Wiley & Sons, New York, (800)879-4539.

Lower end pt	0.66% Strain Range		0.78% Strain Range	
	Interior	Surface	Interior	Surface
3.50–				3
3.55				6
3.60				18
3.65			1	23
3.70		3	3	47
3.75	1	3	1	44
3.80	1	14		57
3.85		15	1	37
3.90		12		20
3.95		2		16
4.00–	1	6	3	12
4.05	1	8	3	4
4.10	1	2	2	3
4.15	1	7	5	10
4.20	1	4	2	3
4.25	1	7	3	4
4.30	4	2	7	7
4.35	4	9	12	4
4.40	13	3	7	2
4.45	12	7	4	3
4.50–	21	4	2	1
4.55	20	3	1	
4.60	25	8	1	
4.65	20	4		
4.70	30	2	1	
4.75	24	2		
4.80	24	2	1	
4.85	13	6		
4.90	10	4		
4.95	2			
5.00–	1	1		
5.05		1		

FIGURE 7.1
Log life by failure cause

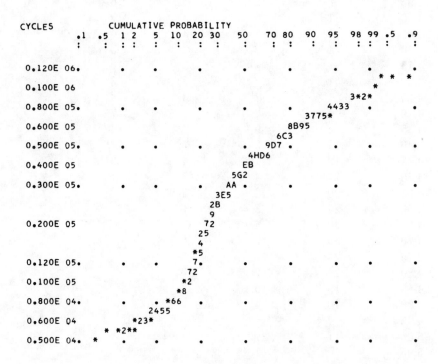

FIGURE 7.2
Lognormal plot of all failures at 0.66% strain range

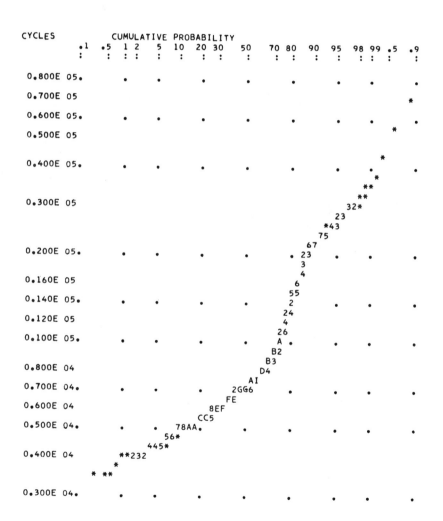

FIGURE 7.3
Lognormal plot of all failures at 0.78% strain range

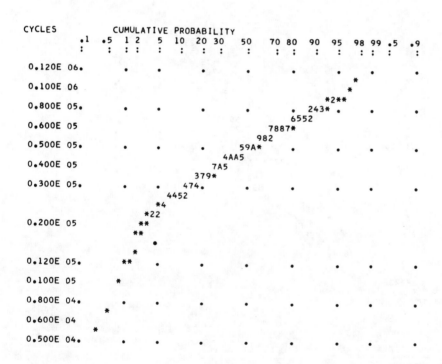

FIGURE 7.4
Lognormal plot of interior failures at 0.66% strain range

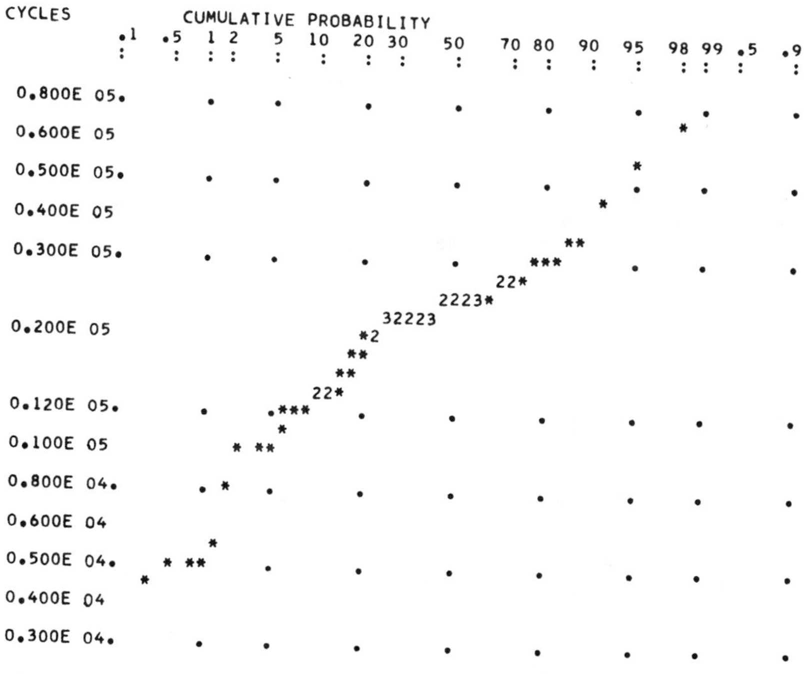

CYCLES   CUMULATIVE PROBABILITY

```
CYCLES              CUMULATIVE PROBABILITY
            .1  .5  1 2   5  10  20 30    50    70 80   90   95   98 99 .5  .9
            :   :   : :   :   :   : :     :     :  :    :    :    : :  :   :

0.800E 05.       •       •        •          •          •       •     •       •

0.600E 05                                                        *

0.500E 05.       •       •        •          •          •      *
                                                              *
0.400E 05                                                   *      •       •
                                                        **
0.300E 05.       •       •        •          •    ***       •       •       •
                                              22*
                                          2223*
                                      32223
0.200E 05                            *2
                                    **
                                    **
                                 22*
0.120E 05.       •    .***    •          •          •       •     •       •
                          *
0.100E 05          *  **

0.800E 04.     •  *    •        •          •          •       •       •
                     *
0.600E 04

0.500E 04.   * **        •          •          •       •     •       •
           *
0.400E 04

0.300E 04.       •       •        •          •          •       •       •
```

FIGURE 7.5

Lognormal plot of interior failures at 0.78% strain range

```
CYCLES            CUMULATIVE PROBABILITY
              .1  .5  1 2   5  10 20 30    50   70 80   90    95   98 99 .5   .9
              :   :   : :   :   :  :  :     :    :  :    :    :    :  :  :    :

0.200E 06.         •      •         •         •         •      •    •      •

0.100E 06                                           •            *
                                          *3*2***
                                             3*
0.500E 05.         •      •         •    33  •         •         •         •
                                        65
0.300E 05                               74
                                        75
0.200E 05.         •      •         54    •           •         •         •
                                    56
                                    55
0.100E 05                           44
                                55674
                            *222334
0.500E 04.  *  *  *.         •           •            •         •         •
```

FIGURE 7.6
Lognormal plot of surface failures at 0.66% strain range

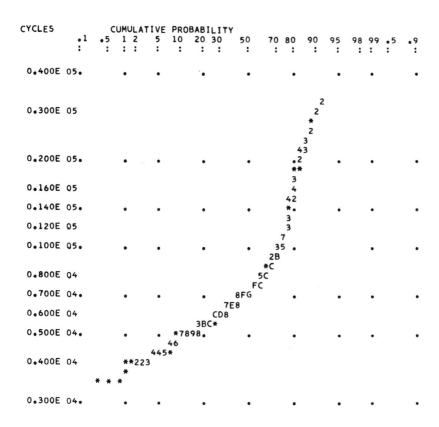

FIGURE 7.7
Lognormal plot of surface failures at 0.78% strain range

8

Role of Field Performance Data and Its Analysis

Kazuyuki Suzuki

The University of Electoro-Communications, Japan

ABSTRACT The definition of "Quality Assurance" in ISO-8402 is identified as to focus on keeping the evidence (data) of a product satisfying the required quality. It is very difficult to confirm the reliability in development stage. Besides reliability testing, the collection of field performance data and its analysis play an important role for confirming reliability. This paper describes importance of field data and its analysis, especially focusing on uses of warranty data.

CONTENTS

8.1 Introduction

8.1.1 Quality Assurance and Reliability

The definition of "Quality Assurance (QA)" in ISO-8402 is "All those planned and systematic actions necessary to provide adequate confidence that a product or service will satisfy given requirements for quality". Here, customer's needs must be reflected on the given requirement. The definition in ISO is known as to focus on keeping the evidence/data of a product satisfying the required quality. It is difficult to confirm the long term customers' satisfaction, at the stage of development or production. This aspect of quality means; "Reliability." To achieve reliability, the detection of failure modes and prevention by prediction in the early stages of new product development using FMEA (Failure Modes and Effect Analysis), FTA(Fault Tree Analysis), design review, accelerated life testing, etc. are important. But the ultimate quality is how well it performs in the field, that is, the collection of the necessary field performance data and their feedback to the quality assurance activities should be done. Through this feedback, the evidence in ISO will become more adequate and it is useful for the prevention against the recurrence of failures/defects.

This paper stresses on the importance of field data and shows some methods for the analysis of field data. For analyzing field reliability data, some difficulties are encountered such as missing information, reporting delays, the need to combine information from different sources, etc. In [9], the authors survey methodologic issues for these problems and illustrate them. This paper focuses on uses of warranty data and application of pseudo-likelihood for them. Section 8.1 describes information sources of field data and their feedback to QA activities. Section 8.2 describes a general way of analysis of field data. Section 8.3 deals with the use of warranty data and pseudo-likelihood. In Section 8.4, a nonparametric approach is shown. Section 8.5 concludes the paper with brief remarks.

8.1.2   Different Types of Field Data Information

The major information sources of field data are the following;

D1) Claim data during the warranty period

D2) Repair record data both during and beyond warranty period

D3) Amount of repair parts which were requested

D4) Data from analyzing actual failure parts/units

D5) Product liability reports (e.g. Commerce Clearing House)

It is not difficult for manufacturers to obtain the above data. Besides them, the following data will be more useful;

D6) Follow-up supplemental information by special monitors

D7) The reports from specially trained reporters on quality and reliability performance

D8) Actual spot survey on the usage and environmental conditions

D9) Periodical preventive maintenance(PM) records

D10) Market survey/researches

D11) Data of customer's need at the time of sales from the salesmen

D12) The useful information of early stages after sales

D13) Publications or surveys from consumers' organizations/groups (e.g. Consumer Reports etc.)

D14) Others

   By D6), a manufacturer can obtain the amount of degradation data about some performances which have much more information than time-to-failure data. In [5], the authors discuss a brake pad wear. Also, the censored data becomes available by follow up study, which is useful for the analysis of warranty data (see §.3). By D7) and D8), a manufacturer can obtain the most important information which affects reliability. The reporters in D7) should include the persons who have experiences on new product development. PM in D9) is limited to some countries, but this information is very vast and fruitful. Market survey is undoubtedly important for new product development to gain the "Attractive Quality Elements" ([6], refer Appendix). This may also be done by D11). The early stage of information after sales must be collected from customers as much as possible. To compare quality and reliability with competitors', publications such as Consumer Reports are important.

   The data sources of D1) to D5) may be limited to the information concerning "Must-be Quality Elements" ([6], see Appendix). Therefore the information such as D6) to D12) becomes important and useful.

8.1.3   Feedback of field information to QA activities

To achieve "Customers' Satisfaction" and enhance it, field performance data has a valuable information. Collecting the field data correctly, and reflecting them into the quality assurance activities are useful. The following feedback is important;

F1) Early warning/detection of wrong design, production process, parts, materials, etc.

F2) Data bank about failure modes and their relation to the environmental and usage condition

F3) The relationship among the test data of development stage, inspection results of the production stage, and the field-performance.

F4) Observe the targets of new product development and supervise; whether the targets were achieved or not ?

8.2   Analysis of Field Performance Data

To implement the feedback actions in Section 8.1.3, it is suggested to apply the following two kinds of analysis.

A1) Quantitative Analysis

A2) Qualitative Analysis

A1) is an analysis based on quantitative data to grasp the over-all figures and narrow the problematic area by Pareto-Principle. A2) focuses on prevention against recurrence based on qualitative data finding true causes and taking countermeasures.

8.2.1   Quantitative Analysis

To grasp the over-all figures, the following information is valuable ;

I1) cumulative number of failures - versus - calendar time plot

I2) Pareto Diagram with a fixed observational period

I3) cumulative failure rate - versus - calendar time (day, week, month) plot

I4) cumulative failure rate - versus - usage time (day, week, month) plot

I5) Weibull plot based on usage time

I6) warranty cost - versus - time plot

I7) number of repair parts and their cost value and etc.

Pareto Diagram in I2) shows the result of a fixed observational period. Therefore, it should be drawn with several different observational periods. In I3), the failure rate is estimated by

$$h(t) = \frac{\text{\# of failures at time } t}{\text{\# of items which are in use beyond time } t}$$

Based on $h(t)$, the cumulative failure rate is estimated by

$$H(t) = \sum_{u < t} h(u), \qquad u < t.$$

Using the above idea, the authors in [4] show the analytical methods of warranty data and examine the properties of their estimates. They focus on the number of failures during a calendar time. But if one is interested in the lifetime distribution measured by usage time, one encounter the following difficulties ; During the warranty period, manufacturer can obtain the data about only the failure products by their claim requests. Therefore, estimating the (cumulative) failure rate becomes very difficult by the unknown denominator of $h(t)$. For this problem, the authors in [13], [14], and [3] stress the importance of supplemental data by follow-up studies and propose pseudo-likelihood estimation methods. Properties of these methods are compared in Section 8.3.

The above information, I1) to I7), can be obtained after the following stratification ;

S1) type of automobiles

S2) parts / units / subsystems

S3) production date

S4) production place(plant) / supplier

S5) dealer

S6) usage conditions, purpose for the use

S7) environmental conditions / area

S8) failure modes

S9) usage time / calendar time until failure

S10) Others

Combinations of the above categories should be done. The main purposes of stratification are early warning and detection of wrong design/material/process; and to make it easy to prevent against the recurrence. Therefore, the traceability listed in S1) to S5) becomes important. By the information of I1) to I7) with stratification, it is easy to identify the problematic area of the Quality Assurance System, as examples;

C1) design problem

C2) wrong material

C3) wrong production process

C4) misusage

C5) unmatched environmental condition, etc.

8.2.2   Qualitative Analysis

After narrowing the problematic area or focusing on critical failure, we must find the true cause of defects/failures and implement a countermeasure. Sometimes we can clarify the true cause by the qualitative analysis, but in the case of reliability problems, we need the engineering approach using the actual failure parts. That is, based on failure modes and examining the stress, we must pursuit the failure mechanism.

It defines as

Failure modes : systematic classification of the states of malfunction

Stress $\left\{ \begin{array}{l} \text{functional stress : stress caused by implementing functions} \\ \text{environmental stress : stress caused by outside environmental factors} \end{array} \right.$

Failure mechanism :the mechanism of failures by which an item fails (Physical, chemical and mechanical process must be clarified).

Manufacturers should construct the data-base system concerning the above three items.

Also, its data-base should be able to be accessed by everybody in the manufacturer.

8.3 Pseudo-Likelihood Approaches for Incomplete Field Data

8.3.1 Notation and Assumptions

The following notation and assumptions are used.

Notation

1. (X_i, Y_i), $i = 1, 2, \ldots, N$, represent independent, identically distributed pairs of random variables, with X_i the variable of interest and Y_i some censoring variable. For example, X_i might be the mileage to the first failure of unit i, and Y_i might be the total mileage of unit i during the warranty period.

2. $f(x)$, $S(x)$: the probability density function (pdf) of the random variable (rv) X and its corresponding survival function.

3. $g(y)$, $G(y)$: the pdf of the rv Y and its corresponding survival function.

4. θ : a vector of unknown parameters, taking on values in the parameter space Ω.

5. (Z_i, δ_i), $i = 1, 2, \ldots, N$: the observed quantities where $Z_i \equiv \min(X_i, Y_i)$, $\delta_i \equiv I[X_i \leq Y_i]$, $i = 1, 2, \ldots, N$. Here $I[\cdot]$ means the indicator function of set $\{\cdot\}$. That is, the random censoring model (e.g., [1]) is applied to the problem. Also in this article, the quantities Z_i in the pairs (Z_i, δ_i) are unobserved for some i's.

6. $D_i \equiv 1$ if the i-th item is followed up; otherwise, $D_i \equiv 0$ $(i = 1, 2, \ldots, N)$. Notice that D_i is a known constant, not a random variable.

In order to facilitate the description in this article, we shall make free use of terminology relative to the automobile example. Thus, we refer to X_i as mileage to the first failure of unit i, and Y_i as mileage in the warranty period of unit i.

7. $n_u \equiv \sum \delta_i$: the number of automobiles that failed in the warranty period. The subscript u on n_u means *uncensored*, and $\sum$ means $\sum_{i=1}^{N}$; this definition of $\sum$ will be used throughout this article.

8. $n_c \equiv \sum(1 - \delta_i)D_i$: the number of automobiles without failure in the warranty period but for which mileage was determined through follow-up. The subscript c on n_c means *censored*.

9. $n_l \equiv \sum(1 - \delta_i)(1 - D_i)$: the number of automobiles without failure that have not been followed up in the warranty period. The mileages for these automobiles have not been observed. The subscript l on n_l means *lost*.

10. $N \equiv n_u + n_c + n_l$: total number of automobiles.

11. $p^* \equiv (1/N)\sum D_i$: percentage of automobiles followed up.

Assumptions

1. X_i and Y_i $(i = 1, 2, \ldots, N)$ are independent for all i; the distributional form of X is known, and that of Y is not specified.

2. The time scale of the rv's X and Y is assumed to be operating time (e.g., mileage, frequency, etc.); whereas the observational period of the study is measured by calendar time (e.g., month, year, etc.).

3. The probability for the failure of an item depends only on its operating time.

4. All failures during the warranty period will be reported to the manufacturer. This is essential for obtaining the mileages of the failures. If there is no failure, the owner will not report the mileage in that period. Consequently, "no record of failure" means there has been no failure.

5. The percentage of follow-ups in the study, p^*, is not equal to zero. Moreover, $n_c \neq 0$ and $n_u \neq 0$.

6. Individual automobiles to be followed up are selected randomly, and the correct mileages of followed-up automobiles are observed with probability 1, even if they have not failed.

8.3.2 Estimators θ^* and $\theta^\#$, and Their Properties

Using the notation and assumptions in Section 8.3.1, the sampling distribution of the observed quantities (Z_i, δ_i), $i = 1, \ldots, N$, is given by

$$\prod_{i=1}^{N} \{f(Z_i)G(Z_i)\}^{\delta_i} \{g(Z_i)S(Z_i)\}^{(1-\delta_i)D_i} \times \{Pr(\delta_i = 0)\}^{(1-\delta_i)(1-D_i)} .$$

Changing subscripts, the likelihood function becomes

$$L = \left[\prod_{i=1}^{n_u} f(\tilde{Z}_i)G(\tilde{Z}_i)\right] \left[\prod_{j=1}^{n_c} g(\bar{Z}_j)S(\bar{Z}_j)\right] \times [Pr(X > Y)]^{n_l} , \qquad (8.3.1)$$

where $\tilde{Z}_i (i = 1, \ldots, n_u)$ is the Z_i conditioned on $X_i \leq Y_i$ ($\delta_i = 1$), and $\bar{Z}_j$ ($j = 1, \ldots, n_c$) is the Z_j conditioned on $X_i > Y_i$ ($\delta_i = 0$) and $D_i = 1$. $Pr(X > Y)$ cannot be expressed in simple closed form except for special distributional cases of X and Y (e.g., both X and Y are exponential; In [11], the author investigated this special case assuming $n_c = 0$). The distribution of the n_l units that did not fail and for which no mileage frequencies are available is the same as that of $\bar{Z}_j$. If we try to estimate the n_l unobserved nonfailure mileages, we should redistribute these n_l observations equally to the n_c observed values of the $\bar{Z}_j$. Therefore, instead of $[Pr(X > Y)]^{n_l}$,

$$\left[\prod_{j=1}^{n_c} g(\bar{Z}_j)S(\bar{Z}_j)\right]^{n_l/n_c}$$

should be applied in maximizing L. Then we have pseudo-likelihood

$$L^{**} = \left[\prod_{i=1}^{n_u} f(\tilde{Z}_i)g(\tilde{Z}_i)\right] \left[\prod_{j=1}^{n_c} \{g(\bar{Z}_j)S(\bar{Z}_j)\}^{1+n_l/n_c}\right] . \qquad (8.3.2)$$

That is, every one of the observed nonfailure data has the additional mass of the n_l/n_c along with its own observed mass of 1. If $G(Z)$ and $g(Z)$ do not involve any parameter of interest, L^{**} can be taken to be

$$L^* = \left[\prod_{i=1}^{n_u} f(\tilde{Z}_i)\right] \left[\prod_{j=1}^{n_c} \{S(\bar{Z}_j)\}^{1+n_l/n_c}\right] . \qquad (8.3.3)$$

The proposed estimator is θ^* in Ω, at which L^* is maximized. θ^* is expected to have properties as good as those of an MLE $\hat{\theta}$ of θ based on (8.3.1). This method is proposed by [14]. A similar approach is done in [3]. They propose the following pseudo-likelihood;

$$L^\# = \left[\prod_{i=1}^{n_u} f(\tilde{Z}_i) \right] \left[\prod_{j=1}^{n_c} \{ S(\bar{Z}_j) \}^{p^*} \right]. \tag{8.3.4}$$

This comes from that $1 + n_l/n_c$ converges to p^*. Let $\theta^\#$ denote the value which maximizes (8.3.4). In the following, properties of θ^* and $\theta^\#$, and comparisons of θ^*, $\theta^\#$ and $\hat{\theta}$ are given. Even when the model for X is parametric and the model for Y is nonparametric, θ^* and $\theta^\#$ are valid.

Example 1
If X follows an exponential distribution $S(x) = \exp(-\lambda x)$, $x > 0$, we get

$$L^* = \left[\prod_{i=1}^{n_u} \lambda \exp(-\lambda \tilde{Z}_i) \right] \times \left[\prod_{j=1}^{n_c} \exp \left\{ -(1 + n_l/n_c)\lambda \bar{Z}_j \right\} \right].$$

From $(\partial/\partial\lambda) \log L^* = 0$, we obtain

$$\lambda^* = n_u \left/ \left\{ \sum_{i=1}^{n_u} \tilde{Z}_i + (1 + n_l/n_c) \sum_{j=1}^{n_c} \bar{Z}_j \right\} \right..$$

$\square$

Similary, we obtain

$$\lambda^\# = n_u \left/ \left\{ \sum_{i=1}^{n_u} \tilde{Z}_i + p^* \sum_{j=1}^{n_c} \bar{Z}_j \right\} \right.,$$

for $L^\#$.

Example 2
If X follows the Weibull distribution $S(x) = \exp(-\lambda x^m)$, $x > 0$, we get

$$L^* = \left[\prod_{i=1}^{n_u} \lambda m \tilde{Z}_i^{m-1} \exp(-\lambda \tilde{Z}_i^m) \right] \times \left[\prod_{j=1}^{n_c} \exp \left\{ -(1 + n_l/n_c)\lambda \bar{Z}_j^m \right\} \right].$$

From $(\partial/\partial m) \log L^* = 0$ and $(\partial/\partial\lambda) \log L^* = 0$, we have

$$1/m + \sum_u \log \tilde{Z}_i / n_u = \frac{\sum_u \left\{ (\log \tilde{Z}_i) \tilde{Z}_i^m \right\} + (1 + n_l/n_c) \sum_c \left\{ (\log \bar{Z}_j) \bar{Z}_j^m \right\}}{\sum_u \tilde{Z}_i^m + (1 + n_l/n_c) \sum_c \bar{Z}_j^m},$$

$$\lambda = n_u \left/ \left[\sum_u \tilde{Z}_i^m + (1 + n_l/n_c) \sum_c \bar{Z}_j^m \right] \right.. \tag{8.3.5}$$

Here $\sum_u$ means $\sum_{i=1}^{n_u}$ and $\sum_c$ means $\sum_{j=1}^{n_c}$. The solution $\theta^* = (m^*, \lambda^*)'$ of the preceding equations can be obtained by using the Newton-Raphson method. Similarly $(1 + n_l/n_c)$ is replaced by p^* for $L^\#$.

The asymptotic properties of θ^* and $\theta^\#$ follow. For proof, refer to [14]. ⬜

Property 1

Asymptotically, the solution θ^* of $(\partial/\partial\theta)\log L^* = 0$ coincides with the solution of θ^* of $(\partial/\partial\theta)\log L_1 = 0$, where L_1 is the full likelihood that every failure and censored value is observed (i.e., $p^* = 1$). Therefore, under regularity conditions ([15], p.194), θ^* is a consistent estimator of θ. $\theta^\#$ has also the above property.

Property 2

Under the regularity conditions,

$$\sqrt{N}(\theta^* - \theta) \to N(0, \ J(\theta)^{-1}I(\theta, p^*)J(\theta)^{-1}) \qquad as \ N \to \infty,$$

where

$$J(\theta) \equiv Pr(X \leq Y)E\left[\frac{\partial^2}{\partial\theta \cdot \partial\theta'}\log f(Z, \ \theta) \mid X \leq Y\right]$$

$$+Pr(X > Y)E\left[\frac{\partial^2}{\partial\theta \cdot \partial\theta'}\log S(Z, \ \theta) \mid X > Y\right] \tag{8.3.6}$$

and

$$I(\theta, \ p^*) \equiv Pr(X \leq.Y)\left\{E\left[\frac{\partial}{\partial\theta}\log f(Z, \ \theta) \times \frac{\partial}{\partial\theta'}\log f(Z, \ \theta) \mid X \leq Y\right]\right.$$

$$\left.-E\left[\frac{\partial}{\partial\theta}\log f(Z, \ \theta) \mid X \leq Y\right] \times E\left[\frac{\partial}{\partial\theta'}\log f(Z, \ \theta) \mid X \leq Y\right]\right\}$$

$$+\frac{Pr(X > Y)}{p^*}\left\{E\left[\frac{\partial}{\partial\theta}\log S(Z, \ \theta) \times \frac{\partial}{\partial\theta'}\log S(Z, \ \theta) \mid X > Y\right]\right.$$

$$\left.-E\left[\frac{\partial}{\partial\theta}\log S(Z, \ \theta) \mid X > Y\right] \times E\left[\frac{\partial}{\partial\theta'}\log S(Z, \ \theta) \mid X > Y\right]\right\}$$

$$-E\left[\frac{\partial}{\partial\theta}\log f(Z, \ \theta) \mid X \leq Y\right] E\left[\frac{\partial}{\partial\theta'}\log S(Z, \ \theta) \mid X > Y\right]. \tag{8.3.7}$$

Here we notice that θ' represents the transposition of a column vector θ,

$$E\left[h(Z) \mid X \leq Y\right] \equiv \int_0^\infty h(Z) \cdot f(Z) \cdot G(Z)/Pr(X \leq Y)dZ,$$

and

$$E\left[h(Z) \mid X > Y\right] \equiv \int_0^\infty h(Z) \cdot g(Z) \cdot S(Z)/Pr(X > Y)dZ.$$

From the law of large numbers, estimates of $J(\theta)$ and $I(\theta, \ p^*)$ are obtainable by substituting the observed $Z's$ and θ^* into (8.3.6) and (8.3.7).

Let $B_p(\theta)$ represent the $100 \times p$ percentile of the lifetime distribution of X, and assume that the partial derivatives $(\partial/\partial\theta)B_p(\theta)$ exist. Then $B_p(\theta^*)$ becomes a consistent estimator of $B_p(\theta)$, and

$$\sqrt{N}(B_p(\theta^*) - B_p(\theta)) \to N(0, \; Q(\theta)' J(\theta)^{-1} I(\theta, \; p^*) \times J(\theta)^{-1} Q(\theta))$$

$$\text{as } N \to \infty$$

where $Q(\theta) = (\partial/\partial\theta)B_p(\theta)$ is the column vector of partial derivatives.
$\sqrt{N}(\theta^\# - \theta)$ follows the same asymptotic distribution of $\sqrt{N}(\theta^* - \theta)$ except $I(\theta, p^*)$ in
(9). For $\sqrt{N}(\theta^\# - \theta)$, $I(\theta, p^*)$ in (8.3.7) is changed into

$$I(\theta, \; p^*) \equiv Pr(X \leq Y) E\left[\frac{\partial}{\partial\theta}\log f(Z, \; \theta) \times \frac{\partial}{\partial\theta'}\log f(Z, \; \theta) \mid X \leq Y\right]$$

$$+ \frac{Pr(X > Y)}{p^*} E\left[\frac{\partial}{\partial\theta}\log S(Z, \; \theta) \times \frac{\partial}{\partial\theta'}\log S(Z, \; \theta) \mid X > Y\right]. \quad (8.3.8)$$

8.3.3 Comparisons of Asymptotic Variances θ^* and $\theta^\#$, and the Effect of Follow-Up Percentage

When X follows the Weibull distribution $S(x) = \exp(-\lambda x^m), x > 0$, it will not be possible for $Pr(X > Y)$ in (3) to be obtained in closed form, except when Y follows the Weibull distribution with the same shape parameter, m. Therefore, we may assume that Y follows the Weibull distribution $G(y) = \exp(-\eta y^m), y > 0$. For comparison, let consider the following estimators.

Usually, the estimation is done using only the follow-up data. In this case, MLE of θ is given by $\tilde{\theta}$ which minimizes

$$\tilde{L} = \left[\prod_{i=1}^{n'_u} f(\tilde{Z}_i)g(\tilde{Z}_i)\right]\left[\prod_{j=1}^{n_c} g(\bar{Z}_j)S(\bar{Z}_j)\right], \quad (8.3.9)$$

where n'_u is the number of observed failure data (censored data) out of the follow-up automobiles. The points that maximize $L^*, L^\#$ and $\tilde{L}$ are represented by $\theta^* = (m^*, \lambda^*)'$, $\theta^\# = (m^\#, \lambda^\#)'$, and $\tilde{\theta} = (\tilde{m}, \tilde{\lambda})'$. Table 8.1 gives their related asymptotic variances. Here we set up $m = 2.0$, $\lambda = \mu = 1.0$, and $p^* = .05,.01,(.1),.5,1.0$. For the Weibull distribution , interest centers on percentiles B_p rather than the distributional parameters m and λ. For example, in the automobile industry of Japan, $B_{.10}$ is the most popular reliability index. From Table 8.1, we know that $\theta^*,\theta^\#$, B_p^* and $B_p^\#$ are superior to $\tilde{\theta}$ and $\tilde{B}_p$. Also, superiority of B_p^* to $B_p^\#$ becomes more clear as p^* becomes small.

Table 8.2 represents the results of $\sqrt{AVar(\cdot)}$ of three estimators $B_{.10}^*$, $B_{.10}^\#$ and $\tilde{B}_{.10}$ for several number of items N and the follow-up percentage p^*. Assume N equals 5,000.

TABLE 8.1
Asymptotic variances of $\theta^* = (m^*, \lambda^*), \theta^\sharp = (m^\sharp, \lambda^\sharp), \widetilde{\theta} = (\widetilde{m}, \widetilde{\lambda}),$
$B_p(\theta^*)(\equiv B_p^*), B_p(\theta^\sharp)(\equiv B_p^\sharp) and B_p(\widetilde{\theta})(\equiv \widetilde{B}_p)(p = .05, .10),$ **Assuming** $S(x) = \exp(-\lambda x^m),$
$G(y) = \exp(-\eta y^m)$

						p^*			
	$AVar(\cdot)$	.01	.05	.1	.2	.3	.4	.5	1.0
m	m^*	99.3	23.0	13.4	8.68	7.09	6.29	5.82	4.86
	$m^\sharp$	99.3	23.0	13.4	8.68	7.09	6.29	5.82	4.86
	$\widetilde{m}$	486	97.2	48.6	24.3	16.2	12.2	9.72	4.86
λ	λ^*	69.6	15.0	8.22	4.82	3.68	3.11	2.77	2.09
	$\lambda^\sharp$	119	24.5	12.7	6.82	4.85	3.86	3.27	2.09
	$\widetilde{\lambda}$	209	41.8	20.9	10.5	6.97	5.23	10.5	2.09
$B_{.05}$	$B_{.05}^*$	.896	.285	.208	.170	.157	.151	.147	.139
	$B_{.05}^\sharp$	1.53	.406	.266	.196	.172	.160	.153	.139
	$\widetilde{B}_{.05}$	13.9	2.78	13.9	.695	.463	.348	.278	.139
$B_{.10}$	$B_{.10}^*$	.780	.294	.233	.203	.192	.187	.184	.178
	$B_{.10}^\sharp$	2.08	.544	.351	.255	.223	.207	.197	.178
	$\widetilde{B}_{.10}$	17.8	3.56	1.78	.890	.593	.445	.356	.178

NOTE : Values of $AVar(\cdot)$ are already multiplied by the number of items.

TABLE 8.2
Comparisons of the three estimation methods for $B_{.10}$ by their asymptotic
variances

(a) $\sqrt{AVar(B_{.10}^*)}$

N	.01	.05	.1	.2	.3	.4	.5	1.0
				p^*				
100	.088	.054	.048	.045	.044	.043	.043	.042
1000	.028	.017	.015	.0143	.0139	.0137	.0136	.0133
2500	.0177	.0108	.0097	.0090	.0088	.0086	.0086	.0084
5000	.0125	.0077	.0068	.0064	.0062	.0061	.0061	.0060

(b) $\sqrt{AVar(B_{.10}^\sharp)}$

N	.01	.05	.1	.2	.3	.4	.5	1.0
				p^*				
100	.144	.074	.059	.051	.047	.046	.044	.042
1000	.046	.023	.019	.016	.0149	.0144	.0140	.0133
2500	.0289	.0148	.0118	.0101	.0094	.0091	.0089	.0084
5000	.0204	.0104	.0084	.0071	.0067	.0064	.0063	.0060

(c) $\sqrt{AVar(\widetilde{B}_{.10})}$

N	.01	.05	.1	.2	.3	.4	.5	1.0
				p^*				
100	.422	.189	.133	.0944	.0770	.0667	.0597	.0422
1000	.133	.0597	.0422	.0298	.0244	.0211	.0189	.0133
2500	.0844	.0377	.0267	.0189	.0154	.0133	.0119	.0084
5000	.0597	.0267	.0189	.0133	.0109	.0094	.0084	.0060

In this case, to attain $\sqrt{AVar(\cdot)} = .0084$, it needs $5,000 \times .5 = 2,500$ items for $\tilde{B}_{.10}^{\#}$. For $B_{.10}^{\#}$, $5,000 \times .1 = 500$ items are necessary. For $B_{.10}^{*}$, $5,000 \times .05 = 250$ is enough because $\sqrt{AVar(B_{.10}^{*})} = .0077 \leq .0084$.

8.4 A Nonparametric Approach for Warranty Data

In this section, a nonparametric approach is done. The generalized maximum likelihood estimator [8] of the survival function of $S(t)$ of the random variable X is given and its statistical properties are described. We assume that X_i and Y_i are independent for all i, and assume that X_i is a discrete random variable taking values $t_1 < \cdots < t_K$. Let t^* denote the minimum of the uppermost support points of $S(\cdot)$ and $G(\cdot)$, and let K^* be the maximum of such that $t_k \leq t^*$. To define an estimator of $S(\cdot)$, use the quantities (Z_i^*, d_i, δ_i^*), $i = 1, \cdots, n^*$ $(n^* \leq n_u)$, instead of (Z_i, δ_i). The Z_i^* are defined as the observed ordered distinct Z_i's; that is, $Z_1^* < \cdots < Z_{n^*}^*$, d_i is the multiplicity of Z_i^* (the number of Z_i's equal to Z_i^*), and δ_i^* is the indicator associated with Z_i^*. If ties $Z_i = Z_j$ for $\delta_i \neq \delta_j$,$i \neq j$ exist, we can break any such ties by treating the censored time as just slightly larger than the failure time. Then δ_i^* is either 0 or 1.

Property 3
The generalized MLE of the survival function $S(t)$ is given by

$$\hat{S}(t) = 1, \qquad\qquad for\ t < Z_1^*$$
$$= \prod_{i:Z_i^* \leq t} (1 - m_i/M_i)^{\delta_i^*}, \ for\ Z_1^* \leq t < Z_{n^*}^*$$
$$= 0 \qquad\qquad if\ \delta_{n^*}^* = 1,\ for\ t \geq Z_{n^*}^*$$
$$= Undefined\ if\ \delta_{n^*}^* = 0,\ for\ t \geq Z_{n^*}^* \tag{8.4.1}$$

where

$$M_i \equiv \textstyle\sum_{j=i}^{n^*} m_j,$$
$$m_i \equiv \begin{array}{ll} d_i, & if\ \delta_i^* = 1 \\ (1 + n_l/n_c)d_i, & if\ \delta_i^* = 0. \end{array}$$

If we apply the approach of the authors in [3], $1 + n_l/n_c$ in (8.4.1) is only changed into p^*. This gives a consistent estimator of $S(\cdot)$.

$\hat{S}(t)$ in (8.4.1) coincides with the estimator used by [7] when $n_l = 0$. This property can be proved in several ways (in details, refer to [13]): one is by Peterson's representation (1977, Theorem 2.1), which relates the subsurvival functions to the survival function $S(\cdot)$. Moreover, the results in [2, Eqs. 6 and 7] can be applied, which permits both subsurvival functions to have jumps in common. The idea of self-consistency, introduced in [1], is also applicable to this problem. Using Efron's definition, we can derive the self-consistent estimator and prove that it coincides with $\hat{S}(\cdot)$ in (8.4.1). Therefore the estimator is described simply via redistribution to the right-algorithm [1].

8.5 Conclusion

Importance of field performance data and some methods of analysis based on warranty data were presented. For warranty data, the authors in [3] propose an estimation method based on "truncated" conditional likelihood. Also, they deal with the case including regression variables. Without non-failure information (censored data), estimation of life-time parameters is difficult, but testing of difference of two populations is possible by only failure data in usage time or numbers of failures. Theoretical investigation should be done for this problem.

Acknowledgements

The author thanks Professor N. Balakrishnan who gave him a chance to write this paper.

Appendix

Two Dimensional Aspects of Quality [6].

A.1 Attractive Quality Element

An element which positively satisfies users if it surpasses a certain level. However, even if it lowers below the level, it does not dissatisfy them as well as it does not satisfy them.

A.2 Must-be Quality Element

An element which strongly dissatisfies users if it lowers below a certain level. However, even if it surpasses the level, it does not satisfy them, instead, it is accepted by users as it must be.

References

1. Efron, B. (1967), The two sample problem with censored data, *Proceedings of the Fifth Berkeley Symposium*, **4**, 831-853.

2. Gill, R.D. (1981), Testing with replacement and the product limit estimator, *Annals of Statistics*, **9**, 853-860.

3. Kalbfleisch, J.D. and Lawless, J.F. (1988), Estimation of reliability from field performance studies (with discussion), *Technometrics*, **30**, 365-388.

4. Kalbfleisch, J.D., Lawless, J.F., and Robinson, J.A. (1991), Methods for the analysis and prediction of warranty claims, *Technometrics*, **33**, 273-285.

5. Kalbfleisch, J.D. and Lawless, J.F. (1992), Some useful statistical methods for truncated data, *Journal of Quality Technology*, **24**, 145-152.

6. Kano, N., Seraku, N., Takahashi, F., and Tuji, S. (1984), Attractive quality and must-be quality, *Journal of the Japanese Society for Quality Control*, **14**, 147-156 (in Japanese).

7. Kaplan, E.L. and Meier, P. (1958), Nonparametric estimation from incomplete observations, *Journal of the American Statistical Association*, **53**, 457-481.

8. Kiefer, J. and Wolfowitz, J. (1956), Consistency of the Maximum likelihood estimator in the presence of infinitely many incidental parameters, *Annals of Mathematical Statistics*, **27**, 887-906.

9. Lawless, J.F. and Kalbfleisch, J.D. (1992), Some issues in the collection and analysis of field reliability data, In *Survival Analysis: State of the Art* (Eds., J.P. Klein and P.K. Goel), 141-152, Kluwer Academic Publishers.

10. Miller, R.G. (1981), *Survival Analysis*, New York: John Wiley & Sons.

11. Miyagawa, M. (1982), Statistical analysis of incomplete data in competing risks model, *Journal of the Japanese Society for Quality Control*, **12**, 23-29 (in Japanese).

12. Peterson, A.V. (1977), Expressing the Kaplan-Meier estimator as a function of empirical subsurvival functions, *Journal of the American Statistical Association*, **72**, 854-858.

13. Suzuki, K. (1985a), Nonparametric estimation of lifetime distribution from a record of failures and follow-ups, *Journal of the American Statistical Association*, **80**, 68-72.

14. Suzuki, K. (1985b), Estimation of lifetime parameters from incomplete field data, *Technometrics*, **27**, 263-271.

15. Zacks, S. (1971), *The Theory of Statistical Inference*, New York: John Wiley & Sons.

9

Analysis of Experiments for Reliability Improvement and Robust Reliability

Michael Hamada

University of Waterloo

ABSTRACT Statistically designed experiments have been used extensively for estimating or demonstrating existing reliability but have seldom been used for improving reliability. Genichi Taguchi has advocated their use not only for improving reliability but also for achieving robust reliability. Robust reliability is part of his robust design philosophy whose aim is to make processes/products insensitive to "noise" factors which are hard or impossible to control such as manufacturing variables that cannot easily be controlled or environmental conditions in which the product is operated. This paper first discusses experimental designs for reliability improvement and robust reliability. Then analysis methods for reliability data are considered. These methods need to be able to handle censored data which commonly occur because all units tested have not failed by the end of the experiment. In analyzing censored data from these experimental designs, some difficulties with standard methods have been encountered and have provided the motivation for recent work. This paper presents an overview of analysis methods which include standard methods, an iterative imputation-based model selection procedure and one which takes a Bayesian approach. Examples of fluorescent lamps, heat exchangers and drill bits are given to illustrate the use of experimental design for improving reliability. The data from these experiments are reanalyzed to show how some of the analysis methods can be applied.

Keywords: Bayesian, Censoring, Complex aliasing patterns, Control and noise factors, Estimability, Fractional factorial, Maximum likelihood estimation, Mixed-level orthogonal array, Plackett-Burman design, Product array, Response surface design, Robust design

CONTENTS

9.1   Introduction

While statistically designed experiments have been used extensively for estimating or demonstrating existing reliability [23], they have seldom been used proactively to improve reliability as advocated by Genichi Taguchi [33, 34], i.e., to identify factors that affect reliability and to recommend factor levels that lead to improved reliability. Taguchi is better known for robust design, whose aim is to make processes/products insensitive to "noise" factors which are hard or impossible to control. Such products/processes are said to be *robust* to the noise factors. Examples of noise factors include manufacturing variables that cannot easily be controlled and environmental conditions in which the product is used. This important paradigm for improving products/processes has attracted much attention in recent years [16] and can be applied to reliability. In order to ensure good stability and adequate life, Taguchi [33, p. 149] recommended that noise factors be considered in any experiment to improve reliability whenever practical to do so.

It is somewhat surprising that statistically designed experiments have not received more attention as a means for improving reliability. Methods for analyzing lifetime data which handle censored data (e.g., arising from units which have not failed by the end of the experiment) existed as early as 1959. See [29] and [41] which discussed, respectively, how maximum likelihood estimation could handle right-censoring (or Type I censoring) and Type II censoring in factorial experiments. Reference [41] considered the effect of two accelerating factors, temperature and voltage, on capacitor lifetime, however, so that the focus was not improvement. The article, "Improve Your Reliability" [15], presented factorial experiments as a way to establish cause-and-effect relationships between factors and a product/process characteristic. Unfortunately, the example presented dealt with quality rather than reliability. While there has been isolated use of such experiments in the years following, the industrial statistical literature has been rather silent on this matter.

It is in the 1980's with North American industry's introduction to Taguchi's quality engineering philosophy and methodology (first [35] and later [33, 34]) that we find a clear message to use designed experiments to improve reliability. In his books, Taguchi provides examples of improving clutch spring and fluorescent lamp reliability. Furthermore, there is documented evidence in the *Symposia on Taguchi Methods* (1984-1993) that industry has heard the message. Reference [32] reported on the improvement of heat-exchanger reliability in a commercial heating system. The improvement of drill bit reliability in a multilayer printed circuit board drilling operation in [22]. Reference [26] also reported an early application of Taguchi's methodology at AT&T which improved router-bit reliability in a printed circuit board cutting operation. The message has recently made its way into textbooks on reliability. The third edition of [25] has a chapter on designed experiments which is apparently influenced by Taguchi. No reliability improvement examples are given, however. In *Reliability Improvement of Design of Experiments* [6] we find the first textbook which is entirely devoted to this subject and in which Taguchi's robust design philosophy figures prominently. In assessing the use of designed experiments for reliability improvement, [6, p. 127] concludes that they have not been used widely by reliability engineers in the past few decades. As for robust design, the book states that it is a potentially powerful tool whose exploitation for reliability improvement is only beginning.

As North American industry began applying Taguchi's methods, his philosophy and methods also attracted the attention of researchers. Various studies were undertaken that generated new lines of research which included improved experimental designs and analysis methods for implementing his philosophy. Regarding experiments to improve reliability, analysis of censored data from them has presented new challenges. Besides right-censored data from units not failing by the end of the experiment, other types of censored data arise when units are inspected periodically for failure. In such situations in which units cannot be monitored continuously, units produce left-censored data if they fail before the first inspection and interval-censored data, otherwise. It is the censored data coupled with the moderately to highly fractionated designs commonly used in industry to study a large number of factors in a small number of runs that causes problems when standard methods are used. In order to overcome these problems, [12] proposed an iterative procedure based on building up a model. Problems with standard methods were further explored by [11], which provided support for the strategy used in the iterative procedure. Reference [10] also explored the analysis of robust reliability experiments with censored data. Recently, [14] proposed a Bayesian approach which overcomes limitations of the iterative procedure and also provides a natural framework for analyzing robust reliability experiments.

This article focuses on the analysis methods for experiments to improve reliability and to achieve robust reliability and is organized as follows. First two classes of experimental designs, fractional factorials and product arrays, are discussed with examples in Sections 9.2 and 9.3, respectively. In Section 9.4, an overview of analysis methods is given. Some of these methods are demonstrated in Sections 9.5 through 9.7 which presents respective analyses of experiments to improve fluorescent lamp and heat exchanger reliability and to achieve robust reliability of drill bits in printed circuit board fabrication. Section 9.8 concludes with a discussion and comments on other types of reliability improvement experiments.

9.2 Experiments for Improving Reliability

Experiments for improving reliability have the following goals: 1) identify the important factors that affect the reliability of a product/process and 2) choose levels of these factors that lead to improved reliability. Like other quality characteristics, the relationship between various factors and reliability can be studied using an appropriately chosen experimental design. Typically in industry, a large number of factors may need to be studied in a relatively small number of runs. Thus, highly fractionated 2^{k-p} designs [3] or non-geometric Plackett-Burman [27] designs such as their 12-run design are often used. In a sequential approach to experimentation, these designs would initially be used for screening out the unimportant factors before conducting a follow-up experiment with more levels so that the the response-factor relationship can modeled in more detail. In reality, the initial experiment may be the only one performed so that a proper assignment of factors for the 2^{k-p} designs can allow some potential interactions to be studied. Reference [13] has also shown that some information on interactions may also be obtained from the Plackett-Burman designs. Taguchi often initially uses designs with more levels. These include the 3^{k-p} designs and mixed-level designs such as the L18(2×3^7), which can be used to study one two-level factor and up to seven three-level factors. In [7] and [37], other mixed-level designs are catalogued. In a sequential experimental approach, however, a design with more levels such as a response surface design [2] which usually is not a fractional factorial would be used in a follow-up experiment.

The applications given in the Introduction illustrate the use of some of these designs. Reference [32] used a 12-run Plackett-Burman design to study ten factors (denoted by A-H,J,K) chosen from many possible product design, material selection and manufacturing

TABLE 9.1
Design and lifetime data for the heat exchanger experiment

					Factor						
F	B	A	C	D	E	G	H	J	K	U	Lifetime
1	1	1	1	1	1	1	1	1	1	1	(93.5, 105)
1	1	1	1	1	2	2	2	2	2	2	(42, 56.5)
1	1	2	2	2	1	1	2	2	2	1	(128, ∞)
1	2	1	2	2	2	2	1	1	2	1	(56.5, 71)
1	2	2	1	2	1	2	1	2	1	2	(56.5, 71)
1	2	2	2	1	2	1	2	1	1	2	(0, 42)
2	1	2	2	1	2	2	1	2	1	1	(56.5, 71)
2	1	2	1	2	2	1	1	1	2	2	(42, 56.5)
2	1	1	2	2	1	2	2	1	1	2	(82, 93.5)
2	2	2	1	1	1	2	2	1	2	1	(82, 93.5)
2	2	1	2	1	1	1	1	2	2	2	(82, 93.5)
2	2	1	1	2	2	1	2	2	1	1	(42, 56.5)

TABLE 9.2
Design and lifetime data for the fluorescent lamp experiment

		Factor				
A	B	C	D	E	Lifetime	
1	1	1	1	1	(14,16)	(20,∞)
1	1	2	2	2	(18,20)	(20,∞)
1	2	1	1	2	(08,10)	(10,12)
1	2	2	2	1	(18,20)	(20,∞)
2	1	1	2	1	(20,∞)	(20,∞)
2	1	2	1	2	(12,14)	(20,∞)
2	2	1	2	2	(16,18)	(20,∞)
2	2	2	1	1	(12,14)	(14,16)

factors to improve heat exchanger reliability. See Table 9.1 which gives the design and interval-censored lifetime data from eight inspections. The interval (93.5,105) indicates that the unit failed between 9350 and 10500 cycles, i.e., between the fifth and sixth inspections. (128,∞) indicates that the unit was still working at 12800 cycles, the last inspection. Note that one unit was still functioning at the last inspection.

Reference [34, p. 930] presented an experiment to improve the lifetime of fluorescent lamps which used a 2^{k-p} design. Five two-level factors denoted by A-E were studied using a twice replicated 8-run experiment with defining relations D=AC and E=BC which was conducted over 20 days with inspections every two days. The design and lifetime data appear in Table 9.2. Besides the main effects, the experimenter also thought that the AB(=DE) interaction might be potentially important. Note that there are right-censored data because seven of the 16 lamps had not failed by the 20 day inspection

Examples of other designs discussed above include the clutch spring experiment in [33, Chapter 9] which used a 3^{10-7} to study seven factors. Reference [26] used a mixed-level 32-run design to study two four-level factors and seven two-level factors. We also know of one study that used a Box-Behnken response surface design [2] as a follow-up experiment to investigate four factors previously identified in a 2^{k-p} screening design.

The data from these designs can be analyzed using a parametric model such as a lognormal or Weibull regression model. These models and their analyses are presented in Section 9.4. Analyses of the fluorescent lamp and heat exchanger experiments will be discussed in Sections 9.5 and 9.6, respectively.

9.3   Experiments for Achieving Robust Reliability

Taguchi's robust design is also referred to as parameter design because its objective is to find levels of engineering parameters (called control factors here) that yield a robust product/process, i.e., that make the product/process insensitive to the variation of hard or impossible to control noise factors. Taguchi's tactics for carrying out robust design are to specify a criterion for assessing noise factor effects and then use experimentation to estimate the criterion. Note that while noise factors are difficult or impractical to control in production or in use, for purposes of the experiment (i.e., to learn about the effect of the noise factors), the noise factors need to be controlled during the experiment. Following the notation used in [40], a criterion for assessing the effect of the noise factors (termed the loss and denoted by $L(\cdot)$) at a particular combination of control factor levels $\mathbf{x}_{control}$ can be defined for a general loss function $l(\cdot)$ as:

$$L(\mathbf{x}_{control}) = \int l(Y(\mathbf{x}_{control}, \mathbf{x}_{noise})) \cdot f(\mathbf{x}_{noise}) d\mathbf{x}_{noise} , \qquad (9.3.1)$$

where $Y(\mathbf{x}_{control}, \mathbf{x}_{noise})$ is the random quality characteristic observed at a particular combination of control and noise factor levels $(\mathbf{x}_{control}, \mathbf{x}_{noise})$ and $f(\cdot)$ is the joint probability density function of the noise factors. In this formulation, the objective of robust design is to find a product/process design $\mathbf{x}_{control}$ with minimum loss. In applying robust design to reliability, Y is the lifetime random variable; some appropriate loss functions will be discussed later in the example given in Section 9.7.

In [34], Taguchi proposed using experimentation to estimate the loss (9.3.1) and modelling the estimated losses in terms of the control factors. Taguchi recommends using specialized experimental plans referred to as product (or crossed) arrays. A product array consists of two plans or arrays, one for the control factors called the "control array" and the other for the noise factors called the "noise array". The product or crossed array design is so named because all the noise factor combinations specified by the noise array are experimented at every combination of the control factors specified by the control array.

As an example, consider an experiment for improving the lifetime of drill bits (i.e., number of holes drilled before breakage) used in fabricating multilayer printed circuit boards [22]. In designing multilayer circuit boards, small diameter holes are desired because they allow more room for the circuitry. The strength of small diameter drill bits is greatly reduced, however, so that breakage becomes a serious problem; broken bits cannot be removed from the boards requiring the boards to be scrapped. A product array consisting of a 16-run control array and an eight run noise array was used to study 11 control factors (A at four levels and B–J and L at two levels) and five noise factors (M–Q at two levels) as displayed in Table 9.3. The control factors were selected material composition and geometric characteristics of drill bits such as the carbide cobalt percentage in a drill bit (factor A) and radial rake (factor F). The noise factors dealt with characteristics of different types of multilayer circuit boards that would be drilled such as board material (factor O) and number of layers in a board (factor P). Thus, 16 different drill bit designs specified by the control factor array were used in the eight different production conditions specified by the noise factor array. Note that testing was

TABLE 9.3
Product array design and lifetime data for the drill bit experiment (experiment ended at 3000 cycles)

Control Array ADBCFGHIEJL	Noise Array							
	1	1	1	1	2	2	2	2 M
	1	1	2	2	1	1	2	2 N
	1	1	2	2	2	2	1	1 O
	1	2	1	2	1	2	1	2 P
	1	2	2	1	2	1	1	2 Q
	Lifetime							
11111111111	1280	44	150	20	60	2	65	25
11111222222	2680	125	120	2	165	100	795	307
12222111122	2670	480	762	130	1422	280	670	130
12222222211	2655	90	7	27	3	15	90	480
21122112212	3000	440	480	10	1260	5	1720	3000
21122221121	2586	6	370	45	2190	36	1030	16
22211112221	3000	2580	20	320	425	85	950	3000
22211221112	800	45	260	250	1650	470	1250	70
31212121211	3000	190	140	2	100	3	450	840
31212212122	3000	638	440	145	690	140	1180	1080
32121212111	3000	180	870	310	2820	240	2190	1100
41221122112	3000	612	1611	625	1720	195	1881	2780
41221211221	3000	1145	1060	198	1340	95	2509	345
32121121222	3000	970	180	220	415	70	2630	3000
42112122121	3000	3000	794	40	160	50	495	3000
42112211212	680	140	809	275	1130	145	2025	125

stopped after 3,000 holes were drilled and 11% of the tested drill bits had not failed by that time.

In [34], Taguchi originally proposed estimating the loss $L(\mathbf{x}_{control})$ for each $\mathbf{x}_{control}$ specified by the control array from the data obtained by varying the noise factors according to the noise array and then modeling it as a function of the control factors. Alternatively, [40] proposed modeling the response Y directly as a function of both the control and noise factors and then evaluating the loss using the estimated response model. Their rationale for the latter approach, termed the response-model approach by [30], was that it would be more likely to find a simple model for the response than one for the much more complicated estimated loss. Examples in [40] and [30] provide evidence for preferring the response-model approach. Reference [30] showed that the approach also provides more information.

For achieving robust reliability, the response-model approach is a natural one because the same parametric regression models mentioned in Section 9.2 can be used. The product array data allows a model to be fit consisting of all C main effects (possibly some $C \times C$ interactions), all $C \times N$ interactions and all N main effects (possibly some $N \times N$ interactions), where C and N denotes control and noise factors, respectively. The $C \times N$ interactions play an important role because the fact that the loss (9.3.1) changes for different control factor combinations means that these interactions must exist. Figure 9.1a displays a simplified relationship between a response Y and one control factor (at two levels) and one noise factor (over an interval) and shows that the effect of the noise factor is substantially smaller at control factor level 1 (C1). Thus, robust design exploits the existence of interactions between control and noise factors. Note that having a $C \times N$ interaction is not sufficient for an opportunity for robustness as is shown in Figure 9.1b where the magnitude of the change over the noise factor interval at both levels of the control factor is the same. Consequently, an N main effect is also needed which explains the inclusion of both $C \times N$ interactions and N main effects in the model. The C main effects and $C \times C$ interactions indicate the general response value about which the response varies as the noise factors vary according to their distribution; the amount of variation depends on the magnitudes of the N main effects and $C \times N$ interactions.

By taking the response-model approach, alternate designs to a product array have also been suggested. For example, [40] proposed using a single plan or array for both the control and noise factors. Reference [30] explored the economic advantages of single arrays over product arrays.

9.4 Analysis Methods for Censored Data

For analyzing the experiments discussed in the previous two sections, we consider the following parametric regression model [18]:

$$y_i = \log(t_i) = \mathbf{x}_i^T \beta + \sigma \epsilon_i, \quad i = 1, \ldots, n, \tag{9.4.1}$$

where the $\{t_i\}$ are the lifetimes, the $\{\mathbf{x}_i\}$ are the corresponding vectors of covariates values, β is the vector of location parameters and σ is the scale parameter. The errors $\{\epsilon_i\}$ are i.i.d. standard extreme-value r.v.'s if the lifetimes have a Weibull distribution and are i.i.d. standard normal r.v.'s if the lifetimes follow a lognormal distribution. The models with these two error distributions are called the lognormal and Weibull regression models, respectively. For the reliability improvement experiments given in Section 9.2, the covariates consist of an intercept, the factor main effects and possibly some interactions. For the robust reliability experiments presented in Section 9.3, the covariates consist of an intercept, the C main effects, possibly some $C \times C$ interactions, the $C \times N$ interactions, the N main effects and possibly some $N \times N$ interactions, where C and N denote control and noise factors, respectively. Because of the typically small amount of data collected in these experiments, there appear to be little qualitative differences between the use of either model; there is not enough data to differentiate between the two error distributions. The lognormal regression model has some advantages, however, because of its connection with the normal regression model and has been exploited in recent work [12, 14]. For complete data, i.e., when all lifetimes observed, the analysis is straightforward using maximum likelihood estimation [18]. As mentioned in the Introduction, the censored data present new challenges in the context of analyzing reliability experiments. Next we give an overview of methods for handling the censored data which highlights these challenges.

9.4.1 A Quick and Dirty Method

A quick and dirty (QD) method which continues to be used in practice treats the right-censoring times as actual failure times and then analyzes them by standard methods for complete data. (For interval-censored data, an interval endpoint or midpoint is used.) Although simple, ignoring the censoring can lead to wrong decisions because the unobserved failure times and right-censoring times may differ greatly depending on the particular factor level combination. A simulation study in [12] showed that the QD method can perform quite poorly. Reference [9] also pointed out that Taguchi's minute accumulating analysis [34] treats the right-censored data similarly.

(a) opportunity for robustness

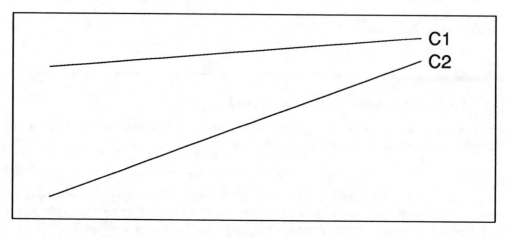

C1

C2

N

(b) no opportunity for robustness

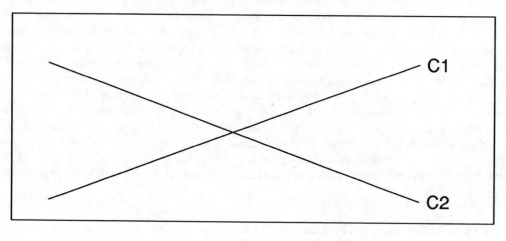

C1

C2

N

FIGURE 9.1
Example response functions and opportunity for robustness

9.4.2 Fitting Saturated Models and Their Submodels

One obvious approach for handling censored data is to specify a saturated model and fit it using maximum likelihood estimation (MLE). The approach has several drawbacks, however. First, the MLEs need not exist, i.e., at least one is infinite, so that testing cannot be done by comparing the MLEs with their standard errors. Reference [31] gave necessary and sufficient conditions for the existence of MLEs for model (9.4.1). In the reliability context, [11] concluded that estimability problems will tend to occur for saturated and nearly saturated submodels.

In [17], an MLE-based forward selection procedure was proposed. Building the model up tends to mitigate estimability problems, but still requires a certain amount of computation; an iterative algorithm is required to obtain the MLEs for each model fit. If a stepwise selection procedure is used instead, the amount of computation required increases substantially. In [19], an efficient algorithm for an all subsets procedure which finds good submodels of a saturated model was proposed. While the MLEs of saturated and nearly saturated models are not likely to exist, the likelihood is still well defined so that sequences of submodels could be fit and compared using appropriate likelihood ratio tests. There are computational difficulties associated with the estimability problems, however, which the software needs to handle as shown in [5]. Nevertheless, the computational cost can be quite high because many possible models may be fit.

For the Plackett-Burman 12-run, mixed-level fractional factorial and 3^{k-p} designs, the computational cost is even higher and may be prohibitive because of the enormous number of possible models. If we consider a comprehensive model for these designs, say containing all main effects and two-factor interactions, the number of effects exceeds the number of runs and therefore cannot be fit. As [14] pointed out, there is no saturated comprehensive model for these designs because of their complex aliasing patterns; e.g., a main effect is partially aliased with several if not many two-factor interactions rather than being completely aliased with a one or a few two-factor interactions as is the case for 2^{k-p} designs. Take for example the 12-run Plackett-Burman design where each factor main-effect is partially aliased with all two-factor interactions not involving the factor. For the heat exchanger experiment with ten factors discussed in Section 9.2, a second-order model would consist of 55 effects plus an intercept. Consequently, the number of possible models to be fit can be enormous, even when restrictions are made on the class of models such as including at least one of the corresponding main effects for any interaction appearing in the model.

As a final comment, the likelihood ratio testing approach which handles the estimability problem may not be entirely useful for robust reliability experiments unless the MLEs of the final model selected exist. In the context of analyzing ordinal data (which applies here as well), [4] pointed out that in choosing control factor levels, finite estimates are needed to evaluate the loss (9.3.1).

9.4.3 Iterative Imputation-Based Model Selection

The computational problems and requirements of the standard MLE based methods presented in the previous section motivated [12] which proposed an iterative model selection procedure based on imputation. By imputing the censored data to obtain "complete normal" or pseudo-complete data, any of the standard model selection techniques can be used at much less computational cost. Note that this procedure can be used for the lognormal regression model (9.4.1) since the log lifetimes are normally distributed.

The iterative model selection procedure consists of a three-step loop:

1. Imputation
2. Model Selection
3. Model Fitting

An initial model chosen by the experimenter which may consist only of main effects for highly fractionated designs is fit using maximum likelihood estimation. In the imputation step, the censored data are replaced by conditional mean lifetimes based on the current model. The next model is then chosen based on these pseudo-complete or pseudo-uncensored data using any standard method such as stepwise selection. The chosen model is then fit using maximum likelihood estimation, the censored data are imputed, and so forth until the next chosen model stops changing.

The procedure exploits the simplicity of a complete data problem to solve an incomplete data problem. While the use of standard methods on pseudo-complete data lacks theoretical justification, the simulation study given in [12] showed that the method performs well and is far superior to the QD method. The procedure also relies on maximum likelihood estimation so that there are potential estimability problems. These tend to be avoided since the procedure builds up the model rather than starting with a saturated model.

In earlier work, [8] used imputation to analyze left-censored yield data from an experiment on a chemical process. Their method starts out by imputing the censored data based on the model obtained by the QD method, i.e., a least-squares fit treating the censoring values as actual observations. Imputing the censored data by their conditional means, the next model is chosen based on the least-squares estimates (LSEs) from the pseudo-complete data. The LSEs of the chosen model, which do not require an additional fitting, are used in the next imputation. A key difference with that proposed by [12] is the reliance on LSEs rather than MLEs; the latter directly incorporates the censoring information in fitting the current chosen model. A rationale for the LSEs is that for a fixed model, iterative LSEs at convergence are nearly equal to MLEs [1] and therefore can be viewed as one-step MLEs. Despite its simplicity, the simulation study in [12] indicates that the LSE based procedure performs worse especially for heavier censoring.

9.4.4 A Bayesian Approach

The estimability problem of the MLE based methods presented in the previous sections motivated the Bayesian approach proposed in [14]. The Bayesian approach is a natural one because important factor effects might be expected to be large but not infinite. By using proper prior distributions, posterior distributions with finite modes result and can be used to obtain finite estimates. Also, the posterior distributions allow the importance of factorial effects to be assessed without using the asymptotic approximations employed by the MLE based methods.

This approach is relatively simple to implement using the recent advances in Bayesian computing. That is, the resampling methodology makes calculating the entire posterior distribution or some characteristic pretty straightforward to do. Reference [39] showed how posteriors for censored data regression models could be calculated using data augmentation [36]. In [38], it was shown how the the posterior maximizer could be calculated without computing the entire posterior by what is called the Monte Carlo EM. In [38, 39], an improper prior was used which [14] showed could be extended to the proper conjugate prior given by [28]. Besides producing well-behaved posteriors, the conjugate prior allows the robustness of the results to be investigated. See [14] for more details.

In analyzing 2^{k-p} designs, a saturated comprehensive can be entertained using a stan-

TABLE 9.4
MLEs and standard errors of lognormal regression model for the fluorescent lamp experiment

Effect	MLE	Std Err
Int	2.939	0.064
A	-0.117	0.062
B	0.201	0.060
AB	-0.049	0.062
C	0.051	0.062
D	-0.273	0.062
E	0.153	0.063
σ	1.590	0.043

dard Bayesian analysis. Thus, for these designs, the iterative model selection procedure of [12] is not needed. For designs with complex aliasing patterns such as the 12-run Plackett-Burman design, however, [14] proposed adapting this iterative model selection procedure. That is, the imputation is based on the posterior maximizer instead of the MLEs. For the same reasons given previously, this adapted iterative model selection procedure requires much less computation than an entirely Bayesian approach.

For analyzing robust design experiments, the Bayesian approach has an additional advantage. In [4], it was shown how the Bayesian approach can easily combine the uncertainty of the model estimates with the variation of the noise factors in choosing control factor levels; previous work had not accounted for the model estimates' uncertainty. This idea can be applied to robust reliability experiments and is currently being implemented.

9.5 Analysis of Fluorescent Lamp Experiment

For the fluorescent lamp experiment presented in Section 9.2, first consider a standard MLE based analysis using the lognormal regression model. Table 9.4 gives the MLEs and standard errors for the five main effects (A-E) and the AB interaction which the experimenters thought might be important. The intercept is denoted by Int. Based on these results, the main effects D, B, E and A are important in the order given.

An additional effect (BD=AE) can be entertained if a saturated model is fit, but the MLEs do not exist for this model since both replicates at the fifth run are right-censored. The Bayesian approach of Section 9.4.4, which circumvents the estimability problem, can be taken using a relatively diffuse prior distribution with mean zero for the effects. The quantiles corresponding to central 0.95 and 0.99 intervals of the posterior distribution are displayed in Table 9.5. These results show that BD is not important and confirm the importance of main effects, D, B, E and A. In [14], it was shown that these results hold for a much sharper prior distribution. The sign of these effects suggests that reliability will improve at recommended settings $A_1 B_2 D_1 E_2$, where the subscript indicates the recommended level.

TABLE 9.5
Posterior quantiles using diffuse prior for the fluorescent lamp experiment

Effect	Quantile			
	.005	.025	.975	.995
Int	2.83	2.84	3.02	3.07
A	-0.24	-0.19	-0.02	-0.00
B	0.07	0.09	0.26	0.31
C	-0.06	-0.05	0.12	0.17
AB	-0.15	-0.10	0.07	0.08
E	0.02	0.04	0.21	0.26
D	-0.40	-0.35	-0.18	-0.17
BD	-0.10	-0.05	0.11	0.13
σ	0.10	0.10	0.20	0.24

TABLE 9.6
Posterior quantiles using diffuse prior for the heat exchanger experiment

Effect	Quantile			
	.005	.025	.975	.995
Int	4.180	4.190	4.262	4.273
F	-0.050	-0.040	0.031	0.041
B	-0.145	-0.134	-0.059	-0.049
A	-0.068	-0.054	0.019	0.029
C	0.034	0.044	0.117	0.123
D	-0.026	-0.016	0.057	0.069
E	-0.340	-0.329	-0.257	-0.247
G	-0.061	-0.049	0.023	0.035
H	-0.058	-0.045	0.029	0.040
J	-0.022	-0.012	0.058	0.068
K	0.028	0.039	0.111	0.121
U	-0.191	-0.176	-0.104	-0.094
σ	0.028	0.028	0.028	0.028

9.6 Analysis of the Heat Exchanger Experiment

In [12], the iterative model selection procedure was used to analyze the heat exchanger experiment presented in Section 9.2. The main-effects model for the ten factors whose MLEs exist was fit and used to impute the censored data. A stepwise selection procedure applied to the pseudo-complete data identified the model (E, EG, EH). The same model was chosen in the next iteration.

Now consider the saturated model consisting of the ten main effects and the effect associated with the unassigned design column U in Table 9.1. Since the MLEs do not exist for this model, the Bayesian approach can be taken. Table 9.6 gives the the quantiles corresponding to central 0.95 and 0.99 intervals of the posterior using a relatively diffuse prior. While B, C, E, K and U appear important, recall that all main effects are aliased with the 36 two-factor interactions not involving the factor; U is aliased with all 45 two-factor interactions between the ten factors. The U effect's importance means that some two-factor interactions are important and can explain the main effects B, C and K. See [12] for details. E is much larger so that E is clearly important. Thus, the Bayesian analysis of the saturated model confirms the importance of E and identifies the presence of some interactions.

9.7 Analysis of the Drill Bit Experiment

Taking the response-model approach, a Weibull regression model (9.4.1) consisting of an intercept, C main effects, one $C \times C$ interaction ($D \times E$), N main effects, two $N \times N$ interactions ($M \times P$, $M \times Q$) and all the $C \times N$ interactions, was fit using the product array data from Table 9.3. Table 9.7 presents only the MLEs and respective standard errors for the important effects. Since factor A has four levels, the main effect is modeled by linear, quadratic and cubic components which are denoted by A_l, A_q and A_c, respectively.

Once the response has been modeled, recommendations for the important control factors settings need to be made. For a simple model with few noise factors, they may be apparent from inspection of the model directly; i.e., by observing what the significant effects are and their signs and magnitudes. In [30], an example is given, but for complicated models, this approach may be tedious if not impossible.

An alternative is to specify some meaningful criterion or loss and use the identified model to evaluate them. The loss (9.3.1) can then be estimated using some distribution for the noise factors. In practice, because it may be difficult to specify the distribution, the criterion can be evaluated over the noise combinations given by the noise array. The same noise array from the experiment does not have to be used, however. For example, instead of a fractional factorial design, the loss could be evaluated using a full factorial design. The noise combinations can also be weighted appropriately to reflect their probabilities of occurrence. Similarly, the loss can be estimated for all possible settings of the control factors.

For achieving robust reliability, besides requiring high reliability on average, as little dependence as possible on the noise factors is desired. This suggests estimating the mean and standard deviation of the mixture of lifetime distributions given by (9.4.1) with the mixture defined by the noise factor distribution. Ideally, there is a control factor setting that simultaneously maximizes both the mean and minimizes the standard deviation; otherwise, tradeoffs between the two need to be made. By taking a worst case approach,

TABLE 9.7
MLEs and standard errors of Weibull regression model for the drill bit experiment

Effect	MLE	Std Err	Effect	MLE	Std Err
Int	6.182	0.047	A_1N	-0.047	0.021
A_1	0.279	0.021	BN	-0.094	0.042
A_q	-0.268	0.043	IN	-0.111	0.044
A_c	0.071	0.018	DO	0.181	0.058
C	-0.194	0.043	BO	-0.136	0.047
D	-0.265	0.043	GO	0.277	0.054
F	0.154	0.042	IO	-0.429	0.054
G	0.132	0.048	EO	-0.376	0.061
H	0.218	0.048	LO	0.294	0.059
I	-0.272	0.044	A_qP	-0.123	0.061
J	-0.231	0.043	DP	0.269	0.059
L	-0.272	0.043	BP	0.213	0.048
DE	-0.225	0.041	CP	-0.119	0.059
M	0.179	0.058	IP	0.195	0.054
N	0.136	0.047	EP	0.143	0.061
O	0.898	0.059	JP	0.156	0.060
P	0.862	0.057	LP	-0.194	0.060
Q	0.548	0.057	MP	0.237	0.057
A_qQ	-0.174	0.061	GM	0.236	0.054
HQ	0.079	0.061	JM	0.149	0.059
IQ	-0.117	0.054	LM	0.123	0.059
LQ	0.202	0.060	σ	0.350	0.030

the minimum mean lifetime over the noise factor distribution provides another criterion which can be maximized as a basis for choosing the control factor settings. Finally, Taguchi's [34] larger-the-better signal-to-noise ratio (LTB S/N ratio) for assessing the effect of noise factors is applicable here and is based on the loss in (9.3.1) using the loss function $l(Y) = 1/Y^2$. Control factor settings with large LTB S/N ratios, defined as $-10log_{10}L(\mathbf{x}_{control})$, can then be identified. Other criteria are possible such as that based on the probability of exceeding a specified time such as a warranty period, but are not considered further.

For the drill bit experiment, the relationship between the response and the control and noise factors as seen from Table 9.7 is too complicated to make control factor level recommendations simply by inspecting the model. Consequently, the criteria discussed above can be estimated, namely the mean, standard deviation, minimum mean and LTB S/N ratio using all possible combinations of noise factors ($32 = 2^5$) and evaluated at all possible combinations of control factor ($4096 = 4 \times 2^{10}$) and then ranked appropriately (out of 4096, with 1 being the best). Table 9.8 presents the best five control factor combinations for each criterion along with the other criteria and their ranks. Several observations can be made: (i) the combination least sensitive to the noise factors ranks rather poorly on other criteria, especially the mean; (ii) the other three criteria identify many of the same combinations; (iii) there is little difference between the top few combinations. Based on Table 9.8, a good choice of factor levels would be $A_4D_2B_2C_2F_1G_2H_1I_2E_1J_2L_2$. Note that this combination is also rather robust to the noise factors.

TABLE 9.8
Best factor settings for the drill bit experiment

settings ADBCFGHIEJL	mean value	mean rank	std dev value	std dev rank	LTB S/N value	LTB S/N rank	min mean value	min mean rank
five largest means								
42121112122	8.877	1	1.084	497	18.776	2	7.122	4
42221112122	8.877	2	1.064	469	18.781	1	7.088	7
42121212122	8.613	3	0.903	205	18.552	4	7.150	3
42221212122	8.613	4	0.681	66	18.618	3	7.116	5
32121112122	8.571	5	1.396	1246	18.312	10	6.128	71
five smallest standard deviations								
42212221121	6.209	2171	0.325	1	15.825	1071	5.540	199
42211221121	6.517	1629	0.325	2	16.249	735	5.848	117
42212211121	6.829	1102	0.325	3	16.658	463	6.160	66
42211211121	7.137	674	0.325	4	17.043	271	6.468	35
12211211121	5.321	3306	0.418	5	14.439	2233	4.414	835
five largest LTB S/N ratios								
42221112122	8.877	2	1.064	469	18.781	1	7.088	7
42121112122	8.877	1	1.084	497	18.776	2	7.122	4
42221212122	8.613	4	0.681	66	18.618	3	7.116	5
42121212122	8.613	3	0.903	205	18.552	4	7.150	3
42222112122	8.569	8	1.064	470	18.461	5	6.780	17
five largest minimum means								
42221111122	8.333	20	0.815	120	18.299	11	7.424	1
42211111122	7.945	88	0.594	32	17.934	41	7.274	2
42121212122	8.613	3	0.903	205	18.552	4	7.150	3
42121112122	8.877	1	1.084	497	18.776	2	7.122	4
42221212122	8.613	4	0.681	66	18.618	3	7.116	5

9.8   Discussion

The paper has presented the use of designed experiments for reliability improvement and for achieving robust reliability. While the experimental designs are the same ones used for improving any quality characteristic, it is the analysis of censored lifetime data from these designs that has provided new challenges. In this context, standard methods need to deal with estimability problems because of the censoring or require a large amount of computation because many possible models may be fitted, especially for designs with complex aliasing patterns. Standard MLE based methods may be used if the censoring is not too severe as was demonstrated in the drill bit experiment, however. A Bayesian approach [14] summarized in the paper addresses both drawbacks. Its current implementation is for the lognormal regression model or for the normal regression model after transforming the lifetime response. The Bayesian approach could be extended to the Weibull regression model.

Other types of reliability improvement experiments need to be explored for highly reliable product where lifetime based experiments discussed in this paper are not feasible. Two possibilities are the use of acceleration factors to speed-up failures [23, 21] and the collection of degradation data, i.e., monitoring the degradation of surrogate characteristics for reliability [20].

There are indications that statistically designed experiments are now being used more often for improving reliability than in the past. It is hoped that the trend continues with more experiments for achieving robust reliability being performed. As other types of reliability improvement experiments are considered, new challenges for analyzing them will arise. Finally, the use of alternate experimental designs needs to be investigated.

References

1. Aitken, M. (1981), A note on the regression analysis of censored data, *Technometrics*, **23**, 161-163.

2. Box, G.E.P. and Draper, N.R. (1987), *Empirical Model-Building and Response Surfaces*, John Wiley & Sons, New York.

3. Box, G.E.P., Hunter, W.G., and Hunter, J.S. (1978), *Statistics for Experimenters*, John Wiley & Sons, New York.

4. Chipman, H. and Hamada, M. (1993), A Bayesian approach for analyzing ordinal data from industrial experiments, *Institute for Improvement in Quality and Productivity Research Report RR-93-06*, University of Waterloo, Waterloo, Canada.

5. Clarkson, D.B. and Jennrich, R.I. (1991), Computing extended maximum likelihood estimates for linear parameter models, *Journal of the Royal Statistical Society, Series B*, **53**, 417-426.

6. Condra, R.I. (1993), *Reliability Improvement with Design of Experiments*, Marcel Dekker, New York.

7. Dey, A. (1985), *Orthogonal Fractional Factorial Designs*, Halsted Press, New York.

8. Hahn, R.I., Morgan, C.B., and Schmee, J. (1981), The analysis of a fractional factorial experiment with censored data using iterative least squares, *Technometrics*, **23**, 33-36.

9. Hamada, M. (1992), An explanation and criticism of minute accumulating analysis, *Journal of Quality Technology*, **24**, 70-77.

10. Hamada, M. (1993), Reliability improvement via Taguchi's robust design, *Quality and Reliability Engineering International*, **9**, 7-13.

11. Hamada, M. and Tse, S.K. (1992), On estimability problems in industrial experiments with censored data, *Statistica Sinica*, **2**, 381-391.

12. Hamada, M. and Wu, C.F.J. (1991), Analysis of censored data from highly fractionated experiments, *Technometrics*, **33**, 25-38.

13. Hamada, M. and Wu, C.F.J. (1992), Analysis of designed experiments with complex aliasing, *Journal of Quality Technology*, **24**, 130-137.

14. Hamada, M. and Wu, C.F.J. (1992), Analysis of censored data from fractionated experiments: a Bayesian approach, *Institute for Improvement in Quality and Productivity Research Report RR-92-11*, University of Waterloo, Waterloo, Canada.

15. Hitzelberger, A.J. (1967), Improve your reliability, *Industrial Quality Control*, **24**, 313-316.

16. Kackar, R.N. (1985), Off-line quality control, parameter design, and the Taguchi method (with discussion), *Journal of Quality Technology*, **17**, 176-209.

17. Krall, J.M., Uthoff, V.A., and Harley, J.B. (1975), A step-up procedure for selecting variables associated with survival, *Biometrics*, **31**, 49-57.

18. Lawless, J.F. (1982), *Statistical Models & Methods for Lifetime Data*, John Wiley & Sons, New York.

19. Lawless, J.F. and Singhal, K. (1978), Efficient screening of nonnormal regression models, *Biometrics*, **34**, 318-327.

20. Lu, C.J. and Meeker, W.Q. (1993), Using degradation measures to estimate a time-to-failure distribution, *Technometrics*, **35**, 161-174.

21. Meeker, W.Q. and Escobar, L.A. (1993), A review of recent research and current issues in accelerated testing, *International Statistical Review*, **61**, 147-168.

22. Montmarquet, F. (1988), Printed circuit drill bit design optimization using Taguchi's methods – .013″ diameter bits, In *Sixth Symposium on Taguchi Methods*, 70-77, American Supplier Institute, Romulus, MI.

23. Nelson, W. (1982), *Applied Life Data Analysis*, John Wiley & Sons, New York.

24. Nelson, W. (1990), *Accelerated Testing - Statistical Models, Test Plans, and Data Analysis*, John Wiley & Sons, New York.

25. O'Connor, P.D.T. (1991), *Practical Reliability Engineering*, Third Edition, John Wiley & Sons, Chichester, England.

26. Phadke, M.S. (1986), Design optimization case studies, *AT&T Technical Journal*, **65**, 51-68.

27. Plackett, R.L. and Burman, J.P. (1946), The design of optimum multifactorial experiments, *Biometrika*, **33**, 305-325.

28. Raiffa, H. and Schlaifer, R. (1961), *Introduction to Statistical Decision Theory*, McGraw-Hill, New York.

29. Sampford, M.R. and Taylor, J. (1959), Censored observations in randomized block experiments, *Journal of the Royal Statistical Society, Series B*, **21**, 214-237.

30. Shoemaker, A.C., Tsui, K.L., and Wu, C.F.J. (1991), Economical experimentation methods for robust design, *Technometrics*, **33**, 415-427.

31. Silvapulle, M.J. and Burridge, J. (1986), Existence of maximum likelihood estimates in regression models for grouped and ungrouped data, *Journal of the Royal Statistical Society, Series B*, **48**, 100-106.

32. Specht, N. (1985), Heat exchanger product design via Taguchi methods, In *Third Symposium on Taguchi Methods*, 302-318, American Supplier Institute, Romulus, MI.

33. Taguchi, G. (1986), *Introduction to Quality Engineering*, Asian Productivity Organisation, Tokyo, Japan.

34. Taguchi, G. (1987), *System of Experimental Design*, Unipub/Kraus International Publications, White Plains, NY.

35. Taguchi, G. and Wu, Y.I. (1980), *Introduction to Off-Line Quality Control*, Japan Quality Control Association, Nagoya, Japan.

36. Tanner, M.A. and Wong, W.H. (1987), The calculation of posterior distributions by data augmentation (with discussion), *Journal of the American Statistical Association*, **82**, 528-540.

37. Wang, J.C. and Wu, C.F.J. (1991), An approach to the construction of asymmetrical orthogonal arrays, *Journal of the American Statistical Association*, **86**, 450-456.

38. Wei, G.C.G. and Tanner, M.A. (1990a), A Monte Carlo implementation of the EM algorithm and the Poor Man's Data augmentation algorithms, *Journal of the American Statistical Association*, **85**, 699-704.

39. Wei, G.C.G. and Tanner, M.A. (1990b), Posterior computations for censored regression data, *Journal of the American Statistical Association*, **85**, 829-839.

40. Welch, W.J., Yu, T.K., Kang, S.M., and Sacks, J. (1990), Computer experiments for quality control by parameter design, *Journal of Quality Technology*, **22**, 15-22.

41. Zelen, M. (1959), Factorial experiments in life testing, *Technometrics*, **1**, 269-288.

10

Unity Values for Bayes Reliability Demonstration Testing

Robert L. Parker, Harry F. Martz, Richard R. Prairie, and William J. Zimmer

University of New Mexico
Los Alamos National Laboratory
University of New Mexico
University of New Mexico

ABSTRACT Since in most component reliability assessment tests there is prior information available from development programs, Bayes reliability demonstration tests (BRDT) are important techniques to consider for assessment programs. In this paper, BRDT's are determined and presented in a concise manner for single fixed sample attribute testing using the Poisson distribution. This is done for several important and useful prior distributions, both non-informative and informative. Tables, graphs and comparisons of the plans are presented and examples are given.

CONTENTS

10.1 Introduction

Reliability demonstration tests (RDT) are usually performed on newly designed components in order to demonstrate whether the component has achieved a required reliability. Test plans generally consist of the amount and type of testing necessary to demonstrate the required reliability at a specified level of assurance. The concept is related to acceptance sampling except that RDT's are often more rigorous than the usual acceptance sampling plans. A RDT may determine the ultimate destiny of a specific type component. For example, the decision of whether to go into production for a component may depend on the result of a RDT, whereas the result of an acceptance sampling plan decision is simply the acceptance or rejection of a particular lot of product. Therefore, RDT's are often more elaborate than acceptance sampling plans.

For many applications RDT's include different types of testing such as accelerated or multiple environmental tests. Regardless of what types of tests are used the need to demonstrate high reliability requires large sample sizes and few failures. For example, with attribute testing, a sample size of $n = 480$ with 0 failures is required to demonstrate a reliability of .995 at 90% confidence. Since this level of testing is often difficult and extremely expensive, other approaches than those which are based exclusively on the sample data are often considered.

In most component assessment programs, data, prior to a RDT, are available from a variety of sources such as development and prototype testing. In addition, the design and development engineers have accumulated a large amount of information about product performance from which they can make discerning judgments about the reliability of the component. An approach to demonstration testing which incorporates, in a structured manner, prior data and engineering knowledge, into the test procedure is the Bayesian approach, Bayes reliability demonstration testing (BRDT). The Bayesian approach can often be used to help assure a reliability requirement with fewer demonstration tests than conventional testing by the formal inclusion of information from a wide range of sources.

In order to study the relationship between the strength of the prior information and the amount of test information needed to demonstrate a required reliability, a simple Bayes model is considered and BRDT plans are presented in a concise, graphical manner. This is accomplished through the use of Poisson unity factors for the test plans and the use of several important and useful prior distributions. The test plans can be easily related to the extent of the prior information as illustrated by one important and intuitively appealing prior parameter.

Also, in practice, the design of BRDT's is often neglected. That is, a pre-posterior analysis is not usually made. This is partly because such a study is, in many cases, a difficult procedure. The approach taken in this paper makes the design of the BRDT and the use of a pre-posterior analysis more available.

In this paper, BRDT's are generated for single fixed sample attribute testing using the Poisson distribution for the distribution of the number of failures. The Pareto and the gamma distributions are used as general priors but, as will be seen, only some special cases of these distributions are of particular interest here. These special cases will include the truncated uniform, the truncated maximum entropy prior distribution, some other versions of non-informative prior distributions and some informative prior distributions. Also, the truncated versions of these prior distributions will be of most interest, because reliability (or unreliability, p, considered here) is bounded.

In the range of interest for most RDT plans, the Poisson approximation to the binomial distribution is sufficiently precise. Usually it is determined that the Poisson is an adequate approximation to the binomial if the value of the parameter p is sufficiently small and the value of the sample size n is sufficiently large. In reliability demonstration testing, the range of p encountered in practice for the types of systems or components of interest, and the sample sizes necessary to demonstrate these reliabilities are such that the Poisson approximation to the binomial is valid.

The unity values R (or operating ratios) are defined as the ratio of the unacceptable unreliability p_r (sometimes referred to as the rejectable quality level) to the acceptable unreliability p_a. The values $R = np_r/np_a$, designated as unity values in the literature, can be used for the construction of sampling plans based on the Poisson distribution. Note that the sample size cancels in the unity value expression. Beginning with [5] and appearing in several tables, single sampling plans for attributes have been tabulated in terms of R. Later, plans were developed for double and multiple sampling using unity values by [7].

In the usual RDT, the demonstration decision is based solely on the results of a single test sample. The BRDT procedure which utilizes prior information and specification of the posterior risks is instinctively appealing in comparison. The posterior risks provide assurance that the quality level is in an appropriate range conditional on the results of the test sample. They are given by [4]

$$\alpha = P\{p \leq p_a | x \in \text{Rej}\} \ , \quad \beta = P\{p \geq p_r | x \in \text{Acc}\} \ , \tag{10.1.1}$$

where Rej and Acc are the rejection and acceptance regions of the test results, respectively. Here α is known as the *producer's posterior risk* and β is known as the *consumer's posterior risk*.

In the non-Bayes case, assurance is given that conforming units will pass the test and non-conforming units will fail the test. No probability statement can be made about the overall design or production process. However, the producer may desire assurance $1 - \alpha$ from the test that the design failing the test is indeed unreliable, and the consumer may desire $1 - \beta$ assurance from the test that the design passing the test is reliable. These assurances are achieved by considering the posterior risks of the BRDT using values of α and β.

Furthermore, these Bayesian posterior risks are influenced by prior information. However, the prior distribution is not only a conduit for the use of prior information in the demonstration test, it can also be used in a way that allows a consideration of unity values in the BRDT. Notice that if the prior information is used to specify a uniform prior distribution upper limit, P, on p, for example, the posterior risks are functions of only the test criteria and P. If the Poisson distribution is used, the posterior risks are functions of only the test criteria in the unity value form and nP.

There are several types of prior distribution that have this facility and these will be examined in this paper. In Section 10.2, the BRDT plans are developed for Pareto type priors and gamma priors and the posterior risks are formulated. In Section 10.3, graphs are developed and presented for the special cases considered here. These facilitate the use of the plans with several types of prior information (or lack of information) and allow for easy comparisons. Section 10.4 contains examples of the use of the graphs. Comparisons and conclusions about the procedure and its robustness are given in Section 10.5.

10.2   The Model, Priors and Posterior Risks

In the case of the usual (classical) RDT, the demonstration decision is based only on the results of the test sample. In order to demonstrate high reliability on the basis of information from only a single sample, it is obvious that the size of such a sample must be very large. For example, consider a test program for a CA 5600 capacitor designed for use in a new system. This capacitor has been custom-designed for the new system and has been allocated an acceptable unreliability of $p_a = 0.01$ in order to meet system requirements. It is determined that an unacceptable reliability is $p_r = .03$. The reliability engineer must devise a reliability demonstration test which will distinguish between capacitors with a 0.01 unreliability and capacitors with an unacceptable unreliability of $p_r = 0.03$. Aside from the very important consideration of test conditions, such as test environments, the number of units that need to be tested to provide the above discrimination is required. If the usual type operating characteristic (OC) curve is used such that $P_A\{p = .01\} = .95$ and $P_A\{p = .03\} = .10$ where P_A is the probability of accepting, then one could use a plan tabled in [2], [3], [6], or in [8] with a unity value of $p_r/p_a = .03/.01 = 3$. Such a plan would require an acceptance number $c = 7$ and a sample size of $n = 398$.

The above plan ignores some important information that is available for the CA 5600 capacitor. The design engineers have experience in designing similar capacitors and considerable data have been collected on the CA 5400 and CA 5500 capacitors, which are earlier versions of the CA 5600 capacitor. Furthermore, it is believed that the test conditions and the capacitors are similar enough that these data and the accumulated judgment should provide information about the CA 5600 capacitor reliability. The problem, then, is to devise a reliability demonstration test that incorporates such information in a structured way that will permit a reduction in sample size.

Because, in RDT, emphasis is placed on the results of a single test, it is important that the test provide assurance that the component design is adequate given the data observed. In terms of the acceptable unreliability p_a and the unacceptable unreliability p_r to be demonstrated by the test, the Bayes posterior risks are those given in (10.1.1). For the plans considered here, the rejection and acceptance regions are defined in terms of an acceptance number c, such that:

$$\text{Rej} = \{x|x > c\} \text{ and Acc} = \{x|x \leq c\} .$$

The desired assurances can then be achieved with small values of α and β. Also, the posterior risks are Bayesian and are influenced by a prior distribution on p. It is the influence of the prior that is important for the possibility of reducing the sample size and it is through the prior that one can incorporate prior information about the unreliability. In some simple cases where the prior information limits either the range of possible values for unreliability or indicates a value of the prior mean unreliability, it is straight-forward and informative to study the relationship between the unity values, the information on the unreliability, and the sample size necessary to support the desired discrimination. For example, assuming that the elicitation of prior information induces a truncated uniform distribution on the unreliability p, the reduction in sample size related to the increasing stringency of the prior information can easily be studied using the unity values.

Let

$$f(x|p) = \frac{e^{-np}(np)^x}{x!}, \qquad x = 0, 1, 2, \ldots \tag{10.2.1}$$

and let $g(p)$ denote the prior distribution. Then the posterior distribution, denoted by $g(p|x)$, is

$$g(p|x) = \frac{f(x|p)g(p)}{f(x)}, \tag{10.2.2}$$

where $f(x) = \int_\rho f(x|p)g(p)dp$, and ρ is the domain for p.

Now the producer's posterior risk is given by:

$$\alpha = P_\alpha\{p \leq p_a|x \in \text{Rej}\} = \int_0^{p_a} g(p|x \in \text{Rej})dp$$

$$= \frac{\int_0^{p_a} [1 - \sum_{x=0}^c f(x|p)] g(p)dp}{P\{x \in \text{Rej}\}}. \tag{10.2.3}$$

Correspondingly, the consumer's posterior risk is given by:

$$\beta = P_\beta\{p \geq p_r|x \in \text{Acc}\}$$

$$= \frac{\int_{p_r}^1 [\sum_{x=0}^c f(x|p)g(p)dp]}{P\{x \in \text{Acc}\}}. \tag{10.2.4}$$

In this paper, two families of prior distributions are used: (1) a truncated Pareto type family of prior distributions and (2) the truncated gamma family of prior distributions. Truncated priors are considered because they allow the practitioners the flexibility to limit the range of the unreliability if that is a part of the prior information. If there is no information to that extent, then the limit can be taken to the appropriate value.

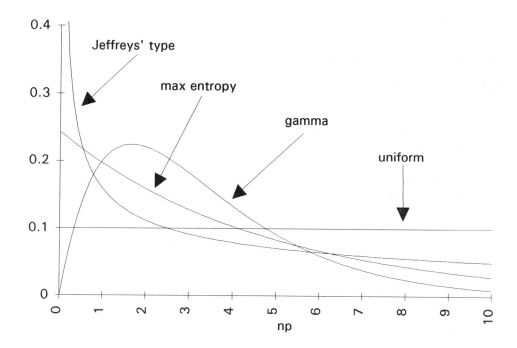

FIGURE 10.1
Examples of Priors

The truncated Pareto type family has two special cases of interest here: (a) the truncated uniform prior and (b) the truncated Jeffreys' type prior. The truncated gamma family also has two special cases of interest here: (a) the truncated maximum entropy prior and (b) the truncated gamma prior with special parameters that allow the use of unity factors.

The first three priors: the truncated uniform, the truncated Jeffreys' type and the truncated maximum entropy are used because of their association with use as non-informative priors in many applications. As will be seen later, these distributions do not appear to act like non-informative priors in the sense that the associated plans are quite different from the classical plans. However, when they are used with partial information, such as the limit on the range of the variable, or knowledge of the prior mean, their usage as non-informative up to the partial prior information is of value to consider and study.

The usual gamma priors are used as informative priors because they are the natural conjugates to the Poisson and they allow the unity values to be computed in a way that permits comparisons.

It is believed that these four priors result in important comparisons of the plans. Also they give the practitioner a range of shapes to consider for the prior information. Most types of prior information that usually is available or is elicited can be represented by one of the four shapes. In Figure 10.1, these four priors are illustrated for a representative member of the family.

These four special cases of prior distributions will be compared through the use of graphs of their unity values in terms of their posterior risk values and parameters.

First, the posterior risks of the general families will be derived and then the special cases will be examined. All of the priors will be derived with inclusion of a truncation point. That is, each distribution will be defined with an upper value on the variable such that exceedence of this value is not possible.

10.2.1   Truncated Pareto Type Prior Distribution

For the prior distributions designated as truncated Pareto type distributions, that is, for:

$$g(p) = K \, p^m, \qquad 0 < p < P, \; m \neq -1 , \tag{10.2.5}$$

where $K = \frac{m+1}{P^{m+1}}$, the producer's and consumer's risks are easily computable in a form that makes for easy use and informative comparisons. With this prior distribution, the prior mean is $\mu = (m+1)P/(m+2)$.

Also with the truncated Pareto type prior form, it is easy to distinguish and determine the risks for a pair of interesting special cases. The truncated uniform and the truncated Jeffreys' type non-informative distributions are both special cases of the truncated Pareto type distribution. These special cases will be presented now; however, the risks for these special cases will be derived as special cases of the risks of the general truncated Pareto type prior.

Truncated Uniform

Note that when $m = 0$, $g(p)$ in (10.2.5) reduces to a truncated uniform:

$$g(p) = 1/P, \; 0 < p < P . \tag{10.2.6}$$

For the truncated uniform prior, the mean is $P/2$.

Truncated Jeffreys' Type Non-Informative

When $m = -1/2$, $g(p)$ becomes the truncated version of the Jeffreys' non-informative prior that is based on the Fisher Information Criterion. That is:

$$g(p) = \frac{P^{-1/2}}{2\sqrt{p}} , \qquad 0 < p < P . \tag{10.2.7}$$

For the truncated Jeffreys' type non-informative prior, the mean is $P/3$.

Posterior Risks of the Pareto Type Prior

In the case of the truncated Pareto type prior distribution, with parameter m, the marginal distribution $f(x)$ becomes:

$$f(x) = \int_0^P f(x|p)g(p)dp$$

$$= \frac{\Gamma(x+m+1)(m+1)\int_0^{nP} \frac{e^{-np}(np)^{x+m} d(np)}{\Gamma(x+m+1)}}{\Gamma(x+1)(nP)^{m+1}}$$

$$= \frac{\Gamma(x+m+1)(m+1)I_{nP}(x+m)}{\Gamma(x+1)(nP^{m+1})} \tag{10.2.8}$$

where $I_y(x) = (1/\Gamma(x+1)) \int_0^y e^{-u} u^x \, du$, the well-known incomplete gamma function. In this case, the posterior distribution can then be written:

$$g(p|x)dp = \frac{e^{-np}(np)^{x+m} d(np)}{\Gamma(x+m+1)I_{nP}(x+m)} \ . \tag{10.2.9}$$

Next, the posterior risks are computed. From before,

$$\alpha = \int_0^{p_a} \frac{P\{x \in \mathrm{Rej}|p\}g(p)dp}{P\{x \in \mathrm{Rej}\}} \ , \tag{10.2.10}$$

and from (10.2.8), it is seen that

$$P\{x \in \mathrm{Rej}\} = 1 - \sum_{x=0}^{c} f(x)$$

$$= 1 - \sum_{x=0}^{c} \frac{(m+1)\Gamma(x+m+1)I_{nP}(x+m)}{(nP)^{m+1}\Gamma(x+1)}$$

and

$$\int_0^{p_a} P\{x \in \mathrm{Rej}|p\}g(p)dp$$

$$= \int_0^{p_a} \left[1 - \sum_{x=0}^{c} f(x|p)\right] g(p)dp$$

$$= \frac{p_a^{m+1}}{P^{m+1}} - \sum_{x=0}^{c} \frac{\Gamma(x+m+1)(m+1)I_{np_a}(x+m)}{\Gamma(x+1)(nP)^{m+1}} \ .$$

It follows that the posterior risks, when the prior is a truncated Pareto type distribution, are ratios of sums of incomplete gamma functions. Thus, the producer's posterior risk is:

$$\alpha = \frac{\frac{p_a^{m+1}}{P^{m+1}} - \sum_{x=0}^{c} \frac{(m+1)\Gamma(x+m+1)I_{np_a}(x+m)}{(nP)^{m+1}\Gamma(x+1)}}{1 - \sum_{x=0}^{c} \frac{(m+1)\Gamma(x+m+1)I_{nP}(x+m)}{(nP)^{m+1}\Gamma(x+1)}}$$

and

$$\alpha = \frac{\frac{(np_a)^{m+1}}{(m+1)} - \sum_{x=0}^{c} \frac{\Gamma(x+m+1)I_{np_a}(x+m)}{\Gamma(x+1)}}{\frac{(nP)^{m+1}}{(m+1)} - \sum_{x=0}^{c} \frac{\Gamma(x+m+1)I_{nP}(x+m)}{\Gamma(x+1)}} \tag{10.2.11}$$

The complement of the consumer's posterior risk is:

$$1 - \beta = \frac{\sum_{x=0}^{c} \frac{\Gamma(x+m+1)I_{np_r}(x+m)}{\Gamma(x+1)}}{\sum_{x=0}^{c} \frac{\Gamma(x+m+1)I_{nP}(x+m)}{\Gamma(x+1)}} \ . \tag{10.2.12}$$

If m is an integer, then (10.2.11) and (10.2.12) can be written as a ratio of sums, since the incomplete gamma function in that case can be written as a sum. Note that the truncated Pareto type prior is invariant with respect to the sample size n. That is, if the prior had been specified or elicited as a Pareto type prior on np, the forms of $f(x)$, $g(np|x)$ and the producer's and the consumer's posterior risks would all have exactly the same forms as above.

Truncated Uniform Posterior Risks

When $g(p)$ is a truncated uniform (i.e., $m = 0$), (10.2.11) and (10.2.12) can be expressed as simple summations. Using the relationships:

$$I_{np}(x) = 1 - \sum_{j=0}^{x} \frac{e^{-np}(np)^j}{j!} \tag{10.2.13}$$

and

$$\sum_{x=0}^{c} \sum_{j=0}^{x} \frac{e^{-np}(np)^j}{j!} = e^{-np} \sum_{x=0}^{c} \frac{(c+1-x)(np)^x}{x!},$$

the producer's posterior risk reduces to:

$$\alpha = \frac{np_a - (c+1) + e^{-np_a}\sum_{x=0}^{c}\frac{(c+1-x)(np_a)^x}{x!}}{nP - (c+1) + e^{-nP}\sum_{x=0}^{c}\frac{(c+1-x)(nP)^x}{x!}} \tag{10.2.14}$$

and the consumer's posterior risk becomes:

$$\beta = 1 - \frac{(c+1) + e^{-np_r}\sum_{x=0}^{c}\frac{(c+1-x)(np_r)^x}{x!}}{(c+1) + e^{-nP}\sum_{x=0}^{c}\frac{(c+1-x)(nP)^x}{x!}} \tag{10.2.15}$$

These relationships are illustrated in Figures 10.2 for $\alpha = .05$, $\beta = .10$ and $c = 0, 1, 2, 3$, and 5. The discussion and an example of the use of these figures will be considered later.

Truncated Jeffreys' Type Posterior Risks

When $g(p)$ is a truncated Jeffreys' type prior (i.e., $m = -1/2$), (10.2.11) and (10.2.12) become:

$$\alpha = \frac{2\sqrt{np_a} - \sum_{x=0}^{c}\frac{\Gamma(x+\frac{1}{2})I_{np_a}(x-\frac{1}{2})}{\Gamma(x+1)}}{2\sqrt{nP} - \sum_{x=0}^{c}\frac{\Gamma(x+\frac{1}{2})I_{nP}(x-\frac{1}{2})}{\Gamma(x+1)}} \tag{10.2.16}$$

and

$$\beta = 1 - \frac{\sum_{x=0}^{c}\frac{\Gamma(x+\frac{1}{2})I_{np_r}(x-\frac{1}{2})}{\Gamma(x+1)}}{\frac{\Gamma(x+\frac{1}{2})I_{nP}(x-\frac{1}{2})}{\Gamma(x+1)}} \tag{10.2.17}$$

These relationships are illustrated in Figure 10.3 for $\alpha = .05$, $\beta = .10$ and $c = 0, 1, 2, 3$, and 5. Use of this figure with an example will also be given later.

Truncated Gamma Type Prior Distribution

Consider prior distributions designated as truncated gamma type distributions, that is,

$$g(p) = \frac{b^a p^{a-1} e^{-bp}}{\Gamma(a) I_{bP}(a-1)} , \qquad 0 < p < P . \qquad (10.2.18)$$

The mean of the truncated gamma distribution is useful for the purposes of this paper and is given by:

$$\mu = \frac{a\, I_{bP}(a)}{b\, I_{bP}(a-1)} . \qquad (10.2.19)$$

With the truncated gamma distribution as a prior, it is also useful to examine the prior variance which is:

$$\sigma^2 = \frac{(a+1)a\, I_{bP}(a+1)}{b^2\, I_{bP}(a-1)} - \frac{a^2 [I_{bP}(a)]^2}{b^2 [I_{bP}(a-1)]^2} . \qquad (10.2.20)$$

Note in (10.2.19) and (10.2.20) that when a is an integer, these forms for μ and σ^2 become ratios of sums using (10.2.13).

With the truncated gamma prior distribution, it is useful for our purposes to specify two special cases: the truncated maximum entropy prior and the truncated informative gamma with mean and variance in a form for easy comparisons to the other distributions we study here.

The maximum entropy prior distribution on $[0,\ P]$ is a special case of the truncated gamma distribution. The maximum entropy prior $g(p)$ for the parameter p, subject to the prior information $E(p) = \mu$, and the restriction on the parameter space $0 < p < P$, is the truncated exponential distribution. That is,

$$g(p) = c_1 e^{-c_2 p} \qquad 0 < p < P \qquad (10.2.21)$$

The constants c_1 and c_2 are determined by

$$E(p) = \int_0^P p\, g(p)\, dp = \mu \qquad (10.2.22)$$

and

$$\int_0^P g(p)\, dp = 1 \qquad (10.2.23)$$

See, for example, [1], p. 92. Substituting (10.2.21) into (10.2.22) and (10.2.23) yields the relations

$$c_1 = c_2 / (1 - e^{-c_2 P}) \qquad (10.2.24)$$

$$\mu = \frac{1 - e^{-c_2 P}(1 + c_2 P)}{c_2 (1 - e^{-c_2 P})} \qquad (10.2.25)$$

Note that when $a = 1$ and $c_2 = b$, (10.2.21) is a specific case of (10.2.18), and (10.2.21) is used with (10.2.25) to specify the truncated maximum entropy prior. Note that, if the mean is not specified, then the maximum entropy prior for the binomial is the uniform. Thus, for the Poisson approximation to the binomial, and with the limit on the value of p, the maximum entropy prior is the truncated uniform which has already been considered.

The other special case of interest here for the comparison of unity values is the informative form of the truncated gamma with special values of the parameters a and b, so that the unity values can be computed. These special cases will now be considered.

Maximum Entropy Prior Distribution

When $a = 1$, the truncated gamma prior reduces to a truncated exponential, which is the maximum entropy prior for the Poisson parameter p for a specified prior mean. The parameter b is chosen such that the mean (10.2.19) with $a = 1$ has the desired value, given by the solution of (10.2.24), with $b = c_2$.

Thus, the maximum entropy prior distribution is a truncated gamma with $a = 1$ and, for our purposes, $b = vn$. Hence

$$g(p)dp = \frac{ve^{-vnp}ndp}{1 - e^{-vnP}} \;,\tag{10.2.26}$$

with

$$\mu = \frac{1 - e^{-vnP}(vnP + 1)}{nv(1 - e^{-vnP})} \;.\tag{10.2.27}$$

Informative Gamma Prior Distribution

The gamma prior is used for comparison purposes as an informative prior of the form (10.2.18) with specified values for parameters a and b. The priors considered will be those with parameters $b = nv$ and a chosen such that the mean and variance of np vary over a range of values of interest.

Posterior Risks for the Truncated Gamma Prior

The risks with the truncated gamma prior are easily computable as ratios of incomplete gamma functions and give rise to the risks of the special cases considered here. The posterior risks will be derived and plotted for both of these special cases of the truncated gamma.

For the truncated gamma prior, the marginal $f(x)$ is:

$$f(x) = \int_0^P f(x|p)g(p)dp = \int_0^P \frac{e^{-p(n+b)}p^{x+a-1}n^x b^a dp}{\Gamma(x+1)\Gamma(a)I_{bP}(a-1)}$$

$$= \left\{ \frac{v^a \Gamma(x+a)}{(v+1)^{x+a}\Gamma(x+1)\Gamma(a)I_{vnP}(a-1)} \right\} \left\{ \int_0^{(v+1)nP} \frac{e^{-u}(u)^{x+a-1}du}{\Gamma(x+a)} \right\}$$

$$= \frac{v^a \Gamma(x+a)I_{(v+1)nP}(x+a-1)}{(v+1)^{x+a}\Gamma(x+1)\Gamma(a)I_{vnP}(a-1)} \;,\tag{10.2.28}$$

where $b = vn$. Also, the posterior distribution $g(p|x)$ is:

$$g(p|x)dp = \frac{e^{-np(v+1)}[(v+1)np]^{x+a-1}d[(v+1)np]}{I_{vnP}(x+a-1)\Gamma(x+a)}.$$ (10.2.29)

The posterior risks are:

$$\alpha = \frac{I_{vnp_a}(a-1) - \sum_{x=0}^{c} \frac{v^a\Gamma(x+a)I_{(v+1)np_a}(x+a-1)}{(v+1)^{x+a}\Gamma(x+1)\Gamma(a)}}{I_{vnP}(a-1) - \sum_{x=0}^{c} \frac{v^a\Gamma(x+a)I_{(v+1)nP}(x+a-1)}{(v+1)^{x+a}\Gamma(x+1)\Gamma(a)}}$$ (10.2.30)

and

$$\beta = 1 - \frac{\sum_{x=0}^{c} \frac{\Gamma(x+a)I_{(v+1)np_r}(x+a-1)}{(v+1)^{x+a}\Gamma(x+1)}}{\sum_{x=0}^{c} \frac{\Gamma(x+a)I_{(v+1)nP}(x+a-1)}{(v+1)^{x+a}\Gamma(x+1)}}$$ (10.2.31)

It is also true here that the posterior risks are invariant with respect to the sample size. That is, the posterior risks are the same whether one places a prior on the parameter p or a corresponding prior on the parameter np, even when the prior mean and/or the prior variance are specified. This fact will be used in computing the unity values for a concise method of comparing the Bayesian plans and examining the robustness of the prior information.

Maximum Entropy Posterior Risks

When $g(p)$ is the maximum entropy prior distribution, the posterior risks (10.2.30) and (10.2.31) become:

$$\alpha = \frac{1 - e^{vnp_a} - \sum_{x=0}^{c} \frac{vI_{(v+1)np_a}(x)}{(v+1)^{x+1}\Gamma(a)}}{1 - e^{vnP} - \sum_{x=0}^{c} \frac{vI_{(v+1)nP}(x)}{(v+1)^{x+1}\Gamma(a)}}$$ (10.2.32)

and

$$\beta = 1 - \frac{\sum_{x=0}^{c} \frac{I_{(v+1)np_r}(x)}{(v+1)^{x+1}}}{\sum_{x=0}^{c} \frac{vI_{(v+1)nP}(x)}{(v+1)^{x+1}}}$$ (10.2.33)

The risks will be computed and plotted for cases when the mean is $nP/3$ (for comparison to the truncated Jeffreys' type prior), and when the mean is $nP/10$ (for comparison to a more informative prior). Note that when the mean is set equal to $nP/2$, the mean of the truncated uniform, the maximum entropy prior becomes the truncated uniform in the limit. To find the appropriate value of v the mean will be equated to the appropriate function of nP and then solved for v as a function of nP. That is,

$$n\mu = \frac{nP}{K} = \frac{1 - e^{-vnP}(vnP+1)}{v(1 - e^{-vnP})},$$

where $K = 3$ or 10. The obtained value of v is then used in the solutions for the posterior risks. The relationship of the posterior risks to the parameters of the plans in terms of the unity values for the truncated maximum entropy prior are illustrated in Figures 10.4 and 10.5.

Posterior Risks for the Truncated Gamma

When $g(p)$ is the general truncated gamma distribution, the posterior risks are given by the equations (10.2.30) and (10.2.31).

For comparison purposes, and for an examination of the robustness of the unity values with respect to the prior distribution, the graphs of the unity values as functions of the nP values will be computed for several values of the gamma prior mean and variance. The values of the prior mean and variance (10.2.19) and (10.2.20), with $b = vn$, can be written as functions of a, v and nP. Furthermore, if v is chosen as an appropriate function of nP and a is an integer , (10.2.19) and (10.2.20) are easily computable functions of nP. In this way, the gamma priors can be chosen in such a way that the means of the gamma priors match the means of the distributions with which they are to be compared. Thus, if a is chosen to be 2, to give a gamma shape, and if $v = 1/nP$, $2/nP$ and $6/np$, the respective gamma priors have means $.6nP$, $.5nP$ and $.3nP$. The means equal to $.5nP$ and $.3nP$ allow for comparisons of the gamma priors to the uniform, Jeffreys' type and maximum entropy priors. The posterior risks as functions of the parameters of the plans, in terms of the unity values, are presented in Figures 10.6, 10.7, and 10.8.

10.3   Unity Graphs

In this section, the graphs resulting from the consideration of the posterior risks in terms of nP, np_a, np_r and R are presented. The graphs are given for $\alpha = .05$ and $\beta = .10$. The graphs are computed for: $1 \leq nP \leq 10$ which should allow consideration of important sample sizes and upper limits on the parameter.

Figure 10.2 is the set of graphs with $c = 0, 1, 2, 3$, and 5 for the truncated uniform priors. Figure 10.3 presents the corresponding graphs for the truncated Jeffreys' type priors. Figure 10.4 gives the graphs for the maximum entropy priors with $\mu = nP/3$ and Figure 10.5 gives the same graphs for $\mu = nP/10$. Figures 10.6, 10.7 and 10.8 present graphs for the gamma priors with $a = 2$ and $v = 1/nP$, $2/nP$ and $6/nP$ respectively.

10.4   Examples

A manufacturer of a product uses several suppliers to provide the many components that are built into the final assembly. Based on contractual agreements it is required that some inspection of the incoming material be made. The material is such that each one of the suppliers provides components that are delivered in lots of size about 1000. The design of the final assembly is such that a component reliability of .99 is acceptable while a reliability of .97 is not. As a result the manufacturer wants to implement an attributes lot acceptance plan that will provide the desired characteristics with regard to lot acceptance. Based upon the requirements, the manufacturer wants a plan such that the probability of accepting lots with reliability $R_a = .99$, $p_a = 1 - R_a = .01$; is $.95$ ($\alpha = .05$) and accepting lots with reliability $R_r = .97$, $p_r = .03$; is $.10$ ($\beta = .10$). The testing of the parts is expensive and the manufacturer would like to use do the smallest amount of testing and yet provide the required protection. Based upon previous experience with this supplier and similar components, the manufacturer believes that one can effectively employ a Bayesian approach to acceptance by defining valid prior distributions on the true lot reliability. Based upon study of the supplier procedures and previous data the manufacturer is confident that one aspect of the prior should be that it have a truncation point. This truncation point represents a value of unreliability that one is certain cannot be exceeded. For any truncation value it is still necessary that a form be specified for the prior distribution. As previously described, four priors for this type situation are of interest. They are the uniform, Jeffreys' type, maximum entropy, and gamma.

It is of interest to determine sampling plans, i.e., truncation values/sample sizes , that are consistent with the desired plan characteristics for each of the types of prior distributions. Use of Figures 10.2-10.8 can be used to determine the plans. Some iteration is required to use the figures. The convention has been used that a plan is chosen that fixes p_a at .01 with $\alpha = .05$ and p_r at .03 with $\beta = .10$, and allows the truncation point to vary.

Uniform: Use of Figure 10.2a with $R = 3$ and $c = 0$ gives a value of $nP = 5.5$. The corresponding values of np_a and np_r are .75 and 2.25, respectively. Fixing p_a at .01 and solving for n gives $n = 75$. For $n = 75$ the exact value of p_r is $2.25/75 = .03$ which is the targeted value. The value of the truncation point is calculated as $P = 5.5/75 = .073$. Thus, if it can be safely assumed that the unreliability is not greater than $P = .073$ and is uniformly distributed a sample of $n = 75$ allowing no failures will provide the desired protection.

Jeffreys' type: Use of Figure 10.3a with $R = 3$ and $c = 0$ gives a value of $nP = 7.2$. The corresponding values of np_a and np_r are .45 and 1.35, respectively. Fixing p_a at .01 and solving for n gives $n = 45$. For $n = 45$ the exact value of p_r is $1.35/45 = .03$ which is the targeted value. The value of the truncation point is calculated as $P = 7.2/45 = .16$. Thus, if it can be safely assumed that the unreliability is not greater than $P = .16$ and the distribution of unreliability follows the Jeffreys' type prior a sample of $n = 45$ allowing no failures will provide the desired protection.

Maximum entropy: Use of Figure 10.4a with $R = 3$ and $c = 0$ gives a value of $nP = 9.2$. The corresponding values of np_a and np_r are .60 and 1.80, respectively. Fixing p_a at .01 and solving for n gives $n = 60$. The value of the truncation point is calculated as $P = 9.2/60 = .15$. Thus, if it can be safely assessed that the unreliability is not greater than $P = .15$ and the unreliability follows the maximum entropy prior a sample of $n = 60$ allowing no failures will provide the desired protection. It is also possible to obtain a plan with the desired characteristics with $c = 1$. Use of Figure 10.4b and $R = 3$ yields $nP = 6$, $np_a = .36$, $np_r = 1.08$ and $n = 36$. The value of the truncation point is calculated as $P = 6/36 = .17$. Thus, a plan with $P = .17$ and $n = 36$ and acceptance number of $c = 1$ will provide the desired characteristics.

Gamma: Use of Figure 10.8a with $R = 3$ and $c = 0$ gives a value of $nP = 9.0$. The corresponding values of np_a and np_r are .77 and 2.32, respectively. Fixing p_a at .01 and solving for n gives $n = 77$. The value of the truncation point is calculated as $P = 9.0/77 = .12$. Thus, if it can be safely assessed that the unreliability is not greater than $P = .12$ and the unreliability follows the gamma$(6/nP)$ prior a sample of $n = 77$ allowing no failures will provide the desired protection.

It is useful to compare the above results with those obtained using classical methods where no assumption about the true unreliability are made. Using [6] with $p_a = .01$ and $p_r = .03$ the values for sample size is $n = 398$ with an acceptance number of $c = 7$. It is apparent that considerable savings in sample size can be had if there is knowledge about unreliability and it can be expressed accurately through a prior distribution.

It is important for implementation that the value for the truncation point is not driven by a target sample size. In practice a truncation point and form for the prior distribution

TABLE 10.1
BRDT plans

prior	n	p
Jeffreys' type ($nP/3$)	45	.16
maximum entropy ($nP/3$)	60	.15
gamma ($.32nP$)	77	.12
uniform	75	.07

should be selected based on technical considerations. With this information the figures can be used to define the appropriate plan and with the potential savings in resources. For some applications it is possible that no plan exists that will have the desired characteristics and selected truncation point with associated form for a prior distribution.

10.5   Summary and Conclusions

In this paper, the unity concept for developing acceptance sampling plans has been expanded to include a formal structure for incorporating prior information via the Bayesian technique, Bayes Reliability Demonstration Testing (BRDT). Charts were prepared that can be used in determining specific plans. For most applications the use of prior information will provide a sampling plan, having the desired risks, with a reduced sample size compared to the classical acceptance sampling plan. As to the comparison of the priors with respect to the consideration of robustness, it can be seen that the plans are not prior-robust; that is, the plans are sensitive to the choice of a prior. This appears to be true even when somewhat non-informative priors are used. Priors which have been suggested as being non-informative in other applications are used here with some prior information (such as a known limit on the range of the parameter or knowledge of the prior mean). However, they are no longer non-informative in this application, even though the limit and the prior mean do not appear to have a large effect on the plans. The plans appear to be most sensitive to the shape of the prior regardless of the other types of prior knowledge.

The examples in the previous section show that the BRDT plans differ substantially depending on the prior which is selected. Those previous examples are summarized here for comparison. In each case, we use $p_a = .01$, $p_r = .03$, with $\alpha = .05$ and $\beta = .10$. Table 10.1 summarizes the resulting plans which meet these specifications for the four classes of priors considered here.

References

1. Berger, J.O. (1985), *Statistical Decision Theory and Bayesian Analysis*, Springer-Verlag, New York.

2. Cameron, J.M. (1952), Tables for constructing and for computing the operating characteristics of single-sampling plans, *Industrial Quality Control*, **9**, 38.

3. Grubbs, F.E. (1949), On designing single sampling inspection, *Annals of Mathematical Statistics*, **20**, 256.

4. Martz, H.F. and Waller, R.A. (1991), *Bayesian Reliability Analysis*, Krieger, Malabar, Florida.

5. Peach, P. and Littauer, S.B. (1946), A Note on sampling inspection, *Annals of Mathematical Statistics*, **17**, 81-84.

6. Shilling, E.G. (1982), *Acceptance Sampling in Quality Control*, Marcel Dekker, New York.

7. Shilling, E.W. and Johnson, L.I. (1980), Tables for the construction of matched single, double and multiple sampling plans with wpplication to MIL-STD 105-D, *Journal of Quality Technology*, **12**, 220-229.

8. Wadsworth, H.M., Stephens, K.S., and Godfrey, A.B. (1986), *Modern Methods for Quality Control and Improvement*, John Wiley & Sons, New York.

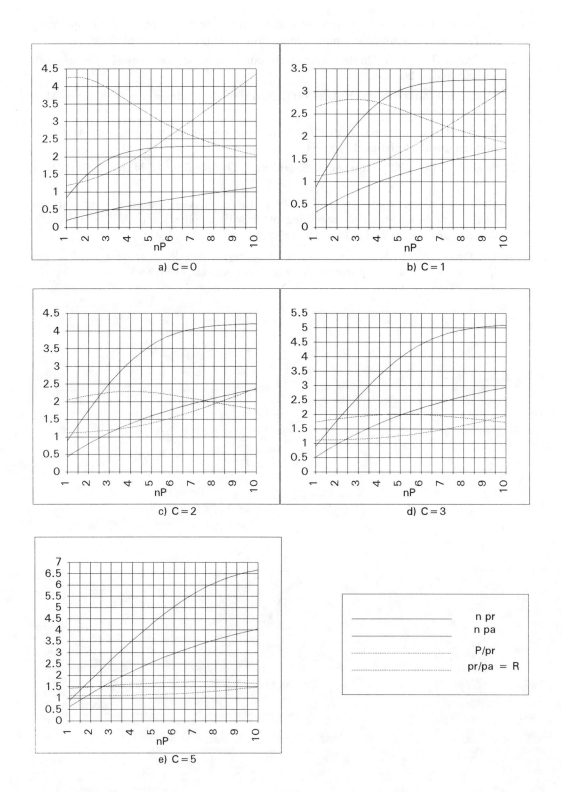

FIGURE 10.2
BRDT plans: uniform plans distribution truncated at nP

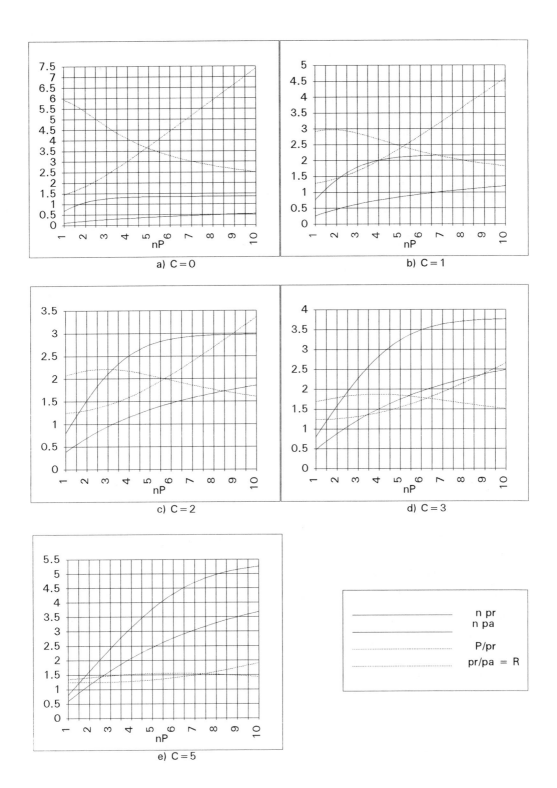

FIGURE 10.3
BRDT plans: Jeffrey's prior distribution truncated at nP

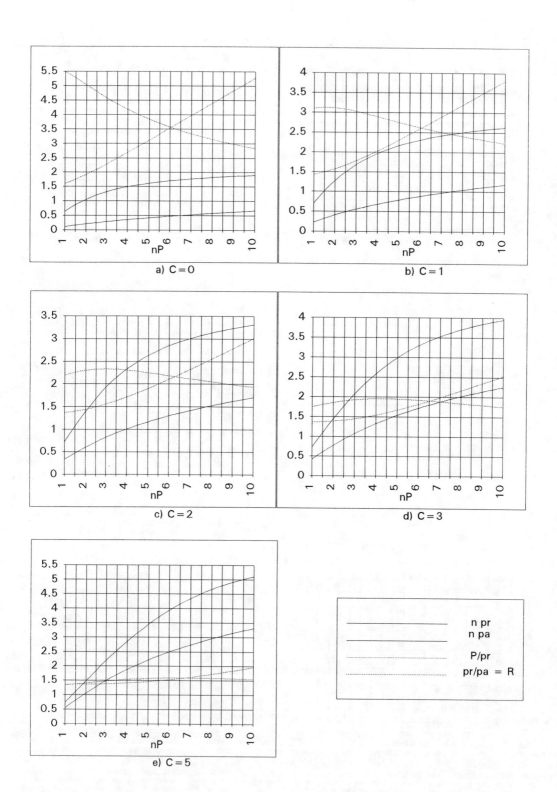

FIGURE 10.4
BRDT plans: maximum entropy (exponential) prior distribution, truncated at nP, mean = $nP/3$

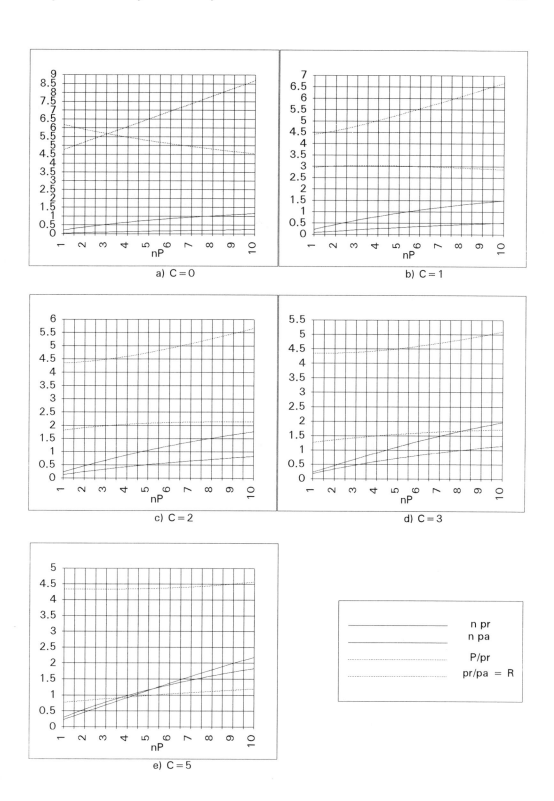

FIGURE 10.5
BRDT plans: maximum entropy (exponential) prior distribution, truncated at nP, mean $= nP/10$

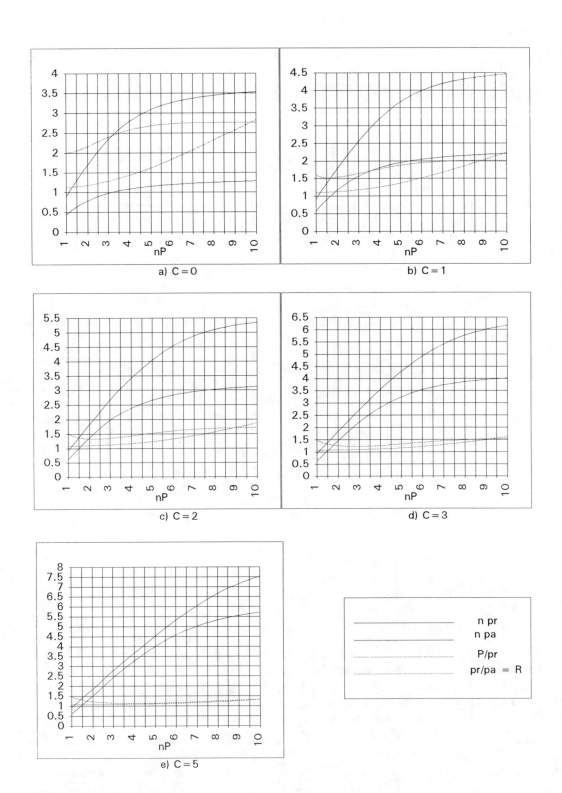

FIGURE 10.6
BRDT plans: gamma prior distribution truncated at nP, $v = 1/nP$, shape
parameter $a = 2$

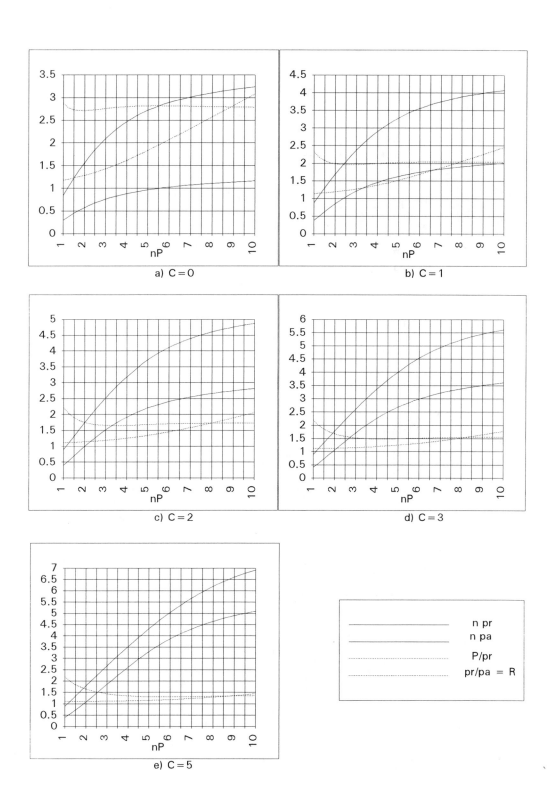

FIGURE 10.7
BRDT plans: gamma prior distribution truncated at nP, $v = 2/nP$, shape
parameter $a = 2$

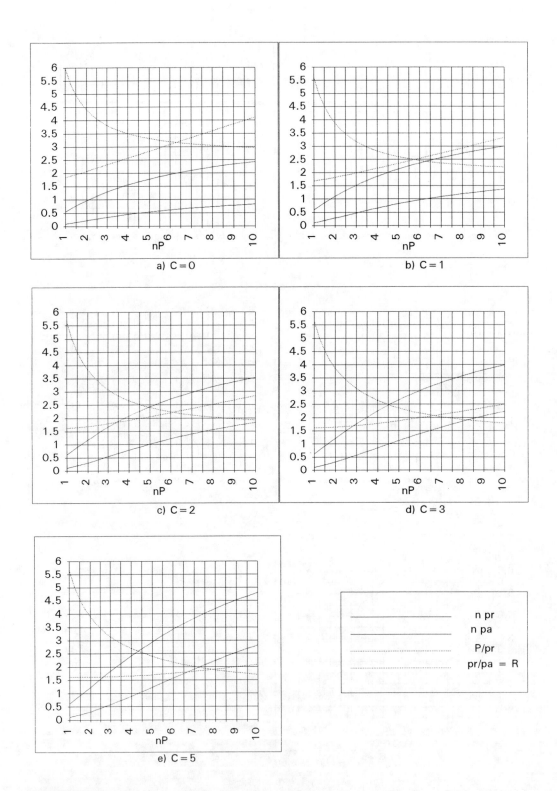

FIGURE 10.8
BRDT plans: gamma prior distribution truncated at nP, $v = 6/nP$, shape parameter $a = 2$

11

Standard Errors for the Mean Number of Repairs on Systems from a Finite Population

Jeffrey A. Robinson

Mathematics Department, General Motors R&D Center

ABSTRACT In [11], Nelson described a simple way to calculate the mean number (or cost) of repairs for data coming from repairable systems. An associated standard error, based on an unbiased variance estimator, is reported in [12]. The standard error is extended here to handle finite populations—a situation that occurs commonly in practice. For either finite or infinite populations, it is shown that the unbiased variance estimates can be negative. The problem is caused by possible negative eigenvalues of the covariance matrix formed from a triangular array of repair frequencies. The triangular shape results from data censoring. Two alternatives, which are always positive, are discussed. One is based on an adjustment to the eigenvalues, and the other is a moment estimator given by [9]. For samples of size 20, simulations indicate that the unbiased variance estimates are rarely negative. The eigenvalue-adjusted estimator has approximately equal mean square error as the unbiased estimator except with very heavy censoring. The moment estimator has smaller mean square error, and its performance improves with the level of censoring. A general Poisson model is described to serve as a basis for the simulations.

CONTENTS

11.1 Introduction

In [11], Nelson proposed an estimator of the accumulated mean number of repairs through age t for a population of repairable systems. He also proposed an appropriate standard error for the estimator. The standard error and a computer program for its calculation are documented in [12]. For the special case when all the systems in the sample have ages greater than t, the mean estimate is equal to the sample average number of repairs through t, and the standard error is the usual one for a sample average. Even in the general case, both the mean estimator and its standard error are nonparametric in the sense that their validity does not depend on the stochastic process that describes the "arrivals" of repairs. However, that validity does require that the choice of systems in the sample and the ages through which these systems are observed not depend on the repair performance of the systems.

There is an extensive literature on point processes that can be applied to data of this type. Some relevant books include [3] , [4], [10], and [7]. Two books that emphasize applications are [1] and [16] . Most of the results are parametric in nature and deal with issues more complex than simply estimating the mean of a population and putting an appropriate standard error on the estimate.

This article considers some extensions and enhancements. The standard error formula is extended to handle finite populations. The analysis of lab and engineering test data often includes the tacit assumption of sampling from a conceptual infinite population. But for field data, such as warranty records, the target of inference is often the units in the field or those that will be eventually. Finite population methods are appropriate for this case. This enhancement is discussed in Section 11.2. The unbiasedness of the mean estimator and the estimator of its variance follow from standard sampling theory. In practice the mean estimate and its standard error are calculated for all ages in the span of the data.

Section 11.3 points out that the unbiased variance estimator can be negative, and discusses why. The difficulty arises because of possible negative eigenvalues of the matrix of sample covariances. This can happen because a particular repairable system contributes to some covariances and not to others because of data censoring.

Two alternative nonparametric variance estimators, which cannot be negative, are also discussed in Section 11.3. The first is derived from adjusting the covariance matrix by setting any negative eigenvalues to zero. The second is a moment estimator proposed by [9] in a more general context. The moment estimator differs from the unbiased estimator only in that variances and covariances are defined with "n" in the denominator rather than "$n-1$."

Section 11.4 discusses a general Poisson model that allows for extra variation. The value of the model is mostly conceptual, but it also provides a framework for a simulation study to compare the three nonparametric variance estimators. The simulation study is discussed in Section 11.5. Section 11.6 is a summary with some additional concluding remarks.

11.2   Background for Finite Populations

Consider a population of N repairable systems from which a sample of size n is taken. Let $X_i(t)$ denote the cumulative number of repairs on the ith system through age t. Let $\mu(t) = \sum_{i=1}^{N} X_i(t)/N$ denote the population mean number of repairs through age t. The goal is to estimate $\mu(t)$ and put a standard error on the estimate.

Suppose the ith system in the sample is observed through age a_i. Without loss of generality assign the indices to the sample in decreasing order of these times on test, so that $a_1 \geq a_2 \geq \ldots \geq a_n$. We will allow for ties among the $a's$, so let $c_1 \geq c_2 \geq \ldots \geq c_{n'}$ denote the distinct test times, where $n' \leq n$. Consider a specific time, t_0, within the observed span of ages $(0, c_1]$. Suppose there are exactly $m-1$ distinct ages less than t_0. Define

$$q_j = \begin{cases} 0, & j = 0; \\ c_{n'+1-j}, & j = 1, ..., m-1; \\ t_0, & j = m. \end{cases}$$

TABLE 11.1
Data array and matrix of covariances for an illustrative example

Data Array			Matrix of Covariances		
j=1	j=2	j=3			
0	0	2	2/3	1	-2
2	2	0	1	4/3	-2
1	0	.	-2	-2	2
1	.	.			
n_j 4	3	2			
$\bar{x}_j$ 1	2/3	1			

The q's are the distinct ages less than t_0 rewritten to be in increasing order with the addition that q_0 is defined to be zero and q_m is defined to be t_0. Figure 11.1 illustrates a simple example for $n = n' = 4$ systems (no ties) having a total of nine repairs.

Let n_j denote the number of systems of age q_j or greater. In this notation, if there are no ties among the ages, then $n_j = n + 1 - j$. Let $x_{ij} = X_i(q_j) - X_i(q_{j-1})$ denote the number of repairs on system i occurring in $(q_{j-1}, q_j]$. The x_{ij}; $j = 1, \ldots, m$, $i = 1, \ldots, n_j$; define a triangular data array.

If $n_m \geq 2$, that is if there are at least two systems of age t_0 or greater, then we can define sample means, variances, and covariances for the first m columns by

$$\bar{x}_j = \sum_{i=1}^{n_j} x_{ij}/n_j, \ 1 \leq j \leq m$$

$$s_{jk} = \sum_{i=1}^{n_k} x_{ij}(x_{ik} - \bar{x}_k)/(n_k - 1), \ 1 \leq j \leq k \leq m \ .$$

In the covariance calculation of s_{jk} $(k > j)$, the entire j^{th} column is not used, but only the first n_k $(< n_j)$ elements. Table 11.1 contains the data array, means, and covariances for the example depicted in Figure 11.1.

The random sampling is assumed not to depend on the ages or the repair histories of the repairable systems. Therefore, from sampling theory, the sample quantities are unbiased estimators of the corresponding population quantities given by

$$\bar{X}_j = \sum_{i=1}^{N} x_{ij}/N \ ,$$

$$S_{jk} = \sum_{i=1}^{N}(x_{ij} - \bar{X}_j)(x_{ik} - \bar{X}_k)/(N - 1) \ ,$$

for all k and j. The common convention is adopted here of denoting population quantities with capital letters, and also of defining population variances and covariances with "$N-1$" in the denominator to simplify the formulas. The population mean number of repairs at time t_0 can now be written as

$$\mu = \sum_{j=1}^{m} \bar{X}_j.$$

An unbiased estimator of the mean number of repairs is

$$\hat{\mu} = \sum_{j=1}^{m} \bar{x}_j,$$

with variance

$$Var(\hat{\mu}) = \sum_{j=1}^{m} var(\bar{x}_j) + 2 \sum_{j=1}^{m-1} \sum_{k=j+1}^{m} cov(\bar{x}_j, \bar{x}_k).$$

From sampling theory [e.g. [2], Chapter 2], $cov(\bar{x}_j, \bar{x}_k) = (1 - n_j/N)S_{jk}/n_j$ for $k \geq j$. (Actually, the covariance result ($k > j$) is a slight generalization of what is usually found in sampling books because $\bar{x}_k$ is an average over fewer terms than $\bar{x}_j$. But its validity is easily established by defining indicator functions that denote whether each unit in the population is chosen in the sample.)

Substituting, the variance of the mean estimate is

$$Var(\hat{\mu}) = \sum_{j=1}^{m}(1 - n_j/N)S_{jj}/n_j + 2 \sum_{j=1}^{m-1}(1 - n_j/N)/n_j \sum_{k=j+1}^{m} S_{jk}.$$

Since the sample variances and covariances are unbiased estimators of their population analogs, the natural unbiased estimator of this variance is

$$\widehat{Var}_u = \sum_{j=1}^{m}(1 - n_j/N)s_{jj}/n_j + 2 \sum_{j=1}^{m-1}(1 - n_j/N)/n_j \sum_{k=j+1}^{m} s_{jk}. \qquad (11.2.1)$$

The square root of (11.2.1) with $N = \infty$ yields Nelson's standard error.

In practice one would calculate the estimate and its standard error at all values of t within the span of the data, not just $t = t_0$. The mean estimate and its standard error are step functions with jumps at the repair times. Several real examples are discussed in [11] and [12]. A straightforward algorithm for updating the means and covariances recursively is discussed in [13].

11.3   The Problem of Negative Variance Estimates

Unfortunately, the variance estimate given by (11.2.1) can be negative in practice. (This was first reported to me by Jason Jones based on some of his simulation work.) For example, calculating from the information in Table 11.1 with $N = \infty$, we find

$$\widehat{Var}_u = \widehat{Var}_u(\bar{x}_1 + \bar{x}_2 + \bar{x}_3) = -2/9.$$

Clearly, negative variance estimates can happen only when there are negative covariances among the columns. Moreover, a nonnegative definite covariance matrix ensures that the variance in (11.2.1) is nonnegative (see Appendix A). Therefore, a negative variance estimate is always associated with negative eigenvalues. Negative eigenvalues can occur because the sample covariances are calculated from a triangular data array.

11.3.1   The Eigenvalue-Adjusted Variance Estimator

A natural solution, then, is to replace any negative eigenvalues by zero. Specifically, denote the sample covariance matrix by

$$\hat{\Sigma} = \begin{pmatrix} s_{11} & s_{12} & \cdots & s_{1m} \\ s_{12} & s_{22} & \cdots & s_{2m} \\ \vdots & \vdots & \ddots & \vdots \\ s_{1m} & s_{2m} & \cdots & s_{mm} \end{pmatrix}.$$

Since it is real and symmetric, it has a spectral decomposition

$$\hat{\Sigma} = QDQ',$$

where $D = \text{diag}(d_1, \ldots, d_m)$ is a diagonal matrix containing the eigenvalues of $\hat{\Sigma}$ and Q is an orthonormal matrix whose columns contain the associated eigenvectors. Define an adjusted variance estimator by replacing the sample variances and covariances in (11.2.1) by

$$[\tilde{s}_{jk}] \equiv \tilde{\Sigma} = Q\tilde{D}Q',$$

where

$$\tilde{D} = \text{diag}[\max(0, d_1), \ldots, \max(0, d_m)].$$

Following the form in (11.2.1), the eigenvalue-adjusted variance estimator is written as

$$\widehat{Var}_a = \sum_{j=1}^{m}(1 - n_j/N)\tilde{s}_{jj}/n_j + 2\sum_{j=1}^{m-1}(1 - n_j/N)/n_j \sum_{k=j+1}^{m} \tilde{s}_{jk}. \qquad (11.3.1)$$

The eigenvalue-adjusted variance estimate is somewhat computationally intensive in practice since it involves solving an eigenvalue problem at each repair time. In [13], the author outlines a method to minimize the effort by recursive calculations. The results rely on the property of interlacing eigenvalues (see [5] and [17]) when adding a row and column to a real symmetric (covariance) matrix.

11.3.2   The Moment Variance Estimator

One could use instead biased sample variances and covariances (sample moments) given by

$$s^*_{jk} = (n_k - 1)s_{jk}/n_k = \sum_{i=1}^{n_k} x_{ij}(x_{ik} - \bar{x}_k)/n_k, \quad 1 \le j \le k \le m.$$

Substituting these into (11.2.1) produces the variance estimator

$$\widehat{Var}_m = \sum_{j=1}^{m}(1 - n_j/N)s^*_{jj}/n_j + 2\sum_{j=1}^{m-1}(1 - n_j/N)/n_j \sum_{k=j+1}^{m} s^*_{jk}. \qquad (11.3.2)$$

This estimator (for infinite populations) was proposed by [9] in a more general context that allows for covariates. Lawless and Nadeau show algebraically that the infinite population version is always positive. For completeness, the argument is duplicated in Appendix B. Furthermore, it is noted that the nonnegativity holds even for finite populations.

TABLE 11.2
Means and standard errors based on the three nonparametric variance estimators

		Mean	Unbiased Std.		Adjusted Std.		Moment Std.	
Age	Order	Est.	Err.	Var.	Err.	Var.	Err.	Var.
1	1	0.25	0.25	0.063	0.25	0.063	0.22	0.047
2	1	0.50	0.29	0.083	0.29	0.083	0.25	0.063
3	1	0.75	0.48	0.229	0.48	0.229	0.41	0.172
4	1	1.00	0.41	0.167	0.41	0.167	0.35	0.125
5	2	1.33	0.73	0.528	0.73	0.528	0.60	0.366
6	2	1.67	1.05	1.111	1.05	1.113	0.87	0.755
7	3	2.17	0.44	0.194	0.70	0.494	0.54	0.296
8	3	2.67	.	-0.222	0.72	0.526	0.30	0.088
9	4	3.67	.	.	.	.	.	.

11.3.3 Estimates for the Artificial Example

Table 11.2 shows the estimates and the three standard errors for the artificial data from Table 11.1. For convenience the repair times have been labeled one through nine. Note that none of the variance estimators are defined for t_0 greater than the next-to-largest repair time since only one system has run beyond that time. Therefore, no standard error can be associated with the mean estimate at the ninth repair time.

As noted earlier, the value of the unbiased variance estimator at the eighth repair time is negative (-0.222). The eigenvalues of the associated three by three covariance matrix shown in Table 11.1 are $-.89$, $.05$, and 4.84. The other variance estimates are always positive.

11.4   A General Poisson Model

The material in this section provides a conceptual framework for a simulation study to compare the three nonparametric standard errors discussed previously. Conceptually, one could think of the repair history for a given repairable system being repeated several times. This does not generally occur in practice since once a system is run, it cannot be put back into its original condition. But even if it could, it would not generate exactly the same repair history because of variation in environmental conditions and the like. We will model this "repeat" variation by a Poisson process. The observed repair history is one random trial from the process.

In general no assumptions are made about the Poisson process. In particular, no specific form for the intensity function is assumed, nor are the intensities assumed to be the same for all systems in the population. This variation over the population will induce sample-to-sample variation on the previously discussed estimators.

Conditioned upon selecting a system in the sample, $X_i(t)$ has a Poisson distribution with mean denoted by $\Lambda_i(t)$. For time $t = t_0$ define the q's as before, and let

$$\lambda_{ij} = \Lambda_i(q_j) - \Lambda_i(q_{j-1}), \quad i = 1, \ldots, n_j, \; j = 1, \ldots, m.$$

The λ's are unobserved. Denote their sample means, variances, and covariances by

$$\bar{\lambda}_j = \sum_{i=1}^{n_j} \lambda_{ij}/n_j \ ,$$

$$\tau_{jk} = \sum_{i=1}^{n_k} \lambda_{ij}(\lambda_{ik} - \bar{\lambda}_k)/(n_k - 1), \text{ for } k \geq j \ .$$

The corresponding population quantities are

$$\bar{\Lambda}_j = \sum_{i=1}^{N} \lambda_{ij}/N \ ,$$

$$T_{jk} = \sum_{i=1}^{N} (\lambda_{ij} - \bar{\Lambda}_j)(\lambda_{ik} - \bar{\Lambda}_k)/(N - 1) \ ,$$

for all k and j. Again, population quantities are denoted with capital letters, and population variances and covariances are defined with "$N - 1$" in the denominator. One can think of the problem now as estimating

$$\mu_* = \sum_{i=1}^{N} \Lambda_i(t_0)/N = \sum_{j=1}^{m} \bar{\Lambda}_j.$$

Let "S" represent the operation of taking a sample of systems, and let "P" represent the conceptual operation of observing the corresponding Poisson process. The mean estimator is still unbiased since

$$E_S[E_P(\sum_{j=1}^{m} \bar{x}_j|S)] = E_S(\sum_{j=1}^{m} \bar{\lambda}_j) = \sum_{j=1}^{m} \bar{\Lambda}_j = \mu_*.$$

The total variance of the mean estimator is

$$Var(\hat{\mu}) = E_S[Var_P\ (\hat{\mu}|S)] + Var_S[E_P\ (\hat{\mu}|S)]$$

$$= E_S[Var_P\ (\sum_{j=1}^{m} \bar{x}_j|S)] + Var_S[E_P\ (\sum_{j=1}^{m} \bar{x}_j|S)]$$

$$= E_S(\sum_{j=1}^{m} \bar{\lambda}_j/n_j) + Var_S(\sum_{j=1}^{m} \bar{\lambda}_j)$$

$$= \sum_{j=1}^{m} \bar{\Lambda}_j/n_j + \sum_{j=1}^{m} (1 - n_j/N)T_{jj}/n_j$$

$$+ 2\sum_{j=1}^{m-1} (1 - n_j/N)/n_j \sum_{k=j+1}^{m} T_{jk} \ . \tag{11.4.1}$$

The independent increments of the Poisson process implies that $Cov_P[(\bar{x}_j, \bar{x}_k)|S] = 0$, for $k \neq j$. Also, the last two terms in (11.4.1) arise by the same argument that produced (11.2.1).

In addition to estimating the parameter μ_*, the mean estimator is a predictor for the ordinary population mean, $\mu = \sum_{j=1}^{m} \bar{X}_j$. The variance associated with the prediction is

$$Var(\hat{\mu} - \mu) = Var(\sum_{j=1}^{m} \bar{x}_j) + Var(\sum_{j=1}^{m} \bar{X}_j) - 2Cov(\sum_{j=1}^{m} \bar{x}_j, \sum_{j=1}^{m} \bar{X}_j).$$

On the right hand side, the first term is (11.4.1), and the second term can be found by substituting $n_j = N$ into (11.4.1). The covariance term (Appendix C) is $-2\sum_{j=1}^{m} \bar{\Lambda}_j/N$. Combining these and doing the algebra yields

$$Var(\hat{\mu} - \mu) = \sum_{j=1}^{m}(1 - n_j/N)(\bar{\Lambda}_j + T_{jj})/n_j$$

$$+2\sum_{j=1}^{m-1}(1 - n_j/N)/n_j \sum_{k=j+1}^{m} T_{jk} . \qquad (11.4.2)$$

The distinction between (11.4.1) and (11.4.2) is the presence of the finite population correction factor, $1 - n_j/N$, with the term $\bar{\Lambda}_j$ in (11.4.2).

The unbiased estimators of the individual terms in (11.4.2) are the appropriate sample covariances. Specifically, (see Appendix D),

$$E(s_{jk}) = \begin{cases} \bar{\Lambda}_j + T_{jj}, & j = k \\ T_{jk}, & j \neq k. \end{cases}$$

Substituting these unbiased estimators into (11.4.2) yields (11.2.1), the finite population version of Nelson's variance estimator. Therefore, Nelson's estimator can also be interpreted as the natural unbiased estimator of the prediction variance in the general Poisson model.

Consider two special cases. For the first case suppose that the Poisson process for each repairable system is the same, that is $\Lambda_i(t) = \Lambda(t)$ for all systems in the population. This assumes that each system in the population is modeled as generating independent sample paths from identical stochastic processes. This assumption implies that the population variances and covariances, T_{jk}, are all zero. The prediction variance in (11.4.2) reduces to

$$Var(\hat{\mu} - \mu) = \sum_{j=1}^{m}(1 - n_j/N)\bar{\Lambda}_j/n_j,$$

which has the unbiased estimator

$$\widehat{Var}_{Pois} = \widehat{Var}(\hat{\mu} - \mu) = \sum_{j=1}^{m}(1 - n_j/N)\bar{x}_j/n_j . \qquad (11.4.3)$$

This estimator is the finite population analog of the usual Poisson estimator. It is guaranteed to be positive but, in this context, is based on rather restrictive assumptions.

A second, less restrictive, model is obtained by taking $\Lambda_i(t) = \beta_i\Lambda(t)$, where $\Lambda(t)$ is some baseline function for the expected cumulative number of repairs and β_i is a positive scale factor unique to the ith system. Without loss of generality, assume $\bar{B} = \sum_{i=1}^{N} \beta_i/N = 1$. The infinite population version of this model has been discussed by [8] with the additional capability of handling covariates. With no covariates this is a mixed Poisson process model as discussed by [14] and [6].

11.5 Simulations

11.5.1 Background

All of the simulations discussed in this section are based on samples of size $n = 20$ from an infinite population. The system repair histories were censored in a uniform fashion; specifically, $a_i = 21 - i$, $i = 1, \ldots, 20$, so that the first system is censored at $t = 20$, the second system is censored at $t = 19$, and so on. This implies that the number of systems contributing to the jth column is $n_j = 21 - j$. With an infinite population the total variance (11.4.1) and the variance of prediction (11.4.2) are the same. This is the true variance about which the variance estimators are compared in the simulation. For $t = q_m$ it is

$$Var_{true} = \sum_{j=1}^{m}(\bar{\Lambda}_j + T_{jj})/(21 - j) + 2\sum_{j=1}^{m-1}(\sum_{k=j+1}^{m} T_{jk})/(21 - j) . \qquad (11.5.1)$$

The true variances were evaluated and stored for $m = 1, \ldots, 19$.

The simulations consist of 10,000 runs. In each run the twenty repair histories were generated by first generating interarrival times from an exponential distribution and then appropriately transforming the time scale [16, Sec. 2.4]. Each variance estimator was evaluated only at the points $m = 1, \ldots, 19$.

Based on the 10,000 simulated samples of repair histories, the mean square error estimates for the variance estimators were calculated as

$$MSE_X = \sum_{i=1}^{10,000} (\widehat{Var}_X^{(i)} - Var_{true})^2/10,000,$$

for each $m = 1, \ldots, 19$, where "X" takes the labels " unbiased," "eigenvalue-adjusted," and "moment." The mean square error

$$RMSE_X = \frac{MSE_X}{MSE_u},$$

relative to the unbiased estimator, is also reported. Similarly, the bias for each estimator is calculated as

$$BIAS_X = \sum_{i=1}^{10,000} \widehat{Var}_X^{(i)}/10,000 - Var_{true},$$

and the bias relative to the true variance is

$$REL\ BIAS_X = \frac{BIAS_X}{VAR_{true}} \cdot 100\%.$$

Three different probability models were employed in the simulations. They are discussed separately in the following subsections.

11.5.2 Model 1: Identical Homogeneous Poisson Process

This is the simple Poisson model discussed in the last section with $\Lambda_i(t) = t, i = 1, \ldots, 20$, that is, each repair history is generated from a common homogeneous Poisson process with a rate of occurrence of failure (ROCOF) equal to one repair per time unit. There is no variation across systems except that induced by the Poisson repeats. In the previous notation, $\lambda_{ij} = \Lambda_i(j) - \Lambda_i(j-1) = 1$ and $\bar{\Lambda}_j = 1$. Since there is no variation across systems $T_{jk} = 0$ for all j and k, and the true variance in (11.5.1) reduces to

$$VAR_1 = \sum_{j=1}^{m} 1/(21 - j),$$

for $m = 1, \ldots, 20$.

Figures 11.2 and 11.3 plot the relative biases and MSE's for this model. The eigenvalue-adjusted estimator has mean square errors that are comparable to the unbiased estimator except with very heavy censoring. The moment estimator is negatively biased, but has smaller mean square errors than either of the other two for all censoring.

One of the practical uses of standard errors is to construct approximate confidence intervals. Two standard error limits are routinely used as approximate 95% confidence limits. Figure 11.4 shows the percent of "two-sigma" confidence intervals that did not cover the true mean in the simulations. The eigenvalue-adjusted estimator was the best at holding nominal 95% confidence levels, but the other two generally have coverage porbabilities that exceed 90%—acceptable for most practical work.

The close match in performance between the adjusted estimator and the unbiased estimator might lead one to believe that the estimate values were rarely different. In fact, the simulations indicated that the two were **always** different for t sufficiently large. Negative eigenvalues occurred for all 10,000 runs. The order of the covariance matrix for which negative eigenvalues first occurred varied between six and thirteen. Despite the existence of negative eigenvalues for every run in the simulation, the unbiased variance estimator was negative only nine times in 10,000 runs. Plots of the nine variance functions that have negative values are shown in Figure 11.5. These are shown along with the true variance function. The negative values only occur at ages corresponding to extremely heavy censoring (three or fewer systems on test out of twenty). The plots of the eigenvalue-adjusted variances for the same nine data sets are shown in Figure 11.6. The improved performance illustrates that the adjustment is reasonable. However, the rarity of negative values and their occurrence only with heavy censoring seems to indicate that the problem of negative variances may be more theoretical than practical.

11.5.3   Model 2: Mixed Homogeneous Poisson Processes

This model sets $\Lambda_i(t) = \beta_i t$, $i = 1, \ldots, 20$. Each repair history is generated from a different homogeneous Poisson process with a ROCOF equal to β_i repairs per time unit. The system-to-system variation is modeled by letting β_i be sampled from a uniform distribution on the interval (0,2). This implies that $\lambda_{ij} \equiv \Lambda_i(j) - \Lambda_i(j-1) = \beta_i$ and $\bar{\Lambda}_j = E(\beta_i) = 1$. Also,

$$T_{jk} \equiv Cov(\lambda_{ij}, \lambda_{ik}) = Cov(\beta_i, \beta_i) = Var(\beta_i) = 1/3.$$

Substituting into (11.5.1) yields the true variance,

$$VAR_2 = \frac{4}{3} \sum_{j=1}^{m} \frac{1}{21 - j} + \frac{2}{3} \sum_{j=1}^{m-1} \frac{m - j}{21 - j}, \quad m = 1, \ldots, 20.$$

The simulation results for this model are shown in Figures 11.7-11.9. They are very similar to the previous case. This time the unbiased estimator was positive for all 10,000 runs, but, again, negative eigenvalues occurred for all 10,000 runs.

11.5.4 Model 3: Mixed Power-Law Processes

This model imposes nonconstant ROCOF's by adopting a power-law form for

$$\Lambda_i(t) \equiv 10(t/10)^{\beta_i},$$

where the β_i's are uniformly distributed on (0,2). In this case

$$\bar{\Lambda}(t) \equiv E[\Lambda_i(t)] = 5 \int_0^2 (\frac{t}{10})^{\beta} d\beta = 10g(\frac{t}{10}),$$

where

$$g(x) = \begin{cases} (x^2 - 1)/(2\ln x), & x \neq 1 \\ 1, & x = 1. \end{cases}$$

Then

$$\lambda_{ij} = \Lambda_i(j) - \Lambda_i(j-1)$$

$$= 10[(\frac{j}{10})^{\beta_i} - (\frac{j-1}{10})^{\beta_i}]$$

and

$$\bar{\Lambda}_j = E(\lambda_{ij}) = 10[g(\frac{j}{10}) - g(\frac{j-1}{10})].$$

Unlike the previous models, the $\bar{\Lambda}_j$ are not identically equal to one. Also,

$$\lambda_{ij}\lambda_{ik} = 100[(\frac{jk}{100})^{\beta_i} - (\frac{j(k-1)}{100})^{\beta_i} - (\frac{(j-1)k}{100})^{\beta_i} + (\frac{(j-1)(k-1)}{100})^{\beta_i}],$$

and

$$T_{jk} = 100[g(\frac{jk}{100}) - g(\frac{j(k-1)}{100}) - g(\frac{(j-1)k}{100}) + g(\frac{(j-1)(k-1)}{100})] - \bar{\Lambda}_j\bar{\Lambda}_k.$$

No attempt is made here to write out the cumbersome expression for the true variance of the mean estimator. In the simulations, a subroutine evaluated $g(\cdot)$, from which $\bar{\Lambda}_j$ and T_{jk} were computed. The true variance was calculated by substituting these values into (11.5.1).

The results of the simulations for this model are shown in Figures 11.10-11.12. They are very similar to the previous two models.

11.5.5 Discussion of Simulations

The simulations indicate that the unbiased estimator is rarely negative. Using it to calculate standard errors should cause few problems in practice. The eigenvalue-adjusted estimator has a mean square error virtually identical to the unbiased estimator except for very heavy censoring. But the moment estimator has the smallest mean square error, and its relative performance improves with the degree of censoring. Coverage probabilities for "two-sigma" confidence limits for the mean number of repairs are closer to 95% for the eigenvalue-adjusted estimator than for the other two. But coverage probabilities for all three estimators should be acceptable for most practical work.

11.6   Summary

This article documented several extensions and enhancements to current methods for estimating the cumulative mean number of repairs and calculating the corresponding standard error. The unbiased variance estimator reported by [12] was extended to handle finite populations. This report points out that the unbiased variance estimate can be negative in practice and explains why. Two nonparametric alternatives, which are always positive, are discussed. Simulations based on samples of size 20 indicated that the unbiased estimator is rarely negative, but the moment estimator reported in [9], which is always positive, performs as well or better.

Appendix A

The variance estimate in (11.2.1) is the sum of all terms in the matrix (a particular quadratic form)

$$R = \begin{pmatrix} a_1 s_{11} & a_1 s_{12} & \cdots & a_1 s_{1m} \\ a_1 s_{12} & a_2 s_{22} & \cdots & a_2 s_{2m} \\ \vdots & \vdots & \ddots & \vdots \\ a_1 s_{1m} & a_2 s_{2m} & \cdots & a_m s_{mm} \end{pmatrix},$$

where $a_j = (1 - n_j/N)/n_j$, j=1,...,m. R can be expressed as

$$a_1 \begin{pmatrix} s_{11} & s_{12} & \cdots & s_{1m} \\ s_{12} & s_{22} & \cdots & s_{2m} \\ \vdots & \vdots & \ddots & \vdots \\ s_{1m} & s_{2m} & \cdots & s_{mm} \end{pmatrix} + (a_2 - a_1) \begin{pmatrix} 0 & 0 & \cdots & 0 \\ 0 & s_{22} & \cdots & s_{2m} \\ \vdots & \vdots & \ddots & \vdots \\ 0 & s_{2m} & \cdots & s_{mm} \end{pmatrix} + \cdots$$

$$\cdots + (a_m - a_{m-1}) \begin{pmatrix} 0 & 0 & \cdots & 0 \\ 0 & 0 & \cdots & 0 \\ \vdots & \vdots & \ddots & \vdots \\ 0 & 0 & \cdots & s_{mm} \end{pmatrix}.$$

If the covariance matrix is nonnegative definite, then so are all submatrices in the above expansion. Furthermore, the a_j's are nondecreasing in j so that the scalar coefficients in the expansion are nonnegative. Any quadratic form in R can be written as a sum of quadratic forms, each term of which is nonnegative. Therefore, nonnegative definiteness of the covariance matrix implies that R also is nonnegative definite and that the variance estimate in (11.2.1) is nonnegative.

Appendix B

For infinite populations ($N = \infty$) and substituting for the s_{jk}^*'s, we have from (11.3.2)

$$\widehat{Var}_m = \sum_{j=1}^{m} \sum_{i=1}^{n_j} (x_{ij} - \bar{x}_j)^2 / n_j^2 + 2 \sum_{j=1}^{m-1} \sum_{k=j+1}^{m} \sum_{i=1}^{n_k} (x_{ij} - \bar{x}_j)(x_{ik} - \bar{x}_k)/(n_j n_k).$$

Introducing the indicator variable

$$\delta_{ij} = \begin{cases} 1, & \text{if } i \le n_j; \\ 0, & \text{if } i > n_j; \end{cases}$$

then

$$\widehat{Var}_m = \sum_{j=1}^{m} \sum_{i=1}^{n} \delta_{ij}^2 (x_{ij} - \bar{x}_j)^2 / n_j^2$$

$$+2 \sum_{j=1}^{m-1} \sum_{k=j+1}^{m} \sum_{i=1}^{n} \delta_{ij} \delta_{ik} (x_{ij} - \bar{x}_j)(x_{ik} - \bar{x}_k)/(n_j n_k)$$

$$= \sum_{i=1}^{n} [\sum_{j=1}^{m} \delta_{ij} (x_{ij} - \bar{x}_j)/n_j]^2 \geq 0 .$$

Nonnegativity for the finite population version holds by essentially the same argument as in Appendix A . This time let

$$R = \begin{pmatrix} a_1 b_{11} & a_1 b_{12} & \cdots & a_1 b_{1m} \\ a_1 b_{12} & a_2 b_{22} & \cdots & a_2 b_{2m} \\ \vdots & \vdots & \ddots & \vdots \\ a_1 b_{1m} & a_2 b_{2m} & \cdots & a_m b_{mm} \end{pmatrix},$$

where $a_j = (1 - n_j/N)$ and $b_{jk} = s_{jk}^*/n_j, j = 1, ..., m$. $\widehat{Var}_m$ is again the sum of the elements in R, and, again, the a_j are nondecreasing in j. Furthermore, the infinite-population result guarantees that the sums of the elements of all submatrices, such as

$$\begin{pmatrix} b_{22} & \cdots & b_{2m} \\ \vdots & \ddots & \vdots \\ b_{2m} & \cdots & b_{mm} \end{pmatrix},$$

are nonnegative. Expanding as in Appendix A shows that $\widehat{Var}_m \geq 0$.

Appendix C

$$Cov(\hat{\mu}, \mu) = Cov_S[E_P(\hat{\mu}|S), E_P(\mu|S)] + E_S[Cov_P(\hat{\mu}, \mu)|S]$$

$$= Cov_S[E_P(\sum_{j=1}^{m} \bar{x}_j|S), E_P(\sum_{j=1}^{m} \bar{X}_j|S)] + E_S[Cov_P(\sum_{j=1}^{m} \bar{x}_j, \sum_{j=1}^{m} \bar{X}_j)|S]$$

$$= Cov_S(\sum_{j=1}^{m} \bar{\lambda}_j, \sum_{j=1}^{m} \bar{\Lambda}_j) + E_S[Cov_P(\sum_{j=1}^{m} \bar{x}_j, \sum_{j=1}^{m} \bar{X}_j)|S] .$$

The first term is zero since $\sum_{j=1}^{m} \bar{\Lambda}_j$ is a constant with respect to the operation of taking samples. Rewriting the covariance within the second term yields

$$Cov(\hat{\mu}, \mu) = E_S\{\sum_{j=1}^{m} [Cov_P(\bar{x}_j, \bar{X}_j)|S] \}$$

$$= E_S \sum_{j=1}^{m} Cov_P\{[\bar{x}_j, n_j \bar{x}_j/N + (N\bar{X}_j - n_j \bar{x}_j)/N]|S\}$$

$$= E_S \sum_{j=1}^{m} [(n_j/N) Var_P(\bar{x}_j|S)]$$

$$= E_S \sum_{j=1}^{m} \bar{\lambda}_j/N = \sum_{j=1}^{m} \bar{\Lambda}_j/N \ .$$

The first equality holds because the Poisson process has independent increments, implying that covariances among different columns are zero. The third equality holds because of the independence among the systems. This implies that $\bar{x}_j$ and $N\bar{X}_j - n_j\bar{x}_j$ are independent since they do not share any common systems. The independence referred to here is over repeats conditioned upon a given sample, not over samples.

Appendix D

$$E(s_{jj}) = E_S[E_P(s_{jj}|S)]$$

$$= E_S[E_P(\frac{\sum_{i=1}^{n_j} x_{ij}^2}{n_j} - \frac{2\sum_{i=1}^{n_j-1}\sum_{l=i+1}^{n_j} x_{ij}x_{lj}}{n_j(n_j-1)})|S]$$

$$= E_S[\frac{\sum_{i=1}^{n_j}(\lambda_{ij}+\lambda_{ij}^2)}{n_j} - \frac{2\sum_{i=1}^{n_j-1}\sum_{l=i+1}^{n_j}\lambda_{ij}\lambda_{lj}}{n_j(n_j-1)}]$$

$$= E_S(\bar{\lambda}_j + \tau_{jj}) = \bar{\Lambda}_j + T_{jj} \ ,$$

$$E(s_{jk}) = E_S[E_P(s_{jk}|S)]$$

$$= E_S[E_P(\frac{\sum_{i=1}^{n_k} x_{ij}(x_{ik}-\bar{x}_k)}{(n_k-1)}|S)]$$

$$= E_S[\frac{\sum_{i=1}^{n_k}\lambda_{ij}(\lambda_{ik}-\bar{\lambda}_k)}{(n_k-1)}]$$

$$= E_S(\tau_{jk}) = T_{jk} \ ,$$

for $k > j$.

References

1. Ascher, H. and Feingold, H. (1984), *Repairable Systems Reliability: Modeling, Inference, Misconceptions and Their Causes*, Marcel Dekker, New York.

2. Cochran, W. G. (1963), *Sampling Techniques, Second Edition*, John Wiley & Sons, New York.

3. Cox, D. R. and Isham, V. (1980), *Point Processes*, Chapman and Hall, London.

4. Cox, D. R. and Lewis, P. A. W. (1966), *The Statistical Analysis of Series of Events*, Methuen, London.

5. Horn, R. A. and Johnson, C. R. (1985), *Matrix Analysis*, Cambridge University

Press, Cambridge.

6. Karr, A. F. (1984), Combined nonparametric inference and state estimation for Mixed Poisson processes, *Zeitschrift für Wahrscheinlichkeitstheorie und Verwande Gebiete*, **66**, 81-96. John Wiley & Sons, New York.

7. Karr, A. F. (1986), *Point Processes and Their Statistical Inference*, Marcel Dekker, New York.

8. Lawless, J. F. (1987), Regression methods for Poisson process data, *Journal of the American Statistical Association*, **82**, 808-814.

9. Lawless, J. F. and Nadeau, J. C. (1993), Some simple robust methods for the analysis of recurrent events, *Technical Report*, Department of Statistics and Actuarial Science, University of Waterloo.

10. Lewis, P. A. W., Ed. (1972), *Stochastic Point Processes*, John Wiley & Sons, New York.

11. Nelson, W. B. (1988), Graphical analysis of system repair data, *Journal of Quality Technology*, **20**, 24-35.

12. Nelson, W. B. and Doganaksoy, N. (1989), A computer program for an estimate and confidence limits for the mean cumulative function for cost or number of repairs of repairable products, *General Electric Research and Development Center Technical Report 89CRD239*.

13. Robinson, J. A. (1990), Standard errors for data from repairable systems, *General Motors Research Publication GMR-6772*.

14. Snyder, D. L. (1975), *Random Point Processes*, John Wiley & Sons, New York.

15. Thompson, W. A., Jr. (1981), On the foundations of reliability, *Technometrics*, **23**, 1-13.

16. Thompson, W. A., Jr. (1988), *Point Process Models with Applications to Safety and Reliability*, Chapman and Hall, New York.

17. Wimmer, H. K. (1989), Generalized singular values and interlacing inequalities, *Journal of Mathematical Analysis and Applications*, **137**, 181-184.

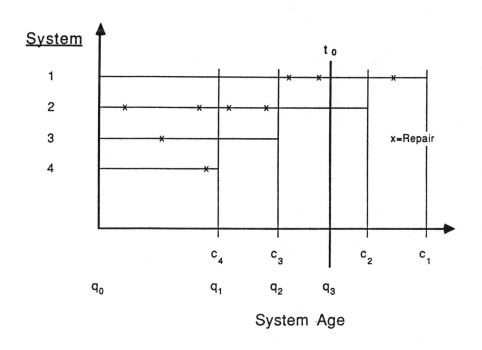

FIGURE 11.1
Illustrative example of repairable systems data

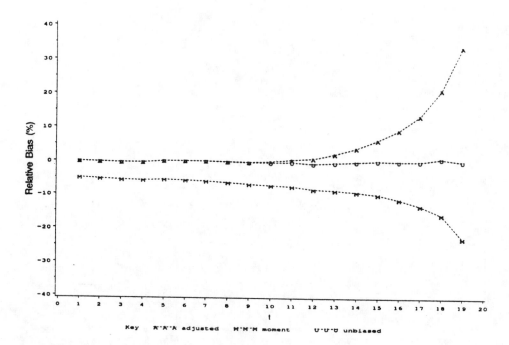

FIGURE 11.2
Bias relative to the true value for Model 1
Identical homogeneous Poisson processes

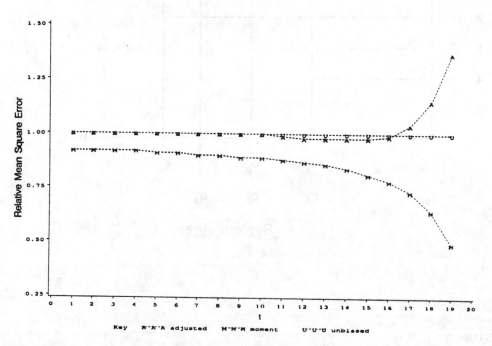

FIGURE 11.3
Relative mean square errors for Model 1
Identical homogeneous Poisson processes

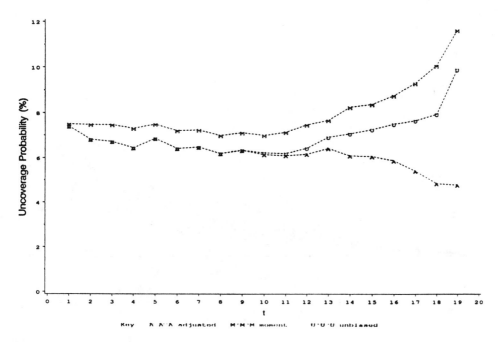

FIGURE 11.4
Two sigma confidence intervals for Model 1
Identical homogeneous Poisson processes

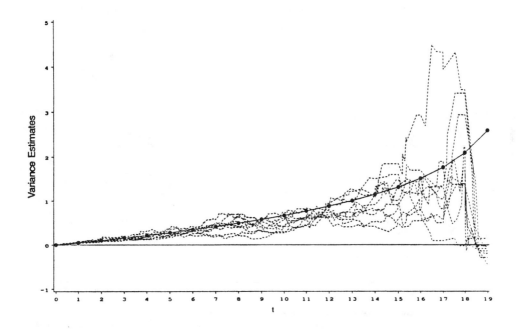

FIGURE 11.5
Nine negative variances for the unbiased estimator
Identical homogeneous Poisson processes

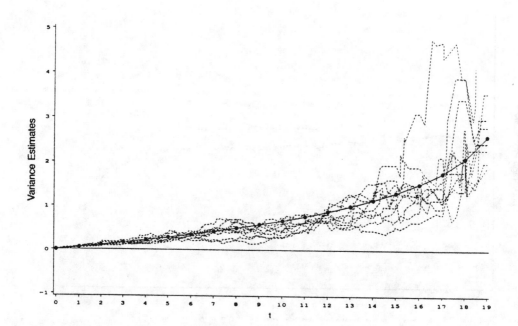

FIGURE 11.6
Nine adjusted variances
Identical homogeneous Poisson processes

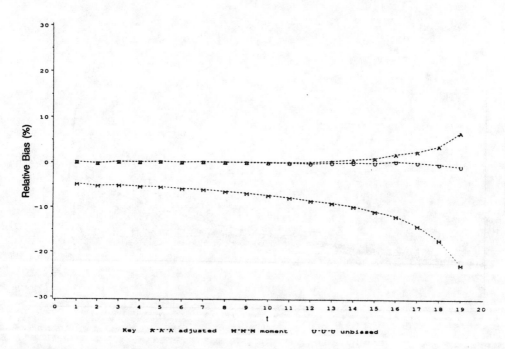

FIGURE 11.7
Bias relative to the true value for Model 2
Mixed homogeneous Poisson processes

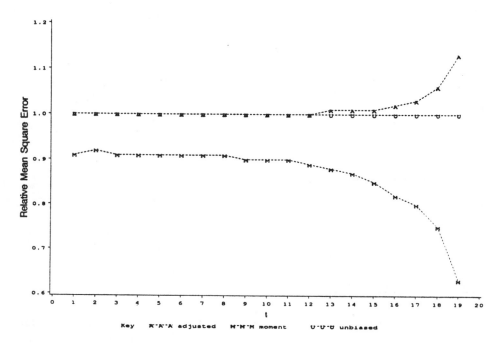

FIGURE 11.8
Relative mean square errors for Model 2
Mixed homogeneous Poisson processes

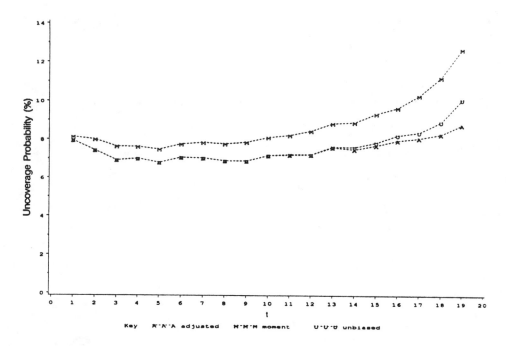

FIGURE 11.9
Two sigma confidence intervals for Model 2
Mixed homogeneous Poisson processes

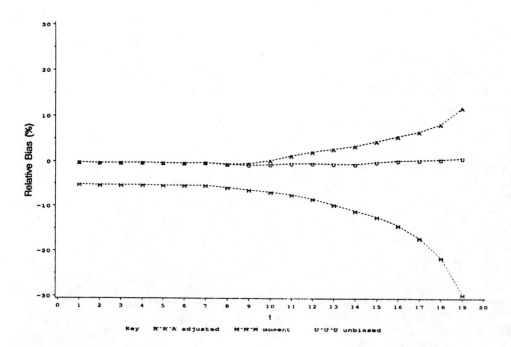

FIGURE 11.10
Bias relative to the true value for Model 3
Mixed power-law processes

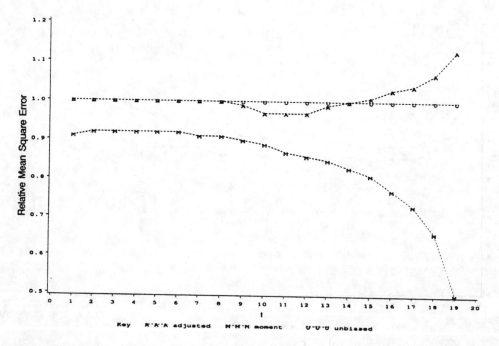

FIGURE 11.11
Relative mean square errors for Model 3
Mixed power-law processes

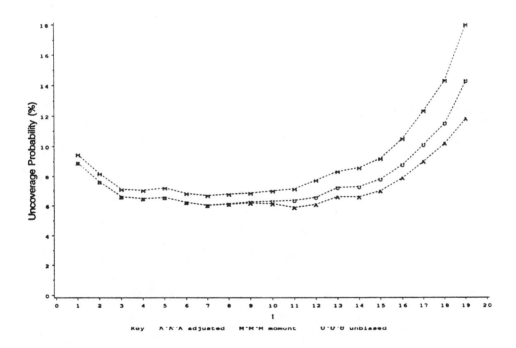

FIGURE 11.12
Two sigma confidence intervals for Model 3
Mixed power-law processes

12

Accelerated Reliability Growth Methodologies and Models

Alec A. Feinberg and Gregory J. Gibson

The Analytic Sciences Corporation, Reading, MA

ABSTRACT Extending reliability growth so that it can be applied in accelerated testing enables the application of all the reliability growth tools and planning advantages to accelerated testing. We describe linking these two areas together into what is termed Accelerated Reliability Growth Testing (ARGT). This work models mathematical equations of ARGT for both iso-stress and step-stress accelerated testing.

Keywords: Iso-stress and step-stress accelerated testing, Reliability growth.

CONTENTS

12.1   Introduction

Using accelerated testing to measure product reliablity is a common practice throughout industry. Historically, it has been applied quite extensively with components and complex assemblies such a semi-conductors, micro-electronics, capacitors, resistors, optoelectronics, lasers, connectors, cells, batteries, lamps, electrical devices, bearings, mechanical components, etc. (see for example, [6], [13], [15], [20], [21], [22], [23], and [24]). Accelerated testing concepts are reviewed in Appendix C.

The need for reliability managements increases with the complexity of the product [9], [10]. This is because of a greater risk in new product development at the system level as compared to the component level. In programs where one wishes to push mature products or complex systems to new reliability milestones, inadequate strategies will be costly. A program manager must know if reliability maturation or growth can be achieved under a program's time and cost constraints. To achieve this, a plan of attack is required for each major subsystem so that the system-level reliability requirement can be met. There may be little time to validate, let alone assess reliability [11]; yet, without some method of assessment, the program itself could be jeopardized. The reliability practitioner is challenged, therefore, to devise a reliability planning methodology for development that incorporates the most time- and cost-effective testing techniques available. The practitioner would like to take advantage of the advances in accelerated testing methodologies, such as test equipment, historical acceleration factors ([15], [21], [22], and [23]), and system-level acceleration factor estimation procedures ([17], [19], and [26]), and incorporate these into reliability growth mathematics as described in MIL HDBK 189 and elsewhere ([1], [2], [9], [11], and [25]).

Today's competitive market requires thorough planning, since system complexity has increased dramatically as competition and technological advances have driven down size and cost. It would be foolish to develop a major system having a handsome price tag without managing its reliability maturation. Properly applied, reliability management methods are powerful. The methodology provides many valuable techniques for system or subsystem levels such as:

- Testing planning
- Growth tracking and assessment
- Fix-effectiveness factor evaluation
- Corrective Action Review Team (CART) set-up and operation
- Test-Analyze-And-Fix (TAAF) strategy.

In this work accelerated testing methodologies (see Appendix C) are combined with the reliability growth techniques of MIL HDBK 189. Mathematical models for combining accelerated testing with the reliability growth models in MIL HDBK 189 are presented. Other investigators have already started to merge these methodologies ([10], [27], and [28]). Our goal is to enhance these technologies so that reliability planning can properly incorporate accelerated testing analysis.

Two kinds of techniques are discussed: Iso-Accelerated Reliability Growth Testing (Iso-ARGT) and Step-Accelerated Reliability Growth Testing (Step-ARGT). Iso-ARGT applies to constant elevated stress testing over time, such a isothermal aging, which uses constant elevated temperature over time ([21], [22]). Step-ARGT applied to step-stress testing of such stresses as temperature, shock, vibration, etc. ([12], [15], [20], [21], and [22]). We describe methodologies for both testing techniques which will allow, we feel, for conservation of time and money that can be wasted when accelerated testing is attempted without a fully planned program.

Accelerated Reliability Growth Methodology

The basic methodology for Accelerated Reliability Growth includes the following ([5] and [10]):

- The design of appropriate accelerated tests to stress expected or unexpected failure modes of the subsystems. These tests should be chosen and designed to stimulate failures at a faster rate. An effective program should include a streamlined root-cause corrective action plan.
- The application of historical acceleration factors and reliability growth "alphas." This will allow estimates of accelerated reliability growth over the program's testing phases to be generated.
- The construction of a reliability management program using idealized accelerated reliability growth equations and curves. These are integral to reliability growth, establishing interim reliability goals, providing target accelerated test times, and

aiding in the estimation of the expected number of failures for each test phase.

- The tracking estimation, and assessment of reliability growth during and after each test phase long with corrective action. Assessment techniques for fix-effectiveness to properly evaluate growth goals/milestones compliance can also be made. Periodic assessments of reliability are essential to assure achievement of planned reliability growth goals.

Once testing has begun, an accelerated test failure should be subjected to standard failure analysis procedures. Failed parts should be traced with a FRACAS (Failure Review and Corrective Action System) to assure adequate post-test examination for failure mechanism identification. The ARGT plan should include guidelines for proper determination of each root-cause and its corrective action. A Correction Action Review Team (CART) should be organized to efficiently review these corrective actions. Once corrective actions have been implemented, reliability growth can be assessed. Assessment, FRACAS, CART and other important processes are methods for management to use to ensure that goals are met.

If tests are not designed correctly (see Appendix C for Guidelines) to initiate all or more of the product's potential failure modes, major problems will result thereby causing delays in the program's progress. Programs sometimes neglect such details as root-cause failure analysis where it is important that corrective actions are properly assigned to relevant failure modes.

Accelerated Reliability Growth (ARG) Curves and Equations

Integral to ARG planning are *idealized accelerated reliability growth* equations and curves that characterize realistic growth of subsystems across major test phases of their development program. They are useful for representing total program growth and determining the sufficiency of testing duration for meeting reliability requirements. They offer important flexibility when planning. For example, growth rates are unknown in the initial program stages. Reliability growth can still be projected, however, by using conservative historical growth "alphas" and initial MTBF subsystem estimates. If accelerated testing is employed at the subsystem level, reliability growth can still be estimated over the testing phases by conservatively estimating time compression. Just as conservative estimates of growth alphas are known, available historical data often enables one to estimates effective "acceleration factors" (see Appendix C for definition) for test planning purposes.

We note that it is not the intention of this article to argue the merit of an effective iso-or step-stress acceleration factor since it is an assumption here. The authors are aware that acceleration factors are failure mode and test strength dependent, and can require creative estimation methods. We refer the reader to a few references on the subject such as [17], [19], and [26] ([16] is for fatigue and fracture mechanics), and also provide some tools that might help guide the reader to an overall approach in the appendix.

In order to properly project reliability growth in an accelerated testing environment, modification of the idealized reliability growth equation [18]

$$
M^\mu(t) =
\begin{cases}
M_I^\mu & , \ 0 < t < t_1^\mu \\[2ex]
\dfrac{M_I^\mu}{1-\alpha} \left(\dfrac{t}{t_1^\mu}\right)^\alpha & , \ t > t_1
\end{cases}
\tag{12.1.1}
$$

is necessary. Here the superscript μ indicates an unstressed or unaccelerated test. $M^\mu(t)$ is the MTBF value at unstressed time t^μ, M_I^μ is the unstressed initial MTBF value, α is the growth rate exponent, and t_1^μ is the initial test time. Equation (12.1.1) becomes modified with the following assumptions:

1. An effective acceleration factor, $\mathbf{A}$, exists and can be estimated.

2. Time is linearly compressed by this factor such that the following transformations are appropriate to modify (12.1.1).

$$t_1^\mu = t_1^s \, A \tag{12.1.2}$$

$$t^\mu = t^s \, A \tag{12.1.3}$$

$$M_I^\mu = M_I^s \, A \tag{12.1.4}$$

3. Reliability growth can be obtained in a compressed time period t^s, as in the uncompressed time period t^μ [see (12.2.3)].

In Assumptions 2 and 3, the superscript s implies the accelerated stress condition.

We note that for N independent accelerated stresses, the total acceleration factor is given by the product rule [20]

$$A_{total} = \prod_{i=1}^{N} A_i \tag{12.1.5}$$

Appendix A provides insight on when this product rule may be applied or when a different rule might be needed to obtain effective system acceleration factor.

12.2   Accelerated Factor Applied at $t = 0$

In this paper we describe modifications of the idealized reliability growth equation for both iso-stress and step-stress testing. We start by considering an accelerated stress for an iso-stress test applied between $t = 0$ and $t = t_f^s$. Referring to Figure 12.1, the accelerated growth curve is given for Iso-ARGT by

$$M^s(t) = \begin{cases} M_I^\mu & , \ 0 < t < t_1^s \\[2mm] \dfrac{M_I^\mu}{1 - \alpha} \left(\dfrac{t}{t_1^s}\right)^\alpha & , \ t_1^s < t < t_f^s \ . \end{cases} \tag{12.2.1}$$

This can be rewritten by using (12.1.2) such that

$$M^A(t) = \begin{cases} M_I & , \ 0 < t < t_1^s \\[2mm] \dfrac{M_I}{1 - \alpha} \left(\dfrac{t}{t_1^\mu}\right)^\alpha A^\alpha & , \ t_1^s < t < t_f^s \ . \end{cases} \tag{12.2.2}$$

Note that this meets the requirement of Assumption 3 above as shown with Eqs. (12.2.1) and (12.2.2) indicating (see Figure 12.1)

$$M^A(t_f^s) = M^\mu(t_f^\mu) \ . \tag{12.2.3}$$

When comparing (12.1.1) with (12.2.2) the stressed environments accelerated testing results in higher reliability growth by a factor of A^α due to effective time compression. For example, at time t_f^s the unaccelerated curve only shows an MTBF of 2.14 AU (arbitrary units) ($3\,\mathrm{AU}/1.4$) compared to the accelerated curve which has an MTBF of $3\,\mathrm{AU}$ where an acceleration factor of 3 is applied. Also illustrated is time compression by a factor of 3 where $t_f^\mu = 3\,t_f^s$ and $t_1^\mu = 3\,t_1^s$. Here the subscripts μ and S refer to unaccelerated and accelerated stress time, respectively.

When a different acceleration factor is applied to each phase we have

$$M_i^A(t) = \frac{M_I}{1-\alpha}\left(\frac{t}{t_1^\mu}\right)^\alpha (A_i)^\alpha , \qquad t_i^s < t < t_{i+1}^s \tag{12.2.4}$$

when $t > t_1^\mu$ and A_i is the ith acceleration factor for the ith phase.

The average number of failures H under stress can be written, referencing MIL HDBK 189 [18], for the ith test phase (see (12.1.3) and (12.1.4)) as

$$H_i^A = \frac{t_i^s - t_{i-1}^s}{M_i^A} . \tag{12.2.5}$$

12.3 Accelerated Factor Applied After t_1

An acceleration factor applied after t_1^μ is illustrated in Figure 12.2. The accelerated growth equation can be written

$$M^A(t) = \begin{cases} M_I & , \ 0 < t < t_1 \\[2ex] \dfrac{M_I}{1-\alpha}\left(\dfrac{t}{t_1}\right)^\alpha F , \ t_1 < t < t_f^s . \end{cases} \tag{12.3.1}$$

Here F is an appropriate multiplication ARGT factor. F is obtained from Assumption **3** where once again we require

$$M^A(t_f^s) = M^\mu(t_f^\mu) . \tag{12.3.2}$$

To find F we first note that accelerated stress is only applied between t_1 and t_f^s. Therefore, we note that

$$(t_f^s - t_1)\,A = (t_f^\mu - t_1) . \tag{12.3.3}$$

The requirement of (12.3.2) yields

$$\frac{M_I}{1-\alpha}\left(\frac{t_f^s}{t_1}\right)^\alpha F = \frac{M_I}{1-\alpha}\left(\frac{t_f^\mu}{t_1}\right)^\alpha \tag{12.3.4}$$

which reduces to

$$F^{1/\alpha}\,t_f^s = t_f^\mu . \tag{12.3.5}$$

Inserting (12.3.3) into (12.3.5), we find that

$$F = A^\alpha \left\{ \frac{t_1}{t_f^s}\left(\frac{1}{A}-1\right)+1 \right\}^\alpha \tag{12.3.6}$$

and when A is much greater than 1 $(A \gg 1)$

$$F \simeq A^\alpha \left\{ 1 - \frac{t_1}{t_f^s} \right\}^\alpha . \tag{12.3.7}$$

Therefore, when an acceleration factor is applied after t_1 with $A \gg 1$, the accelerated growth equation can be written

$$M^A(t) \simeq \begin{cases} M_I & , \ 0 < t < t_1 \\ \dfrac{M_I}{1 - \alpha} \left(\dfrac{t}{t_1} \right)^\alpha A^\alpha \left\{ 1 - \dfrac{t_1}{t_f^s} \right\}^\alpha & , \ t_1 < t < t_f^s . \end{cases} \tag{12.3.8}$$

Comparing (12.3.8) to (12.2.2) we note that (12.2.2) displays more growth, since (12.3.8) is reduced by $\left(1 - \dfrac{t_1}{t_f^s} \right)^\alpha$ when A is applied after t_1. However, this factor turns out to be close to unity for many applications since often $t_f^s \gg t_1$. Figure 12.2 also illustrates an example taken from MIL HDBK 189 [18]. The example is modified using an acceleration factor of 20.4. The ARGT data are displayed in Table 12.1. Note that this acceleration factor leads to an increase in reliability MTBF growth by a factor of 2. In this example, the $\left(1 - \dfrac{t_1}{t_f^s} \right)^\alpha$ factor is about .99 and was treated as unity.

Semiconductor Subsystem Examples

Figures 12.3 through 12.5 illustrate the idealized curves for a semiconductor subsystem using the Iso-ARGT equation. MTBF growth for a test-analyze-and-fix reliability growth program are shown. After each phase, corrective actions are incorporated into the subsystem, yielding a jump in the MTBF growth curve.

Subsystems are divided into electrical and mechanical ARGT planning to help estimate effective acceleration factors. This reflects testing that is usually divided into both electrical and mechanical categories. This supports the practicality of ARGT curves and their flexibility to any testing program. Figures 12.3 and 12.4 display the semiconductor subsystem electrical and mechanical planned ARGT curves respectively. Estimates of electrical and mechanical failure distributions can usually be obtained from historical testing or from their expected failure distributions. Planned electrical and mechanical ARGT may mathematically be combined into an electro-mechanical ARGT curve, as shown in Figure 12.5. Combining electrical and mechanical acceleration factors into an electro-mechanical planned growth curve requires proper analyses, which are described in Appendix A.

These figures are generated by using Iso-ARGT equations similar to those above ([5], [7]). The actual test time may not be known during initial planning. A ratio of the final to initial test time can be assumed if the initial test phase is properly planned.

In Figures 12.3 - 12.5 a comparison is given of the expected reliability growth for an unaccelerated test in the same time frame. The axis on the right allows us to note the ARGT factor from an initial MTBF value. For example, in Figure 12.5 the unaccelerated growth curve indicates a growth factor of about 2.4 as compared to a growth factor of about 7 for the accelerated growth curve. The acceleration factor (AF) in Figure 12.5

is estimated from electrical and mechanical accelerated conditions to be approximately 100, leading to a GRowth Acceleration Factor (GR AF) of 3.2 ($= 100^{.25}$, see (12.2.2)). In these figures a conservative historic growth alpha of 0.25 was used, along with reasonable acceleration factors for typical electrical and mechanical failure modes and test strengths.

12.4 Step Stress Accelerated Reliability Growth Testing (Step-ARGT)

Step-Stress Accelerated Reliability Growth Testing (Step-ARGT) is the application of extreme stress applied at a component, subsystem, or system level. In step-stress testing, the stress is gradually increased in equal time intervals until it reaches a maximum level.

Iso-ARGT is designed so the subsystem will not be over stressed, while Step-ARGT uses a more aggressive step-stress approach. The advantage of Step-ARGT is that it provides rapid feedback on parts that are most likely to fail. The disadvantage is that assessment/validation is difficult. In Iso-ARGT, time compression is easier to estimate facilitating easier linkages to field use conditions for assessment/validation purposes. Step-stress testing, on the other hand is more likely to overstress the subsystem. Overstressing can particularly occur when "common-sense" rules of thumb for stresses are not used (see Appendix C.2) [14]. When engineering judgements are carefully made concerning test results, however, the advantages generally outweight the disadvantages.

An initial phase of step-stress testing should be considered in a development program prior to other testing, such as Iso-ARGT, since it offers many major reliability growth opportunities in highly compressed time periods.

Reliability growth projections are difficult in a step-stress environment. However it is still of interest to find an expression for Step-ARGT in terms of the number of completed step-stress-tests applied. If step-stress testing is applied at $t = 0$, we write

$$M(t^{ss}) = \begin{cases} M_I & , \ 0 < t^{ss} < t_1 \\ \dfrac{M_I}{1-\alpha} \left(\dfrac{A_{eff} \, t^{ss} \, N}{t_1} \right)^{\alpha} & , \ t^{ss} > t_1 \, , \end{cases} \tag{12.4.1}$$

where t^{ss} is the stepping time in the test, A_{eff} is the effective step-stress acceleration factor (see Appendix B for details) such that $t^{ss} A_{eff}$ is the use time for one complete step-stress test, and N is the number of step-stress tests (not the number of steps in the test). Then $N t^{ss} A_{eff}$ is the total step-stress test use time for N completed step-stress tests. Letting

$$M_I \left(\dfrac{A_{eff} \, t^{ss}}{t_1} \right)^{\alpha} = M_I^{ss} \tag{12.4.2}$$

(since t^{ss} is a fixed time interval), then (12.4.1) reads

$$M(N) = \begin{cases} M_I & , \ N = 0 \\ \dfrac{M_I^{ss}}{1-\alpha} N^{\alpha} & , \ N \geq 1 \, . \end{cases} \tag{12.4.3}$$

This expression allows us to write the Step-ARGT equation in terms of the number of planned step-stress tests as a proposed approach for Step-ARGT planning.

Figure 12.6 illustrates an example of a Step-ARGT planning curve. The curve indicates that most of the reliability improvement is achieved in the initial first step-stress test. The following values were used for (12.4.2) and (12.4.3):

$$M_I = 888\,AU\ ,\ A_{eff} = 100\ ,\ \frac{t^{ss}}{t_1} = 3\ ,\ \alpha = .25\ .$$

Assessing Reliability Growth

The practitioner evaluates reliability growth in two ways. The first utilizes assessments (quantitative estimation of current reliability status) that are based on information from the detection of failure sources. This can prove to be pracitcal for ARGT planning, but may be difficult to accomplish under Step-ARGT. The second monitors various activities in the reliability management process to ensure their accomplishment in a timely manner and that the engineering effort and quality comply with the program plan. This latter method may prove practical in Step-ARGT planning where qualitative assessment is easier to make. Each method complements the other when monitoring the growth process. It is always best, however, to make quantitative evaluations if possible.

Method of Assessment

Assessing the effectiveness of engineering fixes for reliability would clearly enhance the ability to plan and manage a reliability growth program. Validation of product progress is a key issue in meeting reliability objectives. However, in practice there are generally not enough failure data both before and after a corrective action to estimate, with reasonable confidence, the effectiveness of a fix and the product's improved failure rate. This will most likely be true even with an accelerated testing program. This lack of information may result in the assignment of unrealistic fix-effectiveness factors, overly optimistic or pessimistic failure rate estimates, and a corresponding incorrect assessment of reliability achievements.

A statistically sound practical methodology for assessing reliability growth from corrective actions is presented in [2] and [11]. This methodology requires proper modification to incorporate the effects of time compression analysis from accelerated testing ([5], [10]). Once properly modified, it is easily implemented during any reliability growth test phase. The effectiveness factor is defined as the percent decrease in a problem failure mode due to a corrective action. Thus, an estimate of the product's improved failure rate can be determined. This is a proposed alternative solution to the prohibitive statistical requirements in a common validation program which usually require large sample sizes and testing times.

12.5   Summary

In this chapter, models were presented that provide the reliability practitioner with a mathematical framework for defining, estimating, or projected accelerated reliability growth over a program's test phases. The models do not perturb the mathematics of Reliability Growth or accelerated testing, but enhance them so that reliability planning can be incorporated into accelerated testing analysis. Assumptions for the model were that an overall effective acceleration factor can be estimated for the subsystem under test and that time is linearly compressed by the measure of this acceleration factor.

Two kinds of ARGT models were presented: Iso-Accelerated Reliability Growth and a model for Step-Accelerated Reliability Growth Testing. Table 12.2 summarizes these modeling results.

This chapter should aid the reader in the proper application of reliability growth methodology to accelerated testing. The methodology described included a general approach for accelerated reliability growth testing. An example was given for designing ARGT planning curves of a semiconductor subsystem. The essential aspect of ARGT planning is to provide an opportunity to discover and eliminate potential failure mechanisms in order to attain accelerated reliability growth goals and to implement a robust design into manufacture in a timely and cost-effective manner. Thus, this chapter stresses the importance of properly managing and planning a complete accelerated reliability growth program, thereby saving considerable time and money.

Appendix A

Effective Acceleration Factor for ARGT

The product rule described by (12.1.5) is valid when a combined stress is used to stress a particular failure mode like temperature and humidity. However, when different failure modes require different acceleration stress test, the effective or total acceleration factor will not obey this product rule for ARGT.

For example, if the system failure rate is given by

$$\lambda_{system} = \lambda_{elec} + \lambda_{mech} \qquad (A.1)$$

and the MTBFs are

$$M^E = 1/\lambda_{elec} , \qquad (A.2)$$

$$M^M = 1/\lambda_{mech} , \qquad (A.3)$$

$$M = 1/\lambda_{sys} . \qquad (A.4)$$

It can be shown that the effective system electromechanical accelerate factor is

$$A^\alpha = \left\{ \frac{A_E^\alpha}{M^E} + \frac{A_M^\alpha}{M^M} \right\} \left\{ \frac{1}{M^E} + \frac{1}{M^M} \right\}^{-1} \qquad (A.5)$$

Appendix B

Effective Acceleration Factor in Step-ARGT

Equation (12.4.1) in the test implies

$$A_{eff}\, t^{ss} = t_{use}$$

where t^{ss} is the stepping time in the test and A_{eff} is the effective overall acceleration factor such that t_{use} is the total effective use-time in a completed step-stress test. To find t_{use} and A_{eff} we note that

$$t_{use} = t^{ss} \sum_{i=1}^{n} A_i$$

where n is the number of steps in the test, and A_i is the i_{th} acceleration factor for the ith step-stress to use conditions, i.e., for an Arrhenius acceleration factor

$$A_i = A_i\,(T_i,\, T_{use}) .$$

Above T_i is the ith stress such as in a temperature step-stress test. Thus, comparing the first two equations, we have identified A_{eff} for (12.4.1) in the test as

$$A_{eff} = \sum_{i=1}^{n} A_i .$$

Appendix C

Concepts of Accelerated Testing

C.1 Accelerated Testing

Accelerated testing is designed to stress potential failure modes and can be designed to test numerous types of failures from temperature sensitive electrical failure to mechanical failures. Figure 12.7 illustrates the concept of stress testing [28]. Failure occurs when the stress exceeds the product's strength. In a product's population, the strength is generally distributed and usually degrades over time. Applying stress simply simulates aging. Increasing stress increases the unreliability overlap area shown in the figure between the strength and stress distributions, and improves the chances for failure occurring. This also means that a smaller sample population of devices can be testing with increase probability of finding failure. Thus stress testing amplifies unreliability so failure can be detected. Accelerated life tests are also used extensively to help make predictions. However predictions can be limited when testing to small samples sizes. Predictions can be erroneously based on the assumption that life test results are representative of the entire populations. Thus it can be difficult to design an efficient experiment that yields enough failures so that the measures of uncertainty in the predictions are not too large. Stresses can also be unrealistic. Fortunately, it is generally rare for an increased stress to cause anomalous failures especially if common sense guidelines are observed.

C.2 Common Sense Guidelines for Preventing Anomalous Failures in Accelerated Testing

Anomalous failures can occur when testing pushes the limits of the material out of the region of intended design capability. Therefore, the natural question is usually asked, what should the guidelines be for designing proper accelerated testing and evaluating failures? The answer is that judgement is required by management and their engineers to make the correct decisions in this regard. To aid such decisions we have provided the following guidelines.

1. Always reference the literature to see what has been done in the area of accelerated testing.

2. Anomalous failures occur when accelerated stress causes "nonlinearities" in the product. For example, material changing phases from solid to liquid is a chemical "nonlinear" phase transition (e.g., solder melting, intermetalic changes, etc.), an electric spark in a material is an electrical nonlinearity, material breakage compared to material flexing is a mechanical nonlinearity. Accelerated stresses causing nonlinear ties should be avoided unless such stresses are plausible in product use conditions.

3. Lastly, tests can be designed in two ways – by avoiding nonlinear stresses, and by allowing higher stress which may or may not cause nonlinear stresses. In the later test design, the CART reviews all failures and decides if the failure is anomalous or not. Then a decision is made whether or not to fix the problem. Conservative decisions may end up fixing some anomalous failures, this of course is not a concern when time and money permit fixing all problems. The problem occurs when normal failures are put in the wrong category as anomalous and no corrective action is taken.

C.3 Acceleration Factor

The acceleration factor may be defined mathematically as

$$A = \frac{t}{t'}$$

where t is the typical life of a failure mode at use condition and t' is the life at accelerated test condition. Since accelerated testing is designed to create failure in a shorter time frame, the life at use condition is usually much longer than the life at test condition, so that A is greater than one. For example, suppose that the acceleration factor is one-hundred, this indicates that the failure mode lasts one-hundred times longer at use conditions than at the accelerated test condition. Thus, acceleration factors as described here, are time acceleration factors. Acceleration factors may also be put in terms of cycle life [16], power [16], parameter life [6], etc. However, the most common application is in the time domain for estimating test time-compression using the time acceleration factor.

Acceleration factors are often modeled using parametric equations. For example, many failure modes effected by temperature, such as chemical processes and diffusion, have what is known as an Arrhenius reaction rate given by

$$\text{Rate} = B \exp\left\{\frac{-E}{KT}\right\} .$$

Here B is a constant that is characteristic of the product failure mechanism and test conditions [21], E is the activation energy (in eV) of the failure mode, T is the temperature (in degrees Kelvin), and K is Boltzmann's constant, 8.617×10^{-5} electron-volts per degree Kelvin. If the kinetics of the failure mechanism are such that failure occurs when a certain amount of the chemical has reacted (or diffused); a simple view of failure is when the critical amount has been consumed [21] e.g.,

$$\text{Critical amount} = \text{rate} \times \text{time-to-failure}$$

or

$$\text{time-to-failure } t = \text{critical amount/rate} .$$

Using this naive approach for the time-to-failure t of the chemical process, the temperature acceleration factor is

$$A_T = \frac{t}{t'} = \exp\left\{\left(\frac{E}{K}\right)\left[\left(\frac{1}{T}\right) - \left(\frac{1}{T'}\right)\right]\right\} .$$

Here T is the temperature at use conditions and T' is test temperature. In order to evaluate the acceleration factor, the value for E must be known or assumed for the failure mode. Often, historical information provides typical values for E, or it may be obtained through analysis of testing data obtained at different temperatures.

C.4 Other Common Models

Other types of acceleration factors can be derived depending on the failure mode and test type. For example, temperature-cycling is a common form of accelerated testing thermal fatigue related failures. Generally a relation of the following form is used to model the number of cycles-to-failure N_f, as

$$N_f \simeq C[\Delta T]^{-m}$$

where C and m are constants and ΔT is the total temperature range during cycling. The temperature-cycle acceleration factor between environments 1 and 2 can then be written

$$A_n = \frac{N_f^{(1)}}{N_f^{(2)}} = \left[\frac{(\Delta T)_2}{(\Delta T)_1}\right]^m .$$

This equation has been used to analyze solder joint failures [4], fracture-intermetallic bond failures [3], chip-out bond failure [3], etc. Roughly, the value of m usually ranges between 4 and 7.

Accelerated testing using voltage stress is also common. Voltage (or current) is a factor in the degradation rate for many of the thermally accelerated failure mechanisms which can affect, for example, integrated circuits. Ionic drift, electromigration, oxide breakdown, etc., are examples of such phenomena. In fact, the failure mechanisms not directly influenced by applied bias are those generally associated with mechanical stress, molecular diffusion or chemical reaction. And these can be indirectly affected by the magnitude of applied bias as a result of local heating. A multiplicative factor for voltage acceleration of the following form is recommended:

$$A_V = \left(\frac{V_T}{V_0}\right)^2$$

where V_T is the supply voltage during life test and V_0 is the maximum normal operating voltage. Total acceleration for the voltage and temperature life testing is then:

$$A = A_T \cdot A_V .$$

Humidity also accelerates failures. Failure rates due to humidity effects have been generally found to be exponential functions of temperature and relative hemidity. In [29], the authors' evaluation of published data on post-molded plastic devices indicate that reaction rate is of the form

$$R :: \exp\left\{-\left[\frac{Ea}{KT} + \frac{B}{RH}\right]\right\}$$

where Ea is an equivalent of activation energy, K is Boltzmann's constant, T is temperature in degree Kelvin, B is a proportionality constant and RH is percent relative humidity. In this view, the acceleration factor resulting from the combination of temperature and humidity is

$$A_{T-H} = A_T \cdot A_H$$

where

$$A_H = \exp\left\{B\left(\frac{1}{RH} - \frac{1}{RH'}\right)\right\} .$$

Other models for A_H have been proposed (see [20]).

Theory of accelerated testing has advanced in a number of testing areas for certain classes of devices, such as in microelectronics. Numerous authors are now working in the area of system-level accelerated testing. When an overall acceleration factor can be estimated (see for example [17], [19], and [26]), then the procedures in this article describe mathematical models for using the estimated value for determining the system's Accelerated Reliability Growth.

References

1. Benton, A.W. and Crow, L. (1989), Integrated reliability growth testing, *Proceedings of the Annual Reliability and Maintainability Symposium*, 160-166.
2. Crow, L.H. (1984), Methods of assessing reliability growth potential, 484-489.
3. Dunn, C.F. and McPherson, J.W. (1991), Temperature-cycling acceleration metallization failure in VLSI applications, *Reliability Physics*, 352-355.
4. Engelmaier, W. (1983), Fatigue life of leadless chip carrier solder joints during power cycling, *IEEE Transactions on Components, Hybrids, and Manufacturing Technology*, **6**, 232-237.
5. Feinberg, A.A. (1992), Fundamental development of equations and methodologies in accelerated reliability growth testing, *Engineering Memorandum, TASC*.
6. Feinberg, A.A. (1992), Gaussian parametric failure-rate model with practical application to quartz crystal device aging, *IEEE Transactions in Reliability*, **41**, 565-571.
7. Feinberg, A.A. (1994), Accelerated reliability growth models, *Journal of the Institute of Environmental Sciences*, 17-23.
8. Feinberg, A.A. and Gibson, G.J. (1993), Accelerated reliability growth, *Proceedings of the Institute of Environmental Sciences*, 102-109.
9. Feinberg, A.A., Gibson, G.J., and Shupe, R.H. (1992), Connecting technology performance maturation levels to reliability growth, *Proceedings of the Institute of Environmental Sciences*, 415-421.
10. Feinberg, A.A. and Stevens, R.A. (1992), Commercial electric vehical propulsion subsystem accelerated reliability growth test plan, *Technical Information Memorandum, TASC*.
11. Gibson, G.J. and Crow, L.H. (1989), Reliability fix effectivness factor estimation, *Proceedings of the Annual Reliability and Maintainability Symposium*, 75-82.
12. Hobbs, G.K. (1992), Highly accelerated life tests, *Proceedings of the Institute of Environmental Sciences*, 377-386.
13. Howes, M.J. and Morgan, D.V. (Eds.) (1981), *Reliability and Degradation of Semiconductor Devices and Circuits*, The Wiley Series in Solid State Devices and Circuits, **6**, New York.
14. Hu, J.M., Baker, D., Dasgupta, A., and Arora, A. (1992), Role of failure-mechanism identification in accelerated testing, *Annual Reliability and Maintainability Symposium*, 181-188.
15. IEEE Index (1988), *The 1988 Index to IEEE Publications*, IEEE Service Center, P.O. Box 1331, Picatinny, NJ 08855, (201) 981-1396 and 9535.
16. Lambert, R. (1993), Accelerated test rationale for damage tolerant critical designs with combined random and mean stress, *Proceedings of the Institute of Environmental Sciences*, 181-190.
17. McKinney, B.T. (1991), Designing through stress, *Proceedings of the the Institute of Environmental Sciences*.
18. MIL-HDBK 198 (1981), *Reliability Growth Management*.
19. Moura, E.C. (1992), A method to estimate the acceleration factor for subassemblies, *IEEE Transactions on Reliability*, **41**, 396-399.
20. Nelson, W. (1980), Accelerated life testing – step-stress models and data analy-

ses, *IEEE Transactions on Reliability*, **R-29**, 105-108; also in *Accelerated Testing: Statistical Models, Test Plans and Data Analyses*, John Wiley & Sons, New York (1990).

21. Nelson, W. (1990), *Accelerated Testing: Statistical Models, Test Plans, and Data Analyses*, John Wiley & Sons, New York.

22. Peck, D.S. and Zierst, C.H. (1974), The reliability of semiconductor devices in the bell system, *Proceedings of the IEEE*, **62**, 260-273.

23. Peck, D.S. and Trap, O.D. (1978), *Accelerated Testing Handbook*, Technology Associated, 51 Hillbrook Dr., Portola Valley, CA 94025, (415) 941-8272, Revised 1987.

24. Reynolds, F.H. (1977), Accelerated test procedures for semiconductor components, *Proceedings of the 15th International Reliability Physics Symposium*, 168-178.

25. Robbins, N.B. (1990), Tracking reliability growth with delayed fixes, *Proceedings of the Institute of Environmental Sciences*, 754-757.

26. Seager, J.D. and Fieselman, C.D. (1988), A method to predict an average activation energy for subassemblies, *IEEE Transactions on Reliability*, **R-37**, 458-461.

27. Schinner, C.E. (1991), Reliability growth through application of accelerated reliability techniques and continued improvement processes, *Proceedings of the Institute of Environmental Sciences*, 347-354.

28. Seusy, C.J. (1987), Achieving phenomental reliability growth, *ASM Conference of Reliability, Key to Industral Success*, 24-26.

29. Striny, K.M. and Schelling, A.W. (1981), Reliability evaluation of aluminum - metallized MOS dynamic RAM in plastic packages in high-humidity and temperature environments, *Proceedings of 31st Electronic Components Conference*. Also in *IEEE Transactions on Components, Hybrids and Manufacturing Technology*, 4, 476-481.

TABLE 12.1
Example corresponding to figure 12.2

Model Input Parameters:			M1 = 50,	α = .23,	AF = 20.36	
			T1 = 1000,	T2 = 2500,	T3 = 5000	
			T4 = 7000,	T5 = 10000		

TIME t	UNACC MTBF M(t)	ACC MTBF M(t)*2	AVE MTBF Mi	Ti	AVE-NUM Fail Hi	PHASE i
.00000	50.0	50.0	50.0	1000.0	20.0	1
500.0	50.0	50.0	50.0	1000.0	20.0	1
1000.0	64.9	129.8	146.4	2500.0	208.7	2
1500.0	71.2	142.5	146.3	2500.0	208.7	2
2000.0	76.1	152.3	146.3	2500.0	208.7	2
2500.0	80.1	160.3	175.0	5000.0	290.8	3
3000.0	83.6	167.2	175.0	5000.0	290.8	3
3500.0	86.6	173.2	175.0	5000.0	290.8	3
4000.0	89.3	178.6	175.0	5000.0	290.8	3
4500.0	91.7	183.5	175.0	5000.0	290.8	3
5000.0	94.0	188.0	195.8	7000.0	207.9	4
5500.0	96.1	192.2	195.8	7000.0	207.9	4
6000.0	98.0	196.1	195.8	7000.0	207.9	4
6500.0	99.8	199.7	195.8	7000.0	207.9	4
7000.0	101.5	203.1	212.1	10000.0	287.9	5
7500.0	103.2	206.4	212.1	10000.0	287.9	5
8000.0	104.7	209.5	212.1	10000.0	287.9	5
9000.0	107.6	215.2	212.1	10000.0	287.9	5
9500.0	108.9	217.9	212.1	10000.0	287.9	5
10000.0	110.2	220.5	212.1	10000.0	287.9	5

AMSAA Crow Parameters:			
PHASE	AVE MTBF	T(I)	CUM FAILURES
1	50.00	1000.0	20.0
2	146.30	2500.0	228.7
3	175.05	5000.0	519.5
4	195.84	7000.0	727.4
5	212.14	10000.0	1015.4
AFα = 2.	AF = 20.36		

*from page 46, MIL HDBK 189.

TABLE 12.2
Summary of ISO-ARGT and step ARGT models

EQUATION	REGION	COMMENT
$M^A(t) = \begin{cases} M_I \\ \dfrac{M_I A^a}{1-a} \left(\dfrac{t}{t_1^u} \right)^a \end{cases}$	$0 < t < t_1^s$ $t > t_1^s$	For A applied at t=0
$M_I^A(t) = \dfrac{M_I}{1-a} \left(\dfrac{t}{t_1^\mu} \right) A_i^a$	$t_i^s < t < t_{i+1}^s$	For A_i applied at t=0 and $t > t_1^s$ with different stresses per phase
$M^A(t) = \begin{cases} M_I \\ \dfrac{M_I A^a}{1-a} \left(\dfrac{t}{t_1} \right)^a \left(1 - \dfrac{t_1}{t_f^s} \right)^a \end{cases}$	$0 < t < t_1$ $t_1 > t_s$	For A applied after t_1 and A>>1.
$M(N) = \begin{cases} M_I \\ \dfrac{M_I^{ss}}{1-a} N^a \end{cases}$	$N = 0$ $N \geq 1$	For step-stress testing where $M_I^{ss} = M_I \left(\dfrac{A_{eff} \, t^{ss}}{t_1} \right)^a$

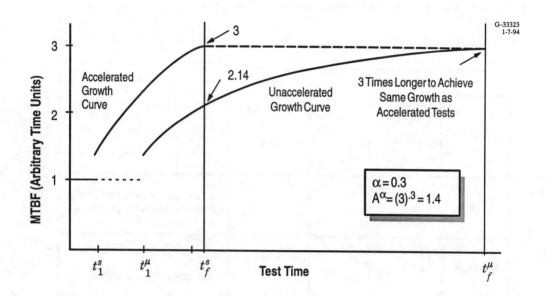

FIGURE 12.1
Accelerated vs. unaccelerated reliability growth

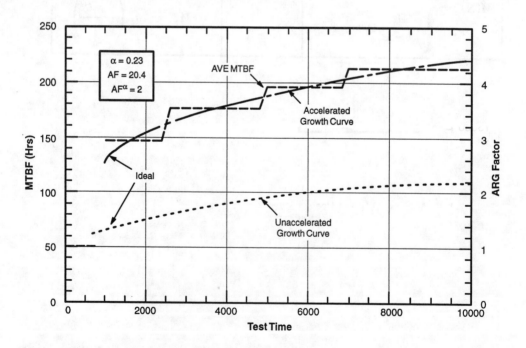

FIGURE 12.2
Illustrates ARGT using the sample in MIL HDBK 189 (page 46) with an
acceleration factor of 20.4. Note that this acceleration factor leads to an increase
in reliability growth by a factor of 2

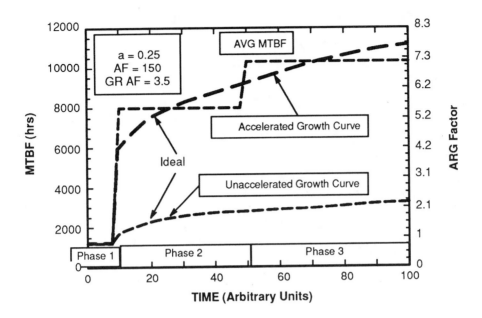

FIGURE 12.3
Idealized accelerated and unaccelerated reliability growth curves for electrical
iso-stress testing of a semiconductor subsystem

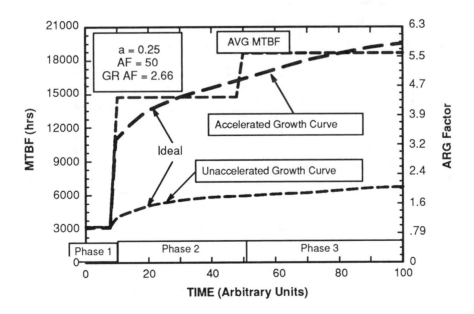

FIGURE 12.4
Idealized accelerated and unaccelerated reliability growth curves for mechanical
iso-stress testing of a semiconductor subsystem

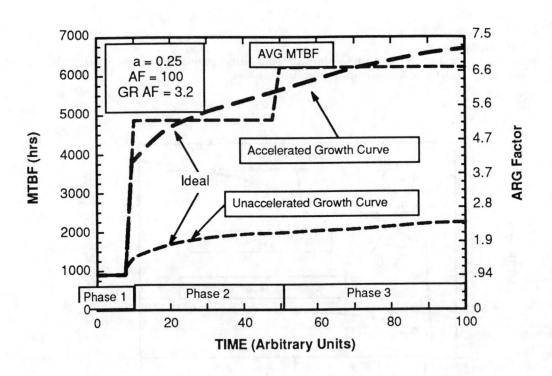

FIGURE 12.5
Composite accelerated and unaccelerated reliability growth curve for electro
mechanical iso-stress testing of a semiconductor subsystem

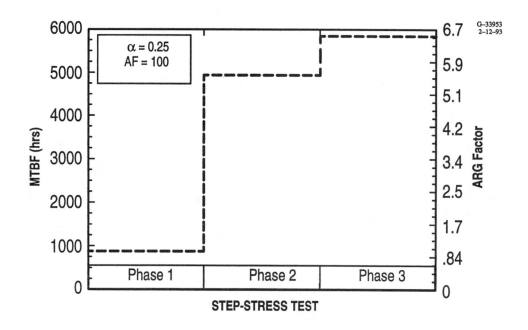

FIGURE 12.6
Step-ARGT figure showing reliability growth as a function of completed step-stress tests 1 and 2

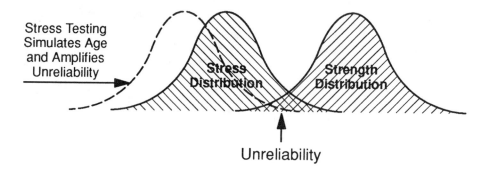

FIGURE 12.7
The accelerated testing principle

13

The Reliability Physics of Thermodynamic Aging

Alec A. Feinberg and Alan Widom

The Analytic Sciences Corporation, Reading, MA
Northeastern University

ABSTRACT This paper describes the reliability physics of thermodynamic aging processes. We present a model where catastrophic phenomena are linked to parametric degradation through thermodynamics for Arrhenius mechanisms. Time dependent kinetics are derived for both parametric and catastrophic Arrhenius aging. It is shown how aging dynamics are dependent upon thermodynamics specific to device reliability physics and that catastrophic phenomena can be correlated to device life dynamics.

Keywords: Aging kinetics, Arrhenius theory, Lognormal failure rates, Log(time) aging, Parametric/catastrophic reliability analysis, Thermodynamic free energy

CONTENTS

13.1 Introduction

The science of reliability is primarily associated with catastrophic failure mechanisms. The notion of catastrophic failure is usually perceived as a sudden failure in time, so that studying related parametric aging kinetics can be overlooked. However, aging kinetics often indicate that failure is precipitated first by parametric degradation. In this case, a measurable parameter changes value. This change can be associated with aging. Such degradation could then lead to catastrophic failure. In this case, the end-of-life phenomenon is continuous, implying that prognostics may be possible.

We will first review the aging process that we describe as primarily due to three types of thermodynamic processes: activation, diffusion, or forcing via an external stress. We provide examples using the corrosion process.

We then describe a model for thermodynamic aging of parametric and catastrophic failure mechanisms using the classical Arrhenius degradation mechanism. Although much work has been done with Arrhenius theory in reliability, very little of it actually applies to parametric reliability analysis. This work provides fundamental details. In the model we first map the reliability physics of this mechanism to parametric degradation. We show how parametric aging can quickly go catastrophic. Furthermore, linking the two

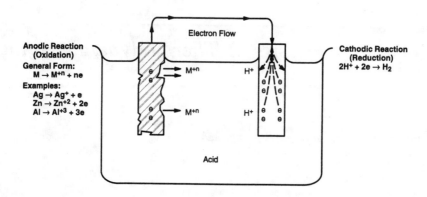

FIGURE 13.1
A simple corrosion cell

phenomena in this way yields a proactive approach to reliability in which catastrophic phenomena are correlated to device life. We then show that life degradation trends can affect catastrophic phenomena. Another conclusion is that Arrhenius degradation behavior can lead to log(time) aging kinetics with failure rates having lognormal form [8].

13.2   The Aging Process

If we define aging simply as an irreversible process, then many theories in physics and thermodynamics can alternatively be phrased in terms of aging. Aging may be then viewed as primarily due to three types of frequently irreversible thermodynamic processes: activation, diffusion, or forced processes via an external stress. Combinations of these processes provide complex forms of aging. Aging depends on the rate controlling process. Any one of these three processes may dominate depending on the failure mode. Alternately, the aging rate of each process may be on the same time scale, making all such mechanisms equally important.

Whatever type of aging is involved, an exchange of matter takes place in the process. As an example, we will examine the corrosion process to illustrate basic principles of aging. Corrosion is important because it is estimated that over 8 billion dollars a year is spent on corrosion problems in the US alone [11]. Like a battery, corrosion requires the presence of four basic elements, metal anode, cathode, conductive liquid (electrolyte), and conductive path. This is illustrated in Figure 13.1. Figure 13.2 illustrates that a similar electrochemical process can be formed with an electrolyte on a simple metal surface.

Small irregularities on the surface can form cathode 'C' and anode 'A' areas usually due to differences in oxygen concentration. The general reaction can be represented by the following equation

$$O + ne < -- > R \tag{13.2.1}$$

where O is the oxidized species, R is the reduced species, and n is the number of electrons involved in electrode process. We are interested in how fast matter is exchanged, e.g. the corrosive reaction rate. The forward rate $(- >)$ can be different from the backwards

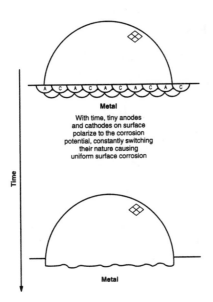

FIGURE 13.2
Uniform electrochemical corrosion depicted on the surface of a metal

reaction rate $(< -)$. According to the law of mass action, at constant temperature, the rate of many reactions is proportional to the concentration of each reactant. The concentrations are those at the electrode surface (O and R) (as opposed to the bulk electrolyte) denoted here as C_O and C_R. In this view, the rates are [18]

$$\text{Forward Rate} = K_f C_O \text{ and Backward Rate} = K_b C_R , \qquad (13.2.2)$$

where K_b and K_f are rate constants. It is of interest to express these rates in term of measurable currents. According to Faraday's law of electrolysis to convert to electrical current requires multiplication of the product nHA, where n is the number of electrons involved in the reaction, A is the electrode area, and H is Faraday's constant. Then the forward (anode) and backward (cathode) reactions currents, I_f and I_b, respectively are [18]

$$I_f = nHAK_f C_O \text{ and } I_b = nHAK_b C_R . \qquad (13.2.3)$$

The actual mass (M) transferred in time (t) in the reaction, which can be thought of as aging the material, is according to Faraday's Law proportional to the net current $I = I_f + I_b$ as

$$M = (A_m/nH) I t \qquad (13.2.4)$$

where A_m is the atomic mass. Thus the corrosion aging rate (dM/dt) is proportional to the current flowing.

The rate constants K_f and K_b are also temperature dependnent with an Arrhenius form [18]

$$K_f = K_{fo} \exp\{-\phi_f/RT\} \text{ and } K_b = K_{bo} \exp\{-\phi_b/RT\} . \qquad (13.2.5)$$

Here R is the gas constant, ϕ_f and ϕ_b represent the standard thermodynamic free energy of the reaction. The magnitude of energy represents a barrier height that must be surmounted to cause a corrosive transition state. The thermodynamic free energy when the forward and reverse reaction rates are equal is [18,21]

$$\phi_f = -\phi_b = |\phi|/2 = |nHE|/2 \qquad (13.2.6)$$

where E is the cell potential. In this case

$$I_f = I_b . \qquad (13.2.7)$$

Substitution of (13.2.3), (13.2.5), and (13.2.6) into (13.2.7) yields the important Nernst equation governing the thermodynamic's of the electrochemical reaction in terms of the concentrations [21]

$$\phi = \phi_o + RT \ln\{C_o/C_R\} \qquad (13.2.8)$$

where $\phi_o = RT \ln\{K_{fo}/K_{bo}\}$. The Nernst equation enables the calculation of the metals electrode potential when concentrations are known. It also can indicate the corrosive tendency of the reaction. When the thermodynamic free energy of the process is negative then there is a spontaneous tendency to corrode. Note that (13.2.8) can be expressed in terms of the reaction's activities [11].

Corrosion is an interesting process because it demonstrates important thermodynamic facets of aging; it shows the importance of the activation process via the Arrhenius relationship, the electropotential external stress, and concentration. Concentration is not only important for electrochemical corrosive aging but also in diffusion. The above equations have illustrated aging due to reaction at the electrode interface.

Diffusion often occurs in mass transport processes. In terms of corrosion, this refers to mass transported in the electrolyte solution to and from the electrodes. Mass transport occurs essentially from three processes. These are (1) convection and stirring, (2) electrical migration due to a field, and (3) diffusion from a concentration gradient [18]. The first two are under our category of forced via an external force. Of these three processes, diffusion is the most important. Diffusion in corrosion can be a rate controlling step. This is often the case in "hot corrosion" or "aqueous corrosion" due to oxidation. Furthermore, many aging processes due to diffusion do not involve electrochemical transitions. To find a diffusion rate associated with aging, we need to first consider the flux associated with the process. A flux of transition particles occurs due to the particles' concentration gradient. The flux 'F' in diffusion is given by Fick's first law as proportional to the concentration gradient where

$$\overline{F} = D\overline{\nabla}C \qquad (13.2.9)$$

where D is the diffusion coefficient, and C is the concentration. To see how this relates to mass transport we can consider the electrochemical cell. The flux in a one dimensional model is associated by a concentration gradient between the electrodes in the cell, can be approximated using (13.2.9) as

$$F = D(C_B - C_E)/d \qquad (13.2.10)$$

where C_B is the bulk concentration, C_E is the electrode concentration, d is an effective thickness of concentration at the electrode's surface. Then using Faraday's law of electrolysis the current is ($I = nHF$)

$$I = nHD(C_B - C_E)/d . \qquad (13.2.11)$$

We note that D, like K (see (13.2.5)) has an Arrhenius temperature dependent form

where

$$D = D_o \exp\{-\phi/RT\} \; . \tag{13.2.12}$$

Another way to measure aging is by considering the change of concentration in time. This is expressed using the transport equation [15]

$$\frac{\partial C}{\partial t} = -\overline{\nabla} \cdot \overline{F} \; . \tag{13.2.13}$$

Substituting in Flick's first law (13.2.9) we arrive at the diffusion equation (Flick's second law) in one-dimension [15]

$$\frac{\partial C}{\partial t} = D \, \frac{\partial^2 C}{\partial x^2} \tag{13.2.14}$$

which expresses the rate of change of concentration with time. Solutions to this equation require that appropriate boundary conditions be imposed. For example if we have initial conditions that the concentration is constant (C_o) at a surface over time, zero everywhere else initially, and falls to zero at infinity, then the solution to the diffusion equation can be shown to be [15]

$$C(x,t) = C_o erfc \left(\frac{x}{2\sqrt{Dt}} \right) \tag{13.2.15}$$

where $erfc$ is the complementary error function. As an interpretation of this equation in terms of aging, we might have a material that is exposed to a concentration C_o at its surface; parametric aging might be defined via the amount of contaminant that penetrates the material over time. The solution indicates that the problem is of course dependent upon the material's diffustivity.

The most generalized diffusion equation expression for aging circumstances includes external forces such as an electric field. Here generalized forces can be included as the gradient of a thermodynamic potential ϕ where [15,21]

$$\frac{\partial C}{\partial t} = D_o e^{\left(\frac{-\phi}{RT} \right)} [\nabla^2 C + \frac{1}{RT} \; \nabla \left([\nabla \phi] \, C \right)] \; . \tag{13.2.16}$$

This expression is fairly complete for our purposes. It unifies these three types of aging processes: diffusion, a forced mechanisms, and the Arrhenius activated processes. All processes are fundamentally thermodynamic in nature. The equation would be extremely difficult to solve if all mechanisms where equally important. However, we can often separate aging into its rate controlling process.

13.3   Activated Aging Processes

When activation is the rate-controlling process, Arrhenius type rate kinetics apply as noted previously in (13.2.5) and (13.2.12). In this section, we will propose a model that can be used to relate Arrhenius aging mechanism to parametric measurable quantities. To do this we first must review the basic idea behind Arrhenius theory. The theory describes the mathematical probability for a trapped particle (e.g., atom or molecule trapped on a surface or in a volume of material) to escape. To escape, the particle requires a certain amount of energy to overcome a potential energy barrier. Such trapped particles are often

described as "Maxwell-Boltzmann" particles [21] in which this probability is associated
with their energy state and is proportional to the factor

$$p \sim e^{\left(\frac{-\phi}{K_B T}\right)} \tag{13.3.1}$$

where T is temperature, ϕ is the potential energy barrier height that we have earlier
identified as the thermodynamic free energy. Note that this term appears in (13.2.5),
(13.2.12), and (13.2.16). In our view, this causes thermal aging by changing the state of
the material. Therefore, this can lead to degradation that is often measurable. To show
how this leads to a measurable quantity, let ν be the characteristic vibrational frequency
of the trapped particles. The probability per unit time that the particles will have enough
thermal energy to pass over this barrier is

$$\frac{dp}{dt} \approx \nu \, e^{\left(\frac{-\phi}{K_B T}\right)} . \tag{13.3.2}$$

Note that in this section we are using Boltzmann's constant K_B and incorporating the
units appropriately into the free energy where $K_B = R/N_o$ (N_o is Avogadro's number).
It is very common to use Boltzmann's constant instead of the gas constant (R) that was
previously used in Section 13.2, since it is independent of the molecular mass.

   We wish to associate the thermodynamic aging kinetics in the material with measurable
parametric changes. Therefore, we model the above as a fractional rate of parametric
change given by

$$\frac{da}{dt} = \nu \, e^{\left(\frac{-\phi}{K_B T}\right)} \tag{13.3.3}$$

where a is a unitless fractional change of the measurable parameter P (e.g., $a = \Delta P/P_o$).
For example, ΔP could be resistance change (or current, see (13.2.3) and (13.2.11)) such
that a is then the fractional resistance (or current) change (possibly in parts per million).

   This model can only be a function of the fractional change if the aging process is closely
related to the parameter change. This implies that the free energy itself will be associated
with the parameter through the thermodynamic work. Thus, ϕ will be a function of a.
It could also be a function of environment factors as well, such as temperature. This
is the basic assumption of the model. We proceed to expand the free energy in terms
of its parametric dependence using a Maclaurin series (with environmental factors held
constant for the moment). The free energy then reads

$$\phi(a) = \phi(0) + a \, y1 + \frac{a^2}{2} \, y2 + \ldots \tag{13.3.4}$$

where $y1$ and $y2$ are given by

$$y1 = \frac{\partial \phi(0)}{\partial a} \; , \;\; y2 = \frac{\partial^2 \phi(0)}{\partial a^2} \; . \tag{13.3.5}$$

13.3.1   Arrhenius Aging Due to Small Parameter Change

When $a \ll 1$, the first and second terms in the Maclaurin series yield

$$\frac{da}{dt} = \nu(T) \, e^{\left(-\frac{a \, y1}{K_B T}\right)} , \tag{13.3.6}$$

where

$$\nu(T) = \nu_o \, e^{\left(-\frac{\phi(0)}{K_B T}\right)} . \tag{13.3.7}$$

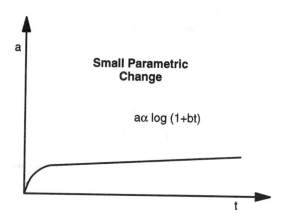

FIGURE 13.3
Log(time) aging for small fractional change of *a* over time *t*

Rearranging terms and solving for *a* as a function of *t* and integrating, we find (13.3.6) reads

$$a = \frac{\Delta P}{P_o} \approx \frac{K_B T}{y1} \ \ln[1 + \frac{\nu(T) \, y1}{K_B T} \, t] \qquad \text{for } a << 1 \, . \tag{13.3.8}$$

Numerous authors have observed similar measurable parametric dependence on aging and kinetic processes [2,3,5,8,9,10,12,14,16,17,19,20,24,25,26,27] that have Arrhenius degradation mechanisms. The key result here is the explicit description of the parameter *y1* that enables one to link the thermodynamics of the physical aging process to this log(time) aging model. Equation (13.3.8) is a general expression for Arrhenius degradation. However, the reliability physics is tied to the specific parametric problem by defining *y1*, which expresses the change in the thermodynamic free energy relative to the parametric change. Explicit examples to aid the reader are provided in the appendix.

Log(time) aging is an extremely important process in reliability physics since the origin of this aging kinetics can mathematically be tied to Arrhenius mechanisms of which numerous examples exist [2,3,5,8,9,10,12,14,16,17,19,20,24,25,26,27]. Figure 13.3 illustrates typical log(time) aging. One notes that aging is highly nonlinear for small time. This curve is representative of many aging and kinetic processes such as crystal frequency aging [8,9,20,27], corrosion of thin films [3,10,11, 25,26], gate oxide stressing [2,19], chemisorption processes [17], early degradation of primary battery life [5,12,16], cold worked metal recrystallization [24], superconducting ring flux leakage (yielding degradation of current in the ring) [14], 1/*f* noise [1] (note that 1/*f* transforms in the time-domain to log(time) dependance and has been shown to have its origin in a Bose-Einstein distribution having similar Arrhenius form [13]), etc.

The significance of parametric log(time) aging can further be put in perspective as it can be tied to catastrophic lognormal failure rates of semiconductors. To understand this hypothesis, first note that it has been shown that log(time) aging leads to parametric lognormal failure rates. This is shown in [8]. Thus, it is possible to presume that such degradation will also lead to lognormal catastrophic failure rates when components degrade slowly towards catastrophic failure via log(time) aging kinetics. That is, if the rate controlling aging process is log(time) aging, and parameters are normally distributed, then catastrophic failure rates will be lognormal. Many papers have shown that semiconductor failure rates are indeed lognormal in time [22, 23]. This could imply that aging is

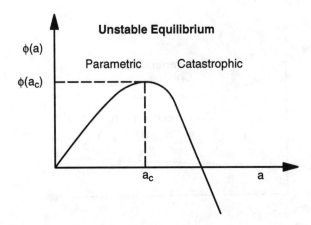

FIGURE 13.4
Thermodynamic potential function showing unstable equilibrium point at critical value a_c

predominantly Arrhenius with log(time) aging kinetics and that devices quickly go catastrophic at some critical threshold. Indeed, such behavior is also explained in the next
section from further expansion of this model.

13.3.2   Arrhenius Aging Due to Large Parameter Change

When a is not much less that 1, we make a Taylor expansion of the free energy potential
function around a critical parametric value a_c. This is shown in Figure 13.4. At this point
aging becomes unstable and catastrophic behavior occurs. Equation (13.3.6) can then be
rewritten

$$\frac{da}{dt} \approx \nu(T)\, e^{\frac{1}{K_B T} \left[\frac{(a-a_c)^2}{2} y2 \right]} \tag{13.3.9}$$

where

$$y1 = 0 = \frac{\partial \phi(a_c)}{\partial a} \;,\; y2 = -\frac{\partial^2 \phi(a_c)}{\partial a^2} \tag{13.3.10}$$

and

$$\nu(T) = v_o\, e^{\left(-\frac{\phi(a_c)}{K_B T} \right)} . \tag{13.3.11}$$

Rearranging terms in (13.3.9) and integrating the Gaussian expression where

$$\int_0^a e^{-\frac{y2}{2K_B T}(a'-a_c)^2} da' \approx \nu(T)\, t \tag{13.3.12}$$

yields

$$Erf\left(\frac{a - a_c}{B}\right) \approx B_1 \nu(T)\, t + C \tag{13.3.13}$$

where

$$B = \sqrt{\frac{2K_B T}{y2}} \;,\; B_1 = \sqrt{\frac{8K_B T}{y2 \pi}}$$

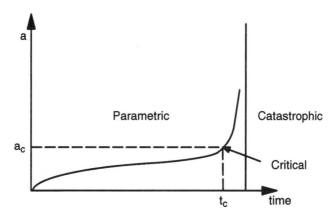

FIGURE 13.5
Aging for large fractional change of a over time t with critical values a_c and t_c occurring prior to catastrophic failure

and

$$C = -Erf\left(\frac{a_c}{B}\right) . \qquad (13.3.14)$$

This unstable function when combined with (13.3.8) depicts an important aging process shown in Figure 13.5. Figure 13.5 shows that aging starts off as to log(time) aging and then quickly goes catastrophic at the critical value a_c corresponding to a critical time t_c. Figure 13.4 illustrates the thermodynamic potential function for the process that provides the source of unstable equilibrium. When the particles reach the critical energy $\phi(a_c)$ the process quickly goes catastrophic. Figure 13.5 illustrates a number of rate processes. For example, batteries [4,5,6,7,12,16,18,] and cold worked metals recrystallizing exhibit forms of this dependence over time [24]. What is interesting in this model is that the rate of initial aging is mathematically connected to its rate of final catastrophic behavior (see (13.3.10)). This suggests that if we truly understand the initial aging process, prognostics is possible! We provide an example in the appendix.

13.4 Conclusions

This work provides fundamental details in the area of reliability physics. We first reviewed the aging process that is described as primarily as due to three types of thermodynamic processes: activation, diffusion, or forcing processes via an external stress. We provided examples using the corrosion process.

We then analyze the Arrhenius activated process in detail and proposed a model where Arrhenius degradation behavior can lead to log(time) aging kinetics with failure rates having lognormal form. We discuss how this helps to explain why the lognormal distribution has been so successfully applied in semiconductor reliability analysis. Log(time) aging is an extremely important process in reliability physics as its origin can mathematically be tied to Arrhenius mechanisms, of which numerous examples exist. Furthermore, this paper ties the reliability physics of the thermodynamic process to the log(time) aging expression. As such, the parametric dependencies of the aging process can be found by analysis of the thermodynamic free energy. Explicit examples to aid the reader are provided in the appendix.

Next we extended this work to the catastrophic case. Here it is shown that parametric aging can mathematically be related to catastrophic phenomena for severe degradation due to Arrhenius mechanisms. Furthermore, linking the two phenomena in this way yields a proactive approach to reliability in which catastrophic phenomena are correlated to device life. The results demonstrate that early life degradation trends provide significant links to eventual catastrophic phenomena.

Appendix

A.1 Crystal Resonator Parametric Aging Law Example

The crystal resonator aging law has the well known form [8,9,20,27]

$$a = \frac{\Delta f}{f_o} = A \ln(1 + bt) \tag{A.1}$$

where $\Delta f = f - f_o$, Δf is the frequency shift due to aging, and f_o is the resonator's resonance frequency. When a resonator ages, its frequency is found to shift in log(time) according to this function where A and b are usually treated as unknown constants. Note that this expression is in agreement with the derived expression in (13.3.8) where according to our model

$$A = \frac{K_B T}{y_1}, \quad b = \frac{\nu(T) y_1}{K_B T} . \tag{A.2}$$

In this case

$$a = \frac{\Delta f}{f_o} , \quad y_1 = \frac{\partial \phi(0)}{\partial a} \approx \frac{\partial \phi(0)}{\partial f} f_o . \tag{A.3}$$

To illustrate the thermodynamic model, we will find the parametric dependencies using $y1$. To do this we will describe the change of resonance frequency due to foreign mass (m) loading or unloading on the crystal's surface. Note this is usually an Arrhenius phenomenon. For example, mass unloading is often the result of moisture evaporating off the surface which increases the resonator's frequency. Conceptually, the molecules receive enough thermal energy to surmount the potential barrier to escape from the surface. To illustrative this we examine the frequency aging of a longitudinal bulk acoustic mode, which is often written by

$$f = \frac{1}{2e} \sqrt{\frac{c}{\rho_q}} \tag{A.4}$$

where e is the thickness of the crystal resonator, c is the longitudinal elastic constant, and ρ_q is the density of the crystal itself. A change in the frequency due to foreign mass loading is often modeled to first order as due to a fractional change in the thickness of the plate due to the foreign mass. In addition, there will be a fractional shift due to density change, however this is negligible by comparison. Then we can write

$$e = e_o + e_m \tag{A.5}$$

where e_o is the thickness of the resonator plate and e_m is the thickness due to the foreign mass loading. Thus, the infinitesimal frequency change due to the foreign mass loading to the plate is

$$df = -\frac{1}{2} \sqrt{\frac{c}{\rho_q}} \frac{1}{e^2} de_m \tag{A.6}$$

but from (A.4) this reads

$$df/dm = -f^2 K_1 \tag{A.7}$$

where

$$K_1 = 2\sqrt{\frac{\rho_q}{c}}\,\frac{1}{\rho_m A} \tag{A.8}$$

and we have let

$$de_m = \frac{1}{\rho_m}\frac{dm}{A}\,. \tag{A.9}$$

Here ρ_m is the density of the foreign mass, A is the surface area of the crystal, and d_m is the infinitesimal change in the foreign mass itself.

We are now in a position to find $y1$, the change in the free energy on the surface with respect to the parametric measurable quantity of frequency shift. We write $y1$ as

$$y1 = \frac{\partial \phi}{\partial f}\,f_o = \frac{\partial \phi}{\partial N}\frac{\partial N}{\partial m}\frac{\partial m}{\partial f}\,f_o\,. \tag{A.10}$$

Choosing ϕ as the Gibbs free energy [21] then

$$\frac{\partial \phi}{\partial N} = \mu \tag{A.11}$$

where μ is the chemical potential, and

$$\frac{\partial N}{\partial m} = \frac{1}{g} \tag{A.12}$$

where g is the gram-molecular weight of the foreign mass (e.g., for $M = Ng$).

With (A.3), (A.7), (A.11), and (A.12), we now have identified the thermodynamic quantity $y1$ as

$$y1 = -\frac{\mu}{g\,K_1 f_o}\,. \tag{A.13}$$

Using physical-based theory and the thermodynamic model, the parameters in (A.1) have been identified. The longitudinal frequency shift of the resonator due to mass loading during aging can be put in terms of parametric measurable quantities, where (A.1) now reads

$$\frac{\Delta f}{f_o} = -K_B T \left(\frac{g\,K_1 f_o}{\mu}\right)\ln\left[1 + \frac{\nu(T)}{K_B T}\left(\frac{\mu}{g\,K_1 f_o}\right)\right]\,. \tag{A.14}$$

This problem illustrates the concepts of the thermodynamic model. The log(time) thermodynamic model links typical log(time) aging for the Arrhenius phenomena and provides a link to the reliability physics of the specific problem through the thermodynamic quantity $y1$. In the case of the crystal filter aging law, we were able to find the dependence of a number of identifiable parameters. For example, the dependence on temperature (T), foreign mass (g) density and elastic constant c (via K_1) and frequency (f) is shown. The frequency dependence shows that aging amplitude increases as the square of the natural resonate frequency of the crystal (e.g., the higher the frequency of the resonator, the larger is its frequency shift over time) due to aging. This result has been observed experimentally and derived by others using bulk acoustic wave theory.

A.2 Catastrophic Aging Example of Primary Battery Aging

Most batteries age primarily due to the Arrhenius mechanism rather than diffusion. Therefore, an expression will be developed here for the voltage degradation parameter over time using the model expressed by (13.3.13). To derive this expression using the catastrophic model of (13.3.13), we start by expanding the thermodynamic free energy in terms of the measurable voltage parameter V about an assumed critical voltage value Vc (representative of a_c in Figure 13.1) e.g.,

$$\phi(a) = \phi(a_c) + a\, y_1(a_c) + a^2\, \frac{y_2(a_c)}{2} \quad \cdots \tag{A.15}$$

Here

$$a = \frac{\Delta V}{V_o} \; , \quad \Delta V = V - V_o \; , \quad a_c = \frac{V_c - V_o}{V_o}$$

and y_1 and y_2 are the first and second derivatives of the thermodynamic potential energy with respect to the voltage (see (13.3.5) and (13.3.10)). This expression for the potential is represented in Figure 13.4 where in the catastrophic case the potential y_2, will be negative and $y_1(a_c)$ vanishes. The expression then simplifies to

$$\phi(a) = \phi(a_c) + a^2\, \frac{y_2(a_c)}{2} \; . \tag{A.16}$$

To evaluate the free energy, we will assume a first order-model for the thermodynamic free energy at constant temperature of

$$\phi(\Delta V) \approx -\frac{1}{2}\, C_p\, \Delta V^2 \tag{A.17}$$

where C_p is a equivalent electrode capacitance of the battery. Using this expression for the value of the free energy about the critical value Vc, and defining the first-order free energy in terms of $\phi(a) = \phi(\Delta V)/(V_0^2)$ then the Taylor expansion about a_c in (A.16) yields

$$\phi(a) = \frac{1}{2}\, C_p \left(\frac{\Delta V_c}{V_o} \right) - \frac{1}{2}\, (a)^2\, C_p \; . \tag{A.18}$$

In this view $y2$ equals C_p and the catastrophic behavior near the critical voltage can now be expressed with the aid of (13.3.13) as

$$Erf \left(\frac{V - V_c}{V_o B} \right) \approx B_1\, f(T)\, t + C_1 \tag{A.19}$$

where now, according to (13.3.14),

$$B = \sqrt{\frac{2\, K_B T}{C_p}} \; , \qquad B_1 = \sqrt{\frac{8\, K_B T}{C_p\, \pi}} \qquad \text{and}$$

$$\tag{A.20}$$

$$C_1 = -Erf \left(\frac{\Delta V_c}{V_o B} \right) \; .$$

The aging expression when combined with (13.3.8) is similar to that shown in Figure 13.5. These equations can be used to fit typical battery aging curves [18,16,17]. For example, (A.19) and (13.3.8) can be combined to generate curves of the form shown in Figure A-1.

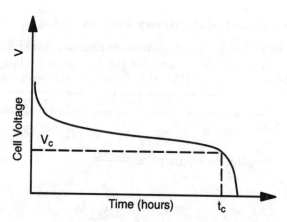

FIGURE A-1
Thermodynamic prediction of primary battery degradation

This displays typical primary battery degradation [16,17,18]. We note that the maximum value of the free energy corresponds to the critical voltage. This maximum value will be a direct consequence of the battery's design according to (A.15). This implies that early aging plays a significant role in battery life through the thermodynamic quantities $y1$ and $y2$.

References

1. Allan, D.W. (1966), Statistics of atomic frequency standards, *Proceedings of the IEEE*, **54**, 221-230.

2. Ash, A.S. and Gorton, H.C. (1989), A practical end of life model for semiconductor devices, *IEEE Transactions on Reliability*, **38**, 485-493.

3. Berry, R.W., Hall, P.M., and Harris, M.T. (1979), *Thin Film Technology*, Krieger, Malabar, Florida.

4. Bogner, R.S. and King, A.M. (1988), Accelerated life testing of intelstat VI batteries, intersoc. *Energy Conversion Engineering Conference*, 477-481.

5. Bro, P. and Levy, S.C. (1990), *Quality and Reliability Methods for Primary Batteries*, John Wiley & Sons, New York.

6. Cataldo, R.L. (1986), Parametric and cycle test of a 40-A-hr bipolar nickel-hydrogen battery, *NASA Technical Memo 88793*.

7. Ewashiaka, J.G. (1984), Results of electric vehicle propulsion system performance on three lead-acid battery systems, *NASA Technical Memo 83657*. Also in *19th Intersociety Energy Conversion Engineering Conference*, **2**, 727-755.

8. Feinberg, A.A. (1992), Gaussian parametric failure-rate model with application to quartz-crystal device aging, *IEEE Transactions on Reliability*, **41**, 565-571.

9. Filler, R.L. (1985), Aging specifications, measurement, and analysis, *7th Quartz Crystal Conference*.

10. Fisher, J.S. (1971), Failure characteristics of thin film capacitors, *Reliability Physics Symposium*, 183-187.

11. Fontana, M.G. and Greene, N.D. (1978), *Corrosion Engineering*, McGraw-Hill, New York.

12. Gabano, J.P. (Ed.) (1983), *Lithium Batteries*, Academic Press, New York.

13. Gersten, J.L. and Levine, R.D. (1981), A mechanism for a $1/f$ power spectrum, *Kinam*, **3**, 43-51.

14. Grassie, A.D.C. (1975), *The Superconducting State*, Sussex Press, Sussex, England.

15. Grove, A.S. (1967), *Physics and Technology of Semiconductor Devices*, John Wiley & Sons, New York.

16. Hobbs, B.S., Keily, T., and Palmer, A.G. (1978), Aspects of nickel-cadmium cells in single cycle application; The effect of long-term storage, *Journal of Applied Electrochemistry*, **10**, 721-727; **8**, 305-311.

17. Lansberg, P.T. (1955), On the logarithmic rate law in chemisorption and oxidation, *Journal of Chemical Physics*, **23**, 1079-1087.

18. Linden, D. (Ed.) (1980), *Handbook of Batteries and Fuel Cells*, McGraw-Hill, New York.

19. McPherson, J.W. and Baglee, D.A. (1985), Acceleration factors of thin gate oxide stressing, *IEEE International Reliability Physics Symposium*, 1-7.

20. Miljovic, M.R., Trifunovic, G.L., and Brajovic, V.J. (1988), Aging prediction of quartz crystal units, *42nd Annual Frequency Control Symposium*, 404-411.

21. Morse, P.M., *Thermal Physics*, Benjamin/Cummings, Reading, Massachusetts.

22. Peck, D.S. and Trapp, O.D. (1981), *Accelerated Testing Handbook*, Technology Association, 51 Hillbrook Dr., Portola Valley, CA 94025, (415) 941-8272, Revised 1987.

23. Peck, D.S. and Zierdt, Jr., C.H. (1974), The reliability of semiconductor device in the Bell system, *Proceedings of the IEEE*, **62**, 260-273.

24. Reed-Hill, R.E. (1973), *Physical Metallurgy Principles*, Van Nostrand, New York.

25. Uhlig, H.H. (1963), *Corrosion and Corrosion Control*, John Wiley & Sons, New York.

26. Vermilyea, D.A. (1953), Then kinetics of formation and structure of anodic oxide films on tantalum, *ACTA Metallurgica*, **1**, 282-294.

27. Warner, A.W., Fraser, D.B., and Stockbridge, C.D. (1965), Fundamental studies of aging in quartz resonators, *IEEE Transactions on Sonics and Ultrasonics*, 52-59.

14

Reliability in Design

Nelson O. Wood

TASC, Midwest City, OK

ABSTRACT Current design practices need to consider reliability from design inception. Strong economic reasons for this include lower per assembly cost, lower cost of manufacturing, and increased market share. Most reliability issues can be identified and addressed through analysis and simulation prior to any breadboarding/prototyping effort. Serious attention to reliability, while perhaps delaying prototyping, will actually shorten development time and lower reliability verification costs. Several key economic impacts of reliability in relation to manufacturing are discussed, as well as many reliability-related analyses which should be performed prior to any hardware realization. The analyses herein described are applicable to electrical, electronic, and mechanical systems.

Keywords: Reliability, SPICE, Stress analysis

CONTENTS

14.1 Introduction

The designer of a product today faces not just local, state, and national competition, but international competition as well. Designers in the United States have no monopoly on innovation or technology. They are under enormous pressures to produce quality products faster, at less cost, and with better performance than competitors who have access to cheaper labor, transportation, natural resources, and newer facilities. Compounding the problem is a legal climate in the United States which is increasingly hostile toward manufacturing and manufacturing error. This paper explores two topics: the first are the key economic reasons a business should strive for a design that is highly reliable from inception; the second addresses many techniques now available to designers to help them realize this goal.

14.2 Economics of Production

Businesses generally exist to make money; they can do this only if they establish, retain, or expand the market share for their product. No business today can be assured that they have an unassailable market niche or market share (note the plight of the U.S. automobile and airline industries). As Dr. W. Edwards Deming has pointed out, one increases the business market share by improving *quality* [2]. Therefore, the goal of any design engineer, in addition to producing a product that satisfies the demand, should be to achieve a reliable product. Quality reflects *reliability over time* [9]. The most cost-effective way of doing this is to incorporate reliability considerations into every phase of design, from concept through production and into product support.

Focusing early design efforts on reliability benefits the manufacturing portion of the business through lower production losses due to component failures, which helps to reduce the cost of producing and supporting the design. It is quite obvious that it costs just as much effort and material to make a defective product as a good product within a particular production lot. What is not obvious, at first glance, is that it costs three to four times *more* to make and fix a bad product as a good one (private discussions, see Note 1). While there are industries where defect rates or inspection rejection may run as high as 50% – for example, gift pottery – it is not typically in the best interest of any business to tolerate such waste. It is much more important for non-throwaway products that the manufacturing defect rate be kept low. Suppose a manufacturer detects, through end-of-line inspections, a bad product prior to shipment. It could be any process but for discussion's sake, let it be a complex electronics printed circuit board. At this point, the manufacturer has two choices: scrap the board or repair (rework) the board.

If the board is scrapped, then the cost of producing that board must be added to the costs of producing the remaining boards in the lot. If, for example, we experience an 8% defect rate, then that 8% must be added to the cost of producing the satisfactory boards, increasing the overall cost of production becomes:

$$P = \frac{n + d}{n} \tag{14.2.1}$$

where n is the cost of the number of good units, and d is the cost of the number of defective units. It is obvious that the article can quickly become prohibitively expensive to produce if scrap rates are high.

Simply scrapping the product never addresses the cause of the bad product. Obviously, if the cause for the failure is not consistent, then neither yield, cost, or profit will be consistent. Until the root cause of the failure is found and corrective action taken, then the number of failures detected in manufacturing should be added to the base failure rate of the product.

The decision to scrap a defective unit increases the total cost to produce beyond the cost of the components of the unit and the labor required. Manufacturing overhead must increase because of the larger inventory of components required to produce a required number of good units. More floor space for inventory, more accounting man-hours, and more man-hours required to handle the inventory are the most obvious labor hours required per good product, and possibly more expensive equipment due to the need for higher capacity to meet anticipated demand.

Rework is not necessarily a viable approach, either. Rework is labor-intensive. A manufacturing facility is geared to producing a quantity of products as a continuous *process*, not towards processing a quantity of units for repair to provide a product for sale. Economy of scale is generally incompatible with a serious volume of rework. In large part this is because, in many industrialized nations, the cost of labor is greater than the cost of the material for many products. Therefore, if we think of the circuit board as being made up of not just components but of components and labor hours, the cost of production equation becomes:

$$P = \frac{n(c + xh_m) + d(c_x h_m + yh_r)}{n(c + xh_m)} \tag{14.2.2}$$

where

$$
\begin{aligned}
n &= \text{number of good units} \\
d &= \text{number of defective units} \\
x, y &= \text{work hours required for product} \\
h_m &= \text{production man-hours cost} \\
h_r &= \text{rework man-hour cost} \\
c &= \text{materials cost}
\end{aligned}
$$

If the manufacturer elects to rework the board, then rework costs must be added to the basic cost of producing the item which, again, may cause the item to be very expensive. These additional costs due to rework are not always appreciated. First, additional components must be purchased, shipped, stocked, and accounted for. This drives up the total inventory costs. Second, additional equipment intended for individual rework rather than large-scale production must be purchased and dedicated to rework. This drives up the capital investment cost. Third, additional workers must be hired or additional man-hours must be funded, which drives up the total item labor cost. Fourth, manufacturing and storage space must be dedicated to rework, driving up the facilities cost. Fifth, when the rework has been accomplished, the units must be retested/reinspected, which drives up the total inspection costs of the product. Finally, all the rework needs to be tracked so that the business knows that the rework has been completed and inspected. This drives up the administrative costs associated with manufacturing the product.

Reliable products also reduce the cost of doing business in other ways. The most apparent way is in reducing the costs associated with warranties. Warranties costs in the U.S. automotive industry, for example, run between $500 Million and $2 Billion annually for each of the "Big Three," depending on the manufacturer and the reliability of the designs that year. The difference between a good year and a bad year for the automotive manufacturer is warranty costs (private discussion, see Note 2). It should be self-evident that warranties subtract from profit – yet annual warranty costs continue to be excessive. The experience of warranty cost is shared by most manufacturers of consumer goods and has resulted in many product lines becoming unprofitable (e.g., the Ford Pinto).

Unreliable products may also open the door to lawsuits. In addition to inherent safety, unreliable performance has been equated with safety for purposes of litigation. A brake failure, for example, is a safety issue. If the failure is caused by unreliable design rather than improper maintenance, the manufacturer is at fault. Major costs in the automotive industry are warranties associated with engine failures. Yet, it is the responsibility of the operator to change the oil regularly. Because of several lawsuits dealing with motor replacements, wherein the manufacturer lost and was forced to replace motors when it was believed the damage was due to improper maintenance, many automobiles now have a particle detection sensor in the oil system to protect the manufacturer from a presumption of manufacturing defect (see Note 3).

It is fair to note that the product service staff tends to grow in inverse proportion to product reliability. After all, the more often something needs to be fixed, the more personnel need to be hired to fix the units – until the customers catch on and quit buying the product altogether.

As with the "scrap" option, overall inventory is increased as both the manufacturer and distributors have to keep larger stocks of components for repair. In addition, shipping costs rise as more parts have to be shipped, as well as facilities costs associated with keeping, tracking, and handling this additional quantity of material due to poor reliability. Finally, the overall product administrative costs rise as more product information has to be tracked, accounted for, verified, and warranty claims processed. This cost will be reflected in both equipment and "support staff" for a product. From the foregoing, it is patent that both scrapping faulty assemblies and reworking faulty assemblies can add significantly to the per-unit cost of the product. The *least costly* product, then, is one that experiences the least reject rate. The product most likely to capture market share is one that is highly reliable. Therefore, it is clear that the most desirable design is one that is highly reliable and easily manufacturable from inception.

14.3   Methods of Designing-In Reliability

Thus, it comes to the design engineer: make it better, faster, more reliable, and cheaper. Fortunately, there is an increasing body of knowledge regarding reliability and how to achieve reliability in the initial design. The remainder of this discussion will concentrate primarily on electronic systems, but the general principles and approaches that help an engineer design in reliability from inception are applicable to any system (see Note 4).

The first requirement for a design engineer intending to develop a really reliable product is to clearly understand the intended use, operational environment, operator expectations, and required life cycle of the needed product. This requirement is directly reflected in the definition of reliability: "The ability or capability of a product to perform the specified function in the designated environment for a minimum length of time or minimum number of cycles or events" [5]. The designer must also be knowledgeable of how the design is to be manufactured and maintained. It would seem very sensible that the design engineer have this basic knowledge; in fact, most designers are faced with ill-defined operational requirements, poorly-identified customer expectations, incomplete or incorrect environmental requirements, and minimal information about manufacturing and maintenance requirements. This lack of up-front system definition can result in poorly-accepted products or misuse of the product and will be reflected in lack of sales and loss of business profitability. While it is not typically the design engineer's responsibility to conduct customer surveys, it is the design engineer's job to both understand what has historically been produced for the particular task at hand and to be able to identify that information which is needed or is in need of clarification for the development to be able to better meet the customer's needs.

Once an acceptable set of requirements has evolved, the design engineer can concentrate on making an initial design for the required functions of the system. In order to have a reliable design, it is as important for the designer to know what has not worked, and the reasons why, as it is to know what has worked, and how successfully, in identical or similar applications. As the design engineer begins defining and outlining the functions, the reliability of the design must be a concern co-equal to that of functionality. There are several methods to help the designer achieve good functionality and highly-reliable designs from the start. In fact, many reliability-related actions can and should be taken before a single component is purchased.

One of the easiest reliability actions during early design is to formulate the design using components of known reliability. Such a parts list identifies not only the component but the manufacturer. Many firms have found it necessary to screen components to not only identify bad parts prior to assembly but also to segregate the manufacturers with the fewest rejects [4]. These manufacturers are then put on a "preferred supplier" list which should be established *without regard to component cost* (within reason). Beware of the "lowest bidder" syndrome so favored by some parts buyers. A good design engineer should insist on parts of known quality from established manufacturers. After all, if there is a problem with the design, management will come looking for the designer not the parts buyer.

In addition to designing using parts of known quality, design using computer simulations, such as SPICE, can cut days to months off project time and allow problems to be spotted long before a single component needs to be purchased or a prototype produced. Despite the ready availability of these programs, a surprising number of design organizations fail to use any simulation programs. A recent conversation with a design engineer responsible for a high-voltage, high-current DC-to-AC power supply illustrates the problem. This individual had spent several years in the aerospace industry before moving to the commercial world. The engineer rather proudly related how he had worked for about six weeks to define and resolve the integral equations defining the proposed design, which finally were expressed on a dozen hand-written sheets of paper. During this time, he worked twelve to fourteen hours a day including weekends. After this was accomplished, he still had to find components capable of supporting his design, lay out the schematic, order the parts, and get a prototype built. It took another two weeks of work to specify the magnetic devices used in the design and to find someone willing to accomplish the proto-typing (private discussion, see Note 5). Use of a simulation, even with nonstandard parts, could have more readily realized the design, provided selfdocumenting schematics and parts lists, and provided reasonably-accurate predictions about what the input and output signal characteristics would be in the proposed design. This approach is not only faster but helps avoid costly design errors that, all too often, are not identified until hardware prototyping or integration. Computer simulation of the proposed system can avoid the expensive (in time and materials) "murder-suicide" pacts between units developed in parallel using different designers. We all have experienced these: one unit puts out an unexpected signal into another unit causing catastrophic failure of the second unit (the murder). The first unit, now outputting into a short, or no load, self-destructs because it is not limited against such possibilities (the suicide). Such test-bench or, worse yet, system integration experiences do not need to be part of any design experience and can best be avoided by doing some mathematical simulations which should identify most, if not all, such catastrophic events before the prototype is built.

In addition to the use of computer simulations and a qualified parts list, several other techniques can help the designer create a good circuit design from the very start. The first of these is an anathema to many designers: reuse reliable circuit designs. For a given set of applications, any designer or design organization will probably have some knowledge of prior designs. The "trap" is using a circuit *without* knowledge of how reliable it proved in use. Find what circuits worked, and use them again even for a design that appears different from any that has gone before. Not only is it costly, re-inventing the circuit every time, but it is another source for design error. A "set" of well-designed circuits which have proven reliable, with precisely-defined performance characteristics, should be part of every designer's "kit." This does not mean the designer is being lazy – far from it. Use of "standard" circuits to achieve required functions not only brings the designs in faster with higher probability of initial operating success, it will free the designer to worry about the tricky parts of the requirement where innovation is *really* needed. Many design engineers fail to appreciate that this is being done for them by suppliers – in the form of ever higher levels of integration in integrated circuits. Use of "standard" circuits for recurring functional requirements does more than just help the design reliability – it can help the designer's reputation. Every business treasures a designer with a reputation for speed, few errors, and innovation. These are the designers who bring in projects ahead of schedule and under budget. It should be pointed out, also, that these "standard" circuits can be kept on file as a macro in a simulation program and in a matter of minutes be integrated into a proposed design for rapid circuit simulation. Standard circuits also help in less obvious ways: higher volume of these parts often results in price breaks for the manufacturer. Maintenance procedures are more easily formulated and failure modes are not only able to be well-defined, but failures should become very infrequent in these parts of the circuit. Additionally, Manufacturing is able to refine the fabrication process so that loss and inefficiencies are minimized for such fabrications (no "learning curve," for example).

A second technique is a good circuit *stress analysis*. This simple analysis is frequently overlooked or ignored by designers convinced of the veracity of their equations and simulations to the exclusion of reality. In current failure analysis experience, at least 90% of all observed component failures are due to a poor design (electrical or mechanical) or poor application rather than a defective or poorly-made part (private discussions, see Note 6). A stress analysis is one tool designed to identify circuit conditions which can drive a device into premature failure.

In conjunction with the stress analysis, a good set of derating guidelines should be used by the designer [9]. Many larger businesses have adopted such guidelines and are having good results. The reason for derating is not meant to imply that the various components are incapable of performing as advertised but a realization of the laws of physics. Devices operated at 100% of capacity fail more frequently than those operated at 50% capacity. A very good example of the positive impact on system reliability of an enforced derating scheme is the electronics in the U.S. Navy's F-18 (Hornet) which have a mean time between failures (MTBF) approximately twice that of comparable systems (private discussion, see note 7). The utility of derating is reflected in the impact of stress on reliability calculations in MIL-HDBK-217 as well as various physics of failure approaches. Identification of stress levels allows the designer to either correct the design to prevent unwanted stresses or choose higher-capacity parts so that premature failures are avoided.

A good stress analysis and derating scheme leads into a design reliability analysis. A great deal of discussion has been, and is continuing to be, generated in regard to statistically-driven prediction methodologies such as MIL-HDBK-217 (-217), Bellcore, and PREL. The point to be made is that the chosen methodology must be applied consistently. Many design engineers dislike reliability analysis because they feel such an analysis inaccurately reflects the frequency of operational failures. This is undoubtedly true, just as it is undoubtedly true that such engineers have misunderstood the purpose, intent, and result of a reliability analysis which are:

(1) Feasibility evaluation

(2) Comparison of competing designs

(3) Identification of potential reliability problems

(4) Provide reliability input for other RIM tasks [3].

The purpose of the reliability analysis during initial design is to gain a rough order of magnitude estimate of how often the design is likely to fail and identify the parts of the design most prone to failure. Unfortunately, probability of failure methodologies such as -217 tend to give a number of hours to failure. This is somewhat deleterious to the best understanding of the analysis inasmuch as it is a human vice to seize upon numbers as if they are immutable. A failure rate, for example, of 5287 hours is meaningless in several senses of the term. First, in relation to required operating hours, it could either be very bad or very good. A failure every 5287 hours, or once every 220 days, would be disastrous to a business such as AT&T, which might use thousands of identical units in a major switching exchange. On the other hand, for an airborne radar such as that carried in a tactical fighter, such a failure rate would be considered phenomenally good. Second, the number reflects a probability, usually with greater than 60% confidence, that most of the items will have failed by that number of operational hours. The important use of the number is in contrasting alternative designs and in identifying those functions or portions of the design which are the reliability drivers. That is, those items which are likely to fail the most often (Power MOSFETs, for example). Once these "drivers" are clearly identified, they can usually be addressed and rectified; and this can be done before any hardware is purchased or assembled.

A final analysis which improves design reliability is a sensitivity or tolerance analysis. That is, what are the effects on circuit performance of allowable component tolerance variation? Some designs, or functions, may prove exceptionally sensitive to small device parameter shifts such as resistance, capacitance, leakage, or gain. The adverse impact of circuit variation on productivity and profit has been well illustrated by Taguchi and others. There are several approaches which have been used through the years: worst case, transpose circuit, root-sum-square. The easiest, and certainly the most useful, is a Monte Carlo simulation of the proposed circuit [9,11]. The output results, when tied to the specific, numerically-generated component variations, can quickly identify undesirable parameter values or combinations. As a result, either tighter tolerances may be in order, or some device optimal parameters may be changed to keep all variation within acceptable limits. This is most easily recognized as the Taguchi method but designers of highly-reliable circuits have been doing this for years. A cautionary note on such optimization: the designer must be wary that the circuit is not optimized to the point of losing the ability to perform as intended. Any optimization which comes up with an "ideal" value such as 1.12 volts for a zener diode needs to be junked [10]. It should also be remembered that this optimization must also be done in conjunction with updated Monte Carlo runs, stress analysis, and a well-thought out set of "what if" inputs to the simulated system to postulate different operational scenarios.

One other analysis should be performed once the designer is satisfied with the functionality and reliability of the design and that is a sneak circuit analysis. A "sneak circuit" is defined as an unexpected electrical pathway that results in an unusual or unintended operation by a system. As electronic designs become increasingly complex, the probabilities of unintended operations increases as well. Software-controlled operations are very prone to these problems. Less frequently, similar unexpected events can take place in hardware that normally performs exactly as expected. The U.S. armed forces have had some bitter experiences with unintended operations due to erroneous equipment settings and have invested a lot of time in designing analysis techniques to identify and allow correction of such sneak paths before equipment is fielded. Given the increasingly litigious climate in the United States, it is prudent for the design engineer to perform a good sneak circuit analysis before releasing new designs [7,8].

At the end of the foregoing analyses, the design engineer should have a very reliable design before any components need to be ordered or assembly needs to be done. Since we now have a reliable design which meets all functional requirements, it is ready to go into prototyping, right? Not if we want a reliable product to be produced the first time into manufacturing.

The design engineer now needs to coordinate the chosen design with several other engineering activities (in addition to getting help from the quality and reliability sections of the organization), foremost of which are the packaging and manufacturing engineers. Packaging, in electronics, refers not only to the component package, but also to the design and layout of the printed circuit boards and larger assemblies. Especially in the case of surface mount technology, the layout of the components forms a critical operation in determining module/system reliability. Three different types of simulations should be performed in support of any proposed package design in order to create a reliable product: Finite Element, Thermal, and EMI/RFI.

A finite element analysis is usually an effective means to identify components or solder joints subject to damage due to mismatches in the coefficient of thermal expansion (CTE) of the components and the substrate before time and resources are wasted on a poor layout or component-substrate matchup [12]. All three axes of movement should be modeled, and should include ambient temperature extremes and thermal shock [1]. Vibration may also be simulated, again to identify potential problems. Several "runs" may have to be made to select the most reliable layout from a physical reliability standpoint. This layout must then be coordinated with the manufacturing engineer to make sure the selected layout is practical and economical from a manufacturability standpoint.

Concurrent with each proposed layout a thermal prediction should be made to identify "hot spots" on the design. These, of course, should be minimized and, if possible, dispersed evenly across the design layout so that heat is generated as uniformly as possible. This is a good time to model possible heat sinks, identify a different substrate, and perform "what if" situations. It should help the designer to identify requirements for forced air, or to redesign the circuit to create less heat for the same functionality (see Note 8). In any case, an extensive simulation of the proposed layout will allow the engineer to determine the most satisfactory design within the physical constraints of the intended system.

Similarly, the wire traces of the proposed circuit can be modeled as transmission lines using SPICE. This should give the designer a gross indication of potential EMI/crosstalk problems and potential RFI sensitivities of the design.

As a result of the above simulations and coordination, changes to the design, layout, and planned manufacturing processes may be made in an iterative fashion until the predicted reliability is as high as possible within the functional requirements of the design. It should be emphasized that, to this point, not a single physical component has been acquired, nor has capital been required to establish inventory or build prototypes, yet the designer can now initiate the prototype phase with high assurance that the design should meet or exceed the reliability and robustness anticipations of the customer. In addition, the wasteful test-and-fix approach is largely avoided and, because a good sensitivity analysis was performed and component values "centered," there should be very little manufacturing waste. In addition, since the performance and reliability is present from the start of the prototyping, the time to market is likely to be less than with more traditional design practices.

14.4 Conclusion

Global competition for market share is placing increased emphasis on product quality and reliability. In order to retain a company's market share, the product must not only meet customer expectations in regard to performance, it needs to exceed those expectations in relation to reliability. The increasingly-litigious climate in the United States makes highly-reliable products essential. Major cost savings and product price advantages accrue from designing products from inception to be very reliable and tolerant of manufacturing variation. Also, there are many computer-supported simulation programs available today to assist the designer. There is an increasingly-large and understandable body of reliability theory and techniques available to assist the designer in the concept and realization of highly-reliable circuits. Early reliability-related analysis and simulations allow the optimal design in terms of functionality, reliability, and manufacturability to be accomplished prior to initial proof-of-concept prototyping. In fact, early, intensive use of available simulations for analysis and improvement should radically reduce design and manufacturing error and, when properly used, significantly reduce rework, scrap, inventory, and warranty costs. The result will be larger market share for the corporation and success for the design engineers.

Notes

1. Numerous private discussions between author and Quality Assurance engineers and managers from Ford, General Motors, Texas Instruments, AT&T, Seagate, Eaton Corporation, and United Design. In particular, Herb Swanson, formerly with Dana Corporation, and George Muzar, formerly with GM, (now private consultants) have provided special insights into this problem.

2. Private conversations with Mr. Joseph Hughlett, Automotive Quality, Ford Motor Company regarding industry-wide warranty data.

3. Oil contamination and engine RPM are, as a result, recorded in non-volatile memory in the engine control computer of most cars today as a result of litigation holding the automobile manufacturer *a priori* at fault for an engine problem within the warranty period.

4. There is nothing magical or unique about these techniques. They have been practiced, to one degree or another, by the military aerospace industry for some time. However, with the exception of those military manufacturers who also do commercial work, most of these techniques have not been used in the commercial world. Moreover, the defense industry has generally been characterized by ineconomies of scale, slow development, and an absence of warranty and litigation pressure. Commercial firms have seen the military requirements as expensive, difficult, time-consuming, and preventing rapid innovation. It is the thesis of this paper that improvements in reliability, when coupled to economics of scale, will result in higher profits and larger market share, and that these historic criticisms are largely invalid.

5. Private communication between the design engineer and author, April 1992.

6. Private communication with many leading reliability engineers and physicists to the author. These include: Mr. Howard Dicken, DM Data, Scottsdale, Arizona; Dr. James Fordemwalt, Arizona State University, Tempe, Arizona; Mr. John Devaney, Hi-Rel Labs, Spokane, Washington, Dr. Ed Hakim, U.S. Army LABCOM; Mr. Charles Leonard, Boeing Aircraft Company; Ms. Barbara Chotras, Intel Corporation Military Products; Dr. Archie Brainard, SCI Corporation, Santa Ana, California; Mr. Charles Murphy, Texas Instruments, Mr. Steve Marting, QA Manager, T.I. Military Products Division, Midland, Texas; Mr. Robert Howell, Motorola Manager, Military Productions Division, Phoenix, Arizona.

7. Private communication with Walt Willoughby, Jr. regarding U.S. Navy derating standards and philosophy. See also AFMC Pamphlet 800-27, "Part Derating Guidelines," available from U.S. Government Printing Office (GPO), Washington, D.C.

8. There are several reasonably-good thermal modeling simulation programs available, such as NISA II from the Engineering Mechanics Research Corporation, SAUNA from Thermal Solutions, Ann Arbor; and the Circuit Board Thermometer, Lakeview Software Corporation. It is not the purpose of this paper to endorse any particular simulation product.

References

1. Bivens, G.A. (1989), *Reliability Assessment Using Finite-Element Techniques*, RADC-TR-89-281.

2. Deming, W.E. (1986), *Out of the Crisis*, Point Five of the Fourteen Points, Cambridge, MA: Massachusetts Institute for Technology, Center for Advanced Engineering Study.

3. Denson, W., Chandler, G., Crowell, W., and Wanner, R. (1991), *Non-Electronic Parts Reliability Data*, Rome, New York: Reliability Analysis Center NPRD-91.

4. Dicken, H. (Ed.) (1990), Plastic package screening, *Semiconductor Reliability News*.

5. Ireson, W.G. and Coombs, Jr., C.F. (1988), *Handbook of Reliability Engineering and Management*, Second edition, McGraw-Hill, New York.

6. MIL-HDBK-217 is published by the Rome Air Development Center, Griffiss AFB, New York and is available from the GPO. The Bellcore reliability prediction program is formally called the ARPP and is available from Bellcore Technology Licensing, Livingston, New Jersey. PREL is a product of the Society for Automotive Engineers (SAE), Detroit, Michigan.

7. Miller, J. (1989), *Sneak Circuit Analysis for the Common Man*, RADC-TR-89-223.

8. Miller, J. (1990), *Integration of Sneak Analysis with Design*, RADC-TR-90-109.

9. O'Connor, P.D.T. (1991), *Practical Reliability Engineering*, Third edition, John Wiley & Sons, New York.

10. Pease, R. (1993), What's all this Taguchi stuff, anyhow? (Part II), *Electronic Design*, 85-92.

11. Spence, R. and Sain, R.S. (1988), *Tolerance Design of Electronic Circuits*, Addison-Wesley, Reading, MA.

12. Wood, N.O. (1992), Predicting SMT solder joint reliability using calculated temperatures, *Electronic Packaging and Production*, 72-74.

15

Bounds and Limit Theorems for Coherent Reliability

Markos V. Koutras, Stavros G. Papastavridis, and Kyriakos I. Petakos

University of Athens, Greece

ABSTRACT In the present paper some reliability bounds are presented for general coherent structures consisting of independent components. Both non-maintained and independently maintained reliability systems (IMRS) are covered. Making use of these bounds, certain extreme value-type limit theorems are developed for the reliability (unreliability) of large systems consisting of components with independent and identically distributed lifetimes and repair times. Finally, the application of the general results to specific reliability structures is discussed.

Keywords: Coherent structures, Consecutive-k-out-of-n systems, Cut sets, Limit theorems, Maintenance, Path sets, Reliability bounds.

CONTENTS

15.1 Introduction

The reliability evaluation of a general coherent system is of special interest for the design and performance improvement of system's characteristics. Since it is not always feasible to develop simple and computationally efficient algorithms for the calculation of the exact reliability value (see [2]; [6]; [30]) it is reasonable to look instead for good approximations, bounds and limit theorems. During the last few years, this promising research area has attracted special attention, as the continuously expanding literature on these subjects indicates.

The earliest approach to the problem of developing efficient reliability bounds for a general coherent structure may be attributed to Esary and Proschan [15]. Due to their simplicity, these bounds have by now been incorporated into the majority of the contemporary reliability textbooks (see for example [5]; [33] etc). Recently the authors in [18] proposed some additional reliability bounds which combined with the classical Esary and Proschan's bounds yield tight approximating intervals. In [24], the authors, using the Chen-Stein method (see [12]; [1]; [3]) established some Poisson-type approximations.

All the above mentioned results refer to the case of non-maintained reliability structures with independent components. For the maintained case, among the most important general reliability bounds seem to be the ones developed in [16] by extending the bounds used in [15]. Various aspects of independently maintained reliability systems (IMRS) have been investigated in [4], [5] , [7], [8], [9], [10], [13], [14], [21], [22], [25], [27], [31], [32] and [34] . In particular, [4, Theorem 2.7] and [13, Theorem 3.5] provide an important characterization theorem for the distribution of time to first system failure starting with all components new at time zero. For the special case of maintained consecutive k-out-of-n systems, the authors in [28], assuming general lifetime distributions and exponential repair times, proved a limit theorem for the time to first failure of both linear and circular such systems.

The present paper is organized as follows. In Section 15.2 we mention without proofs the main results of [24] and [18] for non-maintained coherent structures. In Section 15.3 we establish some reliability bounds for the maintained case along with certain limit theorems, illustrating the asymptotic behavior of system's lifetime. Finally, in Section 15.4 we show how the general theory can be applied to specific systems yielding new results or easy ways of proving well known results.

15.2   Non-Maintained Coherent Structures

15.2.1   Bounds on System Reliability

It is well known that a coherent reliability system is completely determimed by its minimal cut sets or minimal path sets. Let us consider such a system with independent components, and denote by I the set of all its components, $\boldsymbol{C} = \{C_1, C_2, \ldots, C_N\}$ the set of all minimal cut sets and $\boldsymbol{P} = \{P_1, P_2, \ldots, P_M\}$ the set of all minimal path sets. For a system described through its cut sets (resp. path sets) we shall use the term "the system $(I, \boldsymbol{C})$ (resp. the system $(I, \boldsymbol{P})$)". Finally, for any set A we denote by $|A|$ the cardinality of A.

We define the width $\mu(\boldsymbol{C})$ (resp. length $\mu(\boldsymbol{P})$) of any coherent structure as the minimum number of components in a minimum cut set and path set respectively (see [26], [20]), i.e.,

$$\mu(\boldsymbol{C}) = \min\{C_j|, \ 1 \leq j \leq N\} \, ,$$

$$\mu(\boldsymbol{P}) = \min\{|P_j|, \ 1 \leq j \leq M\} \, .$$

The quantities

$$v(\boldsymbol{C}) = \max_{1 \leq i \leq N} |\{C_j \in \boldsymbol{C} : C_j \cap C_i \neq \emptyset\}| \, ,$$

$$v(\boldsymbol{P}) = \max_{1 \leq i \leq M} |\{P_j \in \boldsymbol{P} : P_j \cap P_i \neq \emptyset\}|$$

will also be used in the sequel.

For $i \in I$ we let $q_i(t)$ be the failure distribution of the i-th component (i.e. the probability that component's lifetime is less than or equal to $t \geq 0$), $p_i(t) = 1 - q_i(t)$ be the reliability of the same component, and introduce the quantities

$$q(t) = \max_{i \in I} q_i(t), \qquad p(t) = \max_{i \in I} p_i(t)$$

corresponding to the reliabilities of the worst and best component (at time t) respectively. Also for $A \subseteq I$ we shall use the notation

$$q_A(t) = \prod_{i \in A} q_i(t), \qquad p_A(t) = \prod_{i \in A} p_i(t) \ .$$

In [24], the authors proved the next two theorems which provide bounds for the reliability and failure probability of a coherent system. The first one is based on the minimal cut sets of the structure, and the second on the minimal path sets. More specifically, we have

THEOREM 15.2.1 *(minimal cut bound)*
If T is the time to failure of the coherent system (I, C) then

$$|P[T > t] - e^{-\lambda(t)}| \leq (1 - e^{-\lambda(t)})\{v(\boldsymbol{C})q^{\mu(\boldsymbol{C})}(t) + (v(\boldsymbol{C}) - 1)q(t)\}$$

where $\lambda(t) = \sum_{j=1}^{N} q_{C_j}(t)$

and its dual counterpart

THEOREM 15.2.2 *(minimal path bound)*
If T is the time to failure of the structure $(I, \boldsymbol{P})$ then

$$|P[T \leq t] - e^{-\lambda(t)}| \leq (1 - e^{-\lambda(t)})\{v(\boldsymbol{P})p^{\mu(\boldsymbol{P})}(t) + (v(\boldsymbol{P}) - 1)p(t)\}$$

where $\lambda(t) = \sum_{j=1}^{M} p_{P_j}(t)$.
   In [18], the authors defined an **L-family** *associated with the coherent structure* $(I, \boldsymbol{C})$ as follows: For $j = 1, 2, \ldots, N$, let L_j be a subset of I such that $L_j \cap C_j = \emptyset$ and $L_j \cap C_i \neq \emptyset$ for all $1 \leq i \leq j - 1$ with $C_i \cap C_j \neq \emptyset$. They conventionally set $L_1 = \emptyset$ and if $C_i \cap C_j = \emptyset$ for all $1 \leq i \leq j - 1$ they defined $L_j = \emptyset$. There is always at least one L-family for any coherent structure. For example, given the family $\boldsymbol{C}$ of cut sets, the sequence of sets defined by

$$L_j = \bigcup_{i=1}^{j-1} (C_i \cap C_j')$$

provides a valid L-family for the respective structure. Actually, this choice results in L_j's with "maximum" cardinality, but as we shall see later on, it is in our advantage to find L_j's as "small" as possible. To this end, we note that it is always possible to choose L_j so that

$$|L_j| \leq |\{i : 1 \leq i \leq j - 1 \text{ and } C_i \cap C_j \neq \emptyset\}| \ .$$

To form such an L_j, just choose one element from all $C_i \cap C_j'$ with $1 \leq i \leq j - 1$ and $C_i \cap C_j \neq \emptyset$. It should be emphasized that the definition of an L-family depends upon the selected ordering $C_1, C_2, \ldots, C_N$ of the cut sets. Changing the ordering will in general alter the feasible L-family choices.
   For every cut set C_j, $j = 1, 2, \ldots, N$ we define a random variable T_{C_j} which takes on the value 0 when all components in the minimal cut set C_j have failed in $[0, t]$ and 1 otherwise. In [18], the authors proved the following theorem

THEOREM 15.2.3

If R is the reliability of a coherent system, then

$$\prod_{j=1}^{N}[1 - P(T_{C_j} = 0)] \leq R \leq \prod_{j=1}^{N}\left[1 - \left(\prod_{i \in L_j} P_i\right) P(T_{C_j} = 0)\right] = UB . \quad (15.2.1)$$

As we see, the "smaller" the L_j's are, the better the upper bound is (at least in the i.i.d. case). So it is a challenge, when applying this theorem to specific systems, to find L_j's with as few elements as possible. Another interesting point is that, since the L-families depend upon the assumed ordering of the cut sets, it is to the benefit of the prospective user of this theorem, to find the "best" ordering.

Extensive numerical calculations indicate that for systems with components of *high reliability*, the discrepancy between LB and UB is quite small; Therefore (LB, UB) provides a satisfactory interval estimation of the exact reliability value.

The next theorem, which is the dual counterpart of Theorem 15.2.3, gives lower and upper reliability bounds based on the minimal path sets of the structure. These bounds turn out to be very efficient for the approximation of *low reliability systems*. Let us first define the **K-family** *associated with the structure* (I, P). For $j = 1, 2, \ldots, M$, let K_j be a subset of I such that $K_j \cap P_j = \emptyset$ and $K_j \cap P_i \neq \emptyset$ for all $1 \leq i \leq j-1$ with $P_i \cap P_j \neq \emptyset$. We conventionally set $K_1 = \emptyset$ and if $P_i \cap P_j = \emptyset$ for all $1 \leq i \leq j-1$ we define $K_j = \emptyset$. For every path set P_j, $j = 1, 2, \ldots, M$ we introduce a random variable T_{P_j}, which takes on the value 1 if no component in the minimal path set P_j has failed in $[0, t]$ and 0 otherwise. In [18], the authors proved that

THEOREM 15.2.4

If R is the reliability of a coherent system, then

$$1 - \prod_{j=1}^{M}\left[1 - \left(\prod_{i \in K_j} q_i\right) P(T_{P_j} = 1)\right] \leq R \leq 1 - \prod_{j=1}^{M}[1 - P(T_{P_j} = 1)] .$$

It is evident that the remarks after Theorem 15.2.3 are valid for Theorem 15.2.4 too (with some obvious modifications).

15.2.2   Limit Theorems

In this section we consider a sequence of coherent systems (I_n, C_n) (or (I_n, P_n) with I_n being the set of components, $C_n = \{C_{n1}, C_{n2}, \ldots, C_{nN_n}\}$ the family of the minimal cut sets, $P_n = \{P_{n1}, P_{n2}, \ldots, P_{nM_n}\}$ the family of the minimal path sets and $L_n = \{L_{ni}, i = 1, 2, \ldots, N_n\}$, $K_n = \{K_{ni}, i = 1, 2, \ldots, M_n\}$, the L- and K-families associated with the coherent structure (I_n, C_n) and (I_n, P_n) respectively. Let $t_n \geq 0$ be a sequence of real numbers. In [18], the authors proved the following theorems for the reliability of the system at t_n.

THEOREM 15.2.5

If all the systems of the sequence (I_n, C_n) are series-parallel and

a. $\displaystyle\lim_{n \to +\infty} \sum_{j=1}^{N_n}\left(\prod_{i \in C_{nj}} q_i(t_n)\right) = \lambda$

b. $\displaystyle\lim_{n \to +\infty} \sup_{1 \leq j \leq N_n} \prod_{i \in C_{nj}} q_i(t_n) = 0$

then the reliability $R(t_n)$ tends to $\exp(-\lambda)$ as, $n \to +\infty$ i.e.

$$\lim_{n \to +\infty} R(t_n) = \exp(-\lambda) .$$

THEOREM 15.2.6
If the systems of the sequence (I_n, C_n) are not series-parallel and

a. $\quad \displaystyle\lim_{n \to +\infty} \sum_{j=1}^{N_n} \left(\prod_{i \in C_{nj}} q_i(t_n) \right) = \lambda$

b. $\quad \displaystyle\lim_{n \to +\infty} \left(\max_{1 \le j \le N_n} |L_{nj}| \cdot \max_{i \in I_n} q_i(t_n) \right) = 0$

then the reliability $R(t_n)$ tends to $\exp(-\lambda)$ as $n \to +\infty$, i.e.

$$\lim_{n \to +\infty} R(t_n) = \exp(-\lambda) .$$

There are analogous results for a sequence of coherent systems (I_n, P_n). More specifically we have

THEOREM 15.2.7
If all the systems of the sequence (I_n, P_n) are parallel-series and

a. $\quad \displaystyle\lim_{n \to +\infty} \sum_{j=1}^{M_n} \left(\prod_{i \in P_{nj}} p_i(t_n) \right) = \lambda$

b. $\quad \displaystyle\lim_{n \to +\infty} \sup_{1 \le j \le M_n} \prod_{i \in P_{nj}} p_i(t_n) = 0$

then

$$\lim_{n \to +\infty} R(t_n) = 1 - \exp(-\lambda) .$$

THEOREM 15.2.8
If the systems of the sequence (I_n, P_n) are not parallel-series and

a. $\quad \displaystyle\lim_{n \to +\infty} \sum_{j=1}^{M_n} \left(\prod_{i \in P_{nj}} p_i(t_n) \right) = \lambda$

b. $\quad \displaystyle\lim_{n \to +\infty} \left(\max_{1 \le j \le M_n} |K_{nj}| \cdot \max_{i \in I_n} p_i(t_n) \right) = 0$

then the reliability $R(t_n)$ tends to $1 - \exp(-\lambda)$ as $n \to +\infty$.

15.2.3 Extreme Value Type Theorems

In this section we consider again a sequence of coherent systems (I_n, C_n) (or (I_n, P_n)). Let T_n be the time to failure of the n-th system.

In the following, some interesting limiting results for the special case of equal component reliabilities are established. More specifically, under the assumption that components' times to failure are independent and identically distributed (i.i.d.) with failure distributions

$$q_{ni} = q_n \text{ for every } i \in I_n, \ n \in N$$

proper normalizing constants $a_n > 0$ (scale) and b_n (translation) are given so that the sequence of random variables $(T_n - b_n)/a_n$ converges to a non-degenerate distribution as $n \to +\infty$.

First we are going to introduce some additional terminology, which comes from the area of extreme value theory and will help us to state the results precisely.

Let F, W be distribution functions and $c_n > 0, d_n$ $n = 1, 2, \ldots$ two sequences of real numbers. We shall say that F belongs to the minimum domain of attraction of W with normalizing constants c_n, d_n if and only if

$$\lim_{n \to +\infty} [1 - F(c_n t + d_n)]^n = 1 - W(t)$$

at each point of continuity of W. In other words, if ξ_n is the minimum of n independent identically distributed observations from a common distribution F, then $(\xi_n - d_n)/c_n$ converges in law to W. The distribution function W is called *extreme value distribution*. There exist only three types of nondegenerate extreme value distributions. These are (see [5] ; [11]; [29]) the Weibull, Frechet and Gumbel distributions with respective distribution functions

$$W_{1,a}(t) = 1 - \exp(-t^a), \ t \geq 0 \qquad \text{(Weibull)}$$

$$W_{2,a}(t) = 1 - \exp(-(-t)^{-a}), \ t \leq 0 \qquad \text{(Frechet)}$$

$$W_3(t) = 1 - \exp(-\exp(t)), \ t \in R . \qquad \text{(Gumbel)}$$

The Frechet extreme value distribution $W_{2,a}$ cannot be derived as a limiting distribution of nonnegative random variables, for example life lengths (see [5]). Thus our interest focuses on the Weibull and Gumbel extreme value distributions $W_{1,a}$ and W_3 respectively. Working in the same fashion for maxima instead of minima, the extreme value distribution of maximum domain of attraction can be similarly introduced. More specifically we shall say that F belongs to the maximum domain of attraction of W^* with normalizing constants $c_n > 0$, d_n if and only if

$$\lim_{n \to +\infty} F^n(c_n t + d_n) = W^*(t)$$

at each point of continuity of W^*. The distribution function W^* is called *extreme value distribution*. The respective extreme value distributions for maxima are

$$W_{1,a}^*(x) = \exp(-(-x)^a), \ x \leq 0 \qquad \text{(Weibull)}$$

$$W_{2,1}^*(x) = \exp(-x^{-a}), \ x \geq 0 \qquad \text{(Frechet)}$$

$$W_3^* = \exp(-(e^{-x})), \ x \in R . \qquad \text{(Gumbel)}$$

The Weibull extreme value distribution $W_{1,a}^*$ cannot be derived as a limiting distribution of nonnegative random variables (see [5]) and therefore our interest focuses on the Frechet and Gumbel extreme value distributions $W_{2,a}^*$ and W_3^* respectively. In [24], the authors proved the following

THEOREM 15.2.9
Let $\mu(C_n) = \mu$ be independent of n and define

$$u_n = |\{C_{ni} \in C_n : |C_{ni}| = \mu\}| .$$

If the following conditions hold

1. $\lim_{n \to +\infty} u_n = +\infty$

2. $|\{C_{ni} \in \boldsymbol{C}_n : |C_{ni}| > \mu\}| = |\boldsymbol{C}_n| - u_n = o(u_n^{1+1/\mu})$

3. $\max_{1 \le i \le N_n} |\{C_{nj} \in \boldsymbol{C}_n : C_{nj} \cap C_{ni} \ne \emptyset\}| = o(u_n^{1/\mu})$

4. *the common lifetime distribution of the components belongs to the minimum domain of attraction of W (Weibull or Gumbel) with corresponding normalizing constants $c_n > 0$ and d_n*

then

$$\lim_{n \to +\infty} P[T_n \le a_n t + b_n] = 1 - \exp\{-[-\log(1 - W(t))]^\mu\}$$

where a_n and b_n are given by

$$a_n = c_{[u_n^{1/\mu}]}, \ b_n = d_{[u_n^{1/\mu}]}$$

([t] denotes the integer part of t).

THEOREM 15.2.10
Let $\mu(\boldsymbol{P}_n) = \mu$ be independent of n and define

$$u_n = |\{P_{ni} \in \boldsymbol{P}_n : |P_{ni}| = \mu\}| \ .$$

If the following conditions hold

1. $\lim_{n \to +\infty} u_n = +\infty$

2. $|\{P_{ni} \in \boldsymbol{P}_n : |P_{ni}| > \mu\}| = |\boldsymbol{P}_n| - u_n = o(u_n^{1+1/\mu})$

3. $\max_{1 \le i \le M_n} |\{P_{nj} \in \boldsymbol{P}_n : P_{nj} \cap P_{ni} \ne \emptyset\}| = o(u_n^{1/\mu})$

4. *the common lifetime distribution of the components belongs to the maximum domain of attraction of W^* (Frechet or Gumbel) with corresponding normalizing constants $c_n > 0$ and d_n,*

then

$$\lim_{n \to +\infty} P[T_n \le a_n t + b_n] = \exp\{-[-\log W^*(t)]^\mu\}$$

where a_n and b_n are given by

$$a_n = c_{[u_n^{1/\mu}]}, \ b_n = d_{[u_n^{1/\mu}]} \ .$$

15.3 Independently Maintained Reliability Systems (IMRS)

15.3.1 Bounds on System Reliability

In practice, maintainance policies are followed to reduce the incidence of system failure or to return a failed system to its operating state. Some work has been done yielding results concerning the availability of components and systems of components and concerning various models in which a system is maintained by replacing failed parts by spares and by repairing failed parts so that they may enter the pool of spares. Renewal theory plays an important role in replacement models. There exist also algorithms for constructing optimal spares allocations subject to a budget constraint. In the direction of bounds, among the most important ones seem to be those developed in [16] by extending Esary and Proschan's bounds [15] (see also [17]).

The model we shall study here is the independently maintained reliability system (IMRS). This system is coherent, components are independent and each one is separately maintained and undergoes a perfect repair every time it goes down. Repair starts immediately after a component's breakdown, and the repair time is independent of the other components. We also assume that components' times to first failure and components' repair times after their first failure are mutually independent. For every $i \in I$, let X_i be the time to first failure of the i-th component, Y_i the repair time after the first failure of the i-th component, q_i and r_i the distribution functions of X_i, Y_i respectively and $h_i = q_i * r_i$ the convolution of the random variables X_i, Y_i . We recall that, given a coherent structure (I, C), there always exists an L-family associated with it.

For every $j = 1, 2, \ldots, N$ we denote by $T_{C_j}^{(m)}$ the time to first failure of the maintained cut set C_j considered as a parallel system. Before stating the main theorem of this paragraph we give a lemma which will be proved useful in the sequel.

LEMMA 15.3.1

a. For every $j = 1, 2, \ldots, N$ we have

$$P(T_{C_j}^{(m)} \leq t) \geq \prod_{i \in C_j} [q_i(t) - h_i(t)] \ .$$

b. For every $i \in I$ and every $t > \alpha_i$ we have

$$h_i(t) \, / \, q_i(t) \leq r_i(t - \alpha_i)$$

*where α_i is the **lower end point** of the support of q_i, i.e.*

$$\alpha_i = \inf\{t : q_i(t) > 0\}$$

PROOF   a. It is clear that

$$P(T_{C_j}^{(m)} \leq t) \geq P(X_i \leq t, X_i + Y_i > t \text{ for all } i \in C_j)$$

and since

$$P[X_i \leq t, X_i + Y_i > t \text{ for all } i \in C_j]$$
$$= \prod_{i \in C_j} [P(X_i \leq t) - P(X_i + Y_i \leq t)] = \prod_{i \in C_j} [q_i(t) - h_i(t)]$$

the result follows immediately.

b. For $t > \alpha_i$ we have

$$h_i(t) \, / \, q_i(t) = P(X_i + Y_i \leq t) \, / \, P(X_i \leq t)$$
$$= P(X_i + Y_i - \alpha_i \leq t - \alpha_i) \, / \, P(X_i - \alpha_i \leq t - \alpha_i)$$

and since

$$P(X_i + Y_i - \alpha_i \leq t - \alpha_i, X_i - \alpha_i \leq t - \alpha_i) = P(X_i + Y_i - \alpha_i \leq t - \alpha_i)$$

we may write

$$h_i(t) \, / \, q_i(t) = P(X_i + Y_i - \alpha_i \leq t - \alpha_i \mid X_i - \alpha_i \leq t - \alpha_i)$$

which is less than

$$P(Y_i \leq t - \alpha_i \mid X_i - \alpha_i \leq t - \alpha_i) = r_i(t - \alpha_i)$$

∎

The next theorem provides cut set-based lower and upper bounds for the reliability of an IMRS. Denote by $R^{(m)}$ and R the reliability of the maintained (IMRS) and the non-maintained coherent system respectively.

THEOREM 15.3.2
For the reliability $R^{(m)}$ of an IMRS, we have that

$$\prod_{j=1}^{N} \left[1 - \prod_{i \in C_j} q_i(t) \right] \leq R^{(m)}(t)$$

(15.3.1)

$$\leq \prod_{j=1}^{N} \left[1 - \left(\prod_{i \in L_j} p_i(t) \right) \left(\prod_{i \in C_j} q_i(t) \right) \left(\prod_{i \in C_j} \{1 - r_i(t - \alpha_i)\} \right) \right]$$

PROOF It is clear that $R(t) \leq R^{(m)}(t)$. According to [15] (see also Theorem 15.2.3)

$$\prod_{j=1}^{N} \left[1 - \prod_{i \in C_j} q_i(t) \right] \leq R(t)$$

and the lower bound given in (15.3.1) is immediately obtained. The derivation of the upper bound is based on expressing the reliability $R^{(m)}$ as the probability of intersection of events. To do so, let us denote by S_j, the event that there exists an instance t_0 in the time interval $[0, t]$ such that all components of the cut set C_j are down at t_0 and by S_j' the complementary event, i.e. for every instance τ in $[0, t]$ there exists at least one component of the cut set C_j that functions at τ. It is clear that $R = P\left(\bigcap_{j=1}^{N} S_j' \right)$ and employing the chain rule we obtain

$$R^{(m)}(t) = [1 - P(S_1)] \prod_{j=2}^{N} \left[1 - P\left(S_j \mid \bigcap_{i=1}^{j-1} S_i' \right) \right] = \prod_{j=1}^{N} (1 - \gamma_j) \qquad (15.3.2)$$

where

$$\gamma_1 = P(S_1)$$

$$\gamma_j = P\left(S_j \mid \bigcap_{i=1}^{j-1} S_i' \right), \ j = 2, \dots, N .$$

Let D_j, $j = 1, 2, \dots, N$ be the event that all the components contained in index set L_j are **continuously up** in the time interval $[0, t]$ (If L_j is empty, D_j is considered as an event with probability 1). If $C_i \cap C_j = \emptyset$ for all i such that $1 \leq i \leq j - 1$, then S_j is independent of $\bigcap_{i=1}^{j-1} S_i'$ and therefore

$$\gamma_j = P(S_j) = P(S_j)P(D_j) .$$

When there is at least one i, such that $1 \leq i \leq j-1$ and $C_i \cap C_j \neq \emptyset$, it is obvious that

$$\gamma_j \geq P\left(S_j \mid D_j \cap \left(\bigcap_{i=1}^{j-1} S_i'\right)\right) P\left(D_j \mid \bigcap_{i=1}^{j-1} S_i'\right) = \beta_j \eta_j \qquad (15.3.3)$$

where

$$\beta_j = P\left(S_j \mid D_j \cap \left(\bigcap_{i=1}^{j-1} S_i'\right)\right) , \quad \eta_j = P\left(D_j \mid \bigcap_{i=1}^{j-1} S_i'\right) .$$

We'll show that

$$\beta_j = P(S_j) , \quad \eta_j = P\left(D_j \mid \bigcap_{i=1}^{j-1} S_i'\right) \geq P(D_j) . \qquad (15.3.4)$$

To prove the first equality, observe that the definition of the L-family implies that $D_j \cap S_i' = D_j$ for all i such that $1 \leq i \leq j-1$ and $C_i \cap C_j \neq \emptyset$. Hence,

$$D_j \cap \left(\bigcap_{i=1}^{j-1} S_i'\right) = D_j \cap \left(\bigcap_{\substack{i=1 \\ C_i \cap C_j = \emptyset}}^{j-1} S_i'\right) = A_j . \qquad (15.3.5)$$

Using the right hand side of (15.3.5) and the definition of the L-family, we conclude that the components contained in C_j are not used in event A_j. Therefore events S_j and A_j are independent and

$$\beta_j = P(S_j \mid A_j) = P(S_j) .$$

For the proof of the second part of (15.3.4), we introduce the index set

$$L_j^0 = \{i : C_i \cap L_j = \emptyset, \ 1 \leq i < j\}$$

and we get

$$P(A_j) = P\left(D_j \cap \left(\bigcap_{i \in L_j^0} S_i'\right)\right)$$

$$= P(D_j) P\left(\bigcap_{i \in L_j^0} S_i'\right) \geq P(D_j) P\left(\bigcap_{i=1}^{j-1} S_i'\right)$$

which can be equivalently written as

$$\eta_j = \frac{P(A_j)}{P\left(\bigcap_{i=1}^{j-1} S_i'\right)} \geq P(D_j) .$$

Combining (15.3.2), (15.3.3) and (15.3.4) and taking into consideration that

$$P(S_j) = P(T_{C_j}^{(m)} \leq t) , \qquad P(D_j) = \prod_{i \in L_j} p_i(t)$$

we obtain

$$R^{(m)}(t) \leq \prod_{j=1}^{N} \left[1 - \left(\prod_{i \in L_j} p_i(t)\right) P(T_{C_j}^{(m)} \leq t)\right] . \qquad (15.3.6)$$

On the other hand, Lemma 15.3.1 a. yields that

$$R^m(t) \leq \prod_{j=1}^{N} \left[1 - \left(\prod_{i \in L_j} p_i(t) \right) \left(\prod_{i \in C_j} (q_i(t) - h_i(t)) \right) \right]$$

and the upper bound in (15.3.1) is easily deduced if we write

$$q_i(t) - h_i(t) = q_i(t)(1 - h_i(t) / q_i(t))$$

and take into account Lemma 15.3.1 b. ∎

In the i.i.d. case, i.e. if

$$q_i(t) = q(t), \; r_i(t) = r(t), \; h_i(t) = h(t)$$

for all $i \in I$, the following corollary results immediately from Theorem 15.3.2.

COROLLARY 15.3.3
For the reliability $R^{(m)}$ of an IMRS with identically distributed components, we have that

$$\prod_{j=1}^{N}[1 - q^{|C_j|}(t)] \leq R^{(m)}(t) \leq \prod_{j=1}^{N}[1 - p^{|L_j|}(t)q^{|C_j|}(t)\{1 - r(t - \alpha)\}^{|C_j|}] \quad (15.3.7)$$

15.3.2 Limit Theorems

The problem addressed in this Section can be summarized in the following general question: given a maintained coherent structure (IMRS) and some procedure to increase the number of its components without bound, what are the proper conditions such that the limiting system reliability is non degenerate? The basic tool for the analysis conducted here is provided by Theorem 15.3.2.

Let I_n be the set of components and $C_n = \{C_{n1}, C_{n2}, \ldots, C_{nNn}\}$ the family of the minimal cut sets of the n-th system. For every $i \in I_n$, let $q_{ni}(t)$ be the distribution function of the time to first failure of the i-th component and $r_{ni}(t)$ the distribution function of the repair time after the first failure of the i-th component. Moreover, let $L_n = \{L_{ni}, 1, 2, \ldots, N_n\}$ be the L-family associated with the coherent structure (I_n, C_n). We set

$$a_j(n) = \prod_{i \in C_{nj}} q_{ni}(t_n) ,$$

$$b_j(n) = \left(\prod_{i \in L_{nj}} p_{ni}(t_n) \right) \left(\prod_{i \in C_{nj}} q_{ni}(t_n) \right) \left(\prod_{i \in C_{nj}} \{1 - r_{ni}(t_n - \alpha_{ni})\} \right) ,$$

$$(15.3.8)$$

$$LB_n = \prod_{j=1}^{N_n}[1 - a_j(n)] , \; UB_n = \prod_{j=1}^{N_n}[1 - b_j(n)] ,$$

$$\ell_n = \max\{|L_{nj}|, \; 1 \leq j \leq N_n\} , \; k_n = \max\{|C_{nj}|, \; 1 \leq j \leq N_n\} .$$

It is obvious that $\ell_n \leq k_n$ for every $n \in N$. Denoting by $R_n^{(m)}(t_n)$ the reliability of the

n-th system at time t_n, we may write in virtue of Theorem 15.2.2

$$LB_n \leq R_n^{(m)}(t_n) \leq UB_n, \quad n = 1, 2, \ldots . \tag{15.3.9}$$

The limiting behavior of $R_n(t_n)$ will now be derived by exploring the convergence of LB_n and UB_n as $n \to +\infty$. First, let us state a Lemma without proof, which is needed in the sequel.

LEMMA 15.3.4
If $a_1, a_2, \ldots, a_m$ is a sequence of real numbers with $0 \leq a_j \leq 1, j = 1, 2, \ldots, m$ then,

$$0 \leq \exp\left(-\sum_{j=1}^{m} a_j\right) - \prod_{j=1}^{m}(1 - a_j) \leq \frac{1}{2} \sum_{j=1}^{m} a_j^2 .$$

PROOF   See [36] or [18].   ∎

A first general result referring to the limiting ($n \to +\infty$) behavior of the system's reliability $R_n^{(m)}(t_n)$ is described in the next theorem.

THEOREM 15.3.5
If the following conditions hold

a. $\displaystyle \lim_{n \to +\infty} \sum_{j=1}^{N_n} a_j(n) = \lambda$

b. $\displaystyle \lim_{n \to +\infty} \sum_{j=1}^{N_n} b_j(n) = \lambda$

c. $\displaystyle \lim_{n \to +\infty} \sum_{j=1}^{N_n} a_j^2(n) = 0,$

then the reliability $R_n^{(m)}(t_n)$ of the system tends to $\exp(-\lambda)$ as $n \to +\infty$.

PROOF   Since

$$0 \leq b_j(n) \leq a_j(n), \qquad j = 1, 2, \ldots, N_n$$

we may write

$$\sum_{j=1}^{N_n} b_j^2(n) \leq \sum_{j=1}^{N_n} a_j^2(n)$$

and condition (c) implies

$$\lim_{n \to +\infty} \sum_{j=1}^{N_n} b_j^2(n) = 0 .$$

Applying Lemma 15.3.4 to both sequences $a_j(n)$, $b_j(n)$ we deduce that

$$\lim_{n \to +\infty} LB_n = \lim_{n \to +\infty} \prod_{j=1}^{N_n}[1 - a_j(n)] = \exp(-\lambda)$$

and

$$\lim_{n \to +\infty} UB_n = \lim_{n \to +\infty} \prod_{j=1}^{N_n}[1 - b_j(n)] = \exp(-\lambda) .$$

The assertion of the theorem follows immediately by making use of (15.3.9). ∎

If $q_{ni}(t) = q(t)$ and $r_{ni}(t) = r(t)$, for every $i \in I_n$, $n \in N$ we have the following corollaries.

COROLLARY 15.3.6
If the systems of the sequence are not series-parallel and

a. $\displaystyle \lim_{n \to +\infty} \sum_{j=1}^{N_n} q^{|C_{nj}|}(t_n) = \lambda$

b. $\displaystyle \lim_{n \to +\infty} \ell_n q(t_n) = 0$

c. $\displaystyle \lim_{n \to +\infty} k_n r(t_n - \alpha) = 0$,

then the reliability $R_n^{(m)}(t_n)$ tends to $\exp(-\lambda)$ as $n \to +\infty$.

PROOF From condition (c), it is clear that

$$[1 - q(t_m)]^{\ell_n}[1 - r(t_n - \alpha)]^{k_n} \sum_{j=1}^{N_n} a_j(n) \le \sum_{j=1}^{N_n} b_j(n) \le \sum_{j=1}^{N_n} a_j(n) . \qquad (15.3.10)$$

Since the systems of the sequence are not series-parallel, we have $\ell_n \ge 1$ and therefore condition (b) implies

$$\lim_{n \to +\infty} q(t_n) = 0 . \qquad (15.3.11)$$

Employing conditions (a), (b), (c), and (15.3.10) we get

$$\lim_{n \to +\infty} \sum_{j=1}^{N_n} b_j(n) = \lambda .$$

Using the fact that

$$\sum_{j=1}^{N_n} a_j^2(n) \le q(t_n) \sum_{j=1}^{N_n} a_j(n) ,$$

condition (a) and (15.3.11) yield

$$\lim_{n \to +\infty} \sum_{j=1}^{N_n} a_j^2(n) = 0.$$

The result follows immediately, since all conditions of Theorem 15.3.5 are fulfiled. ∎

COROLLARY 15.3.7
If the systems of the sequence are not series-parallel and

a. $\displaystyle \lim_{n \to +\infty} \sum_{j=1}^{N_n} q^{|C_{nj}|}(t_n) = \lambda$

b. $\displaystyle \lim_{n \to +\infty} \ell_n q(t_n) = 0$

c. *the sequence $|C_{nj}|$, $1 \le j \le N_n$, $n \in N$ is uniformly bounded from above, i.e. $|C_{nj}| \le k$ for all $1 \le j \le N_n$ and $n \in N$,*

then the reliability $R_n^{(m)}(t_n)$ tends to $exp(-\lambda)$ as $n \to +\infty$.

PROOF From condition (c) it is clear that

$$[1 - q(t_n)]^{\ell_n}[1 - r(t_n - \alpha)]^k \sum_{j=1}^{N_n} a_j(n) \leq \sum_{j=1}^{N_n} b_j(n) \leq \sum_{j=1}^{N_n} a_j(n) . \qquad (15.3.12)$$

Since the systems of the sequence are not series-parallel, we have $\ell_n \geq 1$ and therefore condition (b) implies

$$\lim_{n \to +\infty} q(t_n) = 0 . \qquad (15.3.13)$$

Employing conditions (a), (b) and the continuity of $r(\cdot)$ at 0, inequalities (15.3.12) yield

$$\lim_{n \to +\infty} \sum_{j=1}^{N_n} b_j(n) = \lambda .$$

Using the fact that

$$\sum_{j=1}^{N_n} a_j^2(n) \leq q(t_n) \sum_{j=1}^{N_n} a_j(n)$$

condition (a) and (15.3.13) imply

$$\lim_{n \to +\infty} \sum_{j=1}^{N_n} a_j^2(n) = 0$$

and the result is an immediate consequence of Theorem 15.3.5. ∎

In the special case of a sequence of series-parallel systems we have

COROLLARY 15.3.8
If all the systems of the sequence are series-parallel and

a. $\lim_{n \to +\infty} \sum_{j=1}^{N_n} q^{|C_{nj}|}(t_n) = \lambda$

b. $\lim_{n \to +\infty} q(t_n) = 0$

c. $\lim_{n \to +\infty} k_n r(t_n - a) = 0$

or

c'. *the sequence $|C_{nj}|$, $1 \leq j \leq N_n$, $n \in N$ is uniformly bounded from above,*

then the reliability $R_n^{(m)}(t_n)$ tends to $\exp(-\lambda)$ as $n \to +\infty$.

PROOF From condition (c) or (c'), it is clear that

$$[1 - r(t_n - \alpha)]^{k_n} \sum_{j=1}^{N_n} a_j(n) \leq \sum_{j=1}^{N_n} b_j(n) \leq \sum_{j=1}^{N_n} a_j(n)$$

or

$$[1 - r(t_n - \alpha)]^k \sum_{j=1}^{N_n} a_j(n) \leq \sum_{j=1}^{N_n} b_j(n) \leq \sum_{j=1}^{N_n} a_j(n) .$$

The result is obtained working in a similar fashion as in Corollaries 15.3.6 and 15.3.7. ∎

In the following we establish some interesting limiting results for the i.i.d case. Let $T_n^{(m)}$ be the time to first failure of the n-th system. Assuming that components' times to first failure and components' repair times after their first failure are independent and identically distributed, we are going to seek proper normalizing constants $a_n > 0$ (scale) and b_n (translation) such that the limiting distribution of the sequence of random variables $(T_n^{(m)} - b_n)/a_n$ can be calculated as $n \to +\infty$.

We first state two lemmas without proof, that will be needed in the proof of Lemmas 15.3.12, 15.3.13 and our main limit result (Theorem 15.3.15).

LEMMA 15.3.9
In order that

$$\lim_{n \to +\infty} [1 - F(c_n t + d_n)]^n = 1 - W(t)$$

it is necessary and sufficient that

$$\lim_{n \to +\infty} n F(c_n t + d_n) = -\log(1 - W(t))$$

for all continuity points of W such that

$$1 - W(t) \neq 0 .$$

PROOF   See [5], page 241.   ∎

LEMMA 15.3.10
Let $A \subseteq B \subseteq R$ and F, F_0 real functions defined on B. Let also $\{c_n\}$, $\{d_n\}$ be sequences of real numbers such that

$$\lim_{n \to +\infty} n F(c_n t + d_n) = F_0(t)$$

for all $t \in A$. If μ is any positive number, then

$$\lim_{n \to +\infty} n F^\mu(a_n t + b_n) = F_0^\mu(t)$$

for all $t \in A$, where a_n and b_n are defined by

$$a_n = c_{[n^{1/\mu}]}, \quad b_n = d_{[n^{1/\mu}]} .$$

PROOF   See [24].   ∎

In the next proposition we explore the convergence of the corresponding normalizing constants, when q belongs to the minimum domain of attraction of an extreme value distribution (Weibull or Gumbel).

PROPOSITION 15.3.11
Let q be a distribution function with lower end point

$$\alpha = \inf\{t : q(t) > 0\} .$$

If q belongs to the minimum domain of attraction of Weibull or Gumbel distributions, with normalizing constants $c_n > 0$, d_n then

$$\lim_{n \to +\infty} c_n = 0 \text{ and } \lim_{n \to +\infty} d_n = \alpha .$$

PROOF Employing Lemma 15.3.9 and using the fact that F belongs to the minimum domain of attraction of W, we deduce

$$\lim_{n \to +\infty} nF(c_n t + d_n) = -\log(1 - W(t))$$

for all $t > 0$, which yields

$$\lim_{n \to +\infty} F(c_n t + d_n) = 0 \text{ for all } t > 0 .$$

Then $c_n t + d_n \to \alpha$ as $n \to +\infty$ for every t positive. In order to prove that, let's assume that there exists a positive real number t' such that the sequence $(c_n t' + d_n)$, $n \in N$ does not converge to α or equivalently the sequence $c_n t' + d_n - \alpha$ does not converge to 0. Then there exists a subsequence n_k such that

$$\lim_{k \to +\infty} (c_{n_k} t' + d_{n_k} - \alpha) = d$$

with

$$d \in \overline{R} = R \cup \{-\infty, +\infty\} \text{ and } d \neq 0 .$$

If $d \in R$, $d > 0$, the limiting expression

$$\lim_{k \to +\infty} (c_{n_k} t' + d_{n_k}) = d + \alpha$$

guaranties the existence of a $k_0 \in N$ such that

$$d + \alpha - d/2 \leq c_{n_k} t' + d_{n_k} \leq d + \alpha + d/2$$

for every $k \geq k_0$. This implies

$$F(c_{n_k} t' + d_{n_k}) \geq F(d/2 + \alpha) > 0$$

which contradicts the fact that

$$\lim_{k \to +\infty} F(c_{n_k} t' + d_{n_k}) = 0 .$$

If $d = +\infty$, then

$$\lim_{k \to +\infty} (c_{n_k} t' + d_{n_k}) = +\infty$$

and hence

$$\lim_{k \to +\infty} F(c_{n_k} t' + d_{n_k}) = 1$$

which is inept since

$$\lim_{k \to +\infty} F(c_{n_k} t' + d_{n_k}) = 0 .$$

If $d \in R$, $d < 0$, the relation

$$\lim_{k \to +\infty} (c_{n_k} t' + d_{n_k}) = d + \alpha$$

implies that there exists a $k_0' \in N$ such that

$$d + \alpha + d/2 \leq c_{n_k} t' + d_{n_k} \leq d + \alpha - d/2$$

for every $k \geq k_0'$. Then

$$F(c_{n_k} t' + d_{n_k}) \leq F(\alpha + d/2) = 0$$

which yields

$$F(c_{n_k} t' + d_{n_k}) = 0 \text{ for all } k \geq k_0' .$$

Hence from Lemma 15.3.9 and the above relation, we have

$$\lim_{k \to +\infty} n_k F(c_{n_k} t' + d_{n_k}) = 0 = -\log(1 - W(t)) .$$

If F belongs to the minimum domain of attraction of Gumbel, there exists no t such that $-\log(1 - W(t)) = 0$, whereas if F belongs to the minimum domain of attraction of Weibull, the only t's which satisfy the above relation are all the nonpositive ones. This contradicts our initial assumption $t' > 0$.

Finally if $d = -\infty$, then there exists $k_0 \in N$ such that for every $k \geq k_0$,

$$c_{n_k} t' + d_{n_k} - \alpha < -2(\alpha + 1)$$

implying

$$c_{n_k} t' + d_{n_k} < -\alpha - 2$$

and

$$F(c_{n_k} t' + d_{n_k}) = 0 \text{ for all } k \geq k_0 .$$

A contradiction is now easily verified in a similar fashion to the previous case. Since $c_n t' + d_n \to \alpha$ for every positive t (as $n \to +\infty$), it is evident that

$$\lim_{n \to +\infty} c_n = 0 \text{ and } \lim_{n \to +\infty} d_n = \alpha .$$

∎

The following lemma provides the limiting distribution of the reliability of an IMRS provided that the components' time to first failure belongs to the minimum domain of attraction of a certain distribution.

LEMMA 15.3.12
Assume that the quantity $\mu = \min\{|C_{nj}|, \ 1 \leq j \leq N_n\}$ does not depend on n and define

$$u_n = |\{C_{nj} \in C_n : |C_{nj}| = \mu\}| .$$

If the following conditions hold

a. $\lim_{n \to +\infty} u_n = +\infty$

b. $|\{C_{nj} \in C_n : |C_{nj}| > \mu\}| = o(u_n^{1+1/\mu})$

c. $\ell_n = o(u_n^{1/\mu})$

d. *the common distribution function $q(t)$ of the time to first failure of the components belongs to the minimum domain of attraction of W with corresponding normalizing constants $c_n > 0$ and d_n,*

then

$$\lim_{n \to +\infty} P[T_n^{(m)} \leq a_n t + b_n] = 1 - \exp\{-[-\log(1 - W(t))]^\mu\}$$

where a_n and b_n are given by

$$a_n = c_{[u_n^{1/\mu}]}, \qquad b_n = d_{[u_n^{1/\mu}]} .$$

PROOF   Considering the system in the time interval , $[0, t_n]$ with $t_n = a_n t + b_n$, we may write

$$\sum_{j=1}^{N_n} a_j(n) = \sum_{j=1}^{N_n} q^{|C_{nj}|}(t_n) = \sum_{\substack{j=1 \\ |C_{nj}|=\mu}}^{N_n} q^{|C_{nj}|}(t_n) + \sum_{\substack{j=1 \\ |C_{nj}|>\mu}}^{N_n} q^{|C_{nj}|}(t_n) \ .$$

The first summand of $\sum_{j=1}^{N_n} a_j(n)$ equals $u_n q^\mu(t_n)$ which in view of Lemmas 15.3.9 and 15.3.10, converges to

$$[-\log(1 - W(t))]^\mu$$

while for the second one it is not difficult to verify that it is bounded above by

$$o(u_n^{1+1/\mu})q^{\mu+1}(t_n) = [u_n q^\mu(t_n)]^{1+1/\mu} o(1)$$

which converges to 0 as $n \to +\infty$. It is now evident that

$$\lim_{n \to +\infty} \sum_{j=1}^{N_n} a_j(n) = [-\log(1 - W(t))]^\mu \ .$$

We may also write

$$\sum_{j=1}^{N_n} b_j(n) = \sum_{j=1}^{N_n} p^{|L_{nj}|}(t_n) q^{|C_{nj}|}(t_n)\{1 - r(t_n - \alpha)\}^{|C_{nj}|}$$

$$= \sum_{\substack{j=1 \\ |C_{nj}|=\mu}}^{N_n} b_j(n) + \sum_{\substack{j=1 \\ |C_{nj}|>\mu}}^{N_n} b_j(n) \ .$$

For the first summand of $\sum_{j=1}^{N_n} b_j(n)$, it is easy to verify that it is bounded above and below as follows

$$p^{\ell_n}(t_n) u_n q^\mu(t_n)\{1 - r(t_n - \alpha)\}^\mu$$

(15.3.14)

$$\leq \sum_{\substack{j=1 \\ |C_{nj}|=\mu}}^{N_n} b_j(n) \leq u_n q^\mu(t_n)\{1 - r(t_n - \alpha)\}^\mu \ .$$

Condition (c) implies that

$$\ell_n q(t_n) = o(u_n^{\frac{1}{\mu}}) q(t_n) = [u_n q^\mu(t_n)]^{1/\mu} o(1) \to 0 \text{ as } n \to +\infty \ .$$ (15.3.15)

Combining (15.3.15), Lemmas 15.3.9, 15.3.10 and the continuity of $r(\cdot)$ at 0, we deduce that both bounds in (15.3.14) converge to

$$[-\log(1 - W(t))]^\mu \ .$$

It is now clear that

$$\lim_{n \to +\infty} \sum_{j=1}^{N_n} b_j(n) = [-\log(1 - W(t))]^\mu \ .$$

For the sum

$$\sum_{j=1}^{N_n} a_j^2(n) = \sum_{j=1}^{N_n} q^{2|C_{nj}|}(t_n)$$

it is not difficult to verify that it is bounded above by

$$q(t_n) \sum_{j=1}^{N_n} a_j(n) .$$

Moreover, using Lemmas 15.3.9 and 15.3.10 we get

$$\lim_{n \to +\infty} q(t_n) = 0$$

which yields

$$\lim_{n \to +\infty} \sum_{j=1}^{N_n} a_j^2(n) = 0 .$$

We can now see that all conditions of Theorem 15.3.5 are fulfiled and therefore the assertion of the lemma is evident. ∎

The following lemma provides the limiting distribution of the reliability of an IMRS provided that the systems of the sequence (I_n, C_n) are series-parallel and every parallel subsystem has μ components eventually for all $n \in N$.

LEMMA 15.3.13
Assume that the quantity $\mu = \min\{|C_{nj}|, \ 1 \le j \le N_n\}$ does not depend on n and define

$$u_n = |\{C_{nj} \in C_n : |C_{nj}| = \mu\}| .$$

Let the systems of the sequence (I_n, C_n) be series-parallel and every parallel subsystem have μ components eventually for all $n \in N$. If the following conditions hold

a. $\lim_{n \to +\infty} |I_n| = +\infty$

b. $|\{C_{nj} \in C_n : |C_{nj}| = \mu\}| = o(u_n^{1+1/\mu})$

c. $\ell_n = o(u_n^{1/\mu})$

d.   *the common distribution function $q(t)$ of the time to first failure of the components belongs to the minimum domain of attraction of W with corresponding normalizing constants $c_n > 0$ and d_n,*

then

$$\lim_{n \to +\infty} P[T_n^{(m)} \le a_n t + b_n] = 1 - \exp\{-[-\log(1 - W(t))]^\mu\}$$

where a_n and b_n are given by

$$a_n = c_{[u_n^{1/\mu}]}, \ b_n = d_{[u_n^{1/\mu}]} .$$

PROOF From the assumptions of the lemma, it is clear that $u_n = N_n$ eventually for all $n \in N$ where N_n is the cardinality of the cut sets of the n-th system. Condition (a) implies that

$$\lim_{n \to +\infty} \mu N_n = +\infty$$

yielding

$$\lim_{n \to +\infty} N_n = +\infty$$

and therefore

$$\lim_{n \to +\infty} u_n = +\infty .$$

We can see now that all conditions of Lemma 15.3.12 are fulfiled and the assertion of the lemma is immediate. ∎

We present now a lemma concerning the convergence of a sequence of nonnegative integers.

LEMMA 15.3.14
If a sequence (u_n), $n \in N$ of nonnegative integers does not converge to $+\infty$, then it contains a constant subsequence.

PROOF Since u_n does not converge to $+\infty$, there exists an $\varepsilon > 0$ such that for every $n \in N$ there is an $n' \geq n$ such that $u_{n'} \leq \varepsilon$. In this way, we can construct a subsequence (u_{n_k}), $k \in N$ of (u_n), $n \in N$ such that

$$u_{n_k} \leq \varepsilon \text{ for all } k \in N .$$

But (u_{n_k}), $k \in N$ is a sequence of nonnegative integers and due to the above relation, it contains a constant subsequence. ∎

The following theorem provides our main limiting result for the reliability of an IMRS. More specifically we have

THEOREM 15.3.15
Assume that the quantity $\mu = \min\{|C_{nj}|, 1 \leq j \leq N_n\}$ does not depend on n. If the following conditions hold

a. $\lim_{n \to +\infty} |I_n| = +\infty$

b.   *for infinitely many n either $\ell_n \geq 1$ or $\{C_{nj} \in C_n : |C_{nj}| > \mu\} \neq \emptyset$*

c. $|\{C_{nj} \in C_n : |C_{nj}| > \mu\}| = o(u_n^{1+1/\mu})$

d. $\ell_n = o(u_n^{1/\mu})$

e.   *the common distribution function $q(t)$ of the time to first failure of the components belongs to the minimum domain of attraction of W with corresponding normalizing constants , $c_n > 0$ and d_n,*

then

$$\lim_{n \to +\infty} P[T_n^{(m)} \leq a_n t + b_n] = 1 - \exp\{-[-\log(1 - W(t))]^\mu\}$$

where a_n and b_n are given by

$$a_n = c_{[u_n^{1/\mu}]}, \ b_n = d_{[u_n^{1/\mu}]} .$$

PROOF We shall show here that $\lim_{n \to +\infty} u_n = +\infty$ so that all conditions of Lemma

15.3.12 are fulfiled. In order to prove that, let's assume that $\lim\limits_{n \to +\infty} u_n \neq +\infty$. Define

$$A = \{n \in N : \ell_n = 0 \text{ and all the cut sets of the } n\text{-th system have cardinality } \mu\} \ .$$

According to condition (b), either A contains infinitely many but not eventually all n or it is finite.

Let's first examine the case when A contains infinitely many but not eventually all n. From Lemma 15.3.14, there is a constant subsequence (u_{n_k}), $k \in N$ of (u_n), $n \in N$. Since A contains infinitely many n, we can construct a subsequence $(u_{n_{k_\lambda}})$, $\lambda \in N$ of (u_{n_k}), $k \in N$, such that $u_{n_{k_\lambda}} \in A$ for all $\lambda \in N$. By definition of A and condition (a) we get

$$\lim_{\lambda \to +\infty} u_{n_{k_\lambda}} = \lim_{\lambda \to +\infty} N_{n_{k_\lambda}} = +\infty$$

which contradicts the fact that $(u_{n_{k_\lambda}})$, $\lambda \in N$ is constant.

In the second case, i.e. when A is finite, there is again from Lemma 15.3.14 a constant subsequence (u_{n_k}), $k \in N$ of (u_n), $n \in N$. From condition (b), we can construct a subsequence $(u_{n_{k_\lambda}})$, $\lambda \in N$ of the sequence (u_{n_k}), $k \in N$ such that $n_{k_\lambda} \in B = N - A$ (the complement of A with respect to N) for all $\lambda \in N$, i.e., for every $\lambda \in N$, $\ell_{n_{k_\lambda}} \geq 1$ or the set $\{C_{n_{k_\lambda}} \in C_{n_{k_\lambda}} : |C_{n_{k_\lambda}}| > \mu\} \neq \emptyset$. Therefore

$$\max\{\ell_{n_{k_\lambda}}, |\{C_{n_{k_\lambda}j} \in C_{n_{k_\lambda}} : |C_{n_{k_\lambda}j}| > \mu\}|\} \geq 1$$

for all $\lambda \in N$, and the combination of conditions (c) and (d) yields

$$\max\{\ell_{n_{k_\lambda}}, |\{C_{n_{k_\lambda}j} \in C_{n_{k_\lambda}} : |C_{n_{k_\lambda}j}| > \mu\}|\} = o(u^2_{n_{k_\lambda}}) \ .$$

Hence

$$\lim_{\lambda \to +\infty} u^2_{n_{k_\lambda}} = +\infty$$

and

$$\lim_{\lambda \to +\infty} u_{n_{k_\lambda}} = +\infty \ .$$

This is inept since $(u_{n_{k_\lambda}})$, $\lambda \in N$ is constant. $\blacksquare$

The above limit theorem (under a little bit more restrictive conditions) was proved in [24] for the non-maintained case. So the various concrete limit theorems, which appear therein and in [18], have been carried in the maintained case too.

15.4 Applications

In this section we are going to examine some interesting special cases of IMRS systems where the results of the previous sections are applicable.

15.4.1 Consecutive-k-out-of-n: F System

The consecutive-k-out-of-n systems were introduced in [23] . They are of special interest in the study of telecommunication and pipeline networks, vacuum systems in accelerators, computer network structures, space relay stations and several other fields such as quality control and inspection procedures, statistical run tests etc. Such a system fails in the time interval $[0, t]$ if and only if there is an instance t_0 in $[0, t]$ such that at least k consecutive components are down at t_0. In this case, it is easy to verify that

$$C_j = \{j, j+1, \ldots, j+k-1\}, \ 1 \le j \le N$$

where $N = n - k + 1$ and choosing

$$L_j = \{j-1\}, \ 2 \le j \le N ,$$

the validity of the definition of the L-family associated with the coherent structure (I, C) is straightforward. The resulting lower and upper bounds according to Theorem 15.3.2 are

$$LB = \prod_{j=1}^{N} \left[1 - \prod_{i=j}^{j+k-1} q_i(t) \right]$$

$$UB = \prod_{j=1}^{N} \left[1 - p_{j-1}(t) \left(\prod_{i=j}^{j+k-1} q_i(t) \right) \left(\prod_{i=j}^{j+k-1} \{1 - r_i(t - \alpha_i)\} \right) \right] .$$

For the application of the limit Lemma 15.3.12, we observe that

$$\mu = k(\text{independent of } n) ,$$

$$r_n = n - k + 1 \to +\infty \text{ as } n \to +\infty ,$$

$$\ell_n \le 2k - 1 = o((n - k + 1)^{1/k}) ,$$

$$|\{C_{nj} \in C_n : |C_{nj}| > k\}| = 0 = o((n - k + 1)^{1/k}) .$$

So, when q belongs to the minimum domain of attraction of an extreme value distribution W (Weibull or Gumbel), Lemma 15.3.12 yields

$$\lim_{n \to +\infty} P[T_n^{(m)} \le a_n t + b_n] = 1 - \exp(-t^{ak}), \qquad t \ge 0 \tag{15.4.1}$$

for the Weibull distribution and

$$\lim_{n \to +\infty} P[T_n^{(m)} \le a_n t + b_n] = 1 - \exp[-\exp(kt)], \qquad t \in R \tag{15.4.2}$$

for the Gumbel distribution. The normalizing constants a_n and b_n are given in both cases by

$$a_n = c_{[(n-k+1)^{1/k}]}, \qquad b_n = d_{[(n-k+1)^{1/k}]} .$$

The special case

$$q(t) = (\lambda t)^a + o(t^a)$$

under the restrictive assumption of exponential repair was settled and studied by [28]. The assumption $q(t) = (\lambda t)^a + o(t)^a$ implies that q belongs to the minimum domain of attraction of Weibull and by choosing $a_n = n^{-1/ak}$, $b_n = 0$ for every $n \in N$, their result can be obtained by relation (15.4.1).

Another system related to the consecutive-k-out-of-n systems is the r-within-consecutive k-out-of-n system. This system was introduced in [19]. It appears in quality control, in a learning process, in acceptance sampling, in therapeutic trials and in target detection systems. Such a system fails in the time interval $[0, t]$ if and only if there is an instance t_0

in $[0,t]$ such that within k consecutive components there are at least r down at t_0. Our results can easily be applied to this model to derive bounds and limit theorems for the reliability of the system. (See also [24] for the non-maintained case).

15.4.2 Two-Dimensional Consecutive-k-out-of-n: F Model

Let us consider now the 2-dimensional consecutive k-out-of-n system, [35]. This model appears in integrated circuit design, pattern detection and medical diagnostics and is defined as follows: components are arranged on a square grid of size n. The system fails in the time interval $[0,t]$ if and only if there exists an instance t_0 in $[0,t]$ such that there is at least one square grid of size k with all its k^2 components down at t_0 $(1 < k < n)$. Arranging the minimal cut sets

$$C_{ij} = \{(\mu,\nu) : i \le \mu \le i+k-1 \, , j \le \nu \le j+k-1\}, \; 1 \le i,j \le n-k+1$$

in the following way

$$C = \{C_{11}, C_{21}, \ldots, C_{n-k+1,1}, \; C_{12}, C_{22}, \ldots, C_{n-k+1,n-k+1}\}$$

we may deduce the following L-family associated with the coherent structure (I, C)

$$L_{ij} = \{(i-1,j),(i+k-1,j-1)\} \, , \; 2 \le i,j \le n-k+1$$
$$L_{1j} = \{(k,j-1)\} \, , \; 2 \le j \le n-k+1$$
$$L_{i1} = \{(i-1,1)\} \, , \; 2 \le i \le n-k+1$$
$$L_{11} = \emptyset \, .$$

Bounds and limit theorems for the reliability of the above systems can be obtained through the general theory developed in Section 15.3.

15.4.3 Consecutive-2 Graphs

A consecutive-2 graph is a graph $G = (V, E)$ with n vertices and r_n edges. Each vertex is associated with a failure distribution $q(t)$ and the graph fails in the time interval $[0,t]$ if and only if there exists an instance t_0 in $[0,t]$ such that any two adjacent vertices are down at t_0. The consecutive 2-out-of-n: F System is an example of a consecutive-2 graph. The minimal cut sets of the structure are of the form $A = \{u, w\} \in E$ with $u, w \in V$ and therefore

$$\mu = 2(\text{independent of } n), \; r_n = |C_n| = N_n$$

$$|\{C_{nj} \in C_n : |C_{nj}| > 2\}| = 0 \, .$$

So, when $N_n \to +\infty$ as $n \to +\infty$, $\ell_n = o(N_n^{1/2})$ and $q(t)$ belongs to the minimum domain of attraction of an extreme value distribution W (Weibull or Gumbel), Lemma 15.3.12 yields

$$\lim_{n \to +\infty} P[T_n^{(m)} \le a_n t + b_n] = 1 - \exp(-t^{2a}), \qquad t \ge 0$$

for the Weibull distribution and

$$\lim_{n \to +\infty} P[T_n^{(m)} \le a_n t + b_n] = 1 - \exp[-\exp(2t)], \qquad t \in R$$

for the Gumbel distribution. The normalizing constants a_n and b_n are given in both cases by

$$a_n = c_{[N_n^{1/2}]}, \qquad b_n = d_{[N_n^{1/2}]} .$$

Acknowledgements

The authors would like to thank Professor L. Kamarinopoulos, for enlightening discussions pertaining to some of the ideas, leading to this paper.

References

1. Arratia, R., Goldstein, L., and Gordon, L., (1989), Two moments suffice for Poisson approximations: The Chen-Stein method, *Annals of Probability*, **17**, 9-25.

2. Ball, M.O. (1980), Complexity of network reliability computations, *Networks*, **10**, 153-165.

3. Barbour, A.D., Holst, L., and Janson, S. (1991), *Poisson Approximation*, Oxford University Press, New York.

4. Barlow, R.E. and Proschan, F. (1976), Theory of maintained systems: Distribution of time to first system failure, *Mathematics of Operations Research*, **1**, 32-42.

5. Barlow, R.E., and Proschan, F. (1981), *Statistical Theory of Reliability and Life Testing*, To Begin With, Silver Spring, MD,

6. Bodin, L. (1970), Approximations to system reliability using a modular decomposition, *Technometrics*, **12**, 335-344.

7. Brown, M. (1975), The first passage time distribution for a parallel exponential system with repair, In *Reliability and Fault Tree Analysis*, (Eds. R.E. Barlow, J.B. Fussel and N.D. Singpurwalla), 365-396, SIAM, Philadelphia.

8. Brown, M. (1983), Approximating IMRL distributions by exponential distributions with applications to first passage times, *Annals of Probability*, **11**, 419-427.

9. Brown, M. (1984), On the reliability of repairable systems, *Operations Research*, **32**, 607-615.

10. Brown, M. and Chaganty, N.R. (1983), On the first passage time distribution for a class of Markov chains, *Annals of Probability*, **11**, 1000-1008.

11. Castillo, E., (1987), *Extreme Value Theory in Engineering*, Academic Press, San Diego.

12. Chen, L.H.Y. (1975), Poisson approximation for dependent trials, *Annals of Probability*, **3**, 534-545.

13. Chiang, D.T. and Niu, S.C. (1980), On the distribution of time to first system failure, *Journal of Applied Probability*, **17**, 481-489.

14. Chiang, D.T. and Niu, S.C. (1984), On the reliability of repairable systems, *Operations Research*, **32**, 607-615.

15. Esary, J.D. and Proschan, F. (1963), Coherent structures of non identical components, *Technometrics*, **5**, 191-209.

16. Esary, J.D. and Proschan, F. (1970), A reliability bound for systems of maintained interdependent components, *Journal of the American Statistical Association, 65*, 329-338.

17. Esary, J.D., Proschan, F., and Walkup, D.W. (1967), Association of random variables with applications, *Annals of Mathematical Statistics*, **38**, 1466-1474.

18. Fu, J.C. and Koutras, M.V. (1994), Reliability bounds for coherent structures with independent components, *Statistics & Probability Letters* (To appear).

19. Griffith, W.S. (1986), On Consecutive k-out-of-n failure systems and their generalizations, In *Reliability and Quality Control*, (Ed., A.P. Basu), Elsevier, Amsterdam.

20. Kaufmann, A., Grouchko, P., and Cruon R., (1977), *Mathematical Models for the Study of the Reliability of Systems*, Academic Press, San Diego.

21. Keilson, J. (1974), Monotonicity and convexity in system survival functions and metabolic disappearance curve, In *Reliability and Biometry*, (Eds., F. Proschan and R. Serfling), 81-98, SIAM, Philadelphia.

22. Keilson, J. (1975), Systems of independent Markov components and their transient behavior, In *Reliability and Fault Tree Analysis*, (Eds., R.E. Barlow, J.B. Fussel and N.D. Singpurwalla), 351-364, SIAM, Philadelphia.

23. Kontoleon, J.M. (1980), Reliability determination of a r-successive-out-of-n: F system, *IEEE Transactions on Reliability*, **29**, 437.

24. Koutras, M.V. and Papastavridis, S.G. (1993), Application of the Chen-Stein method for bounds and limit theorems in the reliability of coherent structures, *Naval Research Logistics Quarterly*, **40**, 617-631.

25. Miller, D.R. (1979), Almost sure comparison of renewal processes and Poisson processes, with application to reliability theory, *Mathematics of Operations Research*, **4**, 406-413.

26. Nagamochi, H., Sun, Z., and Ibaraki, T. (1991), Counting the number of minimum cuts in undirected multigraphs, *IEEE Transactions on Reliability*, **40**, 610-614.

27. Natvig, B. (1980), Improved bounds for the availability and unavailability in a fixed time interval for systems of maintained interdependent components, *Advances in Applied Probability*, **12**, 200-221.

28. Papastavridis, S.G. and Koutras, M.V. (1992), Consecutive-k-out-of-n systems with maintenance, *Annals of the Institute of Statistical Mathematics*, **44**, 605-612.

29. Resnick, S.I. (1987), *Extreme Values, Regular Variation and Point Processes*, Springer-Verlag, New York.

30. Rosenthal, A. (1977), Computing the reliability of complex networks, *SIAM Journal of Applied Mathematics*, **32**, 384-393.

31. Ross, S.M. (1975), On the calculation of asymptotic system reliability characteristic, In *Reliability and Fault Tree Analysis*, (Eds., R.E. Barlow, J.B. Fussel and N.D. Singpurwalla), 331-350, SIAM, Philadelphia.

32. Ross, S.M. (1976), On the time to first failure in multicomponent exponential reliability systems, *Stochastic Processes and Their Applications*, **4**, 167-173.

33. Ross, S.M. (1985), *Introduction to Probability Models*, Third Edition, Academic Press, San Diego.

34. Ross, S.M. and Schechtman, J. (1979), On the first time a separately maintained parallel system has been down for a fixed time, *Naval Research Logistics Quarterly*, **26**, 285-290.

35. Salvia, A.A. and Lasher, W.C. (1990), 2-Dimensional consecutive-k-out-of-n: F models, *IEEE Transactions on Reliability*, **39**, 382-385.

36. Wang, Y.H. (1993), On the number of successes in independent trials, *Statistica Sinica*, **3**, 295-312.

Part II

Reliability Growth

16

A Reliability Growth Model Under Inherent and Assignable-Cause Failures

Ananda Sen and Gouri K. Bhattacharyya

Oakland University
University of Wisconsin

ABSTRACT A reliability growth (RG) model is an essential tool for tracking the improvement of a system at its developmental stage. In many situations, however, the standard models do not adequately represent the failure mechanism in a real operational setting. In the test-fix-retest setting of observing the failure process of a system, we propose a stochastic model of RG from the consideration of inherent and assignable failure causes. Essentially, this model can be viewed as a two-component series system, where one component pertains to reliability growth following an intervention while the life distribution of the other does not change from one stage to another. For the component exhibiting growth in reliability, a parametric step-intensity structure is assumed which provides a stochastic basis for the Duane plot. Exact and large sample properties of the estimates of the model parameters based on maximum likelihood and nonlinear least squares are investigated. Certain irregularities and pathologies shared by the estimates are pointed out. Extensive simulation study has been carried out and the numerical findings are provided as a supplement to the theoretical results.

Keywords : Consistency, Inherent and assignable-cause failures, Maximum likelihood, Nonlinear least squares, Reliability growth

CONTENTS

16.1 Introduction

Modeling reliability growth has received considerable attention in the statistical, engineering and computer software literature over the past three decades. At the initial stage of many production processes involving complex systems, prototypes are put into life test under a development testing program, corrective actions or design changes are made at the occurrences of failures, and the modified system is tested again. As this test-redesign-retest sequence contributes to an improvement in the system performance, failures become increasingly sparse at the later stages of testing making it more difficult to assess the current reliability. Resolution of this problem is possible if a technique can be devised for use of data from early failures and operating experience while taking proper account of improvements resulting from design changes. Since the basic process involved is one of learning through failures, knowledge of a generally applicable "learning curve" would provide a means of measuring and predicting reliability of the system. This knowledge is often conveyed through a reliability growth (RG) model which provides a structure through which the failure data from the current as well as previous stages of testing could be analyzed in an integrated way.

Reliability growth modeling was given a major thrust by the empirical findings in [8] from examination of the failure data of a variety of systems such as complex hydrome-chanical devices, aircraft generators and jet engines in the course of their development. When plotted on a log-log scale, the cumulative number of failures was typically found to produce a linear pattern of relationship with the cumulative operating time. This phe-nomenon (referred to as the "Duane postulate" in the literature) was given a stochastic basis in [7]. The author modeled the underlying failure process as a nonhomogeneous Poisson process (NHPP) with a specific choice of the intensity function conforming to Duane's observation. This model produces several elegant distributional results but does not reflect the physical phenomenon behind the development process in many real op-erational settings. In the situations where failure occurrences and design changes are synchronized, the continually changing intensity function of an NHPP model is in con-flict with the *test-redesign-retest* course of a development program. As the author in [14] pointed out, "··· some provision needs to be present for altering the process of failures when modifications or corrective actions are applied to the system". In [3], the authors described the NHPP model as an "idealization" of a step-intensity model in the sense that the step structure is approximated by a smooth curve.

In [13] the authors avoided this approximation and constructed a version of the step-intensity model that conforms to the "Duane postulate". They assumed a constant in-tensity at each stage of testing and the rate to be decreasing at each failure following the corrective actions (indicating reliability growth). This translates into the assumption that the times between successive failures are independent exponential random variables with means $1/\lambda_i$, $i = 1, 2, \ldots$. As for the pattern of change of the failure rate, the parametric form

$$\lambda_i = (\mu/\delta)i^{1-\delta}, \quad \mu > 0, \quad \delta > 1 \tag{16.1.1}$$

was chosen to bring the model in line with Duane's observation. In [13], the authors developed inference procedures associated with this *Piecewise Exponential* (PEXP) model.

In this article, we explore another formulation of a continuous-time reliability growth model which generalizes the above PEXP to the context of *inherent* and *assignable-cause* failures. The proposed model essentially draws from the idea of the discrete RG model de-scribed in [2]. Their principal modeling assumption involved a classification of the causes of failure in two categories : "assignable" and "non-assignable (or inherent)". While the latter is an inherent feature of the state-of-the-art and hence is not correctable, the for-mer corresponds to the failure modes which can be rectified through the developmental testing program. In [2], the authors assumed that the probability (q_0) of an inherent failure remains constant throughout the developmental program, while the RG feature manifests itself into a decrease of q_i, the probability of the assignable-cause failure, over successive stages $i = 1, 2, \ldots$. Estimation procedures for a Bayesian version of this model were considered in [15]. Other works dealing with reliability models under different classes of failures include [16], [5], [6], [4], and [7]. In the present article, we would incorporate the concept of inherent and assignable-cause failures into building a new RG model in the continuous-time framework.

16.2 Formulation of the Model

To cast our idea in a physical setting, we conceptualize the system, that is undergoing a development testing program, as one consisting of two components in series. We assume that upon failure of the system, component A can be repaired in order to rectify its defect, while component B cannot be fixed and needs to be replaced by an identical component of the same type. Then under an effective repair scheme, component A, whose failure cause is assignable, undergoes reliability growth, while the clock for the failure process of component B is reset to zero. Essentially, these amount to the assumption that the failure process has two ingredients : one has a step intensity that corresponds to reliability growth following an intervention, while the other pertains to reliability changes (or no change) not associated with the intervention. If we continue observing the process until n failures of the system, the data would consist of the n ordered failure times $0 < T_1 < T_2 < \ldots < T_n$. If T_{1i} and T_{2i} denote the lifetimes of A and B, respectively, at the i-th stage of the developmental program, then the system inter-failure times $T_i - T_{i-1}$ are distributed as $min(T_{1i}, T_{2i})$, $i = 1, \ldots, n$. We further make the following distributional assumptions :

1. For all i, T_{1i} and T_{2i} are independent (independence of the components).

2. T_{1i}'s are independent exponential random variables with parameters λ_i (step-changing pattern) while T_{2i}'s are i.i.d. with failure rate $h(t)$.

The successive failures of the system then arise from a composite intensity function

$$\lambda(t) = \lambda_{N(t)} + h(t), \qquad (16.2.1)$$

where N(t) stands for the number of system failures in the time interval [0,t]. Formulation (16.2.1) encompasses a variety of models as special cases. In [1], p. 385, the authors discussed a model where the failure rate is a polynomial in t. In the case of no reliability growth ($\lambda_{N(t)}$ unchanged over the entire observation period), (16.2.1) yields this polynomial hazard function model if $h(t)$ is taken to be a polynomial. The general model also yields the PEXP of [13] as yet another special case when the $h(t)$ term is deleted from (16.2.1) (i.e. when only component A is present), and $\lambda_{N(t)}$ is taken to be of the form (16.1.1). While a PEXP gives rise to flat steps in the intensity function, a suitable choice of the $h(t)$ function (e.g. Weibull intensity $\lambda \delta t^{\delta-1}$) can yield a curved step-intensity model (the curve may account for such factors as wear out or other contributors to failure that are not affected by design changes). Figure 16.1 illustrates the forms of the composite intensity function for two situations: part(a) corresponds to a superposition of two exponentials, and part(b) results from a sum of an exponential and a Weibull intensity function, the jump points denoting the random failure times.

We now consider a special parametric structure of (16.2.1) where the lifetime of component B is assumed to be exponential with mean θ^{-1}, and component A conforms to the PEXP model in (16.1.1). The system inter-failure times are then independent exponential random variables with failure rates λ_i's given by

$$\lambda_i = \frac{\mu}{\delta i^{\delta-1}} + \theta, \quad \mu > 0, \ \delta > 1, \ \theta > 0. \qquad (16.2.2)$$

Note that the restriction on the parameter δ reflects a growth in reliability. In order for (16.2.2) to attain a relationship to the Duane learning curve property, this property must now be interpreted as applying to the assignable-cause failures only. Henceforth, for the simplicity of notation, we work with the parameterization

$$\theta_1 = \mu/\delta, \quad \theta_2 = 1 - \delta, \quad \theta_3 = \theta, \quad \theta_1 > 0, \ \theta_2 < 0, \ \theta_3 > 0$$

so (16.2.2) takes the form

$$\lambda_i = \theta_1 i^{\theta_2} + \theta_3, \quad \theta_1 > 0, \ \theta_2 < 0, \ \theta_3 > 0. \tag{16.2.3}$$

In the rest of this article, we confine attention to a further special case, namely, when the parameters θ_1 and θ_3 are equal. We study in detail two methods of parameter estimation, namely, the maximum likelihood and the least squares. It turns out that, even in this simple and yet physically meaningful model, one encounters certain irregularities and pathologies. Extensive simulation study has been carried out and the numerical findings are provided as a supplement to the theoretical results. The results we obtain for the special case throw some light into the more general case (16.2.2) and give an indication of some of the pathologies one is expected to encounter there.

16.3 Parameter Estimation – Maximum Likelihood and Least Squares

We consider the failure-truncated sampling scheme under which the successive system failure times $0 < T_1 < T_2 < \ldots < T_n$ are observed up to a prespecified number n of failures. The inter-failure times $Y_i = T_i - T_{i-1}, i = 1, \ldots, n$, are independent exponential random variables with means

$$\lambda_i = \left\{\theta_1(i^{\theta_2} + 1)\right\}^{-1}, \quad \theta_1 > 0, \theta_2 < 0.$$

In this section, we study some finite-sample aspects of maximum likelihood and least squares estimation of the model parameters, and provide some numerical results for their comparison.

16.3.1 Maximum Likelihood

The log-likelihood takes the form

$$\log L = n log\theta_1 + \sum_{i=1}^{n} \log(i^{\theta_2} + 1) - \theta_1 \sum_{i=1}^{n} Y_i(i^{\theta_2} + 1),$$

and the first two derivatives of $\log L$ with respect to θ_1 and θ_2 are :

$$l_{1n}(\boldsymbol{\theta}) \equiv \frac{\partial \log L}{\partial \theta_1} = n\theta_1^{-1} - \sum_{i=1}^{n} Y_i(i^{\theta_2} + 1)$$

$$l_{2n}(\boldsymbol{\theta}) \equiv \frac{\partial \log L}{\partial \theta_2} = \sum_{i=1}^{n} \frac{i^{\theta_2} \log i}{i^{\theta_2} + 1} - \theta_1 \sum_{i=1}^{n} Y_i i^{\theta_2} \log i$$

$$a_{11}(\boldsymbol{\theta}) \equiv -\frac{\partial^2 \log L}{\partial \theta_1^2} = n\theta_1^{-2}$$

$$a_{12}(\boldsymbol{\theta}) \equiv -\frac{\partial^2 \log L}{\partial \theta_1 \partial \theta_2} = \sum_{i=1}^{n} Y_i i^{\theta_2} \log i$$

$$a_{22}(\boldsymbol{\theta}) \equiv -\frac{\partial^2 \log L}{\partial \theta_2^2} = \sum_{i=1}^{n} i^{\theta_2} (\log i)^2 [\theta_1 Y_i - (1 + i^{\theta_2})^{-2}]. \tag{16.3.1}$$

We denote

$$l_n(\boldsymbol{\theta}) \equiv (l_{1n}(\boldsymbol{\theta}), l_{2n}(\boldsymbol{\theta}))', \quad A_n(\boldsymbol{\theta}) \equiv (a_{ij}(\boldsymbol{\theta}))_{i,j=1,2}.$$

For solving the likelihood equations $l_n(\boldsymbol{\theta}) = 0$, an iterative procedure such as the Newton-Raphson method can be used. A reduction to a single-variable iteration, however, is readily available upon simplification. Solving $l_{1n}(\boldsymbol{\theta}) = 0$ for θ_1, we get

$$\hat{\theta}_{1n} = n / \sum_{i=1}^{n} Y_i(i^{\theta_2} + 1)$$

which then reduces the second equation $l_{2n}(\boldsymbol{\theta}) = 0$ to

$$\sum_{i=1}^{n} \frac{i^{\theta_2} \log i}{i^{\theta_2} + 1} = n \left(\sum_{i=1}^{n} Y_i(i^{\theta_2} + 1) \right)^{-1} \sum_{i=1}^{n} Y_i i^{\theta_2} \log i \tag{16.3.2}$$

Finite-sample existence of a root of (16.3.2) is not guaranteed. Also, the possibility of multiple roots is not excluded. Furthermore, a root of (16.3.2), even if it exists, may not correspond to a local maximum of the likelihood. Extensive simulation shows that while the frequency of "non-existence" cases diminish with an increase in the sample size, the problem of multiple roots is still prevalent for large sample sizes (e.g. $n = 50$ or 100).

16.3.2   Nonlinear Least Squares

The difficulties and pathologies encountered in the case of maximum likelihood estimation motivate a search for an alternative, simple estimation procedure. Here, we discuss the nonlinear least squares estimation technique as one such alternative, and compare the properties of the resulting estimates with those of the MLE's.

Note that our basic model is in an exponential regression form, where we observe independent exponential random variables Y_i with means $\{\theta_1(i^{\theta_2} + 1)\}^{-1}$. In terms of $L_i \equiv \log Y_i - \Gamma'(1)$, this reduces to the location model

$$L_i = -\log \theta_1 - \log(i^{\theta_2} + 1) + \epsilon_i, \quad i = 1, \ldots, n, \tag{16.3.3}$$

where ϵ_i's are i.i.d. random variables with mean 0, variance $\pi^2/6$, and $\Gamma'(1) = -0.5772$.

The non-linear least squares estimation procedure involves minimization of the function

$$Q_n(\boldsymbol{\theta}) = \sum_{i=1}^{n} \left[L_i + \log \theta_1 + \log(i^{\theta_2} + 1) \right]^2. \tag{16.3.4}$$

Since the nonlinear function $g_i(\boldsymbol{\theta}) = -\log \left(\theta_1(i^{\theta_2} + 1) \right)$ is continuous in $\boldsymbol{\theta}$ for each i, existence of the least squares estimates (LSE) is ensured in the sense described in [12, p. 367]. If we can assume that the least squares estimates are obtained inside the parameter space, then they would satisfy the equations

$$0 = l_{1n}^*(\boldsymbol{\theta}) \equiv \frac{\partial Q_n(\boldsymbol{\theta})}{\partial \theta_1} = 2\theta_1^{-1} \sum_{i=1}^{n} \left[L_i + \log \theta_1 + \log(i^{\theta_2} + 1) \right],$$

$$0 = l_{2n}^*(\boldsymbol{\theta}) \equiv \frac{\partial Q_n(\boldsymbol{\theta})}{\partial \theta_2} = 2 \sum_{i=1}^{n} \frac{i^{\theta_2} \log i}{i^{\theta_2} + 1} \left[L_i + \log \theta_1 + \log(i^{\theta_2} + 1) \right]. \tag{16.3.5}$$

As in the case of ML estimation, we can reduce the problem of solving equations (16.3.5) to a single variable iteration scheme. Denoting $u_i = L_i + \log(i^{\theta_2} + 1)$, and solving $l_{1n}^*(\boldsymbol{\theta}) = 0$ for θ_1 we get

$$\tilde{\theta}_{1n} = \exp(-\bar{u})$$

which reduces the second equation in (16.3.5) to

$$\sum_{i=1}^{n} \frac{i^{\theta_2} \log i}{i^{\theta_2} + 1}(u_i - \bar{u}) = 0. \tag{16.3.6}$$

Like the MLE's, the possibilities of non-existence of a root as well as existence of multiple roots of (16.3.6) cannot be ruled out. From simulation studies we find that like the MLE's, the frequency of "non-existence" cases decreases with an increase in the sample size, but the frequency of occurrence of multiple roots remains fairly large even for large sample sizes such as $n = 50$ or 100.

Maximum likelihood and least squares estimates were generated from simulated data to study and compare their performance in both small and large samples. Three pairs of $(\theta_{10}, \theta_{20})$ values, $(1.0, -0.3)$, $(2.0, -0.5)$ and $(2.0, -2.0)$ were used for the study, and for each case 500 realizations of the MLE's and the LSE's were obtained with the sample sizes $n=10, 25, 50$ and 100. Using one-variable Newton-Raphson technique, the MLE $\hat{\theta}_{2n}$ and the LSE $\tilde{\theta}_{2n}$ were obtained as roots of equations (16.3.2) and (16.3.6), respectively. The corresponding estimates of θ_1 were then calculated using the closed-form expressions

$$\hat{\theta}_{1n} = n / \sum_{i=1}^{n} Y_i(i^{\theta_2} + 1), \quad \tilde{\theta}_{1n} = \exp(-\bar{u}).$$

Table 16.1 gives the estimated bias and the mean-squared error of these estimates. As far as the estimation of θ_1 is concerned, both the estimates $\hat{\theta}_{1n}$ and $\tilde{\theta}_{1n}$ appear to be quite stable. By contrast, both $\hat{\theta}_{2n}$ and $\tilde{\theta}_{2n}$ exhibit substantial variability (measured by the MSE's) especially for small sample sizes. The simulation does not reveal any definite pattern of superiority for one set of estimates over the other. However, the proportion of cases for which the one-variable Newton-Raphson procedure was non-convergent, turned out to be much higher for the MLE's. Overall, this proportion ranged from a low of 19% to a high of 49% for the MLE's in comparison to the corresponding figures of 7% to 22% for the LSE's.

16.4   Large Sample Properties of the Estimates

We now proceed to derive the asymptotic properties of the maximum likelihood and nonlinear least squares estimates of the parameters θ_1, θ_2. Henceforth, we denote the true parameter vector by $\boldsymbol{\theta}_0 = (\theta_{10}, \theta_{20})$. In the sequel, the MLE $\hat{\boldsymbol{\theta}}_n$ will refer to a root of the likelihood equations $l_n(\boldsymbol{\theta}) = 0$.

16.4.1   MLE asymptotics

To study the asymptotic properties of the MLE's, we start with the Taylor expansion

$$l_n(\boldsymbol{\theta}_0) = \boldsymbol{A}_n(\zeta)(\hat{\theta}_{1n} - \theta_{10}, \hat{\theta}_{2n} - \theta_{20})',$$

where ζ is a point on the line segment joining $\hat{\boldsymbol{\theta}}_n$ and $\boldsymbol{\theta}_0$. For a regular model, the principal steps involve showing that

1. Properly scaled $\boldsymbol{l}_n(\boldsymbol{\theta}_0)$ converges in distribution.
2. Properly scaled $\boldsymbol{A}_n(\boldsymbol{\theta}_0)$ converges in probability to a positive definite matrix $\boldsymbol{\Sigma}$.
3. The difference between the elements of $\boldsymbol{A}_n(\boldsymbol{\theta})$ and $\boldsymbol{A}_n(\boldsymbol{\theta}_0)$, once again appropriately scaled, converges to zero in probability, uniformly in a sequence of neighborhoods $M_n(\boldsymbol{\theta}_0)$ of $\boldsymbol{\theta}_0$.
4. A local maximizer of the log-likelihood function exists in the neighborhood $M_n(\boldsymbol{\theta}_0)$ with probability tending to 1 as $n \longrightarrow \infty$.

In the present situation it turns out that the asymptotics need different treatments for the three cases:

$$-1/2 < \theta_{20} < 0$$
$$\theta_{20} = -1/2$$
$$\theta_{20} < -1/2$$

We will see that for $-1/2 \le \theta_{20} < 0$, the maximum likelihood estimates are asymptotically normal, although the rates of convergence as well as the expressions for the asymptotic variance-covariance matrix are different for the cases $-1/2 < \theta_{20} < 0$, and $\theta_{20} = -1/2$. For $\theta_{20} < -1/2$, we show that no maximum likelihood estimate of θ_2 exists. We now present the results for the three cases separately.

CASE I : $-1/2 < \theta_{20} < 0$
Let

$$\boldsymbol{Z}_n^{(1)} = \left(n^{-1/2}l_{1n}(\boldsymbol{\theta}_0),\; n^{-(\theta_{20}+1/2)}(\log n)^{-1}l_{2n}(\boldsymbol{\theta}_0)\right)'$$

$$\boldsymbol{C}_n^{(1)}(\boldsymbol{\theta}) = n^{-1}\begin{bmatrix} a_{11}(\boldsymbol{\theta}) & a_{12}(\boldsymbol{\theta})/\{(\log n)n^{\theta_{20}}\} \\ a_{12}(\boldsymbol{\theta})/\{(\log n)n^{\theta_{20}}\} & a_{22}(\boldsymbol{\theta})/\{(\log n)^2 n^{2\theta_{20}}\} \end{bmatrix}$$

$$\boldsymbol{W}_n = \left(n^{1/2}(\hat{\theta}_{1n} - \theta_{10}),\; (\log n)n^{\theta_{20}+1/2}(\hat{\theta}_{2n} - \theta_{20})\right). \qquad (16.4.1)$$

A Taylor expansion of $\boldsymbol{l}_n(\boldsymbol{\theta}_0) = 0$ around $\hat{\boldsymbol{\theta}}_n$ yields the relation

$$\boldsymbol{Z}_n^{(1)} = \boldsymbol{C}_n^{(1)}(\boldsymbol{\zeta}_n)\boldsymbol{W}_n, \qquad (16.4.2)$$

where $\boldsymbol{\zeta}_n$ is on the line segment between $\hat{\boldsymbol{\theta}}_n$ and $\boldsymbol{\theta}_0$. We further define the neighborhood $M_n^{(1)}(\boldsymbol{\theta}_0)$ of $\boldsymbol{\theta}_0$ by

$$M_n^{(1)}(\boldsymbol{\theta}_0) = \{(\theta_1, \theta_2) : \theta_1 = \theta_{10} + \tau_1 n^{-\gamma}, \theta_2 = \theta_{20} + \tau_2(\log n)^{-1}n^{-(\gamma+\theta_{20})},\; \|\tau\| \le h\},$$

where γ and h are fixed numbers, $-\theta_{20} < \gamma < 1/2$, $0 < h < \gamma + \theta_{20}$. The main steps for deriving asymptotic normality of $\boldsymbol{W}_n$ are provided in Theorem 16.4.2. The proof of Theorem 16.4.2 requires two auxiliary results concerning sequences of real numbers, which we state in a lemma (proof is omitted).

LEMMA 16.4.1
(a) If $\{a_n\}$, $\{b_n\}$ and $\{h_n\}$ are real sequences such that $a_n \ge 0$ for all n, $h_n \uparrow \infty$, $h_n^{-1}\sum_{i=1}^n a_i \longrightarrow a$ and $b_n \longrightarrow b$. Then $h_n^{-1}\sum_{i=1}^n a_i b_i \longrightarrow ab$.

(b) For any nonnegative integer k,

$$\sum_{i=1}^{n}(\log i)^{k}i^{p} = n^{p+1}(\log n)^{k}(p+1)^{-1} + o(n^{p+1}(\log n)^{k}), \quad -1 < p < 0$$

$$= (k+1)^{-1}(\log n)^{k+1} + o(1), \qquad p = -1$$

$$< \infty, \qquad\qquad\qquad\qquad p < -1.$$

THEOREM 16.4.2

(a) Asymptotically, $Z_n^{(1)}$ has a bivariate normal distribution with mean 0 and variance covariance matrix

$$\Sigma = \begin{bmatrix} \theta_{10}^{-2} & \theta_{10}^{-1}(1+\theta_{20})^{-1} \\ \theta_{10}^{-1}(1+\theta_{20})^{-1} & (1+2\theta_{20})^{-1} \end{bmatrix}.$$

(b) $C_n^{(1)}(\theta_0) \xrightarrow{P} \Sigma$.

(c) The difference $\left[C_n^{(1)}(\theta) - C_n^{(1)}(\theta_0)\right]$ tends to zero in probability uniformly in $\theta \in M_n^{(1)}(\theta_0)$.

(d) With probability tending to 1 as $n \longrightarrow \infty$, there exists a sequence of roots $\hat{\theta}_n \in M_n^{(1)}(\theta_0)$ of the likelihood equations, which correspond to a local maximum of the likelihood function.

PROOF   (a) We prove the asymptotic normality of $Z_n^{(1)}$ using the Cramér-Wold device. For arbitrary real numbers $(c_1, c_2) \neq (0, 0)$, we consider

$$c'Z_n^{(1)} = n^{-1/2}\left[c_1 l_{1n}(\theta_0) + c_2(\log n)^{-1}n^{-\theta_{20}}l_{2n}(\theta_0)\right]$$

$$= \sum_{j=1}^{n}\left[n^{-1/2}c_1/\theta_{10} + \{(\log n)^{-1}n^{-(\theta_{20}+1/2)}c_2\}j^{\theta_{20}}\log j/(j^{\theta_{20}}+1)\right]v_j$$

$$\equiv \sum_{j=1}^{n}d_{nj}v_j,$$

where $1 - v_j = \theta_{10}(j^{\theta_{20}} + 1)Y_j$ are i.i.d. standard exponential random variables. Now, $\sum_{j=1}^{n}d_{nj}^2$ tends to $c'\Sigma c$ which is positive, since Σ is positive definite. Also, using the inequality $(a+b)^2 \leq 2(a^2 + b^2)$ for real numbers a, b, we have

$$\max_{1 \leq j \leq n} d_{nj}^2 \leq \frac{2n^{-1}c_1^2}{\theta_{10}^2} + 2(\log n)^{-2}n^{-(2\theta_{20}+1)}c_2^2 \max_{1 \leq j \leq n}(\log j)^2\left(\frac{j^{\theta_{20}}}{j^{\theta_{20}}+1}\right)^2$$

$$\leq \frac{2n^{-1}c_1^2}{\theta_{10}^2} + 2n^{-(2\theta_{20}+1)}c_2^2$$

which converges to zero as $n \longrightarrow \infty$. The required result then follows from the Noether condition : $\max_{1 \leq j \leq n} d_{nj}^2 / \sum_{j=1}^{n}d_{nj}^2 \longrightarrow 0$ as $n \longrightarrow \infty$, for the asymptotic normality of a linear function of i.i.d. random variables.

   (b) The proof rests on showing that $E(c_{ij}^{(1)}) \longrightarrow \sigma_{ij}$ and $Var(c_{ij}^{(1)}) \longrightarrow 0$ for $i, j = 1, 2$. We shall demonstrate these only for $c_{22}^{(1)}$; the proofs for the remaining two elements are similar. We have from (16.3.1) and the definition of $C_n^{(1)}(\theta_0)$,

$$E(c_{22}^{(1)}(\theta_0)) = (\log n)^{-2}n^{-(1+2\theta_{20})}\sum_{i=1}^{n}(\log i)^2\left(\frac{i^{\theta_{20}}}{i^{\theta_{20}}+1}\right)^2,$$

$$Var(c_{22}^{(1)}(\boldsymbol{\theta}_0)) = (\log n)^{-4} n^{-(2+4\theta_{20})} \sum_{i=1}^{n} (\log i)^4 \left(\frac{i^{\theta_{20}}}{i^{\theta_{20}}+1}\right)^2.$$

By part(b) of Lemma 16.4.1,

$$(\log n)^{-k} n^{-(1+2\theta_{20})} \sum_{i=1}^{n} (\log i)^k i^{2\theta_{20}} \longrightarrow (1+2\theta_{20})^{-1}$$

for any nonnegative integer k. Since $(1 + n^{\theta_{20}})^{-1} \longrightarrow 1$, an application of part(a) of Lemma 16.4.1 yields the required convergences of $E(c_{22}^{(1)}(\boldsymbol{\theta}_0))$ and $Var(c_{22}^{(1)}(\boldsymbol{\theta}_0))$.

(c) Once again, we shall show the derivation only for a typical difference

$$T_{1,2} = |\, c_{12}^{(1)}(\boldsymbol{\theta}) - c_{12}^{(1)}(\boldsymbol{\theta}_0)\,|\,.$$

Since $c_{12}^{(1)}(\boldsymbol{\theta}) = n^{-(1+\theta_{20})}(\log n)^{-1} Y_i i^{\theta_2} \log i$, we have

$$E_{\boldsymbol{\theta}_0}(T_{1,2}) \le \theta_{10}^{-1}(\log n)^{-1} n^{-(1+\theta_{20})} \sum_{i=1}^{n} (\log i)\,|\,i^{\theta_2} - i^{\theta_{20}}\,|\,/(1+i^{\theta_{20}}),$$

which is a consequence of the triangle inequality. Using a Taylor expansion of the right hand side w.r.t. θ_2 around θ_{20}, we obtain

$$E_{\boldsymbol{\theta}_0}(T_{1,2}) \le \theta_{10}^{-1}\,|\,\theta_2 - \theta_{20}\,|\,(\log n)^{-1} n^{-(1+\theta_{20})} \sum_{i=1}^{n} (\log i)^2 i^{\zeta^*}/(1+i^{\theta_{20}})$$

$$\le \theta_{10}^{-1} h(\log n)^{-1} n^{-(\gamma+\theta_{20})}(\log n)^{-1} n^{-(1+\theta_{20})}$$

$$\times \sum_{i=1}^{n} (\log i)^2 i^{\theta_{20}+h(\log n)^{-1} n^{-(\gamma+\theta_{20})}}/(1+i^{\theta_{20}}) \qquad (16.4.3)$$

$$\equiv B_n, \qquad (16.4.4)$$

where ζ^* as usual is an intermediate point between θ_2 and θ_{20}, and the inequality in (16.4.3) uses the bound for θ_2 for $\boldsymbol{\theta} \in M_n(\boldsymbol{\theta}_0)$. Note also that the bound B_n in (16.4.4) does not depend on $\boldsymbol{\theta}$. It follows by a combination of parts (a) and (b) of Lemma 16.4.1 that $B_n = O(n^{-(\gamma+\theta_{20})})$. Since $\gamma + \theta_{20} > 0$, it follows that $E_{\boldsymbol{\theta}_0}(T_{1,2}) \longrightarrow 0$ uniformly in $\boldsymbol{\theta} \in M_n(\boldsymbol{\theta}_0)$. Therefore, the uniform probability convergence of $T_{1,2}$ follows by virtue of the Markov inequality.

The proof of part (d) follows along the same lines of the proof of a corresponding result in [13] and is omitted. ∎

In summary, we have; (i) $\boldsymbol{Z}_n^{(1)} \xrightarrow{d} N(\boldsymbol{0}, \boldsymbol{\Sigma})$ where $\boldsymbol{\Sigma}$ is nonsingular,(ii) $\boldsymbol{C}_n^{(1)}(\boldsymbol{\zeta}_n) \xrightarrow{P} \boldsymbol{\Sigma}$ uniformly in an appropriate neighborhood of $\boldsymbol{\theta}_0$, (iii) a consistent sequence of roots (which correspond to a local maximum of the likelihood function) of the likelihood equations exists. In view of the identity (16.4.2), we arrive at the following result.

THEOREM 16.4.3
For a consistent sequence of roots $\hat{\boldsymbol{\theta}}_n$ of the likelihood equations, $\boldsymbol{W}_n$ as defined in (16.4.1) is asymptotically bivariate normal with mean $\boldsymbol{0}$ and variance-covariance matrix

$$\boldsymbol{D}_1^M \equiv \boldsymbol{\Sigma}^{-1}$$

$$= \begin{bmatrix} \theta_{10}^2(1+\theta_{20}^{-1})^2 & -\theta_{10}\theta_{20}^{-2}(1+\theta_{20})(1+2\theta_{20}) \\ -\theta_{10}\theta_{20}^{-2}(1+\theta_{20})(1+2\theta_{20}) & (1+2\theta_{20})(1+\theta_{20}^{-1})^2 \end{bmatrix}. \qquad (16.4.5)$$

Case II : $\theta_{20} = -1/2$

In this case also, we obtain asymptotic normality of a consistent sequence of roots of the likelihood equations. However, the scalings of the components are different. Also, $\hat{\theta}_{1n}$ and $\hat{\theta}_{2n}$ turn out to be asymptotically independent.

THEOREM 16.4.4
*For a consistent sequence of roots $\widehat{\boldsymbol{\theta}}_n$ of the likelihood equations, the vector $\left(n^{1/2}(\hat{\theta}_{1n} - \theta_{10}),\right.$
$\left.(\log n)^{3/2}(\hat{\theta}_{2n} - \theta_{20})\right)'$, where $\theta_{20} = -1/2$, is asymptotically bivariate normal with mean $\boldsymbol{0}$ and variance-covariance matrix*

$$\boldsymbol{D}_2^M = \begin{bmatrix} \theta_{10}^2 & 0 \\ 0 & 3 \end{bmatrix}. \tag{16.4.6}$$

The line of proof is basically the same as in the previous case and we shall only point out the notational differences without going into the details of the steps. We define
$$\boldsymbol{Z}_n^{(2)} = \left(n^{-1/2} l_{1n}(\boldsymbol{\theta}_0), \ (\log n)^{-3/2} l_{2n}(\boldsymbol{\theta}_0)\right)'$$

$$C_n^{(2)}(\boldsymbol{\theta}) = \begin{bmatrix} a_{11}(\boldsymbol{\theta})/n & a_{12}(\boldsymbol{\theta})/\{n^{1/2}(\log n)^{3/2}\} \\ a_{12}(\boldsymbol{\theta})/\{n^{1/2}(\log n)^{3/2}\} & a_{22}(\boldsymbol{\theta})/(\log n)^3 \end{bmatrix}.$$

Then arguments along the lines of Theorem 16.4.2 yield the results

(i) $\boldsymbol{Z}_n^{(2)} \xrightarrow{d} N_2\left(\boldsymbol{0}, (\boldsymbol{D}_2^M)^{-1}\right)$,

(ii) $C_n^{(2)}(\boldsymbol{\theta}_0) \xrightarrow{P} (\boldsymbol{D}_2^M)^{-1}$
which constitute the crucial steps of the proof.

Case III : $\theta_{20} < -1/2$

In sharp contrast with the previous cases, it turns out that, a consistent sequence of roots of the likelihood equations does not exist in the present case. To pinpoint the nature of the difficulty, we first treat a simpler problem in detail. Specifically, we argue that when θ_1 is a known constant, there is no consistent root of the likelihood equation for the one-dimensional problem. We then indicate the line of proof for inconsistency in the situation when both θ_1 and θ_2 are unknown.

The following lemma concerning independent random variables will be used subsequently to prove distributional convergence of certain random quantities. For a proof of the lemma the reader is referred to [9], p. 259.

LEMMA 16.4.5
Let $X_1, X_2, \ldots$ be independent random variables with mean $E(X_i) = 0$ and variance $Var(X_i) = \sigma_i^2$. If $\sigma^2 \equiv \sum_{i=1}^{\infty} \sigma_i^2 < \infty$, then the distribution G_n of the partial sum $\sum_{i=1}^{n} X_i$ tends to a proper distribution G with zero expectation and variance σ^2.

When θ_1 is a known constant (say θ_{10}), the first two derivatives of the log-likelihood w.r.t. θ_2 is given by

$$\bar{l}_n(\theta_2) \equiv \frac{\partial \log L}{\partial \theta_2} = \sum_{i=1}^{n} \frac{i^{\theta_2} \log i}{i^{\theta_2} + 1} - \theta_{10} \sum_{i=1}^{n} Y_i i^{\theta_2} \log i$$

$$\bar{a}_n(\theta_2) \equiv -\frac{\partial^2 \log L}{\partial \theta_2^2} = \sum_{i=1}^{n} i^{\theta_2} (\log i)^2 [\theta_{10} Y_i - (1 + i^{\theta_2})^{-2}]. \tag{16.4.7}$$

Let $\hat{\theta}_{2n}$ denote a solution of the equation $\bar{l}_n(\theta_2) = 0$. A Taylor expansion of $\bar{l}_n(\theta_{20})$ around $\hat{\theta}_{2n}$ yields the relation

$$\bar{l}_n(\theta_{20}) = \bar{a}_n(\zeta_n)(\hat{\theta}_{2n} - \theta_{20}) \tag{16.4.8}$$

where ζ_n is an intermediate point between $\hat{\theta}_{2n}$ and θ_{20}. The following lemma provides the asymptotic properties of $\bar{l}_n$ and $\bar{a}_n$.

LEMMA 16.4.6
(a) $\bar{l}_n(\theta_{20})$ converges in distribution to a random variable with zero expectation and variance $\sum_{i=1}^{\infty}(\log i)^2 i^{2\theta_{20}}/(1 + i^{\theta_{20}})^2$.
(b) $\bar{a}_n(\theta_{20})$ converges in distribution to a random variable with mean $\sum_{i=1}^{\infty}(\log i)^2 i^{2\theta_{20}}/(1 + i^{\theta_{20}})^2$ and variance $\sum_{i=1}^{\infty}(\log i)^4 i^{2\theta_{20}}/(1 + i^{\theta_{20}})^2$.
(c) For any neighborhood $I(\theta_{20}) = (\theta_{20} - h, \theta_{20} + h)$ of θ_{20} such that $\theta_{20} + h < -1/2$, $[\bar{a}_n(\theta_2) - \bar{a}_n(\theta_{20})]$ is bounded in probability uniformly in $\theta_2 \in I(\theta_{20})$.

PROOF (a) Note from the expressions in (16.4.7) that $\bar{l}_n(\theta_{20})$ can be written in the form

$$\bar{l}_n(\theta_{20}) = \sum_{i=1}^{n} d_{ni}^{(1)} v_i, \quad d_{ni}^{(1)} = \frac{(\log i) i^{\theta_{20}}}{1 + i^{\theta_{20}}}, \quad v_i = 1 - \theta_{10}(1 + i^{\theta_{20}})Y_i.$$

Observe that $1 - v_i$ are i.i.d. standard exponential random variables. The stated result for the asymptotic distribution of $\bar{l}_n(\theta_{20})$ is a direct consequence of Lemma 16.4.5 once we note that $\sum_{i=1}^{\infty}(\log i)^2 i^{2\theta_{20}}/(1 + i^{\theta_{20}})^2$ is finite for $\theta_{20} < -1/2$.
  (b) Referring to (16.4.7), we express $\bar{a}_n(\theta_{20})$ as

$$\bar{a}_n(\theta_{20}) = \sum_{i=1}^{n} i^{\theta_{20}}(\log i)^2 [\theta_{10}Y_i - (1 + i^{\theta_{20}})^{-2}]$$

$$= \sum_{i=1}^{n} \frac{i^{\theta_{20}}(\log i)^2}{(1 + i^{\theta_{20}})}[\theta_{10}(1 + i^{\theta_{20}})Y_i - 1] + \sum_{i=1}^{n}(\log i)^2 i^{2\theta_{20}}/(1 + i^{\theta_{20}})^2$$

$$= \sum_{i=1}^{n} d_{ni}^{(2)} w_i + \sum_{i=1}^{n} d_{ni}^{(3)}$$

where $w_i + 1$ are i.i.d. standard exponential random variables. Also,

$$\sum_{i=1}^{\infty}(d_{ni}^{(2)})^2 = \sum_{i=1}^{\infty}(\log i)^4 i^{2\theta_{20}}/(1 + i^{\theta_{20}})^2 < \infty, \quad \sum_{i=1}^{\infty} d_{ni}^{(3)} = \sum_{i=1}^{\infty}(\log i)^2 i^{2\theta_{20}}/(1 + i^{\theta_{20}})^2 < \infty$$

for $\theta_{20} < -1$ and hence Lemma 16.4.5 yields the required result.
  (c) We write

$$\bar{a}_n(\theta) - \bar{a}_n(\theta_{20}) = \sum_{i=1}^{n}(\log i)^2 Y_i \theta_{10}[i^{\theta_2} - i^{\theta_{20}}] + \sum_{i=1}^{n}(\log i)^2 \left[\frac{i^{\theta_{20}}}{(1 + i^{\theta_{20}})^2} - \frac{i^{\theta_2}}{(1 + i^{\theta_2})^2} \right]$$

$$= \sum_{i=1}^{n} \frac{i^{\theta_{20}}(\log i)^2}{(1 + i^{\theta_{20}})}[i^{\theta_2} - i^{\theta_{20}}][\theta_{10}(1 + i^{\theta_{20}})Y_i - 1] + \sum_{i=1}^{n} \frac{i^{\theta_{20}}(\log i)^2}{(1 + i^{\theta_{20}})}[i^{\theta_2} - i^{\theta_{20}}]$$

$$+ \sum_{i=1}^{n}(\log i)^2 \left[\frac{i^{\theta_{20}}}{(1 + i^{\theta_{20}})^2} - \frac{i^{\theta_2}}{(1 + i^{\theta_2})^2} \right]$$

$$\equiv K_1 + K_2 + K_3 \tag{16.4.9}$$

Since $\theta_{10}(1 + i^{\theta_{20}})Y_i$ are i.i.d. standard exponential random variables, $E(K_1) = 0$. Further,

$$Var(K_1) = \sum_{i=1}^{n} \frac{(\log i)^4}{(1 + i^{\theta_{20}})^2} \left(i^{2\theta_2} + i^{2\theta_{20}} + 2i^{\theta_2 + \theta_{20}} \right). \qquad (16.4.10)$$

Note that for $\theta_2 \in I(\theta_{20})$, $\sum_{i=1}^{\infty}(\log i)^4 i^{2\theta_2}/(1 + i^{\theta_{20}})^2$ is finite. Similar consideration shows that the remaining two terms in (16.4.10) also converge to finite numbers, thus establishing that $K_1 = O_p(1)$. As for the non-random terms K_2 and K_3, in (16.4.9), we observe that after some algebraic manipulations the sum $K_2 + K_3$ can equivalently be expressed as

$$K_2 + K_3 = \sum_{i=1}^{n}(\log i)^2 i^{\theta_2} \left[\frac{i^{\theta_2}}{(1 + i^{\theta_2})} - \frac{i^{\theta_{20}}}{(1 + i^{\theta_{20}})} \right]$$

$$+ \sum_{i=1}^{n}(\log i)^2 \left[\frac{i^{2\theta_{20}}}{(1 + i^{\theta_{20}})^2} - \frac{i^{2\theta_2}}{(1 + i^{\theta_2})^2} \right]. \qquad (16.4.11)$$

Arguing as for the $Var(K_1)$ term above, it can be verified that each term in (16.4.11) (in absolute value) converges to a finite number for all $\theta_2 \in I(\theta_{20})$. The required result now follows. ∎

THEOREM 16.4.7
When $\theta_{20} < -1/2$, there does not exist any consistent sequence of roots of the equations $\bar{l}_n(\theta_2) = 0$.

PROOF   We shall prove the theorem by a contradiction argument. Suppose there exists a consistent sequence of roots $\hat{\theta}_{2n}$ of the equations $\bar{l}_n(\theta_2) = 0$. Rewrite the equation in (16.4.8) as

$$\bar{l}_n(\theta_{20}) = \bar{a}_n(\theta_{20})(\hat{\theta}_{2n} - \theta_{20}) + [\bar{a}_n(\zeta_n) - \bar{a}_n(\theta_{20})](\hat{\theta}_{2n} - \theta_{20}). \qquad (16.4.12)$$

Choose and fix an interval $I(\theta_{20})$ of the type described in part(c) of Lemma 16.4.6. By the consistency of $\hat{\theta}_{2n}$, there exists $n_0 \geq 1$ such that for all $n \geq n_0$, $\hat{\theta}_{2n}$(and hence ζ_n in (16.4.12)) falls inside $I(\theta_{20})$ with an arbitrarily large probability. By part (b) of Lemma 16.4.6, $\bar{a}_n(\theta_{20}) = O_p(1)$ and so the first term on the right of (16.4.12) is $o_p(1)$. Further, part(c) of Lemma 16.4.6 entails that $[\bar{a}_n(\zeta_n) - \bar{a}_n(\theta_{20})]$ is uniformly bounded in probability and so the second term in (16.4.12) is also $o_p(1)$. The relation (16.4.12) now yields a contradiction in view of the fact that $\bar{l}_n(\theta_{20})$ converges in distribution to a nondegenerate random variable. ∎

We now indicate the line of proof for the case when both the parameters θ_1 and θ_2 are unknown. As we have discussed in Section 16.3, the two-variable iteration procedure for solving the likelihood equations $l_n(\theta) = 0$ can be reduced to a single-variable iteration. Specifically, rearranging the terms in (16.3.2), we define

$$g_n(\theta_2) = \left(\frac{1}{n} \sum_{i=1}^{n} \frac{i^{\theta_2} \log i}{i^{\theta_2} + 1} \right) \left(\sum_{i=1}^{n} Y_i(i^{\theta_2} + 1) \right) - \sum_{i=1}^{n} Y_i i^{\theta_2} \log i \qquad (16.4.13)$$

which is a function of θ_2 alone. Denote by $\hat{\theta}_{2n}$ a solution of $g_n(\theta_2) = 0$. A Taylor expansion of $g_n(\theta_{20}) = 0$ around $\hat{\theta}_{2n}$ yields the approximate relation

$$g_n(\theta_{20}) \approx -g'_n(\theta_{20})(\hat{\theta}_{2n} - \theta_{20}). \qquad (16.4.14)$$

Large-sample distributional results for $g_n(\theta_{20})$ and $g'_n(\theta_{20})$ are provided in the following lemma.

LEMMA 16.4.8
(a) $g_n(\theta_{20})$ converges in distribution to a random variable with zero expectation and variance $\theta_{10}^{-2}\sum_{i=1}^{\infty}(\log i)^2 i^{2\theta_{20}}/(1+i^{\theta_{20}})^2$.
(b) $g'_n(\theta_{20})$ converges in distribution random variable with zero expectation and variance $\theta_{10}^{-2}\sum_{i=1}^{\infty}(\log i)^4 i^{2\theta_{20}}/(1+i^{\theta_{20}})^2$.

PROOF (a) Note from the expression in (16.4.13) that $g_n(\theta_{20})$ can be expressed as

$$g_n(\theta_{20}) = \left(\frac{1}{n}\sum_{i=1}^{n}\frac{i^{\theta_{20}}\log i}{i^{\theta_{20}}+1}\right)\left(\sum_{i=1}^{n}Y_i(i^{\theta_{20}}+1)\right) - \sum_{i=1}^{n}Y_i i^{\theta_{20}}\log i$$

$$= \frac{1}{\theta_{10}}\sum_{i=1}^{n}\theta_{10}(1+i^{\theta_{20}})Y_i\left[\frac{1}{n}\sum_{i=1}^{n}\frac{i^{\theta_{20}}\log i}{i^{\theta_{20}}+1} - \frac{i^{\theta_{20}}\log i}{i^{\theta_{20}}+1}\right]$$

$$= \frac{1}{\theta_{10}}\sum_{i=1}^{n}[\theta_{10}(1+i^{\theta_{20}})Y_i - 1]\left[\frac{1}{n}\sum_{i=1}^{n}\frac{i^{\theta_{20}}\log i}{i^{\theta_{20}}+1} - \frac{i^{\theta_{20}}\log i}{i^{\theta_{20}}+1}\right]$$

$$\equiv -\frac{1}{\theta_{10}}\sum_{i=1}^{n}w_i(d_i - \bar{d})$$

where w_i are i.i.d. standard exponential random variables and $d_i = (i^{\theta_{20}}\log i)/(i^{\theta_{20}}+1)$. Now for $\theta_{20} < -1/2$,

$$\sum_{i=1}^{\infty}d_i^2 = \sum_{i=1}^{\infty}\frac{i^{2\theta_{20}}(\log i)^2}{(i^{\theta_{20}}+1)^2} = d < \infty$$

and $\frac{1}{n}\left(\sum_{i=1}^{n}d_i\right)^2 = \frac{1}{n}\left(\sum_{i=1}^{n}\frac{i^{\theta_{20}}(\log i)}{(i^{\theta_{20}}+1)}\right)^2$ converges to zero as $n \longrightarrow \infty$.

Hence, $\sum_{i=1}^{n}(d_i - \bar{d})^2$ tends to d as $n \longrightarrow \infty$, and we have our required result by virtue of Lemma 16.4.5.
   (b) $g'_n(\theta_{20})$ takes the form

$$g'_n(\theta_{20}) = \left(\frac{1}{n}\sum_{i=1}^{n}\frac{i^{\theta_{20}}(\log i)^2}{i^{\theta_{20}}+1}\right)\left(\sum_{i=1}^{n}Y_i(i^{\theta_{20}}+1)\right) + \left(\frac{1}{n}\sum_{i=1}^{n}\frac{i^{\theta_{20}}\log i}{i^{\theta_{20}}+1}\right)\left(\sum_{i=1}^{n}Y_i i^{\theta_{20}}\log i\right)$$

$$- \sum_{i=1}^{n}Y_i i^{\theta_{20}}(\log i)^2$$

which by analogous manipulations to part (a) can be equivalently expressed as the sum

$$g'_n(\theta_{20}) = -\frac{1}{\theta_{10}}\sum_{i=1}^{n}w_i(e_i - \bar{e}) + \left(\frac{1}{n}\sum_{i=1}^{n}\frac{i^{\theta_{20}}\log i}{i^{\theta_{20}}+1}\right)\left(\sum_{i=1}^{n}Y_i i^{\theta_{20}}\log i\right), \qquad (16.4.15)$$

and $e_i = i^{\theta_{20}}(\log i)^2/(i^{\theta_{20}}+1)$. By an expectation-variance argument the second term on the right of (16.4.15) turns out to be $o_p(1)$. By similar considerations as in part (a), it can be shown that $\sum_{i=1}^{n}w_i(e_i - \bar{e})$ converges in distribution to a random variable with zero mean and variance $\theta_{10}^{-2}\sum_{i=1}^{\infty}(\log i)^4 i^{2\theta_{20}}/(1+i^{\theta_{20}})^2$ and so the stated result follows. ∎

If we assume that $\hat{\theta}_{2n}$ consistently estimates θ_2, then in view of Lemma 16.4.8, $g_n(\theta_{20})$ converges to a nondegenerate distribution but $g'_n(\theta_{20})(\hat{\theta}_{2n} - \theta_{20}) = o_p(1)$, thus establishing a contradiction to the relation in (16.4.14). Of course, a finer treatment involving a random point ζ_n (intermediate between $\hat{\theta}_{2n}$ and θ_{20}) is necessary to make the Taylor-expansion proof rigorous. The details follow along similar lines to the "one-parameter known" case and are omitted here.

We now derive the asymptotic results for the nonlinear least squares estimates and compare with those for the MLE's.

16.4.2   LSE asymptotics

Although, there is a substantial amount of literature (e.g. [10], [11], [17]) on the asymptotic properties of nonlinear least squares estimates in a general setup, the theorems involve quite complex sets of regularity conditions not all of which are readily applicable to our model. Thus we take recourse to direct arguments which would be convenient to use in the present situation. Let us denote

$$a_{11}^*(\boldsymbol{\theta}) \equiv \frac{\partial^2 Q_n(\boldsymbol{\theta})}{\partial \theta_1^2} = 2\theta_1^{-2} \left[n - \sum_{i=1}^n \left[L_i + \log \theta_1 + \log(i^{\theta_2} + 1) \right] \right]$$

$$a_{12}^*(\boldsymbol{\theta}) \equiv \frac{\partial^2 Q_n(\boldsymbol{\theta})}{\partial \theta_1 \partial \theta_2} = 2\theta_1^{-1} \sum_{i=1}^n \frac{i^{\theta_2} \log i}{i^{\theta_2} + 1}$$

$$a_{22}^*(\boldsymbol{\theta}) \equiv \frac{\partial^2 Q_n(\boldsymbol{\theta})}{\partial \theta_2^2} = 2 \sum_{i=1}^n \frac{i^{\theta_2}(\log i)^2}{(i^{\theta_2} + 1)^2} \left[L_i + \log \theta_1 + \log(i^{\theta_2} + 1) \right] + 2 \sum_{i=1}^n \left(\frac{i^{\theta_2} \log i}{i^{\theta_2} + 1} \right)^2$$

$$\boldsymbol{A}_n^*(\boldsymbol{\theta}) \equiv (a_{ij}^*(\boldsymbol{\theta}))_{i,j=1,2}. \tag{16.4.16}$$

With the usual Taylor expansion, we have

$$\boldsymbol{l}_n^*(\boldsymbol{\theta}_0) = -\boldsymbol{A}_n^*(\boldsymbol{\zeta}^*)(\tilde{\theta}_{1n} - \theta_{10}, \tilde{\theta}_{2n} - \theta_{20})'$$

where $\boldsymbol{\zeta}^*$ lies on the line segment joining $\tilde{\boldsymbol{\theta}}_n$ and $\boldsymbol{\theta}_0$, and $\tilde{\boldsymbol{\theta}}_n$ is a root of the equation $\boldsymbol{l}_n^*(\boldsymbol{\theta}) = 0$. The line of proof is similar to the MLE case barring a few expressions. We omit the details here and present the main distributional results in the following theorem.

THEOREM 16.4.9
(a) CASE I : $-1/2 < \theta_{20} < 0$
For a consistent sequence of roots $\tilde{\boldsymbol{\theta}}_n$ of the equations $\boldsymbol{l}_n^*(\boldsymbol{\theta}) = 0$, the vector
$\left(n^{1/2}(\tilde{\theta}_{1n} - \theta_{10}), (\log n) n^{\theta_{20} + 1/2}(\tilde{\theta}_{2n} - \theta_{20}) \right)'$ is asymptotically bivariate normal with mean
$\boldsymbol{0}$ and variance-covariance matrix $\boldsymbol{D}_1^L = \pi^2/6 \, \boldsymbol{D}_1^M$, where $\boldsymbol{D}_1^M$ is as given in (16.4.5).
(b) CASE II : $\theta_{20} = -1/2$
For a consistent sequence of roots $\tilde{\boldsymbol{\theta}}_n$ of the equations $\boldsymbol{l}_n^*(\boldsymbol{\theta}) = 0$, the vector
$\left(n^{1/2}(\tilde{\theta}_{1n} - \theta_{10}), (\log n)^{3/2}(\tilde{\theta}_{2n} - \theta_{20}) \right)'$ converges in distribution to a bivariate normal
random variable with mean $\boldsymbol{0}$ and variance-covariance matrix $\boldsymbol{D}_2^L = \pi^2/6 \, \boldsymbol{D}_2^M$, where
$\boldsymbol{D}_2^M$ is as given in (16.4.6).
(c) CASE III : $\theta_{20} < -1/2$
There does not exist any consistent sequence of roots of the equations $\boldsymbol{l}_n^*(\boldsymbol{\theta}) = 0$.

Thus the nonlinear least squares estimates, apart from losing asymptotic efficiency, possess properties identical to the maximum likelihood estimates. For θ_{20} in $[-1/2, 0)$, adequacy of the normal approximation of the LS estimates were investigated through normal scores plots. To reduce the variability in the estimated values, logarithmic transformation on the absolute values of the estimates was used. Under the true parameter values $\theta_{10} = 1.0, \theta_{20} = -0.3$, and for sample sizes $n = 10, 25$, Figure 16.2 shows the normal scores plots for one hundred realizations of $\log \hat{\theta}_{1n}$ and $\log \tilde{\theta}_{1n}$ values. While there is a reasonable agreement to normality for the MLE $\log \hat{\theta}_{1n}$ (see Figure 16.2), for these sample sizes, the corresponding plots for $\log \tilde{\theta}_{1n}$ show a substantial departure from normality. This can be attributed to the greater variability in the LSE's as is revealed from Table 16.1. The distributions of both $\log(-\hat{\theta}_{2n})$ and $\log(-\tilde{\theta}_{2n})$ show a strong departure from normality in terms of normal scores plots, even for large sample sizes such as $n = 50$ or 100. Figures 16.3 and 16.4 show two such sets of plots for the cases $(\theta_{10} = 1.0, \theta_{20} = -0.3)$ and $(\theta_{10} = 1.0, \theta_{20} = -0.5)$, respectively. The inadequacy of normal approximation can be explained by the slow rates $(n^{1/5}(\log n), (\log n)^{3/2}$, respectively, in the cases considered$)$ of convergence. For the case $\theta_{10} = 2.0, \theta_{20} = -2.0$, Figures 16.5 and 16.6 show the histograms (based on 500 realizations) for the simulated values of $\log(\hat{\theta}_{2n}/\theta_{20})$ and $\log(\tilde{\theta}_{2n}/\theta_{20})$, respectively. The plots are provided for four sample sizes $n = 10, 25, 50$ and 100. Figure 16.5 reveals a left-skewed pattern for the distribution of $\log(\hat{\theta}_{2n}/\theta_{20})$. The skewness and spread in the histograms of $\log(\tilde{\theta}_{2n}/\theta_{20})$ (Figure 16.6) are much less prominent in comparison.

It turns out that the inconsistency of both MLE and LSE, though disconcerting, should not be attributed to these methods of estimation. We now establish a general and much stronger result towards that direction.

THEOREM 16.4.10
For our model, there does not exist any consistent estimate of θ_2 in the subspace of Ω defined by $\Omega_s = \{\theta_1 > 0, \theta_2 < -1/2\}$.

To prove the theorem, we will use a result in [17], which we state as the following lemma.

LEMMA 16.4.11
Consider the model $y_i = g_i(\theta) + \eta_i$, where $\theta \in \Theta$, a subset of $\mathcal{R}^p$, $g_i(\theta)$ are real valued functions defined on Θ, η_i's are i.i.d. mean zero random variables with the common distribution F which has a positive (almost everywhere) and absolutely continuous density f with finite Fisher information $I = \int_{-\infty}^{\infty} (f')^2/f$. If there exists an estimate $\bar{\theta}$ such that $\bar{\theta} \xrightarrow{P} \theta$ for all $\theta \in \Theta$, then, as $n \longrightarrow \infty$,

$$D_n(\theta^{(1)}, \theta^{(2)}) \equiv \sum_{i=1}^{n} \left(g_i(\theta^{(1)}) - g_i(\theta^{(2)})\right)^2 \longrightarrow \infty \qquad (16.4.17)$$

for all $\theta^{(1)} \neq \theta^{(2)}$ in Θ.

The lemma gives a simple necessary condition for the existence of a consistent estimate in a general location model. We shall prove Theorem 16.4.10 by showing that the condition (16.4.17) is violated in our model.

PROOF OF THEOREM 16.4.10 To be able to apply Lemma 16.4.11, we first verify the finiteness of the Fisher information I. Note from (16.3.3), that the i.i.d. ϵ_i's here are translated standard extreme value random variables with p.d.f.

$$f(x) = \exp\left(x + \Gamma'(1) - \exp(x + \Gamma'(1))\right), \quad -\infty < x < \infty$$

so the condition of finite Fisher information holds.

To show that condition (16.4.17) fails, let us take any $\boldsymbol{\theta}^{(1)} = (\theta_1^{(1)}, \theta_2^{(1)})$, $\boldsymbol{\theta}^{(2)} = (\theta_1^{(2)}, \theta_2^{(2)})$ in Ω_S, such that they differ only in the second coordinate, i.e. $\theta_1^{(1)} = \theta_1^{(2)}$, $\theta_2^{(1)} \neq \theta_2^{(2)}$. Using Taylor expansion, we then have

$$D_n(\boldsymbol{\theta}^{(1)}, \boldsymbol{\theta}^{(2)}) = \sum_{i=1}^{n} \left[\log(i^{\theta_2^{(2)}} + 1) - \log(i^{\theta_2^{(1)}} + 1)\right]^2$$

$$= (\theta_2^{(2)} - \theta_2^{(1)})^2 \sum_{i=1}^{n} (\log i)^2 i^{2\varsigma} / (1 + i^\varsigma)^2$$

where ς is a point between $\theta_2^{(1)}$ and $\theta_2^{(2)}$. Since $\varsigma < -1/2$, we obtain

$$\sum_{i=1}^{n} (\log i)^2 i^{2\varsigma} / (1 + i^\varsigma)^2 \leq \sum_{i=1}^{n} (\log i)^2 i^{2\varsigma} < \infty.$$

This establishes the nonexistence of consistent estimates for (θ_1, θ_2). In fact, θ_2 is the one which is not consistently estimable. There exist consistent estimates for θ_1 ($\bar{\theta}_{1n} = n/T_n$ is one such). ∎

16.5   Conclusion

In the context of reliability growth analysis for systems undergoing development testing programs, the main thrust of the present article lies in two directions; formulation of a parametric model that captures the essence of certain physical settings, and development of estimation procedures for the model parameters. Cast in the setting of continuous-time monitoring of a system, the proposed model incorporates two sources of failure (i.e. assignable cause failures and inherent failures) and builds in the learning curve concept of RG with regard to the assignable-cause failure rate. It can be viewed as an extension of the Barlow-Scheuer discrete RG model ([2]) to the continuous-time case, and also as an extension of the single component PEXP to one that accounts for a two-component system with one component experiencing reliability growth and the other remaining unchanged.

Maximum likelihood and non-linear least squares estimation of the model parameters are examined under the specific sampling scheme that records the sequence of failure times for a single prototype of the system under the test-redesign-retest cycle. Properties of the estimates and their comparisons are drawn both from a development of necessary asymptotic theory and finite-sample simulation studies. Contrary to the behavior of such estimates in standard parametric models and usual sampling schemes, several irregularities and pathologies surface in the present situation. The most troublesome feature is that no consistent estimates exist in a part of the parameter space. These findings signal the need for a close scrutiny of the routine data analytic methods such as the iterative solutions of the likelihood equations for calculation of the MLE and computation of the estimated Fisher information as well as its use in setting large-sample confidence intervals in the context of RG analysis. Interestingly, the same parametric structure for reliability growth in the absence of inherent-cause failures does not suffer from any such irregularities in the asymptotic behavior of the parameter estimates. The inherent-cause component adds a complexity to the model that makes estimation of parameters difficult from the data of a single system even though testing may be continued over a large number of stages. Evidently, this difficulty can be remedied by allowing sufficient replications in the experiment.

References

1. Bain, L.J. and Engelhardt, M. (1991), *Statistical Analysis of Reliability and Life-Testing Models,* Second edition, Marcel Dekker, New York.

2. Barlow, R. and Scheuer, E. (1966), Reliability growth during a development test program, *Technometrics,* **8**, 53-60.

3. Benton, A.W. and Crow, L.H. (1989), Integrated reliability growth testing, *Proceedings of Annual Reliability and Maintainability Symposium,* 160-166.

4. Berndt, G.D. (1966), Estimating a monotonically changing binomial parameter, Office of Operations Analysis, Strategic Air Command, Offut Air Force Base, Nebraska.

5. Bresenham, J.E. (1964), Reliability growth models, *Technical Report No. 74,* Department of Statistics, Stanford University.

6. Corcoran, W.J., Weingarten, H., and Zehna, P.W. (1964), Estimating reliability after corrective action, *Management Science,* **10**, 786-795.

7. Crow, L.H. (1987), Reliability growth assessments utilizing engineering judgement, *Proceedings of Institute of Environmental Sciences,* 179-183.

8. Duane, J.T. (1964), Learning curve approach to reliability monitoring, *IEEE Transactions on Aerospace,* **2**, 563-566.

9. Feller, W. (1966), *An Introduction to Probability Theory and Its Applications,* Vol. II, John Wiley & Sons, New York.

10. Jenrich, R. (1969), Asymptotic properties of nonlinear least squares estimators, *Annals of Mathematical Statistics,* **40**, 633-643.

11. Malinvaud, E. (1970), The consistency of nonlinear regression, *Annals of Mathematical Statistics,* **41**, 956-969.

12. Rao, B.L.S.P. (1987), *Asymptotic Theory of Statistical Inference,* John Wiley & Sons, New York.

13. Sen, A. and Bhattacharyya, G.K. (1993), A Piecewise exponential model for reliability growth and associated inferences, In *Advances in Reliability,* 331-355, (Ed., A.P. Basu), North-Holland, Amsterdam.

14. Thompson, W.A.(Jr.) (1988), *Point Process Models with Applications to Safety and Reliability,* Chapman and Hall, New York.

15. Weinrich, M. and Gross, A. (1978), The Barlow-Scheuer reliability growth model from a Bayesian viewpoint, *Technometrics,* **20**, 249-254.

16. Wolman, W. (1963), Problems in System Reliability Analysis, In *Statistical Theory of Reliability,* 149-160, (Ed., M. Zelen), University of Wisconsin Press, Madison.

17. Wu, C.F. (1981), Asymptotic theory of nonlinear least squares estimation, *Annals of Statistics,* **9**, 501-513.

TABLE 16.1
Estimated bias and the mean squared error of the ML and the LS estimators

Sample Size (n)	$\theta_{10} = 1.0, \ \theta_{20} = -0.3$							
	MLE				LSE			
	Bias($\hat{\theta}_1$)	MSE($\hat{\theta}_1$)	Bias($\hat{\theta}_2$)	MSE($\hat{\theta}_2$)	Bias($\hat{\theta}_1$)	MSE($\hat{\theta}_1$)	Bias($\hat{\theta}_2$)	MSE($\hat{\theta}_2$)
10	0.416	0.493	−2.502	216.887	0.322	0.711	−2.899	301.318
25	0.233	0.162	−0.545	0.834	0.182	0.181	−1.418	130.412
50	0.134	0.088	−0.631	43.368	0.132	0.106	−1.538	172.176
100	0.073	0.050	−0.189	0.165	0.083	0.052	−0.238	0.332
	$\theta_{10} = 2.0, \ \theta_{20} = -0.5$							
10	0.582	1.201	−1.722	131.375	0.341	1.205	−1.270	87.354
25	0.226	0.349	− 0.686	43.204	0.110	0.404	−1.248	129.269
50	0.037	0.186	−0.463	43.178	−0.028	0.209	−0.106	0.260
100	−0.085	0.153	−0.037	0.158	−0.059	0.134	−0.675	86.593
	$\theta_{10} = 2.0, \ \theta_{20} = -2.0$							
10	0.018	0.515	−1.309	252.335	−0.134	0.779	−0.916	211.396
25	−0.218	0.245	0.709	42.796	−0.269	0.391	−0.352	210.743
50	−0.291	0.257	0.928	44.268	−0.295	0.272	1.184	2.466
100	−0.398	0.284	1.475	2.381	−0.324	0.200	1.420	2.364

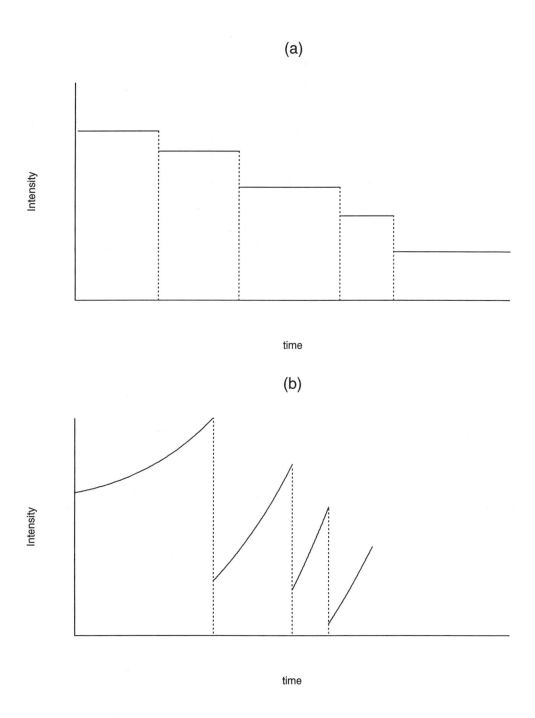

FIGURE 16.1
Forms of the composite intensity functions when (a) both components are
exponential, (b) component A is exponential and component B is Weibull

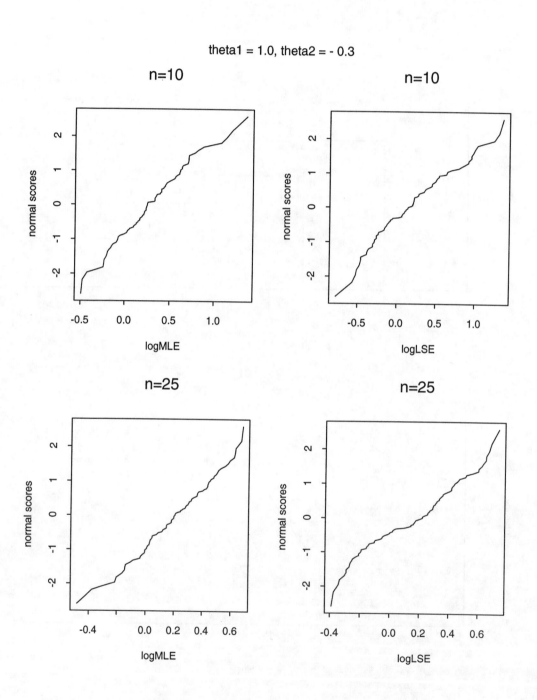

FIGURE 16.2
Normal scores plots for $\log \hat{\theta}_{1n}$ and $\log \tilde{\theta}_{1n}$

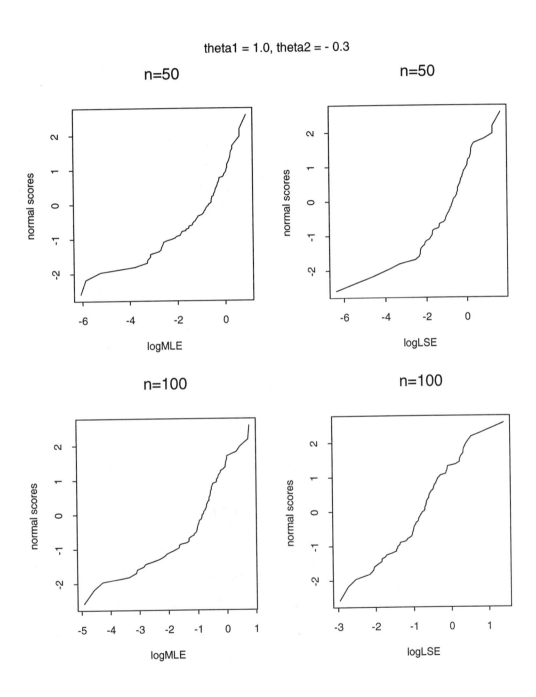

FIGURE 16.3
Normal scores plots for $\log \hat{\theta}_{2n}$ and $\log \tilde{\theta}_{2n}$

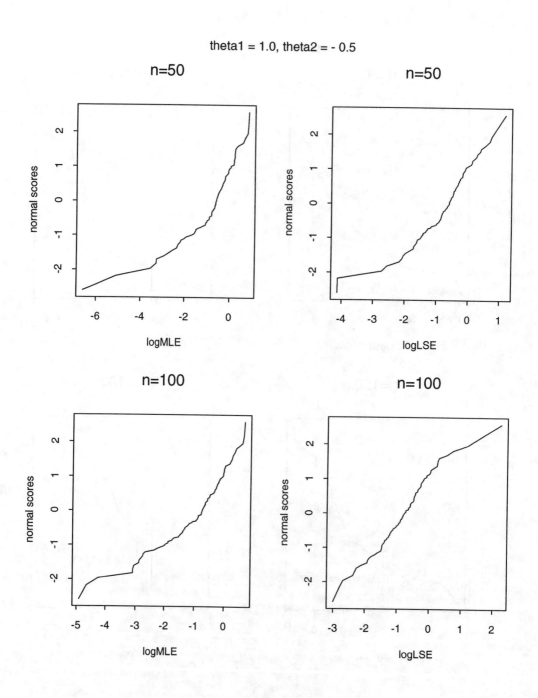

FIGURE 16.4
Normal scores plots for $\log \hat{\theta}_{2n}$ and $\log \tilde{\theta}_{2n}$

theta1 = 2.0, theta2 = - 2.0

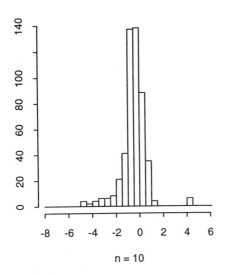

n = 10

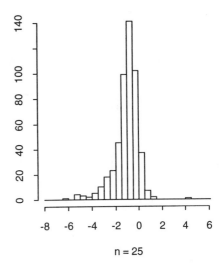

n = 25

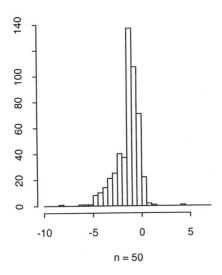

n = 50

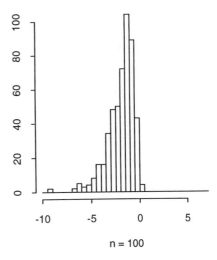

n = 100

FIGURE 16.5
Histograms for the simulated values of $\log(\hat{\theta}_{2n}/\theta_{20})$

theta1 = 2.0, theta2 = - 2.0

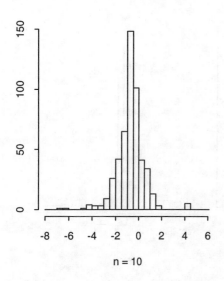

n = 10

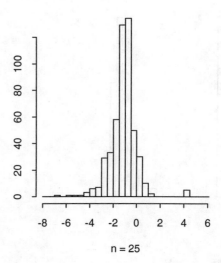

n = 25

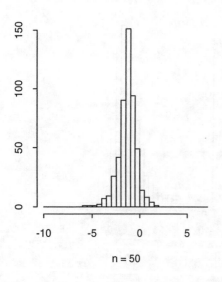

n = 50

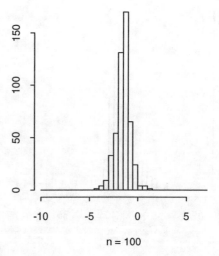

n = 100

FIGURE 16.6
Histograms for the simulated values of $\log(\tilde{\theta}_{2n}/\theta_{20})$

17

Reliability Growth: Nonhomogeneous Poisson Process

Chris P. Tsokos

University of South Florida

ABSTRACT The object of the present study is to discuss some very recent findings in reliability growth modeling. The nonhomogeneous Poisson process has been used to characterize the failure pattern of systems during the development and testing process. One of the key elements in this type of reliability setting is the relationship between the failure intensity function, $V(t)$, and the mean time between failures, $MTBF$. We shall summarize the analytical relationship of $V(t)$ and $MTBF$ and introduce a computational procedure for stastically estimating the intensity function. In addition, we discuss some very recent results in effectively estimating the failure intensity function. Some real data, along with numerical simulations, is used to illustrate the theoretical developments.

CONTENTS

17.1 Introduction

Consider a complex system in the development and testing process. The system is tested until it fails, it is repaired or redesigned, if necessary, and it is tested again until it fails. This process continues until we reach a desirable reliability which would reflect the quality of the final design. This process of testing a system has been referred to as reliability growth. Also, likewise the reliability of a repairable system will improve with time as component defects and flaws are detected, repaired or removed. Initially, [8], observed the consistent system growth patterns in testing several engineering systems. In [8], the author introduced the concept of "learning curve approach", which is a plot illustrating the change of the cumulative failure rate of a system as a function of time. In [6] and [7], the author noticed the approximate linear behavior of the cumulative failure rate to time on log plots and introduced a mathematical model to characterize this failure pattern.

He proposed the nonhomogeneous Poisson process, NHPP, with a failure intensity function given by

$$V(t) = \frac{\beta}{\theta} \left(\frac{t}{\theta} \right)^{\beta-1}, \quad t > 0.$$

0-8493-8972-0/95/$0.00+$.50

This failure intensity function corresponds to the hazard rate function of the Weibull distribution, which is the Weibull process.

The nonhomogeneous Poisson process, NHPP, is an effective approach to analyzing the reliability growth and predicting the failure behavior of a given system. The power-law process often is misleadingly used to describe the NHPP, Weibull process in reliability growth modeling. Some of the recent publications which address some of the fundamental aspects of reliability growth of repairable systems are: [2]; [9] and [10]; [17], [18] and [19]; [1]; [12]; and [4], among others.

In addition to tracking the reliability growth of a system, we can utilize such a modeling for predictions, it is quite important to be able to determine the next failure time after the system has experienced some failures during the developmental process. Being able to have a good estimate as to when the system will fail again is important in strategically structuring maintenance policies. The time difference between the expected next failure time and the current failure time is the **Mean Time Between Failures**, $MTBF$. The reciprocal of the intensity function at current failure time has been used as being equal to $MTBF$. In [20], the author pointed out that such a relation between $V(t)$ and $MTBF$ is only true for the homogeneous process.

Recently, in [15] and [16], the authors have analytically investigated the relationship between $MTBF$ and the reciprocal of the intensity function. In this presentation, a summary of this finding is presented along with some numerical results. In addition, a review of some of the latest findings concerning the estimation of the parameters and failure intensity function that are inherent in the NHPP will be given.

17.2   Nonhomogeneous Poisson Process

The probability of achieving n failures of a given system in the time interval $(0,\ t)$ can be written as

$$P\left(x = n;\ t\right) = \frac{\exp\left\{-\int_0^t v\left(x\right) dx\right\}\left\{\int_0^t v\left(x\right) dx\right\}^n}{n!} \tag{17.2.1}$$

for $t > 0$.

When the failure intensity function $V(t) = \lambda$, expression (17.2.1) reduces into a homogeneous Poisson process,

$$P\left(x = n;\ t\right) = \frac{e^{-\lambda t}\left(\lambda t\right)^n}{n!},\quad 0 < t. \tag{17.2.2}$$

For tracking reliability growth, [3] and [4], proposed that the failure intensity function be

$$V(t) = \frac{\beta}{\theta}\left(\frac{t}{\theta}\right)^{\beta-1},\quad 0 < t,\ \ 0 < \beta,\ \theta \tag{17.2.3}$$

where β and θ are the shape and scale parameters, respectively. Thus, expression (17.2.1) reduces to

$$P\left(x = n;\ t\right) = \frac{1}{n!}\exp\left\{-\frac{t^\beta}{\theta^\beta}\right\}\left(\frac{t}{\theta}\right)^{n\beta}. \tag{17.2.4}$$

The above expression is the nonhomogeneous Poisson, (NHPP) or Weibull process. In reliability growth analysis it is important to be able to determine the next failure time after some failures have already occurred. With respect to this aim, the time difference between the expected failure time and the current failure time or the mean time between failures, $MTBF$, is of significant interest. Until recently, in reliability growth studies, scientists used the reciprocal of the intensity function at current failure time as the $MTBF$. That is,

$$MTBF = \frac{1}{V(t)} \, .$$

However, such a relationship is only true for homogeneous processes. In what follows, we shall discuss the relationship of $MTBF$ and the reciprocal of the intensity function for nonhomogeneous Poisson process which has recently been developed by [15] and [16].

17.2.1 The Relationship of MTBF and $\frac{1}{V(t)}$

Let $T_1, T_2, \ldots, T_n$ denote the first n failure times of the NHPP, where $T_1 < T_2 < \cdots < T_n$. The truncated conditional probability distribution function, $f_i(t \,|t_1, \ldots, t_{i-1})$, in Weibull given by

$$f_i(t \,|t_1, \ldots, t_{i-1}) = \frac{\beta}{\theta}\left(\frac{t}{\theta}\right)^{\beta-1} \exp\left\{-\left(\frac{t}{\theta}\right) + \left(\frac{t_{i-1}}{\theta}\right)^{\beta}\right\}, \quad t_{i-1} < t.$$

The likelihood function for $T_1 = t_1$, $T_2 = t_2$, $\ldots$, $T_n = t_n$ is

$$L(\beta, \theta) = \prod_{i=1}^{n} f_i(t_i \,|t_1, \ldots, t_{i-1})$$

$$= \left(\frac{\beta}{\theta}\right)^{n} \exp\left\{-\left(\frac{t_n}{\theta}\right)^{\beta}\right\} \prod_{i=1}^{n}\left(\frac{t_i}{\theta}\right)^{\beta-1} \, .$$

The maximum likelihood estimates, MLE, for the shape and scale parameters β and θ are

$$\widehat{\beta} = \frac{n}{\sum\limits_{i=1}^{n} \log\left(\frac{t_n}{t_i}\right)} \quad \text{and} \quad \widehat{\theta} = \frac{t_n}{n^{1/\beta}}, \tag{17.2.5}$$

respectively. The MLE of the intensity function and its reciprocal can be approximated using the above estimates of β and θ. In reliability growth modeling, the reliability of a system will improve with time as design flaws are repaired and removed. Thus, we would expect the failure intensity, V_n, to be decreasing with time. This means that $\beta < 1$. However, when $\beta > 1$, this would indicate that the system is wearing out rapidly and would require a higher level of maintenance. Thus, since our aim in this presentation is reliability growth modeling, we need to establish the relationship between $MTBF$ and $\frac{1}{V(t)}$ for $\beta < 1$. For convenience, we shall denote at $t = t_n$, $V(t) = V_n$.

The $MTBF_n$ at $T_n = t_n$ for the Weibull process or NHPP is given by

$$MTBF_n = \int_{t_n}^{\infty} t\, f_{n+1}(t \,|t_1, \ldots, t_n)\, dt - t_n$$

$$= \int_{t_n}^{\infty} t\left(\frac{\beta}{\theta}\right)\left(\frac{t}{\theta}\right)^{\beta-1} e^{-\left(\frac{t}{\theta}\right)^{\beta}+\left(\frac{t_n}{\theta}\right)^{\beta}}\, dt - t_n \, .$$

Case 1: $\beta = 1$

For the special case when $\beta = 1$, we have

$$MTBF_n = \int_{t_n}^{\infty} \frac{t}{\theta}\, e^{-\frac{t}{\theta} + \frac{t_n}{\theta}}\, dt - t_n = \theta \ .$$

Thus, for $\beta = 1$, we have $\dfrac{1}{V_n} = \theta$. The $MTBF_n$ for a general β, we have

$$MTBF_n = \beta e^{\left(\frac{t_n}{\theta}\right)^{\beta}} \int_{t_n}^{\infty} \left(\frac{t}{\theta}\right)^{\beta} e^{-\left(\frac{t}{\theta}\right)^{\beta}}\, dt - t_n$$

$$= \theta \beta e^{\left(\frac{t_n}{\theta}\right)^{\beta}} \int_{\frac{t_n}{\theta}}^{\infty} t^{\beta} e^{-t^{\beta}}\, dt - t_n$$

$$= \theta e^{\left(\frac{t_n}{\theta}\right)^{\beta}} \int_{\frac{t_n}{\theta}}^{\infty} e^{-t^{\beta}}\, dt$$

$$= \frac{\theta}{\beta}\, e^{\left(\frac{t_n}{\theta}\right)^{\beta}} \int_{\left(\frac{t_n}{\theta}\right)^{\beta}}^{\infty} e^{-t}\, t^{\frac{1}{\beta}-1}\, dt \ . \tag{17.2.6}$$

The reciprocal of the intensity function (17.2.3) is

$$\frac{1}{V_n} = \frac{\theta}{\beta}\left(\frac{\theta}{t_n}\right)^{\beta-1}$$

and in relation to $MTBF_n$, we have

$$\frac{1}{V_n} = \frac{\theta}{\beta}\, e^{\left(\frac{t_n}{\theta}\right)^{\beta}} \left(\frac{\theta}{t_n}\right)^{\beta-1} \int_{\left(\frac{t_n}{\theta}\right)^{\beta}}^{\infty} e^{-t}\, dt.$$

Case 2: $\beta > 1$.

When $\beta > 1$, we have $\dfrac{1}{\beta} - 1 < 0$, and

$$\int_{\left(\frac{t_n}{\theta}\right)^{\beta}}^{\infty} e^{-t}\, t^{\frac{1}{\beta}-1}\, dt < \int_{\left(\frac{t_n}{\theta}\right)^{\beta}}^{\infty} e^{-t} \left\{ \left(\frac{t_n}{\theta}\right)^{\beta} \right\}^{\frac{1}{\beta}-1}\, dt$$

$$= \left(\frac{\theta}{t_n}\right)^{\beta-1} \int_{\left(\frac{t_n}{\theta}\right)^{\beta}}^{\infty} e^{-t}\, dt$$

$$= \frac{1}{V_n} \cdot$$

Case 3: $\beta < 1$.

For $\beta < 1$ or $\frac{1}{\beta} - 1 > 0$, we have

$$\int_{\left(\frac{t_n}{\theta}\right)^\beta}^{\infty} e^{-t} \, t^{\frac{1}{\beta}-1} \, dt > \left(\frac{\theta}{t_n}\right)^{\beta-1} \int_{\left(\frac{t_n}{\theta}\right)^\beta}^{\infty} e^{-t} \, dt$$

$$= \frac{1}{V_n} \cdot$$

Thus, the relationship between $MTBF_n$ and $\frac{1}{V_n}$ is summarized by

$$MTBF_n \begin{cases} = \frac{1}{V_n} & \text{for} \quad \beta = 1 \\ > \frac{1}{V_n} & \text{for} \quad \beta < 1 \\ < \frac{1}{V_n} & \text{for} \quad \beta > 1 \end{cases}$$

and for reliability growth modeling, the only relationship of interest is

$$MTBF_n > \frac{1}{V_n} \text{ for } \beta < 1 \ .$$

In [15] and [16], the authors have also developed a computational procedure for estimating $MTBF_n$, V_n and $\frac{1}{V_n}$.

THEOREM 17.2.1
For a nonhomogeneous Poisson process with intensity function given by

$$V_n = \frac{\beta}{\theta} \left(\frac{t_n}{\theta}\right)^{\beta-1} , \quad 0 < t_n$$

we have: for $0 < \beta \le \frac{1}{2}$,

$$\frac{1}{V_n}\left(1 + \frac{\alpha - 1}{h}\right) \le MTBF_n \le \frac{1}{V_n}\left(\frac{1}{1 - \frac{\alpha-1}{h}}\right)$$

and for $\frac{1}{2} < \beta \le 1$,

$$\frac{1}{V_n} \le MTBF_n \le \frac{1}{V_n}\left(1 + \frac{\alpha - 1}{h}\right)$$

where

$$\alpha = \frac{1}{\beta}, \ h = \left(\frac{t_n}{\theta}\right)^\beta \ and \ \frac{1}{V_n} = \frac{\theta}{\beta}\left(\frac{\theta}{t_n}\right)^{\beta-1} \ .$$

PROOF For $\alpha = \dfrac{1}{\beta}$, $h = \left(\dfrac{t_n}{\theta}\right)^{\beta}$ we have $\alpha > 1$ and

$$V_n = \frac{1}{\alpha\theta}\, h^{1-\alpha} \text{ and } \frac{1}{V_n} = \alpha\theta h^{\alpha-1}.$$

Thus, $MTBF_n$ can be written as follows:

$$MTBF_n = \theta e^h\, \alpha \int_h^{\infty} e^{-t}\, t^{\alpha-1}\, dt.$$

For analytical convenience, we have split the behavior of β into two parts. For $\beta \le \dfrac{1}{2}$, we have $\alpha \ge 2$. Assume $\alpha = m + 1 + \delta$, with $m \ge 1$ being an integer and $\delta \in [0, 1)$. For the special case when $\delta = 0$, we have

$$MTBF_n = \theta e^h \alpha e^{-h} \left[h^m + (\alpha - 1)\, h^{m-1} + \cdots + (\alpha - 1)(\alpha - 2) \right.$$

$$\left. \cdots (\alpha - m + 1)\, h + (\alpha - 1)(\alpha - 2)\cdots(\alpha - m) \right]$$

$$= \alpha\theta h^m \left[1 + (\alpha - 1)\frac{1}{h} + \cdots + \frac{(\alpha - 1)\cdots(\alpha - m)}{h^m} \right]$$

$$= \frac{1}{V_n} \left[1 + (\alpha - 1)\frac{1}{h} + \cdots + \frac{(\alpha - 1)\cdots(\alpha - m)}{h^m} \right].$$

For a general δ, $MTBF_n$ is obtained as follows:

$$MTBF_n = \theta\alpha h^{\delta+m} \left[1 + \frac{\alpha - 1}{h} + \cdots + \frac{(\alpha - 1)(\alpha - 2)\cdots(\alpha - m + 1)}{h^{m-1}} \right.$$

$$\left. + \frac{(\alpha - 1)(\alpha - 2)\cdots(\alpha - m)}{h^m} e^h\, h^{-\delta} \int_h^{\infty} t^{\alpha-m-1} e^{-t}\, dt \right]$$

$$= \frac{1}{V_n} \left[1 + \frac{\alpha - 1}{h} + \cdots + \frac{(\alpha - 1)(\alpha - 2)\cdots(\alpha - m + 1)}{h^{m-1}} \right.$$

$$\left. + \frac{(\alpha - 1)(\alpha - 2)\cdots(\alpha - m)}{h^m} e^h\, h^{-\delta} \int_h^{\infty} t^{\alpha-m-1} e^{-t}\, dt \right] \qquad (17.2.7)$$

$$= \frac{1}{V_n} \left[1 + \frac{\alpha - 1}{h} + \cdots + \frac{(\alpha - 1)(\alpha - 2)\cdots(\alpha - m + 1)}{h^{m-1}} \right.$$

$$\left. + \frac{(\alpha - 1)(\alpha - 2)\cdots(\alpha - m)}{h^m} \right.$$

$$+ \frac{(\alpha - 1)(\alpha - 2) \cdots (\alpha - m - 1)}{h^m} \, e^h \, h^{-\delta} \int_h^\infty t^{\alpha - m - 2} \, e^{-t} \, dt \Bigg] \, . \qquad (17.2.8)$$

It is also clear that

$$\alpha - m - 1 = \delta \geq 0$$

and

$$\alpha - m - 2 = \delta - 1 < 0 \, .$$

Furthermore, we have

$$e^h \, h^{-\delta} \int_h^\infty t^{\alpha - m - 1} \, e^{-t} \, dt \geq e^h \, h^{-\delta} \, h^{\alpha - m - 1} \int_h^\infty e^{-t} \, dt = 1 \, .$$

Also,

$$e^h \, h^{-\delta} \int_h^\infty t^{\alpha - m - 2} \, e^{-t} \, dt < e^h \, h^{-\delta} \, h^{\alpha - m - 2} \int_h^\infty e^{-t} \, dt = \frac{1}{h} \, .$$

Thus, from equation (17.2.7) the following inequality holds,

$$MTBF_n \geq \frac{1}{V_n} \left[1 + \frac{\alpha - 1}{h} + \cdots + \frac{(\alpha - 1)(\alpha - 2) \cdots (\alpha - m + 1)}{h^{m-1}} \right.$$

$$\left. + \frac{(\alpha - 1)(\alpha - 2) \cdots (\alpha - m)}{h^m} \right]$$

$$\geq \frac{1}{V_n} \left[1 + \frac{\alpha - 1}{h} \right] \, .$$

Similarly, from equation (17.2.8) we have

$$MTBF_n \leq \frac{1}{V_n} \left[1 + \frac{\alpha - 1}{h} + \cdots + \frac{(\alpha - 1)(\alpha - 2) \cdots (\alpha - m)}{h^m} \right.$$

$$\left. + \frac{(\alpha - 1)(\alpha - 2) \cdots (\alpha - m - 1)}{h^{m+1}} \right]$$

$$\leq \frac{1}{V_n} \left[1 + \frac{\alpha - 1}{h} + \left(\frac{\alpha - 1}{h} \right)^2 + \cdots + \left(\frac{\alpha - 1}{h} \right)^m \right]$$

$$\leq \frac{1}{V_n} \frac{1}{1 - \frac{\alpha - 1}{h}} \, .$$

Therefore, we can conclude that $MTBF_n$ is bounded from below by $\dfrac{1}{V_n} \left(1 + \dfrac{\alpha - 1}{h} \right)$ and above by $\dfrac{1}{V_n} \dfrac{1}{1 - \frac{\alpha - 1}{h}}$, respectively. That is,

$$\frac{1}{V_n} \left(1 + \frac{\alpha - 1}{h} \right) \leq MTBF \, (T_n) \leq \frac{1}{V_n} \frac{1}{1 - \frac{\alpha - 1}{h}}$$

where

$$\alpha = \frac{1}{\beta}, \quad h = \left(\frac{t_n}{\theta}\right)^{\beta}$$

and

$$\frac{1}{V_n} = \frac{\theta}{\beta}\left(\frac{\theta}{t_n}\right)^{\beta-1}.$$

For $\beta > \frac{1}{2}$, we have $1 \le \alpha < 2$. Assume $\alpha = 1 + \delta$, where $0 \le \delta < 1$.
Thus, we can write the $MTBF_n$ as follows:

$$MTBF_n = \theta e^h \, \alpha \int_h^{\infty} e^{-t} t^{\delta} \, dt$$

$$\ge \alpha \theta \, h^{\delta} \, e^{-h} \int_h^{\infty} e^{-t} \, dt$$

$$= \alpha \theta h^{\delta} = \frac{1}{V_n}.$$

At the same time, we can also write

$$MTBF_n = \alpha \theta e^h \left[h^{\delta} e^{-h} + \delta \int_h^{\infty} h^{\delta-1} e^{-x} \, dx \right]$$

$$\le \alpha \theta e^h \left[h^{\delta} e^{-h} + \delta \int_h^{\infty} h^{\delta-1} e^{-t} \, dt \right]$$

$$\le \frac{1}{V_n}\left[1 + \frac{\delta}{h} \right].$$

Therefore, when $\beta > \frac{1}{2}$, $MTBF_n$ is bounded from below and above by $\frac{1}{V_n}$ and $\frac{1}{V_n}\left[1 + \frac{\delta}{h}\right]$, that is,

$$\frac{1}{V_n} \le MTBF_n \le \frac{1}{V_n}\left[1 + \frac{\delta}{h}\right]$$

where $\delta = \frac{1}{\beta} - 1 \in [0, 1)$ and thus Theorem 1. For implementing the above results, we have to utilize estimates of the appropriate parameters that are inherent in V_n. Thus, α, β, θ and V_n must be estimated as follows:

$$\hat{\alpha} = \frac{1}{\hat{\beta}}, \quad \hat{h} = \left(\frac{t_n}{\hat{\theta}}\right)^{\hat{\beta}} \quad \text{and} \quad \frac{1}{\hat{V}_n} = \frac{\hat{\theta}}{\hat{\beta}}\left(\frac{\hat{\theta}}{t_n}\right)^{\hat{\beta}-1}.$$

Using these estimates, we can estimate $MTBF_n$ in accordance with the behavior of β.

For $\beta \le \dfrac{1}{2}$, we have

$$\frac{1}{\widehat{V_n}}\left(1 + \frac{\widehat{\alpha} - 1}{h}\right) \le \widehat{MTBF}_n \le \frac{1}{\widehat{V_n}}\left(\frac{1}{1 - \frac{\widehat{\alpha}-1}{h}}\right)$$

and a recommended estimate, [15] and [16] is given by

$$\widehat{MTBF}_n = \frac{1}{2}\left[\frac{1}{\widehat{V_n}}\left(1 + \frac{\widehat{\alpha} - 1}{h}\right) + \frac{1}{\widehat{V_n}}\left(\frac{1}{1 - \frac{\widehat{\alpha}-1}{h}}\right)\right].$$

For $\beta > \dfrac{1}{2}$, we have

$$\frac{1}{\widehat{V_n}} \le \widehat{MTBF}_n \le \frac{1}{\widehat{V_n}}\left[1 + \frac{\widehat{\delta}}{h}\right], \quad \widehat{\delta} = \frac{1}{\widehat{\beta}} - 1 \in [0,\, 1)$$

and a proposed estimate given by

$$\widehat{MTBF}_n = \frac{1}{2\widehat{V_n}}\left[2 + \frac{\widehat{\delta}}{h}\right].$$

We shall now use Crow's, [6] and [7], failure data of a system undergoing developmental testing to illustrate the results of [15] and [16]. The forty successive failures of the system under development are given below:

Reliability Growth Failure Data

.7	3.7	13.2	17.6	54.5	99.2	112.2	120.9	151.0	163.0
174.5	191.6	282.8	355.2	486.3	490.5	513.3	558.4	678.1	688.0
785.9	887.0	1010.7	1029.1	1034.4	1136.1	1178.9	1259.7	1297.9	1419.7
1571.7	1629.8	1702.4	1928.9	2072.3	2525.2	2928.5	3016.4	3181.0	3256.3

The system failed for the first time at .7 units (ages) of time, $t_1 = .7$ and it failed after the fortieth time at 3256.3 units of time, $t_{40} = 3256.3$. In addition to being able to identify the reliability of the system, we would be interested in knowing the time that the next, fourty-first, failure will occur. Thus, if we locate $MTBF_{40}$, then we will be able to estimate the next time that the system will fail. In addition to being able to calculate the reliability of the system, having a good estimate of the next failure is very important in developing maintenance strategies.

For $n = 40$, using the above estimates, we have

$$\widehat{\beta} = \frac{n}{\sum\limits_{i=1}^{n} \log\left(\frac{t_n}{t_i}\right)} = \frac{40}{\sum\limits_{i=1}^{40} \log\left(\frac{3256.3}{t_i}\right)} \approx .49 < \frac{1}{2},$$

$$\widehat{\theta} = \frac{t_n}{n^{1/\widehat{\beta}}} = \frac{3256.3}{40^{1/.49}} \approx 1.7461,$$

$$\widehat{h} = \left(\frac{t_n}{\widehat{\theta}}\right)^{\widehat{\beta}} = \left(\frac{3256.3}{1.7461}\right)^{.49} \approx 40.25,$$

$$\widehat{V}_n = \frac{\widehat{\beta}t_n^{\widehat{\beta}-1}}{\widehat{\theta}^{\widehat{\beta}}} = \frac{(.49)\,(3256.3)^{-.51}}{1.7461^{.49}} \approx .006$$

and

$$\frac{1}{\widehat{V}_n} \approx 166.$$

Since $\beta < \frac{1}{2}$, we can use the appropriate formulas, recommended above, to obtain lower and upper bound estimates of $MTBF_n$. That is,

$$\text{lower bound} = \frac{1}{\widehat{V}_n}\left(1 + \frac{\widehat{\alpha} - 1}{h}\right)$$

$$= 166\left(1 + \frac{1.0408 - 1}{40.25}\right) \approx 170.235,$$

and

$$\text{upper bound} = \frac{1}{\widehat{V}_n}\,\frac{1}{1 - \frac{\widehat{\alpha}-1}{h}}$$

$$= 166\,\frac{1}{1 - \frac{1.0408}{40.25}} \approx 170.350.$$

Thus, the estimate of the true $MTBF_n$ satisfies the following inequality:

$$170.235 \leq MTBF_n \leq 170.350,$$

and an effective estimate of $MTBF_n$ is

$$MTBF_n = \frac{\text{lower bound} + \text{upper bound}}{2} = 170.3.$$

We can now conclude that an estimate of the next unit of time that the system will fail will be 170.3 ages from the last time the system failed, that is, 3426.6 ages. In [3] and [4], the author had obtained 166 ages from the 40th failure. Note also that his estimate did not fall within the bounds we calculated for $MTBF_n$.

17.3   Parameter Estimation of NHPP

An effective estimate of $MTBF_n$ depends on having a good estimate of V_n, and having a good estimate of the intensity function depends on having good estimates of β and θ, the shape and scale parameters, respectively. We shall review some of the recent developments on the subject area.

You recall the MLE of β and θ are given by

$$\widehat{\beta} = \frac{n}{\sum_{i=1}^{n} \ln\left(\frac{t_n}{t_i}\right)} \quad \text{and} \quad \widehat{\theta} = \frac{t_n}{n^{\frac{1}{\beta}}}.$$

The MLE of the scale parameter, θ, is quite unstable and this problem has not been investigated. In [6], the authors have shown that $\widehat{\beta}$ is chi-squared distributed with $2(n-1)$ degrees of freedom. That is,

$$\widehat{\beta} \sim \frac{2n\beta}{\chi^2_{2(n-1)}}.$$

They have shown that $\widehat{\beta}$ is a biased estimate of β. That is,

$$E\left(\widehat{\beta}\right) = \frac{n}{n-2}\,\beta \text{ and Var}\left(\widehat{\beta}\right) = \frac{n^2}{(n-3)(n-2)^2}\,\beta^2.$$

In [5], the authors have suggested using an unbiased estimate of the shape parameter, that is,

$$\widetilde{\beta} = \frac{n-2}{n}\,\widehat{\beta} = \frac{n-2}{n}\,\frac{1}{\sum\limits_{i=1}^{n} \log \frac{t_n}{t_i}}.$$

Furthermore,

$$E\left(\widetilde{\beta}\right) = \beta, \text{Var}\left(\widetilde{\beta}\right) = \frac{(n-2)^2}{n^2}\,\text{Var}\left(\widehat{\beta}\right) = \frac{1}{n-3}\,\beta^2$$

and

$$MSE\left(\widetilde{\beta}\right) = \text{Var}\left(\widetilde{\beta}\right) = \frac{1}{n-3}\,\beta^2.$$

In [12], the authors have shown that $\dfrac{n\,V_n}{\widehat{V}_n}$ can be approximated by a chi-squared distribution with $(n-1)$ degrees of freedom. That is,

$$\frac{n\,V_n}{\widehat{V}_n} \sim \chi^2_{n-1}.$$

This result was also shown to be quite effective even for small n. Using this result, they obtain approximate unbiased estimates of the intensity function. More precisely, they obtained the following approximate estimates:
   (i) $\widetilde{V}_n$, a biased estimate of V_n under the influence of $\widetilde{\beta}$, which is unbiased:

$$\widetilde{V}_n = \frac{n\widetilde{\beta}}{t_n}$$

   (ii) $\overline{V}_n$, an unbiased estimate of V_n, corrected for biased MLE $\widehat{V}_n$:

$$\overline{V}_n = \frac{(n-3)}{t_n}\,\widehat{\beta}.$$

In [13], the authors have shown that

$$\frac{V_n}{\widetilde{V}_n} \sim \frac{1}{4n^2}\,Z_{2(n-1)}\,S_{2n}$$

where $Z_{2(n-1)}$ and S_{2n} are independent chi-squared random variables with $2(n-1)$ and $2n$ degrees of freedom, respectively. Using this result, [21], have obtained exact unbiased estimates of the intensity function. Let $X = 4n^2\,\dfrac{V_n}{\widetilde{V}_n}$. Then the probability density function of X is given by

$$f(x) = \frac{1}{\Gamma(n)\Gamma(n-1)2^{2n-1}} \int_0^\infty x^{n-2} e^{-\frac{x}{2y}} e^{-\frac{y}{2}} dy, \quad 0 < x < \infty.$$

For $X = ZS$ and $Y = S$ we have

$$f(x,\,y) = \frac{1}{\Gamma(n)\Gamma(n-1)2^{2n-1}} \left(\frac{x}{y}\right)^{n-2} e^{-\frac{x}{2y}} y^{n-1} e^{-\frac{y}{2}} \left(\frac{1}{y}\right)$$

$$= \frac{1}{\Gamma(n)\Gamma(n-1)2^{2n-1}} x^{n-2} e^{-\frac{x}{2y}} e^{-\frac{y}{2}}.$$

Thus,

$$E\left(\widehat{V}_n\right) = E\left(\frac{4n^2 V_n}{x}\right)$$

$$= \frac{4n^2 V_n}{\Gamma(n)\Gamma(n-1)2^{2n-1}} \int_0^\infty \int_0^\infty \frac{1}{x} x^{n-2} e^{-\frac{x}{2y}} e^{-\frac{y}{2}} dy\, dx$$

$$= \frac{4n^2 V_n}{\Gamma(n)\Gamma(n-1)2^{2n-1}} \int_0^\infty e^{-\frac{y}{2}} \int_0^\infty x^{n-3} e^{-\frac{x}{2y}} dx\, dy.$$

Let $u = \frac{x}{y}$. Then the $E\left(\widehat{V}_n\right)$ becomes

$$E\left(\widehat{V}_n\right) = \frac{4n^2 V_n}{\Gamma(n)\Gamma(n-1)2^{2n-1}} \int_0^\infty e^{-\frac{y}{2}} y^{n-2} \int_0^\infty u^{n-3} e^{-\frac{u}{2}} du\, dy$$

$$= \frac{4n^2 V_n}{\Gamma(n)\Gamma(n-1)2^{2n-1}} \Gamma(n-2) 2^{n-2} \Gamma(n-1) 2^{n-1}$$

$$= \frac{n^2}{(n-1)(n-2)} V_n \;.$$

Hence, an unbiased estimate of the intensity function V_n is given by

$$V_n^* = \frac{(n-1)(n-2)}{n^2} \widehat{V}_n = \frac{(n-1)(n-2)}{n^2} \cdot \frac{n\widehat{\beta}}{t_n}$$

$$= \frac{(n-1)(n-2)}{n} \frac{\widehat{\beta}}{t_n} \;.$$

Furthermore,

$$E\left(\widehat{V}_n^2\right) = E\left(\frac{16n^4 V_n^2}{x^2}\right)$$

$$= \frac{16n^4 \, V_n^2}{\Gamma(n) \, \Gamma(n-1) \, 2^{2n-1}} \int_0^\infty e^{-\frac{y}{2}} \, y^{n-3} \int_0^\infty u^{n-4} \, e^{-\frac{u}{2}} \, du \, dy$$

$$= \frac{16n^4 \, V_n^2}{\Gamma(n) \, \Gamma(n-1) \, 2^{2n-1}} \, \Gamma(n-2) \, 2^{n-2} \, \Gamma(n-3) \, 2^{n-3}$$

$$= \frac{n^4 \, V_n^2}{(n-1)(n-2)^2 (n-3)}.$$

Hence,

$$\text{Var}\left(\widehat{V_n}\right) = V_n^2 \, n^4 \left(\frac{1}{(n-1)(n-2)^2 (n-3)} - \frac{1}{(n-1)^2 (n-2)^2} \right)$$

$$= \frac{2n^4 \, V_n^2}{(n-1)^2 (n-2)^2 (n-3)}.$$

Also, the variance of the unbiased estimate of V_n is obtained as follows:

$$\text{Var}\left(\overline{V}_n^*\right) = \frac{(n-1)^2 (n-2)^2}{n^4} \, \text{Var}\left(\widehat{V_n}\right)$$

$$= \frac{2 V_n^2}{n-3}.$$

The MSE of the MLE of V_n is

$$MSE\left(\widehat{V_n}\right) = \text{Var}\left(\widehat{V_n}\right) + \left\{ E\left(\widehat{V_n}\right) - V_n \right\}^2$$

$$= V_n^2 \left\{ \frac{2n^4 + 9n^3 - 39n + 40n - 12}{(n-1)^2 (n-2)^2 (n-3)} \right\}.$$

In summary, the authors in [21], have obtained the following exact estimates of the intensity function:

(i) $\widetilde{V}_n^*$, an exact biased estimate of V_n under the influence of $\widetilde{\beta}$, which is unbiased.

$$\widetilde{V}_n^* = \frac{n-2}{n} \, \widetilde{\beta}$$

(ii) $\overline{V}_n^*$, an exact unbiased estimate of V_n corrected for biased MLE $\widehat{V}_n$.

$$\overline{V}_n^* = \frac{(n-1)(n-2)}{n} \, \frac{\widehat{\beta}}{t_n}$$

In what follows, we shall numerically compute the exact and approximate estimates of the intensity function.

17.4 Numerical Comparison of the Estimates

We shall use the concept of relative efficiency to evaluate the exact and approximate estimates of the intensity function. That is, assume that θ_1 and θ_2 are two possible estimates of the parameter θ. We define the relative efficiency of the estimate θ_1 compared to the estimate θ_2 as follows:

$$\text{Eff}\left(\frac{\theta_1}{\theta_2}\right) = \frac{MSE\,(\theta_2)}{MSE\,(\theta_1)}.$$

If $\text{Eff}\left(\dfrac{\theta_1}{\theta_2}\right) < 1$, we can conclude that the estimate of θ_2 is more efficient than θ_1. Whereas, if $\text{Eff}\left(\dfrac{\theta_1}{\theta_2}\right) > 1$, we conclude that the estimate θ_1 is more efficient than θ_2 with respect to minimizing the MSE. In case the $\text{Eff} = \left(\dfrac{\theta_1}{\theta_2}\right) = 1$, then both estimates are equally effective.

Table 17.1, given below, displays the theoretical efficiencies for the exact and approximate, biased and unbiased estimate of the intensity function.

It is clear that for small sample sizes there is a significant difference between the exact and approximate estimates of the failure intensity function. However, for large n the approximate estimates of V_n converge to the exact estimates.

Table 17.2 contains some numerical results of the relative efficiencies of the exact and approximate estimates of the failure intensity function. For $n = 10$, the exact unbiased estimate of $\overline{V}_n^*$ is about 40% more efficient than the approximate estimate $\overline{V}_n$. The relative efficiency of the exact estimate reduces to about 2% over the approximate estimate of V_n for a sample of $n = 100$.

References

1. Ascher, H.E., Lin, T.T.Y., and Siewiorek, D.P. (1992), Modification of: Error log analysis: Statistical modeling and heuristic trend analysis, *IEEE Transactions on Reliability*, **41**, 599-601.

2. Ascher, H.E. and Feingold, H. (1984), *Repairable Systems Reliability: Modeling, Inference, Misconceptions and Their Causes*, Marcel-Dekker, New York.

3. Bain, L.J. and Engelhardt, M. (1991), *Statistical Analysis of Reliability and Life Testing Models*, Marcel-Dekker, New York.

4. Basin, W.M. (1969), Increasing hazard functions and overhaul policy, *Proceedings of the 1969 Annual Symposium on Reliability*, Chicago, IEEE, **8**, 173-178.

5. Calabria, R., Guida, M., and Pulcini, G. (1988), Some modified maximum likelihood estimators for the Weibull process, *Reliability Engineering and System Safety*, 51-55.

6. Crow, L.H. (1975), Tracking reliability growth, *Proceedings of the Twentieth Conference Design of Experiments*, Report 75-2, U.S. Army Research Office, Research Triangle Park, North Carolina, 741-754.

7. Crow, L.H. (1974), Reliability analysis for complex, repairable systems, In *Reliability and Biometry*, (Eds., F. Proschan and R.J. Serfling), SIAM, Philadelphia, Pennsylvania.

8. Duane, J.T. (1964), Learning curve approach to reliability monitoring, *IEEE Transactions on Aerospace*, **2**, 563-566.

9. Engelhardt, M. and Bain, J.L. (1987), Statistical analysis of a compound power law model for repairable systems, *IEEE Transactions on Reliability*, **R-36**, 392-396.

10. Engelhardt, M. and Bain, J.L. (1978), Prediction intervals for the Weibull process, *Technometrics*, **20**, 167-169.

11. Finkelstein, J.M. (1978), Confidence bounds on the parameters of the Weibull process, *Technometrics*, **18**, 115-117.

12. Higgins, J.J. and Tsokos, C.P. (1981), A quasi-Bayes estimate of the failure intensity of a reliability-growth model, *IEEE Transactions on Reliability*, **R-30**, 471-475.

13. Lee, L. and Lee, S.K. (1978), Some results on inference for the Weibull process, *Technometrics*, **20**, 41-45.

14. Newton, D.W. (1991), Some pitfalls in reliability data analysis, *Reliability Engineering and System Safety*, **34**, 7-21.

15. Qiao, H. and Tsokos, C.P. (1994), Best efficient estimates of the intensity function of the Weibull process, *Submitted for publication*.

16. Qiao, H. and Tsokos, C.P. (1994), Reliability growth: MTBF vs. intensity function of the Weibull process, *Submitted for publication*.

17. Rigdon, S.E. and Basu, A.P. (1990), The effect of assuming a homogeneous Poisson process when the true process is a power law process, *Journal of Quality Technology*, **22**, 111-117.

18. Rigdon, S.E. and Basu, A.P. (1990), The Power law process: a model for the reliability of repairable systems, *Journal of Quality Technology*, **21**, 251-260.

19. Rigdon, S.E. and Basu, A.P. (1990), Estimating the intensity function of a power law process at the current time: Time truncated case, *Communications in Statistics-Simulation and Computation*, **19**, 1079-1104.

20. Thompson, Jr., W.A. (1981), On the foundations of reliability, *Technometrics*, **23**, 1-23.

21. Tsokos, C.P. and Rao, A.N.V. (1994), Estimation of failure intensity for the Weibull process, *Reliability Engineering and System Safety* (To appear).

TABLE 17.1
Theoretical efficiencies

$$\frac{\text{EXACT}}{\text{EXACT}}: \quad \text{Eff}\left(\frac{\overline{V}_n^*}{\widehat{V}_n^*}\right) = \frac{MSE\left(\overline{V}_n^*\right)}{MSE\left(\widehat{V}_n^*\right)} = \frac{2(n-1)(n-2)^2}{2n^3 + 11n^2 - 28n + 12}$$

$$\frac{\text{APPR.}}{\text{APPR.}}: \quad \text{Eff}\left(\frac{\overline{V}_n}{\widehat{V}_n}\right) = \frac{MSE\left(\overline{V}_n\right)}{MSE\left(\widehat{V}_n\right)} = \frac{2(n-3)}{2n + 15}$$

$$\frac{\text{APPR.}}{\text{EXACT}}: \quad \text{Eff}\left(\frac{\widehat{V}_n}{\widehat{V}_n^*}\right) = MSE\left(\frac{\widehat{V}_n}{\widehat{V}*}\right) = \frac{(2n+15)\,(n-1)(n-2)^2}{(2n^3 + 11n^2 - 28n + 12)(n-5)}$$

$$\frac{\text{APPR.}}{\text{EXACT}}: \quad \text{Eff}\left(\frac{\overline{V}_n}{\overline{V}^*}\right) = \frac{MSE\left(\overline{V}_n\right)}{MSE\left(\overline{V}^*\right)} = \frac{n-3}{n-5}$$

TABLE 17.2
Numerical efficiencies

		$n = 5$	$n = 10$	$n = 20$	$n = 30$	$n = 50$	$n = 100$
$\frac{\text{EXACT}}{\text{EXACT}}$:	$\text{Eff}\left(\frac{\overline{V}_n^*}{\widehat{V}_n^*}\right) =$	0.18	0.40	0.62	0.72	0.82	0.90
$\frac{\text{APPR.}}{\text{APPR.}}$:	$\text{Eff}\left(\frac{\overline{V}_n}{\widehat{V}_n}\right) =$	0.16	0.40	0.62	0.72	0.82	0.90
$\frac{\text{APPR.}}{\text{EXACT}}$:	$\text{Eff}\left(\frac{\widehat{V}_n}{\widehat{V}_n^*}\right) =$	——	1.00	1.14	1.08	1.05	1.02
$\frac{\text{APPR.}}{\text{EXACT}}$:	$\text{Eff}\left(\frac{\overline{V}_n}{\overline{V}_n^*}\right) =$	——	1.40	1.13	1.08	1.04	1.02

18

Reliability Growth - Early Models and Some Recent Developments

Philip Prescott

University of Southampton

ABSTRACT During the last thirty years a considerable amount of research has been carried out into the development of reliability growth models. This paper describes the early models proposed for both discrete and continuous measures of reliability, and outlines some of the more recent work on models produced for software reliability and error detection using continuous models based on non-homogeneous Poisson processes. It also discusses continuous analog estimators for discrete single-mission systems using estimators similar in form to those obtained for continuous models.

CONTENTS

18.1 Introduction

Since the introduction in [11], [12] of the reliability growth model for monitoring the changes in the reliability of a product due to design modifications and the corrective actions of testing and fixing errors and faults, and the further development of these ideas in [7], [8], [9] and others, many alternative models, or variants of the Duane model, have been proposed in the literature. Several reviews of this area have appeared including the general survey of reliability growth in [10] and the description of a general framework for learning curve reliability growth models in [18]. Over the last ten years there has been considerable interest in the use of these reliability growth models for assessing reliability in software systems such as operating systems, control programs and large application programs. A review of some of the early work in this area was given in [38]. Further developments are described in [19], [22], [35], [36], [37], [39], [40], and [41].

It is clear from the intense research activity in this area that these models continue to be used widely, perhaps sometimes with confusion over the applicability of the particular models proposed. In [16], the author points out that there is a tendency to forget that reliability growth is not automatic and that any improvement in reliability results from a program of testing-analyzing-and-fixing. In [34], the author asks whether reliability growth is a myth or a mess and recommends more strict definitions of the reliability growth process involved.

Confusion over reliability growth management was only too evident a few years ago when I was asked to look at the statistical aspects of a contract between a large electronics company and a government department. The contract involved the design and manufacture of a complex electronic device, and the company had run into difficulties over meeting the agreed completion time for this project. There was considerable argument between the parties concerned over the level of reliability achieved up to that point in time, and over whether there was any possibility of the company being able to meet the targets specified in the contract. Unfortunately, the contract had been set up under acceptance sampling arrangements which were more suited to the production of a regular stream of large batches of items which could be inspected on an acceptance-rejection basis, with rejected batches being fully inspected for defects and then replaced by other items. This type of quality check on the manufacturing process was completely inappropriate in this case, as these items were large and expensive to make individually, and testing them was virtually a destructive process. It would have been more appropriate to have based the contractual requirements on documents such as Mil-Std-1635 (EC) or Mil-Hdbk-189, [23], [24], since the development program was clearly one of designing and testing small numbers of devices. Reliability growth management methods could then have been used to monitor improvements in reliability, so that the necessary resources could have been allocated to the project in order to reach the specified target of reliability within the contract time. Once a satisfactory design had been achieved, the manufacturing process could then have been monitored with a suitable quality control procedure.

Once it had become clear that an assessment of reliability growth was the appropriate way of looking at the development of the product, it was evident from the discussions about the reliability of these devices that there was still confusion and disagreement about the kind of reliability growth model that could be employed. As with most products in the development stage, there were many types of fault which could occur during the testing program. The equipment was meant to be placed in a strategic location and an aerial was to be deployed prior to a transmitter being switched on to send out signals over a period of time. The operation was basically a two-stage procedure, with the first stage, prior to switch-on, being of a go-nogo type. Reliability during this stage should have been measured in terms of the probability of successful deployment up to the point of switch-on. Reliability in the second stage would be measured in terms of the length of time that a satisfactory signal was transmitted. These two aspects needed to be considered together in a reliability growth program, with initial emphasis being placed on the first stage, since, clearly, a long time to failure in the transmission stage was of no importance if the equipment was not switched on. These two different aspects of reliability lead to different types of reliability growth models, discrete models and continuous models.

18.2   Discrete Growth Models

Amongst the early models proposed for reliability growth based on discrete success-failure data obtained from single-mission systems, using proportions of successes or cumulative proportions of successes, were those discussed in [20]. Useful illustrations of growth curves were given in [1] enabling a variety of possible growth curves to be examined. In [4], the author extended these results. In these models the reliability is assessed at each stage of the test program by the proportion of successes observed in a series of tests, i.e., at the kth out of N tests, involving n_k items of which r_k are successful, the reliability is estimated as $R_k = r_k/n_k$ and these values are modeled by some suitable growth curve.

The Lloyd-Lipow model is $R_k = R_\infty - \alpha/k$, the Bonis model is $R_k = R_\infty - Q\,\beta^{k-1}$, and the exponential model is $R_k = R_\infty(1 - \alpha\,e^{-\beta k})$, where R_∞ is some ultimate anticipated reliability resulting from prolonged testing and re-designing, α and Q are constants and β represents the rate of growth. Suitable plots on log-log or log-linear scales for these growth curves may be used to estimate the parameters and the plots may be extrapolated to estimate the reliability at future stages of the development program. For example, using the Bonis model

$$\log(R_\infty - R_k) = \log(Q/\beta) + k\log(\beta) \qquad (18.2.1)$$

plots as a straight line on log-linear scales. A feasible value for R_∞ may be assumed or this may be estimated using an alternative method such as maximum likelihood, see [4] for details.

Although simple plots can often be useful, more formal estimation procedures are available, for example, [20] give the maximum likelihood and least squares estimates of R_∞ and α for their simple model, together with a lower confidence limit for R_∞. The likelihood equations for $\widehat{R}_\infty$ and $\widehat{\alpha}$, which need to be solved iteratively are:

$$\sum_{k=1}^{N} \frac{kr_k}{D_k n_k} = \widehat{R}_\infty \sum_{k=1}^{N} \frac{k}{D_k} - \widehat{\alpha} \sum_{k=1}^{N} \frac{1}{D_k} \qquad (18.2.2)$$

and

$$\sum_{k=1}^{N} \frac{r_k}{D_k n_k} = \widehat{R}_\infty \sum_{k=1}^{N} \frac{1}{D_k} - \widehat{\alpha} \sum_{k=1}^{N} \frac{1}{kD_k} \,, \qquad (18.2.3)$$

where

$$D_k = k(\widehat{R}_\infty - \widehat{\alpha}/k)(1 - \widehat{R}_\infty + \widehat{\alpha}/k)/n_k \,. \qquad (18.2.4)$$

Initial values for the iterative solution of these equations may be obtained from the least squares estimates which are given by the following:

$$\widehat{\alpha} = \frac{C_1 \sum_{k=1}^{N}(r_k/n_k) - N \sum_{k=1}^{N}(r_k/kn_k)}{N\,C_2 - C_1^2} \qquad (18.2.5)$$

and

$$\widehat{R}_\infty = \frac{C_2 \sum_{k=1}^{N}(r_k/n_k) - C_1 \sum_{k=1}^{N}(r_k/kn_k)}{N\,C_2 - C_1^2} \,, \qquad (18.2.6)$$

where

$$C_1 = \sum_{k=1}^{N} \frac{1}{k} \quad \text{and} \quad C_2 = \sum_{k=1}^{N} \frac{1}{k^2} \,. \qquad (18.2.7)$$

In cases where failures may be classified as either inherent or assignable cause failures, trinomial discrete reliability growth models have been proposed, see for example, [33], [2], and [32]. A time series approach to reliability growth analysis of discrete data was proposed in [29] using Box-Jenkins [5] ARIMA models, which have the advantage that no specific model need be selected in advance. The data available lead to selection of a specific model from the broad and flexible class of ARIMA models. The main disadvantage of this approach is that generally the number of tests required to be carried out is higher than with the other forms of model.

18.3   Continuous Growth Models

With models of this type the measure adopted for assessing reliability growth is time between failures or the number of failures in specified time intervals. The two major sub-categories of continous reliability growth models are those based on non-homogeneous point processes and those requiring the solution of a differential equation. Amongst early examples of the latter type is the IBM model described in [27] based on the solution of a differential equation generated from the assumptions that two types of failure are possible, random failures occurring at constant rate λ and non-random failures due to design or manufacturing defects. The number of non-random failures is fixed but unknown at the beginning of testing. If $N(t)$ is the number of non-random defects remaining at time t, the rate of change of $N(t)$ is proportional to $N(t)$. This is described by the equation $dN(t)/dt = -k_2N(t)$, where k_2 is a constant. The solution to this equation is $N(t) = k_1 \exp(-k_2t)$, where k_1 is the unknown number of non-random failures at time $t = 0$. The expected cumulative number of failures up to time t is then

$$V(t) = \lambda t + k_1 - N(t) = \lambda t + k_1\{1 - \exp(-k_2t)\} \qquad (18.3.1)$$

and the cumulative mean time between failures is $t/[\lambda t + k_1\{1 - \exp(-k_2t)\}]$.

One advantage of this model is its capacity for assessing the fraction of non-random failures that have been removed in the development program. If q represents this fraction, then $q = 1 - \exp(-k_2t)$, since the number of non-random defects removed by time t is

$$N(0) - N(t) = k_1\{1 - \exp(-k_2t)\} \ . \qquad (18.3.2)$$

Once k_2 has been estimated, by $\widehat{k}_2$ say, the time to eliminate $100\,q$ percent of the non-random failures may be estimated by solving for t_q, giving

$$\widehat{t}_q = -\ln(1 - q)/\widehat{k}_2 \ , \qquad (18.3.3)$$

which is particularly useful for planning a testing program. Another important feature of the IBM model is that the number of non-random failures remaining at time t can be estimated from $\widehat{k}_1 \exp(-\widehat{k}_2t)$, and the estimate of λ, $\widehat{\lambda}$ say, gives an estimate of the long-run achievable mean time between failures. Estimates of λ, k_1 and k_2 may be found using iterative least squares. Other differential equation models are described in [20].

Perhaps the most commonly used of all reliability growth models, applicable to the continuous classification, is the Duane model which belongs to the non-homogeneous point process group of models. This model involves the cumulative number of failures F up to time T and is of the form

$$F = F_0T^\beta \ , \qquad (18.3.4)$$

where F_0 and β are constants. The gradient of the curve at any point T is the instantaneous failure rate, i.e., $\lambda(T) = dF/dT = F_0\beta\,T^{\beta-1} = F_0\beta\,T^{-\alpha} = \lambda_0T^{-\alpha}$, where $\alpha = 1-\beta$ and $\lambda_0 = F_0\beta$ is a constant. This gives an alternative form of the model as

$$\lambda(T) = \lambda_0T^{-\alpha} \ . \qquad (18.3.5)$$

If it is assumed that at any stage in the development program the failure rate is $\lambda(T)$ and that this would remain constant if no further improvements were made to the product, then this is equivalent to assuming that the time to failure follows a negative exponential distribution with parameter $\lambda(T)$ and therefore that the mean time to failure is $m(T) = 1/\lambda(T)$.

The model may be expressed in terms of the mean time between failures as

$$m(T) = m_0 T^\alpha \ , \tag{18.3.6}$$

where $m(T)$ is the instantaneous mean time between failures, m_0 is some initial value and α is the growth rate. In a reliability growth program the information consists of a series of failures of components or systems over the development time, giving rise to measures of cumulative test time and cumulative numbers of failures. The ratio $m_c = T/F$, where T is the cumulative test time and F is the cumulative number of failures, represents the cumulative mean time between failures. From this information it is required to assess the change in the instantaneous mean time between failures due to the development program. Using the earlier equations we have that $\lambda_0 = F_0 \beta$ and $m_0 = 1/\lambda_0$, so that $F = T^\beta/(m_0\beta)$, from which

$$m_c = m_0 \beta \, T^{1-\beta} \tag{18.3.7}$$

$$\text{i.e., } m_c = m_0(1-\alpha)T^\alpha \ . \tag{18.3.8}$$

On taking logarithms of (18.3.6) and (18.3.8) we have

$$\log\{m(T)\} = \log(m_0) + \alpha \log(T) \tag{18.3.9}$$

and

$$\log(m_c) = \log(m_0) + \log(1-\alpha) + \alpha \log(T) \ . \tag{18.3.10}$$

A plot of $m_c = T/F$ against T on log-log scales gives a straight line with gradient α for the cumulative mean time between failures. Once α has been estimated, a straight line parallel to this at a distance $\log(1-\alpha)$ from it gives instantaneous mean time between failures. This may be used to estimate $m(T)$. In [12], the author and other such authors as in [28] showed that this simple model fitted a number of practical data sets quite well, and they used graphical procedures or least squares to estimate the parameters involved. In [8], the author formulated the model as a non-homogeneous Poisson process with the Weibull intensity function, and determined maximum likelihood estimators of the parameters of the form

$$\widehat{F}_0 = N/T^{\widehat{\beta}} \text{ and } \widehat{\beta} = N/\sum_{k=1}^{N} \log(T/t_k) \ , \tag{18.3.11}$$

where N is the total number of failures in time T and t_k, $k = 1, \ldots, N$, are the individual times to failure. He also gave the associated confidence intervals based on chi-squared approximations. In a recent paper [42] the authors return to simple graphical analysis of several models based on non-homogeneous Poisson processes, including Duane's model. They recommend that simple plots involving $\log(T)$, $\log(\log(T))$ and $1/T$ be used to investigate the suitability of various models, including power-law (Duane) models, logarithmic models of the form $F = \beta \, \log(\alpha T)$ and $F = \log(1 + \alpha T^\beta)$, log-power models such as $F = \alpha\{\log(T)\}^\beta$ and $F = \alpha\{\log(T+1)\}^\beta$ and the inverse-exponential model $F = \alpha \, \exp(-\beta/T)$. They confirm that these graphical methods provide a useful and effective way of visually validating a model prior to parameter estimation.

Further details of the parameter estimation equations using maximum likelihood for the Duane model and for several other models such as the modified Weibull growth model, the exponential and the gamma growth models, are summarized in [38], who also provide goodness of fit tests based on Chi-squared, Cramér-von Mises and Kolmogorov-Smirnov statistics for the non-homogeneous Poisson process reliability growth models.

In addition, it is pointed out that in some cases the early data do not fit the Duane growth curve very well, possibly because of failure repetitions in the early development stages. In [30], the authors proposed that the reliability growth model should be based on the number of failures after some initial time T_0, leading to a modified model of the form

$$F = F_0(T^\beta - T_0^\beta) \ . \qquad (18.3.12)$$

This phenomenon of curvature for early data of the straight line plots modeled by Duane curves, had been noted elsewhere, see for example [6] in which it is assumed that this was due to inadequate or noisy early data. In [21], the author also reported curvature in the Duane learning curves at the beginning of a testing program and several examples of this are given in [17].

18.4   Application of Non-Homogeneous Poisson to Software Reliability

Although the continuous reliability growth curve models considered above have been widely used in manufacturing industry, particularly in electronics and avionics and in the development programs of computer hardware, perhaps the most active area of recent research has been in the application of reliability growth modeling to software engineering. Software reliability is, according to [40], a key issue in modern software product development, and in recent years they and their colleagues have done much to promote the use of software reliability growth models. In their review paper they discuss several non-homogeneous Poisson process software reliability growth models and classify them according to a reliability growth index and an error detection rate. Maximum likelihood estimates of the parameters of the models and of the software reliability measures are considered and illustrated with practical applications. Their models assume that a software system is subject to software failures at random times caused by software errors and that, each time a failure occurs, the software error which caused it is removed and no new errors are introduced. Using their notation, $N(t)$ represents the cumulative number of errors detected in the time interval $(0, t]$ and the expected value of $N(t)$, the mean value function, is $H(t)$.

Let $a = H(\infty)$ be the expected number of errors eventually detected, i.e. the expected number of errors initially in the software system, and define the error detection rate per error (unit of time) at testing time t to be

$$d(t) = h(t)/[a - H(t)] \ , \qquad (18.4.1)$$

where $h(t)$ is the intensity function of the process given by $h(t) = dH(t)/dt$.

In [15], the authors proposed the exponential software reliability growth model of the form

$$H(t) = a[1 - \exp(-bt)], \qquad b > 0 \ , \qquad (18.4.2)$$

where $d(t) = b$ represents the error detection rate per error at an arbitrary testing time. Clearly, with this model the error detection rate is regarded as a constant. A modification of this model which takes care of different types of errors in the software, depending on their severity, was suggested in [39], [41]. In this model type 1 and type 2 errors are, respectively, easy and difficult to detect, so that the error detection rates b_1 and b_2 are such that $0 < b_2 < b_1 < 1$. The model has mean value function

$$H(t) = a \sum_{i=1}^{2} p_i [1 - \exp(-b_i t)] , \qquad (18.4.3)$$

where p_i, $i = 1, 2$, with $p_1 + p_2 = 1$, are the proportions of type 1 and type 2 errors, respectively. For this model the error detection rate per error at testing time t is given by

$$d(t) = \sum_{i=1}^{2} \left(\frac{p_i b_i \exp(-b_i t)}{p_1 \cdot \exp(-b_1 t) + p_2 \cdot \exp(-b_2 t)} \right) . \qquad (18.4.4)$$

A further modification of this model was proposed in [36] including an extra term which resulted in an S-shaped growth curve. The model has mean value function

$$H(t) = a[1 - (1 + bt) \exp(-bt)], \qquad b > 0 . \qquad (18.4.5)$$

For this model the error detection rate increases with testing time and is given by

$$d(t) = b^2 t / (1 + bt) . \qquad (18.4.6)$$

Another S-shaped software reliability growth model was proposed in [25], called the inflection S-shaped model, describing a software failure detection method in which the more failures found the more undetected failures become detectable. The mean value function for this model is

$$H(t) = a[1 - \exp(-bt)] / [1 + c \cdot \exp(-bt)], \qquad b > 0, \ c > 0 . \qquad (18.4.7)$$

The error detection rate for this model is also montonically increasing with testing time t, and is given by

$$d(t) = b / [1 - c \cdot \exp(-bt)] . \qquad (18.4.8)$$

Recently, in [19], the authors have combined two of these forms of model and suggested a modified S-shaped software reliability growth model 'which reflects the real-life situation more closely by accounting for the types of errors in the software'. Their model allows for errors of different kinds and includes the terms $(1 + b_i t)$, having mean value function

$$H(t) = a \sum_{i=1}^{2} p_i [1 - (1 + b_i t) \exp(-b_i t)] \qquad (18.4.9)$$

and corresponding error detection rate

$$d(t) = \sum_{i=1}^{2} \left(\frac{t \, p_i b_i^2 \exp(-b_i t)}{p_1 (1 + b_1 t) \exp(-b_1 t) + p_2 (1 + b_2 t) \exp(-b_2 t)} \right) . \qquad (18.4.10)$$

For all these models the parameters may be estimated using maximum likelihood, and, provided there is sufficient data, asymptotic normality may be assumed for the distributions of the estimates with variance-covariance matrix given by the inverse of the Fisher information matrix. Confidence intervals may be obtained for the reliability growth curve using these variances and covariances. The maximum likelihood equations are reasonably straight-forward to write down but usually require numerical methods for their solution. As an illustration, the equations for a, b_1 and b_2 in the last model described above are reproduced from [19]. The available data consist of z_i, $i = 1, \ldots, n$, of cumulative numbers of errors detected up to time t_i. The three equations to be solved for $\hat{a}$, $\hat{b}_1$ and $\hat{b}_2$ are as follows:

$$\hat{a} = z_n / \left(\sum_{i=1}^{2} p_i [1 - (1 + \hat{b}_i t_n) \exp(-\hat{b}_i t_n)] \right) \qquad (18.4.11)$$

$$\hat{a}\hat{b}_k t_n^2 \exp(-\hat{b}_k t_n) = \sum_{i=1}^{n} \frac{[(z_i - z_{i-1})\{\hat{b}_k t_i^2 \exp(-\hat{b}_k t_i) - \hat{b}_k t_{i-1}^2 \exp(-\hat{b}_k t_{i-1})\}]}{\sum_{j=1}^{2} p_j [(1 + \hat{b}_j t_{i-1}) \exp(-\hat{b}_j t_{i-1}) - (1 + \hat{b}_j t_i) \exp(-\hat{b}_j t_i)]}$$

$$(18.4.12)$$

for $k = 1, 2$.

In [22], the author considers three different sub-families of exponential order statistic models based on three classes of models for the failure rates of the design flaws initially present in the system. The first class is of a collection of deterministic rates, the second is a finite set of random independently identically distributed rates and the third class is a set of random rates whose joint distribution is that of a non-homogeneous Poisson process. The first class leads to deterministic exponential order statistic models while the other two lead to doubly stochastic exponential order statistic models. Miller shows that this rich source of models contains the power law, logarithmic and many other models as special cases.

18.5   Continuous Analog Estimators for Discrete Reliability Growth Models

The great wealth of results available for the continuous reliability growth models, in particular the development of maximum likelihood estimators and their sampling distributions for the non-homogeneous Poisson process models, has prompted several authors to suggest ways of applying similar methods to the discrete models suitable for estimating reliability growth for single-mission systems.

The assumption underlying the simple logarithmic models for the discrete case, that the mean number of cumulative failures is linearly related to the cumulative test trial number on a log-log scale, is identical to the assumption underlying the derivation of the continuous model, except that the variable time-to-test in the continuous model is replaced by test-trial-number for the discrete model. The resulting model may therefore be viewed as a discrete analog to the non-homogeneous Poisson process model.

In [14], a logarithmic growth model has been proposed, similar to the Duane model, for the reliability of a single-mission system and estimation of the parameters was considered under the premise that one item is tested for each system configuration. The assumption here is that the expected number of failures in M tests is proportional to M^ϕ, so the model is

$$E(Y; M) = q_0 M^\phi , \qquad (18.5.1)$$

where Y is the observed number of failures during the testing program, with reliability growth occurring when $0 < \phi < 1$. The probability of failure for the test number i is then given by

$$q(i) = q_0 [i^\phi - (i - 1)^\phi] \qquad (18.5.2)$$

$$\text{since} \qquad E(Y;M) = \sum_{i=1}^{M} q(i) \ . \tag{18.5.3}$$

The maximum likelihood estimators for ϕ and $q\{M(R)\}$, where $M(R)$ is the number of tests planned to reach a reliability target R, obtained by analogy with those for the parameters of the continuous reliability growth model, are

$$\phi^* = r \ / \ \sum_{i=1}^{r} \log\{M/f(i)\} \text{ and} \tag{18.5.4}$$

$$q^*\{M(R)\} = r/M^{\phi^*}[\{M(R)\}^{\phi^*} - \{M(R) - 1\}^{\phi^*}] \ . \tag{18.5.5}$$

These are not the maximum likelihood estimators of ϕ and q for the discrete model, but are the analogs of those for the Duane model. Simulation methods were used in [14] to show that the continuous analog estimators, produced by paralleling the maximum likelihood estimators from the Duane model in this way, were generally superior to the least squares estimators in that they were less variable and tended to under-estimate the reliability growth, whereas the least squares method tended to over-estimate the reliability.

In [3], the authors extended this idea to the situation where m trials are included in each new configuration of the system, arguing that multiple trials are more common in practice. They show that the properties of continuous analog estimators for the discrete model can be handled analytically just as easily, provided that the number of trials is the same for each of N configurations included in the testing program. Their discrete reliability model has reliability of the system increasing with configuration number i according to

$$R_i = 1 - \lambda[i^{\beta} - (i-1)^{\beta}] \ , \ i = 1, 2, \ldots, N \ , \ 0 < \beta < 1 \ , \ 0 < \lambda < 1 \ . \tag{18.5.6}$$

In this form of the model the parameter λ is the system unreliability for the initial configuration and $(1 - \beta)$ is the growth parameter, with reliability growing if $\beta < 1$. The continuous analog maximum likelihood estimators of β and λ are

$$\beta^* = Y \ / \ \sum_{j=1}^{Y} \log\{N/f(j)\} \tag{18.5.7}$$

$$\text{and} \qquad \lambda^* = Y \ / \ (mN^{\beta^*}) \ , \tag{18.5.8}$$

where Y is the number of observed failures and $f(j)$ is the configuration at which the failure j occurred. Using these estimators, the reliability of the system at configuration N is estimated as

$$R_N^* = 1 - \lambda^*[N^{\beta^*} - (N-1)^{\beta^*}] \ . \tag{18.5.9}$$

The conjectures about the asymptotic properties of consistency and normality of the continuous analog estimators, made in [14], based on the simulation results, were formally established in [3], although the authors warn that the rates of convergence are slow so that 'inference procedures developed for the continuous model should be carefully scrutinized for their appliability to the discrete situation'. The continuous analog estimators were, in general, found to over-estimate β and under-estimate the system reliability. A large sample $(1 - \alpha)$ lower confidence bound for the reliability R_N was shown to be

$$R_N > 1 - (1 - R_N^*) \exp\{z_\alpha (2/Y)^{1/2}\} \ , \tag{18.5.10}$$

where z_α is such that $P(Z < z_\alpha) = 1 - \alpha$, for Z a standard normal variable.

18.6   Concluding Remarks

This meander through the variety of models and estimation methods for assessing reliability growth began by emphasizing the need to consider the kind of model which might
be suitable for the particular development program being undertaken. It was pointed
out that in some development programs different aspects of the way in which a product
functions could mean that assessment of reliability might involve the use of both discrete
and continuous reliability growth models simultaneously. We have looked at some of the
earliest models proposed, which used simple graphical techniques, and considered some of
the more complicated models recently presented. We have not considered nonparametric
methods and Bayesian models, see, for example, [13], [26], and [31].

It is evident that there is no shortage of models and estimation formulae available for fitting the models and for testing the adequacy of the fit. However, we should not lose sight
of the original purpose of implementing a reliability growth program. Its main purpose is
to monitor the progress of a design and development program so that, if the reliability is
falling behind the reliability targets set up to meet the contract specifications, then more
resources can be directed to this project to change the rate of reliability growth. It is possible that, if progress is running ahead of schedule, resources may be re-assigned to other
projects, resulting in a reduction in the reliability growth rate for this program. In this
way the data collected during the design and testing phase of the development of a product are quite likely to have arisen from a number of stages in which the reliability growth
rates were considerably different. Fitting a complicated model to the entire data set is
perhaps not the most appropriate way of using the data to estimate future reliability. For
estimation and extrapolation of reliability, the important data are the most recent values
obtained with the current level of resources allocated to the project. Graphical analyses
using simple plots on suitable scales can provide an easily understood management tool
for monitoring reliability growth. Once the structure of the data is validated, including
any changes due to alterations in the resource allocation, suitable models for prediction
may be fitted using the relevant recent data.

An alternative way of recognizing that the reliability is highly related to the amount of
development resources spent on detecting and correcting faults, is to include in the model
the amount of test-effort expended during the various stages of the development program.
A recent contribution to the area of software reliability growth modeling, allowing for test-
effort, was presented in [35] . In this paper, the authors have described the time-dependent
behavior of the test-effort by a Weibull curve, and assumed that, in a growth model
based on a non-homogeneous Poisson process, the error detection rate is proportional
to the current error content and the proportionality depends on the current test effort.
Illustrations of fitting these adjusted models showed that this approach could be useful,
particularly in assessing software reliability growth where it is reasonable to model test
effort this way.

References

1. Amstadter, B.L. (1971), *Reliability Mathematics*, McGraw-Hill, New York.

2. Barlow, R.E. and Scheuer, E.M. (1966), Reliability growth during a development
    testing program, *Technometrics*, **8**, 53-60.

3. Bhattacharyya, G.K., Fries, A., and Johnson, R.A. (1989), Properties of continuous

analog estimators for a discrete reliability growth model, *IEEE Transactions on Reliability*, **38**, 373-378.

4. Bonis, A.J. (1977), Reliability growth curves for one-shot devices, *Proceedings of the Annual Reliability and Maintainability Symposium*, Philadelphia, 181-185.

5. Box, G.E.P. and Jenkins, G.M. (1970), *Time series analysis: Forecasting and Control*, Holden Day, San Francisco, California.

6. Codier, E.O. (1968), Reliability growth in real life, *Proceedings of the 1968 Annual Symposium on Reliability*, 458-469.

7. Crow, L.H. (1974), Reliability analysis for complex, repairable systems, In *Reliability and Biometry*, (Eds., Proschan and Serfling), SIAM, 379-410.

8. Crow, L.H. (1975), On tracking reliability growth, *Proceedings of the 1975 Annual Reliability and Maintainability Symposium*, Washington, D.C., 438-443.

9. Crow, L.H. (1982), Confidence intervals procedures for the Weibull process with application to reliability growth, *Technometrics*, **24**, 67-72.

10. Dhillon, B.S. (1980), Reliability growth: A survey, *Microelectronics and Reliability*, **20**, 743-751.

11. Duane, J.T. (1962), *Technical Information Series Report DF 62 MD300*, General Electric Company, DCM+G Dept, Erie, Pennsylvania.

12. Duane, J.T. (1964), Learning curve approach to reliability monitoring, *IEEE Transactions on Aerospace*, **2**, 563-566.

13. Fard, N.S. and Dietrich, D.I. (1987), A Bayes reliability growth model for a development testing program, *IEEE Transactions on Reliability*, **R-36**, 568-571.

14. Finkelstein, J.M. (1983), A logarithmic reliability growth model for single mission systems, *IEEE Transactions on Reliability*, **R-32**, 508-511.

15. Goel, A.L. and Okumoto, K. (1980), A time dependent error detection rate model for software performance assessment with applications, *Technical Report*, Syracuse University, New York.

16. Gottfried, P. (1987), Some aspects of reliability growth, *IEEE Transactions on Reliability*, **R-36**, 11-16.

17. Jääskeläinen, P. (1982), Reliability growth and Duane learning curves, *IEEE Transactions on Reliability*, **R-31**, 151-154.

18. Jewell, W.S. (1984), A general framework for learning curve reliability growth models, *Operations Research*, **32**, 547-558.

19. Kareer, N., Kapur, P.K., and Grover, P.S. (1990), An s-shaped software reliability growth model with two types of errors, *Microelectronics and Reliability*, **30**, 1085-1090.

20. Lloyd, D.K. and Lipow, M. (1962), *Reliability: Management Methods and Mathematics*, Prentice-Hall, Englewood Cliffs, New Jersey.

21. Mead, P.H. (1975), Reliability growth of electronic equipment, *Microelectronics and Reliability*, **14**, 439-443.

22. Miller, D.R. (1986), Exponential order statistics models of software reliability growth, *IEEE Transactions on Software Engineering*, **SE-12**, 12-24.

23. MIL-STD-1635(EC) (1978), *Reliability Growth Tests*.

24. MIL-HDBK-189 (1981), *Reliability Growth Management*.

25. Ohba, M. (1984), Software reliability analysis models, *IBM Journal of Research and*

Development, **28**, 428-443.

26. Robinson, D. and Dietrich, D. (1989), A nonparametric-Bayes reliability growth model, *IEEE Transactions on Reliability*, **R-38**, 591-598.

27. Rosner, N. (1961), System analysis - non-linear estimation techniques *Proceedings of the National Symposium on Reliability and Quality Control*, New York, 203-207.

28. Selby, J.D. and Miller, S.G. (1970), *Reliability Planning and Management*, ASQC/SRE Seminar Niagara Falls, Ontario.

29. Singpurwalla, N. (1978), Estimating reliability growth (or deterioration) using time series analysis, *Naval Research Logistics Quarterly*, **25**, 1-14.

30. Smith, S.A. and Oren, S.S. (1980), Reliability growth of repairable systems, *Naval Research Logistics Quarterly*, **27**, 539-547.

31. Sofa, A. and Miller, D.R. (1991), A nonparametric software reliability growth model, *IEEE Transactions on Reliability*, **R-40**, 329-337.

32. Virene, E.P. (1968), Reliability growth and its upper limit, *Proceedings of the Annual Symposium on Reliability*, Boston, Massachusetts, 265-270.

33. Wolman, W. (1963), Problems in system reliability analysis, In *Statistical Theory of Reliability*, (Ed., M. Zelen), Madison, Wisconsin, 149-160.

34. Wong, K.L. (1988), Reliability growth – myth or mess, *IEEE Transactions on Reliability*, **R-37**, 209.

35. Yamada, S., Hishitani, J., and Osaki, S. (1993), Software reliability growth with a Weibull test-effort: A model and application, *IEEE Transactions on Reliability*, **R-42**, 100-105.

36. Yamada, S., Ohba, M., and Osaki, S. (1983), S-shaped reliability growth modeling for software error detection, *IEEE Transactions on Reliability*, **R-32**, 475-478.

37. Yamada, S, Ohba, M., and Osaki, S. (1984), S-shaped software reliability growth models and their applications, *IEEE Transactions on Reliability*, **R-33**, 289-291.

38. Yamada, S. and Osaki, S. (1983), Reliability growth models for hardware and software systems based on non-homogeneous Poisson processes: A survey, *Microelectronics and Reliability*, **23**, 91-112.

39. Yamada, S. and Osaki, S. (1984), Non-homogeneous error detection rate models for software reliability growth, In *Stochastic Models in Reliability Theory*, (Eds., S. Osaki and Y. Hatoyama), Springer-Verlag, Berlin, 120-143.

40. Yamada, S. and Osaki, S. (1985), Software reliability growth modeling: Models and applications, *IEEE Transactions of Software Engineering*, **SE-11**, 1431-1437.

41. Yamada, S. and Osaki, S. (1987), Optimal software release policies with simultaneous cost and reliability requirements, *European Journal of Operational Research*, **31**, 46-51.

42. Xie, M. and Zhao, M. (1993), On some reliability growth models with simple graphical interpretations, *Microelectronics and Reliability*, **33**, 149-167.

19

Recent Advances in Experimental Design Analysis for Reliability Growth Programs

Claudio Benski and Emmanuel Cabau

Merlin Gerin, France

ABSTRACT Corrective maintenance actions in repairable systems strive to lengthen as much as possible the expected time to the next system failure. In order to achieve this goal there are usually several system parameters that can be changed before each new run. The effect of these parameter changes on the system's reliability can best be assessed by specific experimental designs. The response of these experiments will be the actual time until the system fails. Since for a given system only one such time will be available after each corrective maintenance action, the experimental design will be an unreplicated one. This is a typical situation in reliability growth programs, although growth as such will not occur during the experiments. It is therefore important to determine how powerful are the available numerical techniques in identifying significant effects when applied to these unreplicated factorial designs. An extensive Monte Carlo simulation study was undertaken to measure the power of recently published techniques to analyze these experiments. In view of these results, a measure of performance to complexity ratio is also given for some of these techniques.

Keywords: Experimental design, Reliability growth, Fractional factorials

CONTENTS

19.1 Introduction

In reliability growth programs the time pattern of system failures is observed. The usual assumption behind these programs is that, after each system failure, corrective maintenance is applied such that the system's reliability is better after these actions than it was before. This should translate in a (stochastic) increase in times between successive failures. However, in real life, there may be many system parameters that maintenance actions could modify to achieve a reliability improvement and the precise effect of each such parameter on the system's reliability is rarely known. Even less so, are the possible interactions between these parameters.

Consequently, a natural approach to this problem is to use a factorial experimental design to determine the impact of the different system parameters on its reliability. In general, this experimental design will be a fraction of the full factorial. This is so, because high order interactions between the parameters are of little interest and fractional factorial designs allow for a very significant reduction in the number of experiments [5]. Notice that actual reliability growth will only occur after a successful optimizaton of the design parameters following the experiment.

In most cases these designs will also be unreplicated. The reason for this is that, for a single system, the response of these experiments is a single measurement. For example, for each configuration of the factors, only a single measurement of the time to the following failure is obtained. After each change in the system parameters, usually consecutive to a failure, a new value of the system's lifetime is obtained. Note however that, if the modified parameters do have an impact on the system's reliability, this time to failure is no longer issued from the same statistical population as the previous times between failures [1]. The implication here is that there is no independent way to evaluate the residual noise of these measurements and therefore it is not possible to perform either a classical or a non-parametric Analysis of Variance to assess the significance of each factor and eventual interactions. Due to the fractionning feature of these designs, variance pooling is also very restricted [5].

To overcome these limitations, several recent statistical methods have been proposed in the literature, including a Bayesian technique, to detect the presence of significant effects in unreplicated factorials, [3], [6], and [9]. It it recognized however, that these techniques were developed for normally distributed responses. And this may or may not be the case for times between failures. In fact, for Homogeneous Poisson Processes (HPP), these times are exponentially distributed. Still, response data transformations can be applied to these times [4] so that, at least approximately, these procedures can be used. It was therefore considered important to determine how well these different techniques performed in terms of power. The actual details of a fractional factorial design applied in the context of reliability growth are described in Section 19.2 through a specific example. The power comparison results are described in Section 19.3. Some recent results on power testing, never published before, are given in Section 19.4.

19.2   Experimental Designs in Reliability Growth

Failures of a complex system usually result in an intervention by a corrective maintenance team. During system development and after these failures, it is frequently the case that many system parameters (or factors) are considered to be possible candidates for change. As an example, consider the case of an elaborate programmable controller which fails to perform a function it was supposed to perform when a certain external event is present at one of its sensor inputs. This failure can be due to many different causes: a software failure, an incorrect threshold, a thermally induced early failure of a hardware component, poor electromagnetic shielding, etc.

The maintenance team may ponder what system parameters should be changed and in what order. Maybe the hardware component failed due to a bad circuit design or a poor thermal design. Or an electromagnetic glitch, induced by the external event, resulted in a wrong calculation of the triggering threshold. In the Design of Experiments (DoE) parlance, changing a system parameter implies assigning it a different level. Thus, the parameter levels that could be considered include different software versions, the position of a cooling fan, type of grounding of a circuit board, characteristics of a critical component, etc.

Experimenting with one design change at a time is known to be a costly and inefficient way of achieving the sought improvement [5]. Alternatively, using an appropriate experimental design will lead to a more efficient reliability improvement program since several parameters are modified in a single maintenance action. These designs consist in a series of different system configurations in which all the parameter levels are changed at each experiment. Of course, the improvement in system reliability, if any, will not be visible until the end of the whole series of experiments. The statistical analysis of these experiments will suggest the actual configuration of the significant system parameters which will result in the longest expected time to system failure.

Consider, for example, a system in which six design parameters, say A, B, C, D, E and F have been targeted as candidates for having an effect on the system's reliability. With only two levels for each parameter, a complete factorial design will result in 2^6 experiments in which the time to system failure has to be measured for each of the 64 system configurations. This is economically penalizing and usually unnecessary. In fact, if only the effects of each individual parameter and, say, seven of the two parameter interactions are to be estimated, it can be shown [5] that 16 experiments will suffice to obtain the desired result using a fractional factorial design (FFD). At the end of this experimental phase it may well turn out that not all specified factors and interactions have a significant influence on the time to failure. But then, the next phase will only need to concentrate on the significant ones. And here is where important savings occur: less system parameters are modified to obtain a concurrent increase in system reliability. Conversely, without the experimental design, useless and sometimes costly changes will rarely produce reliability growth. And little experience is gained in the process.

To illustrate this problem by a specific example, let us assume that out of a total of 15 two-factor interactions, the seven following ones are suspected as possibly having an effect on the MTBF: AB, AC, BF, CD, CF, DE and DF. The remaining eight interactions are felt to be physically impossible. It is not intuitively obvious which experiments have to be carried out to assess the effects of these six factors and seven interactions. Fortunately, there are now a number of programs that automatically generate the necessary experimental arrays as a function of the specified interactions. With parameter levels labeled − and + , such a design will take the form given in Table 19.1. To each run there will be only on corresponding measurement of the time to system failure. Now, the analysis of FFD's is based on formal or informal Analysis of Variance (ANOVA). This is not always possible, as has been mentioned in the Introduction: there may be no independent estimate of the residual (statistical) noise to which one could compare the estimate of each effect. Of course, a ranking of the effects is an alternative and an arbitrary selection of the two or three largest effects is one possibility [8] . But this solution is a dangerous one: there may be more than three significant effects or perhaps less than two. In both cases, choosing the wrong number of system factors to change can lead to either a sub-optimal reliability improvement or economic waste. The evaluation of the statistical risk incurred in making a wrong choice is not possible by this approach.

In order to illustrate the difficulties associated with the analysis of this type of designs we will assume that the above FFD has resulted in the sixteen times to failure given in Table 19.2 and presented in the same order as the design of Table 19.1.

The natural logarithms of the times to failure are also given since they are the transformed responses that will be used for the analysis for reasons that will become clear later. The times t_i are simulated times between system failures, randomly issued from exponential distributions having different λ parameters depending on the settings of factor B and the interaction AB. The parameter of the exponential distribution used for

TABLE 19.1
Design for 6 main factors and 7 double factor interactions

Run	A	B	C	D	E	F
1	−	−	−	−	−	−
2	+	+	−	−	+	−
3	+	−	−	−	−	+
4	−	+	−	−	+	+
5	+	−	−	+	+	−
6	−	+	−	+	−	−
7	−	−	−	+	+	+
8	+	+	−	+	−	+
9	+	−	+	−	+	−
10	−	+	+	−	−	−
11	−	−	+	−	+	+
12	+	+	+	−	−	+
13	−	−	+	+	−	−
14	+	+	+	+	+	−
15	+	−	+	+	−	+
16	−	+	+	+	+	+

TABLE 19.2
Times to failure and their logarithms for the design in Table 19.1

Run	t_i	$\ln(t_i)$
1	0.4703	−0.754
2	74.279	+4.308
3	0.0322	−3.437
4	1.0980	+0.094
5	0.0005	−7.664
6	0.1884	−1.669
7	0.0176	−4.037
8	8.1720	+2.101
9	0.850	−2.464
10	0.3989	−0.919
11	1.8602	+0.621
12	3.7119	+1.312
13	0.7626	−0.271
14	104.07	+4.645
15	0.0298	−3.512
16	0.0650	−2.733

TABLE 19.3
Calculated effects for the design of Table 19.1 and measurements on Table 19.2

Factor	Effect
A	$+0.62$
B	$+3.58$
C	$+0.97$
D	-1.49
E	-0.01
F	-0.60
AB	$+3.78$
AC	$+0.20$
BF	-0.80
CD	$+1.38$
CF	-0.73
DE	-1.60
DF	-0.20

each run is calculated using the formula:

$$\lambda_i = \lambda_b \, \Pi_B \, \Pi_{AB} \qquad (19.2.1)$$

where λ_b is the base value, arbitrarily set equal to 1, and the Π factors are similar to the ones used by MIL-HDBK 217 for electronic components. This corresponds to a model in which a base λ is multiplied or divided by factors of improvement or degradation depending on whether the Π factors are smaller or greater than 1. For example, if a better cooling system was one of the factor levels, we would expect electronic components to have a reduced failure rate which leads to a smaller intensity of failures for the system. We have used the following numerical values for the Π factors: $\Pi_B = 7$ and $\Pi_{AB} = 5$ giving a thirtyfivefold improvement in MTBF with respect to its base value $1/\lambda$.

Using the logarithms of the times between failures as the response has two effects: the response is now an additive linear model in terms of the active parameters and is approximately normally distributed under the assumption than there aren't any s-significant effects. (The exact distribution is an extreme-value function.)

The goal of the analysis is then to "discover" that, out of 15 potentially active factors or double interactions, only B and AB are significant and to estimate their actual effects. Using Yates algorithm [5] we obtain the calculated effects which are given in Table 19.3, i.e., the six main factors and the seven double factor interactions. Hadn't there been any noise in the simulation, only factors B and AB would have been different from zero. The theoretical expected values of their effects are 3.89 and 3.22 respectively. Applying now the techniques given in [3], [6], and [9] we can assess the statistical significance of the above effects using each one of these techniques.

- The Bayesian technique, in [6] gives as expected the highest posterior probability to effects B and AB: 85% and 88% respectively. All the other are smaller than 20%. The prior probability adopted for this test was a uniform 20% for all effects, a typical assumption in the literature. Real effects are assumed to come from a normal distribution with a scale factor ten times bigger than the noise.

- [9] produces two sets of limits that effects must exceed to be significant. Effects exceeding the outer limits are significant at a 95% confidence level. Effects smaller than the inner limits are not significant at this confidence level. In-between cases "may be significant". This latter situation is the case of effects B and AB since the outer limits are ± 5.7 and the inner limits are ± 2.8. All other effects are smaller than the inner limits and thus are not statistically significant.

- [3] gives a combination of two techniques: a normality test and an outlier detection test. In this technique, effects that are significant must produce a rejection of the hypothesis of normality of all effects *and* be detected as outliers. Only B and AB qualify as such, with the normality test significance level set at 98% and the limits for the outlier test calculated as ± 1.82. No other effects are identified as significant, even when the normality test level is lowered to 60% confidence.

Since in this particular experimental design two degrees of freedom are available to estimate the error variance, a classical ANOVA is also possible for comparison purposes. When this was done, again effects B and AB were identified as significant at a 97% confidence level. According to this ANOVA, three other effects, DE, D and CD, had 86%, 84% and 82% confidence levels. We should point out that an ANOVA would not have been possible, had we been interested in 9 double interactions instead of 7, (except through an afterthought variance pooling.)

The conclusion of all these analyses is that B and AB have been (correctly) identified as being the only significant effects although with a somewhat limited confidence. Setting both, A and B to their $+$ level should therefore maximize the expected time to system failure. The effect of A will only manifest itself via its interaction with B. The four other single factors are irrelevant and can be disregarded as well as all the other two-factor interactions.

To compute the actual improvement that would be possible on the MTBF by setting the significant factors to their $+$ value we must go back to the original time domain. If we denote the effects of B and AB by $\mathcal{B}$ and $\mathcal{AB}$ we would have the formula:

$$\text{MTBF}' = \text{MTBF}_0 \exp(\mathcal{B}/2) \exp(\mathcal{AB}/2) \qquad (19.2.2)$$

where the MTBF$'$ is the improved MTBF and MTBF$_0$ is the baseline MTBF. The $1/2$ factor is due to the fact that in Eq. (19.2.1) the Π factors can multiply or divide the base λ_b. Numerically, the effect of B in lengthening the MTBF is $\exp(3.58/2) = +6$ and the effect of the interaction AB is $\exp(3.78/2) = 6.6$. These numbers are to be compared to the theoretical Π factors of 7 and 5. The agreement is only fair due to the scarcity of the data: we are trying to estimate 13 possible factors using only 16 observations! Overall, we can expect the MTBF of the improved design to be about $6 \times 6.6 \simeq 40$ times better than the original. This compares very well to the theoretical improvement factor of 35 used for the simulation.

A legitimate question arises now as to why the observed times between failures of this experiment are so wildly dispersed, spanning over five orders of magnitude. The answer lies in the exponential nature of these times. We have assumed a HPP and this is a very unrealistic model in this case, in spite of the large amounts of literature using it. Real world systems behave in a more orderly way and times between failures are not so dispersed. For a DoE application, this is good news because the effect of system parameters should be even more easily detectable than this example might otherwise suggest.

TABLE 19.4
Perfect "hits" according to three different techniques

case	Ref. 6	Ref. 8	Ref. 3
0/15	97.53	97.98	92.36
0/31	96.47	97.02	92.65
1/15	68.04	58.76	67.44
1/31	68.98	64.24	66.50
2/15	44.78	36.18	46.35
2/31	48.66	43.28	47.43
3/15	29.25	22.31	33.48
3/31	35.09	29.65	35.52

19.3 Monte-Carlo Power Analysis

Because an experimenter will often find it impossible to perform the variance pooled ANOVA, [3], [6], and [9] will be the only way to assess the statistical significance of the calculated effects. In this section we present a summary of a power study comparing these techniques.

A Monte-Carlo study [2] was carried out in which 15 and 31 normal variates were generated. The noise effects were issued from a normal distribution, $N(0, \sigma)$. This noise distribution was contaminated by simulated "real" effects which were issued from a $N(0, k\sigma)$ distribution with $k = 5$, 10 and 15. The number of contaminants varied from 0 to 3. This corresponds to many typical fractional factorial designs. The number of Monte Carlo samples was always set equal to 10,000 and σ was set equal to 1 without loss of generality. A vectorized machine was used to ensure reasonable response times during these tests. For each technique, we counted the number of times that the correct number of real effects were detected as well as the number of times they detected some but not all of these. Also recorded were the number of times that spurious, (noise), effects were detected either alone or with some of the real effects. This gave a performance measure of each technique. Table 19.4, extracted from this study, gives the percent probabilities that each technique will detect the *exact* number of real effects (0, 1, 2 or 3) out of a total of 15 and 31 possible ones and nothing else. This table corresponds to a simulation in which the prior assumptions of the Bayesian technique were satisfied.

The analysis of this table shows that the Bayesian technique [6] is the most powerful technique, albeit by a small margin.

Many other measures of performance are of course possible. A technique which detects all real effects but also detects lots of non real ones is not very useful. Conversely, it is interesting to assess the ability of a procedure to detect only a large fraction of the real effects without contamination by spurious ones. For example, [2] gives the probabilities of detecting two out of three real effects without adding any other spurious factors. All other combinations are also analyzed.

Another useful characteristic available to compare these techniques is to evaluate their complexity, as measured by the CPU time needed to complete an analysis. This is shown in Table 19.5. The technique described in [8] executed in the shortest time and is used as a reference here by giving it a time of 1.

The technique given in [3], almost as powerful as that in [6], seems to attain a good compromise of desirable statistical characteristics and a straightforward implementation.

TABLE 19.5
Times to complete the analysis according to the three techniques

Ref. [6]	1400
Ref. [8]	1
Ref. [3]	2

TABLE 19.6
Q values as a function of k for three analysis techniques

Reference	Q $(k = 5)$	Q $(k = 10)$
[6]	0.368	0.642
[8]	0.334	0.626
[3]	0.301	0.589

19.4 Recent Results

The Monte-Carlo study mentioned above is only a small sample of an ongoing bench-marking activity. Presently, we have evaluated nine different techniques for the analysis of unreplicated experimental designs. The simulations have been performed for up to 6 real effects out of a total of 15 and 31. In each case 10,000 trials were carried out for all nine techniques. As mentioned in Section 19.4, comparisons are difficult because no single technique was uniformly better than others. Their performance depends on the sample size, number of effects present and actual scale-shift k. In addition, some techniques were best under a particular set of circumstances and worst under others. A comprehensive report on these benchmarks will be published in the near future [7]. We introduce in [7] a figure of merit Q which is defined as follows. Let n^+ be the total number of true active effects detected by a given technique during the 10,000 Monte-Carlo runs. Similarly, let n^- be the total number of false effects detected during those runs. Since we have simu-lated the presence of $0, 1, 2, \ldots, 6$ effects, mixed in samples of size 15 and 31, we had a total of 70,000 samples in which we could count both n^+ and n^-. Q is then naturally defined as:

$$Q = \frac{n^+}{N^+} \left(1 - \frac{n^-}{N^-}\right)$$

where N^+ and N^- are the total maximums of true and false effects that could have been detected. Ideally, the Q values should be as close as possible to 1. Table 19.6 contains the Q values for References [6], [8], and [3] for $k = 5$ and 10. In Reference [7], the Q values are given for all nine techniques examined. These Q values confirm that although Reference [6] has better statistical characteristics, the advantage is small compared to other, simpler methods.

References

1. Ascher, H. and Feingold, H. (1984), Repairable systems reliability, *Lecture Notes in Statistics*, **7**, Marcel Dekker, New York.

2. Benski, C. and Cabau, E. (1992), Significant effects in unreplicated experimental designs, *Proceedings of the Statistical Computing Section of the American Statistical Association*, 98-103.

3. Benski, H.C. (1989), Use of a normality test to identify significant effects in factorial designs, *Journal of Quality Technology*, **21**, 174-178.

4. Box, G.E.P. and Draper, N.D. (1987), *Empirical Model-Building and Response Surfaces*, John Wiley & Sons, New York.

5. Box, G.E.P., Hunter, W.G. and Hunter, J.S. (1978), *Statistics for Experimenters*, John Wiley & Sons, New York.

6. Box, G.E.P. and Meyer, R.D. (1986), An analysis for unreplicated fractional factorials, *Technometrics*, **28**, 11-18.

7. Cabau, E. and Benski, C. (1994), Unreplicated experimental designs in reliability growth programs, *IEEE Transactions on Reliability, Special Issue on Design for Reliability* (submitted for publication).

8. Lenth, R.V. (1989), Quick and easy analysis of unreplicated factorials, *Technometrics*, **31**, 469-473.

9. Kackar, R.N. and Shoemaker, A.C. (1986), Robust design: A cost-effective method for improving manufacturing processes, *AT&T Technical Journal*, **65**, 39-50.

Part III

Parametric Models and Inference

20

Likelihood Ratio Confidence Intervals in Life-Data Analysis

Necip Doganaksoy

GE Corporate Research and Development, Schenectady, NY

ABSTRACT Exact confidence interval estimation for most models used in life-data analysis is impractical or impossible. Consequently, approximate confidence intervals based on the asymptotic normal (AN) theory for the maximum likelihood estimators are widely used in applications. Recent evidence indicates that confidence intervals based on inverting likelihood ratio (LR) tests perform much better than AN confidence intervals in small and censored samples. This paper reviews LR confidence intervals for models commonly used in life-data analysis. This review strongly suggests the LR confidence intervals as the method of choice due to its superior performance in small samples typically encountered in applications.

Keywords: Bartlett correction, Life distribution, Maximum likelihood estimation, Mean and variance correction, Statistical computing

CONTENTS

20.1 Introduction

In statistical applications, confidence interval estimates are often based on large sample theory methods since exact methods for the particular problem are either not available or not feasible. It has been widely demonstrated for many types of models that approximate methods may yield seriously misleading results, especially when the sample size is not large enough to justify their use. Understanding the small sample behavior of large sample methods to identify the circumstances under which they do not perform satisfactorily is of considerable interest. Inferences from life-data analyses often have important implications. A misleading lower confidence limit for the reliability of a life sustaining medical equipment, for example, may lead to the wrong conclusions as to the replacement or maintenance of such equipment.

Some important terms pertaining to confidence intervals are defined here (see [21] for an indepth discussion and review of confidence intervals). The error probability of a confidence interval procedure is defined to be the probability that a random interval does not cover the true value of the parameter. The asymptotic value of an error probability is called the nominal error probability, whereas the error probability achieved in finite samples is called the actual error probability. With exact confidence intervals, the actual error probability equals the nominal error probability. A confidence interval with a combined upper and lower error probability less (greater) than nominal is termed (anti)conservative. In life-data applications one is often interested in one-sided confidence limits; therefore, the lower and upper tail error probabilities should be nearly symmetric (equal).

There is a rich literature on confidence interval estimation using both complete and censored life-data (see, for example, [24], [5], [36], [53]). Except for a few special cases, exact methods for confidence interval estimation are either unavailable or impractical, requiring highly specialized computer programs or extensive tables. Traditionally, large sample confidence intervals based on the asymptotic normal (AN) theory of the maximum likelihood (ML) estimators have been used in applications. AN confidence intervals are relatively easy to compute and are used in most statistical software packages. However, these confidence intervals are usually inaccurate for small and censored samples. The asymptotic chi-square distribution of the likelihood ratio (LR) test statistic provides another approximate method to obtain confidence intervals. Confidence intervals based on inverting LR tests, however, are rarely found in commercial statistical software packages. Nelson [36, Chap. 5] presented a comprehensive overview of software packages for fitting parametric models to life-data.

This paper reviews LR confidence intervals for models commonly used in life-data analysis. Various results indicate that LR confidence intervals are much more accurate for small samples than AN confidence intervals.

Section 20.2 presents a general review of LR confidence intervals. Section 20.3 reviews the performance of LR confidence intervals for common life models. Section 20.4 describes corrections to improve small sample performance of LR confidence intervals. Section 20.5 discusses conclusions, including areas for future research.

20.2   Likelihood Ratio Confidence Intervals

Notation

p	number of unknown parameters.
θ	the parameter of interest.
λ	$(p-1)$ vector of nuisance parameters.
(θ^*, λ^*)	the true parameter values.
$(\hat{\theta}, \hat{\lambda})$	the joint ML estimator of (θ^*, λ^*).
$\tilde{\lambda}$	the restricted ML estimator of λ under a fixed value of θ.
$\sigma_{\hat{\theta}}$	asymptotic standard deviation of $\hat{\theta}$.
$\hat{\sigma}_{\hat{\theta}}$	estimate of $\sigma_{\hat{\theta}}$.
θ_L	a lower confidence limit for θ.
θ_U	an upper confidence limit for θ.
$\ell(\theta, \lambda)$	the log-likelihood function.
$W(\theta)$	$-2[\ell(\theta, \tilde{\lambda}) - \ell(\hat{\theta}, \hat{\lambda})]$, the LR statistic for θ.
z_α	α quantile of the standard normal distribution.
$\chi^2_{(1;\alpha)}$	α quantile of the chi-square distribution with 1 d.f.

The LR statistic for θ is defined by

$$W(\theta) = -2[\ell(\theta, \tilde{\lambda}) - \ell(\hat{\theta}, \hat{\lambda})]. \tag{20.2.1}$$

In many cases, the exact distribution of the LR statistic is intractable. For large samples, under the true parameter values (θ^*, λ^*), W is approximately distributed as a χ^2 random variable with 1 d.f. The $100(1-\alpha)\%$ LR confidence region for θ consists of all parameter values that would not be rejected at the α significance level.

Under usual regularity assumptions on the likelihood function ([4, Chap. 9]), the $100(1-\alpha)\%$ lower and upper LR confidence limits for θ are the two values that satisfy

$$\ell(\theta, \tilde{\lambda}) = \ell(\hat{\theta}, \hat{\lambda}) - (1/2) \chi^2_{(1;1-\alpha)} \tag{20.2.2}$$

with $\theta_L < \hat{\theta}$ and $\theta_U > \hat{\theta}$.

The right side of (20.2.2) is a constant once data are observed. The left side is the profile log-likelihood function for θ. The search for θ_L and θ_U usually requires successive evaluations of the left side of (20.2.2). Section 20.2.2 further discusses computational aspects of LR confidence intervals. Figure 20.1 illustrates LR confidence limits. In this figure, the level of the solid horizontal line is determined by the right side of (20.2.2).

Lawless [24, Appendix E] reviewed the asymptotic distribution theory for the LR statistic. The theory is valid for many types of models and data. A notable exception is models involving threshold parameters.

20.2.1   Remarks

AN and LR confidence intervals are asymptotically equivalent but they may differ considerably in small samples. The following heuristic argument motivates the asymptotic correspondence between AN and LR confidence intervals for the case of a single unknown parameter θ. The basic ideas also extend to those involving nuisance parameters ([18, Chap. 8], [12, Chap. 5]).

The series expansion for the log-likelihood function $\ell(\theta)$ about $\hat{\theta}$ yields

$$\ell(\theta) = \ell(\hat{\theta}) + \ell^{(1)}(\hat{\theta})(\theta - \hat{\theta}) + \ell^{(2)}(\hat{\theta})(\theta - \hat{\theta})^2/2 + \text{smaller terms}. \tag{20.2.3}$$

Here, $\ell^{(i)}(\hat{\theta})$ is the ith derivative of $\ell(\theta)$ with respect to θ evaluated at $\hat{\theta}$. Noting that $\ell^{(1)}(\hat{\theta}) = 0$ and $\ell^{(2)}(\hat{\theta}) = -1/\hat{\sigma}_{\hat{\theta}}^2$, we obtain (ignoring smaller terms)

$$- 2[\ell(\theta) - \ell(\hat{\theta})] = W(\theta) \approx [(\theta - \hat{\theta})/\sigma_{\hat{\theta}}]^2. \tag{20.2.4}$$

Under the AN theory of ML estimators, for large samples, $(\hat{\theta} - \theta)/\sigma_{\hat{\theta}}$ is standard normal distributed when $\theta = \theta^*$. This result also implies the asymptotic χ^2 distribution of $W(\theta^*)$ with 1 d.f. The $100(1 - \alpha)\%$ lower and upper AN confidence limits for θ are

$$\theta_L = \hat{\theta} - z_{(1-\alpha/2)}\hat{\sigma}_{\hat{\theta}} \text{ and } \theta_U = \hat{\theta} + z_{(1-\alpha/2)}\hat{\sigma}_{\hat{\theta}}. \tag{20.2.5}$$

In (20.2.5), $\hat{\sigma}_{\hat{\theta}}$ is usually obtained from the local estimate of the covariance matrix of ML estimators. AN confidence intervals are symmetric in length about the ML estimate (unless the confidence interval endpoints are transformed) and depend on the parametrization of the model. In contrast, LR confidence intervals tend to be asymmetric about the ML estimates in small samples. LR confidence intervals are always in the natural parameter range and are transformation invariant.

The large sample theory for the AN and LR confidence intervals outlined above is based on the asymptotic quadratic form and symmetric shape of the log-likelihood function. The non-quadratic, asymmetric shape of the log-likelihood function in small and censored samples suggests the inadequacy of AN confidence intervals. Considerable research effort has been devoted to improve small sample performance of AN confidence intervals. The most common types of corrections to AN confidence intervals include parameter transformations to normality (see [46] and [47]) and bias corrections for ML estimators (see [7]). Whenever feasible, we advocate a careful examination of the sample profile likelihood as a useful and enlightening data analysis practice. This approach is illustrated in [48], [41], and [29].

LR confidence intervals, unlike the AN confidence intervals, use the asymptotic χ^2 distribution of the LR statistic to calculate interval endpoints from the *observed* asymmetric log-likelihood function. Small sample corrections to LR confidence intervals are reviewed in Section 20.4.

20.2.2   Computational Issues

Except for few trivial cases, closed-form solutions for LR confidence interval endpoints are typically not available and numerical methods must be used to solve (20.2.2).

The basic approach to obtain the LR confidence limits requires successive evaluations of the left hand side of (20.2.2) until the two values of θ satisfying the equation are obtained. The search for θ_L may start at a convenient value of θ, say $\hat{\theta}$. At this value of θ, note that $\tilde{\lambda} = \hat{\lambda}$ and thus (20.2.2) is not satisfied. The value of θ is then decreased by a certain amount (usually proportional to $\hat{\sigma}_{\hat{\theta}}$). Fixing θ at this new value, the log-likelihood function is maximized with respect to λ to obtain $\tilde{\lambda}$. Then the log-likelihood function is evaluated at these values of θ and λ to determine whether or not (20.2.2) is satisfied. The procedure is repeated until two values, say θ_a and θ_b, such that $\ell(\theta_a, \tilde{\lambda}) < \ell(\hat{\theta}, \hat{\lambda}) - (1/2)\chi^2_{(1;1-\alpha)}$ and $\ell(\theta_b, \tilde{\lambda}) > \ell(\hat{\theta}, \hat{\lambda}) - (1/2)\chi^2_{(1;1-\alpha)}$ are obtained. The bisection algorithm then finds the θ value lying between θ_a and θ_b that satisfies (20.2.2). This solution yields θ_L. The procedure to obtain θ_U is similar except that in this case the trial value of θ is increased from one iteration to the next. This approach can be conveniently implemented on most general purpose software packages. Nelson [36, Chap. 5] illustrated this approach with numerical examples.

In [52], Venzon and Moolgavkar suggested a general algorithm to compute LR confidence limits. This algorithm is based on the recognition that the $100(1 - \alpha)\%$ LR confidence limits for θ are the solutions of the system of p equations

$$\ell(\hat{\theta}, \hat{\lambda}) - \ell(\theta, \hat{\lambda}) + (1/2)\chi^2_{(1;1-\alpha)} = 0 \text{ and } \partial\ell/\partial\lambda = 0. \qquad (20.2.6)$$

This approach can be most effectively used in special purpose software developed to calculate LR confidence intervals for particular models.

LR confidence intervals can also be calculated for a function $\phi = \phi(\theta, \lambda)$ of the parameters such as a percentile of a life distribution or survival probability. In some cases, $\phi = \phi(\theta, \lambda)$ can be readily inverted to obtain $\theta = \phi^{-1}(\phi, \lambda)$. One would then reparametrize the log-likelihood function in (20.2.2) in terms of ϕ and proceed to obtain LR confidence intervals for ϕ in the usual manner as described above. For example, Lawless [24, Chap. 4] used this approach to obtain LR confidence intervals for the percentiles of the Weibull distribution. An alternative approach is to maximize the log-likelihood function $\ell(\theta, \lambda)$ subject to the constraint $\phi(\theta, \lambda) = \phi_0$ where ϕ_0 is a trial value of ϕ that varies from one iteration to the next. Such constraint maximization can be carried out using the method of Lagrangian multipliers. The two values of ϕ_0 satisfying (20.2.2) are the lower and upper LR confidence limits for ϕ. This approach is illustrated in [26] and [28].

As the discussion above indicates, robust and powerful numerical algorithms are essential to successful implementation of LR confidence intervals especially when the log-likelihood function is peculiarly shaped. Meeker and Escobar [30] provided useful recommendations to alleviate numerical difficulties encountered in calculating LR confidence intervals. Reparametrization of the log-likelihood function in terms of orthogonal parameters improves its shape and thus reduces the numerical work and increases the numerical stability of calculations for LR confidence intervals ([6], [43], [15]).

20.2.3 Example

This example (adapted from [13]) illustrates the calculation of LR confidence intervals from system failure data containing partially identified failure causes.

Many products have more than one cause of failure. A multi-component system may fail due to failure of any of a number of components. In applications, determination of the exact cause of some of the failures might be impossible or impractical. Table 20.1 presents failure data of a 3-component series system from a sample of $n = 30$ systems. The ith system failure time t_i is the minimum of the component lifelengths. The table gives the system number, system failure time, and identification number of the component that caused the system failure. For some failures, the cause of failure is known (e.g., System 1) and for others the cause is completely unknown (e.g., System 9). In some cases, though the exact cause is unknown, it is possible to isolate the cause to a smaller subset of system components (e.g., System 2). Usher and Hodgson [50] simulated these data assuming independent and exponentially distributed component failure times each with the rate parameter value of 1.

Statistical analyses of system failure data include point and confidence interval estimation of component failure rates. Nelson [36, Chap. 7] demonstrated the use of such estimates in assessing the impact of individual failure causes on system reliability. In the present example, the parameter of interest θ is the rate parameter associated with the first component. λ_1 and λ_2 (nuisance parameters) are the rate parameters for the other two components.

For exponentially distributed component failure-times, the log-likelihood function is (see [50])

$$\ell(\theta, \lambda_1, \lambda_2) = r_1 \ln(\theta) + r_2 \ln(\lambda_1) + r_3 \ln(\lambda_2) + r_{12} \ln(\theta + \lambda_1)$$
$$+ r_{13} \ln(\theta + \lambda_2) + r_{23} \ln(\lambda_1 + \lambda_2) + r_{123} \ln(\theta + \lambda_1 + \lambda_2)$$
$$- (\theta + \lambda_1 + \lambda_2) T, \tag{20.2.7}$$

where r_j is number of system failures due to component j, $r_{k\ell}$ is number of system failures due to the component k or ℓ, r_{123} is number of system failures due to any of the three components, and $T = \sum_{i=1}^{n} t_i$ is the total time on test.

Closed-form expressions for the ML estimators of the parameters θ, λ_1, and λ_2 are not available; therefore, one has to rely on numerical methods to solve the resulting likelihood equations or, numerically maximize (20.2.7). Consequently, exact confidence interval estimation for θ is intractable and large-sample approximate methods are used.

For the data in Table 20.1, we find that $n = 30$, $r_1 = 6$, $r_2 = 6$, $r_3 = 8$, $r_{12} = 3$, $r_{13} = 1$, $r_{23} = 3$, $r_{123} = 3$, and $T = 10.13$. From (20.2.7), the sample log-likelihood function can be written as

$$\ell(\theta, \lambda_1, \lambda_2) = 6 \ln(\theta) + 6 \ln(\lambda_1) + 8 \ln(\lambda_2) + 3 \ln(\theta + \lambda_1) + 3 \ln(\theta + \lambda_2)$$
$$+ 3 \ln(\lambda_1 + \lambda_2) + 3 \ln(\theta + \lambda_1 + \lambda_2) - (\theta + \lambda_1 + \lambda_2) \, 10.13 \, . \tag{20.2.8}$$

TABLE 20.1
Sample system failure data with partially identified failure causes

System i	Failure time t_i	Failing component j	System i	Failure time t_i	Failing component j
1	0.021	2	16	0.281	1
2	0.038	1 or 2	17	0.295	2
3	0.054	3	18	0.310	3
4	0.066	3	19	0.338	3
5	0.076	1 or 2	20	0.341	2
6	0.078	2 or 3	21	0.354	1
7	0.123	3	22	0.358	2
8	0.130	1 or 3	23	0.431	1,2, or 3
9	0.152	1,2, or 3	24	0.457	3
10	0.159	1	25	0.545	1,2, or 3
11	0.199	3	26	0.569	2
12	0.201	1	27	0.677	3
13	0.204	1	28	0.818	2
14	0.215	2 or 3	29	0.946	2 or 3
15	0.218	1 or 2	30	1.486	1

The ML estimates $\hat{\theta} = 0.8588$, $\hat{\lambda}_1 = 0.9890$, and $\hat{\lambda}_2 = 1.1137$ are obtained by directly maximizing (20.2.8). In this illustration, all computations were done on Mathematica [56]. The local estimate of the covariance matrix of ML estimators yields $\hat{\sigma}_{\hat{\theta}} = .3261$. The 95% AN confidence limits for θ are .2196 and 1.4980.

In order to calculate LR confidence limits for θ, we first obtain the maximized value of the log-likelihood function by evaluating (20.2.8) at the ML estimates. This yields $\ell(\hat{\theta}, \hat{\lambda}) = \ell(0.8588, 0.9890, 1.1137) = -22.1101$. For 95% LR confidence limits, we have $\chi^2_{(1;0.95)} = 3.84$. From (20.2.6), the LR confidence interval equations are

$$0 = 6\ln(\theta) + 6\ln(\lambda_1) + 8\ln(\lambda_2) + 3\ln(\theta + \lambda_1) + 3\ln(\theta + \lambda_2)$$
$$+ 3\ln(\lambda_1 + \lambda_2) + 3\ln(\theta + \lambda_1 + \lambda_2)$$
$$- (\theta + \lambda_1 + \lambda_2)10.13 + 24.0301$$

$$0 = [6/\lambda_1] + [3/(\theta + \lambda_1)] + [3/(\lambda_1 + \lambda_2)] + [3/(\theta + \lambda_1 + \lambda_2)] - 10.13$$
$$0 = [8/\lambda_2] + [1/(\theta + \lambda_2)] + [3/(\lambda_1 + \lambda_2)] + [3/(\theta + \lambda_1 + \lambda_2)] - 10.13 . \quad (20.2.9)$$

The two values of θ, .3590 and 1.6442, which satisfy (20.2.9) are the 95% LR confidence limits.

We now illustrate the calculation of LR confidence intervals for system reliability $R(t)$ at $t = 1$. The ML estimate of $R(1) = \exp[-(.8588 + .9890 + 1.1137)] = 0.0517$ with an associated estimate of its standard error .0280. For comparison, the 95% AN confidence limits for $R(1)$ are -.0031 and .1066. A logistic transformation of R would be useful to avoid the negative AN lower confidence limit. To obtain 95% LR confidence limits for $R(1)$, we simply reparametrized (20.2.9) by letting $\theta = -(\ln R + \lambda_1 + \lambda_2)$. The 95% LR confidence interval for $R(1)$ is .0157 and .1321. Based on an extensive Monte Carlo simulation of error probabilities (see Section 20.2.4), Doganaksoy [13] showed that LR confidence intervals are more accurate than AN confidence intervals for this particular type of model and data.

20.2.4 Performance Evaluation of Likelihood Ratio Confidence Intervals

There are several approaches to evaluate the properties of LR confidence intervals (e.g., actual error probability, error symmetry, average length) in small sample situations:

- Comparison with exact confidence intervals. This is possible only in a few specific cases where an exact method for confidence interval estimation is available. For example, Lawless [24, Chap. 4] compared LR confidence intervals from failure censored Weibull data with exact confidence intervals based on the conditional approach.

- Monte Carlo simulation of the sampling distribution of the LR statistic. This approach involves generation of many Monte Carlo samples, typically on the order of several thousands, under different conditions of interest (i.e., various combinations of sample sizes, degrees of censoring, model parameter values). Each Monte Carlo sample is used to evaluate the LR test statistic under the known true value of the parameter of interest. The estimated sampling distribution of the LR statistic is then compared with its asymptotic χ^2 approximation. Myhre and Saunders [33] used this approach to evaluate the LR confidence intervals for system reliability.

- Monte Carlo simulation of error probabilities. This approach is also based on many Monte Carlo samples which are used to check confidence interval violations for the parameters (or their functions) whose true values are known in a simulation study. This can be done by calculating actual confidence interval endpoints or by conducting a LR test to determine whether or not the true value of the parameter would be rejected. The LR test approach, although computationally less demanding, does not provide average confidence interval lengths. Monte Carlo simulation of error probabilities (and confidence interval lengths, in some cases) has been the preferred method of evaluating LR confidence intervals. For example, Ostrouchov and Meeker [38] for interval data and Doganaksoy and Schmee [16] for right censored and complete data used this approach to assess the accuracy of LR confidence intervals computed from Weibull and lognormal samples.

20.3 Likelihood Ratio Confidence Intervals in Life-Data Analysis

The asymptotic χ^2 distribution of the LR statistic was derived by Wilks [54]. Box and Cox [3] used LR confidence intervals extensively in their study of power transformations to normality. Madansky [26] was the first author who used LR confidence intervals in a reliability context. Table 20.2 presents the first summary of past studies which used LR confidence intervals. In this table, the references marked with an asterisk (*) also contain a performance evaluation of the LR confidence intervals for the particular model considered. As the table indicates, LR confidence intervals have been used in a wide range of situations involving reliability and life-data modeling. The comparative studies shown in this table differ from one another considerably with respect to sample size considered, type of data (complete or censored), parameter of interest (a scalar parameter or a function of the parameters) and the approach used in evaluations (see Section 20.2.4). Nevertheless, the patterns that emerge from the findings of these various investigations are very strong:

- AN confidence interval error probabilities are generally anticonservative and asymmetric even for moderately large sample sizes,

- The LR confidence interval error probabilities converge to nominal error probabilities much faster than their AN counterparts,

- LR confidence intervals tend to yield near symmetric error probabilities and thus are more appropriate for one-sided confidence limits,

- The accuracy of LR confidence intervals becomes questionable only for very small samples.

Figure 20.2 provides a specific illustration of these findings for singly time censored samples from Weibull (smallest extreme value) populations. Doganaksoy and Schmee [16] used Monte Carlo simulation to obtain the error probabilities shown in this figure.

20.4   Corrections to Likelihood Ratio Confidence Intervals

During the last decade, much effort has been devoted to small sample corrections of the LR statistic. Barndorff-Nielsen [1] and DiCiccio ([9], [10]) proposed a mean and variance correction to the square root of the LR statistic and the so-called Bartlett correction to improve the convergence of the distribution of the LR statistic to its limiting χ^2 distribution. More recent work explored further aspects of these corrections ([11]). Despite difficulties in implementing these correction factors in general settings, recent advances reviewed here have made their implementation feasible for some important life models.

20.4.1   Mean and Variance Correction

For a scalar parameter θ, the signed square roots of the LR statistic are defined by

$$S(\theta) = \text{sign}(\theta - \hat{\theta})W(\theta)^{1/2}. \tag{20.4.1}$$

Large sample theory shows that the distribution of S under the true parameter value (θ^*, λ^*) tends to the standard normal distribution as the number of complete observations increases. The confidence intervals obtained from S are identical to LR confidence intervals. Barndorff-Nielsen [1] showed that S can be standardized to follow a standard normal distribution to a higher order of accuracy. Let m and s denote the true mean and standard deviation of S, respectively, then $(S - m)/s$ has a distribution closer to a standard normal than the nonstandardized S.

Having obtained the numerical values m and s for the correction factors for a given parameter θ (see Section 20.4.4 for further discussion), the lower and upper confidence limits may be found as the values of θ satisfying $(S - m)/s = -z_{(1-\alpha/2)}$ and $(S - m)/s = z_{(1-\alpha/2)}$. The lower and upper $100(1 - \alpha)\%$ confidence limits are the respective solutions θ_L and θ_U to the equations

$$\ell(\theta, \tilde{\lambda}) = l(\hat{\theta}, \hat{\lambda}) - (1/2)(m - sz_{(1-\alpha/2)})^2 \tag{20.4.2}$$

$$\ell(\theta, \tilde{\lambda}) = \ell(\hat{\theta}, \hat{\lambda}) - (1/2)(m + sz_{(1-\alpha/2)})^2 \tag{20.4.3}$$

such that $\theta_L < \hat{\theta}$ for the lower limit and $\theta_U > \hat{\theta}$ for the upper limit. The right sides of (20.4.2) and (20.4.3) are constants once the data are observed. These equations can be solved by the same method used to solve (20.2.2) for the LR confidence limits. Figure 20.3 shows the effect of this correction. The intersection of the profile log-likelihood with the right sides of (20.4.2) and (20.4.3) which determine the levels of the dashed horizontal lines is at two different levels. The lower and upper confidence limits are individually corrected to obtain more symmetric error probabilities.

TABLE 20.2
Past uses of likelihood ratio confidence intervals in reliability and life-data analysis

MODEL	REFERENCES
LIFE DISTRIBUTIONS	
Exponential	[24, chap. 3]*, [5, chap. 3]
Two-Parameter Exponential	[24, chap. 3], [39]*
(Log)Normal	[24, chap. 5], [38]*, [16]*
Weibull (Smallest Extreme Value)	[24, chap. 4]*, [5, chap. 3], [38]*, [16]*
Three-Parameter Lognormal	[20]
Gamma	[24, chap. 5]
Generalized Log-Gamma	[24, chap. 5], [9]
LIFE DISTRIBUTIONS IN ACCELERATED	
STRESS (REGRESSION) MODELS	
(Log)Normal	[36, chap. 5]
Weibull (Smallest Extreme Value)	[36, chap. 5], [51]*, [14]*
Generalized Log-Gamma	[24, chap. 6]
Logistic	[55]*
COMPONENT LIFE DISTRIBUTIONS	
IN SYSTEMS RELIABILITY MODELS	
Binomial	[26], [33]*, [34], [28]
Exponential	[35], [13]*
Weibull	[22]
Weibull-Exponential	[17]
MIXTURE MODELS	
Limited Failure Population	[29]*
Dead on Arrival Subpopulation	[31]
NONPARAMETRIC ESTIMATION	
Cox's Proportional Hazards Model	[40]*, [32]*
Kaplan-Meier Estimator	[49]*, [5, chap. 4]
VARIOUS MODELS	
Transformations to Normality	[3]
Discriminant Analysis	[8]
Process Monitoring	[19]

(*) indicates the references that evaluate the small sample performance of LR intervals.

TABLE 20.3
Correction factors m, s, and b for the median and .10 quantile of log failure time

Parameter	m	s	b
μ	.0978	1.1093	1.2401
$y_{.10}$	-.2837	1.0737	1.2333

20.4.2   Bartlett Correction

Asymptotically, under the true parameter value (θ^*, λ^*), $E(W) = 1$, the expected value of the limiting χ^2 distribution with 1 d.f. Here, W is the LR statistic for θ. As shown by Cox and Hinkley [4, Chap. 9], for finite samples $E(b^{-1} W)$ is closer to 1 than $E(W)$, where $E(W) = b$. Using b as a bias-correction factor, one can improve the χ^2 approximation to the distribution of $W' = b^{-1} W$. The correction factor b or any asymptotically equivalent quantity is called a Bartlett correction factor. Lawley [25] showed that all moments of W' are closer to those of the χ^2 distribution than the moments of W, even in the presence of nuisance parameters.

Barndorff-Nielsen and Cox [2] presented a general discussion of methods to obtain Bartlett correction factors. Exact analytical expressions for Bartlett correction factors require integration over the sample space and may be difficult to derive. Lawley [25] and Barndorff-Nielsen and Cox [4, Chap. 9] gave approximate formulas for a Bartlett correction factor. Given the numerical value b of a Bartlett correction factor (see Section 20.4.4 for further discussion), the two solutions θ_L and θ_U to

$$\ell(\theta, \tilde{\lambda}) = \ell(\hat{\theta}, \hat{\lambda}) - (b/2) \chi^2_{(1;1-\alpha)} \qquad (20.4.4)$$

are the $100(1-\alpha)\%$ confidence limits, such that $\theta_L < \hat{\theta}$ for the lower confidence limit and $\theta_U > \hat{\theta}$ for the upper confidence limit. Figure 20.4 illustrates the effect of the Bartlett correction. The intersection of the right side of (20.4.4) which determines the level of the dashed horizontal line with the profile log-likelihood is at a different level than the uncorrected intersection of the right side of (20.2.2). The Bartlett correction adjusts the moments of the limiting χ^2 distribution with the aim of achieving error probabilities that are overall closer to the nominal ones.

20.4.3   Example

The data used in this example are from a life test of a Class B insulation for motors at $170°C$ in which 7 of 10 failed ([36, Chap. 5]). The failure times are 1764, 2772, 3444, 3542, 3780, 4860, and 5196 hours. The censoring time is 5448 hours.

The lognormal distribution is fitted. The log failure times are then assumed to be normally distributed $N(\mu, \sigma^2)$. We denote the q quantile of the log failure time as y_q. The parameters of interest are the median (antilog(μ)) and the .10 quantile (antilog($y_{.10}$)) of failure time. The ML estimates based on log_{10} failure times are $\hat{\mu} = 3.6354$, $\hat{\sigma} = .2027$, and $\hat{y}_{0.10} = 3.3759$. The correction factors m, s, and b can be calculated using the formulas of [9]. For the example see Table 20.3. The lower and upper 90% confidence intervals for the median failure time and .10 quantile of failure time are as shown in Table 20.4. The exact 90% confidence limits are calculated using best linear unbiased estimators and the tables of [37]. They are shown here as convenient benchmarks.

In this example, mean and variance corrected LR confidence intervals are almost identical to the exact results; therefore, the correction was successful. Uncorrected and Bartlett

TABLE 20.4
The lower and upper 90% confidence intervals for the median and .10 quantile of
failure time

Method	Parameter Antilog(μ) Lower	Upper	Antilog($y_{.10}$) Lower	Upper
AN	3333	5598	1692	3334
LR	3327	5963	1458	3120
SLR	3278	6444	1200	3049
BLR	3215	6275	1344	3201
Exact	3237	6390	1182	3034

AN: Asymptotic normal theory; LR: Likelihood ratio; SLR: Mean and variance corrected signed
square roots of the LR statistic; BLR: Bartlett corrected LR statistic; Exact: Based on best
linear unbiased estimators.

corrected LR confidence intervals give acceptable results. AN confidence limits are anticonservative. The AN lower confidence limit for the .10 quantile was considerably more
optimistic than the exact result and could have misled the engineers.

20.4.4   Corrected Likelihood Ratio Confidence Intervals in Life-Data Analysis

The corrected LR confidence intervals have been used very little in applications due to
difficulties in obtaining the values of the factors (i.e., m, s, and b) needed for their implementation. The correction factors m and s presented by DiCiccio [9] for the log-generalized
gamma distribution depend only on observed log-likelihood derivatives and apply to many
common life distributions including the Weibull, smallest extreme value, normal, and lognormal distributions. Similar correction factors are also available for location-scale and
linear regression models which are useful in the analysis of data from accelerated tests
([10]). An approximate Bartlett correction factor is $b = m^2 + s^2$. These corrections to LR
confidence intervals have been developed as approximations to conditional procedures.
Their unconditional error probabilities, however, hold asymptotically.

   In [14] and [16] Doganaksoy and Schmee undertook an extensive Monte Carlo study
to evaluate the performance of corrected LR confidence intervals for complete and right
censored samples from the Weibull (smallest extreme value) and (log)normal distributions.
They considered sampling from a single distribution as well as simple linear regression
models which are useful in the analysis of data from accelerated life tests. In general,
their results showed that the uncorrected LR confidence intervals usually performed as
well as the corrected ones. The mean and variance correction, however, considerably
improved the convergence to nominal error probabilities especially for very small samples
(less than 10 complete observations) where LR confidence intervals did not perform as
well. Mean and variance corrected LR confidence intervals were nearly exact in complete
samples. The Bartlett corrected LR confidence intervals tended to yield conservative error
probabilities with small samples.

20.5   Conclusions and Discussion

For a wide class of life-time models, LR confidence intervals are more accurate than the more widely used confidence intervals based on the AN theory. The convergence of AN error probabilities to nominal error probabilities is rather slow. For the LR confidence intervals, actual error probabilities tend to be closer to the nominal ones and are more symmetric. If one has to choose a single method for confidence interval estimation, we recommend LR based confidence interval procedures as the methods of choice, since they are more accurate irrespective of the shape of the sample log-likelihood function.

Presently, only a few general purpose software packages compute LR confidence intervals. With increasing speed of computation and availability of new and efficient algorithms, LR confidence intervals will, no doubt, become more popular in life-data software packages. Some software products have recently incorporated such capability (e.g., SAS product JMP [44] and SAS procedure GENMOD [45]). We expect to see further research to improve efficiency of numerical algorithms to calculate LR confidence intervals especially for models involving many parameters.

The amount of additional computation required to implement the corrected LR confidence intervals is negligible when compared with the overall effort required to calculate LR confidence intervals. With limitations, bootstrapping and other resampling methods may be a way to obtain estimates for correction factors when closed-form expression are not available ([42], [27]).

LR based inferences can also possibly be used beyond hypothesis testing and confidence interval estimation. The prediction problem which is often encountered in life-data modeling is such a potential area for future research ([23]).

Acknowledgements

I would like to thank Gerald J. Hahn, William Q. Meeker, Wayne Nelson, Josef Schmee, and Mark VanDeven for their very useful comments on an earlier version of this paper.

References

1. Barndorff-Nielsen, O.E. (1986), Inference on full or partial parameters based on the standardized signed log likelihood ratio, *Biometrika*, **73**, 307-322.

2. Barndorff-Nielsen, O.E. and Cox, D.R. (1984), Bartlett adjustments to the likelihood ratio statistic and the distribution of the maximum likelihood estimator, *Journal of the Royal Statistical Society, Series B*, **46**, 483-495.

3. Box G.E.P. and Cox, D.R. (1964), An analysis of transformations (with discussion), *Journal of the Royal Statistical Society, Series B*, **26**, 211-252.

4. Cox, D.R. and Hinkley, D.V. (1974), *Theoretical Statistics*, Chapman and Hall, London.

5. Cox, D.R. and Oakes, D. (1984), *Analysis of Survival Data*, Chapman and Hall, London.

6. Cox, D.R. and Reid, N. (1987), Parameter orthogonality and approximate conditional inference (with discussion), *Journal of the Royal Statistical Society, Series B*, **49**, 1-39.

7. Cox, D.R. and Snell, E.J. (1968), A general definition of residuals (with discussion), *Journal of the Royal Statistical Society, Series B*, **30**, 248-275.

8. Critchley, F., Ford, I., and Rijal, O. (1988), Interval estimation based on the profile likelihood: Strong Lagrangian theory with applications to discrimination, *Biometrika*, **75**, 21-28.

9. DiCiccio, T.J. (1987), Approximate inference for the generalized gamma distribution, *Technometrics*, **29**, 33-40.

10. DiCiccio, T.J. (1988), Likelihood inference for linear regression models, *Biometrika*, **75**, 29-34.

11. DiCiccio, T.J. and Martin, M.A. (1990), Simple modifications for signed square roots of likelihood ratio statistics, *Technical Report 353*, Department of Statistics, Stanford University, Stanford, CA.

12. Dobson, A.J. (1983), *An Introduction to Statistical Modelling*, Chapman and Hall, New York.

13. Doganaksoy, N. (1991), Interval estimation from censored and masked system-failure data, *IEEE Transactions on Reliability*, **40**, 280-286.

14. Doganaksoy, N. and Schmee, J. (1991), Comparisons of approximate confidence intervals for the smallest extreme value distribution simple linear regression model under time censoring, *Communications in Statistics - Simulation and Computation*, **20**, 1085-1113.

15. Doganaksoy, N. and Schmee, J. (1993), Orthogonal parameters with censored data, *Communications in Statistics - Theory and Methods*, **22**, 669-685.

16. Doganaksoy, N. and Schmee, J. (1993), Comparisons of approximate confidence intervals for distributions used in life data analysis, *Technometrics*, **35**, 175-184.

17. Dombroski, B.A. and Meeker, W.Q. (1991), Maximum Likelihood Estimation of the Steady State Model, *Technical Report*, Department of Statistics, Iowa State University, Ames, IA.

18. Fraser, D.A.S. (1976), *Probability and Statistics: Theory and Applications*, Duxbury Press, North Scituate, MA.

19. Garrigoux, C. and Meeker, W.Q. (1991), Process Monitoring with Censored Responses, *Technical Report*, Department of Statistics, Iowa State University, Ames, IA.

20. Griffiths, D.A. (1980), Interval estimation for the three-parameter lognormal distribution via the likelihood function, *Applied Statistics*, **29**, 58-68.

21. Hahn, G.J. and Meeker, W.Q. (1991), *Statistical Intervals: A Guide for Practitioners*, John Wiley & Sons, New York.

22. Johnson, C.A. and Tucker, W.T. (1992), Advanced Statistical Concepts of Fracture in Brittle Materials, In *Engineered Materials Handbook, Volume 4: Ceramics and Glasses*, ASM International, Material Park, OH.

23. Kalbfleisch, J.D. (1971), Likelihood Methods of Prediction, In *Foundations of Statistical Inference* (Eds., V.P. Godambe and D.A. Sprott), Holt, Rinehart and Winston, Toronto.

24. Lawless, J.F. (1982), *Statistical Models and Methods for Lifetime Data*, John Wiley & Sons, New York.

25. Lawley, D.N. (1956), A general method for approximating to the distribution of likelihood ratio criteria, *Biometrika*, **43**, 295-303.

26. Madansky, A. (1965), Approximate confidence limits for the reliability of series and parallel systems, *Technometrics*, **7**, 495-503.

27. Martin, M.A. (1990), On bootstrap iteration for coverage correction in confidence interval, *Journal of the American Statistical Association*, **85**, 1105-1118.

28. Matthews, D.E. (1988), Likelihood-based confidence intervals for functions of many parameters, *Biometrika*, **75**, 139-144.

29. Meeker, Jr., W.Q. (1987), Limited failure population life tests: Application to integrated circuit reliability, *Technometrics*, **29**, 51-65.

30. Meeker, W.Q. and Escobar, L.A. (1994), Maximum likelihood methods for fitting parametric statistical models to censored and truncated data, In *Probabilistic and Statistical Methods in the Physical Sciences* (Eds., J. Stanford and S. Vardeman), Academic Press, San Diego.

31. Melroe, S.L. and Meeker, W.Q. (1991), A time-to-failure model when some units are dead on arrival, *Technical Report*, Department of Statistics, Iowa State University, Ames, IA.

32. Moolgavkar, S.H. and Venzon, D.J. (1987), Confidence regions for parameters of the proportional hazards model: A simulation study, *Scandinavian Journal of Statistics*, **14**, 43-56.

33. Myhre, J. and Saunders, S.C. (1968), Comparison of two methods of obtaining approximate confidence intervals for system reliability, *Technometrics*, **10**, 37-49.

34. Myhre, J. and Saunders, S.C. (1968), On confidence limits for the reliability of systems, *Annals of Mathematical Statistics*, **39**, 1463-1472.

35. Myhre, J. and Saunders, S.C. (1971), Approximate confidence limits for complex systems with exponential component lives, *Annals of Mathematical Statistics*, **42**, 342-348.

36. Nelson, W. (1990), *Accelerated Testing: Statistical Models, Data Analyses, and Test Plans*, John Wiley & Sons, New York.

37. Nelson, W. and Schmee, J. (1979), Inference for (log)normal distributions from small singly censored samples and BLUE, *Technometrics*, **21**, 43-54.

38. Ostrouchov, G. and Meeker, Jr., W.Q. (1988), Accuracy of approximate confidence bounds computed from interval censored Weibull or lognormal data, *Journal of Statistical Computation and Simulation*, **29**, 43-76.

39. Piegorsch, W.W. (1987), Performance of likelihood-based interval estimates for two-parameter exponential samples subject to Type I censoring, *Technometrics*, **29**, 41-49.

40. Prentice, R.L. and Mason, M.W. (1986), On the application of linear relative risk models, *Biometrics*, **42**, 109-120.

41. Reiser, B. and Bar Lev, S. (1979), Likelihood inference for life test data, *IEEE Transactions on Reliability*, R-288, 38-43.

42. Rocke, D.M. (1989), Bootstrap Bartlett adjustments in seemingly unrelated regression, *Journal of the American Statistical Association*, **84**, 598-601.

43. Ross, G.J.S. (1990), *Nonlinear Estimation*, Springer-Verlag, New York.

44. SAS Institute Inc. (1993), JMP Statistics and Graphics Guide, Version 3, *SAS Technical Report P-243*, Cary, NC.

45. SAS Institute Inc. (1993), SAS/STAT Software: The GENMOD Procedure, Release 6.09, *SAS Technical Report P-243*, Cary, NC.

46. Sprott, D.A. (1973), Normal likelihoods and their relation to large sample theory of estimation, *Biometrika*, **60**, 457-465.

47. Sprott, D.A. (1980), Maximum likelihood estimation in the presence of nuisance parameters, *Biometrika*, **67**, 515-523.

48. Sprott, D.A. and Kalbfleisch, J.D. (1969), Examples of likelihoods and comparisons with point estimation and large sample approximations, *Journal of the American Statistical Association*, **64**, 468-484.

49. Thomas, D.R. and Grunkemeier, G.L. (1975), Confidence interval estimation of survival probabilities for censored data, *Journal of the American Statistical Association*, **70**, 865-871.

50. Usher, J.A. and Hodgson, T.J. (1988), Maximum likelihood analysis of component reliability using masked system life-test data, *IEEE Transactions on Reliability*, **37**, 550-555.

51. Vander Wiel, S.A. and Meeker, W.Q. (1990), Accuracy of approximate confidence bounds using interval censored Weibull regression data from accelerated life tests, *IEEE Transactions on Reliability*, **39**, 346-351.

52. Venzon, D.J. and Moolgavkar, S.H. (1988), A method for computing profile-likelihood-based confidence intervals, *Applied Statistics*, **37**, 87-94.

53. Viveros, R. and Balakrishnan, N. (1994), Inverval estimation of parameters of life from progressively censored data, *Technometrics*, **36**, 84-91.

54. Wilks, S.S. (1938), The large-sample distribution of the likelihood ratio for testing composite hypotheses, *Annals of Mathematical Statistics*, **9**, 60-62.

55. Williams, D.A. (1986), Interval estimation of the median lethal dose, *Biometrics*, **42**, 641-645.

56. Wolfram, S. (1991), *Mathematica: A System for Doing Mathematics by Computer*, Second Edition, Addison-Wesley, Reading, MA.

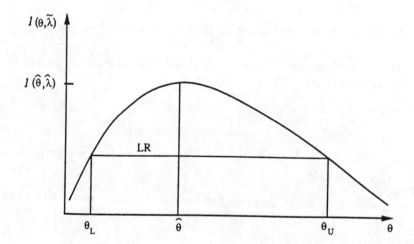

FIGURE 20.1
Confidence limits based on the likelihood ratio statistic

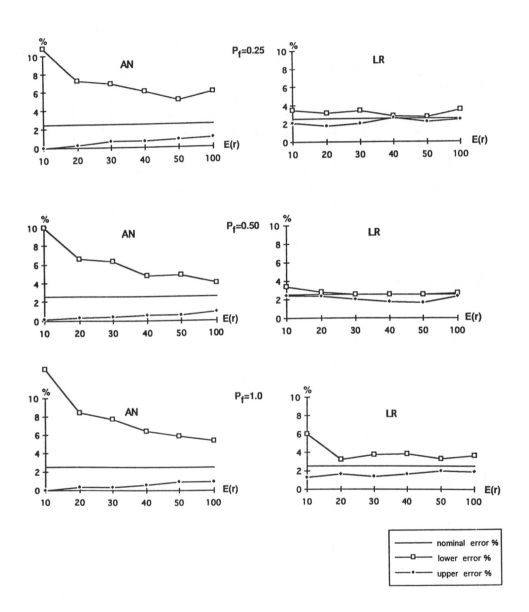

FIGURE 20.2
Actual lower and upper error percentages versus expected number of failures $E(r)$ of approximate confidence intervals for the .01 quantile of the Weibull (smallest extreme value) distribution with the expected proportion of failures P_f

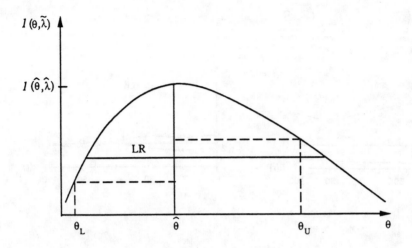

FIGURE 20.3
Confidence limits based on the mean and variance corrected signed square roots of
the likelihood ratio statistic

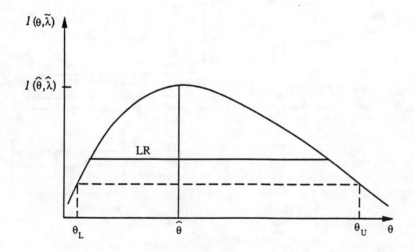

FIGURE 20.4
Confidence limits based on the Bartlett corrected likelihood ratio statistic

21

Statistical Analysis of Life or Strength Data from Specimens of Various Sizes Using the Power-(Log)Normal Model

Wayne Nelson and Necip Doganaksoy

Consultant, Schenectady, NY
GE Corporate Research and Development, Schenectady, NY

ABSTRACT This paper provides the first detailed presentation of properties of the power-normal and -lognormal models, which describe the effect of specimen size on the distribution of life or strength of a product or material. Such a model arises when any specimen can be regarded as a series system of smaller portions, where portions of a certain size have a normal or lognormal life (or strength) distribution. Also, this paper presents maximum likelihood fitting of these distributions and approximate confidence limits, as well as estimates, for model parameters, distribution percentiles, and other quantities of interest. Our work on this model was motivated by an application to data on time to electromigration failure of aluminum conductors for microcircuits. The model was fitted to these data using a new computer program POWNOR.

CONTENTS

21.1 Introduction

Purpose. This paper describes in detail properties of the power-lognormal and related power-normal models and maximum likelihood fitting of them to data. This introduction briefly presents the history of the models, the models, and an overview of this paper. Then the introduction describes electromigration failure data used to illustrate use of the models of Section 21.3 and 21.4.

21.1.1 Background

History. The power-normal and -lognormal models have long been used to represent the effect of specimen size on the distribution of life or strength of products and materials. In an early application [3], Chaplin used the power-normal model to describe the distribution of tensile strength of steel bars of various lengths. In [13, pp. 385-392], Nelson presents models for size effect and applications to cable. In [8], Harter surveys the literature on this and other models for the effect of specimen size for a great variety of applications. Our work on this model was motivated by an application to data on time to electromigration failure of aluminum conductors in microcircuits. A key electromigration reference is [11], whose authors use the model for their data.

The models. The power-normal and power-lognormal models are briefly described here. Their properties appear in detail in Sections 21.3 and 21.4. For specimens of size S, the reliability function of the power-normal distribution is the population fraction above the value y (of life, strength, etc.); namely,

$$\bar{F}(y, S; \mu, \sigma, S_0) = \{\Phi[-(y - \mu)/\sigma]\}^{S/S_0}, \qquad -\infty < y < \infty . \qquad (21.1.1)$$

This is a power of the normal reliability function. Other authors also use the term "power distribution" for the distribution of a power transformations $Y = X^b$ of a random variable (see, for example, [10, p. 389-390]). Here $\Phi[\ ]$ is the standard normal cumulative distribution function and the minus sign converts it into a reliability function. The *location parameter* μ has the range $-\infty < \mu < \infty$. The *scale parameter* σ is positive. The *normal size parameter* S_0 is positive; that is, y has a normal distribution when $S = S_0$. In previous work, people had to assume a value for S_0, which is unknown. The fitting methods described in Section 21.5 provide an estimate of S_0 (and of μ and σ) from data. The ratio $\rho \equiv S/S_0$ is called the *power parameter*. The reliability function of the power-lognormal distribution is the population fraction above the value t; namely,

$$\bar{G}(t, S; \mu, \sigma, S_0) = \{\Phi[-(\ln(t) - \mu)/\sigma]\}^{S/S_0}, \qquad t > 0 . \qquad (21.1.2)$$

The parameters μ and σ have the same names and ranges. Here S_0 is the *lognormal size parameter*; that is, t has a lognormal distribution when $S = S_0$.

Overview. This paper contains:

- Section 21.2 on fitting the power-lognormal model to sample data on time to electromigration failure.
- Section 21.3 on the statistical properties of the power-normal model.
- Section 21.4 on the statistical properties of the related power-lognormal model.
- Section 21.5 on the maximum likelihood theory for fitting either model to data, which may be censored (unfailed runouts) or interval data.
- Section 21.6 on the numerical methods employed to carry out the maximum likelihood calculations.

21.1.2 Example Data

Overview. This section presents electromigration failure data used to illustrate the fitting of a power-(log)normal model to data. The section first gives background on electromigration. Then it describes the data. Finally it states a purpose of fitting a power-lognormal model to the data. The results of fitting appear in Section 21.2.

Electromigration. Aluminum conductors in microcircuits can fail from movement of atoms resulting in an open conductor or in a short (due to an aluminum filament growing between adjoining conductors). Such failures occur more quickly at high current density and high temperature. Microcircuit components and conductors are being built smaller and smaller for higher speed, resulting in higher current densities and temperatures in conductors. The microcircuit industry fears that electromigration failure may limit such

TABLE 21.1
Hours to failure of 400-micrometer electromigration specimens

6.545	9.289	7.543	6.956	6.492	5.459
8.120	4.706	8.687	2.997	8.591	6.129
11.038	5.381	6.958	4.288	6.522	4.137
7.459	7.495	6.573	6.538	5.589	6.087
5.807	6.725	8.532	9.663	6.369	7.024
8.336	9.218	7.945	6.869	6.352	4.700
6.948	9.254	5.009	7.489	7.398	6.033
10.092	7.496	4.531	7.974	8.799	7.683
7.224	7.365	6.923	5.640	5.434	7.937
6.515	6.476	6.071	10.491	5.923	

size reduction. So the industry seeks more knowledge of electromigration, including a model for the distribution of time to failure and for the effect of conductor length. The power-lognormal model has been proposed, but some experts have observed that such time to failure data for various specimen lengths all appear adequately described by just a lognormal distribution. The adjoining data yield information on the adequacy of the power-lognormal and lognormal distributions.

Data. Table 21.1 displays hours to failure of 59 test conductors of length $S = 400$ micrometers. All specimens ran to failure at a certain high temperature and current density. The data come from an interlaboratory comparison reported by [18]. The 59 specimens were all tested under the same temperature and current density. The data appear in a lognormal probability plot in Figure 21.1. The plot is relatively straight. Thus it suggests that a lognormal distribution adequately fits the data and that a power-lognormal distribution fitted to the data is "close" to lognormal; that is, $\rho = S/S_0$ is "close" to 1. Although this example uses complete failure data from specimens of equal length, the computer program POWNOR ([14], [15]) fits the power-(log)normal model to data (including censored and interval life data) from specimens of various sizes.

Purpose. Fitting the power-lognormal distribution to the data serves certain purposes. First, it provides an estimate of the lognormal length S_0, which previously has not been estimated and may lead to physical insight. Second, wide confidence limits for $\rho = S/S_0$ that enclose 1 (the lognormal distribution) would indicate why many data sets appear adequately described by a lognormal distribution. Third, extrapolation to estimate the life distribution of the total length of conductor in microcircuits under normal temperature and current density is of interest. Fourth, the uncertainty of that distribution estimate provides guidance on the appropriate number of test specimens needed to yield estimates of a desired accuracy.

21.2 Fitting the Power-(Log)Normal Model to Electromigration Data

This section presents the maximum likelihood fit of the power-lognormal model to the electromigration data of Section 21.1.2. Although this example uses complete failure data from specimens of equal length, the computer program POWNOR in [14] and [15] is the first that fits the power-(log)normal model to data (including censored and interval like data) from specimens of various sizes. Section 21.5 presents the maximum likelihood theory for fitting the power-(log)normal model to data. The numerical methods that facilitate such fitting are in Section 21.6.

Computer program. The computer program POWNOR in [14] and [15] uses maximum likelihood (ML) methods to fit the model to data. Program output includes ML estimates and approximate confidence limits for model parameters, distribution percentiles, and other quantities of interest. User friendly, the program is written in Fortran 77 and runs on the VAX 11/785 computer under the VMS operating system and on the SUN workstation. It employs IMSL routines. Copies of the program and documentation in [14] may be requested from the Statistical Engineering Division, Admin. Bldg., Room A337, National Institute of Standards and Technology, Gaithersburg, MD 20899, (301) 975-2839.

Output. Figure 21.2 displays the program output from the ML fitting of a power-normal model to the logs of the data on time to electromigration failure of 400-micrometer specimens presented in Section 21.1.2. This output is explained below; the numbers correspond to those in the figure.

I. This output states the number of uncensored and censored observations in the data set.

II. This output shows the ML estimates of the (transformed) model parameters (μ, $\ln \sigma$, $\ln S_0$) and their approximate 95% confidence limits. For example, the estimate of σ is $\hat{\sigma} = \exp(-0.7336) = 0.48$, and its 95% confidence limits are $\underset{\sim}{\sigma} = \exp(-1.6375) =$ 0.19 and $\tilde{\sigma} = \exp(0.1704) = 1.19$. Similarly, the estimate of the lognormal length S_0 is $\hat{S}_0 = \exp(2.6779) = 14.6$ micrometers, and the 95% confidence limits are $\underset{\sim}{S}_0 = \exp(-3.1808) = 0.0415$ micrometers and $\tilde{S}_0 = \exp(8.5366) = 5089$ micrometers. The estimate of the power parameter (for 400-micrometer specimens) is $\hat{\rho} = 14.6/400 = 0.0365$, and the 95% confidence limits are $\underset{\sim}{\rho} = 0.0415/400 = 0.000104$ and $\tilde{\rho} = 5098/400 = 12.7$. For a lognormal distribution, $\rho = 1$. The confidence limits for ρ enclose 1. Thus, the data are consistent with a lognormal distribution. Moreover, the limits are remarkably wide; indeed, the limits differ by a factor of $\tilde{\rho}/\underset{\sim}{\rho} = 12.7/0.000104 = 1.22 \times 10^5$. This large factor (or interval) explains why most electromigration life data are consistent with a lognormal distribution. Thus, the curvature of a lognormal plot of data must be extreme for the data to be inconsistent with a lognormal distribution when ρ is used as a criterion.

III. The local estimates of the information, covariance, and correlation matrices for the (transformed) parameter estimates are shown here. The high correlations (near ± 1) indicate that convergence to the ML estimates may be incomplete and that a better parameterization of the model could avoid possible convergence problems.

IV. This output shows the ML estimates of percentiles (log values) and their 95% confidence limits at each specified length (400 micrometers here). For example, the estimate of median life is $\hat{t}_{.50} = \exp(1.9360) = 6.93$ hours.

21.3 The Power-Normal Distribution

Introduction. This section presents properties of the power-normal distribution in detail. As background, readers need to know basic properties of the normal distribution and normal probability paper. For example, in [12] and [13], Nelson covers these topics and the series-system model (below) in detail. The following topics include its reliability function, cumulative distribution function, the series-system model, percentiles, median, probability density, mode, mean, variance, standard deviation, and hazard function (failure rate).

Reliability function and cdf. The *reliability function* of the power-normal distribution is the population fraction *above* the value y; namely,

$$\bar{F}(y; \mu, \sigma, \rho) \equiv \{\Phi[-(y - \mu)/\sigma]\}^{\rho}, \qquad -\infty < y < \infty. \qquad (21.3.1)$$

Here $-\infty < \mu < \infty$ is the *location parameter*, $\sigma > 0$ is the *scale parameter*, and $\rho > 0$ is the *power parameter* $\rho = S/S_0$ in (21.1.1). Note the minus sign in the standard normal cumulative distribution function $\Phi[\ ]$; it yields the upper tail probability of the normal distribution. For $\rho = 1$, the distribution is normal; only then is μ the mean and median of the distribution, and only then is σ the standard deviation of the distribution. The *cumulative distribution function* (cdf) is the population fraction *below* y; namely,

$$F(y; \mu, \sigma, \rho) = 1 - \{\Phi[-(y - \mu)/\sigma]\}^{\rho}, \qquad -\infty < y < \infty. \qquad (21.3.2)$$

The distribution is simpler in terms of the *standardized deviate* $z \equiv (y - \mu)/\sigma$. Then

$$\bar{F}(y; \mu, \sigma, \rho) = [\Phi(-z)]^{\rho}, \quad F(y; \mu, \sigma, \rho) = 1 - [\Phi(-z)]^{\rho}, \quad -\infty < z < \infty. \quad (21.3.3)$$

In [5, p. 287] and [1, p. 57], the authors list tables of these functions for integer ρ values.

Probability plot. Figure 21.3 displays power-normal cdfs for selected ρ values on normal probability paper. For $\rho = 1$, the cdf is normal and is a straight line. For $\rho > 1$, the cdfs curve concave to the left and shift toward lower y (and z) values as ρ gets larger. For $0 < \rho < 1$, the cdfs curve concave to the right and shift toward higher y (and z) values as ρ gets smaller.

Approximate cdf. For large ρ, the exact cdf formula (21.3.2) requires a table or algorithm for $\Phi(\ )$ that is accurate in the far tails. Asymptotically as $\rho \to \infty$, the power-normal cdf approaches the cdf of a smallest extreme value distribution (convergence in law or in distribution). Bury [2] gives the location and scale parameters of that distribution as a function of μ, σ, and ρ. Convergence to this asymptotic distribution is slow. That is, ρ must be very large (say well above 100) for a satisfactory approximation.

Series-system model. In some applications, product comes in different sizes, and its life, strength, or some property depends on size. For example, time to electromigration failure of conductors in microcircuitry depends on conductor length. Also, strength of metal bars depends on bar length ([3]). Harter [8] surveys models for the effect of size. The following series-system model is a simple, widely-used model for the effect of size. For purposes of presenting the model, reliability terminology and product life are used. The assumptions of the *series-system model* for size effect are:

1. If product is conceptually divided into portions of any size of interest, it is assumed that the times to failure of the portions are statistically independent.

2. The product fails when the first such portion fails.

3. Portions of size S_0 have a reliability function $R_0(t)$, which is the population fraction surviving beyond age t.

It follows from these assumptions that product of size S has a reliability function

$$R_S(t) = [R_0(t)]^{S/S_0}. \qquad (21.3.4)$$

In [12] and [13], Nelson presents this model in more detail and gives various applications. When $R_0(t)$ is the reliability function of a normal distribution, (21.3.4) is the power-normal reliability function (21.3.1). Then the power parameter $\rho \equiv S/S_0$ is the ratio of the product size S to the "normal size" S_0.

Fractiles. The F fractile of the power-normal distribution is

$$\eta_F = \mu + z_{F'}\,\sigma \; ; \tag{21.3.5}$$

here $z_{F'}$ is the $F' = 1 - (1 - F)^{1/\rho}$ fractile of the standard normal distribution. For $\rho \gg 1$, F' usually is very near zero. For ρ near zero, F' usually is very near one. Tables of $z_{F'}$ appear in most statistics books. However, few tables adequately provide $z_{F'}$ values for F' very near zero or one.

Median. The median is the 0.50 fractile $\eta_{0.50}$. It is a "center" of the distribution in that half of the population is below $\eta_{.50}$ and half is above it. For the power-normal distribution,

$$\eta_{0.50} = \mu + z_{F'}\,\sigma \; ; \tag{21.3.6}$$

here $F' = 1 - (0.5)^{1/\rho}$. For $\rho = 1$, the distribution is normal, $F' = 0.50$, $z_{F'} = 0$, and $\eta_{0.50} = \mu$.

Density. The power-normal probability density is the derivative of the cdf (21.3.2); namely,

$$f(y; \mu, \sigma, \rho) = (\rho/\sigma)\phi[(y - \mu)/\sigma]\{\Phi[-(y - \mu)/\sigma]\}^{\rho-1} \; ; \tag{21.3.7}$$

here $\phi(z) \equiv (2\pi)^{-1/2}\exp(-z^2/2)$ is the standard normal probability density. In terms of the standardized deviate $z = (y - \mu)/\sigma$,

$$f(y; \mu, \sigma, \rho) = (\rho/\sigma)\phi(z)[\Phi(-z)]^{\rho-1} \; . \tag{21.3.8}$$

Figure 21.4 displays such probability densities as a function of z for selected ρ values where $\mu = 0$ and $\sigma = 1$. For $\rho = 1$, the density is the standard normal density. These densities are unimodal.

Mode. The mode of a power-normal distribution is the value η_m where the probability density peaks. It is the most probable y value. η_m is the solution of $df(y; \mu, \sigma, \rho)/dy = 0$; namely,

$$\eta_m = \mu + z_m\sigma \; . \tag{21.3.9}$$

Here z_m is a function of just ρ and is the solution of $-z_m\Phi(-z_m)/\phi(z_m) = \rho - 1$. For $\rho = 1$, $z_m = 0$ and the mode $\eta_m = \mu$, a well-known property of the normal distribution.

Mean. The *mean* of the power-normal distribution is

$$EY \equiv \int_{-\infty}^{\infty} y\,f(y; \mu, \sigma, \rho)\,dy = \mu + E_\rho\sigma \; . \tag{21.3.10}$$

Here the mean E_ρ for the standardized distribution ($\mu = 0$ and $\sigma = 1$) is a function of ρ only; namely,

$$E_\rho \equiv \int_{-\infty}^{\infty} z\,\rho\,\phi(z)[\Phi(-z)]^{\rho-1}\,dz \; . \tag{21.3.11}$$

For integer ρ, E_ρ is the mean of the first order statistic of a random sample of size ρ from a standard normal distribution. David [5, p. 291] and Balakrishnan and Cohen [1, p. 57] list tables of E_ρ for integer ρ. E_ρ apparently has not been tabulated for non-integer ρ; then (21.3.11) must be evaluated by numerical quadrature.

Variance and standard deviation. The variance of the power-normal distribution is

$$V(Y) = \int_{-\infty}^{\infty} (y - EY)^2 f(y; \mu, \sigma, \rho) \, dy = V_\rho \sigma^2 ; \qquad (21.3.12)$$

here the variance V_ρ for the standardized distribution ($\mu = 0$ and $\sigma = 1$) is a function of ρ only; namely,

$$V_\rho \equiv \int_{-\infty}^{\infty} (z - E_\rho)^2 \rho \, \phi(z)[\Phi(-z)]^{\rho-1} dz . \qquad (21.3.13)$$

For integer ρ, V_ρ is the variance of the first order statistic of a sample of size ρ from a standard normal distribution. David [5, p. 291] and Balakrishnan and Cohen [1, p. 57] list tables of V_ρ for integer ρ. V_ρ apparently has not been tabulated for non-integer ρ; then (21.3.13) must be evaluated by numerical quadrature. The *standard deviation* of the power-normal distribution is

$$sd(Y) = [V(Y)]^{1/2} = V_\rho^{1/2} \sigma . \qquad (21.3.14)$$

For $\rho = 1$, the distribution is normal, and $V_1 = 1$, $V(Y) = \sigma^2$, and $sd(Y) = \sigma$.

Hazard function. Much used in reliability work, the *hazard function* (also called the *instantaneous failure rate*) of the power-normal distribution is

$$h(y; \mu, \sigma, \rho) \equiv f(y; \mu, \sigma, \rho)/[1 - F(y; \mu, \sigma, \rho)]$$
$$= (\rho/\sigma)\phi[(y - \mu)/\sigma]/\Phi[-(y - \mu)/\sigma]$$
$$= \rho \cdot h[(y - \mu)/\sigma] = \rho \cdot h(z) . \qquad (21.3.15)$$

Here $h(z) \equiv \phi(z)/\Phi(-z)$ is the hazard function of the standard normal distribution and is an increasing function of $z = (y - \mu)/\sigma$. Figure 21.5 depicts $h(z)$. Any power-normal hazard function is merely a multiple ρ of $h(z)$.

21.4 The Power-Lognormal Distribution

Introduction. This section presents properties of the power-lognormal distribution in detail. As background, readers need to know basic properties of the power-normal and normal distributions and lognormal probability paper. For example, in [12] and [13], Nelson covers these topics and the series-system model (below) in detail. The following topics include its reliability and cumulative distribution functions, its relationship to the power-normal distribution, the series-system model, percentiles, median, probability density, mode, mean, variance, standard deviation, and hazard function (failure rate).

Reliability function. The *reliability function* of the power-lognormal distribution is the population fraction surviving beyond age $t > 0$; namely,

$$\bar{G}(t; \mu, \sigma, \rho) \equiv \{\Phi[-(\ln(t) - \mu)/\sigma]\}^\rho , \qquad t > 0 . \qquad (21.4.1)$$

Here $\Phi[\,]$ is the standard normal cumulative distribution function, and $\rho > 0$ is called the *power parameter*. Also μ ($-\infty < \mu < \infty$) is called the *location parameter*, but is better called "mu". Strictly speaking, μ is a location parameter only for the power-normal distribution when $y = \ln(t)$ is used. For the power-*lognormal* distribution, $\tau = e^\mu$ is a scale parameter as shown below. Similarly, $\sigma > 0$ is called the *scale parameter*, but is better called "sigma". Strictly speaking, σ is a scale parameter only for the power-*normal* distribution, when $y = \ln(t)$ is used. For the power-*lognormal* distribution, ρ and σ are actually shape parameters. For $\rho = 1$, the distribution is lognormal; only then is μ the mean of log life $y = \ln(t)$, is e^{μ} the median life, and is σ the standard deviation of log life $y = \ln(t)$. The lognormal distribution is briefly presented in [12] and [13] and in many other statistics books; [19] devotes a book to it.

Cdf. The *cumulative distribution function* (cdf) is the population fraction failing before age t; namely,

$$G(t; \mu, \sigma, \rho) = 1 - \{\Phi[-(\ln(t) - \mu)/\sigma]\}^{\rho} , \qquad t > 0 . \qquad (21.4.2)$$

The distribution is simpler in terms of the *log-transformed standardized deviate* $z \equiv [\ln(t) - \mu]/\sigma$. Then

$$\bar{G}(t; \mu, \sigma, \rho) = [\Phi(-z)]^{\rho} , \qquad G(t; \mu, \sigma, \rho) = 1 - [\Phi(-z)]^{\rho} . \qquad (21.4.3)$$

David [5, p. 287] and Balakrishnan and Cohen [1, p. 57] list tables of these functions for integer ρ values.

Probability plot. Figure 21.6 displays power-lognormal cdfs for selected ρ values on lognormal probability paper. Figure 21.6 is the same as Figure 21.3, except that Figure 21.6 has a data scale (for t) that is logarithmic and Figure 21.3 has a data scale (for y) that is linear. For $\rho = 1$, the cdf is lognormal and is a straight line. For $\rho > 1$, the cdfs curve concave to the left and shift toward lower t (and z) values as ρ gets larger. For $0 < \rho < 1$, the cdfs are curves concave to the right and shift toward higher t (and z) values as ρ gets smaller. Figure 21.7 shows selected cdfs that have common values for the 0.25 and 0.75 fractiles to give these distributions roughly the same spread, as if fitted to the same data set. The figure shows that the lower tails differ appreciably. Thus it is important to know ρ if extrapolating into the lower tail in high reliability applications.

Other parameters. For some purposes, it is useful to express the distribution in terms of its true scale parameter $\tau \equiv \exp(\mu)$. Then

$$\bar{G}(t; \tau, \sigma, \rho) = \{\Phi[-(\ln(t/\tau))/\sigma]\}^{\rho} = \{\Phi[-(\ln(t/\tau)^{1/\sigma})]\}^{\rho} , \qquad (21.4.1')$$

$$G(t; \tau, \sigma, \rho) = 1 - \{\Phi[-(\ln(t/\tau))/\sigma]\}^{\rho} = 1 - \{\Phi[-(\ln(t/\tau)^{1/\sigma})]\}^{\rho} . \qquad (21.4.2')$$

For $\rho = 1$, τ is the distribution median; otherwise, it is the $1 - (0.5)^{\rho}$ fractile.

Approximate cdf. For large ρ, the exact cdf formula (21.4.2) requires a table or algorithm for $\Phi(\ )$ that is accurate in the far tails. Asymptotically as $\rho \to \infty$, the power-lognormal cdf approaches the cdf of a Weibull distribution (convergence in law or in distribution). Bury [2] gives the scale and shape parameters of that Weibull distribution as a function of μ, σ and ρ. Convergence to this asymptotic Weibull distribution is slow. That is, ρ must be very large (say, well above 100) for a satisfactory approximation.

Series-system model. As described in Section 21.3, a product may come in different sizes, and its life may depend on size. The previous assumptions for a series-system model are assumed here. However, if in (21.3.4) the reliability function $R_0(t)$ for product of size S_0 is lognormal, then the reliability function (21.3.4) of product of size S is the power-lognormal one (21.4.1). Then the power parameter $\rho = S/S_0$ is the ratio of the product size S to the "lognormal size" S_0.

Relation to the power-normal. The power-lognormal distribution has a relationship to the simpler power-normal distribution. If t has a power-lognormal distribution with parameters μ, σ, and ρ, then $y = \ln(t)$ has a power-normal distribution with the same values of the parameters μ, σ, and ρ. Thus, one can analyze power-lognormal data by fitting the power-normal distribution to the log data $y = \ln(t)$. The power-normal distribution is simpler in that only ρ is a shape parameter, whereas, for the power-lognormal distribution, both ρ and σ are shape parameters.

Fractiles. The F fractile of the power-lognormal distribution is

$$t_F = \exp(\mu + z_{F'}\sigma) ; \qquad (21.4.4)$$

here $z_{F'}$ is the $F' = 1 - (1-F)^{1/\rho}$ fractile of the standard normal distribution. t_F is the antilog of the F fractile (21.3.5) of the corresponding power-normal distribution.

Median. The median of the power-lognormal distribution is

$$t_{0.50} = \exp(\mu + z_{F'}\sigma) \qquad (21.4.5)$$

where $F' = 1 - (0.5)^{1/\rho}$. For $\rho = 1$ (the lognormal distribution), $z_{F'} = 0$ and $t_{0.50} = \exp(\mu)$.

Density. The power-lognormal probability density is the derivative of the cdf (21.4.2); namely,

$$g(t; \mu, \sigma, \rho) = (\rho/t\sigma)\, \phi\{[\ln(t) - \mu]/\sigma\}\, [\Phi\{-[\ln(t) - \mu]/\sigma\}]^{\rho-1} ; \qquad (21.4.6)$$

here $\phi(z) \equiv (2\pi)^{-1/2}\exp(-z^2/2)$ is the standard normal probability density. Figures 21.8–21.11 display such densities as a function of $r = t/\tau$, where $\tau = \exp(\mu)$. The densities have a variety of shapes depending on the ρ and σ values. These densities are unimodal. For $\rho = 1$, the density is lognormal.

Mode. The mode of a power-lognormal distribution is the value t_m where the probability density peaks. It is the most probable value. t_m is the solution of $dg(t; \mu, \sigma, \rho)/dt = 0$; namely, the solution $z_m \equiv [\ln(t_m) - \mu]/\sigma$ of

$$(z_m + \sigma)\Phi(-z_m) = -(\rho - 1) . \qquad (21.4.7)$$

Here z_m is a function of ρ and σ. For $\rho = 1$ (the lognormal distribution), $z_m = -\sigma$ and $t_m = \exp(\mu - \sigma^2)$.

Mean. The mean of the power-lognormal distribution is

$$ET \equiv \int_0^\infty t\, g(t; \mu, \sigma, \rho)\, dt$$

$$= \int_0^\infty (\rho\sigma)\, \phi\{[\ln(t) - \mu]/\sigma\}\, [\Phi\{-[\ln(t) - \mu]/\sigma\}]^{\rho-1} dt . \qquad (21.4.8)$$

$$= \tau \int_{-\infty}^\infty \rho\, e^{u\sigma}\, \phi(u)\{\Phi(-u)\}^{\rho-1} du \equiv \tau\, E(\sigma, \rho) .$$

Here $\tau = \exp(\mu)$ is the scale parameter, and $E(\sigma, \rho)$ is a function of σ and ρ but not $\mu = \ln(\tau)$. David [5, p. 288] and Balakrishnan and Cohen [1, p. 51] reference tables of $E(\sigma, \rho)$ for integer ρ and $\sigma = 1$. The integral can be evaluated with numerical quadrature.

Variance and standard deviation. The variance of the power-lognormal distribution is

$$V(T) \equiv \int_0^\infty (t - ET)^2\, g(t; \mu, \sigma, \rho)\, dt$$

$$= \int_0^\infty (t - ET)^2 (\rho/\sigma t)\, \phi\{[\ln(t) - \mu]/\sigma\}\, [\Phi\{-[\ln(t) - \mu]/\sigma\}]^{\rho-1} dt , \qquad (21.4.9)$$

$$= \tau^2 \int_{-\infty}^\infty [e^{\sigma u} - E(\sigma, \rho)]^2 \rho\, \phi(u)\, [\Phi(-u)]^{\rho-1} du \equiv \tau^2 V(\sigma, \rho) .$$

Here $\tau = \exp(\mu)$ is the scale parameter, and $V(\sigma,\rho)$ is a function of σ and ρ but not $\mu = \ln(\tau)$. David [5, p. 288] and Balakrishnan and Cohen [1, p. 51] reference tables of the integral for integer ρ and $\sigma = 1$. The integral can be evaluated with numerical quadrature. The standard deviation of the power-lognormal distribution is

$$sd(T) \equiv [V(T)]^{1/2} = \tau[V(\sigma,\rho)]^{1/2} \ . \tag{21.4.10}$$

For $\rho = 1$, the lognormal distribution, $V(\sigma,1) = [\exp(\sigma^2)-1]\exp(\sigma^2)$, $V(T) = \tau^2[\exp(\sigma^2)-1] \times \exp(\sigma^2)$, and

$$sd(T) = \tau[\exp(\sigma^2) - 1]^{1/2} \exp(\sigma^2/2) \ . \tag{21.4.11}$$

Hazard function. The hazard function (or "failure rate") of the power-lognormal distribution is

$$h(t;\mu,\sigma,\rho) \equiv g(t;\mu,\sigma,\rho)/\bar{G}(t;\mu,\sigma,\rho) = \rho \cdot h(t;\mu,\sigma,1) \ ; \tag{21.4.12}$$

here $h(t;\mu,\sigma,1) \equiv (1/\sigma t)\phi\{[\ln(t/\tau)]/\sigma\}/\Phi\{-[\ln(t/\tau)]/\sigma\}$ is the lognormal hazard function and $\tau = \exp(\mu)$ is the scale parameter. Rewritten

$$h(t;\mu,\sigma,\rho) = \rho \cdot h(r;0,\sigma,1) \tag{21.4.13}$$

where $r = t/\tau$. Figure 21.12 shows $h(r;0,\sigma,1)$ for selected values of σ. These are a multiple of the lognormal hazard functions. Thus, they have the same behavior as lognormal hazard functions. In particular, they are all zero at $t = 0$, increase to a maximum, and then decrease back toward zero with increasing time. For most σ values seen in practice, the maximum occurs in the far upper tail. Thus, for most practical purposes, the failure rate increases as the population ages.

21.5   Maximum Likelihood Theory for Fitting the Power-(Log)Normal Model to Data

21.5.1   Introduction

Purpose. This section presents maximum likelihood (ML) theory for fitting the power-(log)normal model to data. The POWNOR program in [14] and [15] implements the ML methods described here. As explained in detail by [12] and [13], the ML method has good properties including

- It is versatile; that is, it applies to most models and types of data including censored and interval data.
- It has good statistical properties; namely, the method yields good estimators and confidence limits.

Overview. The contents of this section are:

- Specimen and sample likelihoods (Section 21.5.2),
- Maximum likelihood estimates (Section 21.5.3),
- Fisher and covariance matrices (Section 21.5.4), and
- Confidence limits (Section 21.5.5).

21.5.2 Likelihoods

Introduction. This section presents specimen and sample (log) likelihoods for the power-(log)normal model and observed, left censored, right censored, and interval data. Terminology for life data is used, but the model may be fitted to strength and other types of data. Needed background is in Section 21.3 on the power-normal distribution. As background, the following paragraph describes types of sample data.

Data. The types of censored life data are defined here. For specimen i with size S_i, failure may occur at an exactly *observed failure age* (life) y_i. If an age t_i is from a power-lognormal distribution, then one uses the log age $y_i = \ln(t_i)$ throughout. If the specimen has run a time y_i without failure, then the failure age is beyond and is *censored on the right* at age y_i; then y_i is called the *survival age* without failure. If the failure occurs at some unknown age *before* a known first inspection age y_i, then the failure age is *censored on the left* at age y_i. If the failure occurs at some unknown age between two adjoining inspection ages $(y_i < y_i')$, then the failure age is *interval censored* and in the interval (y_i, y_i'). For a specimen, each such type of data has a corresponding likelihood for the power-(log)normal model, as described in the following paragraphs.

Observed. The *likelihood L_i* for specimen i of size S_i that has an observed failure at (log) age y_i is the probability density (21.3.7); namely,

$$L_i = f(y_i, S_i; \mu, \sigma, \kappa) = (\kappa S_i / \sigma) \, \phi[(y_i - \mu)/\sigma] \, \{\Phi[-(y_i - \mu)/\sigma]\}^{\kappa S_i - 1} . \qquad (21.5.1)$$

This and other likelihoods are functions of the data y_i and S_i and the model parameters μ, σ, and $\kappa \equiv 1/S_0$ where S_0 is the size of specimens with a (log) normal distribution. The *log likelihood* is

$$\mathcal{L} \equiv \ln(L_i) = \ln(\kappa S_i / \sigma) - \ln(2\pi)^{1/2} - \frac{1}{2}[(y_i - \mu)/\sigma]^2 - (\kappa S_i - 1)H[(y_i - \mu)/\sigma] ;$$

$$(21.5.2)$$

here

$$H(z) \equiv -\ln[\Phi(-z)] \qquad (21.5.3)$$

is the *cumulative hazard function* of the standard normal distribution.

Right censored. The likelihood L_i for specimen i of size S_i that is right censored at (log) age y_i, its current unfailed age, is the survival probability (21.3.1); namely,

$$L_i = \{\Phi[-(y_i - \mu)/\sigma]\}^{\kappa S_i} . \qquad (21.5.4)$$

The log likelihood is

$$\mathcal{L}_i = -\kappa \, S_i H[(y_i - \mu)/\sigma] . \qquad (21.5.5)$$

Left censored. The likelihood for specimen i of size S_i which is left censored at (log) age y_i, its inspection age after its unknown failure age, is the failure probability (21.3.2); namely,

$$L_i = 1 - \{\Phi[-(y_i - \mu)/\sigma]\}^{\kappa S_i} . \qquad (21.5.6)$$

The log likelihood is

$$\mathcal{L}_i = \ln[1 - \{\Phi[-(y_i - \mu)/\sigma]\}^{\kappa S_i}] . \qquad (21.5.7)$$

Interval censored. The likelihood L_i for specimen i of size S_i which is known to have failed in the interval between inspections at ages $y_i < y_i'$ is the probability of failure in that interval; namely,

$$L_i = \{\Phi[-(y_i - \mu)/\sigma]\}^{\kappa S_i} - \{\Phi[-(y_i' - \mu)/\sigma]\}^{\kappa S_i} . \qquad (21.5.8)$$

The log likelihood is $\mathcal{L}_i = \ln(L_i)$.

Sample likelihood. The specimen likelihoods for a sample of n specimens yield the sample likelihood below. The sample log likelihood yields ML estimates and confidence limits for model parameters and other quantities, as described in the following sections. The specimen lifetimes are assumed to be a random sample of statistically *independent* lifetimes from a power-(log)normal distribution. The sample likelihood is the probability of the n specimen outcomes (observed, censored, or interval data). For independent specimen lifetimes, the *sample likelihood* L is the product of the specimen likelihoods; namely,

$$L = L_1 \times L_2 \times \cdots \times L_n . \qquad (21.5.9)$$

Here the sample likelihood is a function of the data values y_i (and y_i') and S_i, the type of data, and the model parameters μ, σ, and $\kappa = 1/S_0$. The *sample log likelihood* $\mathcal{L} \equiv \ln(L)$ is the sum of the specimen log likelihoods; namely,

$$\mathcal{L} = \mathcal{L}_1 + \mathcal{L}_2 + \cdots + \mathcal{L}_n . \qquad (21.5.10)$$

21.5.3   Maximum Likelihood Estimates

Introduction. This section defines maximum likelihood (ML) estimates and presents related theory.

Definition. The *maximum likelihood estimates* for the model parameters μ, σ, and κ are the values $\hat{\mu}$, $\hat{\sigma}$, and $\hat{\kappa}$ that maximize the sample likelihood (21.5.9) (or equivalently that maximize (21.5.10)) over the range of mathematically allowed values of those parameters. For many models and types of data, the maximum likelihood estimates are unique. Also, the ML estimates $(\hat{\mu}, \hat{\sigma}, \hat{\kappa})$ have an asymptotic joint normal distribution with mean vector (μ, σ, κ), the true population values, and covariance matrix (21.5.20) below. In [12] and [13], Nelson references authors who state regularity conditions that the model and data must satisfy to assure this. For some models and types of data, there may be local maxima; then the parameter values at the global maximum are the ML estimates. In many computer programs, the ML estimates are iteratively calculated by direct numerical search for the maximum of (21.5.10); Section 21.6 describes such calculations. An alternative calculation involves the likelihood equations which follow.

Likelihood equations. For some models and types of data, the ML estimates can be found by the calculus method. That is, the parameter values for which the partial derivatives of the sample log likelihood with respect to each parameter all equal zero are the ML estimates. For the power-(log)normal model and *complete* data (all specimens have exactly observed failure ages y_i), the partial derivatives set equal to zero are:

$$0 = \partial \mathcal{L}/\partial \mu = (1/\sigma) \left\{ \Sigma z_i + \Sigma (\kappa S_i - 1) h(z_i) \right\} ,$$

$$0 = \partial \mathcal{L}/\partial \sigma = (1/\sigma) \left\{ -n + \Sigma z_i^2 + \Sigma (\kappa S_i - 1) z_i h(z_i) \right\} , \qquad (21.5.11)$$

$$0 = \partial \mathcal{L}/\partial \kappa = (n/\kappa) - \Sigma S_i H(z_i) .$$

Here the standardized deviate $z_i \equiv (y_i - \mu)/\sigma$ is a function of μ and σ, and each sum runs over all n specimens. These *likelihood equations* cannot be solved explicity for $\hat{\mu}$, $\hat{\sigma}$, and $\hat{\kappa}$. They must be solved iteratively.

Expectations. Mathematical expectations of various quantities appear in the ML theory below. Such expectations are defined here for complete data; in [12] and [13], Nelson defines them for other forms of data. Let $u(y_i)$ denote a given function of the random failure age y_i of specimen i. Then the *expectation* of the random quantity $u(y_i)$ is

$$E\{u(y_i)\} \equiv \int_{-\infty}^{\infty} u(y_i) f(y_i; \mu, \sigma, \kappa) \, dy_i \; ; \tag{21.5.12}$$

here $f(\ )$ is the power-normal probability density (21.3.8), and the parameters μ, σ, and κ have the true population values.

Relationships. A useful general property of the first partial derivatives of the specimen and sample log likelihoods is that their expectations are zero. That is,

$$0 = E\{\partial \mathcal{L}/\partial \mu\} = (1/\sigma) \left[\Sigma E\{z_i\} + \Sigma(\kappa S_i - 1) E\{h(z_i)\}\right] \; ,$$
$$0 = E\{\partial \mathcal{L}/\partial \sigma\} = (1/\sigma) \left[-n + \Sigma E\{z_i^2\} + \Sigma(\kappa S_i - 1) E\{z_i h(z_i)\}\right] \; , \tag{21.5.13}$$
$$0 = E\{\partial \mathcal{L}/\partial \kappa\} = (n/\kappa) - \Sigma S_i E\{H(z_i)\} \; .$$

In these formulas, the y_i are regarded as random quantities and the parameters are equal to their true (unknown) population values. These formulas aid in the evaluation of various expectations later. For example, for a single specimen i, the third (21.5.13) equation yields

$$1/(\kappa S_i) = E\{H(z_i)\} \equiv \int_{-\infty}^{\infty} H(z_i) f(y_i; \mu, \sigma, \kappa S_i) \, dy_i$$

$$= (1/\sigma) \int_{-\infty}^{\infty} H(z_i) f(z_i; 0, 1, \kappa S_i) \, dz_i \; ; \tag{21.5.14}$$

Here $f(z_i; 0, 1, \kappa S_i)$ is the standard power-normal distribution with $\rho_i = \kappa S_i$. Similarly, the first (21.5.13) equation yields

$$E\{h(z_i)\} = E\{z_i\}/(\kappa S_i) = (\kappa S_i - 1)^{-1}(1/\sigma) \int_{-\infty}^{\infty} z_i \, f(z_i, 0, 1, \kappa S_i) \, dz_i \; . \tag{21.5.15}$$

The last integral is the mean of the standard power-normal distribution with $\rho_i = \kappa S_i$. These results are used to evaluate integrals below.

21.5.4 Fisher and Covariance Matrices

Introduction. This section presents the theoretical Fisher information matrix and the theoretical asymptotic covariance matrix of the ML estimates. The subsequent estimates of these matrices are used to obtain approximate confidence limits in Section 21.5.5.

Second derivatives. ML theory employs the second partial derivatives of the sample log likelihood. For the power-normal model and complete data, they are:

$$\partial^2 \mathcal{L}/\partial \mu^2 = (-1/\sigma^2) \{n + \Sigma(\kappa S_i - 1) h(z_i)[h(z_i) - z_i]\} \; ,$$
$$\partial^2 \mathcal{L}/\partial \sigma \partial \mu = (-1/\sigma^2) \{2\Sigma z_i + \Sigma(\kappa S_i - 1) h(z_i)[1 + z_i[h(z_i) - z_i]]\} \; ,$$
$$\partial^2 \mathcal{L}/\partial \kappa \partial \mu = (S_i/\sigma) h(z_i) \; , \tag{21.5.16}$$
$$\partial^2 \mathcal{L}/\partial \sigma^2 = (-1/\sigma^2) \{-n + 3\Sigma z_i^2 + \Sigma(\kappa S_i - 1) h(z_i)[2 + z_i[h(z_i) - z_i]]\} \; ,$$
$$\partial^2 \mathcal{L}/\partial \kappa \partial \sigma = (1/\sigma) \Sigma S_i z_i h(z_i) \; ,$$
$$\partial^2 \mathcal{L}/\partial \kappa^2 = -1/\kappa^2 \; .$$

In these equations, the derivative of the standard normal hazard function satisfies $h'(z_i) = h(z_i)[h(z_i) - z_i]$.

Fisher matrix. The theoretical *Fisher (information) matrix* is the matrix of expectations of the negative second partial derivatives; namely,

$$\mathbf{F} = \begin{bmatrix} E\{-\partial^2 \mathcal{L}/\partial\mu^2\} & E\{-\partial^2 \mathcal{L}/\partial\mu\partial\sigma\} & E\{-\partial^2 \mathcal{L}/\partial\mu\partial\kappa\} \\ E\{-\partial^2 \mathcal{L}/\partial\sigma\partial\mu\} & E\{-\partial^2 \mathcal{L}/\partial\sigma^2\} & E\{-\partial^2 \mathcal{L}/\partial\sigma\partial\kappa\} \\ E\{-\partial^2 \mathcal{L}/\partial\kappa\partial\mu\} & E\{-\partial^2 \mathcal{L}/\partial\kappa\partial\sigma\} & E\{-\partial^2 \mathcal{L}/\partial\kappa^2\} \end{bmatrix} . \qquad (21.5.17)$$

This matrix is symmetric and a function of the parameters, which equal their true (unknown) population values here. The *ML estimate* $\mathbf{F}$ of this matrix is obtained by replacing the parameters with their ML estimates in (21.5.17). Some integrals in (21.5.17) have not been evaluated analytically. For the ML estimate, the integrals can be evaluated by numerical quadrature when the true unknown parameter values are replaced by their ML estimates. For censored data, general formulas for such integrals appear in [12] and [13] and require a censoring time for each specimen, even those that failed. Such a planned censoring time may not be known for field data. The theoretical matrix is useful for theoretically evaluating proposed test plans where all censoring times are specified. Another estimate of $\mathbf{F}$ that does not require all censoring times and is easier to calculate is the *local estimate*

$$\mathbf{F}^* = \begin{bmatrix} E\{-\partial^2 \widehat{\mathcal{L}}/\partial\mu^2\} & E\{-\partial^2 \widehat{\mathcal{L}}/\partial\mu\partial\sigma\} & E\{-\partial^2 \widehat{\mathcal{L}}/\partial\mu\partial\kappa\} \\ E\{-\partial^2 \widehat{\mathcal{L}}/\partial\sigma\partial\mu\} & E\{-\partial^2 \widehat{\mathcal{L}}/\partial\sigma^2\} & E\{-\partial^2 \widehat{\mathcal{L}}/\partial\sigma\partial\kappa\} \\ E\{-\partial^2 \widehat{\mathcal{L}}/\partial\kappa\partial\mu\} & E\{-\partial^2 \widehat{\mathcal{L}}/\partial\kappa\partial\sigma\} & E\{-\partial^2 \widehat{\mathcal{L}}/\partial\kappa^2\} \end{bmatrix} . \qquad (21.5.18)$$

This is the matrix of negative second partial derivatives ((21.5.16) for complete data) of the sample log likelihood where the caret $\widehat{}$ indicates that the parameters are set equal to their ML estimates. For example,

$$-\partial^2 \widehat{\mathcal{L}}/\partial\kappa\partial\mu = -(1/\widehat{\sigma})\Sigma S_i h(\widehat{z}_i) \qquad (21.5.19)$$

where $\widehat{z}_i = (y_i - \widehat{\mu})/\widehat{\sigma}$. In [12], Nelson gives other (equivalent) formulas for the theoretical Fisher information matrix and other estimates for it. This local estimate of the Fisher matrix does not require a censoring time for each specimen, only for the censored specimens. Thus it can always be used, even when the ML estimate $\mathbf{F}$ cannot due to unknown censoring times for failed specimens.

Covariance matrix. The theoretical asymptotic *covariance matrix* $\mathbf{\mathcal{Z}}$ of the ML estimates $\widehat{\mu}$, $\widehat{\sigma}$, and $\widehat{\kappa}$ is the inverse of the theoretical Fisher information matrix; namely,

$$\mathbf{\mathcal{Z}} \equiv \begin{bmatrix} \text{Var}(\widehat{\mu}) & \text{Cov}(\widehat{\mu}) & \text{Cov}(\widehat{\mu}, \widehat{\kappa}) \\ \text{Cov}(\widehat{\sigma}, \widehat{\mu}) & \text{Var}(\widehat{\sigma}) & \text{Cov}(\widehat{\sigma}, \widehat{\kappa}) \\ \text{Cov}(\widehat{\kappa}, \widehat{\mu}) & \text{Cov}(\widehat{\kappa}, \widehat{\sigma}) & \text{Var}(\widehat{\kappa}) \end{bmatrix} = \mathbf{F}^{-1} . \qquad (21.5.20)$$

In [12, pp. 384-386], Nelson gives a heuristic argument for this result. $\mathbf{\mathcal{Z}}$ is a function of the parameters μ, σ, and κ set equal to their true (unknown) population values. The ML estimate of the covariance matrix is the inverse of the ML estimate (21.5.17) of the Fisher matrix; namely, $\mathbf{\mathcal{Z}} = \mathbf{F}^{-1}$. Similarly, the local estimate of the covariance matrix is the inverse of the local estimate (21.5.18) of the Fisher matrix; namely, $\mathbf{\mathcal{Z}}^* = (\mathbf{F}^*)^{-1}$. An estimate of the covariance matrix yields approximate confidence limits for parameters and functions of parameters as described in Section 21.5.5.

21.5.5   Confidence Limits

Introduction. This section presents approximate confidence limits for model parameters and functions of them.

Function estimate. Often one wishes to estimate some function $\theta = \theta(\mu, \sigma, \kappa)$ of the model parameters. For example, the F fractile of the power-normal model for specimens of size S is the function $\eta_F = \mu + z_{F'}\sigma$ where $F' = 1 - (1 - F)^{1/\kappa S}$ and $z_{F'} = \Phi^{-1}(F')$. Also, for example, the "normal size" is the function $S_0 = 1/\kappa$ of just κ. The ML estimate of the true function value $\theta = \theta(\mu, \sigma, \kappa)$ is the function evaluated at the ML estimates of the parameters; namely, $\widehat{\theta} = \theta(\widehat{\mu}, \widehat{\sigma}, \widehat{\kappa})$. For example, $\widehat{F'} = 1 - (1 - F)^{1/\widehat{\kappa} S}$ and $\widehat{\eta}_F = \widehat{\mu} + z_{\widehat{F'}}\widehat{\sigma}$. Under certain regularity conditions that hold for many models and data, the asymptotic cumulative distribution of $\widehat{\theta}$ for "large" samples is close to a normal cumulative distribution with mean θ (the true population value) and variance $\mathrm{Var}(\widehat{\theta})$ given below in (21.5.22). This asymptotic normal distribution yields the approximate confidence limits in the next paragraph.

Confidence limits. Based on the asymptotic normal distribution of $\widehat{\theta}$, lower and upper two-sided approximate $100C\%$ normal confidence limits for the true value θ are

$$\underset{\sim}{\theta} = \widehat{\theta} - z_{C'}[\mathrm{Var}(\widehat{\theta})]^{1/2}, \qquad \widetilde{\theta} = \widehat{\theta} + z_{C'}[\mathrm{Var}(\widehat{\theta})]^{1/2} ; \qquad (21.5.21)$$

here $z_{C'}$ is the standard normal $C' = (1 + C)/2$ fractile, and $\mathrm{Var}(\widehat{\theta})$ appears below in (21.5.22). In practice, one uses in (21.5.21) an estimate of $\mathrm{Var}(\widehat{\theta})$ given below.

Variance. This paragraph presents the formula for the theoretical asymptotic variance $\mathrm{Var}(\widehat{\theta})$. It is assumed that the function $\theta(\mu, \sigma, \kappa)$ has continuous partial derivatives with respect to the parameters μ, σ, and κ. Then

$$\mathrm{Var}(\widehat{\theta}) = \begin{bmatrix} \partial\theta/\partial\mu & \partial\theta/\partial\sigma & \partial\theta/\partial\kappa \end{bmatrix} \Sigma \begin{bmatrix} \partial\theta/\partial\mu \\ \partial\theta/\partial\sigma \\ \partial\theta/\partial\kappa \end{bmatrix} \qquad (21.5.22)$$

$$= (\partial\theta/\partial\mu)^2 \mathrm{Var}(\widehat{\mu}) + 2(\partial\theta/\partial\mu)(\partial\theta/\partial\sigma)\mathrm{Cov}(\widehat{\mu}, \widehat{\sigma})$$

$$+ (\partial\theta/\partial\sigma)^2 \mathrm{Var}(\widehat{\sigma}) + 2(\partial\theta/\partial\sigma)(\partial\theta/\partial\kappa)\mathrm{Cov}(\widehat{\sigma}, \widehat{\kappa})$$

$$+ (\partial\theta/\partial\kappa)^2 \mathrm{Var}(\widehat{\kappa}) + 2(\partial\theta/\partial\kappa)(\partial\theta/\partial\mu)\mathrm{Cov}(\widehat{\mu}, \widehat{\kappa}) . \qquad (21.5.23)$$

Here the partial derivatives and variances and covariances are all evaluated at the true values of the parameters. The local (or maximum likelihood) estimate of this variance is given by (21.5.22) where the partial derivatives are evaluated at the ML estimates of the parameters and the variances and covariances are replaced by their local (or maximum likelihood) estimates. In practice, the local estimate is usually easier to calculate. Either estimate is used to obtain the confidence limits (21.5.21).

Covariance. Suppose that one wishes to estimate another function $\psi = \psi(\mu, \sigma, \kappa)$ of the model parameters. Its ML estimate $\widehat{\psi} = \psi(\widehat{\mu}, \widehat{\sigma}, \widehat{\kappa})$ and $\widehat{\theta} = \theta(\widehat{\mu}, \widehat{\sigma}, \widehat{\kappa})$ have an asymptotic joint normal cumulative distribution for large samples. That joint distribution has a mean vector (θ, ψ), the true population values, variances given by (21.5.22), and covariance

$$\mathrm{Cov}(\widehat{\theta}, \widehat{\psi}) = (\partial\theta/\partial\mu)(\partial\psi/\partial\mu)\mathrm{Var}(\widehat{\mu}) + (\partial\theta/\partial\sigma)(\partial\psi/\partial\sigma)\mathrm{Var}(\widehat{\sigma}) + (\partial\theta/\partial\kappa)(\partial\psi/\partial\kappa)\mathrm{Var}(\widehat{\kappa})$$

$$+ [(\partial\theta/\partial\mu)(\partial\psi/\partial\sigma) + (\partial\theta/\partial\sigma)(\partial\psi/\partial\mu)]\mathrm{Cov}(\widehat{\mu}, \widehat{\sigma})$$

$$+ [(\partial\theta/\partial\sigma)(\partial\psi/\partial\kappa) + (\partial\theta/\partial\kappa)(\partial\psi/\partial\sigma)]\mathrm{Cov}(\widehat{\sigma}, \widehat{\kappa}) \qquad (21.5.24)$$

$$+ [(\partial\theta/\partial\mu)(\partial\psi/\partial\kappa) + (\partial\theta/\partial\kappa)(\partial\psi/\partial\mu)]\mathrm{Cov}(\widehat{\mu}, \widehat{\kappa}) .$$

This result is used in Section 21.6 for transformed parameters used for convenience in place of the natural model parameters μ, σ, and κ.

LR limits. The approximate normal limits (21.5.21) are simple to calculate for most models and types of data, and they are widely used. In practice, such limits usually enclose the corresponding true value with a probability below the stated confidence $100C\%$. Calculation of likelihood ratio (LR) intervals is described by Nelson [13, pp. 297-301]. Such intervals usually have a confidence level closer to the stated $100C\%$. The review paper [7] in this volume provides further discussion on LR intervals and their use in life-data analysis. However, their computation is more complex, and they are not yet available in POWNOR and in most computer programs for ML fitting.

Improved normal limits. The approximate normal confidence limits (21.5.21) can often be improved by working with a transformed value $\zeta = \zeta(\theta)$ and treating its ML estimate $\hat{\zeta} = \zeta(\hat{\theta})$ as normally distributed. Then one calculates the limits

$$\underset{\sim}{\zeta} = \hat{\zeta} - z_{C'}[\text{Var}(\hat{\zeta})]^{1/2} , \qquad \tilde{\zeta} = \hat{\zeta} + z_{C'}[\text{Var}(\hat{\zeta})]^{1/2} \qquad (21.5.25)$$

Here $\text{Var}(\hat{\zeta}) = (\partial\zeta/\partial\theta)^2 \text{Var}(\hat{\theta})$ and is estimated as described above. Then the limits for θ are obtained by inverting the monotone function $\zeta(\ )$ to get $\underset{\sim}{\theta} = \zeta^{-1}(\underset{\sim}{\zeta})$ and $\tilde{\theta} = \zeta^{-1}(\tilde{\zeta})$, if $\zeta(\theta)$ is an increasing function of θ. Otherwise, $\underset{\sim}{\theta} = \zeta^{-1}(\tilde{\zeta})$ and $\tilde{\theta} = \zeta^{-1}(\underset{\sim}{\zeta})$, if $\zeta(\tilde{\theta})$ is a decreasing function of θ. For example, in the power-(log)normal model, a useful transformation is $\zeta = \ln(\sigma)$. This transformation assures that the confidence limits for σ are positive, whereas, (21.5.21) may yield a negative lower limit, which is unacceptable since $\sigma > 0$. Such transformed limits may have a confidence level that is closer to the stated $100C\%$ and more equal error probabilities above and below the interval.

21.6   Numerical Methods

Introduction. This section describes numerical methods that facilitate the calculations involved in the ML theory in Section 21.5. Many formulas and ideas in Section 21.5 are computationally naive and must be altered as shown here. In particular, finding the ML estimates by iteratively maximizing the log likelihood is difficult unless the model is reparametrized. Even the best optimization routine has difficulty optimizing the log likelihood of a poorly parametrized model, and the natural model parameters are often a poor choice. As background, Ross [17] discusses such numerical problems and solutions in detail, and [12, pp. 386-395], [4], and [6] provide further solutions. This section will interest only a few who are concerned with numerical methods.

Starting values. The following method uses percentile estimation to obtain starting values $\hat{\mu}_0$, $\hat{\sigma}_0$, and $\hat{\kappa}_0$ for the ML iteration. The method employs three selected sample order statistics $y_i < y_{i'} < y_{i''}$. with corresponding probability plotting positions $P < P' < P''$. The next paragraph deals with the choice of i, i', and i''. For a complete sample of n specimens of size S, the plotting position of order statistic i is $P_i = (i - 0.5)/n$. The P fractile of the power-normal distribution is

$$\eta_P = \mu + z_\pi \sigma \qquad (21.6.1)$$

where $\pi = 1 - Q^{1/\kappa S}$ and $Q = 1 - P$. For specified

$$Q = 1 - P = (n - i + 0.5)/n , \qquad Q' = 1 - P' = (n - i' + 0.5)/n , \qquad (21.6.2)$$

$$Q'' = 1 - P'' = (n - i'' + 0.5)/n ,$$

the differences of the corresponding fractiles are

$$\eta_{P''} - \eta_{P'} = (z_{\pi''} - z_{\pi'})\sigma , \qquad \eta_{P'} - \eta_P = (z_{\pi'} - z_\pi)\sigma . \qquad (21.6.3)$$

The ratio of these equations yields

$$(\eta_{P''} - \eta_{P'})/(\eta_{P'} - \eta_P) = [z_{1-Q''^{1/\kappa S}} - z_{1-Q'^{1/\kappa S}}]/[z_{1-Q'^{1/\kappa S}} - z_{1-Q^{1/\kappa S}}] . \quad (21.6.4)$$

Substitute the order statistic values y_i, $y_{i'}$, and $y_{i''}$ for the corresponding fractiles η_P, $\eta_{P'}$, and $\eta_{P''}$ on the left side of this equation, and iterate on $b = 1/(\kappa S)$ on the right side to find the value b_0 that satisfies this equation. Then use $\widehat{\kappa}_0 = 1/(b_0 S)$ as the starting value. Then substituting the order statistic values y_i and $y_{i'}$ for η_P and $\eta_{P'}$ into (21.6.3) yields a starting value $\widehat{\sigma}_0$; namely,

$$\widehat{\sigma}_0 = (y_{i'} - y_i)/(z_{\widehat{\pi}'} - z_{\widehat{\pi}}) . \qquad (21.6.5)$$

where $\widehat{\pi} = 1 - Q^{b_0}$ and $\widehat{\pi}' = 1 - Q'^{b_0}$. Finally, substituting the order statistic y_i for η_P in (21.6.1) yields

$$\widehat{\mu}_0 = y_i - z_{\widehat{\pi}}\widehat{\sigma}_0 . \qquad (21.6.6)$$

These starting values are the percentile estimates based on the order statistics y_i, $y_{i'}$, and $y_{i''}$. These starting values can be used to calculate starting values for another parametrization of the model. This method extends readily to multiply (right) censored samples, but must be adapted to interval data. Treat each failure in an interval as if it occurred at the midpoint of its interval, and use the method above. This yields crude starting values.

Order statistics. For a complete sample, the order statistics used in the previous paragraph are chosen as follows. For a (log)normal distribution ($\rho = 1$), the asymptotically "best" (as $n \to \infty$) two order statistics for estimating σ have ranks i and i'' satisfying

$$0.0694 = (i - 0.5)/n , \quad 0.9306 = (i'' - 0.5)/n . \qquad (21.6.7)$$

The solutions are

$$i = 0.0694\,n + 0.5 , \quad i'' = 0.9306\,n + 0.5 , \qquad (21.6.8)$$

which each must be rounded to the nearest integer. Use the rounded values in (21.6.8) to calculate the plotting positions in the previous paragraph. For the middle order statistic use the sample median; that is, $i' = n/2$ for even n and $i' = (n + 1)/2$ for odd n, and $P' = Q' = 0.50$. For a (multiply) censored sample, calculate plotting positions for the failure times as described by [12, Chap. 4], and use one of the early failures P_i, one of the last failures $P_{i''}$ and one about midway between them, $P_{i'} \simeq (P_i + P_{i''})/2$.

Iteration parameters. The theory in Sections 21.3, 21.4, and 21.5 employs the natural model parameters μ, σ, and ρ (or equivalently S_0 or $\kappa = 1/S_0$). These parameters are mathematically convenient and easy to interpret, especially μ and σ, which are well understood as parameters of the normal distribution ($\rho = 1$). For purposes of calculating ML estimates by optimizing a sample log likelihood, other parameters work better. POWNOR uses the parameters μ, $\delta = \ln(\sigma)$, and $\omega = \ln(S_0)$ in the log likelihood. That is, $\exp(\delta)$ replaces σ, and $\exp(\omega)$ replaces S_0. The new parameters all range from $-\infty$ to $+\infty$. Thus the optimization with respect to μ, δ, and ω is unconstrained; whereas iterated values of σ and S_0 would have to be constrained to be positive, which complicates the optimization calculations. Moreover, confidence limits $(\underline{\delta}, \overline{\delta})$ and $(\underline{\omega}, \overline{\omega})$ convert into positive confidence limits $\underset{\sim}{\sigma} = \exp(\underset{\sim}{\delta})$, $\widetilde{\sigma} = \exp(\widetilde{\delta})$, $\underset{\sim}{S}_0 = \exp(\underset{\sim}{\omega})$, and $\widetilde{S}_0 = \exp(\widetilde{\omega})$ avoiding negative lower limits. Also, for various models, such transformations tend to give actual confidence levels that are closer to the intended levels than untransformed parameters [12, p. 393].

Optimization of the log likelihood. The ML estimates of the model parameters are the parameter values that maximize the sample log likelihood, which is calculated as described in Section 21.5.2. The program employs the Powell method [16] to iteratively search for the values of the model parameters that maximize the log likelihood. This method does not use the explicit derivatives of the log likelihood function taken with respect to the parameters. The user must provide the Powell method with starting values for the ML estimates of model parameters. A method to obtain starting values for the ML iteration is given above.

Trimmed standardized deviates. In searching for the parameters that maximize the log likelihood function, POWNOR works with standardized deviates $z_i \equiv (y_i - \hat{\mu})/\hat{\sigma}$ of observations. If the starting values of the model parameters are far from their ML estimates, the standardized deviates are large. Such large values can cause overflow problems and slow down or prevent the convergence of the iterative method. POWNOR avoids this through the use of trimmed standardized deviates, as described in [12, p. 394]. POWNOR uses the trimmed standardized deviates until the iteration is close to converging. Then the program switches to using the (untrimmed) standardized deviates of the observations and completes the ML iteration.

Standard normal cdf and its inverse. As described in Section 21.5, the log likelihood function for a censored value contains the standard normal cdf. POWNOR uses the double-precision IMSL subroutine DNORDF [9] to evaluate the standard normal cdf.

Fisher matrix. After POWNOR obtains the ML estimates, it calculates the local estimate of the Fisher (information) matrix. The elements of this matrix are the negative second partial derivatives of the log likelihood taken with respect to the (transformed) model parameters and evaluated at their ML estimates. POWNOR uses the STATLIB subroutine HESS which obtains these derivatives numerically by a perturbation calculation using only the values of the log likelihood function.

Covariance matrix. The local estimate of the covariance matrix for the ML estimates of the model parameters is the inverse of the local estimate of the Fisher matrix. The program uses the IMSL subroutine DLINDS [9] to invert the local estimate of the Fisher matrix.

Confidence limits. The program calculates approximate confidence limits for μ, $\ln(\sigma)$, and $\ln(S_0)$, using their ML estimates and local estimates of their variances. This assures positive lower limits for σ and S_0. Calculation of confidence limits for a distribution percentile requires the inverse of the standard normal cdf. The program uses the IMSL subroutine DNORIN [9] to evaluate the inverse of the standard normal cdf.

Acknowledgements

The authors gratefully acknowledge the support of this research under their NIST/NSF/ASA Research Fellowship at the National Institute of Standards and Technology (NIST). The National Science Foundation (NSF) generously provided our salary and travel expenses. NIST kindly provided research facilities and colleagues. The American Statistical Association (ASA) handled the Fellowship selection and administration. Dr. Robert Lundegard and Mrs. Ruth Varner of NIST effectively and graciously administered our work at NIST, and provided word processing. Mr. Harry Schafft and Dr. Jim Lechner of NIST provided much valued stimulation for this paper in our collaboration on electromigration research with them. We would also like to thank Prof. William Meeker for his suggestions on the numerical methods in the computer program.

References

1. Balakrishnan, N. and Cohen, A.C. (1991), *Order Statistics and Inference: Estimation Methods*, Academic Press, San Diego.

2. Bury, K.V. (1975), Distribution of smallest log-normal and gamma extremes, *Statistische Hefte*, **16**, 105-114. In English.

3. Chaplin, W.S. (1880), The relation between the tensile strengths of long and short bars, *Van Nostrand Engineering Magazine*, **23**, 441-444.

4. Cox, D.R. and Reid, N. (1987), Parameter orthogonality and approximate conditional inference, *Journal of the Royal Statistical Society, Series B*, **49**, 1-39.

5. David, H.A. (1981), *Order Statistics*, Second edition, John Wiley & Sons, New York.

6. Doganaksoy, N. and Schmee, J. (1993), Orthogonal parameters with censored data, *Communications in Statistics*, **22**, 669-685.

7. Doganaksoy, N. (1994), Likelihood ratio confidence intervals in life-data analysis, In *Recent Advances in Life-Testing and Reliability* (Ed., N. Balakrishnan), CRC Press, Boca Raton, FL.

8. Harter, H.L. (1977), A survey of the literature on the size effect on material strength, *Report AFFDL-TR-77-11, Air Force Flight Dynamics Lab. AFSC*, Wright-Patterson AFB, OH 45433.

9. IMSL (1987), *User's Manual Stat/Library*, Version 1, IMSL, Inc., 2500 ParkWest Tower One, 2500 City West Blvd., Houston, TX 77042-3020, (713)782-6060.

10. Johnson, N.L., Kotz, S., and Balakrishnan, N. (1994), *Continuous Univariate Distributions, Vol. 1*, Second edition, John Wiley & Sons, New York.

11. LaCombe, D.J. and Parks, E.L. (1986), The distribution of electromigration failures, *Proceedings of the 1986 International Reliability Physics Symposium*, 1-6.

12. Nelson, Wayne (1982), *Applied Life Data Analysis*, John Wiley & Sons, New York.

13. Nelson, Wayne (1990), *Accelerated Testing: Statistical Models, Test Plans, and Data Analyses*, John Wiley & Sons, New York, (800)879-4539.

14. Nelson, Wayne and Doganaksoy, N. (1992), A computer program POWNOR for fitting the power-normal and -lognormal models to life or strength data from specimens of various sizes, *NIST Technical Report NISTIR 4760*, Statistical Engineering Div., Adm. Bldg. A337, Gaithersburg, MD 20899.

15. Nelson, Wayne and Doganaksoy, N. (1992), POWNOR, in New Developments in Statistical Computing, *The American Statistician*, **46**, 318-319.

16. Powell, M.J.D. (1964), An efficient method for finding the minimum of a function without derivatives, *Computer Journal*, **7**, 155-162.

17. Ross, G.J.S. (1990), *Nonlinear Estimation*, Springer-Verlag, New York, (800)777-4643.

18. Schafft, H.A., Staton, T.C., Mandel, J., and Shott, J.D. (1987), Reproducibility of electromigration measurements, *IEEE Transactions on Electronic Devices*, **ED-34**, 673-681.

19. Shimizu, K. and Crow, E.L. (Eds.) (1988), *Lognormal Distributions*, Marcel Dekker, New York, (800)228-1160.

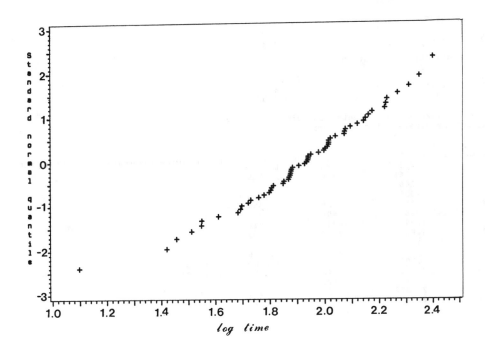

FIGURE 21.1
Lognormal probability plot of 400-micrometer data

POWNOR Version 1.0

Developed by Necip Doganaksoy and Wayne Nelson under 1991 Fellowship
grant from ASA/NSF/NIST. The program is documented in National Inst.
of Standards and Technology Report NISTIR 4760.

I. Total number of data cases= 59
Exact observations   = 59   Right censored observations = 0
Left censored observations= 0   Interval censored observations= 0

    Maximized log-likelihood = 55.8554

II. Maximum likelihood estimates for distribution parameters
with approximate 95% confidence limits

Parameter	ML Estimate	Lower Limit	Upper Limit	Std. Error
1 mu	2.8779	0.8280	4.9279	1.0459
2 ln(sigma)	-0.7336	-1.6375	0.1704	0.4612
3 ln(normal length)	2.6779	-3.1808	8.5366	2.9891

III.   Estimated information matrix

	1	2	3
1	1159.60	-985.06	255.65
2	-985.06	947.61	-200.44
3	255.65	-200.44	59.00

    Estimated covariance matrix

	1	2	3
1	1.09391	0.47809	-3.11576
2	0.47809	0.21270	-1.34900
3	-3.11576	-1.34900	8.93479

    Estimated correlation matrix

	1	2	3
1	1.00000	0.99115	-0.99662
2	0.99115	1.00000	-0.97856
3	-0.99662	-0.97856	1.00000

IV. Maximum likelihood estimates for distribution percentiles
with approximate 95% confidence limits

    Length=   400.

Pct.	ML Estimate	Lower Limit	Upper Limit	Std. Error
0.1	0.9731	0.5994	1.3468	0.1907
0.2	1.0541	0.7256	1.3825	0.1676
0.5	1.1664	0.8968	1.4360	0.1375
1	1.2561	1.0297	1.4825	0.1155
2	1.3507	1.1655	1.5358	0.0945
5	1.4853	1.3496	1.6210	0.0692
10	1.5973	1.4930	1.7016	0.0532
30	1.8074	1.7357	1.8792	0.0366
50	1.9360	1.8719	2.0001	0.0327
70	2.0528	1.9921	2.1135	0.0310
90	2.2044	2.1380	2.2709	0.0339
95	2.2714	2.1944	2.3485	0.0393
98	2.3431	2.2475	2.4387	0.0488
99	2.3890	2.2776	2.5004	0.0568
99.5	2.4298	2.3020	2.5576	0.0652
99.9	2.5108	2.3439	2.6776	0.0851

FIGURE 21.2
POWNOR output for the electromigration data

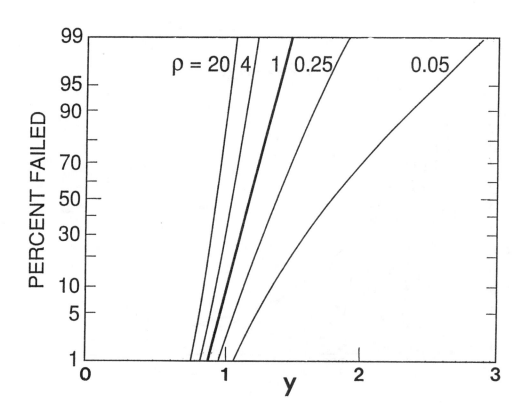

FIGURE 21.3
Power-normal cdfs on normal probability paper

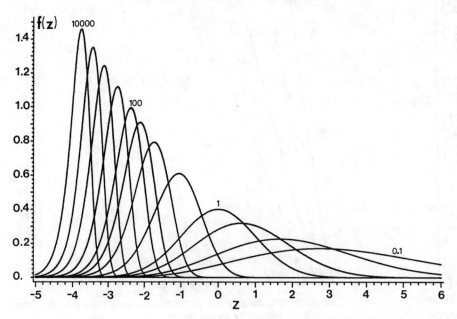

FIGURE 21.4
Power-normal probability densities

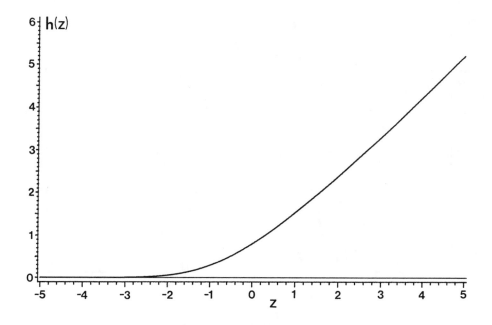

FIGURE 21.5
Standard power-normal hazard function

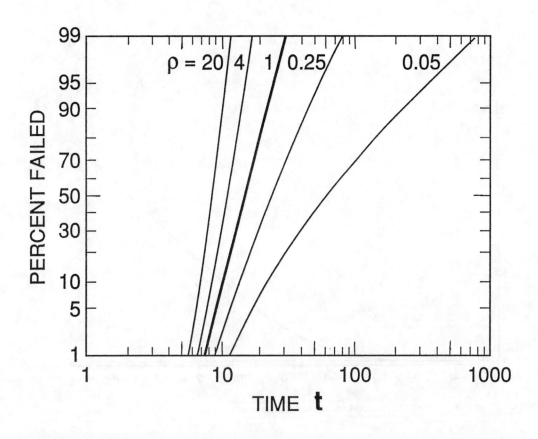

FIGURE 21.6
Power-lognormal cdfs on lognormal probability paper

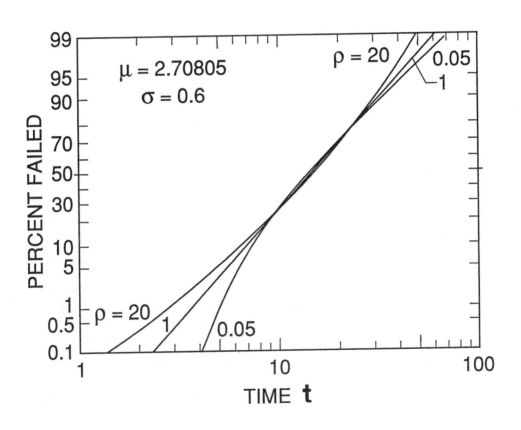

FIGURE 21.7
Power-lognormal cdfs with common 0.25 and 0.75 fractiles

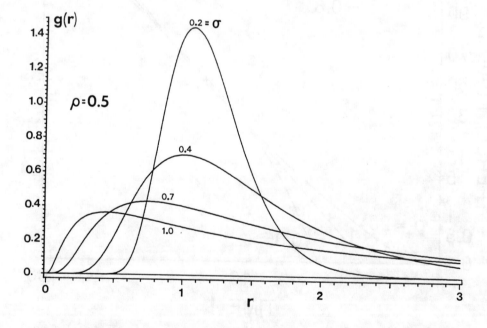

FIGURE 21.8
Power-lognormal probability densities ($\rho = 0.5$)

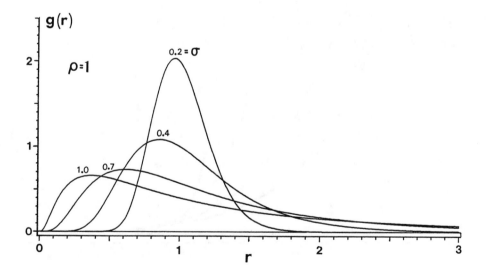

FIGURE 21.9
Power-lognormal probability densities ($\rho = 1$)

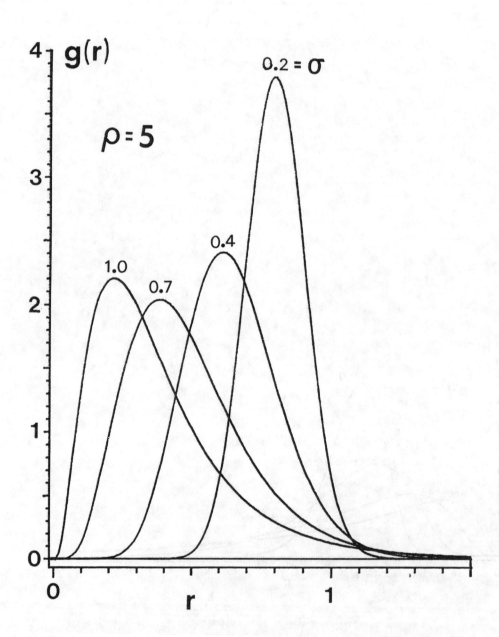

FIGURE 21.10
Power-lognormal probability densities ($\rho = 5$)

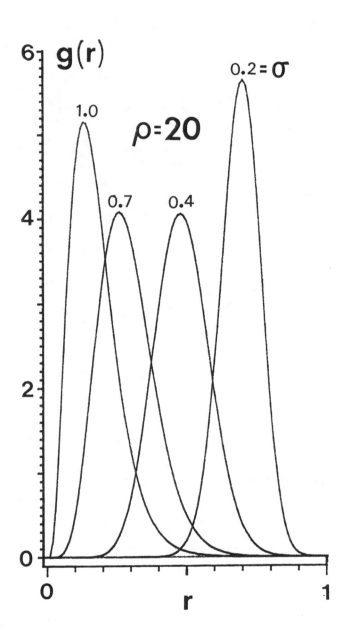

FIGURE 21.11
Power-lognormal probability densities ($\rho = 20$)

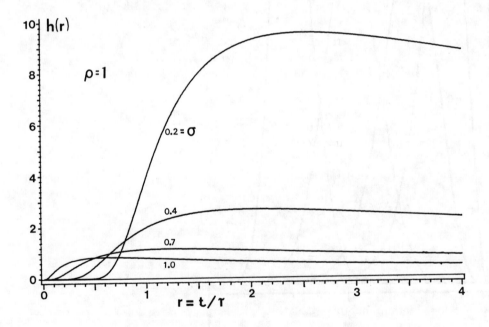

FIGURE 21.12
Power-lognormal hazard functions

22

Maximum Likelihood Estimation for the Log-gamma Distribution under Type-II Censored Samples and Associated Inference

N. Balakrishnan and P.S. Chan

McMaster University
The Chinese University of Hong Kong

ABSTRACT In this paper, we study the bias and variances of the maximum likelihood estimators (MLEs) of the location and scale parameters of the log-gamma distribution under Type II censored samples. We also derive the expected Fisher information matrix through which the asymptotic variances and covariance of the MLEs are tabulated for various proportions of censoring. The small-sample distributions of two pivotal quantities (useful for making inference for the location and scale paramters) are simulated by Monte Carlo method and the large-sample normal approximation is also described. Finally, a life-test data due to [27], and also analysed by [25], [26], is used to illustrate the method of estimation and inference developed in this paper.

Keywords: Asymptotic variance-covariance matrix, Log-gamma distribution, Maximum likelihood estimation, Shape parameter, Type-II censored sample

CONTENTS

22.1 Introduction

Let X be a log-gamma random variable with probability density function

$$f^*(x) = \frac{1}{\Gamma(\kappa)} e^{\kappa x - e^x}, \qquad -\infty < x < \infty, \ \kappa > 0 \qquad (22.1.1)$$

and cumulative distribution function

$$F^*(x) = I_{e^x}(\kappa), \qquad -\infty < x < \infty, \ \kappa > 0, \qquad (22.1.2)$$

where $I_t(\kappa)$ is the incomplete gamma function defined by

$$I_t(\kappa) = \int_0^t \frac{1}{\Gamma(\kappa)} e^{-z} z^{\kappa-1} dz, \qquad 0 < t < \infty, \ \kappa > 0.$$

The special case of $\kappa=1$ is the extreme value distribution which has been discussed in great detail by many authors including [27], [3], [4], [26], [13], and [7].

Note that, for positive integral values of κ, the cumulative distribution function $F^*(x)$ in (22.1.2) can be written as

$$F^*(x) = \sum_{i=\kappa}^{\infty} e^{-e^x} e^{ix} / i!, \qquad -\infty < x < \infty, \ \kappa = 1, 2, \ldots,$$

or, equivalently,

$$1 - F^*(x) = \sum_{i=0}^{\kappa-1} e^{-e^x} e^{ix} / i!, \qquad -\infty < x < \infty, \ \kappa = 1, 2, \ldots. \tag{22.1.3}$$

From (22.1.1), we derive the moment generating function of the log-gamma random variable X to be

$$M_X(t) = E(e^{tX}) = \frac{\Gamma(\kappa + t)}{\Gamma(\kappa)}$$

from which we immediately obtain the mean and variance of X to be

$$E(X) = \frac{\Gamma'(\kappa)}{\Gamma(\kappa)} = \psi(\kappa)$$

and

$$\mathrm{Var}(X) = \frac{\Gamma''(\kappa)}{\Gamma(\kappa)} - \left(\frac{\Gamma'(\kappa)}{\Gamma(\kappa)}\right)^2 = \psi'(\kappa),$$

where

$$\psi(z) = \frac{d}{dz} \ln \Gamma(z) = \frac{\Gamma'(z)}{\Gamma(z)}$$

is the digamma function and

$$\psi'(z) = \frac{d}{dz} \psi(z)$$

is the trigamma function.

As a matter of fact, by using the asymptotic formula that $\psi(\kappa) \sim \log \kappa$ and $\psi'(\kappa) \sim 1/\kappa$ (see for example [1]), in [29], the author suggested a reparametrized model of the density function in (22.1.1) as

$$f(x) = \frac{\kappa^{\kappa-\frac{1}{2}}}{\Gamma(\kappa)} e^{\sqrt{\kappa}x - \kappa e^{x/\sqrt{\kappa}}}, \qquad -\infty < x < \infty, \ \kappa > 0, \tag{22.1.4}$$

and showed the asymptotic normality (as $\kappa \to \infty$) of this family. It can be easily seen that the cumulative distribution function for the density function in (22.1.4) is given by

$$F(x) = I_{\kappa e^x/\sqrt{\kappa}}(\kappa), \qquad -\infty < x < \infty, \ \kappa > 0. \tag{22.1.5}$$

Let Y be a log-gamma random variable with probability density function

$$g(y) = \frac{\kappa^{\kappa-\frac{1}{2}}}{\sigma\Gamma(\kappa)} \exp\left(\sqrt{\kappa}\left(\frac{y-\mu}{\sigma}\right) - \kappa e^{(\frac{y-\mu}{\sigma})/\sqrt{\kappa}}\right),,$$

$$-\infty < y < \infty, \quad -\infty < \mu < \infty, \quad \sigma > 0, \quad \kappa > 0, \qquad (22.1.6)$$

where μ is the location parameter, σ is the scale parameter, and κ is the shape parameter. In [25] and [26], the author has illustrated the usefulness of the log-gamma model in (22.1.6) as a life-test model and discussed the maximum likelihood estimation of the parameters; see also [29]. Recently, in [31] the authors discussed the log-gamma regression model. One may also refer to [25] and [15] for some valuable work on the inference for a related generalized gamma distribution.

The exact best linear unbiased estimation and the asymptotic best linear unbiased estimation of the parameters μ and σ, based on doubly Type II censored samples, have been discussed recently by [5], [6]. In this paper, we study the maximum likelihood estimators of μ and σ (with shape parameter κ known) based on doubly Type-II censored samples. In Section 22.2, we present the likelihood equations for μ and σ and the simulated values of the bias, variances and covariance of the MLEs for various sample sizes, choices of censoring and κ. In Section 22.3, we derive explicit algebraic expressions for the asymptotic variances and covariance of these estimators through the expected Fisher information matrix. Tables of these values are also presented for various choices of censoring and κ. In Section 22.4, we describe an approximation to these MLEs which is in closed form and thus is useful as the initial guess for the numerical iterative algorithm for determining the MLEs. It is also shown in this section that these estimators are themselves nearly as efficient as the MLEs. In Section 22.5, we discuss some inference procedures based on the pivotal quantities $P_1 = \sqrt{n}(\hat{\mu} - \mu)/\hat{\sigma}$ and $P_2 = \sqrt{n}\hat{\sigma}/\sigma$. For application purposes, some percentage points of P_1 and P_2 are determined by Monte Carlo simulation for some selected choices of n, proportions of censoring and κ. The asymptotic normal approximations to the distributions of P_1 and P_2 are also discussed. It is shown that the normal approximation to the distribution of P_2 is fairly good even for a sample of size 40, but the approximation for the distribution of P_1 requires a much larger sample size. Finally, in Section 22.6, we consider the life-test data set given in [27] and discussed further in [25], [26] and illustrate the method of estimation of μ and σ and the construction of confidence intervals for μ and σ discussed in the preceding sections.

It is important to mention here that work of this nature has been carried out in [16], [17], [18], [19], [20], [21], [22], [23], [24], [3], [8], [9], [10], [11], and [12] for a wide range of distributions.

22.2 Maximum Likelihood Estimation

Consider a random sample of size n from a log-gamma population with density function as in (22.1.6), where the shape parameter κ is assumed to be known. Let $Y_{r+1:n} \leq Y_{r+2:n} \leq \cdots \leq Y_{n-s:n}$ be the ordered observations remaining when r smallest and s largest observations have been censored. The likelihood function for μ and σ of the given Type-II censored sample is then

$$L = \frac{n!}{r!s!} \left[G(Y_{r+1:n}) \right]^r \left[1 - G(Y_{n-s:n}) \right]^s \prod_{i=r+1}^{n-s} g(Y_{i:n}) \qquad (22.2.1)$$

or equivalently,

$$L = \frac{n!}{r!s!} \frac{1}{\sigma^A} \left[F(X_{r+1:n}) \right]^r \left[1 - F(X_{n-s:n}) \right]^s \times \prod_{i=r+1}^{n-s} f(X_{i:n}), \qquad (22.2.2)$$

where $A = n - r - s$ and $X = (Y - \mu)/\sigma$ is the standardized variable with the density function $f(\cdot)$ and distribution function $F(\cdot)$ given in (22.1.4) and (22.1.5), respectively. The log-likelihood function is given by

$$\log L = \text{const} - A \log \sigma + r \log \left[F(X_{r+1:n}) \right] + s \log \left[1 - F(X_{n-s:n}) \right]$$

$$+ \sum_{i=r+1}^{n-s} \log f(X_{i:n}). \qquad (22.2.3)$$

Upon differentiating the log-likelihood function with respect to μ and σ, we obtain

$$\frac{\partial \log L}{\partial \mu} = -\frac{1}{\sigma} \left\{ r \frac{f(X_{r+1:n})}{F(X_{r+1:n})} - s \frac{f(X_{n-s:n})}{1 - F(X_{n-s:n})} + A\sqrt{\kappa} - \sqrt{\kappa} \sum_{i=r+1}^{n-s} e^{X_{i:n}/\sqrt{\kappa}} \right\}$$

$$(22.2.4)$$

and

$$\frac{\partial \log L}{\partial \sigma} = -\frac{1}{\sigma} \left\{ A + rX_{r+1:n} \frac{f(X_{r+1:n})}{F(X_{r+1:n})} - sX_{n-s:n} \frac{f(X_{n-s:n})}{1 - F(X_{n-s:n})} \right.$$

$$\left. + \sqrt{\kappa} \sum_{i=r+1}^{n-s} X_{i:n} - \sqrt{\kappa} \sum_{i=r+1}^{n-s} X_{i:n} e^{X_{i:n}/\sqrt{\kappa}} \right\}. \qquad (22.2.5)$$

The MLEs $\hat{\mu}$ and $\hat{\sigma}$ can be obtained by simultaneously solving the equations $\partial \log L/\partial \mu = 0$ and $\partial \log L/\partial \sigma = 0$. Since these two equations cannot be solved analytically, some numerical method must be employed. Newton-Raphson iterative method is one of the ideal procedures as has been demonstrated by many authors. In order to study the performance of these estimators, we generated 3001 pseudorandom samples of size $n = 20$, 25 and 40 from the standardized log-gamma population with chioces of $\kappa = 1.0, 2.0(2.0)10.0$ and employed the Newton-Raphson procedure to solve the likelihood equations. Thus, we determined the bias, variances and covariance of the maximum likelihood estimates $\hat{\mu}$ and $\hat{\sigma}$ for $q_1 = r/n = 0.0(0.1)0.2$ and $q_2 = s/n = 0.0(0.1)0.5$. The values of (1) Bias($\hat{\mu}$)/σ, (2) Bias($\hat{\sigma}$)/σ, (3) Var($\hat{\mu}$)/σ^2, (4) Var($\hat{\sigma}$)/σ^2 and (5) Cov($\hat{\mu}, \hat{\sigma}$)/σ^2 determined through this Monte Carlo process are presented in Tables 22.1-22.6.

22.3   Asymptotic Variances and Covariance

The second partial derivatives of the log-likelihood function in (22.2.3) with respect to μ and σ are obtained to be

$$\frac{\partial^2 \log L}{\partial \mu^2} = \frac{1}{\sigma^2} \left\{ r \frac{f(X_{r+1:n})}{F(X_{r+1:n})} \left[\sqrt{\kappa} - \sqrt{\kappa} e^{X_{r+1:n}/\sqrt{\kappa}} - \frac{f(X_{r+1:n})}{F(X_{r+1:n})} \right] \right.$$

$$\left. - s \frac{f(X_{n-s:n})}{1 - F(X_{n-s:n})} \left[\sqrt{\kappa} - \sqrt{\kappa} e^{X_{n-s:n}/\sqrt{\kappa}} + \frac{f(X_{n-s:n})}{1 - F(X_{n-s:n})} \right] \right.$$

$$- \sum_{i=r+1}^{n-s} e^{X_{i:n}/\sqrt{\kappa}} \Bigg\}, \tag{22.3.1}$$

$$\frac{\partial^2 \log L}{\partial \mu \partial \sigma} = \frac{1}{\sigma^2} \Bigg\{ r \frac{f(X_{r+1:n})}{F(X_{r+1:n})} + r X_{r+1:n} \frac{f(X_{r+1:n})}{F(X_{r+1:n})} \left[\sqrt{\kappa} - \sqrt{\kappa} e^{X_{r+1:n}/\sqrt{\kappa}} - \frac{f(X_{r+1:n})}{F(X_{r+1:n})} \right]$$

$$- s \frac{f(X_{n-s:n})}{1 - F(X_{n-s:n})} - s X_{n-s:n} \frac{f(X_{n-s:n})}{1 - F(X_{n-s:n})}$$

$$\left[\sqrt{\kappa} - \sqrt{\kappa} e^{X_{n-s:n}/\sqrt{\kappa}} + \frac{f(X_{n-s:n})}{1 - F(X_{n-s:n})} \right]$$

$$+ A\sqrt{\kappa} - \sqrt{\kappa} \sum_{i=r+1}^{n-s} e^{X_{i:n}/\sqrt{\kappa}} - \sum_{i=r+1}^{n-s} X_{i:n} e^{X_{i:n}/\sqrt{\kappa}} \Bigg\}, \tag{22.3.2}$$

$$\frac{\partial^2 \log L}{\partial \sigma^2} = \frac{1}{\sigma^2} \Bigg\{ A + 2r X_{r+1:n} \frac{f(X_{r+1:n})}{F(X_{r+1:n})} + r X_{r+1:n}^2 \frac{f(X_{r+1:n})}{F(X_{r+1:n})}$$

$$\left[\sqrt{\kappa} - \sqrt{\kappa} e^{X_{r+1:n}/\sqrt{\kappa}} - \frac{f(X_{r+1:n})}{F(X_{r+1:n})} \right]$$

$$- 2s X_{n-s:n} \frac{f(X_{n-s:n})}{1 - F(X_{n-s:n})} - s X_{n-s:n}^2 \frac{f(X_{n-s:n})}{1 - F(X_{n-s:n})}$$

$$\left[\sqrt{\kappa} - \sqrt{\kappa} e^{X_{n-s:n}/\sqrt{\kappa}} + \frac{f(X_{n-s:n})}{1 - F(X_{n-s:n})} \right] + 2\sqrt{\kappa} \sum_{i=r+1}^{n-s} X_{i:n}$$

$$- 2\sqrt{\kappa} \sum_{i=r+1}^{n-s} X_{i:n} e^{X_{i:n}/\sqrt{\kappa}} - \sum_{i=r+1}^{n-s} X_{i:n}^2 e^{X_{i:n}/\sqrt{\kappa}} \Bigg\}. \tag{22.3.3}$$

Let $q_1 = r/n$, $q_2 = s/n$ and $p = 1 - q_1 - q_2$. As $n \to \infty$ (with q_1 and q_2 fixed), $E(X_{r+1:n}) \to \xi_{q_1}$ where $F(\xi_{q_1}) = q_1$, $E(X_{n-s:n}) \to \xi_{q_2}$ where $F(\xi_{q_2}) = 1 - q_2$,

$$E\left(\sum_{i=r+1}^{n-s} X_{i:n} \right) \to n \int_{\xi_{q_1}}^{\xi_{q_2}} f(t) dt, \tag{22.3.4}$$

$$E\left(\sum_{i=r+1}^{n-s} e^{X_{i:n}/\sqrt{\kappa}} \right) \to n \int_{\xi_{q_1}}^{\xi_{q_2}} e^{t/\sqrt{\kappa}} f(t) dt, \tag{22.3.5}$$

$$E\left(\sum_{i=r+1}^{n-s} X_{i:n} e^{X_{i:n}/\sqrt{\kappa}} \right) \to n \int_{\xi_{q_1}}^{\xi_{q_2}} t e^{t/\sqrt{\kappa}} f(t) dt, \tag{22.3.6}$$

$$E\left(\sum_{i=r+1}^{n-s} X_{i:n}^2 e^{X_{i:n}/\sqrt{\kappa}} \right) \to n \int_{\xi_{q_1}}^{\xi_{q_2}} t^2 e^{t/\sqrt{\kappa}} f(t) dt. \tag{22.3.7}$$

Hence,

$$\lim_{n \to \infty} \left(-\frac{\sigma^2}{n} E \frac{\partial^2 \log L}{\partial \mu^2} \right) = -f(\xi_{q_1}) \left[\sqrt{\kappa} - \sqrt{\kappa} e^{\xi_{q_1}/\sqrt{\kappa}} - \frac{f(\xi_{q_1})}{q_1} \right]$$

$$+f(\xi_{q_2})\left[\sqrt{\kappa}-\sqrt{\kappa}e^{\xi_{q_2}/\sqrt{\kappa}}+\frac{f(\xi_{q_2})}{q_2}\right]+\int_{\xi_{q_1}}^{\xi_{q_2}}e^{t/\sqrt{\kappa}}f(t)dt$$

$$= V^{11},\tag{22.3.8}$$

$$\lim_{n\to\infty}\left(-\frac{\sigma^2}{n}E\frac{\partial^2\log L}{\partial\mu\partial\sigma}\right)=-p\sqrt{\kappa}-f(\xi_{q_1})+f(\xi_{q_2})$$

$$-\xi_{q_1}f(\xi_{q_1})\left[\sqrt{\kappa}-\sqrt{\kappa}e^{\xi_{q_1}/\sqrt{\kappa}}-\frac{f(\xi_{q_1})}{q_1}\right]$$

$$+\xi_{q_2}f(\xi_{q_2})\left[\sqrt{\kappa}-\sqrt{\kappa}e^{\xi_{q_2}/\sqrt{\kappa}}+\frac{f(\xi_{q_2})}{q_2}\right]$$

$$+\sqrt{\kappa}\int_{\xi_{q_1}}^{\xi_{q_2}}e^{t/\sqrt{\kappa}}f(t)dt+\int_{\xi_{q_1}}^{\xi_{q_2}}te^{t/\sqrt{\kappa}}f(t)dt$$

$$= V^{12},\tag{22.3.9}$$

and

$$\lim_{n\to\infty}\left(-\frac{\sigma^2}{n}E\frac{\partial^2\log L}{\partial\sigma^2}\right)=-p-2\xi_{q_1}f(\xi_{q_1})+2\xi_{q_2}f(\xi_{q_2})$$

$$-\xi_{q_1}^2 f(\xi_{q_1})\left[\sqrt{\kappa}-\sqrt{\kappa}e^{\xi_{q_1}/\sqrt{\kappa}}-\frac{f(\xi_{q_1})}{q_1}\right]$$

$$+\xi_{q_2}^2 f(\xi_{q_2})\left[\sqrt{\kappa}-\sqrt{\kappa}e^{\xi_{q_2}/\sqrt{\kappa}}+\frac{f(\xi_{q_2})}{q_2}\right]$$

$$-2\sqrt{\kappa}\int_{\xi_{q_1}}^{\xi_{q_2}}tf(t)dt+2\sqrt{\kappa}\int_{\xi_{q_1}}^{\xi_{q_2}}te^{t/\sqrt{\kappa}}f(t)dt$$

$$+\int_{\xi_{q_1}}^{\xi_{q_2}}t^2 e^{t/\sqrt{\kappa}}f(t)dt$$

$$= V^{22}.\tag{22.3.10}$$

The asymptotic variance-covariance matrix for the maximum likelihood estimators of μ and σ is then given by $\frac{\sigma^2}{n}[V_{ij}]$ where $[V_{ij}]=[V^{ij}]^{-1}$ with V^{ij} being given by Eqs. (22.3.8)–(22.3.10). Note that the computations of V_{ij} involve the four definite integrals given in (22.3.4)–(22.3.7). These four integrals can be evaluated by using the following three lemmas which have been proved in [6].

LEMMA 22.3.1
Let

$$F_\kappa(\lambda)=\int_0^\lambda\frac{1}{\Gamma(\kappa)}t^{\kappa-1}e^{-t}dt.$$

Then

$$F_{\kappa+1}(\lambda)=F_\kappa(\lambda)-\frac{\lambda^\kappa e^{-\lambda}}{\Gamma(\kappa+1)}.$$

LEMMA 22.3.2

Let

$$G_\kappa(\lambda) = \int_0^\lambda \frac{\log t}{\Gamma(\kappa)} t^{\kappa-1} e^{-t} dt.$$

Then

$$G_{\kappa+1}(\lambda) = G_\kappa(\lambda) + \frac{1}{\kappa} F_\kappa(\lambda) - \frac{(\log \lambda)\lambda^\kappa e^{-\lambda}}{\Gamma(\kappa+1)}.$$

LEMMA 22.3.3

Let

$$H_\kappa(\lambda) = \int_0^\lambda \frac{(\log t)^2}{\Gamma(\kappa)} t^{\kappa-1} e^{-t} dt.$$

Then

$$H_{\kappa+1}(\lambda) = H_\kappa(\lambda) + \frac{2}{\kappa} G_\kappa(\lambda) - \frac{(\log \lambda)^2 \lambda^\kappa e^{-\lambda}}{\Gamma(\kappa+1)}.$$

Upon using the above three lemmas, one can rewrite V^{ij} as follows:

$$V^{11} = -f(\xi_{q_1}) \left[\sqrt{\kappa} - \sqrt{\kappa} e^{\xi_{q_1}/\sqrt{\kappa}} - \frac{f(\xi_{q_1})}{q_1} \right] + f(\xi_{q_2}) \left[\sqrt{\kappa} - \sqrt{\kappa} e^{\xi_{q_2}/\sqrt{\kappa}} + \frac{f(\xi_{q_2})}{q_2} \right]$$

$$+ p - \frac{1}{\sqrt{\kappa}} [f(\xi_{q_2}) - f(\xi_{q_1})], \tag{22.3.11}$$

$$V^{12} = \frac{p}{\sqrt{\kappa}} (1 - \kappa \log \kappa) - \xi_{q_1} f(\xi_{q_1}) \left[\sqrt{\kappa} - \sqrt{\kappa} e^{\xi_{q_1}/\sqrt{\kappa}} - \frac{f(\xi_{q_1})}{q_1} - \frac{1}{\sqrt{\kappa}} \right]$$

$$+ \xi_{q_2} f(\xi_{q_2}) \left[\sqrt{\kappa} - \sqrt{\kappa} e^{\xi_{q_2}/\sqrt{\kappa}} + \frac{f(\xi_{n-s})}{q_2} - \frac{1}{\sqrt{\kappa}} \right]$$

$$+ \sqrt{\kappa} \left[G_\kappa(\kappa e^{\xi_{q_2}/\sqrt{\kappa}}) - G_\kappa(\kappa e^{\xi_{q_1}/\sqrt{\kappa}}) \right] \tag{22.3.12}$$

and

$$V^{22} = p - 2p \log \kappa + \kappa p \left(\log \kappa \right)^2 - \xi_{r+1}^2 f(\xi_{r+1}) \left[\sqrt{\kappa} - \sqrt{\kappa} e^{\xi_{r+1}/\sqrt{\kappa}} - \frac{f(\xi_{r+1})}{q_1} - \frac{1}{\sqrt{\kappa}} \right]$$

$$+ \xi_{n-s}^2 f(\xi_{n-s}) \left[\sqrt{\kappa} - \sqrt{\kappa} e^{\xi_{n-s}/\sqrt{\kappa}} + \frac{f(\xi_{n-s})}{q_2} - \frac{1}{\sqrt{\kappa}} \right]$$

$$+ (2 - 2\kappa \log \kappa) \left[G_\kappa(\kappa e^{\xi_{n-s}/\sqrt{\kappa}}) - G_\kappa(\kappa e^{\xi_{r+1}/\sqrt{\kappa}}) \right]$$

$$+ \kappa \left[H_\kappa(\kappa e^{\xi_{n-s}/\sqrt{\kappa}}) - H_\kappa(\kappa e^{\xi_{r+1}/\sqrt{\kappa}}) \right]. \tag{22.3.13}$$

Hence, in order to obtain the values of V^{ij}, only two definite integrals are required. In [6], the authors have tabulated the values of these two functions, G_κ and H_κ, for some choices of κ and several values of λ. Upon using these values, we computed the values of V^{ij} given in (22.3.11)–(22.3.13) for the choices of $q_1 = 0.0(0.1)0.2$, $q_2 = 0.0(0.1)0.5$ and $\kappa = 1.0, 2.0(2.0)10.0$ using which the values of V_{ij} were determined. These values are presented in Table 22.7. From these values, we observe that the asymptotic formulae provide very close approximation to the true values (simulated and presented in Tables 22.1–22.6) even for a sample of size 40.

22.4   First-iterate Likelihood Estimators

The log-likelihood equations in Section 22.2 do not admit explicit solutions. However, by expanding the functions $f(X_{r+1:n})/F(X_{r+1:n})$, $f(X_{n-s:n})/\{1 - F(X_{n-s:n})\}$ and $f'(X_{i:n})/f(X_{i:n})$ in Taylor series around the points ζ_{r+1}, ζ_{n-s} and ζ_i, respectively, where $\zeta_i = F^{-1}(\frac{i}{n+1})$, we may approximate these functions by (see [14] and [2] for reasoning)

$$\frac{f(X_{r+1:n})}{F(X_{r+1:n})} \simeq \alpha_1 - \beta_1 X_{r+1;n}, \tag{22.4.1}$$

$$\frac{f(X_{n-s:n})}{1 - F(X_{n-s:n})} \simeq \alpha_2 + \beta_2 X_{n-s:n}, \tag{22.4.2}$$

and

$$\frac{f'(X_{i:n})}{f(X_{i:n})} = \sqrt{\kappa} - \sqrt{\kappa}e^{X_{i:n}/\sqrt{\kappa}} \simeq \gamma_i - \delta_i X_{i:n}, \tag{22.4.3}$$

where

$$\beta_1 = -\frac{f(\zeta_{r+1})}{F(\zeta_{r+1})}\left[\sqrt{\kappa} - \sqrt{\kappa}e^{\zeta_{r+1}/\sqrt{\kappa}} - \frac{f(\zeta_{r+1})}{F(\zeta_{r+1})}\right], \tag{22.4.4}$$

$$\alpha_1 = \frac{f(\zeta_{r+1})}{F(\zeta_{r+1})} + \beta_1 \zeta_{r+1}, \tag{22.4.5}$$

$$\beta_2 = \frac{f(\zeta_{n-s})}{1 - F(\zeta_{n-s})}\left[\sqrt{\kappa} - \sqrt{\kappa}e^{\zeta_{n-s}/\sqrt{\kappa}} + \frac{f(\zeta_{n-s})}{1 - F(\zeta_{n-s})}\right], \tag{22.4.6}$$

$$\alpha_2 = \frac{f(\zeta_{n-s})}{1 - F(\zeta_{n-s})} - \beta_2 \zeta_{n-s}, \tag{22.4.7}$$

$$\delta_i = e^{\zeta_i/\sqrt{\kappa}}, \tag{22.4.8}$$

$$\gamma_i = \sqrt{\kappa} - \sqrt{\kappa}e^{\zeta_i/\sqrt{\kappa}} + \zeta_i e^{\zeta_i/\sqrt{\kappa}}. \tag{22.4.9}$$

It is easy to see from Eq. (22.4.8) that $\delta_i > 0$, $i = r+1, \cdots, n-s$. In order to show that $\beta_1 \geq 0$, consider

$$f(x) = \int_{-\infty}^{x} f'(t)dt = \int_{-\infty}^{x}\left(\sqrt{\kappa} - \sqrt{\kappa}e^{t/\sqrt{\kappa}}\right)f(t)\,dt\;.$$

Since $\sqrt{\kappa} - \sqrt{\kappa}e^{t/\sqrt{\kappa}}$ is monotonic decreasing function of t,

$$\min_{-\infty < t \leq x}\left(\sqrt{\kappa} - \sqrt{\kappa}e^{t/\sqrt{\kappa}}\right) = \sqrt{\kappa} - \sqrt{\kappa}e^{x/\sqrt{\kappa}}\;.$$

Therefore,

$$f(x) \geq \left(\sqrt{\kappa} - \sqrt{\kappa}e^{x/\sqrt{\kappa}}\right)\int_{-\infty}^{x} f(t)dt \Rightarrow \frac{f(x)}{F(x)} \geq \sqrt{\kappa} - \sqrt{\kappa}e^{x/\sqrt{\kappa}}$$

and thus, $\beta_1 \geq 0$. Similar argument can be used for showing $\beta_2 \geq 0$.

By making use of the linear approximations in (22.4.1)–(22.4.3), we obtain the approximate log-likelihood equations as

$$\frac{\partial \log L}{\partial \mu} \simeq \frac{\partial \log L^*}{\partial \mu}$$

$$= -\frac{1}{\sigma}\left[r(\alpha_1 - \beta_1 X_{r+1:n}) - s(\alpha_2 + \beta_2 X_{n-s:n}) + \sum_{i=r+1}^{n-s}(\gamma_i - \delta_i X_{i:n})\right]$$

$$= 0 \tag{22.4.10}$$

and

$$\frac{\partial \log L}{\partial \sigma} \simeq \frac{\partial \log L^*}{\partial \sigma}$$

$$= -\frac{1}{\sigma}\left[A + r\alpha_1 X_{r+1:n} - s\alpha_2 X_{n-s:n} + \sum_{i=r+1}^{n-s}\gamma_i X_{i:n}\right.$$

$$\left. - r\beta_1 X_{r+1:n}^2 - s\beta_2 X_{n-s:n}^2 - \sum_{i=r+1}^{n-s}\delta_i X_{i:n}^2\right]$$

$$= 0 . \tag{22.4.11}$$

Upon solving Eqs. (22.4.10) and (22.4.11), we derive the first-iterate likelihood estimators of μ and σ as

$$\tilde{\mu} = B - \tilde{\sigma}C \tag{22.4.12}$$

and

$$\tilde{\sigma} = \frac{-D + \sqrt{D^2 + 4AE}}{2A}, \tag{22.4.13}$$

where

$$m = r\beta_1 + s\beta_2 + \sum_{i=r+1}^{n-s}\delta_i,$$

$$B = \frac{1}{m}\left[r\beta_1 Y_{r+1:n} + s\beta_2 Y_{n-s:n} + \sum_{i=r+1}^{n-s}\delta_i Y_{i:n}\right],$$

$$C = \frac{1}{m}\left[r\alpha_1 - s\alpha_2 + \sum_{i=r+1}^{n-s}\gamma_i\right],$$

$$D = r\alpha_1 Y_{r+1:n} - s\alpha_2 Y_{n-s:n} + \sum_{i=r+1}^{n-s}\gamma_i Y_{i:n} - mBC,$$

$$E = r\beta_1 Y_{r+1:n}^2 + s\beta_2 Y_{n-s:n}^2 + \sum_{i=r+1}^{n-s}\delta_i Y_{i:n}^2 - mB^2.$$

It should be mentioned here that upon solving Eq. (22.4.11) for σ we obtain a quadratic equation in σ which has two roots; however, one of them drops out since

$$E = r\beta_1(Y_{r+1:n} - B)^2 + s\beta_2(Y_{n-s:n} - B)^2 + \sum_{i=r+1}^{n-s} \delta_i(Y_{i:n} - B)^2 > 0 \ .$$

It should be mentioned here that such modified (approximate) maximum likelihood estimators have been derived for a wide range of distributions. Interested readers may refer to [30] and [7] for a description of these estimators.

Since the estimators $\tilde{\mu}$ and $\tilde{\sigma}$ are in closed form, they were used as the initial guess (or first-iterate) for the Newton-Raphson procedure to obtain the MLEs discussed in Section 22.2. For the purpose of comparison, we also obtained the bias, variances and covariance of these estimators through Monte Carlo simulations. These simulated values of (6) Bias$(\tilde{\mu})/\sigma$, (7) Bias$(\tilde{\sigma})/\sigma$, (8) Var$(\tilde{\mu})/\sigma^2$, (9) Var$(\tilde{\sigma})/\sigma^2$ and (10) Cov$(\tilde{\mu}, \tilde{\sigma})/\sigma^2$ are presented in Tables 22.1–22.6 with the choices of n, q_1, q_2 and κ considered in the case of MLEs.

22.5 Pivotal Quantities and Inference

In the preceding sections, we discussed the point estimation of the parameters μ and σ (with shape parameter κ being known). In this section, we consider the pivotal quantities

$$P_1 = \frac{\sqrt{n}(\hat{\mu} - \mu)}{\hat{\sigma}} \qquad \text{and} \qquad P_2 = \frac{\sqrt{n}\hat{\sigma}}{\sigma}, \qquad (22.5.1)$$

where $\hat{\mu}$ and $\hat{\sigma}$ are the MLEs of μ and σ. These two pivotal quantities will enable us to construct confidence intervals or carry out tests of hypotheses concerning the parameters μ and σ.

Since the small-sample distributions of P_1 and P_2 are intractable, we simulated the percentage points of P_1 and P_2 (based on 3001 Monte Carlo runs) for sample sizes $n = 20, 25, 40$ and various choices of q_1 and q_2 (proportions of censoring) and κ. These values are presented in Tables 22.8–22.13. Using the appropriate percentage points of P_1, we can develop inference for the parameter μ. Similarly, the simulated percentage points of P_2 may be used for developing inference for the parameter σ.

It may also be noted from the asymptotic theory of the MLEs described in Section 22.3 that

$$P_3 = \frac{\sqrt{n}(\hat{\mu} - \mu)}{\hat{\sigma}\sqrt{V_{11}}} \qquad \text{and} \qquad P_4 = \frac{\sqrt{n}(\hat{\sigma} - \sigma)}{\sigma\sqrt{V_{22}}} \qquad (22.5.2)$$

are both approximate standard normal variables.

These asymptotic results can be used to approximate the percentage points of P_1 and P_2 with the help of the values of V_{11} and V_{22} tabulated in Table 22.7. In order to assess these approximations, we have given in Table 22.14 the simulated values (for $n = 40$) of some percentage points of P_3 and P_4 for some selected choices of q_1, q_2 and κ, and also the corresponding normal quantiles. A comparison of these two sets of values immediately reveals that the normal approximation is quite accurate for the pivotal quantity P_2 (or P_4) even for sample of size 40. However, the normal approximation does not seem to give satisfactory result for the pivotal quantity P_1 (or P_3) even for the sample of size 40. This may be due to the ratio form of the pivotal quantity P_1. It is noted from Table 22.14 that the simulated percentage points of the pivotal quantity P_3 in (22.5.2) (for $n = 40$) are quite close to the corresponding normal quantiles whenever there is only a small proportion of censoring. However, as the proportion of censoring becomes larger the normal approximation needs a much larger sample size to provide close results to the simulated values. If one does not have such a large sample, then one has to resort to a simulation method (as done in constructing Tables 22.8-22.13) for determining the necessary values.

22.6 Illustrative Example

Let us consider here the example of [25], [26], and [27] which gives the number of million revolutions to failure for each of a group of 23 ball bearings in a fatigue test.

The 23 log lifetimes are

$$
\begin{array}{cccccc}
2.884, & 3.365, & 3.497, & 3.726, & 3.741, & 3.820, \\
3.881, & 3.948, & 3.950, & 3.991, & 4.017, & 4.217, \\
4.229, & 4.229, & 4.232, & 4.432, & 4.534, & 4.591, \\
4.655, & 4.662, & 4.851, & 4.852, & 5.156. &
\end{array}
$$

It is of interest to mention here that in [27], the authors assumed a two-parameter Weibull distribution for the original data and hence an extreme value distribution for the log-lifetime data given above. In [25] and [26], the author assumed a generalized gamma distribution for the original data and thus a log-gamma distribution for the log-lifetime. He concluded that the data fit well under this model with the shape parameter $\kappa > 0.4$.

Now, following the same assumptions made in [25] and [26], we present the analysis based on (1) complete sample which has already been discussed by Lawless, and (2) Type-II censored sample with the two largest observations censored. The estimates of μ and σ thus obtained are presented in Tables 22.15 and 22.16, respectively, for various choices of κ.

We also simulated the percentage points of the statistics P_1 and P_2 using which we constructed the 90% confidence intervals for μ and σ. These confidence intervals are given in Table 22.17.

It should be mentioned here that the estimates and the standard errors given for the complete-sample case are the same as those given in [25] and [26]. It is first of all quite interesting to note that the first-iterate estimates $\tilde{\mu}$ and $\tilde{\sigma}$ (and their standard errors) are very close to the corresponding MLEs $\hat{\mu}$ and $\hat{\sigma}$. Further, a comparison of the entries in Tables 22.15 and 22.16 reveal interestingly that the estimates (and the standard errors) based on the right-censored sample are quite close to those based on the complete sample. Finally, we also observe from Tables 22.15 and 22.16 that the estimates of μ and σ (and the standard errors) hardly change for $\kappa > 4$. This, as pointed out earlier by Lawless [25], [26], is due to the fact that the log-gamma distribution fits the complete-sample data whenever $\kappa > 4$ (the relative profile likelihood is presented in Fig. 22.1). The same feature is observed for the right-censored data as well, for which the related profile likelihood is presented in Fig. 22.2. These also explain why the 95% confidence intervals for μ and σ presented in Table 22.17 are quite close for the different choices of κ considered, based on both complete and right-censored samples.

Acknowledgements

The first author would like to thank the Natural Sciences and Engineering Research Council of Canada for funding this research.

References

1. Abramowitz, M. and Stegun, I.A. (Eds.) (1965), *Handbook of Mathematical Functions with Formulas, Graphs, and Mathematical Tables*, Dover Publications, New York.

2. Arnold, B.C. and Balakrishnan, N. (1989), *Relations, Bounds and Approximations For Order Statistics*, Lecture Notes in Statistics No. 53, Springer-Verlag, New York.

3. Bain, L.J. (1972), Inferences based on censored sampling from the Weibull or extreme-value distribution, *Technometrics*, **14**, 693–702.

4. Bain, L.J. (1978), *Statistical Analysis of Reliability and Life-testing Models–Theory and Practice*, Marcel Dekker, New York.

5. Balakrishnan, N. and Chan, P.S. (1994), Log-gamma order statistics and linear estimation of parameters, *Computational Statistics & Data Analysis* (To appear).

6. Balakrishnan, N. and Chan, P.S. (1994), Asymptotic best linear unbiased estimation for the log-gamma distribution, *Sankhyā, Series B* (To appear).

7. Balakrishnan, N. and Cohen, A.C. (1991), *Order Statistics and Inference: Estimation Methods*, Academic Press, Boston.

8. Cohen, A.C. (1951), Estimating parameters of logarithmic normal distributions by maximum likelihood, *Joural of the American Statististical Association*, **46**, 206-212.

9. Cohen, A.C. (1955), Maximum likelihood estimation of the dispersion parameter of a chi-distributed radial error from truncated and censored samples with applications to target analysis, *Journal of the American Statistical Association*, **50**, 884-893.

10. Cohen, A.C. (1959), Simplified estimators for the normal distribution when samples are singly censored or truncated, *Technometrics*, **1**, 217-237.

11. Cohen, A.C. (1965), Maximum likelihood estimation in the Weibull distribution based on complete and on censored samples, *Technometrics*, **7**, 579-588.

12. Cohen, A.C. and Whitten, B.J. (1980), Estimation in the three-parameter lognormal distribution, *Journal of the American Statistical Association*, **75**, 399-404.

13. Cohen, A.C. and Whitten, B.J. (1988), *Parameter Estimation in Reliability and Life Span Models*, Marcel Dekker, New York.

14. David, H.A. (1981), *Order Statistics*, Second edition, John Wiley & Sons, New York.

15. DiCiccio, T.J. (1987), Approximate inference for the generalized gamma distribution, *Technometrics*, **29**, 32-39.

16. Harter, H.L. (1967), Maximum-likelihood estimation of the parameters of a four parameter generalized gamma population from complete and censored samples, *Technometrics*, **9**, 159-165.

17. Harter, H.L. (1970), *Order Statistics and their Use in Testing and Estimation*, Vol. 2., U.S. Government Printing Office, Washington, D.C.

18. Harter, H.L. and Moore, A.H. (1965), Point and interval estimators, based on m order statistics, for the scale parameter of a Weibull population with known shape parameter, *Technometrics*, **7**, 405-422.

19. Harter, H.L. and Moore, A.H. (1965), Maximum-likelihood estimation of the parameters of gamma and Weibull populations from complete and from censored samples, *Technometrics*, **7**, 639-643.

20. Harter, H.L. and Moore, A.H. (1966), Local-maximum likelihood estimation of the parameters of three-parameter lognormal populations from complete and censored samples, *Journal of the American Statistical Association*, **61**, 842-855.

21. Harter, H.L. and Moore, A.H. (1966), Iterative maximum-likelihood estimation of the parameters of normal populations from singly and doubly censored samples, *Biometrika*, **53**, 205-213.

22. Harter, H.L. and Moore, A.H. (1967), Asymptotic variances and covariances of maximum-likelihood estimators, from censored samples, of the parameters of Weibull and gamma populations, *Annals of Mathematical Statistics*, **38**, 557-570.

23. Harter, H.L. and Moore, A.H. (1967), Maximum likelihood estimation, from censored samples, of the parameters of a logistic distribution, *Journal of the American Statistical Association*, **62**, 675-684.

24. Harter, H.L. and Moore, A.H. (1968), Maximum likelihood estimation, from doubly censored samples, of the parameters of the first asymptotic distribution of extreme values, *Journal of the American Statistical Association*, **63**, 889-901.

25. Lawless, J.F. (1980), Inference in the generalized gamma and log-gamma distribution, *Technometrics*, **22**, 67-82.

26. Lawless, J.F. (1982), *Statistical Models & Methods For Lifetime Data*, John Wiley & Sons, New York.

27. Lieblein, J. and Zelen, M. (1956), Statistical investigation of the fatigue life of deep groove ball bearings, *Jounal of Research National Bureau of Standards*, **57**, 273-316.

28. Mann, N.R. and Singpurwalla, N.D. (1974), *Methods for Statistical Analysis of Reliability and Life Data*, John Wiley & Sons, New York.

29. Prentice, R.L. (1974), A log-gamma model and its maximum likelihood estimation, *Biometrika*, **61**, 539-544.

30. Tiku, M.L., Tan, W.Y., and Balakrishnan, N. (1986), *Robust Inference*, Marcel Dekker, New York.

31. Young, D.H. and Bakir, S.T. (1987), Bias correction for a generalized log-gamma regression model, *Technometrics*, **29**, 183-191.

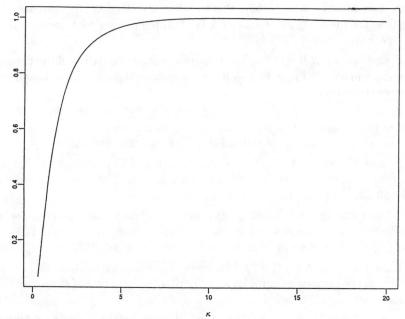

FIGURE 22.1
Relative likelihood for the complete sample

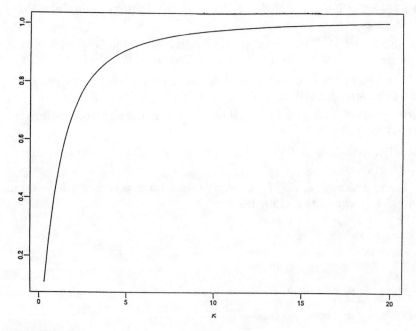

FIGURE 22.2
Relative likelihood for the censored sample

TABLE 22.1
Simulated values of bias, variance and covariance of $(\hat{\mu}, \hat{\sigma})$ and $(\tilde{\mu}, \tilde{\sigma})$ when $\kappa = 1.0$

p	q	(1)	(2)	(3)	(4)	(5)	(6)	(7)	(8)	(9)	(10)
						n= 20					
.0	.0	-.014	-.034	.056	.031	-.012	-.046	-.034	.058	.031	-.013
.0	.1	-.022	-.041	.059	.039	-.008	-.044	-.032	.060	.038	-.008
.0	.2	-.034	-.051	.064	.046	-.001	-.050	-.038	.066	.045	-.002
.0	.3	-.049	-.060	.075	.055	.008	-.061	-.046	.075	.055	.007
.0	.4	-.069	-.070	.093	.066	.022	-.084	-.061	.094	.065	.021
.0	.5	-.106	-.090	.131	.081	.046	-.114	-.077	.128	.079	.043
.1	.0	-.012	-.038	.056	.033	-.013	-.040	-.043	.059	.033	-.014
.1	.1	-.021	-.047	.059	.042	-.009	-.040	-.044	.061	.041	-.009
.1	.2	-.034	-.059	.065	.051	-.002	-.047	-.052	.066	.050	-.003
.1	.3	-.051	-.071	.075	.063	.009	-.061	-.064	.076	.063	.008
.1	.4	-.074	-.087	.094	.078	.026	-.089	-.086	.095	.076	.024
.1	.5	-.120	-.115	.136	.100	.056	-.130	-.112	.133	.096	.052
.2	.0	-.010	-.043	.057	.036	-.015	-.035	-.052	.059	.037	-.016
.2	.1	-.019	-.054	.059	.047	-.010	-.035	-.054	.061	.048	-.011
.2	.2	-.033	-.070	.065	.058	-.003	-.044	-.065	.066	.058	-.004
.2	.3	-.051	-.087	.075	.074	.010	-.060	-.082	.076	.075	.009
.2	.4	-.080	-.110	.095	.095	.030	-.094	-.114	.096	.095	.028
.2	.5	-.138	-.156	.142	.127	.068	-.148	-.157	.139	.127	.065
						n= 25					
.0	.0	-.019	-.032	.045	.025	-.010	-.033	-.032	.044	.024	-.011
.0	.1	-.023	-.035	.046	.029	-.008	-.033	-.032	.046	.029	-.008
.0	.2	-.033	-.042	.051	.037	-.002	-.038	-.037	.049	.036	-.003
.0	.3	-.040	-.046	.057	.042	.004	-.044	-.042	.056	.042	.003
.0	.4	-.062	-.059	.073	.054	.017	-.058	-.050	.073	.052	.016
.0	.5	-.084	-.070	.094	.061	.029	-.075	-.059	.093	.061	.030
.1	.0	-.018	-.035	.045	.026	-.011	-.029	-.039	.044	.025	-.011
.1	.1	-.023	-.039	.047	.031	-.008	-.029	-.040	.046	.030	-.008
.1	.2	-.033	-.047	.051	.040	-.002	-.036	-.047	.049	.038	-.003
.1	.3	-.041	-.052	.057	.046	.004	-.044	-.054	.056	.045	.004
.1	.4	-.065	-.068	.074	.061	.019	-.061	-.067	.074	.057	.019
.1	.5	-.090	-.083	.096	.071	.034	-.083	-.081	.096	.070	.035
.2	.0	-.015	-.040	.045	.029	-.012	-.025	-.046	.045	.028	-.013
.2	.1	-.020	-.046	.047	.034	-.009	-.025	-.049	.046	.035	-.010
.2	.2	-.031	-.058	.051	.046	-.002	-.033	-.059	.050	.045	-.004
.2	.3	-.041	-.065	.057	.055	.005	-.042	-.070	.056	.056	.004
.2	.4	-.071	-.092	.076	.078	.025	-.065	-.091	.075	.075	.023
.2	.5	-.104	-.118	.102	.094	.046	-.096	-.115	.101	.096	.046
						n= 40					
.0	.0	-.010	-.020	.028	.015	-.006	-.024	-.019	.028	.015	-.006
.0	.1	-.014	-.024	.029	.019	-.003	-.024	-.019	.029	.019	-.004
.0	.2	-.019	-.027	.030	.023	-.001	-.027	-.022	.031	.023	-.001
.0	.3	-.026	-.031	.036	.028	.004	-.033	-.026	.036	.028	.003
.0	.4	-.040	-.040	.046	.034	.012	-.043	-.033	.045	.033	.010
.0	.5	-.058	-.049	.064	.041	.023	-.056	-.040	.063	.041	.022
.1	.0	-.010	-.021	.028	.016	-.006	-.022	-.023	.028	.016	-.007
.1	.1	-.014	-.025	.029	.021	-.004	-.022	-.023	.029	.021	-.005
.1	.2	-.019	-.029	.030	.026	-.001	-.026	-.028	.031	.026	-.001
.1	.3	-.027	-.034	.036	.032	.005	-.033	-.034	.036	.031	.004
.1	.4	-.042	-.045	.047	.039	.013	-.045	-.044	.046	.038	.012
.1	.5	-.063	-.058	.066	.050	.028	-.063	-.055	.066	.050	.027
.2	.0	-.009	-.023	.028	.018	-.007	-.020	-.027	.028	.018	-.007
.2	.1	-.013	-.029	.029	.024	-.005	-.020	-.028	.029	.024	-.005
.2	.2	-.018	-.033	.031	.031	-.002	-.024	-.035	.031	.030	-.002
.2	.3	-.027	-.041	.036	.039	.004	-.032	-.043	.036	.037	.004
.2	.4	-.044	-.057	.047	.049	.015	-.047	-.057	.046	.047	.014
.2	.5	-.072	-.078	.069	.066	.034	-.071	-.076	.069	.066	.035

TABLE 22.2
Simulated values of bias, variance and covariance of ($\hat{\mu}, \hat{\sigma}$) and ($\tilde{\mu}, \tilde{\sigma}$) when $\kappa = 2.0$

p	q	(1)	(2)	(3)	(4)	(5)	(6)	(7)	(8)	(9)	(10)
						n= 20					
.0	.0	-.010	-.042	.054	.028	-.007	-.027	-.031	.055	.028	-.010
.0	.1	-.016	-.048	.056	.034	-.004	-.029	-.034	.057	.035	-.007
.0	.2	-.024	-.055	.060	.040	.001	-.035	-.040	.061	.041	-.002
.0	.3	-.034	-.062	.067	.049	.008	-.047	-.051	.068	.049	.005
.0	.4	-.049	-.071	.082	.060	.021	-.063	-.063	.082	.060	.017
.0	.5	-.080	-.090	.108	.074	.039	-.090	-.080	.103	.072	.033
.1	.0	-.008	-.048	.055	.030	-.009	-.021	-.041	.056	.030	-.011
.1	.1	-.014	-.056	.056	.038	-.006	-.023	-.046	.057	.037	-.008
.1	.2	-.023	-.065	.060	.045	.000	-.031	-.056	.061	.045	-.002
.1	.3	-.035	-.075	.067	.056	.008	-.046	-.071	.068	.055	.006
.1	.4	-.052	-.089	.082	.070	.023	-.066	-.089	.082	.069	.020
.1	.5	-.091	-.118	.110	.089	.045	-.102	-.117	.107	.087	.040
.2	.0	-.003	-.057	.056	.033	-.011	-.015	-.049	.057	.033	-.013
.2	.1	-.010	-.068	.057	.043	-.008	-.018	-.056	.058	.043	-.010
.2	.2	-.020	-.081	.060	.053	-.002	-.027	-.070	.062	.053	-.004
.2	.3	-.034	-.096	.067	.067	.008	-.043	-.091	.068	.066	.006
.2	.4	-.056	-.119	.082	.086	.024	-.068	-.118	.083	.085	.022
.2	.5	-.108	-.168	.113	.114	.054	-.117	-.165	.110	.113	.049
						n= 25					
.0	.0	-.014	-.031	.043	.022	-.008	-.024	-.025	.043	.022	-.007
.0	.1	-.018	-.035	.044	.027	-.006	-.025	-.026	.044	.026	-.005
.0	.2	-.026	-.042	.048	.033	-.001	-.030	-.032	.048	.032	.000
.0	.3	-.033	-.047	.052	.038	.004	-.035	-.036	.053	.038	.005
.0	.4	-.054	-.063	.065	.048	.015	-.049	-.046	.066	.048	.017
.0	.5	-.069	-.071	.080	.058	.026	-.066	-.058	.080	.058	.029
.1	.0	-.012	-.035	.044	.024	-.008	-.020	-.032	.043	.024	-.008
.1	.1	-.016	-.040	.045	.029	-.006	-.021	-.034	.044	.028	-.005
.1	.2	-.025	-.049	.048	.036	-.002	-.027	-.042	.048	.036	.000
.1	.3	-.034	-.056	.052	.042	.003	-.034	-.048	.053	.042	.006
.1	.4	-.057	-.076	.065	.054	.016	-.051	-.063	.067	.055	.019
.1	.5	-.075	-.089	.081	.067	.029	-.073	-.079	.082	.067	.033
.2	.0	-.010	-.041	.045	.027	-.010	-.015	-.039	.044	.027	-.009
.2	.1	-.013	-.047	.045	.033	-.008	-.016	-.043	.044	.034	-.007
.2	.2	-.023	-.059	.048	.043	-.003	-.023	-.055	.048	.044	-.001
.2	.3	-.032	-.070	.052	.052	.003	-.031	-.065	.053	.054	.006
.2	.4	-.060	-.100	.066	.070	.018	-.053	-.088	.068	.074	.023
.2	.5	-.086	-.124	.083	.090	.037	-.082	-.117	.086	.095	.043
						n= 40					
.0	.0	-.006	-.018	.026	.014	-.005	-.020	-.016	.027	.014	-.004
.0	.1	-.009	-.019	.027	.017	-.003	-.019	-.014	.027	.017	-.003
.0	.2	-.015	-.025	.029	.020	-.001	-.020	-.015	.029	.020	.000
.0	.3	-.019	-.028	.033	.023	.003	-.024	-.018	.034	.025	.004
.0	.4	-.030	-.036	.039	.029	.009	-.032	-.025	.041	.030	.010
.0	.5	-.046	-.046	.050	.036	.017	-.047	-.035	.054	.038	.020
.1	.0	-.005	-.020	.026	.015	-.005	-.016	-.021	.027	.015	-.005
.1	.1	-.008	-.022	.027	.019	-.004	-.016	-.021	.027	.019	-.003
.1	.2	-.014	-.030	.029	.023	-.001	-.018	-.023	.029	.023	-.001
.1	.3	-.020	-.034	.033	.028	.003	-.023	-.028	.034	.028	.004
.1	.4	-.032	-.045	.040	.035	.010	-.034	-.038	.041	.035	.012
.1	.5	-.051	-.059	.051	.045	.021	-.053	-.053	.056	.046	.024
.2	.0	-.003	-.023	.027	.017	-.006	-.013	-.026	.028	.017	-.006
.2	.1	-.006	-.027	.027	.022	-.004	-.013	-.026	.028	.022	-.005
.2	.2	-.013	-.036	.029	.027	-.001	-.016	-.029	.030	.028	-.002
.2	.3	-.019	-.043	.033	.034	.004	-.022	-.036	.034	.035	.004
.2	.4	-.034	-.059	.040	.045	.013	-.035	-.051	.041	.045	.013
.2	.5	-.059	-.083	.053	.061	.027	-.060	-.075	.057	.063	.029

TABLE 22.3
Simulated values of bias, variance and covariance of ($\hat{\mu}, \hat{\sigma}$) and ($\tilde{\mu}, \tilde{\sigma}$) when $\kappa = 4.0$

p	q	(1)	(2)	(3)	(4)	(5)	(6)	(7)	(8)	(9)	(10)
							n= 20				
.0	.0	-.007	-.037	.050	.027	-.005	-.017	-.035	.054	.025	-.006
.0	.1	-.012	-.042	.052	.032	-.002	-.019	-.037	.056	.030	-.003
.0	.2	-.019	-.048	.055	.037	.001	-.024	-.043	.059	.035	.001
.0	.3	-.030	-.057	.061	.045	.008	-.034	-.052	.065	.042	.007
.0	.4	-.046	-.069	.071	.055	.018	-.049	-.065	.075	.052	.017
.0	.5	-.068	-.083	.090	.068	.034	-.076	-.085	.096	.064	.033
.1	.0	-.005	-.042	.051	.030	-.007	-.010	-.046	.055	.027	-.007
.1	.1	-.010	-.049	.052	.037	-.004	-.013	-.051	.056	.034	-.005
.1	.2	-.017	-.057	.055	.042	.000	-.020	-.060	.059	.040	.000
.1	.3	-.029	-.069	.061	.052	.007	-.032	-.074	.065	.050	.008
.1	.4	-.048	-.086	.071	.066	.018	-.052	-.094	.075	.062	.018
.1	.5	-.076	-.109	.091	.086	.039	-.088	-.126	.097	.078	.038
.2	.0	-.001	-.049	.053	.034	-.010	-.004	-.055	.057	.031	-.010
.2	.1	-.006	-.057	.054	.042	-.007	-.007	-.062	.058	.038	-.007
.2	.2	-.014	-.068	.056	.050	-.003	-.015	-.075	.060	.048	-.003
.2	.3	-.027	-.085	.061	.063	.005	-.028	-.094	.066	.059	.005
.2	.4	-.049	-.111	.071	.085	.019	-.052	-.124	.075	.076	.018
.2	.5	-.086	-.149	.093	.118	.046	-.100	-.177	.098	.100	.041
							n= 25				
.0	.0	-.005	-.031	.041	.020	-.005	-.017	-.028	.043	.022	-.006
.0	.1	-.008	-.035	.042	.024	-.003	-.018	-.031	.045	.025	-.004
.0	.2	-.017	-.045	.045	.030	.001	-.023	-.036	.047	.031	.000
.0	.3	-.024	-.051	.049	.035	.006	-.027	-.040	.052	.036	.005
.0	.4	-.038	-.061	.060	.045	.016	-.041	-.052	.061	.045	.014
.0	.5	-.050	-.070	.071	.053	.026	-.055	-.061	.073	.054	.024
.1	.0	-.003	-.034	.041	.022	-.006	-.013	-.035	.044	.023	-.007
.1	.1	-.007	-.038	.042	.026	-.003	-.015	-.039	.045	.027	-.004
.1	.2	-.016	-.050	.045	.033	.000	-.020	-.045	.047	.033	-.001
.1	.3	-.024	-.058	.049	.039	.005	-.025	-.051	.052	.039	.005
.1	.4	-.039	-.072	.060	.051	.017	-.043	-.068	.061	.050	.015
.1	.5	-.055	-.084	.072	.061	.028	-.060	-.082	.074	.062	.027
.2	.0	.001	-.041	.043	.025	-.008	-.006	-.045	.045	.027	-.009
.2	.1	-.002	-.048	.044	.030	-.006	-.008	-.051	.046	.032	-.007
.2	.2	-.012	-.065	.045	.040	-.002	-.015	-.062	.048	.041	-.003
.2	.3	-.021	-.078	.049	.048	.004	-.022	-.072	.052	.049	.003
.2	.4	-.041	-.102	.060	.066	.017	-.044	-.099	.061	.067	.015
.2	.5	-.064	-.126	.073	.084	.033	-.068	-.127	.075	.085	.031
							n= 40				
.0	.0	-.006	-.020	.026	.013	-.003	-.013	-.020	.025	.013	-.003
.0	.1	-.009	-.023	.027	.017	-.001	-.013	-.021	.026	.016	-.002
.0	.2	-.012	-.026	.028	.020	.001	-.015	-.023	.027	.020	.001
.0	.3	-.016	-.029	.031	.024	.004	-.020	-.028	.030	.023	.004
.0	.4	-.024	-.035	.037	.029	.010	-.024	-.031	.036	.028	.009
.0	.5	-.036	-.044	.047	.036	.018	-.036	-.040	.045	.033	.016
.1	.0	-.004	-.023	.026	.015	-.004	-.009	-.027	.025	.014	-.004
.1	.1	-.007	-.028	.027	.019	-.002	-.010	-.029	.026	.018	-.002
.1	.2	-.011	-.032	.028	.023	.000	-.013	-.033	.028	.022	.001
.1	.3	-.016	-.037	.031	.028	.004	-.018	-.039	.030	.028	.004
.1	.4	-.026	-.047	.037	.035	.010	-.025	-.046	.036	.035	.011
.1	.5	-.042	-.060	.048	.046	.021	-.042	-.061	.046	.042	.019
.2	.0	-.002	-.027	.027	.017	-.005	-.006	-.031	.026	.016	-.005
.2	.1	-.005	-.033	.027	.022	-.003	-.007	-.034	.026	.021	-.004
.2	.2	-.009	-.040	.029	.028	-.001	-.010	-.039	.028	.027	-.001
.2	.3	-.014	-.047	.031	.035	.003	-.017	-.049	.030	.034	.004
.2	.4	-.027	-.061	.037	.045	.011	-.026	-.060	.036	.045	.012
.2	.5	-.048	-.084	.050	.063	.026	-.048	-.084	.047	.060	.024

TABLE 22.4
Simulated values of bias, variance and covariance of $(\hat{\mu}, \hat{\sigma})$ and $(\tilde{\mu}, \tilde{\sigma})$ when $\kappa = 6.0$

p	q	(1)	(2)	(3)	(4)	(5)	(6)	(7)	(8)	(9)	(10)
						n= 20					
.0	.0	-.002	-.040	.051	.025	-.006	-.017	-.038	.055	.025	-.005
.0	.1	-.007	-.046	.053	.030	-.003	-.020	-.042	.056	.030	-.002
.0	.2	-.014	-.053	.055	.037	.001	-.025	-.047	.059	.036	.002
.0	.3	-.025	-.064	.060	.044	.007	-.034	-.056	.064	.042	.007
.0	.4	-.040	-.075	.070	.052	.015	-.047	-.068	.074	.052	.017
.0	.5	-.059	-.089	.089	.066	.032	-.065	-.080	.091	.064	.031
.1	.0	.001	-.046	.052	.029	-.007	-.012	-.047	.055	.029	-.006
.1	.1	-.004	-.054	.053	.034	-.004	-.015	-.053	.056	.035	-.004
.1	.2	-.012	-.064	.055	.043	.000	-.021	-.061	.059	.042	.001
.1	.3	-.024	-.078	.060	.052	.007	-.032	-.074	.064	.051	.007
.1	.4	-.042	-.095	.070	.063	.017	-.049	-.092	.075	.063	.018
.1	.5	-.067	-.117	.091	.085	.038	-.073	-.113	.092	.079	.035
.2	.0	.006	-.053	.053	.033	-.009	-.005	-.057	.057	.033	-.009
.2	.1	.001	-.063	.054	.040	-.007	-.008	-.066	.057	.042	-.007
.2	.2	-.008	-.077	.056	.051	-.003	-.015	-.078	.060	.052	-.002
.2	.3	-.022	-.097	.060	.063	.005	-.028	-.097	.064	.064	.005
.2	.4	-.043	-.123	.070	.080	.017	-.049	-.124	.075	.082	.019
.2	.5	-.077	-.162	.092	.115	.045	-.083	-.162	.093	.108	.040
						n= 25					
.0	.0	-.005	-.034	.041	.020	-.004	-.013	-.030	.041	.021	-.005
.0	.1	-.007	-.036	.042	.024	-.002	-.014	-.031	.042	.025	-.003
.0	.2	-.015	-.044	.045	.029	.001	-.020	-.039	.045	.031	.001
.0	.3	-.021	-.050	.048	.034	.005	-.026	-.045	.048	.035	.005
.0	.4	-.036	-.062	.056	.043	.014	-.039	-.056	.056	.045	.014
.0	.5	-.050	-.072	.067	.049	.022	-.056	-.069	.068	.052	.023
.1	.0	-.003	-.037	.042	.023	-.005	-.010	-.034	.041	.022	-.006
.1	.1	-.006	-.041	.042	.027	-.004	-.011	-.037	.042	.027	-.004
.1	.2	-.014	-.051	.045	.034	.001	-.018	-.046	.045	.034	.000
.1	.3	-.020	-.058	.048	.040	.005	-.024	-.054	.048	.040	.005
.1	.4	-.038	-.074	.057	.051	.015	-.039	-.069	.057	.052	.015
.1	.5	-.054	-.089	.067	.060	.024	-.059	-.087	.069	.062	.026
.2	.0	.001	-.045	.043	.027	-.008	-.005	-.041	.043	.025	-.008
.2	.1	-.001	-.050	.043	.033	-.006	-.006	-.045	.043	.032	-.006
.2	.2	-.010	-.065	.046	.042	-.002	-.013	-.058	.045	.042	-.001
.2	.3	-.017	-.076	.048	.051	.003	-.021	-.070	.049	.049	.004
.2	.4	-.039	-.103	.057	.070	.016	-.039	-.093	.057	.067	.016
.2	.5	-.062	-.130	.068	.087	.030	-.065	-.124	.071	.084	.031
						n= 40					
.0	.0	-.004	-.020	.025	.013	-.002	-.010	-.020	.025	.012	-.003
.0	.1	-.006	-.022	.026	.015	-.001	-.012	-.023	.026	.015	-.001
.0	.2	-.009	-.026	.027	.018	.001	-.015	-.027	.028	.018	.001
.0	.3	-.015	-.031	.031	.022	.004	-.019	-.031	.030	.022	.004
.0	.4	-.024	-.038	.036	.027	.010	-.027	-.037	.035	.027	.009
.0	.5	-.036	-.047	.046	.034	.018	-.035	-.043	.044	.033	.017
.1	.0	-.003	-.022	.026	.014	-.003	-.007	-.025	.025	.014	-.003
.1	.1	-.005	-.025	.026	.018	-.002	-.009	-.029	.026	.017	-.002
.1	.2	-.009	-.029	.027	.021	.000	-.013	-.035	.028	.021	.001
.1	.3	-.015	-.036	.031	.026	.004	-.018	-.040	.030	.026	.004
.1	.4	-.024	-.046	.036	.033	.011	-.027	-.050	.035	.033	.010
.1	.5	-.040	-.059	.047	.044	.021	-.039	-.061	.045	.044	.020
.2	.0	-.001	-.024	.026	.016	-.005	-.004	-.029	.026	.016	-.005
.2	.1	-.003	-.028	.027	.021	-.003	-.006	-.034	.027	.020	-.003
.2	.2	-.007	-.033	.028	.025	-.001	-.010	-.042	.028	.025	-.001
.2	.3	-.014	-.043	.031	.033	.003	-.016	-.051	.030	.032	.003
.2	.4	-.024	-.056	.037	.043	.011	-.027	-.065	.035	.042	.010
.2	.5	-.043	-.078	.048	.062	.026	-.044	-.085	.046	.061	.024

TABLE 22.5

Simulated values of bias, variance and covariance of ($\hat{\mu}, \hat{\sigma}$) and ($\tilde{\mu}, \tilde{\sigma}$) when $\kappa = 8.0$

p	q	(1)	(2)	(3)	(4)	(5)	(6)	(7)	(8)	(9)	(10)
							n= 20				
.0	.0	-.008	-.035	.048	.026	-.004	-.021	-.045	.051	.026	-.006
.0	.1	-.013	-.042	.050	.031	-.001	-.024	-.050	.052	.031	-.003
.0	.2	-.019	-.048	.053	.037	.003	-.029	-.056	.055	.036	.001
.0	.3	-.030	-.059	.059	.043	.009	-.038	-.064	.059	.043	.006
.0	.4	-.045	-.071	.068	.053	.019	-.053	-.078	.068	.051	.015
.0	.5	-.067	-.087	.086	.065	.033	-.071	-.091	.083	.063	.028
.1	.0	-.004	-.041	.049	.029	-.005	-.017	-.051	.052	.030	-.007
.1	.1	-.010	-.050	.050	.035	-.002	-.020	-.059	.053	.036	-.004
.1	.2	-.016	-.058	.053	.043	.003	-.026	-.066	.055	.043	.000
.1	.3	-.029	-.073	.059	.051	.010	-.035	-.078	.059	.052	.005
.1	.4	-.047	-.091	.068	.066	.022	-.054	-.097	.068	.064	.016
.1	.5	-.075	-.117	.089	.084	.041	-.077	-.118	.084	.081	.032
.2	.0	.000	-.048	.050	.033	-.007	-.011	-.060	.053	.033	-.009
.2	.1	-.005	-.059	.051	.041	-.005	-.014	-.070	.054	.041	-.006
.2	.2	-.012	-.071	.053	.052	.000	-.020	-.081	.056	.051	-.002
.2	.3	-.026	-.091	.059	.064	.008	-.031	-.097	.059	.062	.004
.2	.4	-.047	-.118	.069	.086	.023	-.053	-.126	.068	.080	.017
.2	.5	-.084	-.161	.090	.115	.048	-.085	-.164	.086	.109	.039
							n= 25				
.0	.0	-.010	-.031	.040	.021	-.003	-.016	-.030	.040	.021	-.004
.0	.1	-.013	-.035	.040	.025	-.001	-.016	-.031	.041	.025	-.002
.0	.2	-.020	-.041	.043	.030	.003	-.023	-.039	.044	.031	.003
.0	.3	-.025	-.046	.046	.035	.006	-.027	-.043	.047	.035	.006
.0	.4	-.039	-.058	.054	.043	.014	-.039	-.054	.057	.045	.016
.0	.5	-.051	-.067	.063	.051	.023	-.053	-.065	.067	.054	.025
.1	.0	-.009	-.034	.040	.023	-.004	-.012	-.036	.041	.023	-.005
.1	.1	-.012	-.038	.041	.028	-.002	-.013	-.038	.041	.027	-.003
.1	.2	-.019	-.046	.043	.034	.002	-.020	-.049	.044	.035	.002
.1	.3	-.025	-.053	.046	.040	.006	-.025	-.054	.047	.040	.006
.1	.4	-.040	-.068	.054	.050	.015	-.040	-.069	.057	.053	.017
.1	.5	-.054	-.080	.064	.061	.025	-.057	-.084	.068	.065	.028
.2	.0	-.006	-.039	.041	.026	-.005	-.007	-.042	.042	.026	-.007
.2	.1	-.008	-.045	.042	.031	-.004	-.009	-.045	.043	.032	-.005
.2	.2	-.015	-.057	.044	.042	.000	-.016	-.059	.045	.042	.000
.2	.3	-.022	-.066	.046	.049	.005	-.021	-.068	.048	.050	.004
.2	.4	-.041	-.091	.054	.066	.016	-.039	-.091	.057	.069	.018
.2	.5	-.059	-.112	.065	.085	.030	-.061	-.117	.069	.088	.033
							n= 40				
.0	.0	-.007	-.018	.026	.013	-.002	-.008	-.019	.025	.012	-.002
.0	.1	-.009	-.022	.026	.015	-.001	-.010	-.022	.026	.014	-.001
.0	.2	-.012	-.025	.028	.018	.001	-.013	-.027	.027	.018	.001
.0	.3	-.019	-.031	.030	.022	.004	-.017	-.031	.029	.021	.004
.0	.4	-.028	-.039	.035	.027	.009	-.024	-.037	.035	.025	.009
.0	.5	-.036	-.044	.043	.034	.017	-.034	-.044	.042	.032	.016
.1	.0	-.006	-.021	.026	.014	-.003	-.006	-.023	.025	.014	-.003
.1	.1	-.008	-.025	.027	.017	-.002	-.007	-.027	.026	.017	-.002
.1	.2	-.012	-.029	.028	.021	.001	-.011	-.033	.027	.021	.001
.1	.3	-.018	-.036	.030	.026	.004	-.016	-.039	.029	.026	.004
.1	.4	-.028	-.047	.035	.032	.010	-.025	-.048	.035	.032	.010
.1	.5	-.039	-.056	.044	.043	.019	-.037	-.060	.043	.041	.018
.2	.0	-.004	-.023	.027	.017	-.004	-.003	-.027	.026	.016	-.004
.2	.1	-.007	-.028	.027	.021	-.003	-.005	-.032	.026	.020	-.003
.2	.2	-.010	-.033	.028	.026	-.001	-.009	-.040	.027	.026	-.001
.2	.3	-.017	-.043	.031	.033	.003	-.014	-.049	.030	.033	.003
.2	.4	-.028	-.059	.035	.043	.010	-.024	-.062	.035	.042	.010
.2	.5	-.043	-.074	.045	.061	.023	-.041	-.082	.044	.058	.022

TABLE 22.6
Simulated values of bias, variance and covariance of ($\hat{\mu}, \hat{\sigma}$) and ($\tilde{\mu}, \tilde{\sigma}$) when $\kappa = 10.0$

p	q	(1)	(2)	(3)	(4)	(5)	(6)	(7)	(8)	(9)	(10)
						n= 20					
.0	.0	-.005	-.034	.052	.026	-.004	-.016	-.042	.050	.027	-.004
.0	.1	-.009	-.038	.053	.031	-.002	-.019	-.047	.052	.032	-.001
.0	.2	-.015	-.045	.056	.037	.003	-.024	-.053	.054	.038	.002
.0	.3	-.023	-.051	.062	.044	.009	-.033	-.061	.059	.046	.009
.0	.4	-.037	-.063	.071	.054	.018	-.044	-.071	.067	.055	.017
.0	.5	-.056	-.077	.087	.066	.032	-.060	-.082	.083	.067	.031
.1	.0	-.001	-.041	.053	.029	-.005	-.011	-.051	.051	.030	-.005
.1	.1	-.005	-.047	.053	.035	-.004	-.014	-.057	.052	.037	-.002
.1	.2	-.013	-.056	.057	.044	.001	-.020	-.066	.054	.044	.002
.1	.3	-.022	-.066	.062	.053	.008	-.030	-.078	.059	.054	.009
.1	.4	-.038	-.083	.072	.067	.020	-.045	-.094	.068	.067	.019
.1	.5	-.064	-.107	.089	.085	.037	-.066	-.113	.085	.086	.037
.2	.0	.003	-.048	.055	.033	-.008	-.004	-.060	.053	.033	-.008
.2	.1	-.001	-.056	.055	.041	-.006	-.007	-.070	.054	.042	-.005
.2	.2	-.008	-.069	.057	.052	-.001	-.014	-.083	.055	.052	-.001
.2	.3	-.019	-.083	.062	.065	.006	-.026	-.101	.060	.067	.007
.2	.4	-.038	-.109	.072	.085	.020	-.044	-.126	.068	.084	.018
.2	.5	-.072	-.148	.090	.112	.042	-.074	-.161	.086	.113	.041
						n= 25					
.0	.0	-.004	-.031	.041	.020	-.003	-.015	-.026	.041	.020	-.003
.0	.1	-.007	-.036	.042	.023	-.002	-.016	-.027	.042	.024	-.001
.0	.2	-.012	-.041	.044	.028	.002	-.021	-.034	.044	.029	.002
.0	.3	-.018	-.046	.048	.032	.005	-.024	-.037	.047	.034	.006
.0	.4	-.029	-.055	.055	.041	.013	-.034	-.045	.056	.043	.015
.0	.5	-.043	-.066	.065	.048	.021	-.045	-.053	.067	.052	.025
.1	.0	-.002	-.034	.042	.022	-.004	-.010	-.033	.041	.022	-.004
.1	.1	-.005	-.039	.042	.026	-.003	-.012	-.036	.042	.026	-.002
.1	.2	-.011	-.046	.045	.032	.001	-.018	-.045	.044	.033	.002
.1	.3	-.017	-.053	.048	.038	.005	-.022	-.050	.047	.038	.005
.1	.4	-.029	-.064	.055	.049	.014	-.035	-.062	.056	.051	.016
.1	.5	-.046	-.079	.066	.060	.025	-.049	-.075	.067	.062	.027
.2	.0	.004	-.043	.043	.026	-.007	-.005	-.041	.043	.026	-.006
.2	.1	.001	-.051	.044	.031	-.005	-.006	-.045	.043	.031	-.005
.2	.2	-.005	-.061	.046	.040	-.002	-.013	-.057	.045	.040	-.001
.2	.3	-.013	-.073	.048	.048	.003	-.018	-.066	.047	.047	.003
.2	.4	-.030	-.094	.055	.066	.014	-.034	-.087	.056	.066	.016
.2	.5	-.053	-.122	.066	.083	.027	-.053	-.110	.068	.086	.031
						n= 40					
.0	.0	-.007	-.023	.025	.013	-.002	-.012	-.020	.025	.013	-.002
.0	.1	-.009	-.025	.026	.015	-.001	-.012	-.020	.026	.015	-.001
.0	.2	-.011	-.028	.027	.018	.001	-.014	-.022	.027	.018	.001
.0	.3	-.017	-.033	.030	.022	.004	-.019	-.027	.029	.021	.004
.0	.4	-.023	-.038	.034	.027	.009	-.023	-.031	.034	.026	.008
.0	.5	-.035	-.047	.043	.034	.016	-.032	-.037	.041	.032	.015
.1	.0	-.006	-.025	.025	.015	-.003	-.008	-.027	.026	.014	-.003
.1	.1	-.008	-.028	.026	.017	-.002	-.009	-.029	.026	.017	-.002
.1	.2	-.011	-.032	.027	.022	.001	-.011	-.032	.027	.021	.000
.1	.3	-.016	-.039	.030	.027	.004	-.017	-.040	.029	.025	.003
.1	.4	-.024	-.046	.034	.034	.010	-.024	-.047	.034	.032	.008
.1	.5	-.038	-.059	.044	.045	.020	-.037	-.059	.042	.042	.018
.2	.0	-.004	-.028	.026	.017	-.005	-.005	-.031	.026	.017	-.004
.2	.1	-.005	-.032	.027	.021	-.004	-.006	-.034	.027	.021	-.003
.2	.2	-.008	-.037	.028	.027	-.001	-.009	-.038	.028	.026	-.001
.2	.3	-.015	-.047	.030	.035	.003	-.015	-.049	.030	.032	.002
.2	.4	-.023	-.058	.034	.045	.009	-.024	-.061	.034	.042	.009
.2	.5	-.042	-.080	.044	.064	.023	-.041	-.081	.043	.059	.021

TABLE 22.7
Asymptotic values of V_{11}, V_{22} and V_{12} for various proportions of censoring

q_1	q_2	κ	0.5	1.0	1.5	2.0	3.0	4.0	6.0	8.0	10.0	12.0
0.0	0.0	V_{11}	1.181	1.109	1.076	1.059	1.040	1.030	1.021	1.015	1.012	1.010
		V_{22}	0.681	0.608	0.575	0.559	0.540	0.530	0.520	0.515	0.512	0.510
		V_{12}	-0.352	-0.257	-0.210	-0.181	-0.147	-0.127	-0.103	-0.089	-0.080	-0.073
	0.1	V_{11}	1.249	1.152	1.113	1.093	1.071	1.059	1.048	1.042	1.038	1.035
		V_{22}	0.869	0.767	0.721	0.694	0.665	0.650	0.633	0.624	0.619	0.615
		V_{12}	-0.243	-0.176	-0.139	-0.115	-0.087	-0.069	-0.049	-0.037	-0.028	-0.022
	0.2	V_{11}	1.411	1.253	1.197	1.168	1.138	1.122	1.106	1.098	1.092	1.088
		V_{22}	1.051	0.928	0.871	0.837	0.800	0.780	0.757	0.745	0.737	0.732
		V_{12}	-0.074	-0.049	-0.027	-0.012	0.008	0.021	0.036	0.045	0.052	0.057
	0.3	V_{11}	1.727	1.447	1.356	1.311	1.264	1.239	1.214	1.200	1.191	1.185
		V_{22}	1.259	1.122	1.053	1.013	0.966	0.940	0.912	0.896	0.886	0.879
		V_{12}	0.184	0.145	0.143	0.146	0.153	0.158	0.165	0.170	0.173	0.176
	0.4	V_{11}	2.329	1.812	1.652	1.573	1.493	1.452	1.408	1.384	1.369	1.359
		V_{22}	1.526	1.373	1.291	1.241	1.184	1.151	1.115	1.095	1.082	1.073
		V_{12}	0.584	0.447	0.408	0.391	0.376	0.369	0.363	0.361	0.360	0.359
	0.5	V_{11}	3.488	2.510	2.214	2.069	1.923	1.847	1.768	1.725	1.698	1.678
		V_{22}	1.881	1.716	1.618	1.557	1.486	1.445	1.399	1.373	1.357	1.345
		V_{12}	1.226	0.936	0.836	0.786	0.736	0.710	0.683	0.669	0.660	0.654
0.1	0.0	V_{11}	1.195	1.119	1.086	1.070	1.052	1.043	1.034	1.030	1.027	1.026
		V_{22}	0.732	0.655	0.624	0.610	0.596	0.590	0.584	0.582	0.581	0.580
		V_{12}	-0.378	-0.279	-0.232	-0.205	-0.173	-0.154	-0.133	-0.120	-0.112	-0.105
	0.1	V_{11}	1.255	1.157	1.119	1.099	1.078	1.067	1.057	1.051	1.048	1.046
		V_{22}	0.953	0.842	0.798	0.774	0.750	0.738	0.726	0.720	0.717	0.715
		V_{12}	-0.267	-0.196	-0.160	-0.137	-0.111	-0.095	-0.077	-0.067	-0.060	-0.054
	0.2	V_{11}	1.411	1.253	1.198	1.170	1.140	1.125	1.110	1.102	1.097	1.093
		V_{22}	1.177	1.040	0.983	0.953	0.922	0.906	0.889	0.881	0.875	0.872
		V_{12}	-0.083	-0.059	-0.039	-0.025	-0.008	0.003	0.015	0.023	0.028	0.032
	0.3	V_{11}	1.731	1.449	1.357	1.311	1.264	1.240	1.214	1.200	1.192	1.186
		V_{22}	1.444	1.288	1.220	1.183	1.144	1.123	1.101	1.090	1.083	1.078
		V_{12}	0.211	0.161	0.155	0.155	0.158	0.160	0.164	0.166	0.168	0.170
	0.4	V_{11}	2.369	1.834	1.669	1.587	1.504	1.461	1.415	1.391	1.375	1.364
		V_{22}	1.805	1.625	1.544	1.499	1.450	1.425	1.397	1.383	1.373	1.367
		V_{12}	0.690	0.521	0.473	0.450	0.429	0.418	0.408	0.402	0.399	0.396
	0.5	V_{11}	3.674	2.619	2.300	2.144	1.986	1.904	1.818	1.771	1.741	1.720
		V_{22}	2.325	2.125	2.026	1.971	1.911	1.879	1.845	1.826	1.814	1.805
		V_{12}	1.513	1.147	1.024	0.963	0.900	0.867	0.832	0.813	0.800	0.791
0.2	0.0	V_{11}	1.214	1.138	1.108	1.093	1.078	1.071	1.064	1.061	1.060	1.059
		V_{22}	0.797	0.721	0.694	0.685	0.675	0.672	0.670	0.670	0.670	0.670
		V_{12}	-0.413	-0.314	-0.270	-0.246	-0.218	-0.202	-0.183	-0.173	-0.165	-0.160
	0.1	V_{11}	1.265	1.169	1.133	1.115	1.097	1.088	1.080	1.076	1.073	1.072
		V_{22}	1.064	0.952	0.912	0.892	0.873	0.864	0.855	0.851	0.849	0.848
		V_{12}	-0.300	-0.232	-0.200	-0.181	-0.159	-0.146	-0.132	-0.123	-0.117	-0.113
	0.2	V_{11}	1.413	1.257	1.204	1.177	1.150	1.136	1.123	1.116	1.112	1.109
		V_{22}	1.349	1.207	1.156	1.129	1.103	1.090	1.078	1.071	1.068	1.065
		V_{12}	-0.098	-0.083	-0.069	-0.060	-0.049	-0.042	-0.034	-0.029	-0.026	-0.024
	0.3	V_{11}	1.736	1.449	1.358	1.312	1.265	1.241	1.217	1.204	1.196	1.191
		V_{22}	1.709	1.547	1.484	1.451	1.418	1.401	1.384	1.375	1.369	1.366
		V_{12}	0.244	0.172	0.155	0.148	0.142	0.139	0.136	0.134	0.133	0.132
	0.4	V_{11}	2.425	1.859	1.684	1.598	1.510	1.465	1.417	1.392	1.376	1.365
		V_{22}	2.236	2.046	1.969	1.929	1.888	1.866	1.844	1.832	1.825	1.820
		V_{12}	0.846	0.624	0.553	0.518	0.481	0.461	0.439	0.427	0.419	0.413
	0.5	V_{11}	3.976	2.783	2.420	2.242	2.062	1.968	1.869	1.816	1.782	1.758
		V_{22}	3.082	2.870	2.778	2.729	2.677	2.650	2.622	2.606	2.597	2.590
		V_{12}	1.991	1.495	1.324	1.235	1.140	1.088	1.031	0.999	0.978	0.963

TABLE 22.8
Simulated percentage points of the pivotal quantities $P_1 = \sqrt{n}(\hat{\mu} - \mu)/\hat{\sigma}$ and $P_2 = \sqrt{n}\hat{\sigma}/\sigma$

$$\kappa = 1.0$$

				P_1						P_2			
q_1	q_2	.010	.025	.05	.950	.975	.990	.010	.025	.05	.950	.975	.990
							n = 20						
.0	.0	-2.717	-2.292	-1.826	1.858	2.218	2.769	2.701	2.906	3.105	5.688	5.941	6.310
.0	.1	-3.129	-2.478	-2.100	1.854	2.211	2.755	2.515	2.764	2.943	5.853	6.207	6.538
.0	.2	-3.550	-2.861	-2.367	1.822	2.234	2.663	2.284	2.556	2.790	5.907	6.228	6.714
.0	.3	-4.511	-3.566	-2.832	1.817	2.197	2.693	2.122	2.394	2.629	6.072	6.466	7.034
.0	.4	-5.460	-4.399	-3.535	1.801	2.243	2.672	1.896	2.212	2.470	6.163	6.718	7.289
.0	.5	-8.151	-6.280	-4.918	1.940	2.277	2.760	1.606	1.971	2.203	6.425	6.937	7.498
.1	.0	-2.716	-2.300	-1.830	1.885	2.291	2.955	2.648	2.836	3.042	5.693	5.963	6.330
.1	.1	-3.150	-2.501	-2.097	1.873	2.259	2.949	2.495	2.697	2.874	5.893	6.220	6.550
.1	.2	-3.663	-2.945	-2.404	1.843	2.249	2.927	2.174	2.436	2.692	5.980	6.332	6.742
.1	.3	-4.660	-3.644	-2.970	1.877	2.217	2.890	1.943	2.232	2.489	6.156	6.597	7.238
.1	.4	-5.983	-4.857	-3.865	1.818	2.230	2.798	1.733	1.979	2.289	6.267	6.834	7.564
.1	.5	-9.707	-7.009	-5.620	1.926	2.273	2.797	1.407	1.733	1.961	6.563	7.136	7.823
.2	.0	-2.710	-2.274	-1.829	1.932	2.363	3.038	2.510	2.734	2.961	5.758	6.009	6.386
.2	.1	-3.231	-2.517	-2.083	1.922	2.374	3.058	2.322	2.559	2.772	5.961	6.324	6.623
.2	.2	-3.850	-2.932	-2.421	1.859	2.350	3.038	2.029	2.268	2.547	6.038	6.446	6.929
.2	.3	-5.229	-3.997	-3.037	1.892	2.280	3.022	1.744	2.047	2.283	6.259	6.780	7.249
.2	.4	-6.839	-5.366	-4.459	1.846	2.273	2.948	1.427	1.761	2.001	6.388	7.009	7.918
.2	.5	-13.013	-9.536	-7.125	1.925	2.296	2.847	.986	1.298	1.553	6.651	7.338	8.323
							n = 25						
.0	.0	-2.819	-2.325	-1.812	1.832	2.236	2.702	3.129	3.350	3.568	6.177	6.471	6.743
.0	.1	-2.930	-2.435	-2.001	1.808	2.237	2.665	3.050	3.235	3.501	6.292	6.551	6.903
.0	.2	-3.494	-2.783	-2.269	1.829	2.191	2.650	2.801	3.061	3.322	6.454	6.776	7.169
.0	.3	-3.949	-3.112	-2.513	1.809	2.237	2.695	2.669	2.931	3.179	6.542	6.898	7.371
.0	.4	-5.506	-4.178	-3.364	1.817	2.209	2.746	2.366	2.713	2.922	6.779	7.178	7.586
.0	.5	-7.192	-5.429	-4.273	1.876	2.330	2.729	2.209	2.520	2.770	6.809	7.364	7.924
.1	.0	-2.808	-2.323	-1.813	1.876	2.285	2.739	3.074	3.303	3.547	6.221	6.502	6.833
.1	.1	-2.958	-2.450	-2.003	1.876	2.258	2.715	2.985	3.193	3.453	6.329	6.588	6.917
.1	.2	-3.526	-2.838	-2.258	1.811	2.268	2.705	2.684	2.963	3.250	6.496	6.840	7.305
.1	.3	-4.120	-3.168	-2.572	1.852	2.249	2.672	2.525	2.820	3.082	6.612	6.978	7.470
.1	.4	-5.912	-4.377	-3.481	1.833	2.204	2.748	2.221	2.490	2.788	6.832	7.326	7.855
.1	.5	-7.526	-5.937	-4.529	1.876	2.331	2.711	2.010	2.333	2.583	6.932	7.556	7.989
.2	.0	-2.801	-2.315	-1.819	1.912	2.321	2.895	2.946	3.198	3.463	6.226	6.546	6.888
.2	.1	-2.967	-2.446	-1.993	1.885	2.297	2.865	2.854	3.115	3.308	6.369	6.702	7.010
.2	.2	-3.620	-2.872	-2.285	1.861	2.285	2.839	2.513	2.785	3.084	6.607	6.987	7.366
.2	.3	-4.345	-3.274	-2.643	1.867	2.254	2.894	2.261	2.592	2.881	6.788	7.207	7.570
.2	.4	-6.968	-5.395	-3.993	1.843	2.212	2.831	1.768	2.107	2.441	7.028	7.548	8.339
.2	.5	-10.389	-7.526	-5.682	1.882	2.336	2.746	1.562	1.837	2.164	7.202	7.835	8.551
							n = 40						
.0	.0	-2.512	-2.187	-1.847	1.766	2.136	2.506	4.475	4.705	4.967	7.520	7.774	8.058
.0	.1	-2.863	-2.320	-1.978	1.732	2.096	2.506	4.316	4.534	4.797	7.684	7.997	8.252
.0	.2	-3.081	-2.632	-2.180	1.721	2.154	2.497	4.047	4.337	4.625	7.859	8.182	8.538
.0	.3	-3.884	-3.018	-2.489	1.726	2.060	2.499	3.892	4.237	4.514	8.003	8.378	8.825
.0	.4	-4.560	-3.765	-3.156	1.826	2.187	2.572	3.531	3.918	4.319	8.076	8.506	8.905
.0	.5	-5.856	-4.802	-4.038	2.030	2.352	2.740	3.338	3.711	4.022	8.246	8.768	9.347
.1	.0	-2.531	-2.191	-1.858	1.794	2.098	2.559	4.408	4.649	4.888	7.573	7.812	8.155
.1	.1	-2.880	-2.312	-1.980	1.750	2.073	2.513	4.297	4.483	4.700	7.751	8.068	8.365
.1	.2	-3.126	-2.632	-2.156	1.728	2.104	2.516	3.970	4.259	4.511	7.965	8.330	8.730
.1	.3	-3.960	-3.106	-2.498	1.735	2.075	2.579	3.741	4.060	4.371	8.105	8.602	9.060
.1	.4	-4.651	-3.936	-3.264	1.833	2.197	2.579	3.420	3.796	4.105	8.188	8.688	9.279
.1	.5	-6.454	-5.147	-4.294	1.993	2.379	2.740	3.083	3.445	3.762	8.484	9.072	9.688
.2	.0	-2.514	-2.196	-1.855	1.826	2.233	2.639	4.300	4.597	4.807	7.664	7.957	8.173
.2	.1	-2.871	-2.323	-1.974	1.831	2.210	2.636	4.039	4.324	4.584	7.856	8.156	8.558
.2	.2	-3.192	-2.635	-2.174	1.776	2.186	2.611	3.749	4.041	4.400	8.044	8.393	8.953
.2	.3	-3.947	-3.205	-2.592	1.762	2.126	2.618	3.440	3.810	4.146	8.289	8.801	9.191
.2	.4	-5.095	-4.170	-3.456	1.852	2.214	2.662	3.098	3.428	3.807	8.408	8.937	9.452
.2	.5	-7.545	-6.017	-4.833	1.987	2.370	2.702	2.738	3.047	3.393	8.788	9.413	10.218

TABLE 22.9
Simulated percentage points of the pivotal quantities $P_1 = \sqrt{n}(\hat{\mu} - \mu)/\hat{\sigma}$ and $P_2 = \sqrt{n}\hat{\sigma}/\sigma$

$$\kappa = 2.0$$

				P_1							P_2			
q_1	q_2	.010	.025	.05	.950	.975	.990	.010	.025	.05	.950	.975	.990
							$n = 20$						
.0	.0	-2.670	-2.243	-1.873	1.878	2.237	2.727	2.680	2.877	3.112	5.543	5.836	6.099
.0	.1	-3.075	-2.462	-1.998	1.842	2.256	2.720	2.543	2.773	2.953	5.672	5.999	6.345
.0	.2	-3.479	-2.741	-2.228	1.827	2.239	2.673	2.352	2.622	2.848	5.738	6.061	6.577
.0	.3	-3.967	-3.164	-2.526	1.825	2.248	2.676	2.174	2.475	2.661	5.873	6.315	6.879
.0	.4	-5.215	-4.175	-3.255	1.832	2.249	2.708	2.041	2.247	2.532	6.117	6.499	7.185
.0	.5	-6.761	-5.361	-4.281	1.864	2.280	2.766	1.806	2.057	2.312	6.233	6.806	7.438
.1	.0	-2.692	-2.261	-1.873	1.924	2.347	2.799	2.577	2.823	3.035	5.586	5.813	6.226
.1	.1	-3.091	-2.468	-1.998	1.912	2.339	2.811	2.439	2.676	2.869	5.698	6.041	6.391
.1	.2	-3.403	-2.803	-2.293	1.915	2.346	2.783	2.268	2.490	2.730	5.824	6.153	6.612
.1	.3	-4.084	-3.279	-2.582	1.866	2.324	2.725	2.070	2.303	2.519	5.998	6.363	6.968
.1	.4	-5.459	-4.454	-3.489	1.882	2.251	2.681	1.844	2.109	2.304	6.209	6.615	7.251
.1	.5	-8.007	-6.286	-4.956	1.868	2.316	2.706	1.462	1.758	1.996	6.413	6.946	7.661
.2	.0	-2.696	-2.263	-1.872	2.026	2.416	3.084	2.497	2.697	2.914	5.598	5.912	6.194
.2	.1	-3.041	-2.502	-2.024	2.015	2.419	3.086	2.301	2.516	2.731	5.799	6.082	6.519
.2	.2	-3.593	-2.856	-2.305	1.996	2.402	3.067	2.122	2.337	2.552	5.891	6.255	6.633
.2	.3	-4.185	-3.378	-2.711	1.959	2.370	3.042	1.764	2.080	2.325	6.072	6.499	7.090
.2	.4	-6.348	-4.890	-3.715	1.942	2.373	3.024	1.494	1.756	2.045	6.293	6.775	7.473
.2	.5	-10.808	-7.880	-5.890	1.880	2.382	2.903	1.061	1.381	1.611	6.506	7.269	7.916
							$n = 25$						
.0	.0	-2.606	-2.144	-1.780	1.825	2.224	2.765	3.217	3.462	3.654	6.091	6.342	6.661
.0	.1	-2.760	-2.262	-1.910	1.796	2.218	2.745	3.070	3.322	3.530	6.223	6.528	6.828
.0	.2	-3.139	-2.538	-2.095	1.777	2.232	2.756	2.833	3.122	3.399	6.372	6.705	7.041
.0	.3	-3.474	-2.864	-2.336	1.771	2.146	2.695	2.673	2.985	3.239	6.432	6.844	7.302
.0	.4	-4.649	-3.809	-3.099	1.793	2.245	2.718	2.467	2.777	3.056	6.569	7.064	7.623
.0	.5	-5.640	-4.687	-3.788	1.872	2.342	2.759	2.264	2.544	2.865	6.772	7.238	7.970
.1	.0	-2.620	-2.149	-1.787	1.857	2.287	2.853	3.192	3.389	3.584	6.131	6.387	6.618
.1	.1	-2.817	-2.256	-1.922	1.837	2.270	2.830	2.991	3.249	3.469	6.264	6.551	6.828
.1	.2	-3.185	-2.556	-2.118	1.867	2.258	2.811	2.755	3.033	3.286	6.377	6.696	7.082
.1	.3	-3.616	-2.904	-2.383	1.827	2.241	2.787	2.563	2.865	3.139	6.478	6.853	7.374
.1	.4	-4.924	-3.887	-3.249	1.811	2.264	2.742	2.343	2.583	2.845	6.675	7.070	7.557
.1	.5	-6.138	-5.085	-4.145	1.855	2.349	2.813	2.056	2.350	2.666	6.777	7.438	8.003
.2	.0	-2.589	-2.134	-1.778	1.979	2.399	2.901	3.049	3.265	3.464	6.153	6.452	6.745
.2	.1	-2.801	-2.256	-1.917	1.960	2.417	2.868	2.829	3.117	3.330	6.368	6.648	7.008
.2	.2	-3.138	-2.584	-2.158	1.940	2.379	2.861	2.580	2.875	3.087	6.473	6.945	7.383
.2	.3	-3.841	-2.978	-2.414	1.902	2.372	2.868	2.439	2.632	2.905	6.619	7.056	7.632
.2	.4	-5.653	-4.335	-3.488	1.849	2.377	2.925	1.996	2.306	2.567	6.826	7.363	8.009
.2	.5	-8.638	-6.271	-4.628	1.872	2.377	2.915	1.650	1.888	2.184	6.998	7.598	8.854
							$n = 40$						
.0	.0	-2.576	-2.089	-1.725	1.678	2.092	2.539	4.531	4.801	4.984	7.471	7.703	8.020
.0	.1	-2.618	-2.203	-1.814	1.697	2.080	2.459	4.384	4.650	4.856	7.591	7.845	8.218
.0	.2	-2.859	-2.432	-2.045	1.735	2.055	2.535	4.232	4.520	4.775	7.648	7.898	8.269
.0	.3	-3.344	-2.733	-2.229	1.754	2.068	2.533	4.028	4.408	4.647	7.764	8.123	8.594
.0	.4	-4.015	-3.250	-2.674	1.783	2.118	2.550	3.759	4.156	4.419	7.895	8.265	8.873
.0	.5	-4.958	-4.083	-3.343	1.853	2.268	2.640	3.544	3.881	4.234	8.069	8.453	9.004
.1	.0	-2.567	-2.100	-1.718	1.782	2.121	2.521	4.453	4.691	4.929	7.492	7.756	8.063
.1	.1	-2.645	-2.200	-1.808	1.770	2.115	2.492	4.256	4.510	4.754	7.641	7.916	8.225
.1	.2	-2.949	-2.463	-2.050	1.775	2.095	2.515	4.083	4.339	4.607	7.746	8.034	8.367
.1	.3	-3.530	-2.791	-2.285	1.800	2.097	2.443	3.871	4.150	4.432	7.923	8.302	8.644
.1	.4	-4.294	-3.445	-2.852	1.788	2.127	2.576	3.503	3.811	4.155	8.098	8.458	8.927
.1	.5	-5.518	-4.559	-3.597	1.846	2.246	2.639	3.115	3.476	3.864	8.221	8.609	9.258
.2	.0	-2.588	-2.100	-1.717	1.808	2.126	2.592	4.307	4.625	4.846	7.552	7.814	8.082
.2	.1	-2.662	-2.168	-1.813	1.798	2.135	2.574	3.981	4.336	4.629	7.762	8.059	8.391
.2	.2	-3.033	-2.472	-2.032	1.797	2.154	2.560	3.846	4.144	4.469	7.841	8.161	8.601
.2	.3	-3.704	-2.880	-2.305	1.810	2.127	2.543	3.575	3.932	4.232	8.058	8.515	8.949
.2	.4	-4.623	-3.728	-3.018	1.802	2.152	2.600	3.072	3.587	3.903	8.289	8.844	9.420
.2	.5	-6.751	-5.142	-4.163	1.844	2.260	2.619	2.620	3.047	3.469	8.439	9.328	9.876

TABLE 22.10
Simulated percentage points of the pivotal quantities $P_1 = \sqrt{n}(\hat{\mu} - \mu)/\hat{\sigma}$ and $P_2 = \sqrt{n}\hat{\sigma}/\sigma$

$$\kappa = 4.0$$

				P_1						P_2			
q_1	q_2	.010	.025	.05	.950	.975	.990	.010	.025	.05	.950	.975	.990
							$n = 20$						
.0	.0	-2.665	-2.155	-1.775	1.735	2.179	2.578	2.719	2.959	3.144	5.534	5.806	6.051
.0	.1	-2.978	-2.353	-1.890	1.748	2.171	2.548	2.571	2.836	3.032	5.655	5.948	6.319
.0	.2	-3.248	-2.586	-2.091	1.703	2.174	2.558	2.456	2.690	2.903	5.715	6.020	6.393
.0	.3	-3.873	-2.995	-2.398	1.728	2.178	2.554	2.274	2.557	2.748	5.899	6.243	6.637
.0	.4	-4.735	-3.654	-2.976	1.735	2.135	2.499	2.107	2.321	2.583	6.052	6.411	6.906
.0	.5	-5.942	-4.675	-3.811	1.802	2.181	2.546	1.763	2.103	2.352	6.133	6.642	7.273
.1	.0	-2.686	-2.157	-1.786	1.835	2.313	2.765	2.614	2.857	3.038	5.629	5.861	6.121
.1	.1	-2.973	-2.340	-1.892	1.826	2.297	2.741	2.459	2.719	2.918	5.772	6.074	6.348
.1	.2	-3.229	-2.630	-2.130	1.793	2.275	2.722	2.310	2.565	2.770	5.828	6.118	6.545
.1	.3	-4.015	-3.110	-2.446	1.781	2.266	2.686	2.162	2.353	2.596	5.921	6.307	6.693
.1	.4	-5.032	-3.945	-3.090	1.740	2.257	2.574	1.891	2.143	2.349	6.105	6.589	7.035
.1	.5	-7.046	-5.517	-4.274	1.802	2.196	2.630	1.492	1.733	2.064	6.375	6.890	7.409
.2	.0	-2.670	-2.169	-1.800	1.960	2.482	2.867	2.483	2.710	2.948	5.673	5.956	6.294
.2	.1	-2.984	-2.357	-1.885	1.954	2.473	2.849	2.265	2.591	2.810	5.842	6.203	6.549
.2	.2	-3.273	-2.619	-2.156	1.928	2.451	2.897	2.127	2.392	2.633	5.933	6.302	6.647
.2	.3	-4.140	-3.280	-2.569	1.923	2.431	2.872	1.939	2.187	2.408	6.101	6.480	6.941
.2	.4	-5.981	-4.486	-3.449	1.883	2.364	2.815	1.525	1.757	2.068	6.337	6.919	7.479
.2	.5	-9.965	-7.199	-5.205	1.855	2.345	2.848	1.052	1.393	1.642	6.596	7.213	8.164
							$n = 25$						
.0	.0	-2.492	-2.131	-1.739	1.762	2.140	2.551	3.228	3.501	3.705	6.075	6.316	6.605
.0	.1	-2.782	-2.265	-1.860	1.756	2.158	2.577	3.189	3.422	3.625	6.139	6.431	6.811
.0	.2	-3.204	-2.607	-2.077	1.730	2.101	2.585	2.918	3.198	3.426	6.251	6.638	7.042
.0	.3	-3.668	-2.972	-2.317	1.755	2.120	2.567	2.719	3.029	3.294	6.389	6.687	7.202
.0	.4	-4.455	-3.510	-2.863	1.779	2.130	2.712	2.515	2.767	3.058	6.533	6.873	7.535
.0	.5	-5.269	-4.101	-3.479	1.826	2.162	2.683	2.327	2.622	2.890	6.707	7.141	7.724
.1	.0	-2.498	-2.146	-1.737	1.827	2.175	2.611	3.204	3.436	3.650	6.098	6.355	6.657
.1	.1	-2.779	-2.263	-1.851	1.827	2.168	2.635	3.146	3.339	3.553	6.196	6.468	6.777
.1	.2	-3.218	-2.682	-2.065	1.811	2.154	2.594	2.765	3.071	3.340	6.316	6.711	7.053
.1	.3	-3.736	-2.964	-2.350	1.801	2.145	2.598	2.580	2.903	3.144	6.393	6.766	7.266
.1	.4	-4.895	-3.623	-2.890	1.794	2.158	2.693	2.347	2.614	2.922	6.608	7.048	7.531
.1	.5	-5.760	-4.416	-3.592	1.839	2.179	2.731	2.149	2.447	2.678	6.809	7.357	7.871
.2	.0	-2.484	-2.117	-1.744	1.968	2.344	2.930	3.068	3.287	3.514	6.166	6.400	6.696
.2	.1	-2.796	-2.260	-1.849	1.955	2.326	2.958	2.882	3.151	3.389	6.244	6.530	6.862
.2	.2	-3.316	-2.661	-2.086	1.942	2.303	2.971	2.632	2.868	3.120	6.406	6.756	7.272
.2	.3	-3.776	-3.017	-2.395	1.938	2.316	2.968	2.372	2.614	2.914	6.556	6.974	7.473
.2	.4	-5.558	-4.004	-3.141	1.927	2.290	2.961	1.966	2.246	2.539	6.716	7.288	7.938
.2	.5	-7.243	-5.499	-4.132	1.915	2.263	2.904	1.647	1.930	2.215	6.950	7.578	8.310
							$n = 40$						
.0	.0	-2.557	-2.087	-1.708	1.776	2.097	2.514	4.506	4.761	5.017	7.425	7.672	7.899
.0	.1	-2.717	-2.215	-1.817	1.790	2.063	2.516	4.384	4.634	4.861	7.529	7.854	8.217
.0	.2	-2.986	-2.462	-2.017	1.776	2.103	2.516	4.248	4.534	4.725	7.646	7.928	8.400
.0	.3	-3.200	-2.631	-2.168	1.767	2.149	2.454	4.076	4.361	4.597	7.799	8.127	8.478
.0	.4	-4.013	-3.221	-2.569	1.780	2.127	2.517	3.836	4.140	4.412	7.953	8.248	8.700
.0	.5	-4.863	-3.922	-3.122	1.914	2.238	2.629	3.536	3.845	4.165	8.109	8.506	8.942
.1	.0	-2.561	-2.087	-1.687	1.848	2.122	2.470	4.421	4.717	4.930	7.447	7.715	8.011
.1	.1	-2.740	-2.209	-1.808	1.840	2.117	2.490	4.292	4.526	4.784	7.593	7.881	8.307
.1	.2	-3.004	-2.468	-2.012	1.812	2.123	2.515	4.117	4.393	4.632	7.732	8.092	8.424
.1	.3	-3.239	-2.701	-2.218	1.816	2.136	2.518	3.837	4.142	4.425	7.872	8.218	8.658
.1	.4	-4.252	-3.303	-2.671	1.809	2.133	2.544	3.602	3.906	4.202	8.012	8.351	8.852
.1	.5	-5.532	-4.398	-3.442	1.923	2.263	2.628	3.228	3.536	3.875	8.248	8.633	9.288
.2	.0	-2.534	-2.089	-1.673	1.843	2.235	2.642	4.223	4.526	4.800	7.554	7.794	8.049
.2	.1	-2.746	-2.224	-1.799	1.847	2.221	2.601	4.052	4.321	4.602	7.700	8.045	8.425
.2	.2	-3.013	-2.533	-2.005	1.818	2.227	2.609	3.917	4.141	4.394	7.896	8.281	8.659
.2	.3	-3.284	-2.749	-2.272	1.865	2.221	2.616	3.554	3.822	4.152	8.086	8.459	8.891
.2	.4	-4.476	-3.486	-2.818	1.861	2.221	2.628	3.167	3.552	3.867	8.277	8.724	9.268
.2	.5	-6.520	-4.818	-3.808	1.937	2.298	2.655	2.697	3.041	3.432	8.664	9.281	9.942

TABLE 22.11

Simulated percentage points of the pivotal quantities $P_1 = \sqrt{n}(\hat{\mu} - \mu)/\hat{\sigma}$ and $P_2 = \sqrt{n}\hat{\sigma}/\sigma$

$$\kappa = 6.0$$

		P_1						P_2					
q_1	q_2	.010	.025	.05	.950	.975	.990	.010	.025	.05	.950	.975	.990
							$n = 20$						
.0	.0	-2.697	-2.198	-1.803	1.837	2.153	2.605	2.707	2.956	3.159	5.506	5.676	5.991
.0	.1	-2.864	-2.401	-1.923	1.812	2.138	2.601	2.555	2.821	3.032	5.575	5.890	6.132
.0	.2	-3.211	-2.558	-2.121	1.813	2.140	2.527	2.405	2.664	2.884	5.693	6.031	6.339
.0	.3	-4.137	-3.018	-2.311	1.804	2.122	2.558	2.213	2.475	2.760	5.790	6.084	6.521
.0	.4	-4.920	-3.713	-2.882	1.801	2.116	2.547	2.070	2.346	2.604	5.898	6.292	6.713
.0	.5	-6.217	-4.813	-3.878	1.807	2.199	2.608	1.759	2.076	2.349	6.123	6.581	7.011
.1	.0	-2.701	-2.198	-1.810	1.915	2.288	2.766	2.606	2.834	3.048	5.518	5.766	6.023
.1	.1	-2.862	-2.399	-1.970	1.901	2.283	2.727	2.443	2.704	2.914	5.647	5.903	6.271
.1	.2	-3.283	-2.606	-2.149	1.862	2.274	2.721	2.208	2.519	2.736	5.785	6.083	6.409
.1	.3	-4.131	-3.206	-2.413	1.884	2.215	2.678	2.047	2.314	2.527	5.929	6.283	6.636
.1	.4	-5.385	-4.148	-2.978	1.857	2.201	2.676	1.798	2.118	2.358	6.026	6.519	6.917
.1	.5	-7.788	-5.600	-4.360	1.829	2.260	2.646	1.490	1.748	2.033	6.276	6.844	7.334
.2	.0	-2.698	-2.214	-1.812	2.085	2.431	2.941	2.387	2.748	2.940	5.622	5.843	6.217
.2	.1	-2.836	-2.379	-1.990	2.073	2.448	2.957	2.263	2.535	2.784	5.719	6.039	6.395
.2	.2	-3.216	-2.625	-2.170	2.075	2.479	2.941	2.011	2.285	2.515	5.835	6.224	6.678
.2	.3	-4.594	-3.360	-2.485	2.046	2.460	2.964	1.817	2.059	2.320	6.025	6.498	6.932
.2	.4	-6.305	-4.361	-3.257	2.007	2.362	2.939	1.545	1.818	2.067	6.179	6.756	7.348
.2	.5	-10.703	-7.578	-5.439	1.917	2.379	2.879	1.054	1.332	1.596	6.574	7.243	8.144
							$n = 25$						
.0	.0	-2.635	-2.213	-1.798	1.774	2.127	2.580	3.334	3.537	3.710	6.048	6.295	6.559
.0	.1	-2.701	-2.283	-1.874	1.762	2.118	2.569	3.177	3.412	3.595	6.145	6.390	6.713
.0	.2	-3.079	-2.570	-2.112	1.742	2.088	2.562	3.009	3.253	3.410	6.238	6.530	6.754
.0	.3	-3.535	-2.824	-2.296	1.774	2.087	2.492	2.807	3.101	3.339	6.359	6.618	6.931
.0	.4	-4.483	-3.416	-2.717	1.726	2.147	2.492	2.686	2.918	3.136	6.482	6.766	7.178
.0	.5	-5.146	-4.129	-3.269	1.798	2.118	2.515	2.435	2.672	2.977	6.589	6.933	7.418
.1	.0	-2.624	-2.216	-1.817	1.844	2.202	2.605	3.186	3.461	3.606	6.133	6.378	6.699
.1	.1	-2.700	-2.275	-1.878	1.836	2.210	2.593	3.013	3.299	3.527	6.241	6.486	6.871
.1	.2	-3.158	-2.631	-2.114	1.813	2.208	2.556	2.806	3.094	3.306	6.359	6.661	6.976
.1	.3	-3.568	-2.910	-2.301	1.816	2.159	2.545	2.644	2.934	3.197	6.464	6.773	7.140
.1	.4	-4.655	-3.543	-2.814	1.762	2.179	2.554	2.418	2.690	2.960	6.599	7.029	7.472
.1	.5	-5.982	-4.533	-3.531	1.812	2.143	2.523	2.115	2.448	2.733	6.744	7.183	7.657
.2	.0	-2.683	-2.193	-1.815	1.947	2.400	2.807	3.016	3.282	3.455	6.141	6.445	6.803
.2	.1	-2.718	-2.276	-1.870	1.956	2.391	2.813	2.846	3.082	3.307	6.283	6.603	7.011
.2	.2	-3.140	-2.620	-2.131	1.946	2.386	2.795	2.501	2.815	3.099	6.476	6.825	7.199
.2	.3	-3.831	-3.124	-2.331	1.931	2.376	2.843	2.298	2.602	2.873	6.562	6.986	7.440
.2	.4	-5.578	-3.880	-3.057	1.872	2.287	2.831	1.972	2.287	2.547	6.832	7.326	7.912
.2	.5	-7.448	-5.205	-4.220	1.837	2.239	2.776	1.612	1.929	2.235	6.962	7.628	8.288
							$n = 40$						
.0	.0	-2.384	-2.019	-1.663	1.733	2.066	2.533	4.601	4.823	5.059	7.367	7.656	7.911
.0	.1	-2.547	-2.166	-1.709	1.711	2.059	2.522	4.377	4.692	4.934	7.469	7.809	8.097
.0	.2	-2.598	-2.237	-1.912	1.733	2.038	2.486	4.290	4.561	4.801	7.603	7.896	8.309
.0	.3	-3.221	-2.531	-2.148	1.715	2.069	2.529	4.083	4.396	4.640	7.677	8.021	8.454
.0	.4	-3.862	-3.055	-2.477	1.775	2.148	2.569	3.791	4.201	4.469	7.849	8.218	8.723
.0	.5	-4.855	-3.969	-3.092	1.801	2.208	2.585	3.542	3.987	4.258	8.057	8.482	9.012
.1	.0	-2.361	-2.040	-1.664	1.768	2.097	2.571	4.480	4.740	4.977	7.422	7.687	8.031
.1	.1	-2.554	-2.156	-1.705	1.762	2.082	2.577	4.320	4.563	4.825	7.560	7.859	8.186
.1	.2	-2.662	-2.234	-1.918	1.778	2.093	2.553	4.091	4.380	4.685	7.698	8.011	8.439
.1	.3	-3.250	-2.643	-2.163	1.750	2.128	2.470	3.858	4.200	4.490	7.860	8.170	8.636
.1	.4	-4.020	-3.130	-2.561	1.776	2.116	2.574	3.559	3.934	4.226	8.076	8.490	8.918
.1	.5	-5.138	-4.178	-3.333	1.809	2.221	2.568	3.219	3.555	3.928	8.353	8.791	9.508
.2	.0	-2.349	-2.015	-1.660	1.837	2.178	2.685	4.331	4.595	4.862	7.543	7.772	8.132
.2	.1	-2.553	-2.155	-1.708	1.826	2.179	2.708	4.097	4.426	4.678	7.686	8.010	8.255
.2	.2	-2.634	-2.238	-1.913	1.838	2.177	2.678	3.834	4.165	4.499	7.823	8.223	8.590
.2	.3	-3.377	-2.638	-2.198	1.806	2.169	2.668	3.618	3.898	4.225	8.022	8.370	8.960
.2	.4	-4.289	-3.304	-2.697	1.796	2.202	2.662	3.213	3.568	3.921	8.191	8.666	9.335
.2	.5	-6.255	-4.790	-3.713	1.815	2.227	2.589	2.752	3.047	3.435	8.646	9.227	9.980

TABLE 22.12

Simulated percentage points of the pivotal quantities $P_1 = \sqrt{n}(\hat{\mu} - \mu)/\hat{\sigma}$ and $P_2 = \sqrt{n}\hat{\sigma}/\sigma$

$$\kappa = 8.0$$

				P_1						P_2			
q_1	q_2	.010	.025	.05	.950	.975	.990	.010	.025	.05	.950	.975	.990
							$n = 20$						
.0	.0	-2.584	-2.163	-1.811	1.699	2.019	2.477	2.767	2.987	3.173	5.539	5.771	6.004
.0	.1	-2.894	-2.351	-1.925	1.671	2.019	2.460	2.605	2.851	3.026	5.628	5.894	6.130
.0	.2	-3.055	-2.569	-2.117	1.669	2.031	2.421	2.426	2.676	2.904	5.755	6.047	6.341
.0	.3	-3.885	-2.935	-2.390	1.674	2.011	2.372	2.246	2.484	2.725	5.814	6.091	6.459
.0	.4	-4.673	-3.697	-2.894	1.666	1.995	2.392	2.090	2.330	2.558	5.947	6.344	6.722
.0	.5	-6.379	-4.938	-3.799	1.685	2.005	2.511	1.816	2.085	2.336	6.080	6.464	7.006
.1	.0	-2.632	-2.163	-1.789	1.750	2.121	2.513	2.679	2.918	3.097	5.571	5.810	6.158
.1	.1	-2.947	-2.340	-1.926	1.741	2.144	2.502	2.447	2.721	2.937	5.671	5.947	6.197
.1	.2	-3.266	-2.619	-2.179	1.715	2.123	2.511	2.274	2.528	2.758	5.818	6.150	6.463
.1	.3	-4.021	-3.018	-2.487	1.711	2.086	2.512	2.056	2.303	2.552	5.863	6.234	6.648
.1	.4	-5.118	-3.836	-3.095	1.726	2.083	2.483	1.703	2.054	2.333	6.071	6.545	6.873
.1	.5	-7.869	-6.010	-4.537	1.696	2.044	2.545	1.357	1.711	1.984	6.166	6.608	7.399
.2	.0	-2.649	-2.156	-1.797	1.864	2.302	2.850	2.447	2.737	2.989	5.624	5.841	6.258
.2	.1	-2.924	-2.329	-1.915	1.878	2.312	2.866	2.203	2.526	2.763	5.752	6.085	6.338
.2	.2	-3.378	-2.628	-2.194	1.865	2.318	2.899	1.972	2.281	2.551	5.872	6.216	6.591
.2	.3	-4.271	-3.235	-2.589	1.833	2.301	2.897	1.807	2.023	2.323	6.026	6.376	6.871
.2	.4	-5.841	-4.386	-3.468	1.833	2.289	2.886	1.429	1.705	1.992	6.299	6.761	7.355
.2	.5	-10.432	-7.494	-5.622	1.776	2.220	2.888	1.001	1.264	1.522	6.509	6.973	7.805
							$n = 25$						
.0	.0	-2.526	-2.114	-1.798	1.661	2.065	2.529	3.217	3.449	3.678	6.089	6.278	6.578
.0	.1	-2.707	-2.227	-1.890	1.621	2.073	2.563	3.166	3.357	3.572	6.153	6.453	6.687
.0	.2	-2.974	-2.597	-2.062	1.639	2.022	2.501	2.909	3.158	3.394	6.223	6.627	7.001
.0	.3	-3.388	-2.749	-2.239	1.646	2.075	2.549	2.792	3.042	3.293	6.346	6.733	7.148
.0	.4	-4.127	-3.352	-2.725	1.666	2.059	2.607	2.638	2.856	3.072	6.425	6.830	7.265
.0	.5	-5.100	-4.072	-3.243	1.725	2.108	2.606	2.404	2.697	2.950	6.611	7.024	7.562
.1	.0	-2.523	-2.116	-1.806	1.664	2.122	2.577	3.177	3.408	3.587	6.133	6.368	6.634
.1	.1	-2.716	-2.231	-1.894	1.643	2.105	2.593	3.017	3.289	3.504	6.212	6.502	6.809
.1	.2	-3.024	-2.570	-2.059	1.630	2.072	2.580	2.792	3.042	3.289	6.324	6.659	7.120
.1	.3	-3.514	-2.782	-2.246	1.661	2.067	2.558	2.610	2.926	3.151	6.424	6.860	7.328
.1	.4	-4.275	-3.478	-2.777	1.675	2.059	2.597	2.357	2.679	2.915	6.574	6.929	7.516
.1	.5	-5.625	-4.320	-3.505	1.717	2.112	2.610	2.140	2.441	2.733	6.764	7.293	7.808
.2	.0	-2.545	-2.096	-1.803	1.834	2.228	2.842	3.037	3.294	3.512	6.193	6.520	6.835
.2	.1	-2.706	-2.238	-1.875	1.835	2.224	2.833	2.902	3.162	3.382	6.287	6.629	6.978
.2	.2	-3.130	-2.589	-2.072	1.818	2.220	2.831	2.559	2.847	3.111	6.456	6.852	7.259
.2	.3	-3.681	-2.813	-2.288	1.804	2.246	2.829	2.390	2.653	2.919	6.578	6.992	7.453
.2	.4	-4.947	-3.721	-2.988	1.795	2.219	2.807	2.018	2.309	2.628	6.717	7.330	8.050
.2	.5	-6.826	-5.378	-4.001	1.768	2.234	2.811	1.737	2.038	2.307	7.070	7.601	8.520
							$n = 40$						
.0	.0	-2.540	-2.082	-1.769	1.638	1.991	2.411	4.590	4.815	5.050	7.444	7.633	7.940
.0	.1	-2.591	-2.245	-1.818	1.665	2.001	2.338	4.503	4.722	4.951	7.526	7.769	8.039
.0	.2	-2.898	-2.431	-2.023	1.664	1.975	2.320	4.285	4.575	4.822	7.626	7.854	8.125
.0	.3	-3.062	-2.636	-2.218	1.701	1.968	2.314	4.070	4.390	4.623	7.738	8.060	8.386
.0	.4	-3.780	-3.142	-2.598	1.744	2.026	2.346	3.829	4.146	4.407	7.867	8.210	8.571
.0	.5	-4.720	-3.851	-3.152	1.774	2.065	2.466	3.523	3.874	4.196	8.055	8.422	8.856
.1	.0	-2.553	-2.096	-1.761	1.719	2.040	2.426	4.451	4.694	4.980	7.465	7.725	8.004
.1	.1	-2.611	-2.244	-1.806	1.710	2.039	2.406	4.312	4.596	4.842	7.548	7.896	8.115
.1	.2	-2.923	-2.424	-2.018	1.701	2.002	2.368	4.148	4.398	4.675	7.708	7.951	8.309
.1	.3	-3.107	-2.634	-2.239	1.742	2.018	2.392	3.845	4.248	4.499	7.871	8.129	8.473
.1	.4	-3.956	-3.268	-2.659	1.748	2.032	2.388	3.601	3.884	4.231	8.005	8.375	8.788
.1	.5	-5.006	-4.104	-3.361	1.783	2.069	2.506	3.258	3.567	3.915	8.257	8.632	9.096
.2	.0	-2.565	-2.116	-1.766	1.837	2.097	2.483	4.295	4.612	4.864	7.567	7.846	8.145
.2	.1	-2.607	-2.216	-1.810	1.833	2.083	2.461	4.172	4.445	4.728	7.713	8.022	8.433
.2	.2	-3.009	-2.439	-2.011	1.823	2.088	2.448	3.954	4.250	4.520	7.863	8.224	8.685
.2	.3	-3.149	-2.710	-2.282	1.814	2.067	2.443	3.604	3.970	4.293	8.036	8.445	8.939
.2	.4	-4.074	-3.389	-2.759	1.813	2.072	2.447	3.271	3.624	3.942	8.257	8.773	9.324
.2	.5	-5.586	-4.562	-3.688	1.796	2.107	2.539	2.742	3.123	3.510	8.592	9.128	9.901

TABLE 22.13

Simulated percentage points of the pivotal quantities $P_1 = \sqrt{n}(\hat{\mu} - \mu)/\hat{\sigma}$ and $P_2 = \sqrt{n}\hat{\sigma}/\sigma$

$$\kappa = 10.0$$

		P_1						P_2					
q_1	q_2	.010	.025	.05	.950	.975	.990	.010	.025	.05	.950	.975	.990
							$n = 20$						
.0	.0	-2.796	-2.216	-1.799	1.771	2.115	2.534	2.737	2.994	3.162	5.530	5.779	6.101
.0	.1	-2.872	-2.225	-1.849	1.756	2.105	2.522	2.530	2.825	3.075	5.653	5.880	6.220
.0	.2	-3.175	-2.505	-2.029	1.760	2.112	2.538	2.354	2.690	2.920	5.755	6.065	6.430
.0	.3	-3.511	-2.936	-2.373	1.759	2.116	2.593	2.242	2.508	2.762	5.879	6.233	6.603
.0	.4	-4.351	-3.511	-2.865	1.728	2.117	2.624	2.009	2.280	2.578	6.008	6.439	6.827
.0	.5	-6.028	-4.544	-3.525	1.807	2.143	2.572	1.773	2.143	2.397	6.167	6.589	7.095
.1	.0	-2.771	-2.213	-1.787	1.840	2.293	2.739	2.636	2.914	3.084	5.578	5.878	6.241
.1	.1	-2.902	-2.218	-1.869	1.827	2.331	2.777	2.424	2.684	2.950	5.687	6.028	6.432
.1	.2	-3.445	-2.545	-2.059	1.821	2.316	2.811	2.229	2.492	2.751	5.835	6.197	6.690
.1	.3	-3.871	-3.013	-2.385	1.818	2.326	2.782	1.975	2.329	2.612	5.901	6.437	6.901
.1	.4	-4.777	-3.873	-3.046	1.794	2.268	2.787	1.774	2.032	2.329	6.026	6.653	7.204
.1	.5	-7.480	-5.318	-4.065	1.833	2.230	2.612	1.426	1.759	2.028	6.275	6.819	7.612
.2	.0	-2.748	-2.219	-1.787	2.001	2.328	2.960	2.566	2.828	3.019	5.650	5.903	6.288
.2	.1	-2.927	-2.219	-1.858	1.981	2.363	3.000	2.374	2.580	2.824	5.762	6.111	6.613
.2	.2	-3.553	-2.564	-2.056	1.983	2.393	2.984	2.020	2.350	2.605	5.912	6.310	6.793
.2	.3	-4.155	-3.059	-2.473	1.964	2.386	3.008	1.847	2.072	2.342	6.165	6.550	7.151
.2	.4	-6.108	-4.289	-3.175	1.926	2.381	2.976	1.416	1.761	2.054	6.267	6.823	7.547
.2	.5	-8.817	-6.629	-5.029	1.943	2.297	2.864	1.089	1.374	1.674	6.604	7.155	7.846
							$n = 25$						
.0	.0	-2.550	-2.104	-1.764	1.790	2.074	2.422	3.267	3.496	3.713	6.046	6.314	6.562
.0	.1	-2.759	-2.196	-1.842	1.771	2.060	2.431	3.164	3.400	3.607	6.120	6.375	6.669
.0	.2	-3.043	-2.418	-1.940	1.776	2.054	2.399	2.963	3.233	3.461	6.239	6.500	6.855
.0	.3	-3.412	-2.731	-2.234	1.759	2.064	2.410	2.852	3.091	3.358	6.306	6.597	7.015
.0	.4	-4.058	-3.106	-2.591	1.741	2.105	2.445	2.584	2.921	3.158	6.434	6.828	7.230
.0	.5	-4.718	-3.795	-3.113	1.753	2.148	2.479	2.472	2.713	2.984	6.566	6.908	7.481
.1	.0	-2.577	-2.102	-1.745	1.838	2.138	2.481	3.242	3.445	3.646	6.114	6.338	6.658
.1	.1	-2.693	-2.210	-1.845	1.823	2.141	2.471	3.092	3.350	3.539	6.175	6.432	6.782
.1	.2	-3.080	-2.430	-1.974	1.808	2.144	2.458	2.872	3.133	3.346	6.307	6.595	7.012
.1	.3	-3.368	-2.786	-2.263	1.816	2.119	2.460	2.725	2.968	3.241	6.421	6.727	7.235
.1	.4	-4.343	-3.328	-2.669	1.782	2.143	2.438	2.415	2.718	2.934	6.534	7.004	7.518
.1	.5	-5.408	-4.157	-3.353	1.795	2.188	2.490	2.162	2.460	2.702	6.650	7.212	7.857
.2	.0	-2.555	-2.070	-1.756	1.979	2.362	2.820	3.081	3.329	3.515	6.165	6.478	6.854
.2	.1	-2.780	-2.205	-1.847	2.002	2.343	2.776	2.846	3.161	3.366	6.254	6.581	7.030
.2	.2	-3.164	-2.435	-1.986	1.993	2.344	2.850	2.648	2.880	3.150	6.413	6.746	7.260
.2	.3	-3.710	-2.899	-2.358	1.974	2.319	2.816	2.370	2.727	2.976	6.501	6.972	7.536
.2	.4	-4.790	-3.771	-2.916	1.928	2.306	2.794	2.024	2.339	2.620	6.792	7.245	7.790
.2	.5	-6.579	-5.107	-3.866	1.890	2.299	2.785	1.677	1.986	2.317	6.922	7.504	8.376
							$n = 40$						
.0	.0	-2.358	-2.036	-1.712	1.671	2.028	2.403	4.592	4.834	5.033	7.398	7.616	7.959
.0	.1	-2.577	-2.080	-1.758	1.660	2.038	2.407	4.388	4.667	4.940	7.492	7.789	8.038
.0	.2	-2.747	-2.271	-1.920	1.670	2.041	2.457	4.234	4.534	4.781	7.593	7.923	8.296
.0	.3	-3.299	-2.572	-2.106	1.700	1.995	2.469	4.036	4.378	4.616	7.728	8.084	8.425
.0	.4	-3.730	-3.043	-2.490	1.688	2.069	2.519	3.862	4.124	4.409	7.845	8.206	8.629
.0	.5	-4.852	-3.850	-3.042	1.801	2.135	2.564	3.570	3.871	4.145	7.997	8.377	8.913
.1	.0	-2.337	-2.038	-1.698	1.705	2.062	2.500	4.479	4.731	4.947	7.446	7.670	8.000
.1	.1	-2.556	-2.080	-1.764	1.718	2.062	2.490	4.328	4.563	4.805	7.574	7.860	8.178
.1	.2	-2.778	-2.294	-1.925	1.712	2.051	2.475	4.054	4.414	4.669	7.716	8.065	8.478
.1	.3	-3.365	-2.646	-2.136	1.703	2.011	2.517	3.869	4.177	4.484	7.879	8.191	8.641
.1	.4	-3.745	-3.145	-2.570	1.706	2.071	2.569	3.554	3.899	4.212	8.031	8.514	8.870
.1	.5	-5.048	-4.189	-3.306	1.799	2.148	2.566	3.147	3.567	3.838	8.294	8.792	9.298
.2	.0	-2.413	-2.025	-1.714	1.799	2.191	2.683	4.350	4.601	4.807	7.558	7.801	8.048
.2	.1	-2.564	-2.064	-1.750	1.801	2.185	2.693	4.133	4.406	4.632	7.646	8.038	8.386
.2	.2	-2.810	-2.306	-1.920	1.772	2.162	2.694	3.833	4.178	4.428	7.798	8.209	8.705
.2	.3	-3.405	-2.668	-2.169	1.780	2.165	2.674	3.443	3.796	4.149	8.045	8.442	8.869
.2	.4	-3.925	-3.249	-2.579	1.748	2.122	2.707	3.145	3.534	3.833	8.262	8.785	9.247
.2	.5	-5.795	-4.654	-3.759	1.808	2.186	2.661	2.589	2.955	3.340	8.585	9.217	9.952

TABLE 22.14
Comparison of finite-sample (n=40) and asymptotic percentage points of the pivotal quantities $P_3 = \sqrt{n}(\hat{\mu} - \mu)/\hat{\sigma}\sqrt{V_{11}}$ and $P_4 = \sqrt{n}(\hat{\sigma} - \sigma)/\sigma\sqrt{V_{22}}$ when

$$\kappa = 4$$

q_1	q_2	.010	.025	.050	.950	.975	.990	.010	.025	.050	.950	.975	.990
				P_3						P_4			
0.0	0.0	-2.518	-2.056	-1.682	1.750	2.066	2.476	-2.497	-2.147	-1.795	1.511	1.850	2.161
0.0	0.1	-2.639	-2.152	-1.765	1.739	2.004	2.444	-2.407	-2.097	-1.816	1.494	1.897	2.347
0.0	0.2	-2.818	-2.324	-1.904	1.676	1.985	2.375	-2.352	-2.028	-1.811	1.497	1.816	2.351
0.0	0.3	-2.875	-2.363	-1.947	1.587	1.931	2.205	-2.319	-2.025	-1.781	1.521	1.858	2.220
0.0	0.4	-3.331	-2.674	-2.132	1.478	1.765	2.089	-2.319	-2.036	-1.783	1.518	1.793	2.214
0.0	0.5	-3.578	-2.886	-2.297	1.408	1.647	1.934	-2.320	-2.063	-1.797	1.484	1.814	2.178
0.1	0.0	-2.507	-2.043	-1.652	1.810	2.078	2.419	-2.479	-2.093	-1.816	1.461	1.810	2.196
0.1	0.1	-2.653	-2.138	-1.750	1.781	2.049	2.410	-2.366	-2.093	-1.793	1.476	1.812	2.307
0.1	0.2	-2.832	-2.327	-1.897	1.708	2.001	2.371	-2.319	-2.029	-1.778	1.478	1.856	2.205
0.1	0.3	-2.909	-2.426	-1.992	1.631	1.919	2.261	-2.347	-2.060	-1.793	1.460	1.787	2.202
0.1	0.4	-3.518	-2.733	-2.210	1.497	1.765	2.105	-2.281	-2.026	-1.778	1.414	1.697	2.118
0.1	0.5	-4.009	-3.187	-2.494	1.393	1.640	1.905	-3.259	-3.034	-1.787	1.403	1.684	3.162
0.2	0.0	-2.449	-2.019	-1.616	1.781	2.160	2.553	-2.564	-2.194	-1.860	1.500	1.793	2.104
0.2	0.1	-2.633	-2.132	-1.725	1.771	2.130	2.494	-2.446	-2.155	-1.854	1.480	1.851	2.260
0.2	0.2	-2.826	-2.377	-1.881	1.706	2.089	2.448	-2.306	-2.092	-1.848	1.505	1.874	2.235
0.2	0.3	-2.947	-2.467	-2.039	1.673	1.994	2.348	-2.341	-2.114	-1.835	1.488	1.803	2.168
0.2	0.4	-3.698	-2.880	-2.328	1.537	1.835	2.171	-2.311	-2.029	-1.799	1.429	1.756	2.155
0.2	0.5	-4.647	-3.434	-2.714	1.381	1.638	1.893	-2.228	-2.017	-1.777	1.437	1.816	2.222
		-2.326	-1.960	-1.645	1.645	1.960	2.326	-2.326	-1.960	-1.645	1.645	1.960	2.326

TABLE 22.15
The maximum likelihood estimates of μ and σ based on complete sample

κ	$\hat{\mu}(s.e.(\hat{\mu}))$	$\hat{\sigma}(s.e.(\hat{\sigma}))$
3.0	4.300(0.106)	0.498(0.075)
5.0	4.266(0.106)	0.504(0.076)
8.0	4.242(0.107)	0.518(0.076)
10.6	4.230(0.107)	0.510(0.076)
20.0	4.208(0.107)	0.513(0.076)

TABLE 22.16
The maximum likelihood estimates of μ and σ based on censored sample

κ	$\hat{\mu}(s.e.(\hat{\mu}))$	$\hat{\sigma}(s.e.(\hat{\sigma}))$
3.0	4.302(0.108)	0.503(0.083)
5.0	4.270(0.109)	0.512(0.084)
8.0	4.247(0.110)	0.518(0.084)
10.6	4.236(0.110)	0.521(0.084)
20.0	4.215(0.111)	0.526(0.084)

TABLE 22.17
The 95% confidence intervals for μ and σ

	C.I. for μ		C.I. for σ	
	complete sample			
κ	Uncond.	Cond.	Uncond.	Cond.
3	[4.075, 4.527]	[4.073, 4.523]	[0.384, 0.729]	[0.391, 0.726]
4	[4.053, 4.502]	[4.054, 4.504]	[0.390, 0.734]	[0.394, 0.730]
6	[4.028, 4.488]	[4.029, 4.481]	[0.397, 0.730]	[0.398, 0.735]
8	[4.008, 4.473]	[4.015, 4.468]	[0.395, 0.733]	[0.400, 0.738]
10	[4.003, 4.463]	[4.005, 4.459]	[0.404, 0.737]	[0.401, 0.739]
12	[3.995, 4.444]	[3.998, 4.452]	[0.403, 0.730]	[0.402, 0.741]
	cenored sample with $s = 2$			
3	[4.079, 4.544]	[4.075, 4.542]	[0.380, 0.763]	[0.387, 0.763]
4	[4.057, 4.523]	[4.055, 4.525]	[0.390, 0.768]	[0.392, 0.769]
6	[4.029, 4.512]	[4.031, 4.491]	[0.400, 0.770]	[0.398, 0.776]
8	[4.010, 4.495]	[4.018, 4.491]	[0.398, 0.784]	[0.401, 0.781]
10	[4.003, 4.489]	[4.007, 4.482]	[0.406, 0.771]	[0.403, 0.786]
12	[3.999, 4.473]	[4.000, 4.476]	[0.410, 0.776]	[0.405, 0.786]

23

Maximum Likelihood Estimation for the Three-parameter Log-gamma Distribution Under Type-II Censoring

N. Balakrishnan and P.S. Chan

McMaster University
The Chinese University of Hong Kong

ABSTRACT In this paper, we study the maximum likelihood estimation of three parameters of the log-gamma distribution under Type-II censored samples. The score function of the shape parameter and the observed Fisher information are presented. The expected Fisher information matrix is derived through which the asymptotic variances and covariances of the MLEs are tabulated for various proportions of censoring. Finally, a life-test data in [27], and also analysed in [25], [26], and a simulated data set are used to illustrate the method of estimation developed in this paper.

Keywords: Log-gamma distribution; Type-II censored sample; Shape parameter; Maximum likelihood estimation; Asymptotic variance-covariance matrix

CONTENTS

23.1 Introduction

Let X be a log-gamma random variable with probability density function

$$f^*(x) = \frac{1}{\Gamma(\kappa)} e^{\kappa x - e^x}, \qquad -\infty < x < \infty, \ \kappa > 0 \tag{23.1.1}$$

and cumulative distribution function

$$F^*(x) = I_{e^x}(\kappa), \qquad -\infty < x < \infty, \ \kappa > 0, \tag{23.1.2}$$

where $I_t(\kappa)$ is the incomplete gamma function defined by

$$I_t(\kappa) = \int_0^t \frac{1}{\Gamma(\kappa)} e^{-z} z^{\kappa-1} dz, \qquad 0 < t < \infty, \ \kappa > 0.$$

The special case of $\kappa = 1$ is the extreme value distribution which has been discussed in great detail by many authors including [28], [2], [3], [26], [29], [12], and [7].

0-8493-8972-0/95/$0.00+$.50

Note that, for positive integral values of κ, the cumulative distribution function $F^*(x)$ in (23.1.2) can be written as

$$F^*(x) = \sum_{i=\kappa}^{\infty} e^{-e^x} e^{ix}/i!, \qquad -\infty < x < \infty, \ \kappa = 1, 2, \ldots,$$

or, equivalently,

$$1 - F^*(x) = \sum_{i=0}^{\kappa-1} e^{-e^x} e^{ix}/i!, \qquad -\infty < x < \infty, \ \kappa = 1, 2, \ldots . \qquad (23.1.3)$$

From (23.1.1), we derive the moment generating function of the log-gamma random variable X to be

$$M_X(t) = E(e^{tX}) = \frac{\Gamma(\kappa + t)}{\Gamma(\kappa)}$$

from which we immediately obtain the mean and variance of X to be

$$E(X) = \frac{\Gamma'(\kappa)}{\Gamma(\kappa)} = \psi(\kappa)$$

and

$$\mathrm{Var}(X) = \frac{\Gamma''(\kappa)}{\Gamma(\kappa)} - \left(\frac{\Gamma'(\kappa)}{\Gamma(\kappa)}\right)^2 = \psi'(\kappa),$$

where

$$\psi(z) = \frac{d}{dz} \ln \Gamma(z) = \frac{\Gamma'(z)}{\Gamma(z)}$$

is the digamma function, and

$$\psi'(z) = \frac{d}{dz} \psi(z)$$

is the trigamma function.

As a matter of fact, by using the asymptotic formula that $\psi(\kappa) \sim \log \kappa$ and $\psi'(\kappa) \sim 1/\kappa$ (see [1]), in [30], the author suggested a reparametrized model of the density function in (23.1.1) as

$$f(x) = \frac{\kappa^{\kappa - \frac{1}{2}}}{\Gamma(\kappa)} e^{\sqrt{\kappa}x - \kappa e^x/\sqrt{\kappa}}, \qquad -\infty < x < \infty, \ \kappa > 0, \qquad (23.1.4)$$

and showed the asymptotic normality (as $\kappa \to \infty$) of this family. It can be easily seen that the cumulative distribution function for the density function in (23.1.4) is given by

$$F(x) = I_{\kappa e^x/\sqrt{\kappa}}(\kappa), \qquad -\infty < x < \infty, \ \kappa > 0 . \qquad (23.1.5)$$

Let Y be a log-gamma random variable with probability density function

$$g(y) = \frac{\kappa^{\kappa - \frac{1}{2}}}{\sigma \Gamma(\kappa)} \exp\left(\sqrt{\kappa}\left(\frac{y - \mu}{\sigma}\right) - \kappa e^{(\frac{y-\mu}{\sigma})}/\sqrt{\kappa}\right),$$

$$-\infty < y < \infty, -\infty < \mu < \infty, \sigma > 0, \ \kappa > 0, \qquad (23.1.6)$$

where μ is the location parameter, σ is the scale parameter, and κ is the shape parameter. Let $G(y)$ be the corresponding cumulative distribution function. In [25] and [26], the author has illustrated the usefulness of the log-gamma model in (23.1.6) as a life-test model and discussed the maximum likelihood estimation of the parameters; see also [30]. The authors in [15] suggested the use of the relative maximum likelihood method for estimating the shape parameter of this distribution. Recently, in [31], the authors discussed the log-gamma regression model. One may also refer to [25] and [14] for some valuable work on the inference for a related generalized gamma distribution.

The exact best linear unbiased estimation and the asymptotic best linear unbiased estimation of the parameters μ and σ, based on doubly Type II censored samples, have been discussed recently in [4], and [5]. The maximum likelihood estimation of μ and σ (with shape parameter κ known) based on doubly Type-II censored samples has also been studied by [6]. In this paper, we will discuss the maximum likelihood estimation of all three parameters of (23.1.6) based on doubly Type-II censored samples. We first present the likelihood equations for κ, μ and σ. Next, we present the second derivatives of the log-likelihood function with respect to all three parameters so that the Newton-Raphson method can be used to obtain the estimates. In Section 23.3, the asymptotic variance-covariance matrix of these estimators are derived through the expected Fisher information matrix. Finally, we consider the life-test data set given by [27] and a simulated data set to illustrate the estimation method discussed in preceding sections.

It is important to mention here that work of this nature has been carried out in [16], [17], [18], [19], [20], [21], [22], [23], [24], [2], [8], [9], [10], [11], and [12] for a wide range of distributions.

23.2   Maximum Likelihood Estimation

Consider a random sample of size n from a log-gamma population with density function as in (23.1.6). Let $Y_{r+1:n} \leq Y_{r+2:n} \leq \cdots \leq Y_{n-s:n}$ be the ordered observations remaining when r smallest and s largest observations have been censored. The likelihood function for the given Type-II censored sample is

$$L = \frac{n!}{r!s!} [G(Y_{r+1:n})]^r [1 - G(Y_{n-s:n})]^s \prod_{i=r+1}^{n-s} g(Y_{i:n}) \tag{23.2.1}$$

or equivalently,

$$L = \frac{n!}{r!s!} \frac{1}{\sigma^A} [F(X_{r+1:n})]^r [1 - F(X_{n-s:n})]^s \prod_{i=r+1}^{n-s} f(X_{i:n}) , \tag{23.2.2}$$

where $A = n - r - s$ and $X = (Y - \mu)/\sigma$ is the standardized variable with density function $f(\cdot)$ and distribution function $F(\cdot)$ as given in (23.1.4) and (23.1.5), respectively.

The log-likelihood function is then given by

$$\log L = \text{const} - A \log \sigma + r \log [F(X_{r+1:n})]$$

$$+ s \log [1 - F(X_{n-s:n})] + \sum_{i=r+1}^{n-s} \log f(X_{i:n}) . \tag{23.2.3}$$

The derivatives of the log-likelihood function with respect to μ and σ, respectively, have been given in [6] as follows:

$$\frac{\partial \log L}{\partial \mu} = -\frac{1}{\sigma} \left\{ r \frac{f(X_{r+1:n})}{F(X_{r+1:n})} - s \frac{f(X_{n-s:n})}{1 - F(X_{n-s:n})} \right.$$

$$+ A\sqrt{\kappa} - \sqrt{\kappa} \sum_{i=r+1}^{n-s} e^{X_{i:n}/\sqrt{\kappa}} \Bigg\} \qquad (23.2.4)$$

and

$$\frac{\partial \log L}{\partial \sigma} = -\frac{1}{\sigma} \left\{ A + r X_{r+1:n} \frac{f(X_{r+1:n})}{F(X_{r+1:n})} - s X_{n-s:n} \frac{f(X_{n-s:n})}{1 - F(X_{n-s:n})} \right.$$

$$\left. + \sqrt{\kappa} \sum_{i=r+1}^{n-s} X_{i:n} - \sqrt{\kappa} \sum_{i=r+1}^{n-s} X_{i:n} e^{X_{i:n}/\sqrt{\kappa}} \right\}. \qquad (23.2.5)$$

Now, upon differentiating the log-likelihood function in (23.2.3) with respect to κ, we get

$$\frac{\partial \log L}{\partial \kappa} = \frac{r}{F(X_{r+1:n})} \int_{-\infty}^{X_{r+1:n}} \frac{\partial f}{\partial \kappa} dt$$

$$+ \frac{s}{1 - F(X_{n-s:n})} \int_{X_{n-s:n}}^{\infty} \frac{\partial f}{\partial \kappa} dt + \sum_{i=r+1}^{n-s} \frac{\partial \log f(X_{i:n})}{\partial \kappa}. \qquad (23.2.6)$$

Since

$$\frac{\partial f}{\partial \kappa} = \frac{\partial \log f}{\partial \kappa} f$$

and, for the log-gamma density in (23.1.4),

$$\frac{\partial \log f(t)}{\partial \kappa} = \log \kappa + 1 - \frac{1}{2\kappa} - \psi(\kappa) + \frac{t}{2\sqrt{\kappa}} - e^{t/\sqrt{\kappa}} + \frac{t}{2\sqrt{\kappa}} e^{t/\sqrt{\kappa}},$$

Eq. (23.2.6) gives

$$\frac{\partial \log L}{\partial \kappa} = n \left[\log \kappa + 1 - \frac{1}{2\kappa} - \psi(\kappa) \right] + \frac{r}{F(X_{r+1:n})} \int_{-\infty}^{X_{r+1:n}} h(t) f(t) dt$$

$$+ \frac{s}{1 - F(X_{n-s:n})} \int_{X_{n-s:n}}^{\infty} h(t) f(t) dt$$

$$+ \sum_{i=r+1}^{n-s} \left[\frac{X_{i:n}}{2\sqrt{\kappa}} - e^{X_{i:n}/\sqrt{\kappa}} + \frac{X_{i:n}}{2\sqrt{\kappa}} e^{X_{i:n}/\sqrt{\kappa}} \right] \qquad (23.2.7)$$

where

$$h(t) = \frac{t}{2\sqrt{\kappa}} - e^{t/\sqrt{\kappa}} + \frac{t}{2\sqrt{\kappa}} e^{t/\sqrt{\kappa}}.$$

The MLEs $\hat{\kappa}$, $\hat{\mu}$ and $\hat{\sigma}$ can be obtained from (23.2.7), (23.2.4) and (23.2.5) upon solving the equations $\partial \log L/\partial \kappa = 0$, $\partial \log L/\partial \mu = 0$ and $\partial \log L/\partial \sigma = 0$. Since these three equations cannot be solved analytically, some numerical method must be employed. Newton-Raphson method is found to be appropriate here. In order to use the Newton-Raphson method, we need the second derivatives of the log-likelihood function.

The second derivative of the log-likelihood function with respect to κ is

$$\frac{\partial^2 \log L}{\partial \kappa^2} = n \left[\frac{1}{\kappa} + \frac{1}{2\kappa^2} - \psi'(\kappa) \right]$$

$$- \frac{r}{[F(X_{r+1:n})]^2} \left\{ \int_{-\infty}^{X_{r+1:n}} \frac{\partial \log f}{\partial \kappa} f(t) dt \right\} \times \left\{ \int_{-\infty}^{X_{r+1:n}} h(t) f(t) dt \right\}$$

$$-\frac{r}{F(X_{r+1:n})}\int_{-\infty}^{X_{r+1:n}}h'(t)f(t)dt$$

$$+\frac{r}{F(X_{r+1:n})}\int_{-\infty}^{X_{r+1:n}}h(t)\frac{\partial\log f}{\partial\kappa}f(t)dt$$

$$-\frac{s}{[1-F(X_{n-s:n})]^2}\left\{\int_{X_{n-s:n}}^{\infty}\frac{\partial\log f}{\partial\kappa}f(t)dt\right\}\times\left\{\int_{X_{n-s:n}}^{\infty}h(t)f(t)dt\right\}$$

$$-\frac{s}{1-F(X_{n-s:n})}\int_{X_{n-s:n}}^{\infty}h'(t)f(t)dt$$

$$+\frac{s}{1-F(X_{n-s:n})}\int_{X_{n-s:n}}^{\infty}h(t)\frac{\partial\log f}{\partial\kappa}f(t)dt \tag{23.2.8}$$

where

$$h'(t)=\frac{dh(t)}{dt}=\frac{t}{4\kappa\sqrt{\kappa}}-\frac{t}{4\kappa\sqrt{\kappa}}e^{t/\sqrt{\kappa}}-\frac{t^2}{4\kappa^2}e^{t/\sqrt{\kappa}}.$$

The second term on the RHS of (23.2.8) can be rewritten as

$$\frac{r}{[F(X_{r+1:n})]^2}\left\{\int_{-\infty}^{X_{r+1:n}}\frac{\partial\log f}{\partial\kappa}f(t)dt\right\}\left\{\int_{-\infty}^{X_{r+1:n}}h(t)f(t)dt\right\}$$

$$=\frac{r}{F(X_{r+1:n})}\left[\log\kappa+1-\frac{1}{2\kappa}-\psi(\kappa)\right]\int_{-\infty}^{X_{r+1:n}}h(t)f(t)dt$$

$$+\frac{r}{[F(X_{r+1:n})]^2}\left\{\int_{-\infty}^{X_{r+1:n}}h(t)f(t)dt\right\}^2. \tag{23.2.9}$$

The fourth term on the RHS of (23.2.8) can be written as

$$\frac{r}{F(X_{r+1:n})}\int_{-\infty}^{X_{r+1:n}}h(t)\frac{\partial\log f}{\partial\kappa}f(t)dt$$

$$=\frac{r}{F(X_{r+1:n})}\left[\log\kappa+1-\frac{1}{2\kappa}-\psi(\kappa)\right]\int_{-\infty}^{X_{r+1:n}}h(t)f(t)dt$$

$$+\frac{r}{F(X_{r+1:n})}\int_{-\infty}^{X_{r+1:n}}[h(t)]^2f(t)dt. \tag{23.2.10}$$

Using the expressions in (23.2.9) and (23.2.10), we obtain

$$-\frac{r}{[F(X_{r+1:n})]^2}\left\{\int_{-\infty}^{X_{r+1:n}}\frac{\partial\log f}{\partial\kappa}f(t)dt\right\}\left\{\int_{-\infty}^{X_{r+1:n}}h(t)f(t)dt\right\}$$

$$-\frac{r}{F(X_{r+1:n})}\int_{-\infty}^{X_{r+1:n}}h(t)f(t)dt+\frac{r}{F(X_{r+1:n})}\int_{-\infty}^{X_{r+1:n}}h(t)\frac{\partial\log f}{\partial\kappa}f(t)dt$$

$$=-\frac{r}{F(X_{r+1:n})}\int_{-\infty}^{X_{r+1:n}}h'(t)f(t)dt-\frac{r}{[F(X_{r+1:n})]^2}\left\{\int_{-\infty}^{X_{r+1:n}}h(t)f(t)dt\right\}^2$$

$$+\frac{r}{F(X_{r+1:n})}\int_{-\infty}^{X_{r+1:n}}[h(t)]^2f(t)dt. \tag{23.2.11}$$

Similar argument applies to those terms involving $X_{n-s:n}$ in (23.2.8). Finally we get, upon simplification,

$$\frac{\partial^2 \log L}{\partial \kappa^2} = n \left[\frac{1}{\kappa} + \frac{1}{2\kappa^2} - \psi'(\kappa) \right]$$

$$- \sum_{i=r+1}^{n-s} \left[\frac{X_{i:n}}{4\kappa\sqrt{\kappa}} - \frac{X_{i:n}}{4\kappa\sqrt{\kappa}} e^{X_{i:n}/\sqrt{\kappa}} + \frac{X_{i:n}^2}{4\kappa^2} e^{X_{i:n}/\sqrt{\kappa}} \right]$$

$$- \frac{r}{F(X_{r+1:n})} \int_{-\infty}^{X_{r+1:n}} h'(t)f(t)dt - \frac{r}{[F(X_{r+1:n})]^2} \left\{ \int_{-\infty}^{X_{r+1:n}} h(t)f(t)dt \right\}^2$$

$$+ \frac{r}{F(X_{r+1:n})} \int_{-\infty}^{X_{r+1:n}} (h(t))^2 f(t)dt - \frac{s}{1 - F(X_{n-s:n})} \int_{X_{n-s:n}}^{\infty} h'(t)f(t)dt$$

$$- \frac{s}{[1 - F(X_{n-s:n})]^2} \left\{ \int_{X_{n-s:n}}^{\infty} h(t)f(t)dt \right\}^2$$

$$+ \frac{s}{1 - F(X_{r+1:n})} \int_{X_{n-s:n}}^{\infty} (h(t))^2 f(t)dt. \qquad (23.2.12)$$

The second derivatives of the log-likelihood function with respect to μ and σ have been given in [6] as follows:

$$\frac{\partial^2 \log L}{\partial \mu^2} = \frac{1}{\sigma^2} \left\{ r \frac{f(X_{r+1:n})}{F(X_{r+1:n})} \left[\sqrt{\kappa} - \sqrt{\kappa} e^{X_{r+1:n}/\sqrt{\kappa}} - \frac{f(X_{r+1:n})}{F(X_{r+1:n})} \right] \right.$$

$$- s \frac{f(X_{n-s:n})}{1 - F(X_{n-s:n})} \left[\sqrt{\kappa} - \sqrt{\kappa} e^{X_{n-s:n}/\sqrt{\kappa}} + \frac{f(X_{n-s:n})}{1 - F(X_{n-s:n})} \right]$$

$$\left. - \sum_{i=r+1}^{n-s} e^{X_{i:n}/\sqrt{\kappa}} \right\}, \qquad (23.2.13)$$

$$\frac{\partial^2 \log L}{\partial \mu \partial \sigma} = \frac{1}{\sigma^2} \left\{ r \frac{f(X_{r+1:n})}{F(X_{r+1:n})} + r X_{r+1:n} \frac{f(X_{r+1:n})}{F(X_{r+1:n})} \right.$$

$$\times \left[\sqrt{\kappa} - \sqrt{\kappa} e^{X_{r+1:n}/\sqrt{\kappa}} - \frac{f(X_{r+1:n})}{F(X_{r+1:n})} \right]$$

$$- s \frac{f(X_{n-s:n})}{1 - F(X_{n-s:n})} - s X_{n-s:n} \frac{f(X_{n-s:n})}{1 - F(X_{n-s:n})}$$

$$\times \left[\sqrt{\kappa} - \sqrt{\kappa} e^{X_{n-s:n}/\sqrt{\kappa}} + \frac{f(X_{n-s:n})}{1 - F(X_{n-s:n})} \right]$$

$$\left. + A\sqrt{\kappa} - \sqrt{\kappa} \sum_{i=r+1}^{n-s} e^{X_{i:n}/\sqrt{\kappa}} - \sum_{i=r+1}^{n-s} X_{i:n} e^{X_{i:n}/\sqrt{\kappa}} \right\}, \qquad (23.2.14)$$

$$\frac{\partial^2 \log L}{\partial \sigma^2} = \frac{1}{\sigma^2} \left\{ A + 2r X_{r+1:n} \frac{f(X_{r+1:n})}{F(X_{r+1:n})} \right.$$

$$+ r X_{r+1:n}^2 \frac{f(X_{r+1:n})}{F(X_{r+1:n})} \left[\sqrt{\kappa} - \sqrt{\kappa} e^{X_{r+1:n}/\sqrt{\kappa}} - \frac{f(X_{r+1:n})}{F(X_{r+1:n})} \right]$$

$$-2sX_{n-s:n}\frac{f(X_{n-s:n})}{1-F(X_{n-s:n})} - sX_{n-s:n}^2\frac{f(X_{n-s:n})}{1-F(X_{n-s:n})}$$

$$\times\left[\sqrt{\kappa} - \sqrt{\kappa}e^{X_{n-s:n}/\sqrt{\kappa}} + \frac{f(X_{n-s:n})}{1-F(X_{n-s:n})}\right]$$

$$+2\sqrt{\kappa}\sum_{i=r+1}^{n-s}X_{i:n} - 2\sqrt{\kappa}\sum_{i=r+1}^{n-s}X_{i:n}e^{X_{i:n}/\sqrt{\kappa}} - \sum_{i=r+1}^{n-s}X_{i:n}^2e^{X_{i:n}/\sqrt{\kappa}}\bigg\}.$$

$$(23.2.15)$$

The cross derivatives of the log-likelihood function with respect to κ and μ and κ and σ are similarly given by

$$\frac{\partial^2 \log L}{\partial\kappa\partial\mu} = -\frac{1}{\sigma}\left\{-\frac{rf(X_{r+1:n})}{[F(X_{r+1:n})]^2}\int_{-\infty}^{X_{r+1:n}}h(t)f(t)dt + \frac{rf(X_{r+1:n})}{F(X_{r+1:n})}h(X_{r+1:n})\right.$$

$$+\frac{sf(X_{n-s:n})}{[1-F(X_{n-s:n})]^2}\int_{X_{n-s:n}}^{\infty}h(t)f(t)dt - \frac{sf(X_{n-s:n})}{1-F(X_{n-s:n})}h(X_{n-s:n})$$

$$+\sum_{i=r+1}^{n-s}\left[\frac{1}{2\sqrt{\kappa}} - \frac{1}{2\sqrt{\kappa}}e^{X_{i:n}/\sqrt{\kappa}} + \frac{X_{i:n}}{2\kappa}e^{X_{i:n}/\sqrt{\kappa}}\right]\bigg\}. \qquad (23.2.16)$$

and

$$\frac{\partial^2 \log L}{\partial\kappa\partial\sigma} = -\frac{1}{\sigma}\left\{\frac{rX_{r+1:n}f(X_{r+1:n})}{[F(X_{r+1:n})]^2}\int_{-\infty}^{X_{r+1:n}}h(t)f(t)dt\right.$$

$$+\frac{rX_{r+1:n}f(X_{r+1:n})}{F(X_{r+1:n})}h(X_{r+1:n})$$

$$+\frac{sX_{n-s:n}f(X_{n-s:n})}{[1-F(X_{n-s:n})]^2}\int_{X_{n-s:n}}^{\infty}h(t)f(t)dt$$

$$-\frac{sX_{n-s:n}f(X_{n-s:n})}{1-F(X_{n-s:n})}h(X_{n-s:n})$$

$$+\sum_{i=r+1}^{n-s}\left[\frac{X_{i:n}}{2\sqrt{\kappa}} - \frac{X_{i:n}}{2\sqrt{\kappa}}e^{X_{i:n}/\sqrt{\kappa}} + \frac{X_{i:n}^2}{2\kappa}e^{X_{i:n}/\sqrt{\kappa}}\right]. \qquad (23.2.17)$$

Solving the three equations (23.2.7), (23.2.4) and (23.2.5) by using Newton-Raphson method is a straightforward problem (see also [26], Appendix F). A slightly different approach has been used by us here (as compared to the one suggested in [26]) as it was found to be computionally more efficient. The procedure is as follows:

1. Obtain an initial guess of κ.
2. For given κ, compute the MLEs of μ and σ.
3. Substitute these estimates into (23.2.7) and obtain the next κ.
4. Repeat steps (2) and (3) until desired accuracy of $\hat{\kappa}$ is achieved.

23.3   Asymptotic Variances and Covariances

As $n \to \infty$, $r/n \to q_1$, $s/n \to q_2$, $E(X_{r+1:n}) \to \xi_{q_1}$ where $F(\xi_{q_1}) = q_1$, $E(X_{n-s:n}) \to \xi_{q_2}$ where $1 - F(\xi_{q_2}) = q_2$, and

$$\sum_{i=r+1}^{n-s} \left[\frac{X_{i:n}}{4\kappa\sqrt{\kappa}} - \frac{X_{i:n}}{4\kappa\sqrt{\kappa}} e^{X_{i:n}/\sqrt{\kappa}} + \frac{X_{i:n}^2}{4\kappa^2} e^{X_{i:n}/\sqrt{\kappa}} \right]$$

$$\to n \int_{\xi_{q_1}}^{\xi_{q_2}} \left[\frac{t}{4\kappa\sqrt{\kappa}} - \frac{t}{4\kappa\sqrt{\kappa}} e^{t/\sqrt{\kappa}} + \frac{t^2}{4\kappa^2} e^{t/\sqrt{\kappa}} \right] f(t)dt.$$

Hence,

$$\lim_{n\to\infty} \frac{1}{n} E \left(-\frac{\partial^2 \log L}{\partial \kappa^2} \right) = \left[\psi'(\kappa) - \frac{1}{2\kappa^2} - \frac{1}{\kappa} \right]$$

$$+ \frac{1}{4\kappa} \int_{-\infty}^{\infty} \left[\frac{t}{\sqrt{\kappa}} - \frac{t}{\sqrt{\kappa}} e^{t/\sqrt{\kappa}} + \frac{t^2}{\kappa} e^{t/\sqrt{\kappa}} \right] f(t)dt$$

$$- \int_{-\infty}^{\xi_{q_1}} (h(t))^2 f(t)dt + \frac{1}{q_1} \left\{ \int_{-\infty}^{\xi_{q_1}} h(t)f(t)dt \right\}^2$$

$$- \int_{\xi_{q_2}}^{\infty} (h(t))^2 f(t)dt + \frac{1}{q_2} \left\{ \int_{\xi_{q_2}}^{\infty} h(t)f(t)dt \right\}^2 . \quad (23.3.1)$$

Since,

$$\int_{-\infty}^{\infty} \left[\frac{t}{\sqrt{\kappa}} - \frac{t}{\sqrt{\kappa}} e^{t/\sqrt{\kappa}} + \frac{t^2}{\kappa} e^{t/\sqrt{\kappa}} \right] f(t)dt$$

$$= [\psi(\kappa) - \log \kappa] + [\psi'(\kappa + 1) + \psi^2(\kappa + 1)$$

$$+ (\log \kappa)^2 - 2\log \kappa \psi(\kappa + 1)] - [\psi(\kappa + 1) - \log \kappa]$$

$$= \psi'(\kappa) + \psi^2(\kappa) + 2\left(\frac{1}{\kappa} - \log \kappa \right) \psi(\kappa) - \frac{1}{\kappa}(2\log \kappa + 1) + (\log \kappa)^2 , \quad (23.3.2)$$

we can write

$$\lim_{n\to\infty} \frac{1}{n} E \left(-\frac{\partial^2 \log L}{\partial \kappa^2} \right) = C(\kappa) - L_l(\kappa, q_1) - L_r(\kappa, q_2)$$

$$= V^{33} , \quad (23.3.3)$$

where

$$C(\kappa) = \left[\psi'(\kappa) - \frac{1}{2\kappa^2} - \frac{1}{\kappa} \right] + \frac{1}{4\kappa} \left[\psi'(\kappa) + \psi^2(\kappa) \right.$$

$$\left. + 2(\frac{1}{\kappa} - \log \kappa)\psi(\kappa) - \frac{1}{\kappa}(2\log \kappa + 1) + (\log \kappa)^2 \right] , \quad (23.3.4)$$

$$L_l(\kappa, q_1) = \int_{-\infty}^{\xi_{q_1}} (h(t))^2 f(t)dt - \frac{1}{q_1} \left\{ \int_{-\infty}^{\xi_{q_1}} h(t)f(t)dt \right\}^2 , \quad (23.3.5)$$

and

$$L_r(\kappa, q_2) = \int_{\xi_{q_2}}^{\infty} (h(t))^2 f(t)dt - \frac{1}{q_2}\left\{\int_{\xi_{q_2}}^{\infty} h(t)f(t)dt\right\}^2 . \tag{23.3.6}$$

Similarly, we obtain

$$\lim_{n\to\infty} -\frac{\sigma}{n}E\left(\frac{\partial^2 \log L}{\partial\kappa\partial\mu}\right) = -\frac{1}{q_1}f(\xi_{q_1})\int_{-\infty}^{\xi_{q_1}} h(t)f(t)dt + f(\xi_{q_1})h(\xi_{q_1})$$

$$+\frac{1}{q_2}f(\xi_{q_2})\int_{\xi_{q_2}}^{\infty} h(t)f(t)dt - f(\xi_{q_2})h(\xi_{q_2})$$

$$+\frac{1}{2\sqrt{\kappa}}(1 - q_1 - q_2)$$

$$+\frac{1}{2\sqrt{\kappa}}\int_{\xi_{q_1}}^{\xi_{q_2}}\left[\frac{t}{\sqrt{\kappa}}e^{t/\sqrt{\kappa}} - e^{t/\sqrt{\kappa}}\right]f(t)dt$$

$$= V^{13} , \tag{23.3.7}$$

and

$$\lim_{n\to\infty} -\frac{\sigma}{n}E\left(\frac{\partial^2 \log L}{\partial\kappa\partial\sigma}\right) = -\frac{\xi_{q_1}}{q_1}f(\xi_{q_1})\int_{-\infty}^{\xi_{q_1}} h(t)f(t)dt + \xi_{q_1}f(\xi_{q_1})h(\xi_{q_1})$$

$$+\frac{\xi_{q_2}}{q_2}f(\xi_{q_2})\int_{\xi_{q_2}}^{\infty} h(t)f(t)dt - \xi_{q_2}f(\xi_{q_2})h(\xi_{q_2})$$

$$+\frac{1}{2}\int_{\xi_{q_1}}^{\xi_{q_2}} h(t)f(t)dt$$

$$= V^{23} . \tag{23.3.8}$$

Similar expressions for V^{11}, V^{22} and V^{12}, as given by [6], are

$$\lim_{n\to\infty}\left(-\frac{\sigma^2}{n}E\frac{\partial^2 \log L}{\partial\mu^2}\right) = -f(\xi_{q_1})\left[\sqrt{\kappa} - \sqrt{\kappa}e^{\xi_{q_1}/\sqrt{\kappa}} - \frac{f(\xi_{q_1})}{q_1}\right]$$

$$+f(\xi_{q_2})\left[\sqrt{\kappa} - \sqrt{\kappa}e^{\xi_{q_2}/\sqrt{\kappa}} + \frac{f(\xi_{q_2})}{q_2}\right]$$

$$+\int_{\xi_{q_1}}^{\xi_{q_2}} e^{t/\sqrt{\kappa}}f(t)dt$$

$$= V^{11} , \tag{23.3.9}$$

$$\lim_{n\to\infty}\left(-\frac{\sigma^2}{n}E\frac{\partial^2 \log L}{\partial\mu\partial\sigma}\right) = -p\sqrt{\kappa} - f(\xi_{q_1}) + f(\xi_{q_2})$$

$$-\xi_{q_1}f(\xi_{q_1})\left[\sqrt{\kappa} - \sqrt{\kappa}e^{\xi_{q_1}/\sqrt{\kappa}} - \frac{f(\xi_{q_1})}{q_1}\right]$$

$$+\xi_{q_2}f(\xi_{q_2})\left[\sqrt{\kappa} - \sqrt{\kappa}e^{\xi_{q_2}/\sqrt{\kappa}} + \frac{f(\xi_{q_2})}{q_2}\right]$$

$$+ \int_{\xi_{q_1}}^{\xi_{q_2}} e^{t/\sqrt{\kappa}} f(t)dt + \int_{\xi_{q_1}}^{\xi_{q_2}} t e^{t/\sqrt{\kappa}} f(t)dt$$

$$= V^{12} , \tag{23.3.10}$$

and

$$\lim_{n \to \infty} \left(-\frac{\sigma^2}{n} E \frac{\partial^2 \log L}{\partial \sigma^2} \right) = -p - 2\xi_{q_1} f(\xi_{q_1}) + 2\xi_{q_2} f(\xi_{q_2})$$

$$-\xi_{q_1}^2 f(\xi_{q_1}) \left[\sqrt{\kappa} - \sqrt{\kappa} e^{\xi_{q_1}/\sqrt{\kappa}} - \frac{f(\xi_{q_1})}{q_1} \right]$$

$$+\xi_{q_2}^2 f(\xi_{q_2}) \left[\sqrt{\kappa} - \sqrt{\kappa} e^{\xi_{q_2}/\sqrt{\kappa}} + \frac{f(\xi_{q_2})}{q_2} \right]$$

$$-2\sqrt{\kappa} \int_{\xi_{q_1}}^{\xi_{q_2}} t f(t)dt + 2\sqrt{\kappa} \int_{\xi_{q_1}}^{\xi_{q_2}} t e^{t/\sqrt{\kappa}} f(t)dt$$

$$+ \int_{\xi_{q_1}}^{\xi_{q_2}} t^2 e^{t/\sqrt{\kappa}} f(t)dt$$

$$= V^{22} . \tag{23.3.11}$$

These values of V^{11}, V^{22}, V^{33}, V^{12}, V^{13} and V^{23} are presented in Tables 23.1 and 23.2 for various choices of censoring proportions. The asymptotic variance-covariance matrix of the MLEs $\hat{\mu}$, $\hat{\sigma}$ and $\hat{\kappa}$ is simply

$$\frac{n}{\sigma^2} \text{Var} \begin{pmatrix} \hat{\mu} \\ \hat{\sigma} \\ \hat{\kappa} \end{pmatrix} = \begin{pmatrix} V_{11} & V_{12} & V_{13} \\ V_{12} & V_{22} & V_{23} \\ V_{13} & V_{23} & V_{33} \end{pmatrix} = \begin{pmatrix} V^{11} & V^{12} & \sigma V^{13} \\ V^{12} & V^{22} & \sigma V^{23} \\ \sigma V^{13} & \sigma V^{23} & \sigma^2 V^{33} \end{pmatrix}^{-1}$$

23.4   Illustrative Examples

Example 1

Let us consider here the examples in [27] and [25], [26] which gives the number of million revolutions to failure for each of a group of 23 ball bearings in a fatigue test.

The 23 log lifetimes are

$$\begin{array}{llllll}
2.884, & 3.365, & 3.497, & 3.726, & 3.741, & 3.820, \\
3.881, & 3.948, & 3.950, & 3.991, & 4.017, & 4.217, \\
4.229, & 4.229, & 4.232, & 4.432, & 4.534, & 4.591, \\
4.655, & 4.662, & 4.851, & 4.852, & 5.156. &
\end{array}$$

It is of interest to mention here that in [27] assumed a two-parameter Weibull distribution for the original data and hence an extreme value distribution for the log-lifetime data given above. In [25] and [26], the author assumed a generalized gamma distribution for the original data and thus a log-gamma distribution for the log-lifetime. He concluded that the data fit well under this model with the shape parameter $\kappa > 0.4$.

Now, following the same assumptions made in [25] and [26], we fit the data into the three parameter log-gamma model and obtained the MLEs and their standard errors (obtained from the observed information matrix) to be

$$\hat{\kappa} = 10.596 \ , \quad \text{S.E.}(\hat{\kappa}) = 2.206,$$
$$\hat{\mu} = 4.230 \ , \quad \text{S.E.}(\hat{\mu}) = 0.107,$$
$$\hat{\sigma} = 0.510 \ , \quad \text{S.E.}(\hat{\sigma}) = 0.076.$$

These estimates agree with the results in [25] and [26] where the authors determined the estimates by the relative maximum likelihood method. If we consider the censored sample for this data set, for example, with the two largest observations censored, then the normal distribution ($\kappa \to \infty$) gives the best fit for the data set among the family of distributions in (23.1.6). ▯

Example 2

For the purpose of illustration, we simulated a data set from the log-gamma distribution with $\kappa = 3$, $\mu = 5$ and $\sigma = 3$, which is as follows:

$$
\begin{array}{ccccc}
-1.9974, & -1.6676, & -0.4294, & 1.3450, & 1.6295, \\
1.7602, & 1.7948, & 1.8090, & 2.5278, & 2.6610, \\
2.7733, & 2.9744, & 2.9868, & 3.0018, & 3.3436, \\
3.4982, & 3.8417, & 4.1522, & 4.4195, & 4.4853, \\
4.5749, & 4.7166, & 4.4853, & 5.3266, & 5.6463, \\
5.9627, & 6.2254, & 6.2864, & 6.3596, & 6.4510, \\
6.4602, & 6.8362, & 6.9387, & 7.4741, & 7.5384, \\
7.6678, & 7.8987, & 8.1963, & 8.9105, & 9.1561
\end{array}
$$

By considering the complete as well as some right-censored samples, we determined the MLEs and their standard errors. These results are presented in the following table.

Maximum Likelihood Estimates for the Simulated Data

Obs. right-censored	$\hat{\kappa}$	$\hat{\mu}$	$\hat{\sigma}$
0	3.217(0.507)	5.198(0.410)	2.521(0.294)
2	4.601(0.726)	5.086(0.418)	2.580(0.312)
3	5.800(0.915)	5.030(0.424)	2.619(0.322)
4	7.919(1.255)	4.967(0.432)	2.668(0.333)

▯

Acknowledgements

The first author would like to thank the Natural Sciences and Engineering Research Council of Canada for funding this research.

References

1. Abramowitz, M. and Stegun, I.A. (Eds.) (1965), *Handbook of Mathematical Functions with Formulas, Graphs, and Mathematical Tables*, Dover Publications, New York.

2. Bain, L.J. (1972), Inferences based on censored sampling from the Weibull or extreme-value distribution, *Technometrics*, **14**, 693–702.

3. Bain, L.J. (1978), *Statistical Analysis of Reliability and Life-testing Models–Theory and Practice*, Marcel Dekker, New York.

4. Balakrishnan, N. and Chan, P.S. (1994), Log-gamma order statistics and linear estimation of parameters, *Computational Statistic & Data Analysis* (To appear).

5. Balakrishnan, N. and Chan, P.S. (1994), Asymptotic best linear unbiased estimation for the log-gamma distribution, *Sankhyā, Series B* (To appear).

6. Balakrishnan, N. and Chan, P.S. (1994), Maximum likelihood estimation for the log-gamma distribution under Type-II censored samples and associated inference, *In this volume*.

7. Balakrishnan, N. and Cohen, A.C. (1991), *Order Statistics and Inference: Estimation Methods*, Academic Press, San Diego.

8. Cohen, A.C. (1951), Estimating parameters of logarithmic normal distributions by maximum likelihood, *Journal of the American Statistical Association*, **46**, 206-212.

9. Cohen, A.C. (1955), Maximum likelihood estimation of the dispersion parameter of a chi-distributed radial error from truncated and censored samples with applications to target analysis, *Journal of the American Statistical Association*, **50**, 884-893.

10. Cohen, A.C. (1959), Simplified estimators for the normal distribution when samples are singly censored or truncated, *Technometrics*, **1**, 217-237.

11. Cohen, A.C. (1965), Maximum likelihood estimation in the Weibull distribution based on complete and on censored samples, *Technometrics*, **7**, 579-588.

12. Cohen, A.C. and Whitten, B.J. (1980), Estimation in the three-parameter lognormal distribution, *Journal American Statistical Association*, **75**, 399-404.

13. Cohen, A.C. and Whitten, B.J. (1988), *Parameter Estiamtion in Reliability and Life Span Models*, Marcel Dekker, New York.

14. DiCiccio, T.J. (1987), Approximate inference for the generalized gamma distribution, *Technometrics*, **29**, 32-39.

15. Farewell, V.T. and Prentice, R.L. (1977), A study of distributional shape in life testing, *Technometrics*, **19**, 69-75.

16. Harter, H.L. (1967), Maximum-likelihood estimation of the parameters of a four parameter generalized gamma population from complete and censored samples, *Technometrics*, **9**, 159-165.

17. Harter, H.L. (1970), *Order Statistics and their Use in Testing and Estimation*, Vol. 2, U.S. Government Printing Office, Washington, D.C.

18. Harter, H.L. and Moore, A.H. (1965), Point and interval estimators, based on m order statistics, for the scale parameter of a Weibull population with known shape parameter, *Technometrics*, **7**, 405-422.

19. Harter, H.L. and Moore, A.H. (1965), Maximum-likelihood estimation of the parameters of gamma and Weibull populations from complete and from censored samples,

Technometrics, **7**, 639-643.

20. Harter, H.L. and Moore, A.H. (1966), Local-maximum likelihood estimation of the parameters of three-parameter lognormal populations from complete and censored samples, *Journal of the American Statistical Association*, **61**, 842-855.

21. Harter, H.L. and Moore, A.H. (1966), Iterative maximum-likelihood estimation of the parameters of normal populations from singly and doubly censored samples, *Biometrika*, **53**, 205-213.

22. Harter, H.L. and Moore, A.H. (1967), Asymptotic variances and covariances of maximum-likelihood estimators, from censored samples, of the parameters of Weibull and gamma populations, *Annals of Mathematical Statistics*, **38**, 557-570.

23. Harter, H.L. and Moore, A.H. (1967b), Maximum likelihood estimation, from censored samples, of the parameters of a logistic distribution, *Journal of the American Statistical Association*, **62**, 675-684.

24. Harter, H.L. and Moore, A.H. (1968), Maximum likelihood estimation, from doubly censored samples, of the parameters of the first asymptotic distribution of extreme values, *Journal of the American Statistical Association*, **63**, 889-901.

25. Lawless, J.F. (1980), Inference in the generalized gamma and log-gamma distribution, *Technometrics*, **22**, 67-82.

26. Lawless, J.F. (1982), *Statistical Models & Methods For Lifetime Data*, John Wiley & Sons, New York.

27. Lieblein, J. and Zelen, M. (1956), Statistical investigation of the fatigue life of deep groove ball bearings, *Journal of Research of National Bureau of Standards*, **57**, 273-316.

28. Mann, N.R., Schafer, R.E. and Singpurwalla, N.D. (1974), *Methods for Statistical Analysis of Reliability and Life Data*, John Wiley & Sons, New York.

29. Nelson, W. (1982), *Applied Life Data Analysis*, John Wiley & Sons, New York.

30. Prentice, R.L. (1974), A log-gamma model and its maximum likelihood estimation, *Biometrika*, **61**, 539-544.

31. Young, D.H. and Bakir, S.T. (1987), Bias correction for a generalized log-gamma regression model, *Technometrics*, **29**, 183-191.

TABLE 23.1
Values of V^{ij} for various proportions of censoring

$$q_1 = 0.0$$

| | $q_2 = 0.0$ | | | | | |
κ	V^{11}	V^{22}	V^{33}	V^{12}	V^{13}	V^{23}
0.5	1.0000001	1.7335861	0.6683886	0.5159315	0.5159345	-0.2664115
1.0	1.0000000	1.8236809	0.1008542	0.4227844	0.2113931	-0.0881596
1.5	1.0000000	1.8727876	0.0312985	0.3645961	0.1215327	-0.0438437
2.0	1.0000001	1.8953346	0.0133925	0.3247559	0.0811896	-0.0261660
3.0	1.0000002	1.9258927	0.0039866	0.2728073	0.0454680	-0.0123506
4.0	1.0000004	1.9427230	0.0016780	0.2396468	0.0299554	-0.0071607
6.0	1.0000004	1.9606611	0.0004942	0.1984702	0.0165386	-0.0032778
8.0	1.0000001	1.9700623	0.0002076	0.1730991	0.0108186	-0.0018716
10.0	0.9999994	1.9758335	0.0001059	0.1554808	0.0077739	-0.0012078
12.0	0.9999999	1.9797515	0.0000611	0.1423339	0.0059304	-0.0008440
	$q_2 = 0.1$					
0.5	0.8469548	1.2168899	0.5868417	0.2368289	0.4134781	-0.4585211
1.0	0.9000032	1.3513100	0.0890996	0.2069933	0.1780764	-0.1621230
1.5	0.9204613	1.4218307	0.0273734	0.1774937	0.1043913	-0.0853868
2.0	0.9316354	1.4659097	0.0116135	0.1547024	0.0705022	-0.0535813
3.0	0.9438385	1.5186669	0.0034121	0.1227607	0.0399650	-0.0275215
4.0	0.9505447	1.5496202	0.0014222	0.1013418	0.0265087	-0.0170998
6.0	0.9579356	1.5851392	0.0004148	0.0738454	0.0147494	-0.0087403
8.0	0.9620407	1.6054173	0.0001715	0.0564732	0.0096364	-0.0055299
10.0	0.9647122	1.6187707	0.0000844	0.0442352	0.0070061	-0.0037276
12.0	0.9666156	1.6283264	0.0000454	0.0350189	0.0053437	-0.0027908
	$q_2 = 0.2$					
0.5	0.7115470	0.9548941	0.5654768	0.0497803	0.3594608	-0.5330899
1.0	0.8000021	1.0796177	0.0851163	0.0424811	0.1582469	-0.1948962
1.5	0.8360103	1.1493113	0.0259835	0.0260817	0.0936230	-0.1048299
2.0	0.8561950	1.1944059	0.0109669	0.0118632	0.0635633	-0.0668119
3.0	0.8787095	1.2498850	0.0031972	-0.0092883	0.0362572	-0.0351001
4.0	0.8913238	1.2832537	0.0013245	-0.0240004	0.0241357	-0.0221617
6.0	0.9054545	1.3224058	0.0003801	-0.0433285	0.0134764	-0.0115950
8.0	0.9134312	1.3452722	0.0001590	-0.0557461	0.0088599	-0.0073502
10.0	0.9186721	1.3605155	0.0000759	-0.0645618	0.0063774	-0.0051856
12.0	0.9224368	1.3716316	0.0000430	-0.0712471	0.0049231	-0.0038195

TABLE 23.2
Values of V^{ij} for various proportions of censoring

$$q_1 = 0.0$$

	$q_2 = 0.3$					
κ	V^{11}	V^{22}	V^{33}	V^{12}	V^{13}	V^{23}
0.5	0.5880740	0.8069553	0.5630185	-0.0861584	0.3250130	-0.5729850
1.0	0.7000003	0.9025642	0.0831750	-0.0903190	0.1443623	-0.2135279
1.5	0.7479448	0.9631187	0.0252782	-0.1017279	0.0857784	-0.1162826
2.0	0.7754514	1.0038033	0.0106320	-0.1119624	0.0583883	-0.0747959
3.0	0.8067095	1.0551630	0.0030831	-0.1274736	0.0334013	-0.0398085
4.0	0.8245167	1.0866870	0.0012730	-0.1383869	0.0222875	-0.0253536
6.0	0.8447360	1.1242714	0.0003660	-0.1528171	0.0124756	-0.0134149
8.0	0.8562862	1.1465111	0.0001520	-0.1621227	0.0082303	-0.0085397
10.0	0.8639401	1.1614933	0.0000738	-0.1687436	0.0059150	-0.0060672
12.0	0.8694672	1.1724396	0.0000392	-0.1737676	0.0045657	-0.0045105
	$q_2 = 0.4$					
0.5	0.4750694	0.7250530	0.5573572	-0.1819010	0.3012624	-0.5922272
1.0	0.5999997	0.7919512	0.0821310	-0.1951966	0.1341760	-0.2242975
1.5	0.6564119	0.8403503	0.0248897	-0.2074626	0.0798380	-0.1231816
2.0	0.6895463	0.8742913	0.0104445	-0.2171831	0.0543922	-0.0797203
3.0	0.7278970	0.9182880	0.0030185	-0.2310927	0.0311533	-0.0427859
4.0	0.7500919	0.9458020	0.0012435	-0.2405492	0.0207928	-0.0274139
6.0	0.7756193	0.9790245	0.0003547	-0.2527804	0.0116534	-0.0146127
8.0	0.7903676	0.9988675	0.0001437	-0.2605377	0.0076771	-0.0093680
10.0	0.8002154	1.0124223	0.0000787	-0.2660246	0.0055522	-0.0066280
12.0	0.8073594	1.0221905	0.0000345	-0.2701311	0.0042355	-0.0050266
	$q_2 = 0.5$					
0.5	0.3719149	0.6895328	0.5557018	-0.2423914	0.2825195	-0.6027499
1.0	0.5000000	0.7313436	0.0814867	-0.2726308	0.1261600	-0.2304894
1.5	0.5612642	0.7678922	0.0246530	-0.2901152	0.0751010	-0.1273264
2.0	0.5981678	0.7947267	0.0103303	-0.3020909	0.0511721	-0.0827335
3.0	0.6417186	0.8304226	0.0029786	-0.3177749	0.0293099	-0.0446521
4.0	0.6673363	0.8531106	0.0012253	-0.3278076	0.0195643	-0.0287188
6.0	0.6971888	0.8807766	0.0003491	-0.3402518	0.0109676	-0.0153835
8.0	0.7146337	0.8974335	0.0001439	-0.3478976	0.0072365	-0.0098798
10.0	0.7263613	0.9087579	0.0000705	-0.3531845	0.0052269	-0.0070154
12.0	0.7349194	0.9171276	0.0000412	-0.3571259	0.0039894	-0.0053249

24

Likelihood and Modified Likelihood Estimation for Distributions with Unknown Endpoints

Richard L. Smith

University of North Carolina

CONTENTS

24.1 Introduction

A recurrent theme in reliability research is the analysis of distributions with unknown endpoints. Specific examples are the three-parameter Weibull, lognormal, gamma and inverse Gaussian distributions. Each of these distributions has the property of being concentrated on an interval (γ, ∞), where γ, typically called an endpoint or threshold parameter, is unknown. In most cases this is non-negative and is interpreted as the minimum possible lifetime of the system, a parameter of obvious importance in reliability analysis. In such families it has been known for a long time that there are difficulties with maximum likelihood estimation.

Clifford Cohen was one of the first to consider maximum likelihood estimation (MLE) for distributions of this type, and over the years has made many contributions. However, he has also been one of the pioneers in considering alternative methods. He has developed a number of variants of a method which he called modified maximum likelihood estimation (MMLE), and in recent years has also worked on modified moments estimators (MME). The main motivation for developing these alternative methods has apparently been ease of numerical implementation. However, it has also been suggested in a number of simulation studies that they may have superior statistical properties as well, in the sense of having smaller bias or variance in small or moderate-sized samples. In this paper I want to explore these aspects further, with particular attention to a theoretical comparison of the MLE and MMLE techniques.

The paper is organised as follows. In Section 24.2, I review the fields of MLE and MMLE for distributions with unknown endpoints, with particular attention to Cohen's own work on these techniques. Then in Section 24.3, I review some known results on asymptotics for MLE, and develop some new approximations for MMLE. Finally, Section 24.4 presents some numerical and simulation results.

0-8493-8972-0/95/$0.00+$.50

24.2   Estimation of Distributions With Unknown Endpoints

Historically, the first of these distributions to be studied was the three-parameter log-normal distribution. Cohen's paper [6] was one of the first papers to consider maximum likelihood estimation for this family (though not the first: the paper by [28] preceded him). Starting with the density

$$f(x; \gamma, \delta, \beta) = \frac{1}{\sqrt{2\pi}\delta(x - \gamma)} \exp\left[-\frac{\log^2\{(x - \gamma)/\beta\}}{2\delta^2}\right], \qquad x > \gamma, \qquad (24.2.1)$$

he developed the maximum likelihood estimating equations and proposed a graphical method of numerical solution. He also calculated the theoretical Fisher information matrix for the unknown parameter vector (γ, δ, β).

However, in this paper he also introduced modified maximum likelihood. In this, the likelihood equation for γ was replaced by the equation

$$x_0 - \gamma = \beta e^{\delta t_0} . \qquad (24.2.2)$$

Here, $x_0 = X_1 + \frac{1}{2}\eta$, X_1 being the smallest observation in the sample, η is the precision or measurement accuracy of the observations, and t_0 is given by

$$\Phi(t_0) = \frac{k}{n} \qquad (24.2.3)$$

where Φ is the standard normal distribution function, n the total sample size and k the number of observations recorded as being equal to X_1. This formulation allows for the possibility of rounded data; if the data were recorded to a very high precision so that there were effectively no ties, then one would presumably take $x_0 = X_1$ and $k = 1$. Thus it can be seen that the maximum likelihood equation for γ is replaced by a moments equation based on the approximate mean of the smallest order statistic. The usual maximum likelihood estimating equations for the other parameters δ and β are retained, thus giving a system of three nonlinear equations in the three unknown parameters. His main argument for presenting this approach was computational simplicity, but he also hinted that it may have superior statistical properties as well.

In work subsequent to Cohen's original paper on the lognormal distribution, [20] pointed out that the solutions to the likelihood equations – that is, setting the derivatives of the likelihood function equal to 0 – in fact gave only a local and not a global maximum of the likelihood function, the global maximum being $+\infty$ attained as $\gamma \uparrow x_1$. Nevertheless, he argued in favor of the local maximum likelihood estimator, essentially by analogy with the corresponding Bayesian solution. In [19], Harter and Moore also advocated using the local maximum, extending the discussion to include censored data, and proposed an alternative modified scheme based on treating X_1 as a censored value when the local maximum does not exist. In [18], Griffiths expanded further on the theme of local maximum likelihood estimation by proposing interval estimation procedures based on the maximized likelihood. Meanwhile, in [10], Cohen and Whitten gave a more detailed review of the problem presenting a number of different forms of modified maximum likelihood and moments procedures.

Possibly to an even greater extent than the lognormal distribution, the three-parameter Weibull is widely studied and hotly debated. In [8], Cohen defined this family by the density f and distribution function F given by

$$f(x; \gamma, \delta, \theta) = \frac{\delta}{\theta}(x - \gamma)^{\delta-1} \exp\left\{-\frac{(x - \gamma)^\delta}{\theta}\right\},$$

$$F(x; \gamma, \delta, \theta) = 1 - \exp\left\{-\frac{(x-\gamma)^\delta}{\theta}\right\}, \tag{24.2.4}$$

both expressions being valid on $x > \gamma$, and considered the multi-censored likelihood function

$$L = \prod_{i=1}^{n} f(X_i; \gamma, \delta, \theta) \prod_{j=1}^{k} \{1 - F(T_j; \gamma, \delta, \theta)\}^{r_j}$$

which is based on n uncensored observations at $X_1, \ldots, X_n$ together with a group of r_j right-censored observations at T_j for each of k censoring points $T_1, \ldots, T_k$. An earlier development for the two-parameter Weibull (in which γ is known) was given in [7].

For the three-parameter case with $\delta > 1$, he advocated direct application of the maximum likelihood method. The above parametrization in fact allows one to eliminate θ analytically, so the equations actually reduce to a set of two nonlinear equations in two unknowns. For $\delta < 1$, these equations have no solution since the likelihood function becomes unbounded when $\gamma \uparrow X_1$, where we assume $X_1 = \min(X_1, \ldots, X_n)$. In this case he advocated replacing the likelihood equation for the endpoint γ by $\hat{\gamma} = X_1 - \frac{1}{2}\eta$, where η is the sampling precision, or else by the more sophisticated estimating equation

$$X_1 = \gamma + \left(\frac{\theta}{N}\right)^{1/\delta} \Gamma\left(1 + \frac{1}{\delta}\right) \tag{24.2.5}$$

where N is the total sample size $n + r_1 + \ldots + r_k$ and $\Gamma(\cdot)$ the gamma function. Equation (24.2.5) amounts to equating the smallest order statistic X_1 to its expected value under the assumed model, and thus is very much in the same spirit as Cohen's earlier suggestion [6] for the lognormal distribution.

In [12], Cohen and Whitten proposed a number of alternative forms for the modified maximum likelihood idea. In this paper they used the parametrization

$$f(x; \gamma, \delta, \beta) = \frac{\delta}{\beta^\delta}(x-\gamma)^{\delta-1} \exp\left\{-\left(\frac{x-\gamma}{\beta}\right)^\delta\right\}, \qquad x > \gamma. \tag{24.2.6}$$

Based on simulations of the bias and standard deviation of the estimators in the complete sample case, they actually advocated maximum likelihood only when $\delta > 2.2$, and discussed five forms of modified maximum likelihood estimator (MMLE) for use in other cases:

I Set $F(X_r) = r/(n+1)$, where X_r is the r'th smallest order statistic in a sample of size n and F is the cumulative distribution function. Detailed discussion was restricted to $r = 1$ though they also pointed out that $r > 1$ might be desirable for robustness purposes.

II Set $X_1 = \gamma + n^{-1/\delta}\beta\Gamma(1 + 1/\delta)$, as in (24.2.5) with $N = n$.

III Equate the sample and population means, i.e., $\bar{X} = \gamma + \beta\Gamma(1 + 1/\delta)$.

IV Equate the sample and population variances, i.e., $s_X^2 = \beta^2\{\Gamma(1+2/\delta) - \Gamma^2(1+1/\delta)\}$.

V Equate the sample and population medians, i.e. $X_{\text{med}} = \gamma + \beta(\ln 2)^{1/\delta}$.

In each case the new moment equation replaced the likelihood equation for γ, $d\ln L/d\gamma = 0$, the likelihood equations for β and δ being retained in their usual form. MMLE of types I and II had been considered in the lognormal case in [10]. Other related papers were [11] for the three-parameter gamma distribution, and [3] for the three-parameter inverse Gaussian distribution. They also considered a family of "modified moment estimators" in which the usual equations for the first two moments, III and IV above, are supplemented by one of I, II or V instead of the more traditional third-moment equation.

For the three-parameter Weibull case, the authors in [12] carried out a simulation study based on samples of sizes $n = 10$, 25 and 100 and six different values of δ. They suggested that ordinary maximum likelihood estimation works well when $\delta > 2.2$ but that versions I and II of MMLE would work well over the whole range of δ. They also computed the asymptotic variance of the maximum likelihood estimates when $\delta > 2$ but pointed out that for $\delta \leq 2$ the standard asymptotic theory fails – we consider this aspect in more detail in Section 24.3. They also gave some attention to the MME approach and this was developed further in subsequent papers: [16], [17], [13], and [14] , respectively for the Weibull, lognormal, inverse Gaussian and gamma cases.

Cohen has also studied truncated distributions. In [5], a truncated two-parameter Weibull distribution was considered, of the form

$$f_{LT}(x; \alpha, \delta, \theta) = \frac{f(x; \delta, \theta)}{1 - F(\alpha; \delta, \theta)}, \qquad x > \alpha , \tag{24.2.7}$$

f and F denoting the density and distribution function of the two-parameter Weibull distribution,

$$f(x; \delta, \theta) = \frac{\delta}{\theta} x^{\delta-1} \exp\left(-\frac{x^\delta}{\theta}\right), \ F(x; \delta, \theta) = 1 - \exp\left(-\frac{x^\delta}{\theta}\right), \qquad x > 0 . \tag{24.2.8}$$

In this case maximum likelihood estimation results in $\widehat{\alpha} = x_1$ but modifications similar to I and II above may also be considered to reduce the obvious bias in this. Right-truncation was also considered.

Cohen's books ([15] and [9]) have reviewed all of these estimation techniques, and have extended them to numerous parametric families beyond the ones discussed here. Throughout Cohen's work, there has been a strong emphasis on efficient computational implementation of the procedures, and it would appear that this aspect was his own motivation for developing alternatives to MLE. Nevertheless, a recurring theme developed in simulations of these estimators is that in small samples MMLE or MME may have superior statistical properties as well. I believe this aspect is ultimately more important than issues of computational convenience, and this is what I want to discuss in the remainder of the paper. I concentrate on MMLE as I believe this is the more interesting and general technique – for example, it extends easily to the case of censored data – but the general approach taken may be relevant for MME as well.

24.3   Theory of MLE and MMLE

In this section I review the known maximum likelihood theory for three-parameter Weibull and related distributions, and develop some new results for modified maximum likelihood estimation. The MMLEs will always be version II in the classification of Section 24.1, in which the smallest order statistic X_1 is equated to its expected value under the model, together with the usual likelihood estimating equations for β and δ. I restrict detailed discussion to the three-parameter Weibull distribution with density defined by (24.2.6). However, the general concepts and qualitative results are also applicable to many of the other estimators and distributions considered in Section 24.1.

Maximum likelihood estimation for the three-parameter Weibull distribution was considered by a number of authors including [29], [25], [22], and of course [8]. When $\delta > 2$

the Fisher information is finite and the maximum likelihood estimators obey all the usual asymptotic properties, such as asymptotic normality and asymptotic efficiency, of regular parametric problems. Moreover, it is appropriate to seek a local maximum of the log likelihood function and to ignore the singularity in the likelihood that occurs as $\gamma \uparrow X_1$ when $\delta < 1$. Rigorous proofs of these statements were provided in [26]. There is no guarantee that a local maximum of the likelihood function exists, but in [25] the authors conjectured that when a local maximum does exist, it is unique. As far as I know, this conjecture has never been proved, but I have never encountered a counterexample and believe the conjecture to be true for all practical purposes.

The Fisher information matrix has been given by numerous authors; here I follow Section 3.8 of [15] by writing the Fisher information matrix $A = A(\gamma, \delta, \beta) = (a_{ij})$, $i, j = 1, 2, 3$ (so $a_{11} = \mathrm{E}\{-\partial^2 \log f(X; \gamma, \delta, \beta)/\partial \gamma^2\}$, etc.) in the form

$$a_{11} = \frac{C}{\beta^2}, \qquad a_{22} = \frac{K}{\delta^2}, \qquad a_{33} = \frac{\delta^2}{\beta^2},$$

$$a_{12} = a_{21} = \frac{J}{\beta}, \quad a_{13} = a_{31} = \frac{\delta^2}{\beta^2}\Gamma\left(2 - \frac{1}{\delta}\right), \quad a_{23} = a_{32} = -\frac{\Psi(2)}{\beta}, \qquad (24.3.1)$$

where

$$C = (\delta - 1)\left\{\Gamma\left(1 - \frac{2}{\delta}\right) + \delta\Gamma\left(2 - \frac{2}{\delta}\right)\right\} = (\delta - 1)^2\Gamma\left(1 - \frac{2}{\delta}\right)$$

$$J = \Gamma\left(1 - \frac{1}{\delta}\right) - \left\{1 + \Psi\left(2 - \frac{1}{\delta}\right)\right\}\Gamma\left(2 - \frac{1}{\delta}\right),$$

$$K = \Psi'(1) + \Psi^2(2) .$$

Here $\Psi(\cdot)$ is the digamma function (the derivative of $\log \Gamma(\cdot)$) and $\Psi'(\cdot)$, the derivative of $\Psi(\cdot)$, is the trigamma function. These functions are tabulated in [1].

Provided $\delta > 2$, the (local) maximum likelihood estimates are asymptotically unbiased and normally distributed, with variance-covariance matrix given by $n^{-1}A^{-1}$, where n is the sample size. However, as $\delta \downarrow 2$, $C \uparrow \infty$, so this theory breaks down when $\delta \leq 2$.

The case $\delta \leq 2$ was considered in detail in [26]. If δ and β are known, so that the only unknown parameter is γ itself, then there are several different cases depending on the exact value of δ. For $\delta = 2$ the maximum likelihood estimator $\hat{\gamma}$ is asymptotically normal but $\hat{\gamma} - \gamma$ is $O_p((n \log n)^{-1/2})$, instead of the usual $O_p(n^{-1/2})$ [30]. For $1 < \delta < 2$, $\hat{\gamma}$ has an extremely complicated non-normal distribution obtained in [31]; here $\hat{\gamma} - \gamma = O_p(n^{-1/\delta})$. For $\delta \leq 1$ no local maximum of the likelihood function exists. However, for $\delta < 2$, in [2], Akahira showed that $O_p(n^{-1/\delta})$ is the optimal rate of convergence for any estimator and in [21], Ibragimov and Has'minskii obtained asymptotically optimal estimates, though these optimality results are also very complicated and not easy to apply in practice. A different approach is based on the smallest order statistic X_1, since if we define $Z_1 = n^{1/\delta}\beta^{-1}(X_1 - \gamma)$, we have

$$\Pr\{Z_1 \leq z\} = 1 - \exp(-z^\delta), \qquad z \geq 0 , \qquad (24.3.2)$$

so if X_1 is treated as an estimator of γ, it has an extremely simple distributional form with an error of $O_p(n^{-1/\delta})$. Akahira's result shows that this is the optimal *rate* of convergence whenever $\delta < 2$, though not for $\delta > 2$ (since in this case the MLE converges at rate $O_p(n^{-1/2})$).

In this context, the MMLE proposed by Cohen reduces to

$$\bar{\gamma} = X_1 - n^{-1/\delta}\beta\Gamma(1 + 1/\delta) \qquad (24.3.3)$$

and we see that equation (24.3.2) gives the exact distribution of this estimator; we still have $\bar{\gamma} - \gamma = O_p(n^{-1/\delta})$ but the subtracted term in (24.3.3) amounts to a bias correction which improves its properties as a point estimator. The estimator therefore appears to be a good estimator when $\delta < 2$, made more attractive by the simplicity of (24.3.2). However for $\delta > 2$ the maximum likelihood estimator is asymptotically more efficient.

Now let us turn to the case where all three parameters are unknown, and $\delta < 2$. In this case, I showed (in [26]) that there is an asymptotic independence property between estimates of γ and those of the other parameters: in large samples, ignorance about δ and β does not affect one's ability to estimate γ, and conversely, if one tries to estimate δ and β using an estimated value of γ, then one gets exactly the same asymptotic distribution as if γ were known. The intuitive reason for this is that efficient estimates of γ converge at rate $O_p(n^{-1/\delta})$, which is smaller than the error $O_p(n^{-1/2})$ in $\hat{\delta}$ and $\hat{\beta}$, so asymptotically, the error in estimating γ has no influence on the solution.

In pursuit of a specific procedure with these properties, I proposed estimating γ by the sample minimum X_1, and then constructing the usual likelihood equations for δ and β based on the differences $X_2 - X_1, \ldots, X_n - X_1$. When $\delta < 2$, these estimators have the same asymptotic properties as the regular two-parameter Weibull estimates when γ is known, but when $\delta > 2$, they converge only at the rate $O_p(n^{-1/\delta})$ and are therefore inferior to the three-parameter maximum likelihood estimates [26, Theorem 4].

Although this procedure has the right asymptotic properties, it is questionable whether it is in fact the best way to proceed in small samples. In a recent paper [23], Lockhart and Stephens have proposed an "iterative bias reduction" procedure very similar to Cohen's MMLE, and claimed on the basis of Monte Carlo studies that it gave a better fit to the Weibull distribution than estimates based on X_1. It therefore seems worthwhile to re-examine the whole idea of MMLE estimation in the light of these asymptotic results.

First, an elementary but important result:

THEOREM 24.3.1

Suppose an estimator $\tilde{\gamma}$ is defined by the right hand side of (24.3.3), but with estimators $\tilde{\delta}, \tilde{\beta}$ substituted for the true values δ and β. Suppose these estimators satisfy

$$\tilde{\beta} - \beta \to_p 0, \qquad \log n \, (\tilde{\delta} - \delta) \to_p 0 \qquad \text{as } n \to \infty \ . \tag{24.3.4}$$

Here $\to_p$ denotes convergence in probability. Then $\tilde{\gamma}$ has the same asymptotic properties as $\bar{\gamma}$; in particular, the distribution of

$$\tilde{Z}_1 = n^{-1/\delta}\frac{\tilde{\gamma} - \gamma}{\beta} + \Gamma(1 + \frac{1}{\delta})$$

converges to the Weibull random variable Z_1 given by (24.3.2).

We should also note that $\Gamma(1 + 1/\delta)$ is the mean of Z_1, so this result also makes precise the claim that $\tilde{\gamma}$ is asymptotically unbiased. We note also that this result applies for all values of δ, not just $\delta < 2$.

PROOF Taylor expansion shows that

$$\tilde{\gamma} - \bar{\gamma} \approx (\tilde{\delta} - \delta)\frac{\partial \bar{\gamma}}{\partial \delta} + (\tilde{\beta} - \beta)\frac{\partial \bar{\gamma}}{\partial \beta}$$

$$= (\tilde{\delta} - \delta)n^{-1/\delta}\frac{\beta}{\delta^2}\left\{\Gamma'\left(1 + \frac{1}{\delta}\right) - \log n \, \Gamma\left(1 + \frac{1}{\delta}\right)\right\} - (\tilde{\beta} - \beta)n^{-1/\delta}\Gamma\left(1 + \frac{1}{\delta}\right) \ .$$

$$\tag{24.3.5}$$

Under the assumption (24.3.4), the right hand side of (24.3.5) is $o_p(n^{-1/\delta})$, so $\tilde{Z}_1 - Z_1 = o_p(1)$, and the result follows. ∎

Now, let us consider a corresponding result for the estimation of δ and β:

THEOREM 24.3.2
Suppose γ is estimated by an estimator $\tilde{\gamma}$ such that $\tilde{\gamma} - \gamma = O_p(n^{-1/\delta})$. Suppose the estimators $\tilde{\delta}$ and $\tilde{\beta}$ are obtained by maximizing the two-parameter log likelihood for (δ, β), but with $\tilde{\gamma}$ substituted for γ. Also let $(\bar{\delta}, \bar{\beta})$ denote the corresponding estimators computed from the same data using the true value of γ. Then $\tilde{\delta} - \bar{\delta}$ and $\tilde{\beta} - \bar{\beta}$ are each of $O_p(n^{-1/\delta})$ if $\delta > 1$, $O_p(n^{-1} \log n)$ if $\delta \leq 1$.

The idea of the proof is the same as [26, Theorem 4]; see also [27, Theorem 3]. We omit the details.

For $\delta < 2$, $\tilde{\delta} - \bar{\delta}$ and $\tilde{\beta} - \bar{\beta}$ are of smaller order of magnitude than $\bar{\delta} - \delta$ and $\bar{\beta} - \beta$, the latter being of $O_p(n^{-1/2})$, so in this case Theorem 24.3.2 makes precise the notion that $\tilde{\delta}$ and $\tilde{\beta}$ are asymptotically unaffected by γ being unknown. For $\delta > 2$, however, the $O_p(n^{-1/\delta})$ error dominates, so this implies that the MMLE procedure is inefficient in this case.

So far, I have confined my discussion entirely to asymptotic theory. However, there are a number of respects in which these results do not ring true for finite samples. For example, consider Figure 24.1. In this figure I have computed n times the asymptotic variance of $\hat{\gamma}$, assuming $\beta = 1$ and for a range of values of $\delta > 2$, where (a) $\hat{\gamma}$ is the MLE for γ assuming δ and β unknown (this is curve I on Figure 24.1), (b) $\hat{\gamma}$ is the MLE for γ assuming δ and β known (curve II), (c) $\hat{\gamma}$ is the MMLE $\tilde{\gamma}$ (curve III). In this case I assume sample size $n = 100$, and the plotted curves are based on (a) the top left hand entry of A^{-1}, where the matrix $A = (a_{ij})$ is defined by (24.3.1), (b) $1/a_{11}$, again from (24.3.1), (c) $n^{1-2/\delta} \text{Var}(Z_1) = n^{1-2/\delta} \{ \Gamma(1 + 2/\delta) - \Gamma^2(1 + 1/\delta) \}$ where Z_1 has the Weibull distribution (24.3.2). It can be seen that curve III always lies above curve II but, for $\delta > 2.45$, it lies below curve I. The interpretation, taken literally, is that for $\delta > 2.45$ we would prefer to use MMLE, despite the asymptotic inefficiency as $n \to \infty$. The comparison is not so good for larger n – for $n = 1000$, the crossover between curves I and III occurs at $\delta = 2.96$ – but in most reliability applications we have $n < 100$ so comparisons based on $n = 100$ seem meaningful in practice. However, the sharp contrast between curves I and II for MLE suggests there should be a similar contrast for MMLE, but our asymptotic results do not capture that. Similarly, for $\delta < 2$ and the estimation of δ and β, the asymptotic results say that one can ignore the effect of estimating γ, but this does not seem very realistic either. The remaining curve IV will be developed in the later part of this secion to represent an improved approximation to the curve. An additional point is that when $\delta > 2$, the Cramer-Rao inequality is valid, and this indicates that curve I is a lower bound to the variance of any unbiased estimator of γ. This casts even further doubt on the usefulness of the approximation III.

For the remainder of this section, then, I discuss an alternative approach to characterizing the variance-covariance matrix of the MMLE, with more attention to practical approximation than theoretical asymptotics.

The estimating equations for MMLE may be written in the form $h_n(\gamma, \delta, \beta) = 0$, where the suffix n is written to make explicit the dependence on the sample size n, and

$$h_n = \begin{bmatrix} h_n^{(1)} \\ h_n^{(2)} \\ h_n^{(3)} \end{bmatrix}$$

$$= \begin{bmatrix} -(X_1 - \gamma) + n^{-1/\delta}\beta\Gamma\left(1 + \frac{1}{\delta}\right) \\ -\frac{1}{\delta} - \frac{1}{n}\sum \log\left(\frac{X_i - \gamma}{\beta}\right) + \frac{1}{n}\sum\left(\frac{X_i - \gamma}{\beta}\right)^\delta \log\left(\frac{X_i - \gamma}{\beta}\right) \\ \frac{\delta}{\beta} - \frac{\delta}{\beta}\frac{1}{n}\sum\left(\frac{X_i - \gamma}{\beta}\right)^\delta \end{bmatrix}$$

Define the matrix

$$H_n = \begin{bmatrix} \frac{\partial h_n^{(1)}}{\partial \gamma} & \frac{\partial h_n^{(1)}}{\partial \delta} & \frac{\partial h_n^{(1)}}{\partial \beta} \\ \frac{\partial h_n^{(2)}}{\partial \gamma} & \frac{\partial h_n^{(2)}}{\partial \delta} & \frac{\partial h_n^{(2)}}{\partial \beta} \\ \frac{\partial h_n^{(3)}}{\partial \gamma} & \frac{\partial h_n^{(3)}}{\partial \delta} & \frac{\partial h_n^{(3)}}{\partial \beta} \end{bmatrix}. \tag{24.3.6}$$

Assuming consistent solutions $(\tilde{\gamma}_n, \tilde{\delta}_n, \tilde{\beta}_n)$ exist, we will have

$$h_n(\tilde{\gamma}_n, \tilde{\delta}_n, \tilde{\beta}_n) - h_n(\gamma, \delta, \beta) = H_n^* \cdot \begin{bmatrix} \tilde{\gamma}_n - \gamma \\ \tilde{\delta}_n - \delta \\ \tilde{\beta}_n - \beta \end{bmatrix}$$

where H_n^* is H_n evaluated at some $(\gamma_n^*, \delta_n^*, \beta_n^*)$ on the line joining $(\tilde{\gamma}_n, \tilde{\delta}_n, \tilde{\beta}_n)$ to (γ, δ, β). Noting that $h_n(\tilde{\gamma}_n, \tilde{\delta}_n, \tilde{\beta}_n) = 0$, we then have

$$\begin{bmatrix} \tilde{\gamma}_n - \gamma \\ \tilde{\delta}_n - \delta \\ \tilde{\beta}_n - \beta \end{bmatrix} = -(H_n^*)^{-1} \cdot h_n \tag{24.3.7}$$

where h_n (here and in all subsequent appearances) is evaluated at the true parameter values (γ, δ, β).

For large n and for $\delta > 1$, the entries of H_n converge to asymptotic values given by the law of large numbers. (There is a difficulty when $\delta \le 1$, which is explained below). Moreover, the entries are continuous functions of the parameters (γ, δ, β). Thus for approximate calculations we may replace H_n^* by a matrix $\tilde{H}_n$, defined as the limiting form of H_n at the true parameter values. This gives the approximation

$$\begin{bmatrix} \tilde{\gamma}_n - \gamma \\ \tilde{\delta}_n - \delta \\ \tilde{\beta}_n - \beta \end{bmatrix} \approx -(\tilde{H}_n)^{-1} \cdot h_n, \tag{24.3.8}$$

which is easier to handle than (24.3.7) because $\tilde{H}_n$ is now a fixed (non-random) matrix.

Using this approximation, if the variance-covariance matrix of h_n is denoted V_n, we have that the limiting variance-covariance matrix of $(\tilde{\gamma}_n, \tilde{\delta}_n, \tilde{\beta}_n)$ is given by

$$W_n = (\tilde{H}_n)^{-1} V_n (\tilde{H}_n^T)^{-1} .$$

Now let us look more closely at the structure of $\tilde{H}_n$ and V_n. We may write

$$\tilde{H}_n = \begin{bmatrix} 1 & b_n^T \\ c & A_1 \end{bmatrix}, \qquad V_n = \begin{bmatrix} d_n & f_n^T \\ f_n & n^{-1}A_1 \end{bmatrix},$$

where b_n and c are (column) vectors of constants, $d_n = n^{-2/\delta}\{\Gamma(1 + 2/\delta) - \Gamma^2(1 + 1/\delta)\}$ and A_1 is the lower right 2×2 submatrix of A, where A was defined in (3.1). The vector f_n will be explained further below. It should be pointed out that if we had a more general

p-dimensional problem, in which the first parameter was an endpoint parameter estimated by an equation similar to (3.3) and the remaining $p-1$ parameters were estimated from their likelihood equations, then the same general structure, and the manipulations to follow, would be valid. Thus the theory being outlined here could be taken as a first attempt at a general theory for MMLE in problems involving endpoint estimation, though I am confining my detailed calculations to the three-parameter Weibull case.

If we define

$$U_n = (A_1 - cb_n^T)^{-1},$$

then we readily check that

$$\tilde{H}_n^{-1} = \begin{bmatrix} v_n & -b_n^T U_n \\ -U_n c & U_n \end{bmatrix}$$

where $v_n = 1 + b_n^T U_n c$. Consequently,

$$W_n = \begin{bmatrix} w_{n1} & w_{n2}^T \\ w_{n2} & W_{n3} \end{bmatrix}$$

where the scalar w_{n1}, the 2×1 vector w_{n2} and the 2×2 matrix W_{n3} are given by

$$w_{n1} = d_n v_n^2 - 2v_n f_n^T U_n^T b_n + \frac{1}{n} b_n^T U_n A_1 U_n^T b_n,$$

$$w_{n2} = -d_n v_n U_n c + U_n c f_n^T U_n^T b_n + v_n U_n f_n - \frac{1}{n} U_n A_1 U_n^T b_n,$$

$$W_{n3} = d_n U_n cc^T U_n^T - U_n (c f_n^T + f_n c^T) U_n^T + \frac{1}{n} U_n A_1 U_n^T.$$

These expressions may be expressed directly in terms of the elements of $\tilde{H}_n$ and V_n if we write

$$U_n = A_1^{-1} + v_n A_1^{-1} cb_n^T A_1^{-1}, \qquad v_n = (1 - b_n^T A_1^{-1} c)^{-1}$$

from which it also follows that

$$b_n^T U_n = v_n b_n^T A_1^{-1}, \qquad U_n c = v_n A_1^{-1} c.$$

We then have

$$w_{n1} = v_n^2 \left(d_n - 2f_n^T A_1^{-1} b_n + \frac{b_n^T A_1^{-1} b_n}{n} \right),$$

$$w_{n2} = -v_n^2 \left(d_n - 2f_n^T A_1^{-1} b_n + \frac{b_n^T A_1^{-1} b_n}{n} \right) A_1^{-1} c - v_n A_1^{-1} \left(\frac{b_n}{n} - f_n \right),$$

$$W_{n3} = v_n^2 \left(d_n - 2f_n^T A_1^{-1} b_n + \frac{b_n^T A_1^{-1} b_n}{n} \right) A_1^{-1} cc^T A_1^{-1}$$

$$+ v_n A_1^{-1} \left\{ c \left(\frac{b_n}{n} - f_n \right)^T + \left(\frac{b_n}{n} - f_n \right) c^T \right\} A_1^{-1} + \frac{1}{n} A_1^{-1}. \tag{24.3.9}$$

To evaluate these expressions, we still need to know b_n, c and f_n. However, from (24.3.6) we have $c = (a_{12} \ a_{13})^T$ where a_{12} and a_{13} are given by (24.3.1). Also,

$$b_n = \begin{bmatrix} n^{-1/\delta} \left(\frac{\beta}{\delta^2} \right) \{ \Gamma \left(1 + \frac{1}{\delta} \right) \log n - \Gamma' \left(1 + \frac{1}{\delta} \right) \} \\ n^{-1/\delta} \Gamma \left(1 + \frac{1}{\delta} \right) \end{bmatrix}.$$

To evaluate f_n, we need to calculate

$$\text{Cov}\left\{X_1, \frac{1}{n}\sum\left(\frac{X_j - \gamma}{\beta}\right)^\theta\right\}$$

for arbitrary index $\theta \geq 0$.

To simplify the calculations, we assume without loss of generality that $\gamma = 0$, $\beta = 1$. Consider the following scheme for generating the sample $(X_1, \ldots, X_n)$: set $X_1 = n^{-1/\delta}Z_1$, where Z_1 has c.d.f. (24.3.2), then generate $X_2, \ldots, X_n$ conditionally independently from the density

$$\exp(X_1^\delta)\delta x^{\delta-1}\exp(-x^\delta), \qquad x > X_1 .$$

This is equivalent to drawing an independent sample $(X_1, \ldots, X_n)$ and reordering so that X_1 is the smallest value (with the others ordered randomly).

Now, for $j > 1$,

$$E\{X_j^\theta|X_1\} = \exp(X_1^\delta)\int_{X_1}^\infty \delta x^{\theta+\delta-1}\exp(-x^\delta)dx$$

$$= \exp(X_1^\delta)\Gamma\left(1+\frac{\theta}{\delta}\right)\left\{1 - \int_0^{X_1}\delta x^{\theta+\delta-1}\exp(-x^\delta)dx\right\}$$

$$= \Gamma\left(1+\frac{\theta}{\delta}\right)\left(1 + X_1^\delta + \frac{1}{2}X_1^{2\delta} + \cdots\right)\left(1 - \frac{\delta}{\theta+\delta}X_1^{\theta+\delta} + \frac{\delta}{\theta+2\delta}X_1^{\theta+2\delta} - \cdots\right)$$

$$= \Gamma\left(1+\frac{\theta}{\delta}\right)\left(1 + X_1^\delta - \frac{\delta}{\theta+\delta}X_1^{\theta+\delta} + \cdots\right)$$

where we have retained only the three leading terms and dropped the rest. It follows that

$$E\{X_1 X_j^\theta\} = \Gamma\left(1+\frac{\theta}{\delta}\right)E\left\{X_1 + X_1^{\delta+1} - \frac{\delta}{\theta+\delta}X_1^{\theta+\delta+1} + \cdots\right\}$$

$$= n^{-1/\delta}\Gamma\left(1+\frac{\theta}{\delta}\right)\Gamma\left(1+\frac{1}{\delta}\right) + n^{-1-1/\delta}\Gamma\left(1+\frac{\theta}{\delta}\right)\Gamma\left(2+\frac{1}{\delta}\right)$$

$$-n^{-1-1/\delta-\theta/\delta}\frac{\delta}{\theta+\delta}\Gamma\left(1+\frac{\theta}{\delta}\right)\Gamma\left(2+\frac{1}{\delta}+\frac{\theta}{\delta}\right) + \cdots$$

Hence

$$E\left\{X_1 \cdot \frac{1}{n}\sum_2^n X_j^\theta\right\} = \left(1-\frac{1}{n}\right)\left\{n^{-1/\delta}\Gamma\left(1+\frac{\theta}{\delta}\right)\Gamma\left(1+\frac{1}{\delta}\right) + \cdots\right\} .$$

We also have

$$E\left\{\frac{1}{n}X_1^{\theta+1}\right\} = n^{-1-1/\delta-\theta/\delta}\Gamma\left(1+\frac{1}{\delta}+\frac{\theta}{\delta}\right)$$

and that

$$E\{X_1\} = n^{-1/\delta}\Gamma\left(1+\frac{1}{\delta}\right), \qquad E\left\{\frac{1}{n}\sum_1^n X_j^\theta\right\} = \Gamma\left(1+\frac{\theta}{\delta}\right) .$$

Hence we deduce

$$\mathrm{Cov}\left\{X_1, \frac{1}{n}\sum_1^n X_j^\theta\right\} \approx$$

$$-n^{-1-1/\delta}\Gamma\left(1+\frac{\theta}{\delta}\right)\Gamma\left(1+\frac{1}{\delta}\right) + n^{-1-1/\delta}\Gamma\left(1+\frac{\theta}{\delta}\right)\Gamma\left(2+\frac{1}{\delta}\right)$$

$$-n^{-1-1/\delta-\theta/\delta}\frac{\delta}{\theta+\delta}\Gamma\left(1+\frac{\theta}{\delta}\right)\Gamma\left(2+\frac{1}{\delta}+\frac{\theta}{\delta}\right) + n^{-1-1/\delta-\theta/\delta}\Gamma\left(1+\frac{1}{\delta}+\frac{\theta}{\delta}\right)$$

$$= n^{-1-1/\delta}\cdot\frac{1}{\delta}\Gamma\left(1+\frac{\theta}{\delta}\right)\Gamma\left(1+\frac{1}{\delta}\right)$$

$$-n^{-1-1/\delta-\theta/\delta}\Gamma\left(1+\frac{1}{\delta}+\frac{\theta}{\delta}\right)\left\{\left(1+\frac{1}{\theta+\delta}\right)\Gamma\left(1+\frac{\theta}{\delta}\right)-1\right\}$$

$$= R_n(\theta) \qquad \text{say}. \qquad\qquad (24.3.10)$$

We may also formally differentiate with respect to θ in (24.3.10) to deduce

$$\mathrm{Cov}\left\{X_1, \frac{1}{n}\sum_1^n X_j^\theta \log X_j\right\} \approx R_n'(\theta)$$

$$= n^{-1-1/\delta}\cdot\frac{1}{\delta^2}\Gamma'\left(1+\frac{\theta}{\delta}\right)\Gamma\left(1+\frac{1}{\delta}\right)$$

$$+n^{-1-1/\delta-\theta/\delta}\frac{\log n}{\delta}\Gamma\left(1+\frac{1}{\delta}+\frac{\theta}{\delta}\right)\left\{\left(1+\frac{1}{\theta+\delta}\right)\Gamma\left(1+\frac{\theta}{\delta}\right)-1\right\}$$

$$-n^{-1-1/\delta-\theta/\delta}\frac{1}{\delta}\Gamma'\left(1+\frac{1}{\delta}+\frac{\theta}{\delta}\right)\left\{\left(1+\frac{1}{\theta+\delta}\right)\Gamma\left(1+\frac{\theta}{\delta}\right)-1\right\}$$

$$-n^{-1-1/\delta-\theta/\delta}\Gamma\left(1+\frac{1}{\delta}+\frac{\theta}{\delta}\right)\left\{\frac{1}{\delta}\left(1+\frac{1}{\theta+\delta}\right)\Gamma'\left(1+\frac{\theta}{\delta}\right)-\frac{1}{(\theta+\delta)^2}\Gamma\left(1+\frac{\theta}{\delta}\right)\right\}$$

Based on these approximations, and reinserting the scale parameter β, we deduce that

$$f_n \approx \left[\frac{\beta\{R_n'(0)-R_n'(\delta)\}}{\delta R_n(\delta)}\right]. \qquad\qquad (24.3.11)$$

However $f_n = O(n^{-1-1/\delta}\log n)$, so it seems safe to drop terms in which the power of n is $-2-1/\delta$. With this further simplification for $R_n(\delta)$ and $R_n'(\delta)$, we have

$$R_n(\delta) = n^{-1-1/\delta}\cdot\frac{1}{\delta}\Gamma\left(1+\frac{1}{\delta}\right),$$

$$R_n'(0)-R_n'(\delta) = n^{-1-1/\delta}\cdot\frac{1}{\delta^2}\Gamma\left(1+\frac{1}{\delta}\right)$$

$$\cdot\left\{\log n - \Psi\left(1+\frac{1}{\delta}\right)-(\delta+1)\Psi(1)\right\}. \qquad\qquad (24.3.12)$$

Equations (24.3.11) and (24.3.12) define our final approximation for f_n, and complete our derivation of the approximate variances and covariances in (24.3.9).

We are now in a position to consider the implications of these results. For $\delta < 2$ we have $w_{n1} \sim d_n$ and $W_{n3} \sim n^{-1}A_1^{-1}$. Also $w_{n2} = o(n^{-1/\delta-1/2}\log n)$, so that the correlations between $\tilde{\gamma}_n$ and $(\tilde{\delta}_n, \tilde{\beta}_n)$ are asymptotically negligible in this case. Thus for $\delta < 2$, the asymptotic form of W_n is

$$\begin{bmatrix} d_n & 0 \\ 0 & n^{-1}A_1^{-1} \end{bmatrix}$$

confirming our earlier result that estimation of γ and the pair (δ, β) are effectively independent in this case. On the other hand, for $\delta > 2$ the matrix W_n is asymptotic to

$$d_n \begin{bmatrix} 1 & -c^T A_1^{-1} \\ -A_1^{-1}c & A_1^{-1}cc^T A_1^{-1} \end{bmatrix}$$

which is of rank 1; in other words, in this case the variability of $\tilde{\gamma}$ effectively dominates the whole problem.

It is of more interest, however, to examine the individual terms in relation to their asymptotic values. In the case when δ and β are known, the variance of $\tilde{\gamma}$ is d_n; in w_{n1}, the main change is to multiply this by a factor v_n^2 which tends to 1 at rate $O(n^{-1/\delta} \log n)$ as $n \to \infty$. In fact we often have $v_n > 2$; for instance, at $n = 100$ this is true for all $\delta > 2.5$. Thus in practice the difference between the two variances is not negligible and the calculations we have made are essential to obtain realistic approximations. This will be supported by simulations in Section 24.4.

If we look instead at W_{n3}, it can be seen that this is written as the sum of three terms, the first of which measures the influence of the variation in $\tilde{\gamma}$, the last is the asymptotic covariance matrix when γ is known, and the second represents a cross-correlation between these two effects. The first term is of $O(n^{-2/\delta})$ and the third of $O(n^{-1})$; although asymptotically one of these terms always dominates the other (except when $\delta = 2$), from the point of view of practical calculation it would seem much more sensible to retain both of them.

Before leaving this section we return briefly to the case $\delta \leq 1$, for which the results are not valid. The difficulty is that in this case $\partial h_n^{(2)}/\partial \gamma$ does not obey the law of large numbers, so that c is infinite. In fact $\partial h_n^{(2)}/\partial \gamma = O_p(n^{1/\delta - 1})$ [26, Lemma 4.II(iii)] so we could bound c by an increasing sequence c_n of this (or very slightly larger) order of magnitude. However in this case the direct interpretability of the expression for W_{n3} is lost.

Another point to make is that, although I have calculated an approximate covariance matrix for the estimators, it does not follow that they are approximately normally distributed. Indeed from (24.3.8), in which $h_n^{(2)}$ and $h_n^{(3)}$ are asymptotically bivariate normal but $h_n^{(1)}$ has a Weibull distribution, it will follow that the true limiting distribution is a complicated mixture of Weibull and normal random variables. However, I have not attempted to work out the details of this.

24.4   Numerical Results and Simulations

Curve IV on Figure 24.1 plots n times the approximate variance of $\tilde{\gamma}$, when $n = 100$, based on (24.3.9). It can be seen that it lies well above the first-order approximation based on Theorem 24.3.1, and still somewhat above curve I, which is for three-parameter maximum likelihood estimation. So we do not actually claim that MMLE is more efficient than MLE, though the difference is not too great when based on the new approximations for MMLE. In fact, if the curves in Figure 24.1 are continued for larger values of δ, they cross over (curve IV lies below curve I) at about $\delta = 15$, but this is most likely a numerical artifact; we cannot expect the approximations in Section 24.3 to work well for very large δ.

In Figures 24.2 and 24.3, similar curves are plotted for n times the variances of the estimators of δ and β respectively. Here curve I is for the three-parameter MLE, curve II for the two-parameter MLE when γ is known, and curve III is the new approximation based on (24.3.9). When $\delta \gg 2$, curve I lies well above curve II; again, curve III is above both of them, but the difference between curve I and curve III is not too great. For $\delta \leq 2$, based on the results of [26], curves I and II merge into a single curve, but still curve III lies considerably above them, reflecting the real influence of γ being unknown. Only as δ approaches 1 does the new approximation appear to break down; I conjecture that this is because of the instability of the parameter a_{12}, which becomes infinite as $\delta \downarrow 1$.

To examine how well these results reflect the true variances of the estimators, a simulation study was performed. The simulation was based on samples of size 100 for each of four values of δ, namely 4.0, 2.5, 1.75 and 1.2, and 1000 replications were performed of each experiment. For each sample, the MLE $(\hat{\gamma}, \hat{\delta}, \hat{\beta})$ was calculated using the optimization routine DFPMIN from [24] . This routine assumes that function values and first-order derivatives are available, but I have approximated the first-order derivatives of the log likelihood by using a simple differencing procedure, so effectively converting it into a derivative-free optimizer. For the MMLE, the objective is to find a value $\tilde{\gamma}$ such that (24.3.3) holds when δ and β are replaced by their MLEs $\tilde{\delta}$ and $\tilde{\beta}$ under the two-parameter model with $\gamma = \tilde{\gamma}$ fixed. This was done by starting with two trial values of $\tilde{\gamma}$ and linearly interpolating to obtain a third value, the process being iterated to convergence. This procedure, which has been advocated by Cohen in a number of his papers, usually converges in fewer than ten iterations. In contrast, the authors in [23] advocated direct iteration between (24.3.3) and the maximum likelihood solution for (δ, γ); this can be a very slow procedure. There are two known circumstances under which MLE can fail to converge. The first has already been pointed out, and arises when $\delta \leq 1$ and $\gamma \uparrow X_1$. The second occurs for large δ and has been called *the embedded model problem* in [4]. This occurs because, under some conditions involving $\delta \to \infty$, $\beta \to \infty$ and $\gamma \to -\infty$, the three-parameter Weibull distribution converges to a quite different model (the Gumbel or two-parameter extreme value distribution) which may fit the data better than any finite case of the three-parameter Weibull. If this happens, then the estimates will appear to diverge. In principle, this could happen with MMLE as well, though in fact in only one of our 4000 simulations, involving the MLE with $\delta = 4.0$, did this actually happen. A more serious point, however, is that there were several instances in which both δ and $X_1 - \tilde{\gamma}$ were very large, and this distorted the variances of the estimators. Some trial runs with $n = 25$ and $\delta = 4.0$ suggested that for this smaller value of n, there were many more instances when the procedures did not converge at all, but these are not reproduced here.

As an assessment of the performance of the estimators, both variances and mean squared errors (MSEs) of the estimators were calculated, ignoring any instances where the estimator did not converge. In most cases the MSE was not much larger than the variance, showing that bias is not a serious issue at this sample size. However, as a robust alternative to the standard deviation, I also computed 0.741 times the interquartile range (IQR) of the simulated estimates. In this case, non-converging cases were not deleted as these may legitimately be regarded as outliers of the sampling distribution. The multiplier 0.741 is based on the fact that the standard deviation is 0.741 IQR for a normal distribution, though as previously pointed out, the estimators are not exactly normally distributed (even asymptotically) for this problem. In all cases, the quantity actually tabulated is n times the estimated variance of the distribution.

For each value of δ, the four rows of Table 24.1 represent (i) the sample variance of the simulations, (ii) the MSE, (iii) the IQR-based method just described, (iv) the theoretical value obtained from the information matrix in the case of MLE, and (24.3.9) for MMLE. With MLE for $\delta < 2$, no asymptotic value is shown for $\hat{\gamma}$, while the estimates for $\hat{\delta}$ and $\hat{\beta}$ are those for the two-parameter Weibull, since this is the right asymptotic result when $\delta < 2$. Also shown in parentheses under the MMLE for γ is nd_n, which is the value that would be obtained under the first-order approximation given by Theorem 24.3.1 of Section 24.3.

For $\delta = 4$, the differences between the variance (or MSE) and the IQR-based estimates are drastic; however the variance is dominated by the effect of a few outlying terms, and for this reason I believe that the IQR is a much more realistic estimate of the variability of this procedure. For the other values of δ, this difference does not appear to matter.

In comparing theoretical and simulated (IQR-based) variances, it can be seen that with $\delta = 4$ or 2.5, the theoretical results for both MLE and MMLE somewhat underestimate the true value, but the difference is not too bad, and certainly much better than for the uncorrected estimated for the variance of $\tilde{\gamma}$, which is out by a factor of up to 10. When $\delta = 1.75$, the approximation for the MMLE is very good, and in the case of $\tilde{\delta}$ and $\tilde{\beta}$ certainly better than the uncorrected approximations, which are the same as for MLE. Only for $\delta = 1.2$ are the new results somewhat unconvincing, but I have already given a plausible explanation of this.

In an attempt to see just how well the results are capable of doing, the whole experiment was repeated with 500 replications of sample size 500, with results in Table 24.2. This is too large a sample size for most reliability experiments, but the simulation is included mainly to demonstrate that for a value of n that is large but not extremely large, the new approximations do perform well and much better than the first-order asymptotics. In this case, such a conclusion seems substantiated for all cases expect $\delta = 1.2$; here there is not much to choose between the two methods of approximation.

As far as the comparison between MLE and MMLE is concerned — well, there are no cases when MMLE does clearly better than MLE on the simulations, except when judged by raw variances in the $n = 100$, $\delta = 4$ case. It may be that such a result would be seen for smaller sample sizes, but in that case even our new approximations are not going to be good enough to explain what is going on theoretically. The computational simplicity of MMLE, which Cohen has demonstrated on numerous occasions, remains one feature in its favor. Apart from that, I believe the main advantage of MMLE is that its behavior is reasonably well understood, and certainly better understood than the multi-parameter MLE, for the regime when δ is near to or less than 2, which is when the standard asymptotics of MLE start to break down. This is very much consistent with the conclusions Cohen himself has reached from his extensive numerical studies. It is hoped that the results given here will serve to stimulate further theoretical research into these issues.

Acknowledgements

This research was partially supported by NSF grant DMS-9205112, and by the National Institute of Statistical Sciences.

References

1. Abramowitz, M. and Stegun, I.A. (1964), *Handbook of Mathematical Functions.* National Bureau of Standards, Washington D.C., reprinted by Dover, New York.

2. Akahira, M. (1975), Asymptotic theory for estimation of location in non-regular cases. I: Order of convergence of consistent estimators. II: Bounds of asymptotic distributions of consistent estimators, *Rep. Statist. Appl. Res. Union Jap. Sci. Eng.,* **22**, 8-26 and 99-115.

3. Chan, M., Cohen, A.C., and Whitten, B.J. (1984), Modified maximum likelihood and modified moment estimators for the three-parameter inverse Gaussian distribution, *Comm. Stat. Simul. Comp.,* **13**, 47-68.

4. Cheng, R.C.H. and Iles, T.C. (1989), Embedded models in three parameter distributions and their estimation, *J.R. Statist. Soc. B.,* **52**, 135-149.

5. Charernkavanich, D. and Cohen, A.C. (1984), Estimation in the singly truncated Weibull distribution with unknown truncation point, *Comm. Stat. Theor. Meth.,* **A13**, 843-857.

6. Cohen, A.C. (1951), Estimating parameters of logarithmic-normal distributions by the method of maximum likelihood, *Journal of the American Statistical Association,* **46**, 206-212.

7. Cohen, A.C. (1965), Maximum likelihood estimation in the Weibull distribution based on complete and on censored samples, *Technometrics,* **7**, 579-588.

8. Cohen, A.C. (1975), Multi-censored sampling in the three parameter Weibull distribution, *Technometrics,* **17**, 347-351.

9. Cohen, A.C. (1991), *Truncated and Censored Samples: Theory and Applications,* Marcel Dekker, New York.

10. Cohen, A.C. and Whitten, B.J. (1980), Estimation in the three-parameter lognormal distribution, *Journal of the American Statistical Association,* **75**, 399-404.

11. Cohen, A.C. and Whitten, B.J. (1982a), Modified moment and maximum likelihood estimators for parameters of the three-parameter gamma distribution, *Comm. Stat. Simul. Comp.,* **11**, 197-214.

12. Cohen, A.C. and Whitten, B.J. (1982b), Modified maximum likelihood and modified moment estimators for the three-parameter Weibull distribution, *Comm. Stat. Theor. Meth.,* **11**, 2631-2656.

13. Cohen, A.C. and Whitten, B.J. (1985), Modified moment estimation for the three-parameter inverse Gaussian distribution, *J. Qual. Tech.,* **17**, 147-154.

14. Cohen, A.C. and Whitten, B.J. (1986), Modified moment estimation for the three-parameter gamma distribution, *J. Qual. Tech.,* **18**, 53-62.

15. Cohen, A.C. and Whitten, B.J. (1988), *Parameter Estimation in Reliability and Life Span Models,* Marcel Dekker, New York.

16. Cohen, A.C., Whitten, B.J., and Ding, Y. (1984), Modified moment estimation for the three-parameter Weibull distribution, *J. Qual. Tech.,* **16**, 159-167.

17. Cohen, A.C., Whitten, B.J., and Ding, Y. (1985), Modified moment estimation for the three-parameter lognormal distribution, *J. Qual. Tech.,* **17**, 92-99.

18. Griffiths, D.A. (1980), Interval estimation for the three-parameter lognormal distributions via the likelihood equations, *Applied Statistics,* **29**, 58-68.

19. Harter, H.L. and Moore, A.M. (1966), Local maximum likelihood estimation of the parameters of three-parameter log-normal populations from complete and censored samples, *Journal of the American Statistical Association*, **61**, 842-851.

20. Hill, B.M. (1963), The three-parameter log-normal distribution and Bayesian analysis of a point-source epidemic, *Journal of the American Statistical Association*, **58**, 72-84.

21. Ibragimov, I.A. and Has'minskii, R.Z. (1981), *Statistical Estimation*, Springer-Verlag, Berlin.

22. Lemon, G.H. (1975), Maximum likelihood estimation for the three-parameter Weibull distribution based on censored samples, *Technometrics*, **17**, 247-254.

23. Lockhart, R.A. and Stephens, M.A. (1994), Estimation and tests of fit for the three-parameter Weibull distribution, *Journal of the Royal Statistical Society, Series B*, **56**, 491-500.

24. Press, W.H., Flannery, B.P., Teukolsky, S.A., and Vetterling, W.T. (1986), *Numerical Recipes: The Art of Scientific Computation*, Cambridge University Press, Cambridge.

25. Rockette, H., Antle, C., and Klimko, L.A. (1974), Maximum likelihood estimation with the Weibull model, *Journal of the American Statistical Association*, **69**, 246-249.

26. Smith, R.L. (1985), Maximum likelihood estimation in a class of nonregular cases, *Biometrika*, **72**, 67-90.

27. Smith, R.L. (1994), Nonregular regression, *Biometrika*, **81**, 173-183.

28. Wilson, E.B. and Worcester, J. (1945), The normal logarithmic transform, *Rev. Econ. Statist.*, **27**, 17-22.

29. Wingo, D.R. (1973), Solution of the three-parameter Weibull equations by constrained modified quasi-linearization (progressively censored samples), *IEEE Transactions on Reliability*, **R-22**, 96-102.

30. Woodroofe, M. (1972), Maximum likelihood estimation of a translation parameter of a truncated distribution, *Annals Math. Stat.*, **43**, 113-122.

31. Woodroofe, M. (1974), Maximum likelihood estimation of translation parameter of truncated distribution II, *Ann. Statist.*, **2**, 474-488.

TABLE 24.1
Simulation results and approximations for $n = 100$

Number	MLE $\widehat{\gamma}$	$\widehat{\delta}$	$\widehat{\beta}$	Number	MMLE $\tilde{\gamma}$	$\tilde{\delta}$	$\tilde{\beta}$
			(a) $\delta = 4$				
999	35.7	825	36.4	1000	17.7	461	18.5
	35.7	828	36.4		18.4	479	19.1
	3.85	88.2	4.29		6.68	143	7.33
	3.21	76.1	3.54		4.84	95.0	5.21
					(0.647)		
			(b) $\delta = 2.5$				
1000	1.18	18.9	1.67	1000	1.40	21.0	1.95
	1.25	19.1	1.77		1.43	22.0	1.99
	0.890	13.1	1.26		1.31	19.0	2.00
	0.396	8.04	0.670		1.08	13.1	1.47
					(0.362)		
			(c) $\delta = 1.75$				
1000	0.254	3.78	0.722	1000	0.292	4.17	0.794
	0.332	4.10	0.870		0.294	4.29	0.795
	0.219	3.26	0.753		0.282	4.11	0.894
	–	1.86	0.362		0.267	3.62	0.774
					(0.143)		
			(d) $\delta = 1.2$				
921	0.0305	1.01	0.725	1000	0.0407	1.31	0.829
	0.0556	1.07	0.810		0.0412	1.38	0.829
	0.0207	1.17	0.695		0.0296	1.25	0.753
	–	0.875	0.770		0.050	1.75	1.00
					(0.029)		

This table lists n times the variance of the MLEs and MMLEs for simulated samples of size $n = 100$ and four values of δ (together with $\gamma = 0$, $\beta = 1$). The results are based on 1000 replications and "Number" denotes the number of those replications for which the estimators were successfully computed. For each δ, the four rows of numbers are based on 1. the sample variance of the simulations, 2. the mean squared deviation from the true value (sample variance plus squared bias), 3. variance estimator computed from the IQR, 4. theoretical result as obtained from the Fisher information matrix in the case of MLE or (24.3.9) in the case of MMLE. The fifth number in parentheses in the column for $\tilde{\gamma}$ is nd_n, i.e. the crude variance approximation based on direct application of Theorem 24.3.1.

TABLE 24.2
Simulation results and approximations for $n = 500$

		MLE				MMLE	
Number	$\widehat{\gamma}$	$\widehat{\delta}$	$\widehat{\beta}$	Number	$\tilde{\gamma}$	$\tilde{\delta}$	$\tilde{\beta}$
			(a) $\delta = 4$				
500	4.11	91.8	4.39	500	6.92	148	7.32
	4.11	91.8	4.40		6.98	150	7.38
	3.69	85.6	4.13		7.22	152	7.42
	3.21	76.1	3.54		6.53	138	7.09
					(1.45)		
			(b) $\delta = 2.5$				
500	0.790	12.1	1.14	500	1.06	14.8	1.49
	0.808	12.3	1.16		1.07	14.9	1.50
	0.695	11.1	1.10		1.16	16.0	1.66
	0.396	8.04	0.670		1.02	13.7	1.41
					(0.500)		
			(c) $\delta = 1.75$				
500	0.126	2.97	0.611	500	0.146	3.17	0.641
	0.167	3.24	0.695		0.146	3.17	0.642
	0.108	3.23	0.612		0.134	3.24	0.604
	–	1.86	0.362		0.158	3.06	0.615
					(0.114)		
			(d) $\delta = 1.2$				
500	0.0099	0.974	0.883	500	0.0109	1.01	0.904
	0.0181	1.04	0.933		0.0109	1.02	0.905
	0.0069	0.883	0.946		0.0080	0.880	0.943
	–	0.875	0.770		0.0117	1.09	0.825
					(0.0098)		

This table is computed in the same way as for Table 24.1, except that it is for 500 replications based on samples of size $n = 500$.

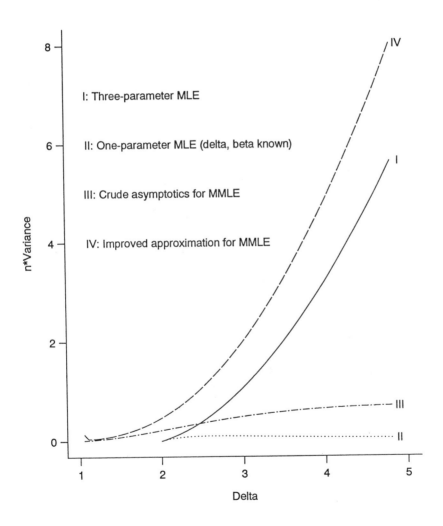

FIGURE 24.1
Variance approximations for estimating Gamma (sample size $n = 100$)

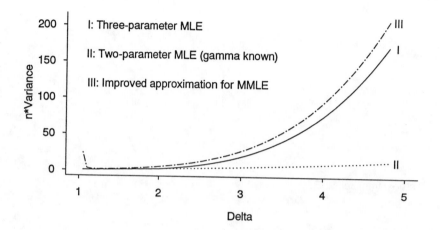

FIGURE 24.2
Variance approximations for estimating Delta (sample size $n = 100$)

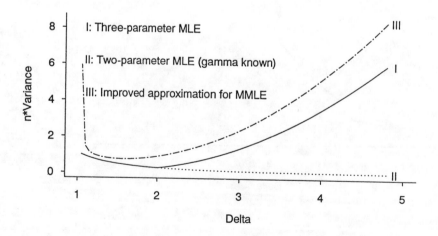

FIGURE 24.3
Variance approximations for estimating Beta (sample size $n = 100$)

25

Some Tests for Discriminating Between Lognormal and Weibull Distributions - An Application to Emissions Data

Gary C. McDonald, Lonnie C. Vance, and Diane I. Gibbons

Consumer and Operations Research Department, General Motors R & D Center
Cadillac Luxury Car Division, General Motors Corporation
Mathematics Department, General Motors R & D Center

ABSTRACT This article discusses statistical tests for discriminating between the two-parameter lognormal and the two-parameter Weibull distributions. Critical values and the power of these tests are derived using Monte Carlo methods. The tail probabilities and failure rates of the lognormal and Weibull distributions are shown in order to illustrate some important differences between the two models. These tests are applied to emissions data from an experimental car. The importance of discrimination conclusions is highlighted by determining the mean emissions required to achieve a specified probability of failing to meet a previously proposed hydrocarbon (HC) emission standard.

Keywords: Coefficient of variation, Failure rate, Goodness-of-fit, Power, Tail probability

CONTENTS

25.1 Introduction

The lognormal and the Weibull distributions are often considered for situations in which a skewed distribution for a nonnegative random variable is needed. Both the lognormal and the Weibull distributions have been used to describe various types of pollution data. For example, in [11] and [12], the author has conjectured, by means of a large number of data comparisons, that observed air pollution concentration distributions are closely approximated by the lognormal function. Emission data have been approximated by the lognormal and the Weibull distributions. As an aid in selecting one of these distributions various statistical tests may be used. In a recent review of the book **Lognormal Distributions: Theory and Applications** [6], the need to add a chapter on model discrimination (when to use the lognormal distribution and when not) was noted [18]. This article, hopefully, will help fill that need.

There is a wealth of literature concerning what are known as goodness-of-fit tests. There are, for example, the chi-square test ([24], [25]), the Kolmogorov-Smirnov test ([14], [15]), and the Srinivasan test ([23]). In [2], the authors provide a review of model selection procedures relevant to the Weibull distribution. They review probability plotting, correlation type goodness-of-fit tests, chi-square tests, as well as selection between Weibull and lognormal models. In [26], the authors consider the goodness-of-fit statistics based on the empirical distribution function and provide percentile values for test statistics appropriate for the two-parameter Weibull distribution when parameters are estimated by maximum likelihood. In [4], the author utilizes a statistic based on the stabilized probability plot for the assessment of goodness-of-fit of a two-parameter Weibull distribution.

In [7], the authors have shown that if the distribution in question has only a location and/or a scale parameter, goodness-of-fit tests, such as those listed above, which depend upon the probability integral transformation, are independent of the true parameter values when "proper" estimates are substituted for them.

For testing the two-parameter lognormal distribution versus the two-parameter Weibull distribution, in addition to the tests given above, we shall consider the ratio of maximum likelihoods (RML) test ([1] and [8]), the Shapiro-Wilk test [21], and the Mann-Scheuer-Fertig test [17].

Suppose an experimenter has observed $x_1, x_2, \ldots, x_n$ and on the basis of these observations wishes to choose either the two-parameter lognormal density,

$$f(x; \mu, \sigma) = \frac{1}{x\sqrt{2\pi\sigma^2}} \exp[-(\ln x - \mu)^2/2\sigma^2], \qquad x > 0, \ \sigma > 0, \qquad (25.1.1)$$

or the two-parameter Weibull density,

$$g(x; b, c) = \frac{c}{b}\left(\frac{x}{b}\right)^{c-1} \exp\left[-\left(\frac{x}{b}\right)^c\right], \qquad x > 0, \ b > 0, \ c > 0, \qquad (25.1.2)$$

as a model. Since $\ln X$ is distributed as a normal random variable with mean μ and variance σ^2 when X has a lognormal distribution, any of the common methods for testing for normality may be applied. If X is distributed as a Weibull random variable, then $\ln X$ has a type 1 extreme value (for minimums) distribution. For example, if X has the distribution in (25.1.2) then $\ln X = Y$ has as its density

$$h(y; u, d) = \frac{1}{d}\exp\left(\frac{y-u}{d}\right) \times \exp\left[-\exp\left(\frac{y-u}{d}\right)\right], \qquad (25.1.3)$$

where $u = \ln b$ and $d = \frac{1}{c}$.

If the parameters in the model are estimated from the data, as will no doubt be the case, then the critical values need to be adjusted accordingly. Furthermore, the same critical values presented for testing the lognormal versus Weibull distributions may be used directly for testing the normal versus type 1 extreme value (for minimums) distributions.

In the next section of this article, we discuss statistical tests for deciding on the appropriateness of the two-parameter lognormal or the two-parameter Weibull models for a set of data. We also consider tests designed to discriminate between the two model choices. The importance of this discrimination problem is illustrated in Section 25.3 by comparing the tail probabilities and failure rates of the lognormal and Weibull distributions. Section 25.4 presents critical values and the power of these tests which are derived using Monte Carlo methods. In Section 25.5, these tests are applied to emissions data from an experimental car. The importance of discrimination conclusions is highlighted by determining the mean emissions required to achieve a specified probability of failing to meet a previously proposed hydrocarbon (HC) emission standard. A summary and conclusions section comprises the final section of this article.

25.2 Description of Tests

Chi-square Test

It is well-known [9, pp. 293-298] that the classical Pearson chi-square goodness-of-fit test can be modified to fit the case when the parameters are unspecified. The distribution of the chi-square test statistic using the maximum likelihood estimators does not depend upon the true values of the nuisance parameters if the model is a location and scale parameter distribution or one such as the Weibull or lognormal obtained from a location and scale parameter distribution. When there are unknown parameters in the model being considered, the critical values need to be adjusted appropriately.

Although the chi-square test may be useful for determining the appropriateness of a given model for the data at hand, there are several objections to this test such as the arbitrariness in the choice of class intervals and its validity when the sample size is small. In comparison with other tests the chi-square test is not very powerful. An excellent treatment of the chi-square goodness-of-fit test is given in [5], [24], and [25].

Kolmogorov-Smirnov Test

The Kolmogorov-Smirnov test, like the chi-square test, is a test of fit with no particular alternative specified. Generally speaking it has at least two major advantages over the chi-square test; namely (1) it can be used with small sample sizes, and (2) it appears to be a more powerful test than the chi-square test for any sample size.

When certain parameters of the distribution must be estimated from the sample, the commonly tabulated critical points for using the Kolmogorov-Smirnov test no longer apply. The critical values and power of the test can be determined by Monte Carlo methods.

As mentioned previously, when X is distributed as a lognormal random variable then $ln\, X = Y$ is distributed as a normal random variable. Consequently, given a sample of n observations, we can determine

$$\hat{D}_n = \sup_y |S_n(y) - F_0(y; \hat{\mu}, \hat{\sigma}^2)|$$

where $S_n(y)$ is the sample cumulative distribution function and $F_0(y; \hat{\mu}, \hat{\sigma}^2)$ is the hypothesized normal cumulative distribution function with μ replaced by $\hat{\mu} = \bar{y}$, the sample mean, and σ^2 replaced by $\hat{\sigma}^2 = \frac{1}{n} \sum_{i=1}^{n} (y_i - \bar{y})^2$.

In [14] and [15], the author has shown that for the normal and exponential cases, the test based on $\hat{D}_n$ is more powerful than the chi-square test for any sample size against certain alternatives. He has obtained the critical values for various sample sizes and commonly used levels of significance.

Srinivasan Test[1]

Another modification of the Kolmogorov-Smirnov test statistic based on a different estimator of the hypothesized cumulative distribution function F_0 was given by [23]. A brief account of the general method is given.

---

[1] The critical values and power calculations given by [23] are incorrect ([20]).

Suppose that $F_0(x; \theta_1, \ldots, \theta_k)$ is such that $\tilde{\theta}_1, \ldots, \tilde{\theta}_k$ are joint complete sufficient statistics for the parameters $\theta_1, \ldots, \theta_k$. For a fixed real number u define the random variable Z as follows:

$$z = \begin{cases} 1, \text{ if } x_1 \leq u \\ 0, \text{ otherwise.} \end{cases}$$

Then it is obvious that Z is an unbiased estimator of $F_0(u; \theta_1, \ldots, \theta_k)$ under H_0. It now follows from the general theory of sufficient statistics ([3], [13], [19]) that the statistic

$$\tilde{F}_0(u; \theta_1, \ldots, \theta_k) = E(Z|\tilde{\theta}_1, \ldots, \tilde{\theta}_k)$$

is the unique minimum variance unbiased estimator of $F_0(u; \theta_1, \ldots, \theta_k)$. Define the statistic $\tilde{D}_n$ as

$$\tilde{D}_n = \sup_x |S_n(x) - \tilde{F}_0(x; \theta_1, \ldots, \theta_k)| \ .$$

If the distribution of $\tilde{D}_n$ does not depend on $\theta_1, \ldots, \theta_k$, then it would serve as an appropriate statistic for testing a composite null hypothesis H_0.

In [23], the author discusses this test for exponentiality. He also considers the case of normal distributions. In both cases Monte Carlo methods are used to obtain critical points of the test statistics for various sample sizes and commonly used levels of significance. Also in each case a Monte Carlo power comparison is made with the corresponding test developed by Lilliefors in [14] and [15].

Shapiro-Wilk Test

As stated previously any test for normality may be used when testing lognormal versus Weibull. In [21], the authors have proposed a useful test for normality. They propose that one use the ratio of the best linear unbiased estimate (BLUE) of σ to the maximum likelihood estimate of σ as a statistic to test the lack of fit of the normal model. This W (Shapiro-Wilk) statistic has been shown to be an effective measure of normality even for small samples ($n < 20$) against a wide spectrum of non-normal alternatives.

As a test for the normality of complete samples, the W statistic has several good features – namely, that it may be used as a test of the composite hypothesis and that it is very simple to compute once the table of linear coefficients, which are required to compute the BLUE of σ, is available.

A drawback of the W test is that for large sample sizes it may prove awkward to tabulate or approximate the necessary values of the coefficients in the numerator of the statistic. Also, it may be difficult for large sample sizes to determine percentage points of its distribution. In [21], the authors give coefficients for the W test for normality for $n = 2(1)50$. They also give percentage points for $n = 3(1)50$.

In [22], the authors summarize some of the results of an empirical sampling study of the comparable sensitivities of nine statistical procedures for evaluating the supposed normality of a complete sample. Forty-five alternative distributions in twelve families and five sample sizes ($n = 10, 15, 20, 35, 50$) were studied. Their general findings include that the W statistic provides a generally superior omnibus measure of non-normality.

Ratio of Maximum Likelihoods (RML) Test

Instead of testing the lack of fit of a particular model, the author in [1] proposed a test which enables the experimenter to choose from two possible models where the two possible models are restricted to be location and scale parameter distributions with unknown parameters. This would include the situation where the model is either of the lognormal or Weibull type (with unknown scale and shape parameters). The RML test differs from the tests mentioned previously because they are tests of fit with no particular alternative, whereas the RML test requires and uses a particular alternative.

To illustrate the RML test, with H_0: lognormal and H_1: Weibull, let $f_0(x; \mu, \sigma)$ be the lognormal density and let $f_1(x; b, c)$ be the Weibull density. The maximum likelihood estimates $\hat{b}$ and $\hat{c}$ of b and c must be obtained by iterative methods.

Then

$$RML = \frac{\max_{b,c} \prod_1^n f_1(x_i; b, c)}{\max_{\mu,\sigma} \prod_1^n f_0(x_i; \mu, \sigma)},$$

so

$$(RML)^{\frac{1}{n}} = (2\pi e \hat{\sigma}^2)^{\frac{1}{2}} \left[\prod_1^n x_i f_1(x_i; b, c) \right]^{\frac{1}{n}},$$

where $\hat{\sigma}^2 = \frac{1}{n} \sum_1^n (\ln x_i - \hat{\mu})^2$ and $\hat{\mu} = \frac{1}{n} \sum_1^n \ln x_i$.

We reject the lognormal in favor of the Weibull distribution whenever $(RML)^{1/n} > (RML)_c^{1/n}$ where $(RML)_c^{1/n}$ denotes the critical values for $(RML)^{1/n}$. The critical points and the power of the test are obtained by simulation. In [1] and [8], the authors have obtained critical values and the power of the RML test for testing H_0: lognormal against H_1: Weibull for sample sizes $n = 20(10)50$ and levels of significance $\alpha = .01, .05, .10,$ and .20. The RML test appears to be clearly superior to the chi-square test, the Kolmogorov-Smirnov test developed by Lilliefors, and the Srinivasan test for discriminating between the lognormal and the Weibull distributions.

Mann-Scheuer-Fertig Test

In [17], the authors developed a test of fit to the type 1 extreme value distribution with unknown parameters. To make a test of fit to the Weibull distribution, one first takes natural logarithms of the supposed Weibull data. This test is shown to have desirable power properties, relative to analogues of certain classical tests, against two important classes of alternatives. The test statistic is easy to calculate and can be used for censored samples. Mann, Scheuer, and Fertig have provided percentage points and certain expected values for sample sizes $n = 3(1)25$ which are needed to implement the test. The closeness of these percentage points to Beta percentage points, even for small n, and the existence of tables of the needed expected values for $n = 1(1)50(5)100$ make it possible to use the test in the range $25 < n < 100$ as well.

The statistic on which their test was based is

$$S = \frac{\sum_{i=[\frac{m}{2}]+1}^{m-1} (X_{i+1,n} - X_{i,n})/[E(Y_{i+1,n}) - E(Y_{i,n})]}{\sum_{i=1}^{m-1} (X_{i+1,n} - X_{i,n})/[E(Y_{i+1,n}) - E(Y_{i,n})]}$$

where $X_{i,n}$ is the ith ordered sample variate from the extreme value distribution given by (25.1.3), m is the value at which the sample is censored, $Y_{i,n} = \frac{X_{i,n} - u}{d}$, and $[\frac{m}{2}]$ is the largest integer contained in $\frac{m}{2}$.

They compared the power of several different versions of their S statistic with analogues (using best linear invariant estimates in place of parameters) of the Kolmogorov-Smirnov test, the Cramer-von Mises test, the Anderson-Darling test, and the Kuiper test. These power comparisons were made via Monte Carlo simulation.

They considered various three-parameter Weibulls as alternatives to the hypothesis of a two-parameter Weibull, and a normal distribution as an alternative to the hypothesis

TABLE 25.1
Descriptive measures for the lognormal and Weibull distributions

| Descriptive | Distribution | |
Measure	Lognormal	Weibull
Mean	$\exp(\mu + \sigma^2/2)$	$b\Gamma(1 + 1/c)$
Median	$\exp(\mu)$	$b[(ln\,2)^{1/c}]$
Mode	$\exp(\mu - \sigma^2)$	$b(1 - 1/c)^{1/c}$
Variance	$\exp(2\mu + \sigma^2)[\exp(\sigma^2) - 1]$	$b^2[\Gamma(1 + 2/c) - \Gamma^2(1 + 1/c)]$
Percentile	$\exp(\mu + \nu_q\sigma)$	$b[-ln(1 - q)]^{1/c}$

of an extreme value distribution. (The latter comparison is equivalent to comparing a two-parameter Weibull distribution with a two-parameter lognormal distribution.)

25.3 A Comparison of the Tail Probabilities and Failure Rates for the Two-Parameter Lognormal and Weibull Distributions

A Comparison of the Tails

One justification for using the lognormal distribution as the preferred distribution in an hypothesis testing context is that the tail of this distribution is longer (or heavier) than the corresponding tail of a Weibull distribution. (Unless otherwise stated, it will be assumed that the family of distributions to which reference is made is the two-parameter family.) If an analyst is entertaining the possibility of fitting empirical data with either the lognormal family or the Weibull family, the "conservative" approach would be to specify the (composite) null hypothesis to be H_0: X_i, $i = 1, ..., n$ are independent identically distributed with a lognormal distribution. The (composite) alternative hypothesis would then be a similar statement with "lognormal" replaced by "Weibull". This approach would seem to be particularly attractive if the goal of the analysis is a tail probability statement, or an inference based on such computations. Most statistical analyses of failure or fatigue data do involve these tail probabilities. In other words, the acceptance of a lognormal hypothesis leads to larger estimates of tail probabilities than does the acceptance of a Weibull hypothesis. In this sense, the analyst is conservative. However, if the sample data is sufficiently in favor of the Weibull distribution, then the lognormal hypothesis would be rejected.

To clarify these assertions, let $\Lambda(\mu, \sigma^2)$ and $W(b, c)$ refer to the lognormal and Weibull distributions respectively, where the densities are specified in (25.1.1) and (25.1.2). Let ν_q be the quantile of order q of a standard normal random variable and let the gamma function be denoted by $\Gamma(x) = \int_0^\infty z^{x-1} \exp(-z)\, dz$. The expressions for several descriptive measures of the distributions are given in Table 25.1.

A comparison of the tail of the Weibull distribution to that of the lognormal distribution can be made by matching the median and .95 quantile. Taking $\mu = 0$ and $\sigma = 1$, the median and .95 quantile of the standard lognormal distribution is 1 and $\exp(\nu_{.95}) = \exp(1.645)$, respectively. The parameters b and c can be determined from the equations:

$$\begin{cases} 1 = b(ln\,2)^{1/c} \\ e^{1.645} = b[-ln(1 - .95)]^{1/c}, \end{cases}$$

TABLE 25.2
Ratio of probabilities of Weibull and lognormal distributions (each distribution has the same median and .95 quantile)

x	$P(X \geq x \mid X \sim W(1.51, .89))/P(X \geq x \mid X \sim \Lambda(0, 1))$
1.0	1.00000
2.0	1.13406
3.0	1.16535
4.0	1.11765
5.0	1.02101
6.0	.89985
7.0	.77155
8.0	.64731
9.0	.53360
10.0	.43352

which yield

$$\begin{cases} c = \frac{1}{1.645} \, ln(ln\,20/ln\,2) \doteq .89 \\ b = (ln\,2)^{-1/c} \doteq 1.51. \end{cases}$$

In Figure 25.1, the $\Lambda(0, 1)$ and W(1.51, .89) densities are drawn for $x \geq 5.181$. The common .95 quantile for these distributions is approximately 5.181, so the cumulative area under these densities to this point is .95. Furthermore, it can be shown that

$$P(X \geq x \mid X \sim W(1.51, .89)) < P(X \geq x \mid X \sim \Lambda(0, 1)), \quad x > 5.2,$$

i.e., the tail (or reliability) of the Weibull distribution is less than the tail of a lognormal distribution which is matched at the median and .95 quantile. The ratio of the two probabilities is given in Table 25.2. A similar comparison for the densities of five distributions is provided in [10, p. 44].

A Comparison of the Failure Rate

If the distribution function F has a density f, the **failure rate** function $r(t)$ is defined for those values of t for which $F(t) < 1$ by

$$r(t) = f(t)/[1 - F(t)].$$

The quantity $r(t)dt$ represents the probability that an item fails in the interval $[t, t + dt]$, given the item is working at time t. If X has a $\Lambda(\mu, \sigma^2)$ distribution, the failure rate function increases initially and then decreases for all choices of μ and σ^2. However, if X has a W(b,c) distribution $r(t)$ is strictly increasing if $c > 1$, constant if $c = 1$, or strictly decreasing if $c < 1$. Figure 25.2 presents the failure rates for the $\Lambda(0, 1)$ and $W(1.51, .89)$ distributions which were previously compared in terms of the tail probabilities.

25.4 Critical Values and Power of the Tests for H_0: Lognormal and H_1: Weibull

Table 25.3 presents the critical values of the various tests of H_0: lognormal against H_1: Weibull and their power (P(reject $H_0 \mid H_0$ is false)). The values are the result of simulations

TABLE 25.3
Critical values and power of the tests (sample size $n = 20$)

	$\alpha = .20$	$\alpha = .10$	$\alpha = .05$	$\alpha = .01$
		Critical Values		
Chi-square	4.0	5.5	6.5	9.5
Kolmogorov-Smirnov	.159	.175	.191	.220
Srinivasan	.156	.174	.188	.218
Shapiro-Wilk	.937	.922	.901	.862
RML	1.006	1.038	1.067	1.140
		Power		
Chi-square	.282	.166	.117	.041
Kolmogorov-Smirnov	.448	.313	.212	.092
Srinivasan	.471	.330	.224	.096
Shapiro-Wilk	.568	.433	.306	.150
RML	.731	.605	.479	.238

in which 1,000 samples were used for each level. For the chi-square test, five cells were chosen so that the fitted model (under H_0) gives a probability of .2 to each cell. It is easily seen that the RML test is the most powerful test.

25.5   Emission Examples

Experimental Car Emission Example

The data given in Table 25.4 is experimental car emission data. This data has been chosen for illustrative purposes and does not represent recent model year production cars. The first five tests described above were used to test the hypothesis that the observations were from a lognormal distribution. The RML test and the Mann-Scheuer-Fertig test were used to test the null hypothesis that the samples were drawn from a Weibull distribution. For the chi-square test, four cells were chosen so that the fitted model under the null hypothesis gives a probability of .25 to each cell.

The results of the hypothesis testing situations for each of three emissions with their corresponding mileage conditions are given in Table 25.5. It appears that hydrocarbon (HC) emission data at 0 miles and 4,000 miles follows a Weibull distribution while HC at 24,000 miles (before and after maintenance) follows a lognormal distribution. The lognormal seems to be the more appropriate distribution for carbon monoxide (CO) over the various mileage ranges. For nitrogen oxides (NO_x), the lognormal seems to be the more appropriate distribution, except possibly at 0 miles. These results are consistent with those of [16] who found that the lognormal distribution provided a reasonable approximation to the distribution of HC and CO emission data for a typical engine configuration.

Emissions Standards

Directly related to the problem of discriminating between the two-parameter lognormal and the two-parameter Weibull distributions are questions which pertain to meeting emissions standards. One such question which might be posed is what is the mean of

TABLE 25.4
Experimental car emission data[2]

	0 Miles			4,000 Miles			24,000 Miles (before maintenance)			24,000 Miles (after maintenance)		
Car	HC	CO	NO_x	HC	CO	NO_x	HC	CO	NO_x	HC	CO	NO_x
1	.16	2.89	2.21	.26	1.16	1.99	.23	2.64	2.18	.36	3.15	1.79
2	.38	2.17	1.75	.48	1.75	1.90	.41	2.43	1.59	.40	3.74	1.81
3	.20	1.56	1.11	.40	1.64	1.89	.35	2.20	1.99	.26	2.36	1.68
4	.18	3.49	2.55	.38	1.54	2.45	.26	1.88	2.29	.26	2.47	2.58
5	.33	3.10	1.79	.31	1.45	1.54	.43	2.58	1.95	.35	3.81	1.92
6	.34	1.61	1.88	.49	2.59	2.01	.48	4.08	2.21	.65	4.88	2.22
7	.27	1.14	2.20	.25	1.39	1.95	.41	2.49	2.51	.40	2.82	2.51
8	.30	2.50	2.46	.23	1.26	2.17	.36	2.23	1.88	.30	2.79	2.07
9	.41	2.22	1.77	.39	2.72	1.93	.41	4.76	2.48	.45	3.59	2.87
10	.31	2.33	2.60	.21	2.23	2.58	.26	3.73	2.70	.30	3.78	2.68
11	.15	2.68	2.12	.22	3.94	2.12	.58	2.48	2.32	.52	3.94	2.61
12	.36	1.63	2.34	.45	1.88	1.80	.70	3.10	2.18	.60	3.41	2.23
13	.33	1.58	1.76	.39	1.49	1.46	.48	2.64	1.69	.44	2.44	1.76
14	.19	1.54	2.07	.36	1.81	1.89	.33	2.99	2.35	.31	2.97	2.37
15	.23	1.75	1.59	.44	2.90	1.85	.48	3.04	1.79	.44	3.90	1.71
16	.16	1.47	2.25	.22	1.16	2.21	.45	3.78	2.03	.47	2.42	2.04

[2] Data are in grams/mile.

the lognormal distribution or the mean of the Weibull distribution necessary in order that only one car in 10,000 fails to meet a previously proposed standard for HC of .41 grams/mile.

The assumptions made are (1) that the HC emissions are distributed either as lognormal or Weibull, and (2) that a linear relationship exists between the mean and the coefficient of variation (CV), where CV represents the ratio of the standard deviation to the mean. The linear relationship was determined by using the estimated means and estimated CVs determined from end-of-line emission data and the experimental car HC emission data at 0 miles from Table 25.4 under the lognormal and Weibull model assumptions. The estimates are computed by first calculating the estimated means and standard deviations from the appropriate formulas in Table 25.1 with the population parameters replaced by the maximum likelihood estimates. The estimated CV is taken as the ratio of the estimated standard deviation to the estimated mean.

Assuming a lognormal distribution, the estimated mean and CV for the end-of-line HC emission data are 1.705 grams/mile and .190, respectively; and the corresponding quantities for the experimental car are .269 grams/mile and .343, respectively. Assuming a Weibull distribution, the estimated mean and CV for the end-of-line HC emission data are 1.639 grams/mile and .224, respectively; and the corresponding quantities for the experimental car are .270 grams/mile and .306, respectively. These linear relationships (Figure 25.3) are assumed to exist for the population parameters applicable to potential production vehicles. Using Figure 25.3 and the condition that $P(X > .41) = .0001$ where $X \sim \Lambda(\mu, \sigma^2)$ or $X \sim W(b,c)$, it is possible to solve for the respective means of the lognormal and the Weibull distributions.

For example, for the lognormal distribution the probability requirement $P(X > .41) = .0001$ implies the following relationship between μ and σ:

$$\mu = ln(.41) - \Phi^{-1}(.9999)\sigma, \tag{25.5.1}$$

where Φ^{-1} is the inverse of the standard normal cumulative distribution function. Since the mean of the lognormal distribution is $\exp(\mu + \sigma^2/2)$ and the CV of the lognormal distribution is $\sqrt{\exp(\sigma^2) - 1}$ (see Table 25.1), the linear relationship in Figure 25.3 provides

TABLE 25.5
Conclusions of the tests

	0	4,000	24,000 (before maint.)	24,000 (after maint.)
			Mileage	

HC

H_0: lognormal vs. H_1: Weibull

	0	4,000	24,000 (before maint.)	24,000 (after maint.)
Chi-square	$R^\dagger$	A	R°	A
K-S	$R^\dagger$	$R^\dagger$	A	A
Srinivasan	$R^\dagger$	$R^\dagger$	A	A
Shapiro-Wilk	$R^\dagger$	R°	A	A
RML	$R^\dagger$	$R^\dagger$	A	A

H_0: Weibull vs. H_1: lognormal

	0	4,000	24,000 (before maint.)	24,000 (after maint.)
RML	A	A	$R^\dagger$	R°
Mann-Schuer-Fertig	A	A	$R^\dagger$	$R^\dagger$

CO

H_0: lognormal vs. H_1: Weibull

	0	4,000	24,000 (before maint.)	24,000 (after maint.)
Chi-square	A	A	A	A
K-S	$R^\dagger$	A	$R^\dagger$	A
Srinivasan	$R^\dagger$	A	$R^\dagger$	A
Shapiro-Wilk	A	A	A	A
RML	A	A	A	A

H_0: Weibull vs. H_1: lognormal

	0	4,000	24,000 (before maint.)	24,000 (after maint.)
RML	$R^\dagger$	R^{**}	R^*	$R^\dagger$
Mann-Schuer-Fertig	A	R^*	R^{**}	A

NO$_x$

H_0: lognormal vs. H_1: Weibull

	0	4,000	24,000 (before maint.)	24,000 (after maint.)
Chi-square	A	A	A	A
K-S	A	A	A	A
Srinivasan	A	A	A	A
Shapiro-Wilk	R°	A	A	A
RML	R^*	A	A	A

H_0: Weibull vs. H_1: lognormal

	0	4,000	24,000 (before maint.)	24,000 (after maint.)
RML	A	$R^\dagger$	A	$R^\dagger$
Mann-Schuer-Fertig	A	R^*	A	A

A	Accept null hypothesis
$R^\dagger$	Reject null hypothesis at the $\alpha = .20$ level
R°	Reject null hypothesis at the $\alpha = .10$ level
R^*	Reject null hypothesis at the $\alpha = .05$ level
R^{**}	Reject null hypothesis at the $\alpha = .01$ level

the second equation in the two quantities μ and σ:

$$.372 - .107 \exp(\mu + \sigma^2/2) = \sqrt{\exp(\sigma^2) - 1}. \qquad (25.5.2)$$

Substituting (25.5.1) in (25.5.2) allows us to solve for $\sigma(= .348)$, resulting in the lognormal mean estimate of .119 grams/mile and a lognormal CV estimate of .359.

Following a similar approach, the estimated Weibull mean satisfying the probability requirement is .198 grams/mile, about 66% larger than the corresponding quantity based on the lognormal assumptions. The Weibull assumption allows for a larger mean to meet the probability requirement because the tail is smaller than that of the lognormal model as shown in Figure 25.1. Note the mean values obtained to meet the probability requirement are considerably smaller than the values realized by the experimental car under either distributional assumption.

25.6   Summary and Conclusions

For discriminating between the two-parameter lognormal and the two-parameter Weibull distributions the ratio of maximum likelihoods (RML) test appears to be the most powerful test and consequently the preferred test to use for this problem. Applying the statistical tests to various sets of experimental car emission data it appears that some emissions follow the lognormal distribution while others follow the Weibull distribution. For example, HC emission data at 0 miles and 4,000 miles appears to follow a Weibull distribution while HC at 24,000 miles (before and after maintenance) follows a lognormal distribution. This seems to indicate that one of these families of distributions is not sufficient to describe the stochastic behavior of all emissions over the operating life of the vehicle. The lognormal distribution seems to be the more appropriate distribution for CO over the range of mileage investigated. With respect to NO_x, the lognormal distribution appears to be the more appropriate, except possibly at 0 miles.

Assuming that a linear relation exists between the coefficient of variation and the mean based on end-of-line HC emission data and experimental car (0 miles) HC emission data, and that emissions are either distributed as a two-parameter lognormal or as a two-parameter Weibull, the mean for the lognormal and the mean for the Weibull are determined in order to meet a specified reliability condition. The mean values so determined differ substantially depending on the underlying distributional assumptions, thus reinforcing the importance of effective discrimination between lognormal and Weibull models.

References

1. Antle, C.E. (1972), Choice of model for reliability studies and related topics, *Technical Report TR 72-0108*, Aerospace Research Laboratories, Wright-Patterson Air Force Base, Dayton.

2. Bain, L.J. and Engelhardt, M. (1983), A review of model selection procedures relevant to the Weibull distribution, *Communications in Statistics – Theory and Methods*, **12**, 589-609.

3. Blackwell, D. (1947), Conditional expectation and unbiased sequential estimation,

Annals of Mathematical Statistics, **18**, 105-110.

4. Coles, S.G. (1989), On goodness-of-fit tests for the two-parameter Weibull distribution derived from the stabilized probability plot, *Biometrika*, **76**, 593-598.

5. Chernoff, H. and Lehmann, E.L. (1954), The use of maximum likelihood estimates in χ^2 tests for goodness of fit, *Annals of Mathematical Statistics*, **25**, 579-586.

6. Crow, E.L. and Shimizu, K. (Eds.) (1988), *Lognormal Distributions: Theory and Applications*, Marcel Dekker, New York.

7. David, F.N. and Johnson, N.L. (1948), The probability integral transformation when parameters are estimated from the sample, *Biometrika*, **35**, 182-190.

8. Dumonceaux, R. and Antle, C.E. (1973), Discrimination between the log-normal and Weibull distributions, *Technometrics*, **15**, 923-926.

9. Guttman, I., Wilks, S.S., and Hunter, J.S. (1971), *Introductory Engineering Statistics*, John Wiley & Sons, New York.

10. Hájek, J. (1969), *A Course in Nonparametric Statistics*, Holden-Day, San Francisco.

11. Larsen, R.I. (1970), Relating air pollutant effects to concentration and control, *Journal of the Air Pollution Control Association*, **20**, 214-224.

12. Larsen, R.I. (1971), A mathematical model for relating air quality measurements to air quality standards, *Office of Air Program Publication No. AP-89*, U. S. Environmental Protection Agency.

13. Lehmann, E.L. and Scheffé, H. (1950), Completeness, similar regions, and unbiased estimation, Part I, *Sankhyā*, **10**, 305-340.

14. Lilliefors, H.W. (1967), On the Kolmogorov-Smirnov test for normality with mean and variance unknown, *Journal of the American Statistical Association*, **62**, 399-402.

15. Lilliefors, H.W. (1969), On the Kolmogorov-Smirnov test for the exponential distribution with mean unknown, *Journal of the American Statistical Association*, **64**, 387-389.

16. Lorenzen, T.J. (1980), Determining statistical characteristics of a vehicle emissions audit procedure, *Technometrics*, **22**, 483-493.

17. Mann, N.R., Scheuer, E.M., and Fertig, K.W. (1973), A new goodness-of-fit test for the two-parameter Weibull or extreme-value distribution with unknown parameters, *Communications in Statistics*, **2**, 383-400.

18. McDonald, G.C. (1989), Book Review of Lognormal Distributions: Theory and Applications, *Journal of the American Statistical Association*, **84**, 845-846.

19. Rao, C.R. (1952), *Advanced Statistical Methods in Biometric Research*, John Wiley & Sons, New York.

20. Schafer, R.E., Finkelstein, J.M., and Collins, J. (1972), On a goodness-of-fit test for the exponential distribution with mean unknown, *Biometrika*, **59**, 222-224.

21. Shapiro, S.S. and Wilk, M.B. (1965), An analysis of variance test for normality (complete samples), *Biometrika*, **52**, 591-611.

22. Shapiro, S.S., Wilk, M.B., and Chen, H.J. (1968), A comparative study of various tests for normality, *Journal of the American Statistical Association*, **63**, 1343-1372.

23. Srinivasan, R. (1970), An approach to testing the goodness of fit of incompletely specified distributions, *Biometrika*, **57**, 605-611.

24. Watson, G.S. (1957), The χ^2 goodness-of-fit test for normal distributions, *Biometrika*, **44**, 336-348.

25. Watson, G.S. (1958), On chi-square goodness-of-fit tests for continuous distributions, *Journal of the Royal Statistical Society, Series B*, **20**, 44-72.

26. Wozniak, P.J. and Li, X. (1990), Goodness-of-fit for the two-parameter Weibull distribution with estimated parameters, *Journal of Statistical Computation and Simulation*, **34**, 133-143.

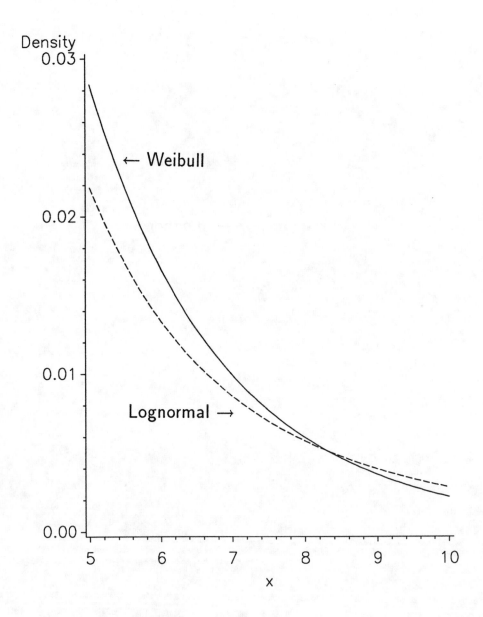

FIGURE 25.1
Tails of the Lognormal and Weibull Densities (Each distribution has the same
median and .95 quantile)

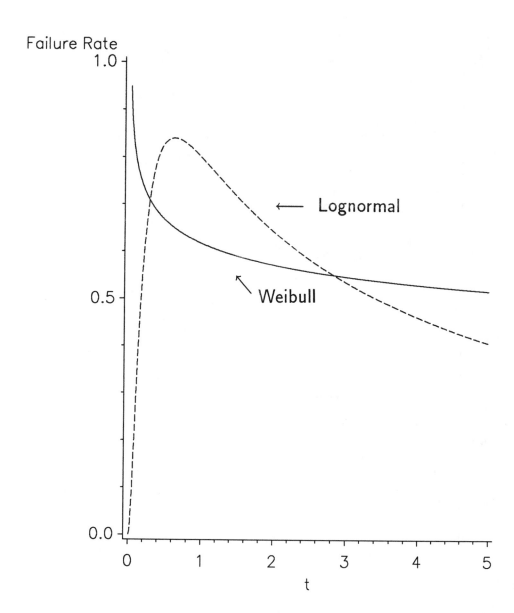

FIGURE 25.2
Failure Rates of the Lognormal and Weibull Distributions (Each distribution has
the same median and .95 quantile)

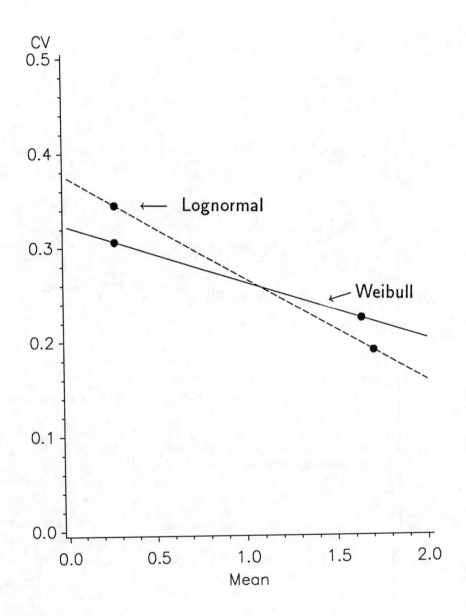

FIGURE 25.3
Linear Relationship Between the Mean and the Coefficient of Variation (CV)

26

Choosing a Model from Among Four Families of Distributions

K.B. Kulasekera and Peter R. Nelson

Clemson University

ABSTRACT We consider several methods of choosing from among the three-parameter Weibull, lognormal, inverse Gaussian, and gamma families based on parameter estimates obtained using modified moment estimators. The best method turns out to be a calculating Kolmogorov-Smirnov type statistics for each family and choosing the family with the smallest statistic value.

CONTENTS

26.1 Introduction

In a series of papers that appeared in the *Journal of Quality Technology* the authors in [2], [3], [4], and [5] discussed modified moment estimation for four families of three parameter distributions: Weibull, lognormal, inverse Gaussian, and gamma. In each paper the modified moment estimators (MME's) were derived for the particular distribution and their properties were investigated (mostly numerically). In all four papers the authors demonstrated their estimation methods using the same two data sets; see also [1] for a detailed comparative analysis based on these four distributions to the two data sets. The question thus naturally arises as to which model is the most appropriate for each of these two data sets. The purpose of this paper is to try to answer that question by considering the more general problem of model selection from among these four families of three-parameter distributions.

26.2 Notation

We will use the same parametrizations for each distribution as was used in [1], [2], [3], [4], and [5]. Specifically, for the Weibull distribution the threshold, shape, and scale parameters will be denoted by γ, δ, and β, respectively; and the Weibull density function

0-8493-8972-0/95/$0.00+$.50

is

$$f(x;\gamma,\delta,\beta) = \frac{\delta}{\beta^\delta}(x-\gamma)^{\delta-1}\exp\left\{-\left[(x-\gamma)/\beta\right]^\delta\right\} \tag{26.2.1}$$

for $\gamma < x < \infty$, $\delta > 0$, and $\beta > 0$; and zero otherwise. For a lognormal distribution with location parameter γ based on a normal distribution with mean μ and variance σ^2 the density function is

$$f(x;\gamma,\mu,\sigma) = \frac{1}{\sigma\sqrt{2\pi}(x-\gamma)}\exp\left\{\frac{-[\ell n(x-\delta)-\mu]^2}{2\sigma^2}\right\} \tag{26.2.2}$$

for $\gamma < x < \infty$ and $\sigma > 0$, and zero otherwise. The density function for the three-parameter inverse Gaussian distribution is

$$f(x;\gamma,\mu,\sigma) = \frac{\mu}{\sigma}\sqrt{\frac{\mu}{2\pi(x-\gamma)^3}}\exp\left[\frac{-\mu\{(x-\gamma)-\mu\}^2}{2\sigma^2(x-\gamma)}\right] \tag{26.2.3}$$

for $\gamma < x < \infty$, $\mu > 0$, and $\sigma > 0$; and zero otherwise. For a gamma distribution with threshold, shape, and scale parameters denoted by γ, ρ, and β, respectively; the density function is

$$f(x;\gamma,\rho,\beta) = \frac{(x-\gamma)^{\rho-1}}{\Gamma(\rho)\beta^\rho}\exp\left[-(x-\gamma)/\beta\right] \tag{26.2.4}$$

for $\gamma < x < \infty$, $\rho > 0$, and $\beta > 0$; and zero otherwise.

26.3   Model Selection

In the event that a complete data set is available, one can estimate the parameters for each model using the MME's suggested in [1], [2], [3], [4], and [5]. We consider two methods for choosing the best model from among the four distributions.

26.3.1   Method 1

The first method is the counterpart to choosing a distribution from among the Pearson family. Namely, one computes estimates of the skewness and kurtosis nonparametrically using

$$\hat{\alpha}_3 = \frac{m_3}{s^3} \text{ and } \hat{\alpha}_4 = \frac{m_4}{s^4} \tag{26.3.1}$$

where

$$m_j = \frac{1}{n}\sum_1^n (X_i - \bar{X})^j$$

$$s = \sqrt{m_2}$$

and compares these values with

$$\tilde{\alpha}_3 = \frac{\hat{\mu}_3}{\hat{\sigma}^3} \text{ and } \tilde{\alpha}_4 = \frac{\hat{\mu}_4}{\hat{\sigma}^4}$$

where for each distribution $\hat{\mu}_3$, $\hat{\mu}_4$, and $\hat{\sigma}$ are computed based on their relationship to the MME's of the distribution's parameters. We considered two ways of measuring the distance between the nonparametric estimates and the parametric estimates, namely, the Euclidean distance in the (skewness, kurtosis) plane and the maximum of the absolute differences of the parametric and nonparametric estimates of skewness and kurtosis. For both measures the family with the minimum distance is selected.

Weibull Distribution

Since the location and scale parameters (i.e., γ and β) do not effect the skewness or kurtosis, one need only consider the case $\gamma = 0$ and $\beta = 1$. In that case the rth moment about the origin is given by (see, e.g., [7])

$$\Gamma_r = \mu'_r = \Gamma\left(r/\delta + 1\right).$$

It follows that

$$\sigma^2 = E\left[(X - \Gamma_1)^2\right] = \Gamma_2 - \Gamma_1^2$$

$$\mu_3 = E\left[(X - \Gamma_1)^3\right] = \Gamma_3 - 3\Gamma_1\Gamma_2 + 2\Gamma_1^3$$

$$\mu_4 = E\left[(X - \Gamma_1)^4\right] = \Gamma_4 - 4\Gamma_1\Gamma_3 + 6\Gamma_2\Gamma_1^2 - \Gamma_1^4\,.$$

Therefore,

$$\tilde{\alpha}_3 = \frac{\widehat{\Gamma}_3 - 3\widehat{\Gamma}_1\widehat{\Gamma}_2 + 2\widehat{\Gamma}_1^3}{\left(\widehat{\Gamma}_2 - \widehat{\Gamma}_1^2\right)^{3/2}}$$

and

$$\tilde{\alpha}_4 = \frac{\widehat{\Gamma}_4 - 4\widehat{\Gamma}_1\widehat{\Gamma}_3 + 6\widehat{\Gamma}_2\widehat{\Gamma}_1^2 - 3\widehat{\Gamma}_1^4}{\left(\widehat{\Gamma}_2 - \widehat{\Gamma}_1^2\right)^2}$$

where

$$\widehat{\Gamma}_r = \Gamma\left(r/\widehat{\delta} + 1\right)$$

and $\widehat{\delta}$ is the solution to (see [4])

$$\frac{s^2}{\left(\bar{x} - x_{(1)}\right)^2} = \frac{\widehat{\Gamma}_2 - \widehat{\Gamma}_1^2}{\left[\left(1 - n^{-1/\widehat{\delta}}\right)\widehat{\Gamma}_1\right]^2}\,.$$

Lognormal Distribution

For the lognormal distribution the skewness and kurtosis are given by (see, e.g., [7])

$$\alpha_3 = (\omega + 2)\sqrt{\omega - 1}$$

and

$$\alpha_4 = \omega^4 + 2\omega^3 + 3\omega^2 - 3$$

where $\omega = \exp(\sigma^2)$. Therefore,

$$\tilde{\alpha}_3 = (\hat{\omega} + 2)\sqrt{\hat{\omega} - 1}$$

and

$$\tilde{\alpha}_4 = \hat{\omega}^4 + 2\hat{\omega}^3 + 3\hat{\omega}^2 - 3$$

where $\hat{\omega}$ is the solution to (see [5])

$$\frac{s^2}{(\bar{x} - x_{(1)})^2} = \frac{\hat{\omega}(\hat{\omega} - 1)}{\left[\sqrt{\hat{\omega}} - \exp\left\{\sqrt{\ell n\hat{\omega}}E\left(Z_{(1),n}\right)\right\}\right]^2}$$

and $Z_{(1),n}$ is the first order statistic for a sample of size n from a standard normal distribution.

Inverse Gaussian Distribution

For the inverse Gaussian distribution the skewness and kurtosis are (see, e.g., [7])

$$\alpha_3 = \frac{3\sigma}{\mu}$$

and

$$\alpha_4 = 3 + 15\left(\frac{\sigma}{\mu}\right)^2 = 3 + 15\left(\frac{\alpha_3}{3}\right)^2 .$$

In this case α_3 is estimated in [2] directly as the solution to

$$G\left(z_1; 0, 1, \tilde{\alpha}_3\right) = \frac{1}{n+1} \tag{26.3.2}$$

where

$$G(z; 0, 1, \alpha_3) = \Phi\left[\frac{z}{\sqrt{1 + (\alpha_3/3)z}}\right] + \exp\left(18/\alpha_3^2\right)\Phi\left[\frac{-(z + 6/\alpha_3)}{\sqrt{1 + (\alpha_3/3)z}}\right]$$

$z_1 = \left(x_{(1)} - \bar{x}\right)/s$, and $\Phi(\cdot)$ is the cumulative standard normal distribution function. It follows immediately that

$$\tilde{\alpha}_4 = 3 + 15\left(\frac{\tilde{\alpha}_3}{3}\right)^2 .$$

Gamma Distribution

For the gamma distribution the skewness and kurtosis are (see, e.g., [7])

$$\alpha_3 = \frac{2}{\sqrt{\rho}}$$

and

$$\alpha_4 = 3 + \frac{6}{\rho} = 3 + \frac{24}{\alpha_3^2} .$$

In this case also α_3 is estimated in [3] directly as the solution to equation (26.3.2), where in this case

$$G(z; 0, 1, \alpha_3) = \int_{-2/\alpha_3}^{z} g(t; 0, 1, \alpha_3) dt$$

$$g(z; 0, 1, \alpha_3) = \frac{a^{a^2-1}}{\Gamma(a^2)} (z + a)^{a^2-1} \exp(-a(z + a))$$

and $a = 2/\alpha_3$. It follows immediately that

$$\tilde{\alpha}_4 = 3 + \frac{24}{\tilde{\alpha}_3^2}.$$

26.3.2 Method 2

The second method consists of calculating Kolmogorov-Smirnov type statistics for each of the four families and choosing the family with the smallest statistic value. More specifically, one computes the MME's of the parameters for each distribution and compares the estimated cumulative distribution function for each family with the empirical distribution function. The family for which the maximum difference between the estimated function and the empirical function is smallest is declared to be the best model.

For the Weibull family one computes

$$S_1 = \sup_x \left| 1 - \exp\left\{ -\left[(x - \hat{\gamma})/\hat{\beta} \right]^{\hat{\delta}} \right\} - F_n(x) \right|$$

where $F_n(x)$ is the empirical distribution function. For the other three families values of the cumulative distribution function must be obtained numerically. Letting $\hat{F}(\cdot)$ represent the estimated cumulative distribution function, one computes

$$S = \sup_x \left| \hat{F}(x) - F_n(x) \right|$$

Thus, for the lognormal family one computes

$$S_2 = \sup_x \left| F(x; \hat{\gamma}, \hat{\mu}, \hat{\sigma}) - F_n(x) \right|$$

where $F(\cdot)$ is the integral of (26.2.2). Similarly one computes S_3 and S_4 for the inverse Gaussian family and the gamma family using the integrals of (26.2.3) and (26.2.4), respectively.

26.4 The Probability of Correct Selection

A simulation study was conducted to evaluate the probability of correct selection for each of these methods. A separate simulation study was conducted for each family. For samples of size $n = 100, 200, 500$, and 1000 data sets were generated, the parameters were estimated for each of the four distributions, and for each of the two methods the proportion of correct selections was recorded. Four different sets of parameter combinations were used for each distribution. The location parameter γ for all the distributions and the scale parameter for the Weibull were arbitrarily chosen to be 10 and 16, respectively. Then four shape parameters were chosen for the Weibull, and the parameters for the remaining distributions were chosen to match the means and variances of the four Weibull distributions. For each sample size and parameter combination pair 1000 trials were run. The results are reported in Tables 26.1-26.4 for the Weibull, lognormal, inverse Gaussian, and gamma families, respectively. The random deviates were generated using [6]. More specifically, the lognormal and gamma random deviates were obtained directly using IMSL generators, the Weibull deviates were obtained by converting IMSL uniform deviates, and the inverse Gaussian deviates were obtained using the algorithm of [8], where the required chi-squared deviates were obtained from IMSL normal deviates. With 1000 trials the standard deviations associated with the probabilities in Tables 26.1-26.4 are less than $\sqrt{0.001/4} = 0.0158$.

From looking at these tables two things are immediately obvious. First, the two distance measures for the skewness/kurtosis method lead to essentially the same probabilities of correct selection for all the cases studied. Second, which of the two methods results in the higher probability of correct selection depends on the underlying distribution, and in the case of the Weibull, it also depends on the parameter values.

26.4.1   Combining the Two Methods

In order to see if the probability of correct selection could be improved and to provide a definitive selection rule, we considered several ways of combining the two methods. Clearly, if both methods indicate the same distribution, then that distribution should be selected. If the two methods do not agree, then the combined selection rules given in Table 26.5 appeared to give the best results. These rules for selection can be summarized as follows. If either method indicates the Weibull distribution, then choose Weibull. If neither method indicates the Weibull but one of them indicates the gamma distribution, then choose the gamma. If only the lognormal or the inverse Gaussian are indicated, then choose the distribution selected by the Kolmogorov-Smirnov type statistic.

While this method of combining the two statistics works well for the Weibull and gamma distributions, the probabilities of correct selection for the log normal are slightly worse then just using the Kolmogorov-Smirnov type statistics and for the inverse Gaussian the probabilities of correct selection are abysmal (see Tables 26.1-26.4).

An investigation into why the procedure performed so poorly with the inverse Gaussian distributions revealed that this particular set of inverse Gaussian distributions had very large kurtosis values (138, 84.4, 59.2, and 35.7), and with large kurtosis values the non-parametric estimate of kurtosis $\hat{\alpha}_4$ (see (26.3.1)) was seriously underestimating the true value. We tried two different methods to compensate for this. First, we tried to reduce the impact of the kurtosis by weighting the skewness and kurtosis differences. Specifically, we tried replacing the Euclidean distance with

$$\sqrt{\left[\frac{\hat{\alpha}_3 - \tilde{\alpha}_3}{\tilde{\alpha}_3}\right]^2 + \left[\frac{\hat{\alpha}_4 - \tilde{\alpha}_4}{\tilde{\alpha}_4}\right]^2}. \tag{26.4.1}$$

This did not produce satisfactory results (see Tables 26.6-26.9), so we tried considering the skewness/kurtosis results only when the the parametric estimate of kurtosis was less than ten. Unfortunately, that didn't work much better. Almost all of the probabilities of correct selection were within $\sigma = 0.16$ of the probabilities obtained using the weighted measure (26.4.1). The only probabilities with differences greater than σ were obtained with inverse Gaussian data and are given in Table 26.7.

Taking all this into account, it seems that to select from among the four families one does best by just considering the Kolmogorov-Smirnov type statistic. If however, one can rule out the inverse Gaussian and is interested in selecting from among the remaining three distributions, then one should use the skewness/kurtosis measure (26.4.1) and the selection rules in Table 26.5.

26.5 Examples

The two data sets used in [1], [2], [3], [4], and [5] contained only 20 and 10 values, and with so few values one would not expect to be able to identify the underlying three-parameter distribution with very high probability. However, as examples we have computed the proposed statistics for these data sets. The results are given in Tables 26.10 and 26.11 for the flood data and the fatigue data, respectively. From Table 26.10 one sees that the minimum Kolmogorov-Smirnov type statistic is 0.125, which is associated with the inverse Gaussian distribution; and the minimum (weighted) moment statistic is 0.29, which is associated with the Weibull distribution. Thus, one would conclude that for the flood data the underlying population is inverse Gaussian. Since our procedure selects the distribution that appears to fit the best, and there is no evaluation of the goodness of this fit, it is a good idea to check on the reasonableness of the selection. A chi-square goodness-of-fit test (applied using five equiprobable categories) results in $\chi^2 = 2.8$ and presents no evidence to contradict the conclusion that the population is inverse Gaussian.

From Table 26.11 one sees that the minimum Kolmogorov-Smirnov type statistic is 0.153, which is associated with the lognormal distribution; and the minimum (weighted) moment statistic is 0.60, which is associated with the gamma distribution. Thus, one would conclude that for the fatigue data the underlying population is lognormal. A probability plot of the fatigue data (obtained by transforming the data so it should be normal) is shown in Figure 26.1 and is consistent with the conclusion that the underlying population is lognormal.

References

1. Balakrishnan, N. and Cohen, A.C. (1991), *Order Statistics and Inference: Estimation Methods*, Academic Press, San Diego.

2. Cohen, A.C. and Whitten, B.J. (1985), Modified moment estimation for the three-parameter inverse Gaussian distribution, *Journal of Quality Technology*, **17**, 147-154.

3. Cohen, A.C. and Whitten, B.J. (1986), Modified moment estimation for the three-parameter gamma distribution, *Journal of Quality Technology*, **18**, 53-62.

4. Cohen, A.C., Whitten, B.J., and Ding, Y. (1984), Modified moment estimation for the three-parameter Weibull distribution, *Journal of Quality Technology*, **16**, 159-167.

5. Cohen, A.C., Whitten, B.J., and Ding, Y. (1985), Modified moment estimation for the three-parameter lognormal distribution, *Journal of Quality Technology*, **17**, 92-99.

6. IMSL (1991), International Mathematical & Statistical Libraries. Houston, TX.

7. Johnson, N.L., Kotz, S., and Balakrishnan, N. (1994), *Continuous Univariate Distributions-1*, Second Edition, John Wiley & Sons, New York.

8. Michael, J.R., Schucany, W.R., and Haas, R.W. (1976), Generating random variates using transformations with multiple roots, *The American Statistician*, **30**, 89.

TABLE 26.1
Selection probabilities for Weibull data. The first entry is for the
Kolmogorov-Smirnov Method, the second entry is for the minimum Euclidean
distance, the third entry is the maximum absolute difference, and the fourth entry
is the combination method given in Table 26.5

$W(\delta, \beta, \gamma)$	$n = 100$	$n = 200$	$n = 500$	$n = 1000$
$W(1, 16, 10)$	0.615	0.681	0.867	0.961
	0.388	0.310	0.300	0.326
	0.389	0.310	0.301	0.325
	0.457	0.387	0.359	0.380
$W(1.25, 16, 10)$	0.615	0.626	0.695	0.773
	0.801	0.804	0.760	0.762
	0.803	0.804	0.759	0.757
	0.817	0.815	0.769	0.790
$W(1.5, 16, 10)$	0.635	0.680	0.784	0.853
	0.717	0.796	0.798	0.867
	0.732	0.812	0.820	0.866
	0.841	0.855	0.857	0.869
$W(2, 16, 10)$	0.659	0.713	0.851	0.942
	0.811	0.866	0.904	0.973
	0.810	0.867	0.905	0.969
	0.842	0.884	0.909	0.973

TABLE 26.2
Selection probabilities for inverse Gaussian data. The first entry is for the
Kolmogorov-Smirnov method, the second entry is for the minimum Euclidean
distance, the third entry is the maximum absolute difference, and the fourth entry
is the combination method given in Table 26.5

$IG(\delta, \beta, \gamma)$	$n = 100$	$n = 200$	$n = 500$	$n = 1000$
$IG(16,16,10)$	0.568	0.684	0.839	0.866
	0.189	0.316	0.474	0.642
	0.187	0.313	0.473	0.642
	0.064	0.155	0.340	0.506
$IG(14.9,24.7,10)$	0.407	0.552	0.728	0.806
	0.211	0.327	0.473	0.630
	0.208	0.324	0.469	0.623
	0.071	0.141	0.302	0.476
$IG(14.4,34.7,10)$	0.288	0.413	0.593	0.724
	0.194	0.284	0.440	0.601
	0.190	0.281	0.435	0.596
	0.059	0.129	0.270	0.392
$IG(14.2,58.6,10)$	0.196	0.266	0.400	0.517
	0.169	0.234	0.343	0.478
	0.164	0.233	0.341	0.472
	0.031	0.112	0.202	0.301

TABLE 26.3
Selection probabilities for Gamma data. The first entry is for the
Kolmogorov-Smirnov method, the second entry is for the minimum Euclidean
distance, the third entry is the maximum absolute difference, and the fourth entry
is the combination method given in Table 26.5

$G(\rho, \beta, \gamma)$	$n = 100$	$n = 200$	$n = 500$	$n = 1000$
$G(1, 16, 10)$	0.615	0.697	0.872	0.972
	0.388	0.310	0.300	0.326
	0.389	0.310	0.301	0.325
	0.792	0.817	0.853	0.884
$G(4.84, 3.09, 10)$	0.443	0.495	0.608	0.627
	0.733	0.718	0.701	0.737
	0.733	0.716	0.705	0.744
	0.784	0.784	0.811	0.844
$G(5.59, 2.58, 10)$	0.441	0.502	0.532	0.602
	0.702	0.701	0.685	0.717
	0.688	0.708	0.681	0.717
	0.753	0.779	0.769	0.822
$G(7.20, 1.97, 10)$	0.424	0.482	0.504	0.517
	0.661	0.670	0.645	0.650
	0.665	0.661	0.642	0.650
	0.712	0.741	0.742	0.735

TABLE 26.4
Selection probabilities for lognormal data. The first entry is for the
Kolmogorov-Smirnov method, the second entry is for the minimum Euclidean
distance, the third entry is the maximum absolute difference, and the fourth entry
is the combination method given in Table 26.5

$LN(\rho, \beta, \gamma)$	$n = 100$	$n = 200$	$n = 500$	$n = 1000$
LN(2.42,0.69,10)	0.466	0.545	0.614	0.709
	0.061	0.147	0.234	0.329
	0.062	0.146	0.227	0.320
	0.306	0.430	0.594	0.697
LN(2.45,0.50,10)	0.490	0.553	0.594	0.664
	0.155	0.235	0.292	0.345
	0.149	0.229	0.282	0.335
	0.265	0.348	0.482	0.543
LN(2.50,0.38,10)	0.471	0.543	0.567	0.607
	0.213	0.323	0.384	0.454
	0.215	0.338	0.381	0.457
	0.245	0.357	0.480	0.556
LN(2.53,0.24,10)	0.401	0.465	0.499	0.565
	0.264	0.378	0.440	0.494
	0.268	0.382	0.438	0.486
	0.236	0.310	0.428	0.480

TABLE 26.5
Selection Rules for Choosing from Among Weibull (W), Lognormal (L), Inverse
Gaussian (I), and Gamma (G) Distributions When the Kolmogorov-Smirnov Type
Statistic and the Statistic Based on Moments Do Not Agree

K-S	Moments	Choose
W	G	W
W	L	W
W	I	W
G	W	W
G	L	G
G	I	G
I	W	W
I	G	G
I	L	I
L	W	W
L	G	G
L	I	L

TABLE 26.6
Selection probabilities for Weibull data. The first entry is for the
Kolmogorov-Smirnov method, the second entry is for the minimum weighted
Euclidean distance, the third entry is the maximum weighted absolute difference,
and the fourth entry is the combination method given in Table 26.5

$W(\delta, \beta, \gamma)$	$n = 100$	$n = 200$	$n = 500$	$n = 1000$
$W(1, 16, 10)$	0.615	0.681	0.867	0.961
	0.392	0.322	0.297	0.376
	0.418	0.338	0.298	0.367
	0.405	0.383	0.379	0.455
$W(1.25, 16, 10)$	0.619	0.609	0.820	0.829
	0.817	0.735	0.738	0.747
	0.820	0.744	0.743	0.757
	0.829	0.746	0.720	0.733
$W(1.5, 16, 10)$	0.672	0.676	0.779	0.853
	0.822	0.785	0.821	0.859
	0.822	0.789	0.822	0.847
	0.831	0.804	0.824	0.860
$W(2, 16, 10)$	0.671	0.706	0.861	0.952
	0.752	0.781	0.904	0.942
	0.748	0.770	0.898	0.942
	0.768	0.793	0.903	0.942

TABLE 26.7
Selection probabilities for inverse Gaussian data. The first entry is for the
Kolmogorov-Smirnov method, the second entry is for the minimum weighted
Euclidean distance, the third entry is the maximum weighted absolute difference,
and the fourth entry is the combination method given in Table 26.5

$IG(\delta, \beta, \gamma)$	$n = 100$	$n = 200$	$n = 500$	$n = 1000$
IG(16,16,10)	0.582	0.690	0.826	0.884
	0.260	0.425	0.611	0.772
	0.237	0.401	0.573	0.730
	0.098	0.240	0.488	0.706
	(0.248)	(0.323)	(0.517)	
IG(14.9,24.7,10)	0.391	0.546	0.750	0.800
	0.256	0.402	0.563	0.718
	0.247	0.383	0.529	0.690
	0.065	0.212	0.433	0.627
				(0.657)
IG(14.4,34.7,10)	0.282	0.400	0.630	0.723
	0.246	0.335	0.545	0.696
	0.244	0.327	0.536	0.675
	0.072	0.157	0.375	0.501
				(0.554)
IG(14.2,58.6,10)	0.149	0.246	0.404	0.528
	0.205	0.259	0.425	0.560
	0.209	0.261	0.425	0.564
	0.040	0.101	0.213	0.376

TABLE 26.8
Selection probabilities for Gamma data. The first entry is for the
Kolmogorov-Smirnov method, the second entry is for the minimum weighted
Euclidean distance, the third entry is the maximum weighted absolute difference,
and the fourth entry is the combination method given in Table 26.5

$G(\rho, \beta, \gamma)$	$n = 100$	$n = 200$	$n = 500$	$n = 1000$
$G(1, 16, 10)$	0.767	0.884	0.970	0.994
	0.890	0.897	0.877	0.887
	0.897	0.902	0.878	0.883
	0.932	0.956	0.981	0.997
$G(4.84, 3.09, 10)$	0.444	0.486	0.578	0.633
	0.627	0.626	0.654	0.686
	0.640	0.631	0.656	0.683
	0.684	0.704	0.751	0.801
$G(5.59, 2.58, 10)$	0.441	0.497	0.546	0.608
	0.608	0.601	0.619	0.648
	0.629	0.615	0.619	0.647
	0.665	0.686	0.728	0.759
$G(7.20, 1.97, 10)$	0.426	0.454	0.488	0.521
	0.568	0.587	0.539	0.586
	0.579	0.594	0.551	0.592
	0.626	0.644	0.647	0.676

TABLE 26.9
Selection probabilities for lognormal data. The first entry is for the
Kolmogorov-Smirnov method, the second entry is for the minimum weighted
Euclidean distance, the third entry is the maximum weighted absolute difference,
and the fourth entry is the combination method given in Table 26.5

$LN(\rho, \beta, \gamma)$	$n = 100$	$n = 200$	$n = 500$	$n = 1000$
LN(2.42,0.69,10)	0.474	0.521	0.622	0.697
	0.145	0.233	0.334	0.423
	0.131	0.210	0.319	0.404
	0.346	0.470	0.616	0.687
LN(2.45,0.50,10)	0.510	0.555	0.590	0.648
	0.220	0.297	0.369	0.413
	0.210	0.290	0.356	0.391
	0.315	0.417	0.511	0.589
LN(2.50,0.38,10)	0.492	0.542	0.584	0.617
	0.276	0.348	0.427	0.472
	0.277	0.349	0.427	0.469
	0.319	0.405	0.488	0.562
LN(2.53,0.24,10)	0.425	0.441	0.520	0.548
	0.356	0.403	0.498	0.531
	0.342	0.385	0.484	0.528
	0.309	0.339	0.472	0.513

TABLE 26.10
Kolmogorov-Smirnov type statistics and moment statistics (Eq. (26.4.1)) for the
flood data

Distribution	Kolmogorov-Smirnov Type	Moment
W	0.135	0.29
L	0.131	0.71
I	0.125	0.47
G	0.129	0.39

TABLE 26.11
Kolmogorov-Smirnov type statistics and moments statistics (Eq. (26.4.1)) for the
fatigue data

Distribution	Kolmogorov-Smirnov Type	Moment
W	0.213	0.61
L	0.153	0.99
I	0.183	0.74
G	0.216	0.60

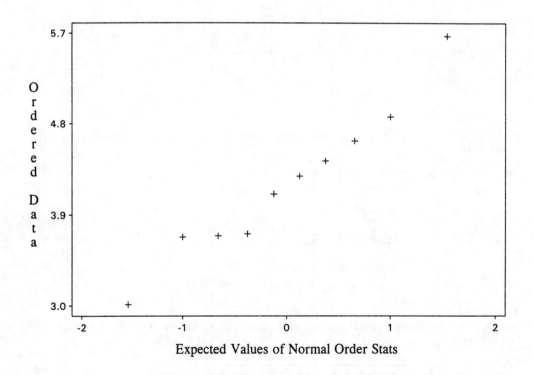

FIGURE 26.1
Probability plot of fatigue data

27

Predictive Distributions of Order Statistics

Richard A. Johnson, James Evans, and David Green

University of Wisconsin-Madison
U.S. Forest Products Lab., Madison, Wisconsin
U.S. Forest Products Lab., Madison, Wisconsin

ABSTRACT We derive and calculate predictive distributions for order statistics based on random samples from normal, log normal and Weibull distributions. The practical application which motivated this work is described and examples are given.

Keywords : Order statistics, Predictive distribution

CONTENTS

27.1   Introduction

This paper develops posterior distributions for percentiles and predictive distributions for order statistics for possible use in monitoring the strength properties of lumber.

There is a need to establish procedures for economically and efficiently evaluating the strength of lumber populations on a continuing basis. Such a procedure would be used by the grading agencies to monitor the mechanical properties of visually graded dimension lumber. The procedure could also be used by the same agencies, or by companies, to monitor the strength of lumber being produced at an individual mill. The establishment of such a monitoring scheme is essential if the lumber producers, consumers, and engineering code authorities are to have confidence in published design values. Bayesian methods can make efficient use of data collected from a large scale national study called the in-grade testing program.

The use of prior information, such as that obtained from the in-grade study, can greatly reduce the number of specimens that must be destructively tested in order to maintain an accurate lower bound on a quality criterion. Competing with this sample reducing feature is the need to check that the model for the prior data is still valid for the future data. We address this issue by deriving predictive distributions and studying their power to detect changes. Occasionally, relatively large samples may be required for this purpose.

Whereas most classical quality control procedures apply to monitoring population means, United States lumber standards are set in terms of a lower percentile. Therefore it was necessary to develop Bayesian procedures for making inferences about population percentiles. Another major difference is that future sampling of dimension lumber for monitoring purposes is apt to be done (i) only at irregular sampling periods and (ii) only for a few species/size/grade combinations and even those combinations selected may change from year to year. Another point that influenced our formulation is that wood products engineers cannot provide a well defined alternative distribution for any specific strength property. It may be a shift of the distribution upwards but more likely downwards. In fact it is unlikely to be a shift of the whole distribution but more of a drooping of the lower tail towards smaller values. This latter behavior has been observed in actual moisture content studies.

Based on previous data summarized by Forest Products personnel and industry usage, it was decided that the normal, log-normal and Weibull distributions were all candidate models for some strength or stiffness property. Bayesian procedures for the normal and log-normal cases are presented in Section 27.2. The two parameter Weibull case is treated in Section 27.3. The procedures are applied to predict the FPL 64 data in Section 27.4.

27.2 Bayesian Procedures for Normal or Log-Normal Strength Distributions

We treat two aspects of the monitoring problem. First, we present expressions for updating an informative prior distribution, for a population percentile, when future data become available. Then, in Section 27.2.2, we develop predictive distributions for checking for changes from the prior model. The predictive distribution for an order statistic from a future sample must be evaluated numerically. We describe our simulation procedure for approximating these predictive distributions and then evaluate their ability to detect changes in the population percentile. Tables of detection probabilities are presented in Section 27.2.2.

We follow the normal theory treatment presented in [2] as summarized in [6]. If the data are log-normal, we recommend applying the normal theory results to the logarithms of the observations. If $\ln(X)$ is normal with $100q$-th percentile η_q, then X has $100q$-th percentile $\exp(\eta_q)$. Further, if $\underline{\eta}_q$ is a lower bound for $\eta_q, \exp(\underline{\eta}_q)$ is a lower bound for $\exp(\eta_q)$.

27.2.1 The Posterior Distribution of a Normal Percentile

Current knowledge concerning a normal distribution for a strength property can be summarized in the form of an informative prior on the mean μ and standard deviation σ.

$$g(\mu|\sigma^2)g(\sigma^2)$$

$$= \frac{\sqrt{n_0}}{\sqrt{2\pi}\,\sigma} \exp\left(-\frac{n_0(\mu - \overline{x}_0)^2}{2\sigma^2}\right) \frac{k}{\sigma^{(n_0+1)}} \exp\left(\frac{-(n_0 - 1)^2 s_0^2}{2\sigma^2}\right) \qquad (27.2.1)$$

The present prior would result from applying Bayes theorem to combine in-grade data $n_0, \overline{x}_0, s_0$ with the non-informative prior $\sigma^{-2} d\mu d\sigma^2$. It should be noted that some authors obtain a slightly more general informative conjugate prior than (27.2.1) by allowing an arbitrary power on σ. It is well known that, apriori,

$$\frac{(n_0 - 1)s_0^2}{\sigma^2} \sim \chi^2_{n_0-1}$$

$$\mu \text{ given } \sigma \sim N\left(\overline{x}_0, \frac{\sigma^2}{n_0}\right)$$

and, unconditionally,

$$\frac{n_0^{1/2}(\mu - \overline{x}_0)}{s_0} \sim t_{n_0 - 1} \tag{27.2.2}$$

A future sample $x_1, x_2, \ldots, x_n$ of size n has likelihood

$$\prod_{i=1}^{n} \frac{1}{\sqrt{2\pi}\sigma} \exp\left(\frac{(x_i - \mu)^2}{2\sigma^2}\right) = \left(\frac{1}{2\pi\sigma^2}\right)^{\frac{n}{2}} \exp\left(\frac{-n(\overline{x} - \mu)^2 - (n-1)s^2}{2\sigma^2}\right)$$

Since (27.2.1) is a conjugate prior distribution, the posterior distribution is of the same form but with updated parameters

$$n_0 + n$$

$$\frac{n_0 \overline{x}_0 + n\overline{x}}{(n_0 + n)}$$

$$(n_0 - 1)s_0^2 + (n-1)s^2 + n_0 n(\overline{x}_0 - \overline{x})^2/(n_0 + n)$$

That is,

(i) $\dfrac{V}{\sigma^2} = \dfrac{(n_0 - 1)s_0^2 + (n-1)s^2 + n_0 n(\overline{x}_0 - \overline{x})^2/(n_0 + n)}{\sigma^2}$

$$\sim \chi^2_{n_0 + n - 1} \tag{27.2.3}$$

(ii) μ given $\sigma^2 \sim N\left(\dfrac{n_0 \overline{x}_0 + n\overline{x}}{n_0 + n}, \dfrac{\sigma^2}{n_0 + n}\right)$

(iii) unconditionally,

$$\frac{\mu - \frac{n_0 \overline{x}_0 + n\overline{x}}{n_0 + n}}{\sqrt{\frac{V}{n_0 + n}}} \sim t_{n_0 + n - 1}$$

Consequently, the Bayesian $100(1 - \gamma)\%$ lower limit on the population 5-th percentile $\eta_{.05} = \mu + \sigma z_{.05}, z_{.05} = -1.645$, is given by

$$\frac{n_0 \overline{x}_0 + n\overline{x}}{n_0 + n} + \frac{1}{\sqrt{n_0 + n}} \sqrt{\frac{V}{n_0 + n - 1}} t_{n_0 + n - 1, 1 - \gamma}[(n_0 + n)^{1/2} z_{.05}] \tag{27.2.4}$$

Here $t_{n_0 + n - 1, 1 - \gamma}$ is the non-central t-value with probability $1 - \gamma$ to the right.

$$t = \frac{N(0, 1) + (n_0 + n)z_{.05}}{\sqrt{\frac{\chi^2_{n_0 + n - 1}}{n_0 + n - 1}}}$$

27.2.2 Predictive Distributions for Normal Order Statistics

If we are to observe $X_1, X_2, \ldots, X_n$ the normal assumption $X_i | \mu, \sigma^2$ are distributed $N(\mu, \sigma^2)$ can be checked by making normal plots, calculating the modified Shapiro-Wilk statistic, or applying any other goodness-of-fit procedure.

Predictive distributions can be used to check if the new data conform to the prior model. In [3], [4], and [5], the author describes the general procedure. In the case of normal likelihoods, we wanted to obtain predictive distributions for:

$$\text{sample mean} \quad \overline{x}$$

$$\text{sample standard deviation } s = \sqrt{\frac{\sum_{i=1}^{n}(x_i - \bar{x})^2}{n-1}}$$

$$\text{sample } \alpha \text{ percentile} \qquad \bar{x} + z_\alpha s \tag{27.2.5}$$

Although the predictive distributions for $\bar{x}$ and s are essentially determined by [4] and [5], those for the percentile estimates are considerably more difficult to obtain analytically. The predictive cumulative distribution for the r-th order statistic is given by

$$\sum_{j=r}^{n} \binom{n}{j} \int \int \Phi^j \left(\frac{y-\mu}{\sigma}\right) \left[1 - \Phi\left(\frac{y-\mu}{\sigma}\right)\right]^{n-j} g(\mu|\sigma^2)g(\sigma^2)d\mu \, d\sigma^2$$

Our key idea is to obtain good approximations to the predictive distributions using simulation techniques. In [8], the author proposed the same idea. For the normal theory model, the steps are

(1) (a) Generate W distributed as Chi-square $(n_0 - 1)$. Then

$$\sigma^2 = \frac{(n_0 - 1)s_0^2}{W}$$

(b) Given σ^2, generate Z distributed as $N(0,1)$ and set

$$\mu = \bar{x}_0 + \sigma \frac{Z}{\sqrt{n_0}}$$

(2) Generate $X_1, X_2, \ldots, X_n$ independently distributed as $N(\mu, \sigma^2)$

(3) Calculate $\bar{x}, s, \bar{x} + z_{.05}s$, sample 5-th and other percentiles.

Repeat steps (1) - (3) 2000 times.

We used prior data that were close to those for Select Structural Hem Fir 2×6 dried to 15% moisture content. In particular, we took $\bar{x}_0 = 9.870$ and $s_0 = 2.490$.

All twelve combinations of

$$\begin{aligned} \text{prior sample size:} \quad & n_0 = 120, 360, 1000 \\ \text{future sample size:} \quad & n = 51, 201, 350, 501 \end{aligned}$$

were run. The results included empirical estimates of the predictive cdf's for each of the statistics (27.2.5).

We will summarize the ability of both estimates of the 5-th percentile

$$\text{sample 5-th percentile and } \bar{x} - 1.645\,s$$

and both estimates of the median

$$\text{sample 50-th percentile and } \bar{x}$$

in detecting model changes. Our analysis is restricted to detecting a decrease in the population percentile although the procedure is clearly applicable to increases or two-sided alternatives. We decide to say that a decrease in the population percentile has occurred if the corresponding order statistics in the future sample falls below the 5-th percentile of its predictive distribution.

A difference of $D = .250$ (250 lbs./in^2), in the location of the population or one of its percentiles, was specified as the smallest amount likely to be of technical interest. Therefore, we considered shifts that are multiples of .250 units. Tables 27.1-27.4 pertain to alternatives that contain two sources of variation. First, (i) the mean and standard deviation for the cumulative distribution function are selected according to the informative prior (27.2.1). This cumulative distribution is then shifted downward by the specified amount of shift. Secondly, (ii) a random sample of size n is drawn, from this shifted normal distribution function, and the statistics (27.2.5) obtained. Equivalently, the predictive distribution of the order statistic can be shifted. The shift can also be accomplished by decreasing the prior mean parameter, $\overline{x}_0$, by the amount of shift.

In summary, suppose the process generating the new observations were to produce values kD units below those of the in-grade data. Tables 27.1-27.4 provide the amount of predictive probability, calculated from a shifted predictive distribution to model new data, that lies below the 5-th percentile signal value determined from the in-grade data based predictive distribution.

Tables 27.1 and 27.2 reveal that the parametric estimator of the fifth percentile has higher detection probabilities than the sample 5-th percentile. Except for the smallest future sample size $n = 51$, this dominance is restricted to the shifts of sizes .250 to 1.250. Both statistics do very well at detecting larger shifts. The few entries that do not follow the dominance pattern are most likely caused by the randomness in the procedure used to approximate the predictive cdf's.

A comparison of Tables 27.3 and 27.4 shows that the parametric estimator (sample mean) of the population 50-th percentile dominates the sample 50-th percentile (median). The range of shifts where the differences are large is .250 to 1.000, except for the future sample size $n = 51$.

When Tables 27.1 and 27.2 are contrasted with Tables 27.3 and 27.4, it is apparent that the estimates of the population 50-th percentile (median) have higher detection probabilities than the estimates of the population 5-th percentile, for a given amount of shift. The simulations give an indication of the properties in a case that is realistic for monitoring lumber.

The parametric estimate $\overline{x} - 1.645s$, of the 5-th percentile, uses data from both tails to estimate μ and σ and then the percentile. When the future data have a normal distribution, this is clearly more efficient than estimating the 5-th percentile by using a single order statistic. However, if only the lower tail of the distribution is normal, it is the sample 5-th percentile that will retain its properties while the predictive distribution for the parametric estimator will no longer remain valid. That is, less assumptions are required for using the sample 5-th percentile than are needed for using the normal theory parametric estimator.

27.3 Bayesian Procedures for Weibull Strength Distributions

This study is restricted to the two parameter Weibull distribution.

27.3.1 Posterior Distributions for Population Percentiles

No conjugate prior is known. However, since our informative prior will be based on data, a reasonable choice is available. Let

$$x_{10}, x_{20}, \ldots, x_{n_0 0}$$

be in-grade data that we combine, according to Bayes theorem, with the initial non-informative distribution

$$\frac{1}{\theta p} \, d\theta^p \, dp = \frac{p}{\theta} \, d\theta \, dp \, .$$

The resulting informative prior has the form

$$g(p, \theta) = \frac{p^{n_0}}{(\Sigma x_{i0}^p)^{n_0}} \prod_{i=1}^{n_0} x_{i0}^{p-1} \frac{1}{c_0 \Gamma(n_0)} \left(\frac{\Sigma x_{i0}^p}{\theta^p} \right)^{n_0} \exp\left(-\frac{\Sigma x_{i0}^p}{\theta^p} \right) \frac{p}{\theta} \tag{27.3.1}$$

Restricting p to be greater than 1, the constant

$$c_0 = \int_1^\infty \frac{p^{n_0} \prod x_{i0}^{p-1}}{(\Sigma x_{i0}^p)^{n_0}} \, dp$$

is evaluated by numerical integration. From (27.3.1), given the shape parameter p,

$$\Sigma \frac{x_{i0}^p}{\theta^p} \text{ is distributed as Gamma } (n_0)$$

and p has marginal distribution

$$g(p) = p^{n_0} \frac{\prod x_{i0}^{p-1}}{c_0 (\Sigma x_{i0}^p)^{n_0}}$$

The posterior distribution, given $x_1, x_2, \ldots, x_n$ has the same form but with n_0 updated to $n_0 + n$ and the sum and product terms expanded to include the new data.

$$g(p, \theta | \underline{x}) = p^{(n_0+n)} \prod_{i=1}^{n_0} x_{i0}^{p-1} \prod_{i=1}^{n} x_i^{p-1} \Big/ (\Sigma x_{i0}^p + \Sigma x_i^p)^{(n_0+n)}$$

$$\frac{1}{c_1 \Gamma(n_0 + n)} \left(\frac{\Sigma x_{i0}^p + \Sigma x_i^p}{\theta^p} \right)^{(n_0+n)} \exp\left\{ -\left(\frac{\Sigma x_{i0}^p + \Sigma x_i^p}{\theta^p} \right) \right\} \frac{p}{\theta}$$

$$\tag{27.3.2}$$

Here

$$c_1 = \int_1^\infty \frac{p^{(n_0+n)} \prod x_{i0}^{p-1} \prod x_i^{p-1}}{(\Sigma x_{i0}^p + \Sigma x_i^p)^{(n_0+n)}} \, dp$$

is again evaluated by numerical intergration. The population 100 q-th percentile is related to the scale and location parameters. The posterior distribution of the percentile

$$\eta_q = \theta[-\ln(1 - \theta)]^{1/p}$$

can be expressed as a one-dimensional integration by first conditioning on the shape parameter p. Let $\underline{x} = (x_{10}, \ldots, x_{n_0 0}, x_1, \ldots, x_n)'$, so

$$P[\eta_q \leq z | \underline{x}] = \int_1^\infty P[\eta_q \leq z | p, \underline{x}] \frac{\prod x_{i0}^{p-1} \prod x_i^{p-1}}{c_1 (\Sigma x_{i0}^p + \Sigma x_i^p)^{(n_0+n)}} p^{(n_0+n+1)} \, dp$$

where

$$P[\eta_q \leq z | p, \underline{x}] = P[\theta^p(-\ln(1 - q)) \leq z^p \, | p, \underline{x}]$$

$$= P[-\ln(1 - q) \frac{\Sigma x_{i0}^p + \Sigma x_i^p}{z^p} \leq \frac{\Sigma x_i^p + \Sigma x_i^p}{\theta^p} \, | p, \underline{x}]$$

$$= \int_{a(q,z,\underline{x})}^\infty \frac{1}{\Gamma(n_0 + n)} v^{(n_0+n)-1} e^{-v} \, dv$$

where,

$$a(q, z, \boldsymbol{x}) = -\ln(1-q) \, \frac{\Sigma \, x_{i0}^p + \Sigma \, x_i^p}{z^p}$$

which is an incomplete gamma function. This can be evaluated as a continued fraction (see [1], relation (6.5.31) and also (6.1.41) for an expansion of the log-gamma function.) Consequently, we have reduced the computation of the posterior cdf of η_q to a single numerical integration.

27.3.2 Predictive Distributions for the Weibull Case

The assumption of a Weibull likelihood for future observations $X_1, X_2, \ldots, X_n$ can be checked by plotting $\ln(x_i)$ versus $i/(n+1)$ or calculating goodness-of-fit statistics.

Predictive distributions can be used to check if the new data conform to the prior model. It was desired to obtain predictive distributions for the statistics:

maximum likelihood estimator of shape: $\qquad\qquad \hat{p}$

maximum likelihood estimator of scale: $\qquad\qquad \hat{\theta}$

maximum likelihood estimator of the $100q$-th percentile: $\quad \hat{\eta}_q = \hat{\theta}[-\ln(1-q)]^{1/\hat{p}}$

$$\text{(27.3.3)}$$

where $q = .05, .10, .25, .50, .75$.

In order to obtain approximations to the predictive distributions for the statistics (27.3.3), we followed a simulation procedure similar to that for the normal case. The steps for the Weibull are:

Step (1) a) Use numerical integration to evaluate the marginal cdf

$$F(z) = \int_1^\infty \frac{p^{n_0} \prod x_{i0}^{p-1}}{c_0 (\Sigma \, x_{i0}^p)^{n_0}} \, dp$$

at grid of points 1.80 (.01) 4.81. (We used the IMSL integration subroutine DE-CARDE.) Linearly connect these values of $F(z)$ to obtain a cdf $\overline{F}(z)$ that closely approximates $F(z)$. Generate a uniform $(0,1)$ variable, $U = u$, and then determine p so that $u = \overline{F}(p)$.

b) Given p, generate V distributed as Gamma (n_0) and calculate

$$\theta = \frac{V^{1/p}}{(\Sigma \, x_{i0}^p)^{1/p}}$$

Step (2) Generate $X_1, X_2, \ldots, X_n$ as independent Weibull (p, θ). This can be accomplished by generating independent uniform $U_1, U_2, \ldots, U_n$ and then calculating

$$X_i = \theta(-\ln(U_i))^{1/p}$$

Step (3) Calculate the statistics (27.3.3)

Repeat steps (1) - (3) 2000 times and obtain approximations to the predictive cdf's.

In our implementation of this procedure, we used the IMSL routine GGUO to generate an ordered sample of n_0 uniform random variables and then capitalized on the ordering to invert $\overline{F}(.)$ and obtain an ordered sample of values for the shape parameter p.

We took as prior data

$$x_{i0} = \theta_0(-\ln(1 - i/(n_0 + 1))^{1/p_0} \qquad\qquad \text{(27.3.4)}$$

where x_{i0} is approximately equal to the expected value of the i-th order statistic in a sample of size n_0 from a Weibull (p_0, θ_0) population. Here $\theta_0 = 10.803$ and $p_0 = 4.726$. These parameter values are those obtained from fitting strength data on Hem fir Select Structural 2×6's dried to 15% moisture content.

The detection probabilities, obtained from shifting the predictive distribution in steps of .250, are given in Tables 27.5-27.8 for both the maximum likelihood and sample percentile estimates of the population 5-th percentile and the population 50-th percentile. Each signal value is again taken as the 5-th percentile of the un-shifted predictive distribution.

Comparing Tables 27.5 and 27.6, we see that the parametric estimator, $\hat{\theta}[-\ln(1-q)]^{1/\hat{p}}$, has higher detection probabilities than the sample 5-th percentile. The range of shift sizes when the differences are large is .250 to 1.250, except for the smallest future sample size $n = 51$ where the range extends to 1.750. The same interpretation holds for Tables 27.7 and 27.8, except that larger differences are confined to the range .250 to 1.000. As is the case with the normal theory procedures, for a given amount of shift, the estimates of the population 50-th percentile have higher detection probabilities than those of the population 5-th percentile.

27.4   An Example

We illustrate our predictive procedure using data on southern pine. The prior data, collected in 1980 by the Forest Products Laboratory, consist of $n_0 = 597$ values of MOR for 2×8 specimens of southern pine that are on grade. After adjusting these data to 12% moisture content, the maximum likelihood estimates for the two Weibull parameters from these prior data are

$$\text{mle shape}: \hat{p} = 3.39$$

$$\text{mle scale}: \hat{\theta} = 7.36$$

As an example, we predict the values of order statistics from the sample of size $n = 99$ specimens of 2×8 southern pine included in the FPL64 study.

The results of the predictions are summarized in Table 27.9 where the predictive probability to the left of the observed order statistic is given in the body of the table. For instance, the sample 5-th percentile from the FPL64 data is the fifth order statistic $x_{(5)} = 2.847$.

From Table 27.9 it appears that the FPL64 2×8 southern pine MOR values are lower than that of the 1980 data. This is particularly true of the middle of the distribution. The sample median $X_{(50)} = 5.697$ lies in the extreme lower tail of its predictive distribution. A similar pattern holds for the maximum likelihood estimate of the percentiles.

The maximum likelihood estimates for the FPL64 data are

$$\text{mle shape}: \hat{p} = 3.05$$

$$\text{mle scale}: \hat{\theta} = 6.76$$

The scale estimate is less than the 1-st percentile of its predictive distribution while the shape estimate is at about the 10-th percentile of its predictive distribution. Also, the FPL64 data do not seem to fit the two parameter Weibull distribution particularly well. Our procedure indicates that a change has occured at least in the middle of the distribution.

Acknowledgements

Richard Johnson would like to thank the Wisconsin Alumni Research Foundation for support.

References

1. Abramowitz, M. and Stegun, I. (Eds.) (1964), *Handbook of Mathematical Functions*, National Bureau of Standards, Applied Mathematics Series No. 55, Dover, New York.

2. Aitchinson, J. (1964), Bayesian tolerance regions, *Journal of the Royal Statistical Society, Series B*, **26**, 161-175.

3. Box, G. (1980), Sampling inference, Bayes' inference and robustness in the advancement of learning, In *Bayesian Statistics* (Eds., DeGroot, M.H., Bernardo, J.M., Lindley, D.V., and Smith, A.F.M.), 366-381, Proceedings of the First International Meeting held in Valencia, Spain, University Press, Valencia.

4. Box, G. (1980), Sampling and Bayes' inference in scientific modeling and robustness (with discussion), *Journal of the Royal Statistical Society, Series A*, **143**, 383-430.

5. Box, G. (1983), An apology for ecumenism in statistics, In *Scientific Inference, Data Analysis, and Robustness* (Eds., Box, G.E.P., Leonard, T., and Wu, C.-F.), 51-84, Academic Press, New York.

6. Guttman, I. (1970), *Statistical Tolerance Regions: Classical and Bayesian*, Charles Griffin, London.

7. Rubin, D. (1984), Bayesian justifiable and relevant frequency calculations for the applied statistician, *Annals of Statistics*, **12**, 1151-1172.

TABLE 27.1
Detection probabilities: Predictive distributions for sample 5-th percentiles from normal populations

(Shifted Predictive Distribution)

The prior is a normal-gamma distribution, with $\bar{x}_0 = 9.87$ and $s_0 = 2.49$.

Future sample size	Shift in Predictive Distribution								
	.000	.250	.500	.750	1.000	1.250	1.500	1.750	2.000
Prior size = 120									
51	.050	.084	.134	.197	.284	.387	.504	.617	.746
201	.050	.109	.232	.406	.607	.779	.916	.973	.993
351	.050	.124	.269	.468	.691	.864	.956	.992	.999
501	.050	.134	.301	.510	.740	.909	.971	.997	1.000
Prior size = 360									
51	.050	.077	.128	.196	.290	.425	.547	.693	.802
201	.050	.148	.318	.527	.747	.911	.980	.996	.999
351	.050	.186	.421	.705	.896	.981	.998	1.000	1.000
501	.050	.195	.474	.787	.953	.996	1.000	1.000	1.000
Prior size = 1000									
51	.050	.083	.133	.210	.308	.430	.563	.701	.805
201	.050	.140	.319	.574	.813	.946	.988	.999	1.000
351	.050	.209	.507	.804	.956	.995	1.000	1.000	1.000
501	.050	.212	.549	.870	.987	.999	1.000	1.000	1.000

TABLE 27.2
Detection probabilities: Predictive distributions for the parametric estimate of the 5-th percentile of a normal population

(Shifted Predictive Distribution)

The prior is a normal-gamma distribution, with $\bar{x}_0 = 9.87$ and $s_0 = 2.49$.

	Future sample size	Shift in Predictive Distribution								
		.000	.250	.500	.750	1.000	1.250	1.500	1.750	2.000
Prior size = 120										
	51	.050	.115	.222	.396	.594	.788	.910	.975	.994
	201	.050	.125	.308	.568	.794	.932	.988	.999	1.000
	351	.050	.132	.314	.575	.805	.947	.992	1.000	1.000
	501	.050	.155	.350	.600	.835	.955	.993	1.000	1.000
Prior size = 360										
	51	.050	.149	.331	.575	.795	.928	.986	.998	.999
	201	.050	.220	.555	.870	.979	1.000	1.000	1.000	1.000
	351	.050	.232	.622	.916	.992	1.000	1.000	1.000	1.000
	501	.050	.270	.698	.952	.998	1.000	1.000	1.000	1.000
Prior size = 1000										
	51	.050	.162	.359	.625	.851	.957	.992	1.000	1.000
	201	.050	.304	.750	.975	1.000	1.000	1.000	1.000	1.000
	351	.050	.382	.880	.994	1.000	1.000	1.000	1.000	1.000
	501	.050	.451	.915	.998	1.000	1.000	1.000	1.000	1.000

TABLE 27.3
Detection probabilities: Predictive distributions for sample 50-th percentiles from normal populations

(Shifted Predictive Distribution)

The prior is a normal-gamma distribution, with $\bar{x}_0 = 9.87$ and $s_0 = 2.49$.

Future sample size	Shift in Predictive Distribution								
	.000	.250	.500	.750	1.000	1.250	1.500	1.750	2.000
Prior size = 120									
51	.050	.112	.241	.425	.629	.796	.911	.964	.990
201	.050	.195	.478	.765	.935	.991	.999	1.000	1.000
351	.050	.222	.561	.856	.972	.998	1.000	1.000	1.000
501	.050	.258	.604	.871	.983	.999	1.000	1.000	1.000
Prior size = 360									
51	.050	.128	.282	.491	.699	.852	.948	.987	.996
201	.050	.257	.621	.909	.989	1.000	1.000	1.000	1.000
351	.050	.344	.786	.975	.999	1.000	1.000	1.000	1.000
501	.050	.415	.869	.993	1.000	1.000	1.000	1.000	1.000
Prior size = 1000									
51	.050	.136	.304	.513	.733	.888	.959	9.92	.999
201	.050	.302	.711	.950	.996	1.000	1.000	1.000	1.000
351	.050	.383	.843	.991	1.000	1.000	1.000	1.000	1.000
501	.050	.485	.935	.999	1.000	1.000	1.000	1.000	1.000

TABLE 27.4
Detection probabilities: Predictive distributions for the parametric estimate of the 50-th percentile of a normal population

(Shifted Predictive Distribution)

The prior is a normal-gamma distribution, with $\bar{x}_0 = 9.87$ and $s_0 = 2.49$.

	Future sample size	Shift in Predictive Distribution								
		.000	.250	.500	.750	1.000	1.250	1.500	1.750	2.000
Prior size = 120										
	51	.050	.139	.304	.547	.755	.897	.970	.992	.997
	201	.050	.232	.574	.848	.968	.997	1.000	1.000	1.000
	351	.050	.258	.624	.892	.986	.999	1.000	1.000	1.000
	501	.050	.281	.644	.898	.993	1.000	1.000	1.000	1.000
Prior size = 360										
	51	.050	.163	.377	.646	.858	.957	.990	.999	1.000
	201	.050	.313	.747	.968	.999	1.000	1.000	1.000	1.000
	351	.050	.408	.865	.993	1.000	1.000	1.000	1.000	1.000
	501	.050	.422	.894	.999	1.000	1.000	1.000	1.000	1.000
Prior size = 1000										
	51	.050	.162	.382	.662	.867	.968	.994	1.000	1.000
	201	.050	.360	.836	.992	1.000	1.000	1.000	1.000	1.000
	351	.050	.521	.950	.999	1.000	1.000	1.000	1.000	1.000
	501	.050	.575	.976	1.000	1.000	1.000	1.000	1.000	1.000

TABLE 27.5
Detection probabilities: Predictive distributions for sample 5-th percentiles from Weibull populations

(Shifted Predictive Distribution)

The prior is developed from expected observations based on a
Weibull with shape = 4.726 and scale = 10.803.

	Future sample size	Shift in Predictive Distribution								
		.000	.250	.500	.750	1.000	1.250	1.500	1.750	2.000
Prior size = 120										
	51	.050	.084	.136	.221	.314	.428	.546	.660	.771
	201	.050	.132	.256	.438	.640	.514	.934	.974	.996
	351	.050	.140	.310	.550	.757	.901	.974	.994	.998
	501	.050	.158	.358	.600	.807	.937	.986	.997	1.000
Prior size = 360										
	51	.050	.080	.134	.221	.318	.437	.565	.698	.806
	201	.050	.166	.342	.554	.776	.926	.978	.996	.999
	351	.050	.169	.405	.691	.896	.978	.995	1.000	1.000
	501	.050	.194	.484	.770	.945	.998	1.000	1.000	1.000
Prior size = 1000										
	51	.050	.096	.142	.232	.342	.468	.588	.720	.888
	201	.050	.139	.330	.572	.805	.937	.986	.998	1.000
	351	.050	.172	.460	.757	.936	.992	.999	1.000	1.000
	501	.050	.261	.612	.896	.989	.999	1.000	1.000	1.000

TABLE 27.6
Detection probabilities: Predictive distributions for the MLE of the 5-th percentile of a Weibull population

(Shifted Predictive Distribution)

The prior is developed from expected observations based on a
Weibull with shape = 4.726 and scale = 10.803.

	Future sample size	Shift in Predictive Distribution								
		.000	.250	.500	.750	1.000	1.250	1.500	1.750	2.000
Prior size = 120										
	51	.050	.109	.217	.367	.534	.691	.825	.923	.968
	201	.050	.141	.340	.588	.815	.939	.985	.996	1.000
	351	.050	.178	.398	.663	.862	.968	.995	1.000	1.000
	501	.050	.184	.431	.701	.894	.976	.996	1.000	1.000
Prior size = 360										
	51	.050	.127	.261	.434	.616	.792	.888	.953	.987
	201	.050	.214	.515	.804	.953	.993	.999	1.000	1.000
	351	.050	.260	.617	.901	.987	.999	1.000	1.000	1.000
	501	.050	.273	.645	.924	.998	1.000	1.000	1.000	1.000
Prior size = 1000										
	51	.050	.136	.264	.443	.637	.819	.919	.972	.993
	201	.050	.228	.600	.882	.980	.998	1.000	1.000	1.000
	351	.050	.298	.751	.968	.998	1.000	1.000	1.000	1.000
	501	.050	.395	.873	.993	1.000	1.000	1.000	1.000	1.000

TABLE 27.7
Detection probabilities: Predictive distributions for sample 50-th percentiles from Weibull populations

(Shifted Predictive Distribution)

The prior is developed from expected observations based on a
Weibull with shape = 4.726 and scale = 10.803.

Future sample size	Shift in Predictive Distribution								
	.000	.250	.500	.750	1.000	1.250	1.500	1.750	2.000
Prior size = 120									
51	.050	.128	.265	.470	.684	.857	.949	.984	1.000
201	.050	.195	.500	.808	.958	.994	1.000	1.000	1.000
351	.050	.227	.580	.867	.980	1.000	1.000	1.000	1.000
501	.050	.254	.623	.900	.986	.999	1.000	1.000	1.000
Prior size = 360									
51	.050	.142	.291	.506	.715	.876	.957	.991	.999
201	.050	.263	.637	.924	.994	1.000	1.000	1.000	1.000
351	.050	.339	.770	.977	1.000	1.000	1.000	1.000	1.000
501	.050	.377	.844	.991	1.000	1.000	1.000	1.000	1.000
Prior size = 1000									
51	.050	.144	.314	.537	.759	.905	.969	1.000	.999
201	.050	.301	.726	.959	.998	1.000	1.000	1.000	1.000
351	.050	.357	.860	.993	1.000	1.000	1.000	1.000	1.000
501	.050	.485	.940	1.000	1.000	1.000	1.000	1.000	1.000

TABLE 27.8
Detection probabilities: Predictive distributions for the MLE of the 50-th percentile of a Weibull population

(Shifted Predictive Distribution)

The prior is developed from expected observations based on a
Weibull with shape = 4.726 and scale = 10.803.

	Future sample size	Shift in Predictive Distribution								
		.000	.250	.500	.750	1.000	1.250	1.500	1.750	2.000
Prior size = 120										
	51	.050	.109	.217	.367	.534	.691	.825	.923	1.000
	201	.050	.141	.340	.588	.815	.939	1.000	1.000	1.000
	351	.050	.178	.398	.663	.862	.968	1.000	1.000	1.000
	501	.050	.184	.431	.701	.894	.976	1.000	1.000	1.000
Prior size = 360										
	51	.050	.127	.261	.434	.616	.792	.888	.953	1.000
	201	.050	.214	.515	.804	.953	1.000	1.000	1.000	1.000
	351	.050	.260	.617	.901	1.000	1.000	1.000	1.000	1.000
	501	.050	.273	.645	.924	1.000	1.000	1.000	1.000	1.000
Prior size = 1000										
	51	.050	.136	.264	.443	.637	.819	.919	1.000	1.000
	201	.050	.228	.600	.882	1.000	1.000	1.000	1.000	1.000
	351	.050	.298	.751	.968	1.000	1.000	1.000	1.000	1.000
	501	.050	.395	.873	1.000	1.000	1.000	1.000	1.000	1.000

TABLE 27.9
Cumulative predictive probabilities evaluated at observed order statistics and Weibull MLE from FPL64

Observed			Cumulative Probability
$x(5)$	=	2.847	.365
$\hat{\eta}_{(.05)}$	=	2.553	.222
$x(10)$	=	3.286	.109
$\hat{\eta}_{(.10)}$	=	3.232	.107
$x(25)$	=	4.184	.004
$\hat{\eta}_{(.25)}$	=	4.493	.030
$x(50)$	=	5.697	.001
$\hat{\eta}_{(.50)}$	=	5.995	.017
$x(75)$	=	8.011	.369
$\hat{\eta}_{(.75)}$	=	7.524	.029

28

Residuals from Type II Censored Samples

Edsel A. Peña

Bowling Green State University

ABSTRACT Exact and asymptotic distributional properties of hazard-based residuals arising from Type II censored samples are studied. These residuals are presently used in diagnostics of reliability models by checking their approximate unit exponentiality property. It is shown that, under Type II censoring, this unit exponentiality property may not be viable, and residual-based diagnostic procedures which revolve on checking this property may be misleading. An alternative approach utilizing exact and asymptotic distributional properties is discussed.

Keywords: Goodness-of-fit, Hazard function, Model diagnostics, Order statistics, Spacings, Total-time-on-test, Type II Censoring, Unit exponentiality property

CONTENTS

28.1 Introduction and Summary

Model diagnostics is an important component of reliability studies since the validity of statistical procedures depend on whether the assumed reliability model is viable or not. Hazard-based residuals, which are obtained by evaluating an estimator of the cumulative hazard function at the failure or censored times, have been used in model diagnostics in Reliability and Survival Analysis. For instance, in the reliability setting, Lawless [18, p. 813] suggests their use in checking a Poisson process assumption; while in the discussion of Arjas' paper [2], Anderson [1] has also alluded to using these residuals for model diagnostics, although Arjas cautioned that there might be problems in their use especially for small samples. In the survival analysis setting, residuals have been used in checking the Cox proportional hazards model [7] such as in [9], [15], [16, p. 150], [14, p. 96], [17, p. 93], [8, p. 89; pp. 107-109], [11, Chap. 4], and [2, Sec. VII.3]. Their use in model validation revolves on checking whether the residual vector is approximately a random sample from a unit exponential distribution, and this is typically done via graphical methods such as hazard plots or the total-time-on-test plot of the residuals, or by performing formal tests of exponentiality.

However, since the residuals are obtained through the evaluation of an estimator of the cumulative hazard function, it is not clear whether the above-mentioned procedures in the use of residuals for model validation are actually appropriate. There are very few papers which have so far examined formally the distributional properties of these residuals. In [12], the authors have studied the asymptotic properties of statistics when applied to residuals arising from the Cox proportional hazards model. Their paper is important since it showed that the asymptotic variance is different from what is expected under the unit exponentiality property. Just recently, in [4], it was shown that one needs to be very careful in using the above-mentioned procedures when dealing with residuals from randomly right-censored data, since the critical assumption that the residuals should be approximately a random sample from the unit exponential distribution, when the model is true, may actually be far from being true. The present paper is related to paper [4]. Distributional properties of hazard-based residuals arising from Type II censored data are obtained, and the exact properties are compared to what is expected under the unit exponentiality property. The exact properties are obtained under the assumption that the failure times are exponentially distributed, both for the scale and location-scale models of the exponential distribution. Asymptotic properties of a class of test statistics when applied to the residuals are also ascertained. It will be seen that the exact and asymptotic variances of statistics when applied to the Type II residuals is at most that of the asymptotic variance when the statistic is applied to a Type II censored sample from the unit exponential distribution. It will be shown that some test statistics are totally noninformative when applied to the residual vector in the sense that their asymptotic variance is zero, while other test statistics can be utilized for model validation using the residuals. Whereas randomly right-censored samples were considered in [4], we restrict our attention in this paper to Type II censored samples. This type of censoring is prevalent in reliability studies due to the desire to control the number of failed test units. To obtain our exact results we restrict to exponentially distributed failure times (scale and location-scale models), while for the asymptotic results we only impose regularity conditions similar to those in [6], which dealt with asymptotics of a class of statistics for Type II censored samples.

We now outline the contents of this paper. Section 28.2 will present notations and background information. Exact properties of the Type II censored residuals will be presented in Section 28.3, with Subsection 28.3.1 containing results under a scale exponential model for the failure times, and Subsection 28.3.2 containing results under a location-scale exponential model. Section 28.4 presents asymptotic results and some concrete examples. In Section 28.5 we discuss how the exact and asymptotic results can be used in constructing goodness-of-fit procedures based on Type II censored residuals.

28.2   Residuals, Type II Censoring and Notations

Let $T_1, \ldots, T_n$ be independent and identically distributed (i.i.d.) nonnegative failure-time variables from an absolutely continuous distribution function F. We denote by $\bar{F} = 1 - F$ the associated survivor function. It is assumed that F belongs to a parametric family of distribution functions $\mathcal{F} = \{F(\cdot; \boldsymbol{\theta}) : \boldsymbol{\theta} \in \Theta\}$, where Θ is an open subset of M-dimensional Euclidean space $\Re^M$. We let $f(\cdot; \boldsymbol{\theta})$ denote the density function and $\lambda(\cdot; \boldsymbol{\theta})$ denote the hazard function of $F(\cdot; \boldsymbol{\theta})$. We recall that the hazard function is defined via

$\lambda(t;\boldsymbol{\theta}) = f(t;\boldsymbol{\theta})/\bar{F}(t;\boldsymbol{\theta})$. The cumulative hazard function is denoted by

$$\Lambda(t;\boldsymbol{\theta}) = -\ln\bar{F}(t;\boldsymbol{\theta}) = \int_0^t \lambda(u;\boldsymbol{\theta})\,du.$$

The true, but unknown, parameter value will be denoted by $\boldsymbol{\theta}_0$.

If $T_1,\dots,T_n$ are i.i.d. $F(\cdot;\boldsymbol{\theta}_0)$, then the 'true' residuals $\boldsymbol{R}_n^0 = (R_1^0,\dots,R_n^0)$, where $R_i^0 = \Lambda(T_i;\boldsymbol{\theta}_0)$, $i=1,\dots,n$, form a random sample from the unit exponential distribution. This is the so-called *unit exponentiality property* (U.E.P.) of $\boldsymbol{R}_n^0$. However, since the exact value of $\boldsymbol{\theta}_0$ is unknown, one obtains instead the 'observed' residuals $\boldsymbol{R}_n = (R_1,\dots,R_n)$, where $R_i = \Lambda(T_i;\hat{\boldsymbol{\theta}})$, $i=1,\dots,n$, with $\hat{\boldsymbol{\theta}}$ being some estimator of $\boldsymbol{\theta}$ based on $T_1,\dots,T_n$. In this paper, we restrict our treatment to the situation where $\hat{\boldsymbol{\theta}}$ is the maximum likelihood estimator (MLE). The use of $\boldsymbol{R}_n$ in model diagnostics revolves on the expectation, or more appropriately the belief, that since $\boldsymbol{R}_n^0$ has the U.E.P. when the model holds then so should $\boldsymbol{R}_n$, at least approximately.

In the Type II censorship model, which is the prevalent form of censoring in reliability life-testing studies due to the desire to control the number of failed test units, n test units are placed on test simultaneously at time zero and the study is terminated at the instant d units have failed, where $d \in \{1,2,\dots,n\}$ is a prespecified integer. Henceforth, this censoring model will be shortened to Type $\mathrm{II}_{[d:n]}$. The observable random variables are therefore

$$\boldsymbol{T}_{[d:n]} = (T_{1:n}, T_{2:n}, \dots, T_{d:n}), \tag{28.2.1}$$

where $T_{1:n} < T_{2:n} < \dots < T_{d:n}$ are the first d order statistics. A realization of $\boldsymbol{T}_{[d:n]}$ will be denoted by $\boldsymbol{t}_{[d:n]} = (t_{1:n}, t_{2:n},\dots,t_{d:n})$, and in the sequel we follow this convention by letting lower-case characters represent realizations of the associated 'upper-case' random entities. The failure times of the remaining $n-d$ test units are only known to exceed $T_{d:n}$, and these are the censored values. The relevant portion of the likelihood function associated with the realization $\boldsymbol{T}_{[d:n]} = \boldsymbol{t}_{[d:n]}$ of (28.2.1) is (cf., [10])

$$\mathcal{L}(\boldsymbol{\theta};\boldsymbol{t}_{[d:n]}) = \mathcal{L}(\boldsymbol{\theta}) = \left[\prod_{i=1}^{d} f(t_{i:n};\boldsymbol{\theta})\right]\left[\bar{F}(t_{d:n};\boldsymbol{\theta})\right]^{n-d}. \tag{28.2.2}$$

From this likelihood function one can obtain the MLE $\hat{\boldsymbol{\theta}}$ of $\boldsymbol{\theta}$. The associated Type $\mathrm{II}_{[d:n]}$ censored 'true' residuals are $\boldsymbol{R}_{[d:n]}^0 = (R_{1:n}^0, R_{2:n}^0, \dots, R_{d:n}^0)$, where $R_{i:n}^0 = \Lambda(T_{i:n};\boldsymbol{\theta}_0)$, $i = 1,\dots,d$. The 'observed' Type $\mathrm{II}_{[d:n]}$ censored residuals are

$$\boldsymbol{R}_{[d:n]} = (R_{1:n}, R_{2:n}, \dots, R_{d:n}),$$

where $R_{i:n} = \Lambda(T_{i:n};\hat{\boldsymbol{\theta}})$, $i = 1,\dots,d$.

For our notation we let, for $i = 1,2,\dots,n$, and $k = 1,2$,

$$c_{i:n} = 1/(n-i+1) \quad \text{and} \quad C_{i:n}^{(k)} = \sum_{j=1}^{i} c_{j:n}^k,$$

We also let $\mathrm{Exp}(\beta)$ denote the exponential distribution with scale parameter β; $\mathrm{Dir}_{ns}(\boldsymbol{\alpha})$ and $\mathrm{Dir}_{si}(\boldsymbol{\alpha})$ denote the nonsingular and singular Dirichlet distributions with parameter $\boldsymbol{\alpha}$, respectively; $\mathrm{Be}(\alpha,\beta)$ denote the beta distribution with parameters α and β; and by $\mathrm{Ga}(\alpha,\beta)$ denote the gamma distribution with shape parameter α and scale parameter β. The 'equal-in-distribution' relation will be denoted by '$\overset{\text{st}}{=}$', while '$\vee$' and '$\wedge$' will denote the binary operations of taking maximum and minimum, respectively. The indicator function of set A will be denoted by $I_A(\cdot)$. The symbols '$\xrightarrow{\mathcal{L}}$', '$\xrightarrow{\text{pr}}$' and '$\xrightarrow{\text{wp1}}$' will denote convergence in distribution, probability, and almost surely, respectively.

28.3 Exact Distributional Properties of Residuals

28.3.1 Scale Exponential Failure Times

In this subsection we assume that the failure time distribution is scale exponential, so that $f(t;\theta) = \theta\exp\{-\theta t\}I_{[0,\infty)}(t)$ for $\theta \in \Theta = (0,\infty)$. The survivor function is $\bar{F}(t;\theta) = \exp\{-\theta t\}I_{[0,\infty)}(t)$, and the hazard function and cumulative hazard function are $\lambda(t;\theta) = \theta I_{[0,\infty)}(t)$ and $\Lambda(t;\theta) = \theta t I_{[0,\infty)}(t)$, respectively. From the likelihood function in (28.2.2) it is seen that

$$S(\boldsymbol{T}_{[d:n]}) = \sum_{i=1}^{d} T_{i:n} + (n-d)T_{d:n}, \tag{28.3.1}$$

the total-time-on-test (TTOT) statistic, is sufficient for θ, and since S has a gamma distribution with parameters d and θ, then it is complete. The MLE of θ is

$$\hat{\theta} = d/S(\boldsymbol{T}_{[d:n]}), \tag{28.3.2}$$

so the Type $\text{II}_{[d:n]}$ censored residuals are $\boldsymbol{R}_{[d:n]} = (R_{1:n}, R_{2:n}, \ldots, R_{d:n})$, where

$$R_{i:n} = \frac{dT_{i:n}}{S(\boldsymbol{T}_{[d:n]})}, \quad i = 1, 2, \ldots, d. \tag{28.3.3}$$

We now present our first result concerning exact properties of $\boldsymbol{R}_{[d:n]}$.

THEOREM 28.3.1
If $f(t;\theta) = \theta\exp\{-\theta t\}I_{[0,\infty)}(t)$, then

$$\boldsymbol{R}_{[d:n]} \overset{\text{st}}{=} d\left(\sum_{j=1}^{i} c_{j:n}V_j, \ i = 1, 2, \ldots, d\right),$$

where $(V_1, V_2, \ldots, V_d) \sim \text{Dir}_{si}(\boldsymbol{1}_d)$ and $\boldsymbol{1}_d = (1, 1, \ldots, 1)$ is a $1 \times d$ vector of 1's.

PROOF Since

$$T_{i:n} = \sum_{j=1}^{i}(T_{j:n} - T_{j-1:n}) = \sum_{j=1}^{i} c_{j:n}W_j$$

and

$$S(\boldsymbol{T}_{[d:n]}) = \sum_{j=1}^{d} T_{j:n} + (n-d)T_{d:n} = \sum_{j=1}^{d} W_j,$$

where $W_j = (n-i+1)(T_{j:n} - T_{j-1:n})$, $j = 1, 2, \ldots, d$, it follows from (28.3.3),

$$\boldsymbol{R}_{[d:n]} = d\left(\sum_{j=1}^{i} c_{j:n}\left[\frac{W_j}{\sum_{l=1}^{d} W_l}\right], \ i = 1, 2, \ldots, d\right).$$

Since the normalized spacings statistics $W_1, W_2, \ldots, W_d$ are i.i.d. Exp(θ) (cf., [5]), then $(V_1, V_2, \ldots, V_d) = (W_1, W_2, \ldots, W_d)/\sum_{l=1}^{d} W_l$ has a singular Dirichlet distribution with parameter $\mathbf{1}_d = (1, 1, \ldots, 1)$. ∎

COROLLARY 28.3.2
Under the conditions of Theorem 28.3.1, for each $i, j = 1, 2, \ldots, d$ with $i \neq j$,

(i) $R_{i:n} \overset{st}{=} d\sum_{l=1}^{i} c_{l:n} V_l;$

(ii) $(R_{i:n}, R_{j:n}) \overset{st}{=} d(\sum_{l=1}^{i} c_{l:n} V_l, \sum_{l=1}^{j} c_{l:n} V_l);$

(iii) $\mathcal{E}\{R_{i:n}\} = C_{i:n}^{(1)};$

(iv) $Var\{R_{i:n}\} = \frac{1}{d+1}[dC_{i:n}^{(2)} - \{C_{i:n}^{(1)}\}^2];$ *and*

(v) $Cov\{R_{i:n}, R_{j:n}\} = Var\{R_{i \wedge j:n}\} - \sum_{k=1}^{i \wedge j} \sum_{l=i \wedge j+1}^{i \vee j} c_{k:n} c_{l:n}.$

PROOF   Results (i) and (ii) follow immediately from Theorem 28.3.1, while (iii), (iv) and (v) are routinely obtained from the facts that $V_j \sim Be(1, d-1)$, $(V_i, V_j) \sim Dir_{ns}(1, 1, d-2)$, $(i \neq j)$, and using properties of moments of the Dirichlet random vectors. (See the proofs in [4] for similar details.) ∎

Let $E_1, E_2, \ldots, E_n$ be i.i.d. Exp(1) random variables. It is clear that the associated Type II$_{[d:n]}$ censored sample satisfies

$$\mathbf{E}_{[d:n]} \equiv (E_{1:n}, E_{2:n}, \ldots, E_{d:n}) \overset{st}{=} \left(\sum_{j=1}^{i} c_{j:n} E_j, \ i = 1, 2, \ldots, d \right). \qquad (28.3.4)$$

Under the U.E.P., this is the distributional property that Type II$_{[d:n]}$ censored residuals are expected to approximately possess, hence one could compare and contrast the results in Theorem 28.3.1 and Corollary 28.3.2 with that of (28.3.4) to see how the U.E.P. is approximated by the Type II$_{[d:n]}$ censored residuals, at least under this scale exponential model.

Whereas, under the U.E.P, the random variables E_i's in the linear combinations in (28.3.4) are i.i.d. with unbounded support, the random variables in the linear combinations for the Type II$_{[d:n]}$ residuals in Theorem 28.3.1, which are the dV_j's, are dependent but exchangeable and have bounded supports. Since d is usually small relative to n, then $\mathbf{R}_{[d:n]}$ could have a joint distribution which is quite different from what is expected under the U.E.P. This indicates that model validation procedures which revolves on checking the U.E.P. for $\mathbf{R}_{[d:n]}$ may be misleading under Type II censoring.

Let us examine further the moments. Under the U.E.P., $\mathcal{E}\{E_{i:n}\} = C_{i:n}^{(1)}$, so from Corollary 28.3.2, the means of $R_{i:n}$ and $E_{i:n}$, $i = 1, \ldots, d$, coincide. With regards to the variance, we have that $Var\{E_{i:n}\} = C_{i:n}^{(2)}$, so that

$$\frac{Var\{R_{i:n}\}}{Var\{E_{i:n}\}} = \frac{d}{d+1} \left\{ 1 - \frac{1}{d} \left[\frac{1}{CV(E_{i:n})} \right]^2 \right\}, \quad i = 1, 2, \ldots, d,$$

where $[CV(E_{i:n})]^2 = C_{i:n}^{(2)}/[C_{i:n}^{(1)}]^2$ is the coefficient of variation of $E_{i:n}$. Letting $n \to \infty$ with $d/n \to \rho \in (0, 1]$ and $i/n \to \gamma$ with $\gamma \leq \rho$, we obtain

$$\lim_{n \to \infty; \ d/n \to \rho; \ i/n \to \gamma} \frac{Var\{R_{i:n}\}}{Var\{E_{i:n}\}} = h_\rho(\gamma) \equiv 1 - \frac{1}{\rho} \left(\frac{1-\gamma}{\gamma} \right) [\ln(1-\gamma)]^2.$$

This limiting expression is slightly different from that obtained under the random censorship model considered in [4] [see equation (3.7) in that paper]. Observe that $\lim_{\gamma \downarrow 0} h_\rho(\gamma) = \lim_{\gamma \uparrow 1} h_\rho(\gamma) = 1$, and since $g(\gamma) = [(1-\gamma)/\gamma][\ln(1-\gamma)]^2$ achieves its maximum value at γ_0 satisfying the equation $\exp\{-2\gamma_0\} = 1 - \gamma_0$, which is approximately equal to $\gamma_0 = 0.7968$, then $h_\rho(\gamma)$ decreases on $[0, \gamma_0]$ and increases on $[\gamma_0, 1]$. Since $0 \leq \gamma \leq \rho$, then

$$M(\rho) \equiv \min_{\gamma \leq \rho} h_\rho(\gamma) = 1 - \frac{1}{\rho} \left[\frac{1 - \rho \wedge \gamma_0}{\rho \wedge \gamma_0} \right] [\ln(1 - \rho \wedge \gamma_0)]^2 . \tag{28.3.5}$$

Figure 28.1 displays a graph of ρ versus $M(\rho)$ as ρ ranges over $[0, 1]$. Thus, even though the means of $E_{i:n}$ and $R_{i:n}$ are identical, there is a big discrepancy between $Var\{E_{i:n}\}$ and $Var\{R_{i:n}\}$, which is vividly illustrated in Figure 28.1. The discrepancy becomes more pronounced as ρ, the limiting proportion of uncensored values, decreases to zero, with the variance of $R_{i:n}$ becoming smaller relative to $E_{i:n}$. This discrepancy is a manifestation of the effect of the dependence of the V_j's mentioned earlier, and this has implications on the behavior of test statistics when applied to $R_{[d:n]}$.

To amplify, consider the complete and sufficient statistic $S(T_{[d:n]})$ in (28.3.1) associated with the Type II$_{[d:n]}$ censored data. If we apply this statistic to $E_{[d:n]}$, we obtain

$$S(E_{[d:n]}) = \sum_{i=1}^{d} E_{i:n} + (n - d)E_{d:n}$$

$$= \sum_{i=1}^{d} (n - i + 1)(E_{i:n} - E_{i-1:n})$$

$$\stackrel{\text{st}}{=} \sum_{i=1}^{d} (n - i + 1)c_{i:n} E_i \quad \text{using (28.3.4)}$$

$$= \sum_{i=1}^{d} E_i$$

which has a gamma distribution with shape parameter d and scale parameter 1. In stark constrast, if we apply the statistic S to the Type II$_{[d:n]}$ censored residuals $R_{[d:n]}$, we obtain

$$S(R_{[d:n]}) = \sum_{i=1}^{d} R_{i:n} + (n - d)R_{d:n}$$

$$= \sum_{i=1}^{d} (n - i + 1)(R_{i:n} - R_{i-1:n})$$

$$\stackrel{\text{st}}{=} \sum_{i=1}^{d} (n - i + 1)c_{i:n} \, dV_i \quad \text{using Theorem 28.3.1}$$

$$= d \sum_{i=1}^{d} V_i$$

$$= d \quad \text{since} \quad \sum_{i=1}^{d} V_i = 1,$$

which is degenerate! Consequently, statistics for testing exponentiality which depends on the complete and sufficient statistic S under random sampling from the exponential distribution becomes totally noninformative when applied to the Type $\mathrm{II}_{[d:n]}$ censored residuals $\boldsymbol{R}_{[d:n]}$. More seriously, if one is unaware of the above degeneracy and tries to verify the approximate U.E.P. of $\boldsymbol{R}_{[d:n]}$ using such a test statistic, then one is bound to always conclude that the U.E.P. is satisfied by $\boldsymbol{R}_{[d:n]}$, and hence may erroneously conclude that the model is viable when infact it is not.

More generally, consider an $M \times 1$ vector of linear statistics for Type $\mathrm{II}_{[d:n]}$ censored data of form

$$S(\boldsymbol{E}_{[d:n]}) = \sum_{i=1}^{d} \boldsymbol{a}_i E_{i:n} \tag{28.3.6}$$

where $\boldsymbol{a}_i$, $i = 1, \ldots, d$, are $M \times 1$ real vectors. By (28.3.4) it follows that

$$S(\boldsymbol{E}_{[d:n]}) = \sum_{i=1}^{d} \boldsymbol{a}_i E_{i:n} \stackrel{st}{=} \sum_{i=1}^{d} \boldsymbol{a}_i \left(\sum_{j=1}^{i} c_{j:n} E_j \right) = \sum_{j=1}^{d} \bar{\boldsymbol{A}}_j c_{j:n} E_j$$

where $\bar{\boldsymbol{A}}_j = \sum_{i=j}^{d} \boldsymbol{a}_i$, $j = 1, 2, \ldots, d$. Consequently, it follows routinely that

$$\mathcal{E}\{S(\boldsymbol{E}_{[d:n]})\} = \sum_{j=1}^{d} \bar{\boldsymbol{A}}_j c_{j:n} \quad \text{and} \quad Cov\{S(\boldsymbol{E}_{[d:n]})\} = \sum_{j=1}^{d} c_{j:n}^2 \bar{\boldsymbol{A}}_j^{\otimes 2},$$

where, for a matrix $\boldsymbol{A}$, $\boldsymbol{A}^{\otimes 2} = \boldsymbol{A}\boldsymbol{A}^*$, with "*" denoting transpose. When the linear statistic in (28.3.6) is applied to $\boldsymbol{R}_{[d:n]}$, we obtain

$$S(\boldsymbol{R}_{[d:n]}) = \sum_{i=1}^{d} \boldsymbol{a}_i R_{i:n} \stackrel{st}{=} \sum_{i=1}^{d} \boldsymbol{a}_i \left(d \sum_{j=1}^{i} c_{j:n} V_j \right) = \sum_{j=1}^{d} \bar{\boldsymbol{A}}_j c_{j:n} \, dV_j.$$

Consequently, $\mathcal{E}\{S(\boldsymbol{R}_{[d:n]})\} = \sum_{j=1}^{d} \bar{\boldsymbol{A}}_j c_{j:n}$ which coincides with $\mathcal{E}\{S(\boldsymbol{E}_{[d:n]})\}$. On the otherhand, the covariance matrix is

$$Cov\{S(\boldsymbol{R}_{[d:n]})\} = Cov\{\sum_{j=1}^{d} \bar{\boldsymbol{A}}_j c_{j:n} \, dV_j, \sum_{l=1}^{d} \bar{\boldsymbol{A}}_l c_{l:n} \, dV_l\}$$

$$= \sum_{j=1}^{d} \sum_{l=1}^{d} c_{j:n} c_{l:n} \bar{\boldsymbol{A}}_j \bar{\boldsymbol{A}}_l^* Cov\{dV_j, dV_l\}$$

$$= \sum_{j=1}^{d} c_{j:n}^2 \bar{\boldsymbol{A}}_j^{\otimes 2} Var\{dV_j\} + \sum_{j \neq l} c_{j:n} c_{l:n} \bar{\boldsymbol{A}}_j \bar{\boldsymbol{A}}_l^* Cov\{dV_j, dV_l\}$$

$$= \frac{d-1}{d+1} \sum_{j=1}^{d} c_{j:n}^2 \bar{\boldsymbol{A}}_j^{\otimes 2} - \frac{1}{d+1} \sum_{j \neq l} c_{j:n} c_{l:n} \bar{\boldsymbol{A}}_j \bar{\boldsymbol{A}}_l^*$$

$$= \sum_{j=1}^{d} c_{j:n}^2 \bar{\boldsymbol{A}}_j^{\otimes 2} - \frac{2}{d+1} \sum_{j=1}^{d} c_{j:n}^2 \bar{\boldsymbol{A}}_j^{\otimes 2} - \frac{1}{d+1} \sum_{j \neq l} c_{j:n} c_{l:n} \bar{\boldsymbol{A}}_j \bar{\boldsymbol{A}}_l^*$$

$$= Cov\{S(\boldsymbol{E}_{[d:n]})\} - \frac{1}{d+1} \left\{ \sum_{j=1}^{d} c_{j:n}^2 \bar{\boldsymbol{A}}_j^{\otimes 2} + \left(\sum_{j=1}^{d} c_{j:n} \bar{\boldsymbol{A}}_j \right)^{\otimes 2} \right\}.$$

Therefore,

$$Cov\{S(E_{[d:n]})\} - Cov\{S(R_{[d:n]})\} = \frac{1}{d+1}\left\{\sum_{j=1}^{d} c_{j:n}^2 \bar{A}_j^{\otimes 2} + \left(\sum_{j=1}^{d} c_{j:n}\bar{A}_j\right)^{\otimes 2}\right\}.$$

Since both matrices in the right-hand side are nonnegative definite, then $S(R_{[d:n]})$ is *never* more variable than $S(E_{[d:n]})$.

28.3.2   Location-Scale Exponential Failure Times

In this subsection we assume that $T_1^\star, T_2^\star, \ldots, T_n^\star$ are i.i.d. from the location-scale exponential distribution which has density function

$$f(t^\star; \theta, \alpha) = \theta \exp\{-\theta(t^\star - \alpha)\} I_{[\alpha,\infty)}(t^\star),$$

where $\theta > 0$, $\alpha > 0$. We denote this distribution by $\mathrm{Exp}(\theta, \alpha)$. [This is also called the guaranteed exponential density function with the parameter α serving as the guaranteed minimum lifetime.] The cumulative hazard function associated with this distribution function is

$$\Lambda(t^\star; \theta, \alpha) = \theta(t^\star - \alpha) I_{[\alpha,\infty)}(t^\star).$$

In order to make meaningful statistical inferences about the parameters under Type II censoring, it is required that d should be at least 2. Using the likelihood function in (28.2.2), it is easy to see that the minimal sufficient statistics for (θ, α), given a Type $\mathrm{II}_{[d:n]}$ censored data $T_{[d:n]}^\star = (T_{1:n}^\star, \ldots, T_{d:n}^\star)$, is the vector of statistics $[T_{1:n}^\star, S^\star(T_{[d:n]}^\star)]$, where

$$S^\star(T_{[d:n]}^\star) = \sum_{i=2}^{d} (T_{i:n}^\star - T_{1:n}^\star) + (n-d)(T_{d:n}^\star - T_{1:n}^\star).$$

The maximum likelihood estimators of θ and α are given, respectively, by

$$\hat{\theta} = d/S^\star(T_{[d:n]}^\star) \quad \text{and} \quad \hat{\alpha} = T_{1:n}^\star.$$

Consequently, the relevant Type $\mathrm{II}_{[d:n]}$ censored residuals under this location-scale model are $R_{[d:n]}^\star = (R_{2:n}^\star, \ldots, R_{d:n}^\star)$, where

$$R_{i:n}^\star = \Lambda(T_{i:n}^\star; \hat{\theta}, \hat{\alpha}) = d(T_{i:n}^\star - T_{1:n}^\star)/S^\star(T_{[d:n]}^\star), \quad i = 1, \ldots, d. \qquad (28.3.7)$$

Note that $R_{1:n}^\star = 0$.

THEOREM 28.3.3
Let $T_1^\star, \ldots, T_n^\star \overset{\text{i.i.d.}}{\sim} \mathrm{Exp}(\theta, \alpha)$ and let $R_{[d:n]}^\star = (R_{2:n}^\star, \ldots, R_{d:n}^\star)$ be the associated Type $\mathrm{II}_{[d:n]}$ censored residuals. Then

$$R_{[d:n]}^\star \overset{\text{st}}{=} \left(\frac{d}{d-1}\right) R_{[d-1:n-1]},$$

where $R_{[d-1:n-1]} = (R_{1:n-1}, \ldots, R_{d-1:n-1})$ is the Type $\mathrm{II}_{[d-1:n-1]}$ censored residuals associated with $T_1, \ldots, T_{n-1} \overset{\text{i.i.d.}}{\sim} \mathrm{Exp}(\theta)$.

PROOF   It is well-known that

$$(T_{2:n}^\star - T_{1:n}^\star, \ldots, T_{d:n}^\star - T_{1:n}^\star) \overset{\text{st}}{=} (T_{1:n-1}, \ldots, T_{d-1:n-1}),$$

where $T_{j:n-1}$ is the jth order statistic of a random sample of size $n-1$ from $\text{Exp}(\theta)$. Since

$$S^\star(\boldsymbol{T}^\star_{[d:n]}) = \sum_{i=2}^{d}(T^\star_{i:n} - T^\star_{1:n}) + (n-d)(T^\star_{d:n} - T^\star_{1:n})$$

$$\overset{st}{=} \sum_{i=1}^{d-1} T_{i:n-1} + [(n-1) - (d-1)]T_{d-1:n-1} = S(\boldsymbol{T}_{[d-1:n-1]}),$$

then, from (28.3.7), $\boldsymbol{R}^\star_{[d:n]} \overset{st}{=} d\left(T_{i:n-1}/S(\boldsymbol{T}_{[d-1:n-1]})\right)$, $i = 1, 2, \ldots, d-1)$
$= \left(\frac{d}{d-1}\right)\boldsymbol{R}_{[d-1:n-1]}$. ∎

From Corollary 28.3.2 and Theorem 28.3.3, the following results follow immediately.

COROLLARY 28.3.4
Under the conditions of Theorem 28.3.3, for each $i, j = 2, \ldots, d$ with $i \neq j$,

(i) $R^\star_{i:n} \overset{st}{=} d\sum_{l=1}^{i-1} c_{l:n-1}V_j$;

(ii) $(R^\star_{i:n}, R^\star_{j:n}) \overset{st}{=} d(\sum_{l=1}^{i-1} c_{l:n-1}V_l, \sum_{l=1}^{j-1} c_{l:n-1}V_l)$;

(iii) $\mathcal{E}\{R^\star_{i:n}\} = \left(\frac{d}{d-1}\right)C^{(1)}_{i-1:n-1}$;

(iv) $Var\{R^\star_{i:n}\} = \frac{d}{(d-1)^2}\left\{(d-1)C^{(2)}_{i-1:n-1} - [C^{(1)}_{i-1:n-1}]^2\right\}$; and

(v) $Cov\{R^\star_{i:n}, R^\star_{j:n}\} = \left(\frac{d}{d-1}\right)^2\{Cov\{R_{i-1:n-1}, R_{j-1:n-1}\}\}$,

where $(V_1, V_2, \ldots, V_{d-1}) \sim \text{Dir}_{si}(\mathbf{1}_{d-1})$.

In contrast to the case where the failure-time distribution is the scale exponential model considered in subsection 28.3.1 , from Corollary 28.3.4(iii) we see that in the location-scale exponential model, $\mathcal{E}\{R^\star_{i:n}\} \neq \mathcal{E}\{E_{i:n}\}$. Indeed, $R^\star_{1:n}$ is degenerate at zero, while $E_{1:n}$ has an exponential distribution with parameter n. Thus, there could be a substantial difference between the distributions of $(E_{1:n}, \ldots, E_{d:n})$, which possesses the U.E.P., and $(R^\star_{1:n}, \ldots, R^\star_{d:n})$, the Type II$_{[d:n]}$ censored residuals. This distributional difference may make model diagnostic procedures, which revolves on checking the approximate U.E.P. of the Type II$_{[d:n]}$ censored residuals, misleading. What one ought to do is to make use of the exact distributional results in Theorems 28.3.1 and 28.3.3 and Corollaries 28.3.2 and 28.3.4 or the asymptotic results in Section 28.4 to develop appropriate model diagnostic procedures. This approach will be discussed in Section 28.5.

28.4   Asymptotic Properties of Statistics Applied to Residuals

In this section we assume that $T_1, T_2, \ldots, T_n$ are i.i.d. from a density function $f(t; \boldsymbol{\theta})$ where $\boldsymbol{\theta} = (\theta_1, \ldots, \theta_K)^* \in \Theta$, with Θ being an open subset of $\mathfrak{R}^K$. We denote by $\boldsymbol{\theta}_0$ the unknown true value of the parameter, and we assume that $\boldsymbol{\theta}_0$ is an interior point of Θ. We adopt the convention that when a function of $\boldsymbol{\theta}$ is evaluated at $\boldsymbol{\theta}_0$, we suppress writing $\boldsymbol{\theta}_0$, for example, $\rho(x; \boldsymbol{\theta}_0) = \rho(x)$. For a differentiable function $S : \Theta \to \mathfrak{R}$, we let

$$\dot{\boldsymbol{S}}(\boldsymbol{\theta}) = \left(\frac{\partial S}{\partial \theta_1}, \ldots, \frac{\partial S}{\partial \theta_K}\right)^*;$$

while for an $M \times 1$ vector of differentiable functions $\boldsymbol{S} : \Theta \rightarrow \Re^M$, we let

$$
\dot{\boldsymbol{S}}(\boldsymbol{\theta}) = \begin{bmatrix} \frac{\partial S_1}{\partial \theta_1} & \cdots & \frac{\partial S_1}{\partial \theta_K} \\ \vdots & \ddots & \vdots \\ \frac{\partial S_M}{\partial \theta_1} & \cdots & \frac{\partial S_M}{\partial \theta_K} \end{bmatrix}.
$$

Thus, note that if $S : \Theta \rightarrow \Re$ is twice-differentiable with respect to the θ_i's, then

$$
\ddot{\boldsymbol{S}}(\boldsymbol{\theta}) = \left(\frac{\partial^2 S}{\partial \theta_i \partial \theta_j} \right)_{i,j=1,\dots,K}.
$$

We shall assume that the censoring parameter d satisfies $d = [np]$ for some $0 < p \leq 1$, and where $[\cdot]$ means 'integer part'. We denote by ξ the pth quantile of $f(\cdot) \equiv f(\cdot; \boldsymbol{\theta}_0)$, and so $F(\xi) = p$ and $\Lambda(\xi) = -\ln q \equiv \gamma$ where $q = 1 - p$.

28.4.1   Asymptotics

Suppose that we have an $M \times 1$ vector of statistics given by

$$
\boldsymbol{S}_n(\boldsymbol{E}_{[d:n]}) = \frac{1}{n} \left\{ \sum_{i=1}^{d} \boldsymbol{A}(E_{i:n}) + (n-d)\boldsymbol{B}(E_{d:n}) \right\} \tag{28.4.1}
$$

which may be used to test the U.E.P. given a Type II$_{[d:n]}$ censored data $\boldsymbol{E}_{[d:n]}$. The jth elements of $\boldsymbol{A}$ and $\boldsymbol{B}$ will be denoted by A_j and B_j, respectively. Our interest is to determine the asymptotic distribution of $\boldsymbol{S}_n$ when applied to the Type II$_{[d:n]}$ censored residuals $\boldsymbol{R}_{[d:n]} = (R_{1:n}, \dots, R_{d:n})$, where $R_{i:n} = \Lambda(T_{i:n}; \hat{\boldsymbol{\theta}}_n)$, $\hat{\boldsymbol{\theta}}_n$ being the MLE of $\boldsymbol{\theta}$ based on $\boldsymbol{T}_{[d:n]}$. To shorten our notation we let

$$
\psi(t; \boldsymbol{\theta}) = \ln f(t; \boldsymbol{\theta}) \quad \text{and} \quad \rho(t; \boldsymbol{\theta}) = \ln \bar{F}(t; \boldsymbol{\theta}).
$$

From (28.2.2) the log-likelihood function is therefore

$$
l_n(\boldsymbol{\theta}) = \sum_{i=1}^{d} \psi(T_{i:n}; \boldsymbol{\theta}) + (n-d)\rho(T_{d:n}; \boldsymbol{\theta}),
$$

and the MLE $\hat{\boldsymbol{\theta}}_n$ satisfies

$$
\dot{l}_n(\hat{\boldsymbol{\theta}}_n) = \sum_{i=1}^{d} \dot{\psi}(T_{i:n}; \hat{\boldsymbol{\theta}}_n) + (n-d)\dot{\rho}(T_{d:n}; \hat{\boldsymbol{\theta}}_n) = \mathbf{0}. \tag{28.4.2}
$$

We now enumerate the regularity conditions needed for the asymptotic result.
Condition I: For each $j = 1, 2, \dots, M$,

1. $\int_0^{\gamma} [A_j(u)]^2 e^{-u} \, du < \infty$;
2. $A_j(t)$ and $B_j(t)$ are continuously differentiable, with derivatives $A_j'(t)$ and $B_j'(t)$.

Condition II: For each $j = 1, 2, \dots, M$, and for each $x > 0$,

1. $\frac{\partial}{\partial \boldsymbol{\theta}} \int_0^x f(t; \boldsymbol{\theta}) \, dt = \int_0^x \dot{f}(t; \boldsymbol{\theta}) \, dt$;
2. $\frac{\partial}{\partial \boldsymbol{\theta}} \int_0^x \dot{f}(t; \boldsymbol{\theta}) \, dt = \int_0^x \ddot{f}(t; \boldsymbol{\theta}) \, dt$;
3. $\frac{\partial}{\partial \boldsymbol{\theta}} \int_0^x A_j[\Lambda(t; \boldsymbol{\theta})]f(t; \boldsymbol{\theta}) \, dt = \int_0^x \frac{\partial}{\partial \boldsymbol{\theta}} \{A_j[\Lambda(t; \boldsymbol{\theta})]f(t; \boldsymbol{\theta})\} \, dt$.

Condition III: There exist a compact neighborhood Θ_1 of θ_0 and a positive number ϵ_1 such that for each j, l,

1. $A'_j[\Lambda(t;\boldsymbol{\theta})]\dot{\Lambda}_l(t;\boldsymbol{\theta})$ is continuous in $\boldsymbol{\theta} \in \Theta_1$ for each t;
2. there exists a $C_1(t)$ such that for each $\boldsymbol{\theta} \in \Theta_1$ and for each t, $| A'_j[\Lambda(t;\boldsymbol{\theta})]\dot{\Lambda}_l(t;\boldsymbol{\theta}) | \leq C_1(t)$ and $\int_0^\infty C_1(t)f(t)\ dt < \infty$;
3. $B'_j[\Lambda(t;\boldsymbol{\theta})]\dot{\Lambda}_l(t;\boldsymbol{\theta})$ is continuous in $[\xi - \epsilon_1, \xi + \epsilon_1] \times \Theta_1$.

Condition IV: There exist a compact neighborhood Θ_2 of $\boldsymbol{\theta}_0$ and a positive number ϵ_2 such that for each j, l,

1. $\ddot{\psi}_{jl}(t;\boldsymbol{\theta})$ is continuous in $\boldsymbol{\theta} \in \Theta_2$ for each t;
2. there exists a $C_2(t)$ such that for each $\boldsymbol{\theta} \in \Theta_2$ and for each t, $| \ddot{\psi}_{jl}(t;\boldsymbol{\theta}) | \leq C_2(t)$ and $\int_0^\infty C_2(t)f(t)\ dt < \infty$;
3. $\ddot{\rho}_{jl}(t;\boldsymbol{\theta})$ is continuous in $[\xi - \epsilon_2, \xi + \epsilon_2] \times \Theta_2$.

THEOREM 28.4.1
Let $T_1, \ldots, T_n \overset{\text{i.i.d.}}{\sim} f(t;\boldsymbol{\theta})$, $\boldsymbol{\theta} \in \Theta$, and let $\boldsymbol{R}_{[d:n]} = (R_{1:n}, \ldots, R_{d:n})$ be the Type II$_{[d:n]}$ censored residuals. If conditions (I)–(IV) are satisfied, then as $n \to \infty$ with $d = [np]$ and under $\boldsymbol{\theta} = \boldsymbol{\theta}_0$,

$$\sqrt{n}\,[S_n(\boldsymbol{R}_{[d:n]}) - \boldsymbol{\mu}] \overset{\mathcal{L}}{\to} N(\mathbf{o},\ \boldsymbol{\Xi} \equiv \boldsymbol{\Sigma}_{11} - \boldsymbol{\Upsilon}\boldsymbol{\Sigma}_{22}^{-1}\boldsymbol{\Upsilon}^*),$$

where

$$\boldsymbol{\mu} = \int_0^\gamma \boldsymbol{A}(u)e^{-u}\ du + q\boldsymbol{B}(\gamma); \tag{28.4.3}$$

$$\boldsymbol{\Sigma}_{11} = \int_0^\gamma [\boldsymbol{A}(u)]^{\otimes 2}e^{-u}\ du - \frac{1}{p}\left(\int_0^\gamma \boldsymbol{A}(u)e^{-u}\ du\right)^{\otimes 2}$$

$$+pq\left\{\boldsymbol{A}(\gamma) + \boldsymbol{B}'(\gamma) - \frac{1}{p}\int_0^\gamma \boldsymbol{A}(u)e^{-u}\ du\right\}^{\otimes 2}; \tag{28.4.4}$$

$$\boldsymbol{\Sigma}_{22} = \int_0^\xi [\boldsymbol{\psi}(t)]^{\otimes 2}f(t)\ dt + \frac{1}{q}[\dot{\boldsymbol{F}}(\xi)]^{\otimes 2}; \tag{28.4.5}$$

$$\boldsymbol{\Upsilon} = \int_0^\xi \boldsymbol{A}'[\Lambda(t)][\dot{\boldsymbol{\Lambda}}(t)]^*f(t)\ dt + q\boldsymbol{B}'(\gamma)[\dot{\boldsymbol{\Lambda}}(\xi)]^*. \tag{28.4.6}$$

PROOF   Let $\boldsymbol{Z}_n(\boldsymbol{\theta}) = [\boldsymbol{Z}_{1n}(\boldsymbol{\theta})^*\ \boldsymbol{Z}_{2n}(\boldsymbol{\theta})^*]^*$, where

$$\boldsymbol{Z}_{1n}(\boldsymbol{\theta}) = \frac{1}{n}\left\{\sum_{i=1}^d \boldsymbol{A}[\Lambda(T_{i:n};\boldsymbol{\theta})] + (n-d)\boldsymbol{B}[\Lambda(T_{d:n};\boldsymbol{\theta})]\right\}$$

and

$$\boldsymbol{Z}_{2n}(\boldsymbol{\theta}) = \frac{1}{n}\left\{\sum_{i=1}^d \boldsymbol{\psi}(T_{i:n};\boldsymbol{\theta}) + (n-d)\dot{\rho}(T_{d:n};\boldsymbol{\theta})\right\}.$$

It follows from Theorem 1 of [6] that under $\boldsymbol{\theta} = \boldsymbol{\theta}_0$ and conditions (I), (II) and (IV),

$$\sqrt{n}[\boldsymbol{Z}_n - \boldsymbol{\nu}] \overset{\mathcal{L}}{\to} N_{M+K}(\mathbf{o}, \boldsymbol{\Sigma}),$$

where

$$\boldsymbol{\nu} = \begin{pmatrix} \boldsymbol{\mu} \\ \mathbf{o} \end{pmatrix} \quad \text{and} \quad \boldsymbol{\Sigma} = \begin{bmatrix} \boldsymbol{\Sigma}_{11} & \boldsymbol{\Sigma}_{12} \\ \boldsymbol{\Sigma}_{12}^* & \boldsymbol{\Sigma}_{22} \end{bmatrix},$$

with

$$\Sigma_{12} = \int_0^\xi A[\Lambda(t)][\dot{f}(t)]^* \, dt - [A(\gamma) + B'(\gamma)][\dot{F}(\xi)]^*.$$

By Taylor's Theorem we have that

$$S_n(R_{[d:n]}) = S_n(\Lambda(T_{1:n}, \ldots, \Lambda(T_{d:n})) + [\Upsilon_n(\theta_n^\dagger)](\hat{\theta}_n - \theta_0),$$

where

$$\Upsilon_n(\theta) = \frac{1}{n} \left\{ \sum_{i=1}^d A'[\Lambda(T_{i:n}; \theta)][\dot{\Lambda}(T_{i:n}; \theta)]^* + (n - d)B'[\Lambda(T_{d:n}; \theta)][\dot{\Lambda}(T_{i:n}; \theta)]^* \right\}$$

and $\theta_n^\dagger$ is in the line segment connecting $\hat{\theta}_n$ and θ_0. By condition (III) and Theorem 2 of [6], under $\theta = \theta_0$, $\sup_{\theta \in \Theta_1} \| \Upsilon_n(\theta) - \Upsilon(\theta) \| \overset{\text{wp1}}{\to} 0$, where $\| C \| = \max_{i,j} | C_{ij} |$, and

$$\Upsilon(\theta) = \int_0^\xi A'[\Lambda(t; \theta)][\dot{\Lambda}(t; \theta)]^* f(t; \theta) \, dt + qB'[\Lambda(t; \theta)][\dot{\Lambda}(t; \theta)]^*.$$

Furthermore, from [6],

$$\sqrt{n}(\hat{\theta}_n - \theta_0) = \left[-\frac{1}{n} \ddot{i}_n(\theta_n^\ddagger) \right]^{-1} \frac{1}{\sqrt{n}} \dot{i}_n,$$

with $\theta_n^\ddagger$ lying in the line segment joining $\hat{\theta}_n$ and θ_0. Letting

$$J(\theta) = -\int_0^\xi \ddot{\psi}(t; \theta) f(t; \theta) \, dt - q\ddot{\rho}(\xi; \theta),$$

then, under condition (IV) and $\theta = \theta_0$, $\sup_{\theta \in \Theta_2} \| -n^{-1} \ddot{i}_n(\theta) - J(\theta) \| \overset{\text{wp1}}{\to} 0$. Since $\| \hat{\theta}_n - \theta_0 \| \overset{\text{wp1}}{\to} 0$, it follows that

$$\sqrt{n}[S_n(R_{[d:n]}) - \mu] = \sqrt{n}[S_n(\Lambda(T_{1:n}), \ldots, \Lambda(T_{d:n})) - \mu] + \Upsilon J^{-1} \frac{1}{\sqrt{n}} \dot{i}_n + o_p(1).$$

By noting that $Z_{1n} = S_n(\Lambda(T_{1:n}, \ldots, \Lambda(T_{d:n}))$ and $Z_{2n} = n^{-1} \dot{i}_n$, it follows that

$$\sqrt{n}[S_n(R_{[d:n]}) - \mu] \overset{\mathcal{L}}{\to} N_M(0, \Xi \equiv \Sigma_{11} + \Upsilon J^{-1} \Sigma_{22} J^{-1} \Upsilon^* + 2\Upsilon J^{-1} \Upsilon^*).$$

It is routine to show that $\Sigma_{22} = J$ (cf., [6]), so the 2nd term in the limiting covariance matrix equals $\Upsilon \Sigma_{22}^{-1} \Upsilon^*$. On the otherhand, since $q\dot{\Lambda}(\xi) = \dot{F}(\xi)$, and by using condition (II),

$$A(\gamma)[\dot{F}(\xi)]^* = \frac{\partial}{\partial \theta} \int_0^{\Lambda(\xi; \theta)} A(u) e^{-u} \, du \mid_{\theta = \theta_0}$$

$$= \frac{\partial}{\partial \theta} \int_0^\xi A[\Lambda(t; \theta)] f(t; \theta) \, dt \mid_{\theta = \theta_0}$$

$$= \int_0^\xi A'[\Lambda(t)][\dot{\Lambda}(t)]^* f(t) \, dt + \int_0^\xi A[\Lambda(t)][\dot{f}(t)]^* \, dt,$$

then

$$\Upsilon = -\int_0^\xi A[\Lambda(t)][\dot{f}(t)]^* \, dt + A(\gamma)[\dot{F}(\xi)]^* + B'(\gamma)[\dot{F}(\xi)]^* = -\Sigma_{12}.$$

Consequently, the 3rd term in the limiting covariance matrix equals $-2\boldsymbol{\Upsilon}\boldsymbol{\Sigma}_{22}^{-1}\boldsymbol{\Upsilon}^*$, implying that $\boldsymbol{\Xi} = \boldsymbol{\Sigma}_{11} - \boldsymbol{\Upsilon}\boldsymbol{\Sigma}_{22}^{-1}\boldsymbol{\Upsilon}^*$. ∎

REMARK In the course of the proof of Theorem 28.4.1, we note the two alternative computational formulas for $\boldsymbol{\Sigma}_{22}$ and $\boldsymbol{\Upsilon}$ given by

$$\boldsymbol{\Sigma}_{22} = -\int_0^\xi \ddot{\psi}(t) f(t) \; dt - q\ddot{\rho}(\xi), \tag{28.4.7}$$

and

$$\boldsymbol{\Upsilon} = -\int_0^\xi \boldsymbol{A}[\Lambda(t)][\dot{\boldsymbol{f}}(t)]^* \; dt + [\boldsymbol{A}(\gamma) + \boldsymbol{B}'(\gamma)][\dot{\boldsymbol{F}}(\xi)]^*. \tag{28.4.8}$$

∎

REMARK The asymptotic mean vector $\boldsymbol{\mu}$ and the 1st term $\boldsymbol{\Sigma}_{11}$ of $\boldsymbol{\Xi}$ are the asymptotic mean vector and covariance matrix of $S_n(\boldsymbol{E}_{[d:n]})$. Thus, the effect of estimating θ by $\hat{\theta}_n$ to obtain the residuals is contained in the 2nd term $\boldsymbol{\Upsilon}\boldsymbol{\Sigma}_{22}^{-1}\boldsymbol{\Upsilon}^*$ of $\boldsymbol{\Xi}$. Note that $\boldsymbol{\Sigma}_{22}^{-1}$ is the asymptotic covariance matrix of the MLE $\hat{\theta}_n$. ∎

28.4.2   Some Examples

In this subsection we apply Theorem 28.4.1 on some concrete situations in order to illustrate the change in asymptotic variance when a statistic is applied to Type II$_{[d:n]}$ censored residuals.

Example 1

Let $f(t;\theta) = \theta \exp\{-\theta t\} I_{[0,\infty)}(t)$ for $\theta > 0$, and let $A(u) = B(u) = u$, so the statistic is

$$S_1(\boldsymbol{E}_{[d:n]}) = \frac{1}{n}\left\{\sum_{i=1}^d E_{i:n} + (n-d)E_{d:n}\right\}, \tag{28.4.9}$$

which is a version of the TTOT statistic. By routine calculations, it follows that

$$\mu = \int_0^\gamma u e^{-u} \; du + q\gamma = p; \tag{28.4.10}$$

$$\Sigma_{11} = \int_0^\gamma u^2 e^{-u} \; du - \frac{1}{p}\left(\int_0^\gamma u e^{-u} \; du\right)^2 + pq\left(\gamma + 1 - \frac{1}{p}\int_0^\gamma u e^{-u} \; du\right)^2 = p; \tag{28.4.11}$$

$$\Sigma_{22} = \xi^2\left(\frac{1}{\gamma^2}\int_0^\gamma (1-u)^2 e^{-u} \; du + q\right) = \frac{\xi^2}{\gamma^2}p; \tag{28.4.12}$$

and

$$\Upsilon = \xi\left(\frac{1}{\gamma}\int_0^\gamma u e^{-u} \; du + q\right) = \frac{\xi}{\gamma}p;$$

so that

$$\Xi = p - \frac{(\xi p/\gamma)^2}{\xi^2 p/\gamma^2} = 0.$$

Thus, $\sqrt{n}\,[S_1(\boldsymbol{R}_{[d:n]}) - p] \xrightarrow{\text{pr}} 0$, a result which is consistent with the exact results obtained in Subsection 28.3.1. ☐

Example 2
Let us still consider the exponential distribution in Example 1, but now let $A(u) = 2\exp\{-u\}$ and $B(u) = \exp\{-u\}$. The resulting statistic is

$$S_2(\boldsymbol{E}_{[d:n]}) = \frac{1}{n}\left\{2\sum_{i=1}^{d} e^{-E_{i:n}} + (n-d)e^{-E_{d:n}}\right\}, \qquad (28.4.13)$$

which was considered in [12]. In this case it follows, by straightforward manipulations, that

$$\mu = \int_0^\gamma 2e^{-u}e^{-u}\,du + qe^{-\gamma} = 1; \qquad (28.4.14)$$

$$\Sigma_{11} = \int_0^\gamma 4e^{-2u}e^{-u}\,du - \frac{1}{p}\left(\int_0^\gamma 2e^{-u}e^{-u}\,du\right)^2$$

$$+pq\left(2e^{-u} - e^{-u} - \frac{1}{p}\int_0^\gamma 2e^{-u}e^{-u}\,du\right)^2 = \frac{1}{3}(1-q^3); \qquad (28.4.15)$$

$$\Upsilon = \int_0^\xi (-2e^{-\theta_0 t})t\theta_0 e^{-\theta_0 t}\,dt + q(-e^{-\gamma})\xi = -\frac{\xi}{2\gamma}(1-q^2).$$

Note that Σ_{22} is given by (28.4.12). Consequently, the asymptotic variance becomes

$$\Xi = \frac{1}{3}(1-q^3) - \frac{\{-\xi(1-q^2)/(2\gamma)\}^2}{\xi^2 p/\gamma^2} = \frac{1}{12}(1-q)^3 = \frac{p^3}{12}.$$

☐

We summarize this result more formally as follows:

COROLLARY 28.4.2
If $T_1, \ldots, T_n \overset{\text{i.i.d.}}{\sim} f(t;\theta) = \theta e^{-\theta t}I_{[0,\infty)}(t)$ for $\theta > 0$, and $\boldsymbol{R}_{[d:n]} = (R_{1:n}, \ldots, R_{d:n})$ are the associated Type II$_{[d:n]}$ censored residuals, then as $n \to \infty$ with $d = [np]$ where $0 < p \le 1$,

$$\sqrt{n}\,[S_2(\boldsymbol{R}_{[d:n]}) - 1] \xrightarrow{\mathcal{L}} N\left(0, \frac{p^3}{12}\right).$$

REMARK   When the Horowitz and Neumann [12] statistic is applied to $\boldsymbol{E}_{[d:n]}$, we obtain

$$\sqrt{n}\,[S_2(\boldsymbol{E}_{[d:n]}) - 1] \xrightarrow{\mathcal{L}} N\left(0, \frac{1}{3}(1-q^3)\right),$$

so, relative to the asymptotic variance under U.E.P., the asymptotic variance of S_2, when applied to the Type II$_{[d:n]}$ censored residuals $\boldsymbol{R}_{[d:n]}$, is

$$\text{Ratio}(p) = \frac{Avar[S_2(\boldsymbol{R}_{[d:n]})]}{Avar[S_2(\boldsymbol{E}_{[d:n]})]} = \frac{1}{4}\frac{(1-q)^3}{1-q^3},$$

which takes values of $1/4, 27/252, 1/28$, and $1/148$ for q-values of $0, 1/4, 1/2$, and $3/4$, respectively. This indicates that the variance of a statistic when applied to the residuals can be substantially smaller compared to the variance under the U.E.P. ∎

For the examples concerning the Weibull density function below, we will utilize the incomplete gamma function defined via $\Gamma(\kappa; x) = \int_0^x w^{\kappa-1}e^{-w}\,dw$ where $x > 0$ and $\kappa > 0$. We denote by $\dot{\Gamma}(\kappa; x)$ and $\ddot{\Gamma}(\kappa; x)$ the first and second partial derivatives of $\Gamma(\kappa; x)$ with respect to κ, so that

$$\dot{\Gamma}(\kappa; x) = \int_0^x w^{\kappa-1}(\ln w)e^{-w}\,dw \quad \text{and} \quad \ddot{\Gamma}(\kappa; x) = \int_0^x w^{\kappa-1}(\ln w)^2 e^{-w}\,dw. \quad (28.4.16)$$

Example 3

Let $f(t; \alpha) = \alpha t^{\alpha-1} \exp\{-t^\alpha\}I_{[0,\infty]}(t)$ for $\alpha > 0$, which is the Weibull density function with shape parameter α and scale parameter 1. It follows that $\bar{F}(t; \alpha) = \exp(-t^\alpha)$, $\lambda(t; \alpha) = \alpha t^{\alpha-1}$, and $\Lambda(t; \alpha) = t^\alpha$. Let us consider the TTOT statistic S_1 in (28.4.9). Then, as in Example 1, $\mu = p$ and $\Sigma_{11} = p$. Using formula (28.4.7),

$$\Sigma_{22} = -\int_0^\xi \left[-\frac{1}{\alpha_0^2} - t^{\alpha_0}(\ln t)^2\right]\alpha_0 t^{\alpha_0-1} \exp\{-t^{\alpha_0}\}\,dt - q(-\xi^{\alpha_0})(\ln \xi)^2$$

$$= \frac{1}{\alpha_0^2}\left\{\int_0^\gamma [1 + w(\ln w)^2]\,e^{-w}\,dw + q\gamma(\ln \gamma)^2\right\}$$

$$= \frac{1}{\alpha_0^2}\left\{p + \int_0^\gamma w(\ln w)^2 e^{-w}\,dw + q\gamma(\ln \gamma)^2\right\},$$

where α_0 is the true value of α, $q = 1 - p = \exp\{-\xi^{\alpha_0}\}$, $\gamma = \xi^{\alpha_0} = -\ln q$, and ξ is the pth quantile of $f(t; \alpha_0)$. Using (28.4.6), we obtain

$$\Upsilon = \int_0^\xi (t^{\alpha_0} \ln t)\alpha_0 t^{\alpha_0-1} \exp\{-t^{\alpha_0}\}\,dt + q(\xi^{\alpha_0} \ln \xi)$$

$$= \frac{1}{\alpha_0}\left\{\int_0^\gamma (w \ln w)e^{-w}\,dw + q\gamma \ln \gamma\right\}.$$

Therefore, the limiting variance of $\sqrt{n}\,[S_1(\boldsymbol{R}_{[d:n]}) - p]$ is

$$\Xi = p - \frac{\left\{\int_0^\gamma (w \ln w)e^{-w}\,dw + q\gamma \ln \gamma\right\}^2}{\left\{p + \int_0^\gamma w(\ln w)^2 e^{-w}\,dw + q\gamma(\ln \gamma)^2\right\}},$$

which in terms of the functions in (28.4.16) becomes

$$\Xi = p - \frac{\{\dot{\Gamma}(2; \gamma) + q\gamma \ln \gamma\}^2}{\{p + \ddot{\Gamma}(2; \gamma) + q\gamma(\ln \gamma)^2\}}. \quad (28.4.17)$$

To see graphically the change in asymptotic variance when S_1 is applied to $\boldsymbol{R}_{[d:n]}$ relative to the asymptotic variance under the U.E.P., we plotted in Figure 28.2 p versus Ξ/p as p ranges over $(0, 1)$. The functions in (28.4.16) were numerically evaluated using the subroutine QDAGS of the IMSL [13] library. Notice that there is a substantial difference when the proportion of uncensored values is small. ∎

Example 4

Still assuming the Weibull density function in Example 3, we now consider the statistic S_2 in (28.4.13). As in Example 2, $\mu = 1$ and $\Sigma_{11} = (1 - q^3)/3$, while Σ_{22} is as in Example 3. Applying (28.4.6), we obtain

$$\Upsilon = \int_0^\xi (-2\exp\{-t^{\alpha_0}\})(t^{\alpha_0}\ln t)\alpha_0 t^{\alpha_0-1}\exp\{-t^{\alpha_0}\}\,dt + q(-e^{-\gamma})(\xi^{\alpha_0}\ln\xi)$$

$$= -\frac{1}{\alpha_0}\left\{2\int_0^\gamma (w\ln w)e^{-2w}\,dw + q^2\gamma\ln\gamma\right\}.$$

Consequently, the limiting variance of $\sqrt{n}\,[S_2(\boldsymbol{R}_{[d:n]}) - 1]$ is

$$\Xi = \frac{1}{3}(1 - q^3) - \frac{\left\{2\int_0^\gamma(w\ln w)e^{-2w}\,dw + q^2\gamma\ln\gamma\right\}^2}{\left\{p + \int_0^\gamma w(\ln w)^2 e^{-w}\,dw + q\gamma(\ln\gamma)^2\right\}}.$$

Expressed in terms of the functions in (28.4.16), this becomes

$$\Xi = \frac{1}{3}(1 - q^3) - \frac{\{\frac{1}{2}\dot\Gamma(2;2\gamma) + [\gamma q^2 - \frac{1}{2}(1 - q^2)]\ln 2 + q^2\gamma\ln\gamma\}^2}{\left\{p + \ddot\Gamma(2;\gamma) + q\gamma(\ln\gamma)^2\right\}}. \qquad (28.4.18)$$

As in Example 3, we also present in Figure 28.3 a plot of p versus $\Xi/[(1 - q^3)/3]$ as p ranges over $(0, 1)$ in order to illustrate the change in asymptotic variance of S_2 when applied to the residuals relative to that under the U.E.P. Again notice the substantial difference when p is small. ▯

28.5   Goodness-of-Fit Based on Residuals

The exact and asymptotic results in Sections 28.3 and 28.4 indicate that model diagnostic procedures based on a Type II$_{[d:n]}$ censored sample which rely on checking the (approximate) U.E.P. of the associated Type II$_{[d:n]}$ censored residuals may lead to misleading conclusions. To amplify, suppose that a test statistic S is such that $\sqrt{n}\,[S(\boldsymbol{E}_{[d:n]}) - \mu]$ converges in distribution to a normal random variable with mean zero and variance unity. From Theorem 28.4.1, under certain regularity conditions, $\sqrt{n}\,[S(\boldsymbol{R}_{[d:n]}) - \mu]$ will converge in distribution to a normal random variable with mean zero and variance σ^2, where $\sigma^2 < 1$. Indeed, it is possible as in Example 1, that $\sigma^2 = 0$. Consequently, if one is to decide that the model is inadequate if and only if $\sqrt{n}\,|\,S(\boldsymbol{R}_{[d:n]}) - \mu\,|$ exceeds $z_{\alpha/2}$, then such a diagnostic procedure is conservative, and it can actually be *unacceptably* conservative.

A plausible approach for use of the Type II$_{[d:n]}$ censored residuals in model diagnostics is to utilize the exact distributional and/or asymptotic results such as those in Sections 28.3 and 28.4. For example, suppose it is possible to estimate consistently the limiting covariance matrix $\boldsymbol{\Xi}$ in Theorem 28.4.1 by an estimator $\hat{\boldsymbol{\Xi}}$ which depends only on the Type II$_{[d:n]}$ censored data $\boldsymbol{T}_{[d:n]}$, and suppose that the asymptotic mean $\boldsymbol{\mu}$ is fully known as in the case of the statistic S_2 in Examples 2 and 4. Then an asymptotic α-level omnibus diagnostic procedure is to reject the viability of the hypothesized model if the quadratic form $n[S(\boldsymbol{R}_{[d:n]}) - \mu]^* \hat{\boldsymbol{\Xi}}^{-1} [S(\boldsymbol{R}_{[d:n]}) - \mu]$ exceeds $\chi^2_{M;\,\alpha}$.

A concrete example of this approach is the problem of testing the null hypothesis that $T_1, T_2, \ldots, T_n \overset{\text{i.i.d.}}{\sim} \text{Exp}(\theta)$ where $\theta > 0$. With S_2 being the Horowitz and Neumann statistic, Corollary 28.4.2 shows that $\sqrt{n}\,[S_2(\boldsymbol{R}_{[d:n]}) - 1]$ converges in distribution to a normal random variable with variance $p^3/12$. Since a consistent estimator of p is $\hat{p} = 1 - \exp\{-R_{d:n}\}$, then a possible test is to reject the hypothesis if the test statistic

$$Q = 12n\,\frac{[S_2(\boldsymbol{R}_{[d:n]}) - 1]^2}{[1 - \exp\{-R_{d:n}\}]^3}$$

exceeds $\chi^2_{1;\,\alpha}$. Analogous diagnostic procedures can be concocted for testing the Weibull model based on the results of Examples 3 and 4. Procedures based on residuals of the above variety still need to be developed and carefully examined. It is clear though that such procedures will improve on diagnostic tests which simply try to verify the U.E.P. of the Type II$_{[d:n]}$ censored residuals, since this property does not seem to hold for $\boldsymbol{R}_{[d:n]}$ even when the model is true under Type II censoring.

Acknowledgements

The author wishes to thank Professor N. Balakrishnan for the invitation to contribute in this Volume in honor of Professor A. Clifford Cohen, and for his persistence and patience in eliciting this contribution; and Inmaculada Baltazar-Aban for helpful discussions concerning hazard-based residuals in failure-time models.

References

1. Andersen, P. (1989), Discussion on survival models and martingale dynamics, by E. Arjas, *Scandinavian Journal of Statistics*, **16**, 177-225.

2. Andersen, P., Borgan, O., Gill, R. and Keiding, N. (1993), *Statistical Models Based on Counting Processes*, Springer-Verlag, New York.

3. Arjas, E. (1989), Survival models and martingale dynamics (with discussions), *Scandinavian Journal of Statistics*, **16**, 177-225.

4. Baltazar-Aban, I. and Peña, E. (1994), Properties of hazard-based residuals and implications in model diagnostics, *Journal of the American Statistical Association* (To appear).

5. Barlow, R. and Proschan, F. (1981), *Statistical Theory of Reliability and Life Testing*, Holt, Rinehart and Winston, New York.

6. Bhattacharyya, G. (1985), The asymptotics of maximum likelihood and related estimators based on Type II censored data, *Journal of the American Statistical Association*, **80**, 398-404.

7. Cox, D. (1972), Regression models and life tables (with discussion), *Journal of the Royal Statistical Society, Series B*, **34**, 187-220.

8. Cox, D. and Oakes, D. (1984), *Analysis of Survival Data*, Chapman and Hall, London.

9. Crowley, J. and Hu, M. (1977), Covariance analysis of heart transplant survival data, *Journal of the American Statistical Association*, **72**, 27-36.

10. Epstein, B. and Sobel, M. (1953), Life testing, *Journal of the American Statistical Association*, **48**, 486-502.

11. Fleming, T. and Harrington, D. (1991), *Counting Processes & Survival Analysis*, John Wiley & Sons, New York.

12. Horowitz, J. and Neumann, G. (1992), A generalized moments specification test of the proportional hazards model, *Journal of the American Statistical Association*, **87**, 234-240.

13. IMSL, Inc. (1987), *Math/Library*, Houston, Texas.

14. Kalbfleisch, J. and Prentice, R. (1980), *The Statistical Analysis of Failure-Time Data*, John Wiley & Sons, New York.

15. Kay, R. (1977), Proportional hazard regression models and the analysis of censored survival data, *Applied Statistics*, **26**, 227-237.

16. Lagakos, S. (1979), General right censoring and its impact on the analysis of survival data, *Biometrics*, **35**, 139-156.

17. Lagakos, S. (1981), The graphical evaluation of explanatory variables in proportional hazards regression models, *Biometrika*, **68**, 93-98.

18. Lawless, J. (1987), Regression methods for Poisson process data, *Journal of the American Statistical Association*, **82**, 808-815.

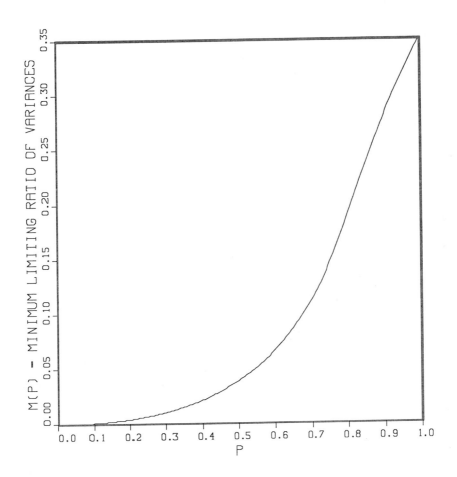

FIGURE 28.1
Graph of ρ versus $M(\rho)$ with $M(\rho)$ defined in (28.3.5). This graph shows the minimum limiting ratios between the variance of the residuals and the variance under the unit exponentiality property as the censoring parameter ρ varies, when the failure-times are exponentially-distributed

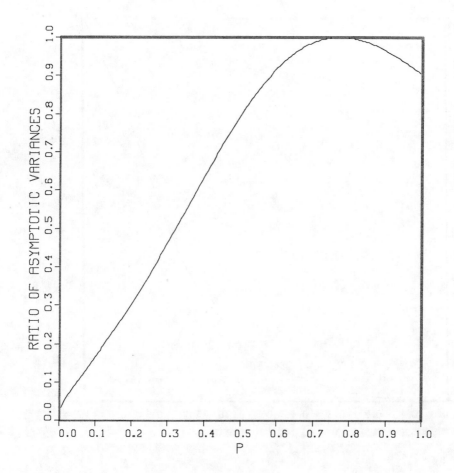

FIGURE 28.2
Ratio of the asymptotic variances of the total-time-on-test statistic S_1 when applied to the residuals and under the unit exponentiality property. The failure time distribution is a Weibull distribution with unknown shape parameter

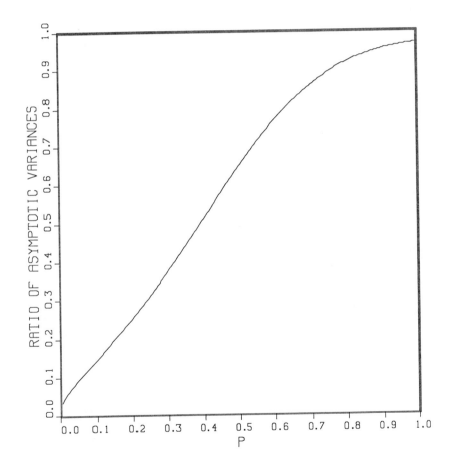

FIGURE 28.3
Ratio of the asymptotic variances of the Horowitz and Neumann statistic S_2 when applied to the residuals and under the unit exponentiality property. The failure time distribution is a Weibull distribution with unknown shape parameter

29

The Use of L-Moments in the Analysis of Censored Data

J.R.M. Hosking

IBM Research Division, T.J. Watson Research Center

ABSTRACT *L*-moments are summary statistics for probability distributions. They are measures of distributional shape—mean, dispersion, coefficient of variation, skewness, kurtosis, etc.—that are in many ways preferable to the conventional moments. Their statistical applications to complete data samples are described in [9]. Here we discuss the application of *L*-moments to the analysis of censored data.

Keywords: Estimation, Identification, Moments

CONTENTS

29.1 Introduction

In [9], the author defined *L*-moments, measures of location, scale and shape for probability distributions and data samples based on expected values of linear combinations of order statistics. The discussion and examples in [9] dealt with complete data samples. Here we consider how *L*-moments can be used when only a censored sample of data is available.

An outline of the chapter is as follows. Section 29.2 summarizes the definition and applications of *L*-moments for complete samples. The definitions are extended to censored samples in Section 29.3. Sampling distributions of *L*-moments for censored samples are discussed in Section 29.4. The next three sections illustrate the uses of *L*-moments in the analysis of censored data: as summary statistics of data samples, in Section 29.5; for choosing an appropriate distribution to fit to a data set, in Section 29.6; for estimating parameters of distributions, in Section 29.7. Conclusions are given in Section 29.8. The Appendix contains algebraic expressions for *L*-moments of censored probability distributions.

We use the following notation for censored samples. In Type I censoring, the sample size is n, of which m values are observed and $n - m$ are censored above a known threshold T; m is a random variable with a binomial distribution. In Type II censoring, the sample size is n but only the smallest m sample values, the order statistics $X_{1:n}, X_{2:n}, \ldots, X_{m:n}$, are observed; here m is fixed and the largest $n - m$ values are censored above the random threshold $X_{m:n}$. These definitions are for right singly censored samples, which are all that we shall consider. The methods described here can be easily extended to left singly censored samples and to doubly censored samples. Some other censoring patterns arising in hydrology, where a set of measured values of streamflow data may be augmented by "historical information", estimates of large values that occurred before stream gauging commenced, have been discussed in [5] and [18]. More complicated censoring patterns than these are of course possible, but inference procedures for them based on L-moments have not yet been developed.

29.2　L-Moments for Complete Samples

For a random variable X with cumulative distribution function F, the quantities

$$\beta_r = E[X\{F(X)\}^r] \tag{29.2.1}$$

are probability weighted moments (PWMs), defined in [7] and used therein to estimate the parameters of probability distributions. In [8] and [9], the author defined L-moments to be linear combinations of probability weighted moments:

$$\lambda_1 = \beta_0, \tag{29.2.2}$$

$$\lambda_2 = 2\beta_1 - \beta_0, \tag{29.2.3}$$

$$\lambda_3 = 6\beta_2 - 6\beta_1 + \beta_0, \tag{29.2.4}$$

$$\lambda_4 = 20\beta_3 - 30\beta_2 + 12\beta_1 - \beta_0, \tag{29.2.5}$$

and in general

$$\lambda_{r+1} = \sum_{k=0}^{r} p_{r,k}^* \beta_k, \tag{29.2.6}$$

where

$$p_{r,k}^* = (-1)^{r-k} \binom{r}{k} \binom{r+k}{k} = \frac{(-1)^{r-k}(r+k)!}{(k!)^2 (r-k)!}. \tag{29.2.7}$$

L-moment ratios are the dimensionless quantities

$$\tau_r = \lambda_r / \lambda_2, \qquad r = 3, 4, \ldots. \tag{29.2.8}$$

L-moments are similar to but more convenient than probability weighted moments, because they are more easily interpretable as measures of distributional shape. In particular λ_1 is the mean of the distribution, a measure of location; λ_2 is a measure of dispersion; τ_3 and τ_4 are measures of skewness and kurtosis respectively. The L-CV, $\tau = \lambda_2/\lambda_1$, is analogous to the usual coefficient of variation.

The foregoing quantities are defined for a probability distribution, but in practice must often be estimated from a finite sample. Let $X_{1:n} \leq \cdots \leq X_{n:n}$ be the order statistics of a sample of size n. Landwehr et al. [12] have shown that

$$b_r = n^{-1} \sum_{j=1}^{n} \frac{(j-1)(j-2)\ldots(j-r)}{(n-1)(n-2)\ldots(n-r)} X_{j:n} \qquad (29.2.9)$$

is an unbiased estimator of β_r. Analogously to (29.2.6), the *sample L-moments* are defined by

$$\ell_{r+1} = \sum_{k=0}^{r} p^*_{r,k} b_k, \qquad r = 0, 1, \ldots, n-1, \qquad (29.2.10)$$

the coefficients $p^*_{r,k}$ being as defined in (29.2.7). The sample L-moment ℓ_r is an unbiased estimator of λ_r. The estimators $t_r = \ell_r/\ell_2$ of τ_r and $t = \ell_2/\ell_1$ of τ are consistent but not unbiased.

The quantities ℓ_1, ℓ_2, t_3 and t_4 are useful summary statistics of a sample of data. In particular the skewness and kurtosis measures t_3 and t_4 avoid two problems arising with conventional moment measures of skewness and kurtosis: mathematically constrained bounds—the conventional skewness of a sample cannot exceed $(n-2)/\sqrt{n-1}$, no matter how large the skewness of the underlying distribution—and severe bias in small samples. L-moments can be used to identify the distribution from which a sample was drawn. They can also be used to estimate parameters when fitting a distribution to a sample, by equating the sample and population L-moments. Further details and examples are given in [9] and [14].

29.3 PWMs and L-Moments for Censored Samples

There are two natural ways of extending the definitions of PWMs and L-moments to singly censored samples. What we term the "A"-type PWMs and L-moments are simply the PWMs and L-moments of the uncensored sample of m observed values. Thus "A"-type PWMs are defined by

$$b_r^A = m^{-1} \sum_{j=1}^{m} \frac{(j-1)(j-2)\ldots(j-r)}{(m-1)(m-2)\ldots(m-r)} X_{j:n}. \qquad (29.3.1)$$

and "A"-type L-moments are defined in terms of the b_r^A, analogously to (29.2.10).

"B"-type PWMs and L-moments are calculated from the "completed sample", in which the $n-m$ censored values are replaced by the censoring threshold T. Thus "B"-type PWMs are defined by

$$b_r^B = n^{-1} \left\{ \sum_{j=1}^{m} \frac{(j-1)(j-2)\ldots(j-r)}{(n-1)(n-2)\ldots(n-r)} X_{j:n} \right.$$

$$\left. + \left(\sum_{j=m+1}^{n} \frac{(j-1)(j-2)\ldots(j-r)}{(n-1)(n-2)\ldots(n-r)} \right) T \right\}, \qquad (29.3.2)$$

and "B"-type L-moments are defined in terms of the b_r^B, analogously to (29.2.10).

When there are more than a few censored values, "B"-type PWMs are most conveniently calculated by first calculating the "A"-type PWMs and then using the expression

$$b_r^B = Zb_r^A + \frac{1-Z}{r+1}T,$$ (29.3.3)

where

$$Z = \frac{m(m-1)\ldots(m-r)}{n(n-1)\ldots(n-r)}.$$ (29.3.4)

When a large proportion of the sample values are censored, the "B"-type *L*-moments are dominated by the contribution from the $n - m$ copies of T in the completed sample. Therefore, for summarizing a sample we prefer to use the "A"-type *L*-moments, even though they ignore the information contained in the censored values.

For estimating parameters of distributions, however, "B"-type *L*-moments seem preferable, because they make use of the information in the censored values. As an example, consider estimation of the exponential distribution with lower bound zero and cumulative distribution function $F(x) = 1 - e^{-x/\alpha}$. Under type II censoring, the maximum-likelihood estimate of the scale parameter α is a multiple of the "B"-type mean:

$$\hat{\alpha}^{\mathrm{ML}} = \hat{\alpha}^B = m^{-1}\left\{\sum_{j=1}^{m} X_{j:n} + (n-m)X_{m:n}\right\}.$$ (29.3.5)

The unbiased estimator of α based on the "A"-type mean is

$$\hat{\alpha}^A = \sum_{j=1}^{m} X_{j:n} \bigg/ \sum_{j=1}^{m} \frac{m-j+1}{n-j+1},$$ (29.3.6)

and is less efficient: its asymptotic efficiency lies in the range $[0.75, 1]$.

Wang [17] defined "partial probability weighted moments", which are equivalent to the "A"-type PWMs defined above, differing from them by a factor m/n. We prefer our "A"-type PWMs, because they yield *L*-moments that can be interpreted the same way as for complete samples: for example, the second *L*-moment, $\ell_2^A = 2b_1^A - b_0^A$, is always positive and is a measure of the dispersion of the uncensored data values.

29.4   Sampling Distributions of L-Moments for Censored Samples

In type I censoring from a distribution with cumulative distribution function $F(x)$ and quantile function $x(u)$, let the threshold T satisfy $F(T) = \zeta$. Conditional on the achieved value of m, the uncensored values are a random sample of size m from a distribution with quantile function

$$y^A(u) = x(u\zeta), \qquad 0 < u < 1.$$ (29.4.1)

The "completed sample" is a random sample of size n from a distribution with quantile function

$$y^B(u) = \begin{cases} x(u), & 0 < u < \zeta, \\ x(\zeta), & \zeta \le u < 1. \end{cases}$$ (29.4.2)

The population PWMs for these distributions are respectively

$$\beta_r^A = \int_0^1 u^r \, y^A(u) \, du$$

$$= \frac{1}{\zeta^{r+1}} \int_0^\zeta u^r \, x(u) \, du \tag{29.4.3}$$

$$= \frac{1}{\{F(T)\}^{r+1}} \int_{-\infty}^T x\{F(x)\}^r \, dF(x) \tag{29.4.4}$$

and

$$\beta_r^B = \int_0^1 u^r \, y^B(u) \, du$$

$$= \int_0^\zeta u^r \, x(u) \, du + \frac{1 - \zeta^{r+1}}{r+1} \, x(\zeta) \tag{29.4.5}$$

$$= \int_{-\infty}^T x\{F(x)\}^r \, dF(x) + \frac{T[1 - \{F(T)\}^{r+1}]}{r+1}. \tag{29.4.6}$$

The population L-moments λ_r^A and λ_r^B are given in terms of the β_r^A and β_r^B by (29.2.6). These quantities are evaluated in the Appendix for several common distributions.

The sampling distributions of L-moments for complete samples apply immediately to these censored samples. Sample PWMs and L-moments are unbiased estimators of the corresponding population quantities. Asymptotically (as $n \to \infty$ for "B"-type quantities; conditionally on m, as $m \to \infty$ for "A"-type quantities) they are Normally distributed with asymptotic covariances that can be calculated from the expressions in [9, Theorem 3].

Sampling distributions under type II censoring are more difficult to derive. For example, the mean of the uncensored values has expectation

$$m^{-1} \sum_{j=1}^m EX_{j:n}, \tag{29.4.7}$$

and this does not have a simple algebraic form even when the sample is drawn from an exponential distribution. Asymptotic distributions, as $n \to \infty$ with $m = [n\zeta]$, can be obtained. For "A"-type PWMs and L-moments, asymptotic distributions follow from the results of [16] for weighted sums of order statistics. For example, the sample PWM b_r^A is asymptotically equivalent to the sum

$$S_n = n^{-1} \sum_{i=1}^n w\left(\frac{i}{n+1}\right) X_{i:n} \tag{29.4.8}$$

where the weight function w is

$$w(u) = \begin{cases} u^r/\zeta, & 0 < u < \zeta, \\ 0, & \zeta \le u < 1. \end{cases} \tag{29.4.9}$$

Provided that $EX^2 < \infty$ and the quantile $x(\zeta)$ is uniquely defined, S_n satisfies the conditions of [16] (Theorem 6) and it follows that b_r^A is asymptotically Normal with $n^{1/2}(Eb_r^A - \beta_r^A) \to 0$ as $n \to \infty$.

Asymptotic distributions of "B"-type PWMs under type II censoring can be obtained from results given in [15] (section 8.2.4) for statistical functionals. Let $T(F)$ be the statistical functional

$$T(F) = \int_0^1 J(u) F^{-1}(u) \, du + c \, F^{-1}(p). \tag{29.4.10}$$

Its sample analogue is $T(F_n)$, where F_n is the empirical distribution function. By choosing

$$J(u) = \begin{cases} u^r, & 0 < u < \zeta, \\ 0, & \zeta \le u < 1. \end{cases} \qquad (29.4.11)$$

$c = (1 - \zeta^{r+1})/(r + 1)$ and $p = \zeta$, we obtain a functional $T(.)$ such that $T(F_n)$ is asymptotically equivalent to b_r^B and $T(F) = \beta_r^B$. Under the conditions

$$\int_{-\infty}^{\infty} [F(x)\{1 - F(x)\}]^{1/2-\delta} \, dx < \infty, \qquad (29.4.12)$$

for some $0 < \delta < \frac{1}{2}$, and $F(x)$ differentiable at $x = T$, Theorem B of [15, p. 283] implies that $n^{1/2}\{T(F_n) - T(F)\}$ is asymptotically Normal with mean zero and finite variance.

29.5 Summarizing a Sample

As with complete samples, a few L-moments summarize the salient features of the data: location, scale, skewness, kurtosis. Compared with ordinary moments, L-moments are less affected by outlying observations and have numerical values that are easier to interpret because they lie within finite intervals: for example, L-skewness satisfies $-1 \le t_3 \le 1$ and the L-CV of a set of nonnegative numbers satisfies $0 \le t \le 1$.

As an example, we consider a data set consisting of 100 measurements of the stress-rupture life, at 80% stress level, of Kevlar 49/Epoxy spherical pressure vessels. The data were analysed in [4] and are included in [2, p. 183]. Figure 29.1 plots the data against expected exponential order statistics. An exponential distribution, which would plot as a straight line on the graph, would fit most of the data well, but in both tails of the data there is some deviation from a straight-line plot. Though in this case a complete sample is available, we artificially censor the sample by removing the upper 20%, 50% and 80% of the data and observe the effect on the L-moments. Table 29.1 shows the L-moments for data samples with different amounts of censoring. The effect of censoring on the "B"-type L-moments is predictable. As the censoring fraction increases, the completed sample contains an increasing number of repetitions of the largest uncensored data point. As this fraction increases, the mean approaches this largest uncensored data value, the L-CV decreases almost proportionally to m, and the L-skewness approaches -1, its minimal value for a negatively skew sample. The "A"-type moments change in less predictable ways: t and t_3 decrease as the censoring fraction increases, but not monotonically. These irregularities can be interpreted as reflecting the sample's deviation from an exponential distribution. For example, with 50% censoring the sample L-skewness, t_3, is negative: this is inconsistent with the L-skewness values of censored exponential distributions.

29.6 Identification of Distributions

L-moments of probability distributions can be plotted on an "L-moment ratio diagram". Sample L-moments can then be compared with the population curves to suggest an appropriate distribution to fit to the data. This is a simple graphical method that gives an intuitive feeling for which distributions are appropriate. It is effective because L-moment ratios have low bias (much smaller than the bias of conventional moment ratios, for example). Population L-moments for several common distributions are given in the Appendix.

TABLE 29.1
L-moments of the stress-rupture life data

Censoring fraction	m	ℓ_1	t	t_3	t_4
			"A"-type		
0	100	209.2	0.466	0.334	0.222
20%	80	131.6	0.363	0.078	0.089
50%	50	79.4	0.345	−0.086	−0.026
80%	20	29.3	0.467	0.147	−0.095
			"B"-type		
0	100	209.2	0.466	0.334	0.222
20%	80	166.1	0.351	0.006	−0.035
50%	50	114.3	0.213	−0.437	0.058
80%	20	61.6	0.114	−0.747	0.450

TABLE 29.2
Lifetimes, in weeks, of 34 transistors in an accelerated life test, after [19]. Three of the times, denoted by asterisks, are censored at 52 weeks

3, 4, 5, 6, 6, 7, 8, 8, 9, 9, 9, 10, 10, 11, 11, 11, 13, 13, 13, 13, 13, 17, 19, 19, 25, 29, 33, 42, 42, 52, 52*, 52*, 52*

They can be used to construct L-moment ratio diagrams corresponding to different censoring proportions. Figure 29.2 is an example, for censoring fraction 50%. The diagram includes both "A"- and "B"-type L-moment ratios, but it is clear that the "B"-type population curves lie so close to each other that they offer little prospect of distinguishing between distributions. Thus L-moment ratio diagrams are most useful when constructed from "A"-type quantities.

For distributions with a lower bound of zero, an L-moment ratio diagram of L-skewness and L-CV can be useful. As an example, Table 29.2 shows a data set consisting of the lifetimes of 34 transistors in an accelerated life test. Three of the lifetimes are censored, so the censoring fraction is 3/34 or 8.8%. The data were given in [19], where it is stated that "there is reason, from past experience, to expect that the gamma distribution might reasonably approximate the failure time distribution". In [19], and also in [13, p. 208], a gamma distribution was fitted to the data without questioning whether the distribution gave a good fit to the data. The "A"-type mean, L-CV and L-skewness of the data are $\ell_1 = 15.7$, $t = 0.389$ and $t_3 = 0.393$. In Figure 29.3, t_3 and t are plotted on an L-moment ratio diagram of L-skewness and L-CV for distributions with lower bound zero and 8.8% censoring. These sample L-moment ratios lie far away from the L-skewness–L-CV relation for gamma distributions, suggesting that a gamma distribution is not appropriate for this data set. Indeed, the L-moments suggest that none of the distributions in Figure 29.3 is appropriate for this data set.

Further analysis suggests that the upper part of the distribution is reasonably well fitted by a generalized Pareto distribution. Using the method of L-moments, analogously to the procedures described in section 29.7 below, a generalized Pareto distribution with fixed lower bound $\xi = X_{7:34} = 8$ was fitted to the data from the eighth order statistic upwards; the estimated parameters were $\alpha = 7.34$, $k = -0.89$. The fit of the distribution to the data is shown in Figure 29.4. This distribution has a very heavy tail: it has probability density $f(x)$ asymptotically proportional to $x^{-1.12}$ as $x \to \infty$, so one might say that the mean of the distribution is "almost infinite". However, the lower tail of the sample is deficient in observations compared to this Pareto distribution. A Wakeby distribution ([11], [1]) would be capable of modeling this data set, but we hesitate to fit this five-parameter distribution when only 31 uncensored observations are available.

As a further example, we compare generalized extreme-value (GEV), Weibull and generalized Pareto distributions. The distributions are 50% censored and the distribution parameters are chosen so that the uncensored part of each distribution has L-skewness $\tau_3^A = 0.02$. The distributions are illustrated in Figure 29.5. Fifty samples of size $n = 100$ with 50% type I censoring were simulated from each distribution. The "A"-type L-moment ratios were calculated for each sample and the L-skewness and L-kurtosis values were plotted on a graph, together with the L-skewness–L-kurtosis relations for the three families of distributions. The resulting graph is Figure 29.6. The plotted points from samples from the different distributions form clouds that overlap each other to a large extent, indicating that it is difficult to distinguish these distributions on the basis of samples of this size.

As the sample size increases, discrimination becomes easier. Figure 29.7 was obtained in the same way as Figure 29.6, but with samples of size $n = 1000$. Even here the clouds of points from the Weibull and generalized Pareto samples overlap to some extent, but the GEV samples have clearly different L-kurtosis from the others.

In this example, discrimination between distributions on the basis of their L-moment ratios requires very large samples. This is despite the fact that the complete distributions are very different: the generalized Pareto has a finite upper bound, the Weibull has no upper bound but has all moments finite, and the GEV has infinite variance (its shape parameter is $k = -0.73$).

29.7   Estimation

To estimate parameters of distributions, we can equate sample and population L-moments. This is the "method of L-moments," and is analogous to the usual method of moments. With censored samples, as noted previously, "B"-type L-moments seem preferable. For a distribution with a quantile function $x(\,.\,;\theta)$ that is a function of a vector θ of p unknown parameters, the simplest approach is to estimate θ by setting $\zeta = m/n$ and b_r^B equal to the corresponding population PWM given in (29.4.5). This yields an estimator $\hat{\theta}$ of θ that satisfies equations of the form

$$b_r^B = \int_0^\zeta u^r\, x(u;\hat{\theta})\, du + \frac{1 - \zeta^{r+1}}{r+1}\, x(\zeta;\hat{\theta}), \qquad r = 0,\ldots,p-1. \qquad (29.7.1)$$

This approach has some theoretical disadvantages. In type I censoring, it does not use the known constraint $T = x(\zeta;\theta)$. In type II censoring, the right side of (29.7.1) is only asymptotically the mean of b_r^B. But these disadvantages become negligible for large samples, and to overcome them would require much more complicated calculations. Although finite-samples corrections may in some circumstances be worthwhile, we shall here consider the asymptotically consistent estimators defined by (29.7.1).

TABLE 29.3
Data set of [13, p. 152]

−2.982, −2.849, −2.546, −2.350, −1.983, −1.492, −1.443, −1.394, −1.386, −1.269,
−1.195, −1.174, −0.845, −0.620, −0.576, −0.548, −0.247, −0.195, −0.056, −0.013,
0.006, 0.033, 0.037, 0.046, 0.084, 0.221, 0.245, 0.296

Estimation for the reverse Gumbel distribution

We shall consider in more detail parameter estimation for the reverse Gumbel distribution, with quantile function $x(u) = \xi + \alpha \log\{-\log(1-u)\}$. This distribution is important in the analysis of censored data, for it is the distribution of a logarithmically transformed two-parameter Weibull distribution, a common model in reliability applications. L-moments of the censored distribution are given in the Appendix, equations (A.11), (A.12). From them we obtain parameter estimates

$$\hat{\alpha} = \ell_2^B / \{\log 2 + \text{Ei}(-2\log(1-\zeta)) - \text{Ei}(-\log(1-\zeta))\}, \qquad (29.7.2)$$

$$\hat{\xi} = \ell_1^B + \hat{\alpha}\{\gamma + \text{Ei}(-\log(1-\zeta))\}. \qquad (29.7.3)$$

As an example, we fit a reverse Gumbel distribution to a data set used in [13, p. 152]. Table 29.3 contains the data, which form a type II censored sample with $n = 40$ and $m = 28$. Thus we take $\zeta = m/n = 0.7$ and calculate $\text{Ei}(-\log(1-\zeta)) = 0.1574$, $\text{Ei}(-2\log(1-\zeta)) = 0.0281$. The first two sample L-moments are $\ell_1^B = -0.5161$, $\ell_2^B = 0.5217$. Then from (29.7.2) and (29.7.3) we have $\hat{\alpha} = 0.9252$, $\hat{\xi} = 0.1636$. For comparison, the maximum-likelihood estimates for this sample are $\hat{\alpha}^{ML} = 0.9104$, $\hat{\xi}^{ML} = 0.1563$.

A simulation study was performed to compare the L-moment-based estimators with the maximum-likelihood estimators. Maximum-likelihood estimation for the reverse Gumbel distribution requires an iterative procedure, described in [12, pp. 143,169]. The L-moment estimators can be computed with no iteration, but require the computation of the exponential-integral function (this is available in standard Fortran subroutine libraries such as IMSL and NAG). In complete samples the L-moment estimators are equivalent to Downton's [6] "linear estimates with polynomial coefficients"; they are unbiased and have asymptotic efficiencies 76% for $\hat{\alpha}$ and 99.6% for $\hat{\xi}$; see [3, pp. 109-119].

The simulation study used samples of size n ranging from 20 to 200, and censoring fractions $1 - \zeta$ between 0 and 80%. For each combination of n and $1 - \zeta$, 10 000 simulated samples were generated using both type I and type II censoring. Without loss of generality, the distribution parameters used to generate the samples were $\xi = 0$ and $\alpha = 1$. Parameters were estimated for each simulated sample by the methods of L-moments and maximum-likelihood, and the bias and root mean square error (RMSE) of the estimates of ξ and α were computed.

Table 29.4 summarizes the results for type I censoring. It can be seen that the L-moment-based estimators are only a little less accurate than the maximum-likelihood estimators. As the censoring fraction increases, L-moment estimators become more competitive. For samples with 50% or more censoring, there is little to choose between the two methods, and the method of L-moments may be preferred because of its noniterative nature.

Table 29.5 summarizes the results for type II censoring. The RMSE values are quite similar to those for type I censoring, but the biases are different. Again, maximum-likelihood appears preferable when the censoring fraction is small, but L-moments are competitive

TABLE 29.4
Bias and root mean square error of estimators of parameters of the reverse
Gumbel distribution with type I censoring. Estimation methods are L-moments
(L) and maximum-likelihood (ML)

		Censoring fraction, $1 - \zeta$							
		0		20%		50%		80%	
n	Method	Bias	RMSE	Bias	RMSE	Bias	RMSE	Bias	RMSE
		Estimation of ξ							
20	L	−.003	.239	−.021	.263	−.061	.429	−.159	1.319
	ML	.016	.240	.000	.260	−.030	.414	−.099	1.270
50	L	.000	.149	−.007	.160	−.021	.240	−.069	.673
	ML	.008	.149	.007	.160	−.009	.236	−.048	.663
100	L	.001	.106	−.002	.113	−.010	.165	−.038	.440
	ML	.005	.106	.002	.113	−.004	.164	−.027	.436
200	L	.000	.075	−.002	.080	−.005	.114	−.017	.297
	ML	.002	.075	.001	.080	−.002	.114	−.011	.296
		Estimation of α							
20	L	−.001	.205	.001	.233	.020	.318	.027	.571
	ML	−.040	.179	−.016	.217	−.006	.307	−.002	.553
50	L	−.001	.128	.003	.144	.006	.193	.010	.329
	ML	−.017	.111	−.007	.136	−.004	.190	−.001	.325
100	L	.000	.090	.002	.101	.004	.134	.008	.225
	ML	−.007	.078	−.003	.095	−.001	.132	.003	.224
200	L	.000	.064	.001	.071	.002	.094	.003	.157
	ML	−.004	.055	−.002	.068	−.001	.093	.000	.156

when the censoring fraction is high. Indeed, when the censoring fraction is 50% or more,
the L-moment estimators have a little less bias than the maximum-likelihood estimators
and effectively the same RMSE. As the sample size increases, the difference between
type I and type II censoring becomes less important. This is particularly noticeable in
the similarity of the RMSE entries for $n = 200$ in Tables 29.4 and 29.5.

Estimation for the GEV distribution

The GEV distribution, with quantile function $x(u) = \xi + \alpha\{1 - (-\log u)^k\}/k$, is of interest
because it includes as special cases several widely used distributions such as the Gumbel
and Weibull. Because ξ and α are location and scale parameters, the L-skewness τ_3 is a
function only of the shape parameter k. From (A.4) and (A.1) we obtain

$$\tau_3^B = \frac{\zeta(1 - \zeta)(1 - 2\zeta)(-\log \zeta)^k - g_1 + 3g_2 - 2g_3}{-\zeta(1 - \zeta)(-\log \zeta)^k + g_1 - g_2} \qquad (29.7.4)$$

where

$$g_r = r^{-k}\{\Gamma(1 + k) - \Gamma(1 + k, -r\log \zeta)\}; \qquad (29.7.5)$$

here $\Gamma(.)$ is the gamma function and $\Gamma(.\,,.)$ is the incomplete gamma function defined
in (A.2) below. This equation can be solved for k by an iterative method such as interval

TABLE 29.5
Bias and root mean square error of estimators of parameters of the reverse Gumbel distribution with type II censoring

		Censoring fraction, $1 - \zeta$							
		0		20%		50%		80%	
n	Method	Bias	RMSE	Bias	RMSE	Bias	RMSE	Bias	RMSE
					Estimation of ξ				
20	L	−.003	.240	.014	.257	.075	.374	.473	.996
	ML	.016	.241	.034	.258	.105	.377	.521	1.001
50	L	.000	.149	.007	.160	.030	.229	.177	.601
	ML	.008	.149	.015	.162	.043	.230	.200	.603
100	L	.001	.106	.004	.112	.013	.160	.088	.413
	ML	.005	.106	.008	.112	.020	.160	.100	.413
200	L	.000	.075	.002	.080	.007	.113	.047	.292
	ML	.002	.075	.004	.080	.010	.113	.054	.293
					Estimation of α				
20	L	−.001	.205	−.014	.231	−.052	.303	−.215	.493
	ML	−.040	.179	−.052	.219	−.089	.299	−.246	.492
50	L	−.001	.128	−.007	.145	−.023	.189	−.081	.313
	ML	−.017	.111	−.022	.138	−.038	.187	−.095	.312
100	L	.000	.090	−.002	.102	−.011	.133	−.042	.220
	ML	−.007	.078	−.010	.098	−.018	.132	−.049	.220
200	L	.000	.064	−.001	.071	−.004	.094	−.022	.156
	ML	−.004	.055	−.005	.068	−.008	.093	−.026	.156

bisection. Once k is given, α and ξ can be found directly: from (A.4) and (A.1) we obtain

$$\alpha = \lambda_2^B k / \{-\zeta(1 - \zeta)(-\log \zeta)^k + g_1 - g_2\}, \tag{29.7.6}$$

$$\xi = \lambda_1^B - \alpha\{1 - (1 - \zeta)(-\log \zeta)^k - g_1\}/k. \tag{29.7.7}$$

As an example, we fit a GEV distribution to the data of Table 29.3, multiplied by -1. This is equivalent to fitting a reverse GEV distribution, a generalization of the reverse Gumbel distribution, to the original data. The sample L-moments are $\ell_1^B = 0.5161$, $\ell_2^B = 0.5217$, $t_3^B = 0.3679$. These yield estimated parameters $\hat{k} = -0.0191$, $\hat{\alpha} = 0.9027$, $\hat{\xi} = -0.1629$. The estimate of k is very close to zero, indicating that the Gumbel distribution (the special case $k = 0$ of the GEV distribution) provides an adequate fit to the data.

For complete samples, Hosking et al. [10] investigated estimates of GEV parameters and quantiles based on PWMs (equivalent to L-moments), for samples of size up to 100 from GEV distributions with shape parameters between -0.5 and 0.5. They found that the PWM estimators were in many cases superior to the maximum-likelihood estimators, particularly for estimating quantiles in the upper tail of the distribution. Preliminary results from a simulation study using censored samples indicate that the L-moment-based estimators continue to perform well in this case. One problem that arises with censored samples concerns estimation of the mean of the complete distribution, $\mu = \xi + \alpha\{1 - \Gamma(1 + k)\}/k$. The complete distribution has infinite mean if $k \leq -1$. Even when the population distribution has $k > -1$, censored samples can (and often do) yield estimates of k less than -1. This is the case for both L-moment and maximum-likelihood estimation methods. Such samples give rise to an infinite estimated mean and very large estimated upper quantiles of the GEV distribution. Method of L-moments estimates of k tend to be positively biased, and large negative estimates of k occur less often than with maximum-likelihood, so L-moments are less subject to the "infinite estimated mean" problem.

29.8   Conclusions

L-moments are known to have good properties in complete samples. We have shown that several complete-sample techniques based on L-moments can be used with censored samples too.

In particular, the L-moment ratio diagram is a useful graphical tool that is easily adapted for use with censored data. It can give a visual indication of which distributions are candidates for giving a good fit to a given data set.

L-moments can be used to estimate parameters of distributions fitted to a data set. For this application their performance can be competitive with computationally more complex methods such as maximum-likelihood.

Appendix: Population L-Moments of Censored Distributions

Introduction

Population PWMs for censored distributions can be calculated from (29.4.3) and (29.4.5) (or from (29.4.4) and (29.4.6)). "A"- and "B"-type PWMs can be computed from each other by the relation

$$r\beta^{B}_{r-1} = \zeta^{r} r\beta^{A}_{r-1} + (1 - \zeta^{r})x(\zeta), \tag{A.1}$$

which follows from (29.4.3) and (29.4.5). *L*-moments are obtained from PWMs via (29.2.6). Since these quantities are easily calculated from one another, we give here whichever of the forms β^{A}_{r}, β^{B}_{r} or λ^{B}_{r} has the simplest algebraic expression. The expressions are rarely as simple as for complete distributions and numerical integration is sometimes necessary. The critical integral is $\int_{0}^{\zeta} u^{r} x(u)\, du$. For several of the distributions discussed below, it can be expressed in terms of the incomplete gamma function

$$\Gamma(\alpha, x) = \int_{0}^{x} t^{\alpha-1} e^{-t}\, dt. \tag{A.2}$$

Generalized extreme-value (GEV) distribution

The GEV distribution has quantile function $x(u) = \xi + \alpha\{1 - (-\log u)^{k}\}/k$. It contains the extreme-value distributions of types I, II and III (Gumbel, Fréchet, Weibull) as special cases. See [10] for more details. In [17], it is shown that

$$\int_{0}^{\zeta} r\, u^{r-1}(-\log u)^{k}\, du = r^{-k}\{\Gamma(1 + k) - \Gamma(1 + k, -r\log \zeta)\}. \tag{A.3}$$

It follows from (29.4.3) that

$$r\beta^{A}_{r-1} = \xi + \frac{\alpha}{k} - \frac{\alpha r^{-k}}{k\zeta^{r}}\{\Gamma(1 + k) - \Gamma(1 + k, -r\log \zeta)\}. \tag{A.4}$$

Weibull distribution

The three-parameter Weibull distribution has cumulative distribution function

$$F(x) = 1 - \exp\left\{-\left(\frac{x - \xi}{\alpha}\right)^{\delta}\right\}, \qquad x \geq \xi. \tag{A.5}$$

It is a reverse GEV distribution: i.e., $1 - F(-x)$ is the cumulative distribution function of a GEV distribution. Using Wang's [18] results for GEV distributions, we can show that

$$r\beta^{A}_{r-1} = \xi\{1 - (1 - \zeta)^{r}\} + \alpha r^{-1/\delta}\Gamma(1 + 1/\delta, -r\log(1 - \zeta)). \tag{A.6}$$

Reverse Gumbel distribution

The reverse Gumbel distribution has quantile function

$$x(u) = \xi + \alpha \log(-\log(1-u)). \tag{A.7}$$

It is the distribution of $\log X$ when X has a two-parameter Weibull distribution (i.e., a three-parameter Weibull distribution with lower bound zero). Rather than $\int_0^\zeta u^r x(u)\, du$, for this distribution it is easier to evaluate

$$\int_\zeta^1 r(1-u)^{r-1} \log(-\log(1-u))\, du$$

$$= \int_{-\log(1-\zeta)}^\infty r e^{-rt} \log t\, dt$$

$$= (1-\zeta)^r \log(-\log(1-\zeta)) + \mathrm{Ei}(-r \log(1-\zeta)) \tag{A.8}$$

(the last step follows from integration by parts), where

$$\mathrm{Ei}(x) = \int_x^\infty t^{-1} e^{-t}\, dt \tag{A.9}$$

is the exponential-integral function. The expression corresponding to (A.8) for the complete (uncensored) Gumbel distribution is

$$\int_0^1 r(1-u)^{r-1} \log(-\log(1-u))\, du = -\gamma - \log r \tag{A.10}$$

([8], [9]), where $\gamma = 0.5772...$ is Euler's constant. Using (A.8) and (A.10) it is straightforward to evaluate PWMs and L-moments of the censored reverse Gumbel distribution. For example, the first two "B"-type L-moments are

$$\lambda_1^B = \xi - \alpha\gamma - \alpha \, \mathrm{Ei}(-\log(1-\zeta)), \tag{A.11}$$

$$\lambda_2^B = \alpha\{\log 2 + \mathrm{Ei}(-2\log(1-\zeta)) - \mathrm{Ei}(-\log(1-\zeta))\}. \tag{A.12}$$

Generalized Pareto distribution

The generalized Pareto distribution has quantile function $x(u) = \xi + \alpha\{1 - (1-u)^k\}/k$. It is straightforward to show that

$$\int_0^\zeta r(1-u)^{r-1} x(u)\, du + (1-\zeta)^r x(\zeta) = \xi + \alpha m_r \tag{A.13}$$

where $m_r = \{1 - (1-\zeta)^{r+k}\}/(r+k)$. PWMs and L-moments can be easily evaluated, using (A.13): for example, we have

$$\lambda_1^B = \xi + \alpha m_1, \tag{A.14}$$

$$\lambda_2^B = \alpha(m_1 - m_2), \tag{A.15}$$

$$\lambda_3^B = \alpha(m_1 - 3m_2 + 2m_3), \tag{A.16}$$

$$\lambda_4^B = \alpha(m_1 - 6m_2 + 10m_3 - 5m_4). \tag{A.17}$$

Gamma distribution

The gamma distribution has cumulative distribution function $F(x) = \Gamma(\alpha, x/\beta)/\Gamma(\alpha)$, $x \geq 0$. In [8] (Appendix A.12), the author evaluated the PWMs of the complete distribution using integration by parts of the integral $\int_0^\infty x\{F(x)\}^r \, dF(x)$. For the censored distribution the upper limit of integration is the threshold T (regarded as a function of ζ defined implicitly by $\zeta = F(T)$) rather than ∞, and Hosking's method is effective only for the first two PWMs. These are

$$\beta_0^A = \alpha\beta - \frac{T^\alpha e^{-T/\beta}}{\beta^{\alpha-1}\zeta\Gamma(\alpha)}, \tag{A.18}$$

$$\beta_1^A = \tfrac{1}{2}\alpha\beta - \frac{T^\alpha e^{-T/\beta}}{\beta^{\alpha-1}\zeta\Gamma(\alpha)} + \frac{\beta\Gamma(2\alpha, 2T/\beta)}{2^{2\alpha}\zeta^2\{\Gamma(\alpha)\}^2}. \tag{A.19}$$

The L-skewness values plotted in Figure 29.3 were obtained by simulation: for each of the L-CV values $\tau = 0.2, 0.25, \ldots, 0.6$, 10 samples of size $100\,000$ were generated from a censored gamma distribution with the given L-CV; the population L-skewness was taken to be the average of the L-skewnesses of these simulated samples; the "Gamma" curve on Figure 29.3 is a spline fit to these nine points.

References

1. anon. (1988), Wakeby distributions, In *Encyclopedia of Statistical Sciences, Vol. 9*, (Eds., N. L. Johnson, S. Kotz, and C. B. Read), John Wiley & Sons, New York.

2. Andrews, D.F. and Herzberg, A.M. (1985), *Data: a collection of problems from many fields for the student and research worker*, Springer-Verlag, New York.

3. Balakrishnan, N. and Cohen, A.C. (1991), *Order Statistics and Inference: Estimation Methods*, Academic Press, San Diego.

4. Barlow, R.E., Toland, R.H., and Freeman, T. (1984), A Bayesian analysis of stress-rupture life of Kevlar/epoxy spherical pressure vessels, In *Proceedings of the Canadian Conference in Applied Statistics, 1981*, (Ed., T. D. Dwivedi), Marcel Dekker, New York.

5. Ding J. and Yang, R. (1988), The determination of probability weighted moments with the incorporation of extraordinary values into sample data and their application to estimating parameters for the Pearson type three distribution, *Journal of Hydrology*, **101**, 63-81.

6. Downton, F. (1966), Linear estimates of parameters in the extreme value distribution, *Technometrics*, **8**, 3-17.

7. Greenwood, J.A., Landwehr, J.M., Matalas, N.C., and Wallis, J.R. (1979), Probability weighted moments: definition and relation to parameters of several distributions expressable in inverse form, *Water Resources Research*, **15**, 1049-1054.

8. Hosking, J.R.M. (1986), The theory of probability weighted moments, *Research Report RC12210*, IBM Research Division, Yorktown Heights, N.Y.

9. Hosking, J.R.M. (1990), *L*-moments: analysis and estimation of distributions using linear combinations of order statistics, *Journal of the Royal Statistical Society, Series B*, **52**, 105-124.

10. Hosking, J.R.M., Wallis, J.R., and Wood, E.F. (1985), Estimation of the generalized extreme-value distribution by the method of probability-weighted moments, *Technometrics*, **27**, 251-261.

11. Houghton, J.C. (1978), Birth of a parent: the Wakeby distribution for modeling flood flows, *Water Resources Research*, **14**, 1105-1109.

12. Landwehr, J.M., Matalas, N.C., and Wallis, J.R. (1979), Probability weighted moments compared with some traditional techniques in estimating Gumbel parameters and quantiles, *Water Resources Research*, **15**, 1055-1064.

13. Lawless, J.F. (1982), *Statistical Models & Methods for Lifetime Data*, John Wiley & Sons, New York.

14. Royston, P. (1992), Which measures of skewness and kurtosis are best? *Statistics in Medicine*, **11**, 333-343.

15. Serfling, R.J. (1980), *Approximation Theorems of Mathematical Statistics*, John Wiley & Sons, New York.

16. Stigler, S.M. (1974), Linear functions of order statistics with smooth weight functions, *Annals of Statistics*, **2**, 676-693. Correction: *Annals of Statistics*, **7** (1979), 466.

17. Wang, Q.J. (1990), Estimation of the GEV distribution from censored samples by method of partial probability weighted moments, *Journal of Hydrology*, **120**, 103-114.

18. Wang, Q.J. (1990), Unbiased estimation of probability weighted moments and partial probability weighted moments from systematic and historical flood information and their application to estimating the GEV distribution, *Journal of Hydrology*, **120**, 115-124.

19. Wilk, M.B., Gnanadesikan, R., and Huyett, M.J. (1962), Estimation of parameters of the gamma distribution using order statistics, *Biometrika*, **49**, 525-545.

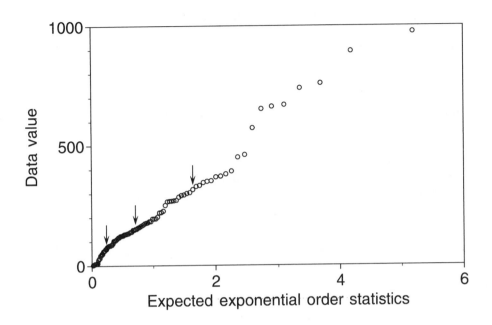

FIGURE 29.1
Stress-rupture life data from [4], plotted against expected exponential order
statistics. Arrows mark the 20th, 50th and 80th order statistics

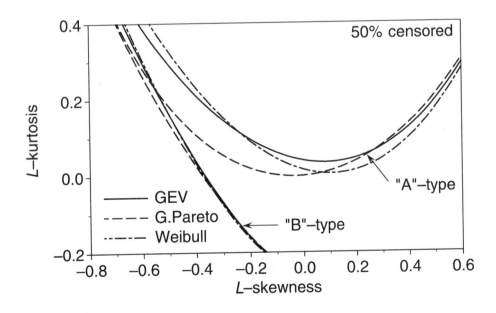

FIGURE 29.2
L-moment ratio diagram for censoring proportion 50%

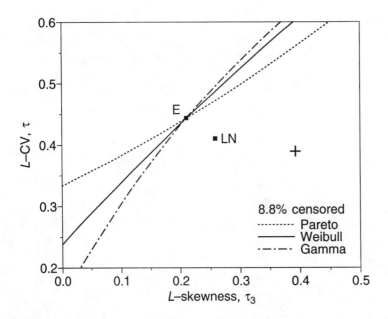

FIGURE 29.3
L-moment ratio diagram for distributions with lower bound zero and 8.8%
censoring. Continuous curves are the L-skewness–L-CV relations for Pareto,
Weibull and gamma distributions with lower bound zero. The points E and LN are
L-skewness and L-CV values for the exponential and lognormal distributions,
respectively. The "+" mark indicates the sample L-skewness and L-CV for the
transistor lifetime data in Table 29.2

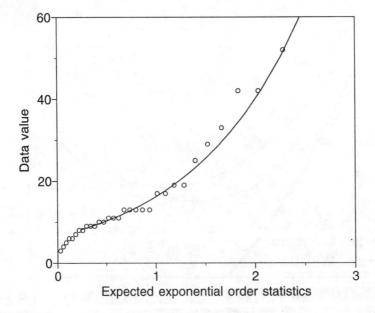

FIGURE 29.4
Transistor lifetime data from [19], plotted against expected exponential order
statistics. Solid line is a generalized Pareto distribution fitted to the order
statistics $X_{j:34}$, $j = 8, \ldots, 31$

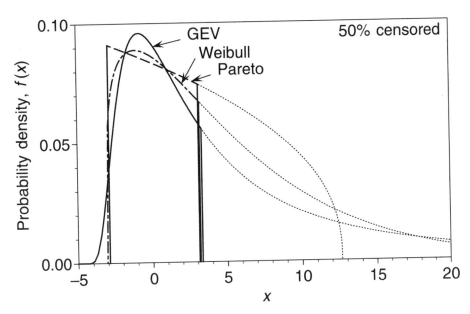

FIGURE 29.5
Three distributions with L-skewness $\tau_3^A = 0.02$ under 50% censoring. Heavy lines indicate the uncensored part, and light dotted lines the censored part, of each distribution

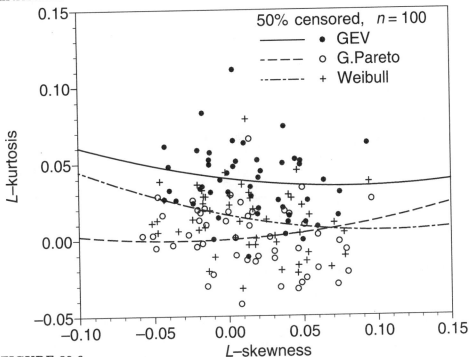

FIGURE 29.6
L-skewness and L-kurtosis of censored samples from the three distributions in Figure 29.5 for sample size $n = 100$. Continuous lines are L-skewness–L-kurtosis relations for GEV, Weibull and generalized Pareto distributions

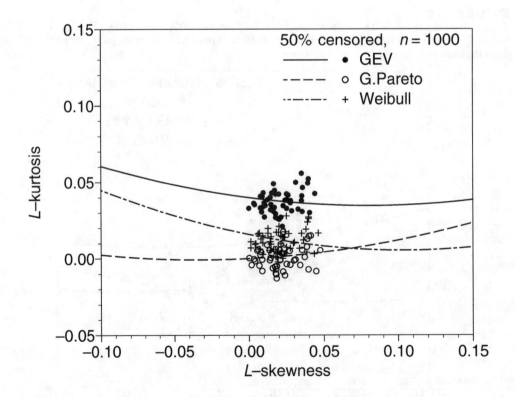

FIGURE 29.7
L-skewness and *L*-kurtosis of censored samples from the three distributions in
Figure 29.5 for sample size $n = 1000$. Continuous lines are *L*-skewness–*L*-kurtosis
relations for GEV, Weibull and generalized Pareto distributions

30

The Infeasibility of Probability Weighted Moments Estimation of Some Generalized Distributions

Gemai Chen and N. Balakrishnan

University of Waterloo
McMaster University

ABSTRACT The method of probability weighted moments is a very useful method for estimating parameters of continuous distributions. Its good robust property, small bias and rapid convergence to the normal distribution, all make this method attractive. However, when estimating the parameters of some generalized distributions, such as the generalized extreme value, Pareto and logistic distributions, the method of probability weighted moments can lead to infeasible estimates in the sense that the estimated distribution has an upper or lower bound and one or more of the data values lie outside this bound. In this paper, we investigate the infeasibility of the parameter estimates obtained by the method of probability weighted moments for the parameters of the above mentioned generalized distributions.

Keywords : Generalized extreme value distribution, Generalized logistic distribution, Generalized Pareto distribution, Parameter estimation, Probability weighted moments

CONTENTS

30.1 Introduction

Probability weighted moments (PWMs) are expectations of some functions of a random variable with finite mean. In [1], the authors introduced PWMs primarily for the purpose of estimating the parameters of the Wakeby distribution. Since then, PWMs have been found very useful in (1) characterization of distributions, (2) description and summarization of sample data, (3) estimation of parameters and quantiles, and (4) hypothesis tests. [2] contains a comprehensive treatment of the above topics mainly from the point of view of PWMs, while [3] covers the same topics, but from the point of view of L-moments, which are expectations of some linear combinations of order statistics. Because PWMs can be expressed as linear combinations of L-moments and vice versa, procedures based on PWMs and on L-moments are equivalent. We use PWMs in this paper and the interested reader should consult [2], [3] and the references therein for more details.

PWMs have several advantages over the conventional moments: PWMs are more robust to outliers in the data, possess smaller bias in parameter estimation, and converge more

0-8493-8972-0/95/$0.00+$.50

rapidly to the asymptotic normal distribution. These advantages enable more secure inferences to be made about the underlying distribution from small samples. However, when estimating the parameters of some generalized distributions, such as the generalized extreme value, Pareto and logistic, the method of probability weighted moments can lead to infeasible estimates in the sense that the estimated distribution has an upper or lower bound and one or more of the data values lie outside this bound. In this paper, we investigate the infeasibility of the parameter estimates in estimating the above mentioned distributions when the method of probability weighted moments is used.

An introduction to the method of probability weighted moments is given in Section 30.2, along with the parameter estimation procedures for the three generalized distributions. A simulation study is described in Section 30.3 to investigate the infeasibility of the parameter estimates, and some conclusions are drawn in Section 30.4.

30.2   The Method of Probability Weighted Moments

Let X be a random variable with distribution function F. The probability weighted moments of X, as defined in [1] , are the quantities

$$M_{p,r,s} = E\left[X^p\{F(X)\}^r\{1 - F(X)\}^s\right], \qquad (30.2.1)$$

where p, r and s are non-negative real numbers and $E|X|^p$ exists.

When $r = s = 0$, $M_{p,0,0}$ are just the conventional moments. However, the key idea of PWMs is to enter X linearly into the definition of PWMs, namely, $M_{1,r,s}$, and to allow r and s to take non-negative integer values. The following special PWMs are the ones usually used to develop statistical inference procedures in practice:

$$\alpha_s = M_{1,0,s} = E\left[X\{1 - F(X)\}^s\right], \quad s = 0, 1, \ldots, \qquad (30.2.2)$$

$$\beta_r = M_{1,r,0} = E\left[X\{F(X)\}^r\right], \quad r = 0, 1, \ldots. \qquad (30.2.3)$$

Because β_r can be expressed as a linear combination of α_s and vice versa, procedures based on β_r and on α_s are also equivalent. We work with β_r in this paper.

Let $x_{1:n} \leq \cdots \leq x_{n:n}$ be an ordered random sample of size n from a distribution F with finite mean, and define

$$b_r = \frac{1}{n}\sum_{j=1}^{n} \frac{(j-1)(j-2)\cdots(j-r)}{(n-1)(n-2)\cdots(n-r)} x_{j:n}, \qquad (30.2.4)$$

where $n > r$ and $b_0 = \bar{x}$ is the sample mean. Then b_r is an unbiased estimator of β_r; see [6]. Alternatively, we can estimate β_r by

$$\hat{\beta}_r[p_{j:n}] = \frac{1}{n}\sum_{j=1}^{n} p_{j:n}^r x_{j:n}, \qquad (30.2.5)$$

where $p_{j:n}$ is a distribution-free estimate of $F(x_{j:n})$, called a plotting position. A common plotting position is $p_{j:n} = (j+\gamma)/(n+\delta)$ for $\delta > \gamma > -1$. Plotting position estimators of β_r are in general biased for finite n, but they become equivalent to b_r as $n \to \infty$, and hence are consistent estimators of β_r.

For a distribution whose specification involves a finite number, m, of unknown parameters, the method of probability weighted moments can be used to estimate the unknown parameters by equating the (first m) sample PWM estimators to the corresponding theoretical PWMs and solving for the unknown parameters in terms of the sample PWM estimators. In the following, we give this estimation procedure for three generalized distributions; more details can be found in [2].

30.2.1 Generalized Extreme Value Distribution

For real ξ, real k and real $\alpha > 0$, the distribution function is given by

$$F(x) = \begin{cases} \exp[-\{1 - k(x - \xi)/\alpha\}^{1/k}], & \xi + \alpha/k \leq x < \infty, & \text{if } k < 0, \\ \exp[-\exp\{-(x - \xi)/\alpha\}], & -\infty < x < \infty, & \text{if } k = 0, \\ \exp[-\{1 - k(x - \xi)/\alpha\}^{1/k}], & -\infty < x \leq \xi + \alpha/k, & \text{if } k > 0. \end{cases} \tag{30.2.6}$$

The PWMs estimators for the parameters are given by

$$\hat{k}, \qquad \hat{\alpha} = \frac{(2b_1 - b_0)\hat{k}}{\Gamma(1 + \hat{k})(1 - 2^{-\hat{k}})}, \qquad \hat{\xi} = b_0 - \frac{\hat{\alpha}\{1 - \Gamma(1 + \hat{k})\}}{\hat{k}}, \tag{30.2.7}$$

where $\hat{k}$ is the solution to $(3b_2 - b_0)/(2b_1 - b_0) = (1 - 3^{-\hat{k}})(1 - 2^{-\hat{k}})$ and can be approximated by

$$\hat{k} = 7.8590c + 2.9554c^2, \qquad c = \frac{2b_1 - b_0}{3b_2 - b_0} - \frac{\ln 2}{\ln 3}, \tag{30.2.8}$$

and $\Gamma(\cdot)$ is the Gamma function.

30.2.2 Generalized Pareto Distribution

For real k and real $\alpha > 0$, the distribution function is given by

$$F(x) = \begin{cases} 1 - (1 - kx/\alpha)^{1/k}, & 0 \leq x < \infty, & \text{if } k < 0, \\ 1 - \exp(-x/\alpha), & 0 \leq x < \infty, & \text{if } k = 0, \\ 1 - (1 - kx/\alpha)^{1/k}, & 0 \leq x < \alpha/k, & \text{if } k > 0. \end{cases} \tag{30.2.9}$$

The PWMs estimators for the parameters are given by

$$\hat{k} = \frac{b_0}{2b_1 - b_0} - 2, \qquad \hat{\alpha} = \frac{2b_0(b_0 - b_1)}{2b_1 - b_0}. \tag{30.2.10}$$

30.2.3 Generalized Logistic Distribution

For real ξ, real k and real $\alpha > 0$, the distribution function is given by

$$F(x) = 1/(1 + y),$$

$$\text{where } y = \begin{cases} \{1 - k(x - \xi)/\alpha\}^{1/k}, & \xi + \alpha/k \leq x < \infty, & \text{if } k < 0, \\ \exp\{-(x - \xi)/\alpha\}, & -\infty < x < \infty, & \text{if } k = 0, \\ \{1 - k(x - \xi)/\alpha\}^{1/k}, & -\infty < x \leq \xi + \alpha/k, & \text{if } k > 0. \end{cases} \tag{30.2.11}$$

The PWMs estimators for the parameters are given by

$$\hat{k} = -\frac{6b_2 - 6b_1 + b_0}{2b_1 - b_0}, \qquad \hat{\alpha} = \frac{2b_1 - b_0}{\Gamma(1 - \hat{k})\Gamma(1 + \hat{k})}, \qquad \hat{\xi} = b_0 - \frac{\hat{\alpha}\{1 - \Gamma(1 - \hat{k})\Gamma(1 + \hat{k})\}}{\hat{k}}. \tag{30.2.12}$$

The above discussion uses the unbiased sample PWMs b_r. However, as discussed in [2, pp. 32-33], there is no general reason to use b_r or any other sample PWMs. Nevertheless, a reasonable choice can be based on computational convenience or the accuracy of the estimators in finite samples. For example, the use of the plotting position estimators with $p_{j:n} = (j - 0.35)/n$ has been found to give good results for generalized extreme value and generalized Pareto distributions ([4]; [5]).

30.3 The Infeasibility of PWM Estimators

When the range of a distribution is bounded by a function of the parameters, such as the cases we have here, we would ideally like to obtain parameter estimators such that the data used to find the estimators are also bounded by the same function of these estimators. This, unfortunately, does not always happen when the method of probability weighted moments is used to estimate the parameters of the three generalized distributions discussed in the previous section.

Because analytical expressions for the probabilities of obtaining infeasible parameter estimates are intractable, we use Monte Carlo simulations to study the frequencies at which infeasible parameter estimates occur for small to moderate sample sizes. Following [4] and [5], we restrict our attention to the case where $-0.5 < k < 0.5$, since this range of k values are most often encountered in practice, and the usual asymptotic theory holds for these values.

Without loss of generality, we set $\xi = 0$ and $\alpha = 1$ in our simulation study. This is because the method of probability weighted moments is invariant under linear transformations of the data. For each distribution and each k value ($k = -0.45$ to 0.45, with increment 0.10; $k \neq 0$), we generate $10,000$ random samples and estimate the relevant parameters using (i) the unbiased PWM estimators b_r, and (ii) the plotting position PWM estimators $\hat{\beta}_r[p_{j:n}]$ with $p_{j:n} = (j - 0.35)/n$. For every set of $10,000$ random samples, we record the proportion of times infeasible parameter estimates occur. Table 30.1 contains the simulation results.

We see that for small samples ($n = 15$ or 25), the probabilities of obtaining infeasible parameter estimates are almost always greater than 5% (there are six exceptions out of 96 cases) and can be as high as 40% ($n = 15$, $k = .15$ for the generalized Pareto distribution). As the sample size becomes larger, we see that the probabilities become smaller for the generalized extreme value and Pareto distributions except when $k = .45$, that the probabilities become smaller for the generalized logistic distribution for small k values ($|k| \leq .25$) and become larger for large k values ($|k| \geq .35$). In most cases, using $\hat{\beta}_r[p_{j:n}]$ with $p_{j:n} = (j - 0.35)/n$ leads to smaller probabilities than using b_r for the generalized extreme value distribution, but the pattern is not clear for the other two generalized distributions. If we take 5% as a tolerance level, then the problem of having infeasible parameter estimates can be ignored for the generalized extreme value distribution only when $k \leq .25$ and $n \geq 100$; for the generalized Pareto distribution only when $k \leq .15$ and $n \geq 500$; and for the generalized logistic distribution only when $|k| \leq .25$ and $n \geq 100$.

In [2], the author observed the infeasibility problem for the generalized extreme value and generalized logistic distributions, and offered a method to overcome the problem. Let x denote $x_{1:n}$ or $x_{n:n}$; if the boundary condition is found to be violated by the PWM estimators of the parameters, in [2], the author suggested equating x to $\xi + \alpha/k$ and solving for k. This leads to $\hat{k} = -\ln\{(2b_1 - x)/(b_0 - x)\}/\ln 2$ for the generalized extreme value distribution, and $\hat{k} = (2b_1 - b_0)/(x - b_0)$ for the generalized logistic distribution, and the other parameters are estimated as before. When the same idea is applied to the generalized Pareto distribution, we find $\hat{k} = b_0/(x - b_0)$ to be the modification. Table 30.2 contains the simulation results when the above modifications are incorporated. As can be seen, the probabilities of obtaining infeasible parameter estimates are greatly reduced, except for small samples ($n = 15$ or 25), the probabilities remain sizable.

The effect of having infeasible parameter estimates can be seen from Table 30.3 in terms of bias and root mean square error (RMSE). For small to moderate samples ($n = 15, 25, 50$ and 100), the authors in [4] find that the PWM estimators for the parameters of the generalized extreme value distribution have in general smaller standard deviation and larger bias than the maximum likelihood estimators, and in [5] the authors find that the PWM estimators for the parameters of the generalized Pareto distribution have smaller bias in general and larger RMSE than the maximum likelihood estimators and the conventional moment estimators when $k > 0$ (the conventional moment estimators have the smallest RMSE). Our simulation results in Table 30.3 show that the RMSE of the PWM estimators with modifications are the smallest compared to those of the maximum likelihood estimators and the conventional moment estimators, and the bias of the PWM estimators with modifications are slightly smaller than those of the PWM estimators without modifications in the case of the generalized extreme value distribution, and are slightly larger than those of the PWM estimators without modifications in the case of the generalized Pareto distribution.

30.4 Conclusions

When the method of probability weighted moments is used to estimate the parameters of the generalized extreme value, Pareto and logistic distributions, the problem of having infeasible parameter estimates cannot be ignored. By making suitable modifications, this problem can be greatly reduced, except for small samples ($n = 15$ or 25). In general, the PWM estimators with modifications have smaller RMSE and comparable bias when compared to the maximum likelihood estimators and the conventional moment estimators. We recommend that routine check be carried out to see whether the problem of infeasible parameter estimates occurs, and use modified PWM estimators if the problem does occur.

References

1. Greenwood, J.A., Landwehr, J.M., Matalas, N.C., and Wallis, J.R. (1979), Probability weighted moments: Definitions and relation to parameters of several distributions expressible in inverse form, *Water Resources Research*, **15**, 1049-1054.

2. Hosking, J.R.M. (1986), The theory of probability weighted moments, *Research Report PC12210*, IBM Research, Reissued with corrections, 3 April, 1989.

3. Hosking, J.R.M. (1990), L-moments: Analysis and estimation of distributions using linear combinations of order statistics, *Journal of the Royal Statistical Society, Series B*, **52**, 105-124.

4. Hosking, J.R.M., Wallis, J.R., and Wood, E.F. (1985), Estimation of the generalized extreme-value distribution by the method of probability weighted moments, *Technometrics*, **27**, 251-261.

5. Hosking, J.R.M. and Wallis, J.R. (1987), Parameter and quantile estimation for the generalized pareto distribution, *Technometrics*, **29**, 339-349.

6. Landwehr, J.M., Matalas, N.C., and Wallis, J.R. (1979), Probability weighted moments compared with some traditional techniques in estimating Gumbel parameters and quantiles, *Water Resources Research*, **15**, 1055-1064.

TABLE 30.1
Proportions of times infeasible parameter estimates occur when the method of probability weighted moments is used to estimate the parameters of the generalized extreme value, Pareto and logistic distributions. Each case is based on 10,000 random samples. For each pair of entries, the upper entry is the proportion when the unbiased estimator b_r is used; the lower entry is the proportion when the plotting position estimator $\hat{\beta}_r[p_{j:n}]$ with $p_{j:n} = (j - 0.35)/n$ is used

n	k							
	-.45	-.35	-.25	-.15	.15	.25	.35	.45
Generalized Extreme Value Distribution								
15	.099	.150	.217	.313	.225	.131	.087	.083
	.059	.102	.159	.230	.314	.186	.095	.050
25	.039	.071	.133	.240	.159	.070	.060	.090
	.024	.050	.096	.187	.214	.081	.036	.029
50	.007	.015	.044	.140	.073	.039	.070	.115
	.004	.011	.034	.111	.095	.027	.037	.059
100	.004	.002	.007	.050	.023	.034	.078	.132
	.003	.001	.006	.042	.024	.021	.053	.087
200	.003	.001	.000	.011	.007	.028	.079	.142
	.002	.001	.000	.010	.005	.022	.061	.109
500	.002	.000	.000	.000	.004	.023	.077	.151
	.002	.000	.000	.000	.003	.020	.066	.130
Generalized Pareto Distribution								
15					.409	.338	.290	.264
					.359	.294	.266	.251
25					.348	.270	.226	.227
					.312	.253	.224	.235
50					.254	.187	.186	.213
					.235	.187	.197	.230
100					.156	.123	.167	.231
					.150	.129	.181	.247
200					.071	.103	.172	.251
					.071	.110	.180	.261
500					.025	.080	.166	.245
					.026	.084	.172	.252
Generalized Logistic Distribution								
15	.079	.082	.118	.202	.200	.119	.085	.081
	.066	.064	.071	.112	.345	.193	.101	.049
25	.090	.070	.064	.133	.135	.066	.072	.086
	.091	.078	.060	.083	.205	.077	.034	.019
50	.118	.085	.043	.053	.049	.048	.087	.115
	.124	.096	.051	.040	.069	.030	.048	.063
100	.149	.099	.045	.013	.016	.046	.102	.147
	.157	.109	.052	.014	.015	.031	.070	.103
200	.172	.108	.043	.006	.007	.041	.109	.174
	.177	.117	.049	.007	.004	.030	.088	.144
500	.199	.113	.037	.003	.003	.030	.119	.199
	.202	.119	.040	.003	.003	.026	.104	.178

TABLE 30.2
Proportions of times infeasible parameter estimates occur after modification has
been made, when the method of probability weighted moments is used to estimate
the parameters of the generalized extreme value, Pareto and logistic distributions.
Each case is based on 10,000 random samples, and the plotting position estimator
$\hat{\beta}_r[p_{j:n}]$ with $p_{j:n} = (j - 0.35)/n$ is used

n	k							
	-.45	-.35	-.25	-.15	.15	.25	.35	.45
	Generalized Extreme Value Distribution							
15	.027	.049	.072	.108	.117	.067	.036	.020
25	.011	.022	.041	.082	.087	.031	.015	.009
50	.002	.004	.015	.050	.039	.010	.014	.020
100	.002	.001	.003	.018	.010	.008	.021	.029
200	.001	.000	.000	.004	.002	.009	.023	.036
500	.001	.000	.000	.000	.001	.008	.026	.044
	Generalized Pareto Distribution							
15					.069	.049	.044	.052
25					.056	.040	.037	.048
50					.047	.037	.028	.047
100					.032	.029	.026	.045
200					.014	.025	.027	.041
500					.004	.021	.028	.031
	Generalized Logistic Distribution							
15	.021	.016	.015	.024	.079	.043	.026	.012
25	.031	.022	.013	.019	.046	.018	.009	.006
50	.038	.025	.013	.008	.016	.008	.013	.022
100	.047	.027	.013	.003	.002	.009	.018	.032
200	.045	.027	.012	.002	.001	.007	.024	.039
500	.052	.030	.010	.001	.000	.007	.026	.046

TABLE 30.3
Comparison of bias and root mean square error (RMSE). For every triplet of entries, the first entry is for the method of probability weighted moments without modification, the second entry is for the method of probability weighted moments with modification, and the third entry is for the maximum likelihood method in the case of the generalized extreme value distribution and for the conventional method of moments in the case of the generalized Pareto distribution. The PWM estimators using $\hat{\beta}_r[p_{j:n}]$ is used, where $p_{j:n} = (j - 0.35)/n$

		Bias			RMSE		
n	k	$\hat{\xi}$	$\hat{\alpha}$	$\hat{k}$	$\hat{\xi}$	$\hat{\alpha}$	$\hat{k}$
		Generalized Extreme Value Distribution					
15	-.4	.10 .08 .03	.01 -.03 -.07	.11 .06 -.04	.33 .32 .32	.33 .36 .28	.23 .18 .36
	-.2	.05 .02 .03	-.07 -.17 -.07	.03 -.07 -.02	.30 .29 .32	.26 .34 .25	.20 .17 .32
	.2	.00 .05 .05	-.11 -.03 -.06	-.08 .05 .04	.28 .28 .30	.23 .20 .22	.19 .17 .27
	.4	-.03 -.01 .04	-.12 -.10 -.07	-.12 -.09 .03	.28 .27 .28	.23 .21 .21	.22 .19 .23
25	-.4	.06 .06 .01	.01 -.01 -.04	.08 .06 -.02	.25 .24 .24	.25 .26 .21	.19 .17 .24
	-.2	.04 .01 .02	-.04 -.11 -.04	.02 -.05 -.01	.23 .22 .24	.20 .26 .19	.16 .15 .21
	.2	.00 .02 .03	-.07 -.03 -.03	-.04 .01 .04	.22 .21 .23	.17 .16 .16	.15 .13 .18
	.4	-.02 -.02 .04	-.07 -.07 -.03	-.07 -.07 .05	.22 .21 .22	.17 .17 .17	.17 .16 .17
50	-.4	.04 .04 .01	.01 .01 -.02	.05 .04 -.01	.17 .17 .17	.17 .18 .15	.15 .15 .15
	-.2	.02 .01 .01	-.02 -.05 -.02	.01 -.01 .00	.16 .16 .16	.14 .17 .13	.12 .12 .13
	.2	.00 .00 .02	-.03 -.03 -.02	-.02 -.01 .02	.16 .16 .16	.12 .11 .11	.10 .10 .11
	.4	-.01 -.01 .02	-.04 -.04 -.01	-.04 -.04 .03	.16 .16 .16	.12 .12 .11	.11 .11 .11
100	-.4	.02 .02 .00	.01 .01 -.01	.03 .03 -.01	.12 .12 .12	.12 .12 .10	.11 .11 .10
	-.2	.01 .01 .00	-.01 -.02 -.01	.01 .00 .00	.12 .12 .11	.10 .11 .09	.09 .09 .09
	.2	.00 .00 .01	-.02 -.02 -.01	-.01 -.01 .01	.11 .11 .11	.08 .08 .08	.07 .07 .07
	.4	.00 -.01 .01	-.02 -.02 .00	-.02 -.02 .02	.11 .11 .11	.08 .08 .08	.08 .08 .07
		Generalized Pareto Distribution					
15	.2		.09 .16 .10	.08 .18 .10		.45 .39 .43	.37 .30 .34
	.4		.08 .08 .10	.08 .09 .10		.45 .38 .46	.42 .31 .41
25	.2		.05 .10 .06	.05 .11 .06		.33 .29 .31	.27 .22 .24
	.4		.05 .04 .06	.05 .04 .05		.33 .28 .32	.31 .23 .28
50	.2		.03 .05 .03	.02 .05 .03		.22 .20 .21	.18 .15 .15
	.4		.02 .01 .03	.02 .00 .02		.22 .19 .21	.21 .17 .18
100	.2		.02 .02 .02	.01 .02 .01		.15 .14 .14	.13 .11 .10
	.4		.01 .00 .01	.01 -.01 .01		.16 .14 .15	.14 .12 .12

A Class of Tests for Exponentiality Against Decreasing Bivariate Mean Residual Life Alternatives

Dipankar Bandyopadhyay and Asit P. Basu

AT&T, Bedminster, NJ
University of Missouri - Columbia

ABSTRACT A class of tests is proposed for testing bivariate exponentiality against the Decreasing Bivariate Mean Residual Life (DBMRL) class of probability distributions. These tests are consistent and asymptotically unbiased against all continuous DBMRL alternatives. They are U-statistics and hence asymptotically normally distributed. Small sample powers are studied based on Monte Carlo simulation.

Keywords : Bivariate exponential distribution, DMRL distributions, DBMRL distributions, Nonparametric test, Reliability, Tests for exponentiality, Unbiasedness, U-statistics

CONTENTS

31.1 Introduction

The class of Decreasing Mean Residual Life (DMRL) distributions holds an important position in the statistical theory of Reliability. These distributions, also, arise naturally in Survival Analysis, where testing for, and estimation of, mean residual life is of great importance. In [2], the authors find that these distributions are especially useful in studying empirical data. These distributions are readily interpretable, and can easily be applied to physical models in actual experimental settings. These distributions also arise as life distributions from various useful shock models.

Univariate mean residual life concept has been found very useful in many areas including medical study and other physical sciences. Multivariate extensions of univariate mean residual life concept would be useful in modelling and analyzing multivariate lifetime data when there is a lack of independence among the components of a system.

For the bivariate case, let X and Y be the lifetimes of two devices having a joint distribution function $F_2(x, y)$. The joint survival function of these two devices is given by $\overline{F}_2(x, y) = P[X > x, Y > y]$. It is assumed that $\overline{F}_2(0, 0) = 1$. We have the following:

DEFINITION 31.1.1 F_2 belongs to the class of Decreasing Bivariate Mean Residual Life - I (DBMRL-I) distributions if and only if for $c = 1, 2$

$$m_c(x + t_1, \ y + t_2) \leq m_c(x, y) \tag{31.1.1}$$

for all (x, y) and $(t_1, t_2) \in \mathcal{R}_2^+$ and $m_c(\cdot)$ is the conditional mean residual life function.

DEFINITION 31.1.2 F_2 belongs to the class of Decreasing Bivariate Mean Residual Life - II (DBMRL-II) distributions if and only if

$$m_1(x + t, y) \leq m_1(x, y) \tag{31.1.2}$$

and

$$m_2(x, y + t) \leq m_2(x, y) \tag{31.1.3}$$

for all $(x, y) \in \mathcal{R}_2^+$ and all $t \in \mathcal{R}^+$. We get the dual classes of distributions by reversing the direction of inequality in (31.1.1), (31.1.2), and (31.1.3) simultaneously.

Result 1

Equality in (31.1.1) $\Leftrightarrow X$ and Y are independent.

PROOF   Proof of X and Y independent $\Rightarrow$ equality in (31.1.1) is easy. So let us assume that for $c = 1$ and 2, $m_c(x + t_1, \ y + t_2) = m_c(x, y)$ for all (x, y) and $(t_1, t_2) \in \mathcal{R}_2^+$. We can prove the following very easily.

$$\frac{\partial}{\partial x} [m_1(x, y)] = r_1(x, y) \cdot m_1(x, y) - 1 \tag{31.1.4}$$

where $r_1(x, y)$ is the first component of the hazard gradient. Now, by assumption we have

$$m_1(x + t_1, \ y + t_2) = m_1(x, y) \Rightarrow \frac{\partial}{\partial x} [m_1(x + t_1, \ y + t_2)] = \frac{\partial}{\partial x} [m_1(x, y)]$$

$$\Rightarrow r_1(x + t_1, \ y + t_2) \cdot m_1(x + t_1, \ y + t_2) - 1 = r_1(x, y) \cdot m_1(x, y) - 1 \qquad \text{[by (31.1.4)]}$$

$$\Rightarrow r_1(x + t_1, \ y + t_2) = r_1(x, y) \qquad \text{[by assumption]}.$$

The above equality is true for all (x, y) and $(t_1, t_2) \in \mathcal{R}_2^+$. So let us set $y = t_1 = 0$ to get for all x and $t_2 \in \mathcal{R}^+$

$$\Rightarrow r_1(x, t_2) = r_1(x, 0) \Rightarrow -\frac{\partial}{\partial x} \log \overline{F}_2(x, t_2) = -\frac{\partial}{\partial x} \log \overline{F}_2(x, 0)$$

$$\Rightarrow -\frac{\partial}{\partial x} [\log \overline{F}_2(x, t_2) - \log \overline{F}_2(x, 0)] = 0$$

Therefore $[\log \overline{F}_2(x, t_2) - \log \overline{F}_2(x, 0)]$ must be stationary with respect to x. In particular if we set $x = 0$, we get

$$\log \overline{F}_2(x, t_2) - \log \overline{F}_2(x, 0) = \log \overline{F}_2(0, t_2) - \log \overline{F}_2(0, 0) = \log \overline{F}_2(0, t_2)$$

$$\Rightarrow \log \overline{F}_2(x, t_2) = \log \overline{F}_2(x, 0) + \log \overline{F}_2(0, t_2)$$

$$\Rightarrow \overline{F}_2(x, t_2) = \overline{F}_2(x, 0) \cdot \overline{F}_2(0, t_2) \qquad \text{for all } x \text{ and } t_2 \in \mathcal{R}^+ .$$

$$\Rightarrow X \text{ and } Y \text{ are independent.}$$

∎

From the above result it is clear that condition (31.1.1) is too stringent to yield any meaningful boundary distributions between DBMRL-I and its dual class. So let us concentrate on DBMRL-II and its dual class. We have the following result:

Result 2
Equality in (31.1.2) and (31.1.3) ⇒ $\overline{F}_2$ *follows Bivariate Exponential distribution of Gumbel (1960).*

PROOF   Let $\overline{F}_2$ be Gumbel's (1960) Bivariate Exponential distribution. That is

$$\overline{F}_2(x,y) = \exp\{-\lambda_1 x - \lambda_2 y - \delta xy\} , \ (x,y) \in \mathcal{R}_2^+ \ ; \ 0 \le \delta \le 1 \ ; \lambda_1, \lambda_2 > 0 \ . \ (31.1.5)$$

Direct calculations will show that

$$m_1(x,y) = (\lambda_1 + \delta y)^{-1} \text{ and } m_2(x,y) = (\lambda_2 + \delta x)^{-1} \ .$$

$$m_1(x+t,y) = (\lambda_1 + \delta y)^{-1} = m_1(x,y) \text{ and } m_2(x,y+t) = (\lambda_2 + \delta x)^{-1}$$

$$= m_2(x,y) \ .$$

Hence we have equality in (31.1.2) and (31.1.3). Conversely, suppose that we have equality in (31.1.2) and (31.1.3). So for all $(x,y) \in \mathcal{R}_2^+$ and all $t \in \mathcal{R}^+$,

$$m_1(x+t,y) = m_1(x,y) \Rightarrow \frac{\partial}{\partial x}[m_1(x+t,y)] = \frac{\partial}{\partial x}[m_1(x,y)]$$

$$\Rightarrow r_1(x+t,y) \cdot m_1(x+t,y) - 1 = r_1(x,y) \cdot m_1(x,y) - 1 \qquad \text{[by (31.1.4)]}$$

$$\Rightarrow r_1(x+t,y) = r_1(x,y) \ . \qquad \text{[by assumption]}$$

The above equality is true for all $(x,y) \in \mathcal{R}_2^+$ and all $t \in \mathcal{R}^+$. So let us set $x = 0$ to get for all y and $t \in \mathcal{R}^+$

$$r_1(t,y) = r_1(0,y) \ .$$

Similarly, we will get from the other equality, for all x and $t \in \mathcal{R}^+$

$$r_2(x,t) = r_2(x,0) \ .$$

Hence the first (second) component of the hazard gradient corresponding to $\overline{F}_2$ is stationary in x (y). Therefore, using the result of [3, theorem 5.4.11, p. 129], $\overline{F}_2$ must be the survival function corresponding to a bivariate exponential distribution of Gumbel (1960). ∎

Although properties of DBMRL (IBMRL) distributions have been studied by many authors, testing for bivariate exponentiality against bivariate DMRL (IMRL) distributions has not been considered in the literature. In this paper, we describe some results obtained for a testing procedure for Bivariate Exponentiality against Decreasing BIvariate Mean Residual Life (DBMRL) alternatives. In Section 31.2, we define two new classes of statistics which are seen to be U-statistics [4]. In Section 31.3, we prove asymptotic normality of these statistics and establish consistency and asymptotic unbiasedness. In Section 31.4, one numerical example has been worked out to illustrate the testing procedure. Finally, in Section 31.5, we study small sample power of our test based on Monte Carlo simulation.

31.2   The Proposed Classes of Test Statistics

Let $\{(X_1, Y_1), (X_2, Y_2), \ldots, (X_n, Y_n)\}$ be a random sample from F_2. Suppose we are interested in testing the following hypotheses:

$$H_0 : \overline{F}_2(x, y) = \exp\{-\lambda_1 x - \lambda_2 y - \delta x y\}, \qquad \in \mathcal{R}_2^+ ; \; 0 \leq \delta \leq 1 ;$$

$$\lambda_1, \lambda_2 > 0 \text{ and } \lambda_1, \lambda_2 \text{ and } \delta \text{ are unspecified}, \qquad (31.2.1)$$

against

$$H_1 : \overline{F}_2(x, y) \text{ is DBMRL-II and NOT as given in } H_0 ,$$

on the basis of the random sample. We observe that we can modify the definition of D(I)BMRL-II as follows:

DEFINITION 31.2.1 $\overline{F}_2$ *belongs to the class of D(I)BMRL-II distribution if and only if for all* $\kappa_1, \kappa_2 \in [0, 1)$ *and* $(x, y) \in \mathcal{R}_2^+$

$$m_1(\kappa_1 \cdot x, y) \geq (\leq) m_1(x, y) \qquad (31.2.2)$$

and

$$m_2(x, \kappa_2 \cdot y) \geq (\leq) m_2(x, y) . \qquad (31.2.3)$$

Let us, now, define the following functions:

$$D_1(x, y; \kappa_1; F2) = \overline{F}_2(\kappa_1 \cdot x, y) \cdot \overline{F}_2(x, y)[m_1(\kappa_1 \cdot x, y) - m_1(x, y)] ,$$
$$D_2(x, y; \kappa_2; F2) = \overline{F}_2(x, \kappa_2 \cdot y) \cdot \overline{F}_2(x, y)[m_2(x, \kappa_2 \cdot y) - m_2(x, y)] .$$

Therefore, $\overline{F}_2$ belongs to the class of D(I)BMRL-II distributions if and only if $D_i(x, y; \kappa_i; F_2) \geq (\leq) 0$ for all $\kappa_i \in [0, 1)$; $i = 1, 2$ and $(x, y) \in \mathcal{R}_2^+$. We propose the following two measures of deviations from H_0 towards H_1:

$$\Delta_i(\kappa_i; F_2) = \int_0^\infty \int_0^\infty D_i(x, y; \kappa_i; F_2) \, dF_2(x, y)$$

and

$$\Gamma_i(\kappa_i; F_2) = \int_0^\infty \int_0^\infty D_i(x, y; \kappa_i; F_2) \, dF(x) \, dG(y) .$$

Here F and G are the marginal distributions of X and Y respecitvely. Let us define

$$\Delta(\kappa_1, \kappa_2; F_2) = \frac{1}{2}[\Delta_1(\kappa_1; F_2) + \Delta_2(\kappa_2; F_2)] ,$$

and

$$\Gamma(\kappa_1, \kappa_2; F_2) = \frac{1}{2}[\Gamma_1(\kappa_1; F_2) + \Gamma_2(\kappa_2; F_2)] .$$

$\Delta(\kappa_1, \kappa_2; F_2)$ and $\Gamma(\kappa_1, \kappa_2; F_2)$ are two measures of deviation from H_0 towards H_1. Under H_0, $\Delta(\kappa_1, \kappa_2; F_2)[\Gamma(\kappa_1, \kappa_2; F_2)] \equiv 0$ and under H_1, $\Delta(\kappa_1, \kappa_2; F_2)[\Gamma(\kappa_1, \kappa_2; F_2)] > 0$. Hence viewing $\Delta(\kappa_1, \kappa_2; F_2)[\Gamma(\kappa_1, \kappa_2; F_2)]$ as a measure of deviation of $\overline{F}_2$ from H_0, the classical non-parametric approach of replacing F_2 by the empirical c.d.f. F_{2n} in the above integrals suggests rejecting H_0 in favor of H_1 if $\Delta(\kappa_1, \kappa_2; F_{2n})[\Gamma(\kappa_1, \kappa_2; F_{2n})]$ is significantly large, where $F_{2n}(x, y) = N(x, y)/n$, that is, F_{2n} is the empirical c.d.f. with $N(x, y)$

being the number of (X_i, Y_i)'s in the sample such that $X_i \leq x$ and $Y_i \leq y$. Let $\overline{F}_{2N}$ denote the corresponding empirical survival function, that is, $\overline{F}_{2n}(x, y) = N^*(x, y)/n$, with $N^*(x, y)$ being the number of (X_i, Y_i)'s in the sample such that $X_i > x$ and $Y_i > y$. $\Delta(\kappa_1, \kappa_2; F_{2n})[\Gamma(\kappa_1, \kappa_2; F_{2n})]$ is the sample analogue of $\Delta(\kappa_1, \kappa_2; F_2)[\Gamma(\kappa_1, \kappa_2; F_2)]$ obtained when we replace F_2 by F_{2n} and $\overline{F}_2$ by $\overline{F}_{2n}$ in the right hand side of the integrals. Let us, now, define the following functions:

$$a_1[(x_j, y_j); 1 \leq j \leq 3; \kappa_1]$$
$$= (x_1 - \kappa_1 \cdot x_3) \cdot I(x_1 > \kappa_1 \cdot x_3) \cdot I(x_2 > x_3) \cdot I(y_1 > y_3) \cdot I(y_2 > y_3) ,$$

$$b_1[(x_j, y_j); 1 \leq j \leq 3; \kappa_1]$$
$$= (x_1 - x_3) \cdot I(x_1 > x_3) \cdot I(x_2 > \kappa_1 \cdot x_3) \cdot I(y_1 > y_3) \cdot I(y_2 > y_3) ,$$

$$a_2[(x_j, y_j); 1 \leq j \leq 3; \kappa_2]$$
$$= (y_1 - \kappa_2 \cdot y_3) \cdot I(y_1 > \kappa_2 \cdot y_3) \cdot I(y_2 > y_3) \cdot I(x_1 > x_3) \cdot I(x_2 > x_3) ,$$

$$b_2[(x_j, y_j); 1 \leq j \leq 3; \kappa_2]$$
$$= (y_1 - y_3) \cdot I(y_1 > y_3) \cdot I(y_2 > \kappa_2 \cdot y_3) \cdot I(x_1 > x_3) \cdot I(x_2 > x_3) ,$$

where the indicator function $I(a > b) = 1$ or 0 according as $a > b$ or not. We have the following result:

LEMMA 31.2.2
$\phi_u[(X_j, Y_j); 1 \leq j \leq 3; \kappa_u] = \{a_u[(X_j, Y_j); 1 \leq j \leq 3; \kappa_u] - b_u[(X_j, Y_j); 1 \leq j \leq 3; \kappa_u]\}$, $u = 1, 2$ with $(X_j, Y_j); 1 \leq j \leq 3$ i.i.d. copies of (X, Y), is unbiased estimator for $\Delta_u(\kappa_u; F_2)$.

PROOF The result follows easily by conditioning on (X_3, Y_3) ans using mutual independence between the pairs. ∎

We thus have an unbiased kernel for $\Delta_u(\kappa_u; F_2)$, namely, the kernel $\phi_u[(X_j, Y_j); 1 \leq j \leq 3; \kappa_u]$, $u = 1, 2$. One might use $\Delta(\kappa_1, \kappa_2; F_{2n})$ as the test statistic for testing H_0 against H_1, but it is more convenient to reject H_0 in favor of H_1 for significantly large values of $V_n^*(\kappa_1, \kappa_2)$ where

$$\delta_{u,n}(\kappa_u) = [^n P_3]^{-1} \sum_{p3} \phi_u[(X_{i_j}, Y_{i_j}); 1 \leq j \leq 3; \kappa_u]; \quad u = 1, 2 \text{ and}$$

$$V_n^*(\kappa_1, \kappa_2) = \frac{1}{2} [\delta_{1,n}(\kappa_1) + \delta_{2,n}(\kappa_2)]$$

with $\sum_{p3}$ denoting summation over all possible permutations of the distinct subscripts $\{i_1, i_2, i_3\}$ chosen from $\{1, 2, \ldots, n\}$. Clearly, then, $V_n^*(\kappa_1, \kappa_2)$ is an unbiased estimator for $\Delta(\kappa_1, \kappa_2; F_2)$. We note that the statistic $V_n^*(\kappa_1, \kappa_2)$ is a U-statistic while $\Delta(\kappa_1, \kappa_2; F_{2n})$ is a von Mises' (1947) differentiable statistical functional corresponding to $\Delta(\kappa_1, \kappa_2; F_2)$. They are known to be asymptotically equivalent. The distribution of $V_n^*(\kappa_1, \kappa_2)$ depends on λ_1, λ_2 and thus is not scale invariant. We might use the following as our scale invariant test statistic:

$$V_n(\kappa_1, \kappa_2) = \frac{1}{2} \left[\frac{\delta_{1,n}(\kappa_1)}{\overline{X}_n} + \frac{\delta_{2,n}(\kappa_1)}{\overline{Y}_n} \right] .$$

Here $\overline{X}_n$ and $\overline{Y}_n$ are the marginal sample means. As for the other measure, $\Gamma(\kappa_1, \kappa_2; F_2)$, the kernels will be defined as follows:

$$f_1[(x_j, y_j); 1 \leq j \leq 4; \kappa_1]$$
$$= (x_1 - \kappa_1 \cdot x_4) \cdot I(x_1 > \kappa_1 \cdot x_4) \cdot I(x_2 > x_4) \cdot I(y_1 > y_3) \cdot I(y_2 > y_3) ,$$

$$g_1[(x_j, y_j); 1 \leq j \leq 4; \kappa_1]$$
$$= (x_1 - x_4) \cdot I(x_1 > x_4) \cdot I(x_2 > \kappa_1 \cdot x_4) \cdot I(y_1 > y_3) \cdot I(y_2 > y_3) ,$$

$$f_2[(x_j, y_j); 1 \leq j \leq 4; \kappa_2]$$
$$= (y_1 - \kappa_2 \cdot y_4) \cdot I(y_1 > \kappa_2 \cdot y_4) \cdot I(y_2 > y_4) \cdot I(x_1 > x_3) \cdot I(x_2 > x_3) ,$$

$$g_2[(x_j, y_j); 1 \leq j \leq 4; \kappa_2]$$
$$= (y_1 - y_4) \cdot I(y_1 > y_4) \cdot I(y_2 > \kappa_2 \cdot y_4) \cdot I(x_1 > x_3) \cdot I(x_2 > x_3) .$$

We will have the following result:

LEMMA 31.2.3
$\Psi_u[(X_j, Y_j); 1 \leq j \leq 4; \kappa_u] = \{f_u[(X_j, Y_j); 1 \leq j \leq 4; \kappa_u] - g_u[(X_j, Y_j); 1 \leq j \leq 4; \kappa_u]\}$, $u = 1, 2$ with $(X_j, Y_j); 1 \leq j \leq 4$ *i.i.d. copies of* (X, Y), *is unbiased for* $\Gamma_u(\kappa_u; F_2)$.

PROOF   Follows similarly to that of Lemma 31.2.2.   ∎

Therefore, we have an unbiased kernel for $\Gamma_u(\kappa_u; F_2)$, namely, the kernel $\Psi_u[(X_j, Y_j)$; $1 \leq j \leq 4; \kappa_u]$, $u = 1, 2$. One might use $\Gamma(\kappa_1, \kappa_2; F_{2n})$ as the test statistic for testing H_0 against H_1, but it is more convenient to reject H_0 in favor of H_1 for significantly large values of $W_n^*(\kappa_1, \kappa_2)$ where

$$\gamma_{u,n}(\kappa_u) = [{}^n P_4]^{-1} \sum_{p4} \Psi_u[(X_{i_j}, Y_{i_j}); 1 \leq j \leq 4; \kappa_u]; \ u = 1, 2 \text{ and}$$

$$W_n^*(\kappa_1, \kappa_2) = \frac{1}{2} \left[\gamma_{1,n}(\kappa_1) + \gamma_{2,n}(\kappa_2) \right]$$

with $\sum_{p4}$ denoting the summation over all possible permutations of the distinct subscripts $\{i_1, i_2, i_3, i_4\}$ chosen from $\{1, 2 \ldots, n\}$. Clearly, then $W_n^*(\kappa_1, \kappa_2)$ is an unbiased estimator for $\Gamma(\kappa_1, \kappa_2; F_2)$. We note that the statistics $W_n^*(\kappa_1, \kappa_2)$ is a U-statistic while $\Gamma(\kappa_1, \kappa_2; F_{2n})$ is a von Mises' (1947) differentiable statistical functional corresponding to $\Gamma(\kappa_1, \kappa_2; F_2)$. They are known to be asymptotically equivalent. The distribution of $W_n^*(\kappa_1, \kappa_2)$ depends on λ_1, λ_2 and thus is not scale invariant. We might use the following as our scale invariant test statistic:

$$W_n(\kappa_1, \kappa_2) = \frac{1}{2} \left[\frac{\gamma_{1,n}(\kappa_1)}{\overline{X}_n} + \frac{\gamma_{2,n}(\kappa_2)}{\overline{Y}_n} \right] .$$

31.3   Properties of the Test Statistics

We shall, now, study the asymptotic properties of the test statistic $\Delta(\kappa_1, \kappa_2, F_{2n})$. We know $\Delta(\kappa_1, \kappa_2, F_{2n})$ is asymptotically equivalent to $V^*(\kappa_1, \kappa_2)$ and $V_n(\kappa_1, \kappa_2)$ is a function of U-statistics. So, we can apply the results from [4] on U-statistics. Before that let us introduce the following notations:

$$\phi_u^*[(x_j, y_j); 1 \le j \le 3; \kappa_u] = [3!] \Sigma' \phi_u[(x_{i_j}, y_{i_j}); 1 \le j \le 3; \kappa_u]; \ u = 1, 2 \ ,$$

with Σ' denoting summation over all the 3! permutations $\{i_1, i_2, i_3\}$ of $\{1, 2, 3\}$. Therefore, ϕ_u^* is a symmetric version of ϕ_u; $u = 1, 2$. Let us define for $u = 1, 2$

$$\phi_u^{**}(x, y; \kappa_u) = E\{\phi_u^*[(X_j, Y_j); 1 \le j \le 3; \kappa_u] \,|\, (X_1, Y_1) = (x, y)\} \ ,$$

and

$$\sigma_u^2(\kappa_u) = \text{var}[\phi_u^{**}(X, Y; \kappa_u)];$$
$$\sigma_{12}(\kappa_1, \kappa_2) = \text{cov}[\phi_1^{**}(X, Y; \kappa_1), \phi_2^{**}(X, Y; \kappa_2)].$$

We asusme $\sigma_1^2(\kappa_1)$ and $\sigma_2^2(\kappa_2)$ exist. Now, we can state the following result:

Result 3
Under H_0, the limiting distribution of $\sqrt{n} V_n(\kappa_1, \kappa_2)$ is normal with mean zero and variance $\sigma^2(\kappa_1, \kappa_2) = \frac{9}{4} [\sigma_1^2(\kappa_1) + \sigma_1^2(\kappa_2) + 2 \sigma_{12}(\kappa_1, \kappa_2)]$.

PROOF   From the result from [4] about U-statistics one can prove that under H_0,

$$\frac{\sqrt{n} \delta_{1,n}(\kappa_1)}{\overline{X}_n} \sim AN(0, 9 \sigma_1^2(\kappa_1)) \ , \qquad \frac{\sqrt{n} \delta_{2,n}(\kappa_2)}{\overline{Y}_n} \sim AN(0, 9 \sigma_2^2(\kappa_2)) \ ,$$

and

$$\lim \text{cov}\left[\frac{\sqrt{n} \delta_{1,n}(\kappa_1)}{\overline{X}_n}, \ \frac{\sqrt{n} \delta_{2,n}(\kappa_2)}{\overline{Y}_n}\right] = 9 \sigma_{12}(\kappa_1, \kappa_2) \qquad \text{as } n \to \infty \ .$$

Now, using Slutsky's theorem, the results follows.   ∎

One can find explicit algebraic expressions for $\sigma_1^2(\kappa_1)$, $\sigma_2^2(\kappa_2)$, and $\sigma_{12}(\kappa_1, \kappa_2)$ in terms of variances and covariances of the original kernels $a_u(\kappa_u)$ and $b_u(\kappa_u)$; $u = 1, 2$ and these, in turn, can be evaluated under H_0. However, it is a very tediuos job. We can use jackknifing to avoid going through the complicated process of evaluating actual variance. Jackknifing would not only reduce the bias but also enable us to estimate the variance of $\sqrt{n} V_n(\kappa_1, \kappa_2)$. The jackknifed estimator of $\text{var}[\sqrt{n} V_n(\kappa_1, \kappa_2)]$ is

$$\widehat{\text{var}}[\sqrt{n} V_n(\kappa_1, \kappa_2)] = \frac{n}{n-1} \sum_{i=1}^{n} [V_n^{(i)}(\kappa_1, \kappa_2) - \overline{V}_n(\kappa_1, \kappa_2)]^2 \ ,$$

where $V_n^{(i)}(\kappa_1, \kappa_2) = V_{n-1}(\kappa_1, \kappa_2)$ based on the sample $\{(X_1, Y_1), \ldots, (X_{i-1}, Y_{i-1}), (X_{i+1}, Y_{i+1}), \ldots, (X_n, Y_n)\}$; $1 \le i \le n$, and

$$\overline{V}_n(\kappa_1, \kappa_2) = \frac{1}{n} \sum_{i=1}^{n} V_n^{(i)}(\kappa_1, \kappa_2) \ .$$

Result 4
Under H_0, the limiting distribution of $\dfrac{\sqrt{n} V_n(\kappa_1, \kappa_2)}{\widehat{\text{var}}^{1/2}[\sqrt{n} V_n(\kappa_1, \kappa_2)]}$ is normal with mean zero and variance one.

PROOF   Follows from [5] and Slutsky's theorem.   ∎

We, now, state our approximate level α test of H_0 vs H_1 as follows:

Reject H_0 in favor of H_1 if $\dfrac{\sqrt{n}V_n(\kappa_1,\kappa_2)}{\widehat{\mathrm{var}}^{1/2}[\sqrt{n}V_n(\kappa_1,\kappa_2)]} \geq z_\alpha$, where z_α is the upper α-percentile point of the standard normal distribution.

REMARK 1. We can carry out similar calculations for the other test statistic $W_n(\kappa_1,\kappa_2)$. The only difference would be in Result 3. The formula for variance would be

$$\sigma^2(\kappa_1,\kappa_2) = 4[\sigma_1^2(\kappa_1) + \sigma_2^2(\kappa_2) + 2\sigma_{12}(\kappa_1,\kappa_2)] .$$

∎

REMARK 2. Using the fact that $V_n((\kappa_1,\kappa_2)$ $[W_n(\kappa_1,\kappa_2)]$ is asymptotically normal [Result 3], one can prove that the test based on $V_n((\kappa_1,\kappa_2)$ $[W_n(\kappa_1,\kappa_2)]$ is consistent and asymptotically unbiased against continuous DBMRL-II alternatives. ∎

REMARK 3. These results can be extended for IBMRL-II alternatives. ∎

REMARK 4. These tests can be used as tests for BNBUE-A or BNBUE-B alternatives since DBMRL-II implies both classes of distributions. In fact, for $\kappa_1 = \kappa_2 = 0$, equations (31.2.2) and (31.2.3) reduce to the condition of F_2 being bivariate NBUE-A. Hence, $V - n(0,0)$ $[W_n(0,0)]$ should be a test statistic for testing against BNBUE-A. Recently the authors in [6] developed two tests for BNBUE-A. We observe that $V^*(0,0)$ and $W^*(0,0)$ are essentially those tests. ∎

31.4   An Example

In this section, we consider an example to illustrate the testing procedures developed in the previous sections.

Example 1
Let us suppose $\overline{F}_2$ be given as follows:

$$\overline{F}_2(x,y) = \exp[-\lambda_1 x - \lambda_2 y - \lambda_{12}\max\{x,y\}] : (x,y) \in \mathcal{R}_2^+ ;$$

$$\text{all } \lambda\text{'s} \geq 0 \text{ such that } \lambda_i + \lambda_{12} > 0; \; i = 1,2 . \quad (31.4.1)$$

⧠

LEMMA 31.4.1
$\overline{F}_2$ [Marshall and Olkin's (1967) BVE($\lambda_1,\lambda_2,\lambda_{12}$)], given above, is a Decreasing Bivariate Mean Residual Life (DBMRL) distribution.

PROOF   For the given $\overline{F}_2$, one can show by direct calculations that $m_1(x,y)$ is of the following form:

$$m_1(x,y) = \begin{cases} \dfrac{1}{\lambda_1} - \dfrac{\lambda_{12}}{\lambda_1(\lambda_1+\lambda_{12})} \cdot \dfrac{\overline{F}(y,y)}{\overline{F}(x,y)} & \text{if } x < y \\ \dfrac{1}{\lambda_1+\lambda_{12}} & \text{if } x \geq y \end{cases} \quad (31.4.2)$$

Now, for all x, y and $t \geq 0$, we can have the following cases:

(a) $x \geq y \ (\Rightarrow x + t \geq y)$, $m_1(x + t, y) = (\lambda_1 + \lambda_{12})^{-1} = m_1(x, y)$, [using (31.4.2)].

(b) $x < y$ and $x + t < y$, $m_1(x + t, y) \leq (\geq) m_1(x, y) \Leftrightarrow \overline{F}_2(x + t, y) \leq (\geq) \overline{F}_2(x, y)$, [using (31.4.2].

Now for all x, y and $t \geq 0$, $\overline{F}_2(x + t, y) \leq \overline{F}_2(x, y)$, hence $m_1(x + t, y) \leq m_1(x, y)$.

(c) $x < y$ and $x + t \geq y$, $m_1(x + t, y) \leq (\geq) m_1(x, y) \Leftrightarrow \overline{F}_2(y, y) \leq (\geq) \overline{F}_2(x, y)$, [using (31.4.2)].

Now for all $x < y$, $\overline{F}_2(y, y) \leq \overline{F}_2(x, y)$, hence $m_1(x + t, y) \leq m_1(x, y)$.

For all x, y and $t \geq 0$, we have $m_1(x + t, y) \leq m_1(x, y)$. Similarly, by symmetry, one can show that $m_2(x, y + t) \leq m_2(x, y)$ for all x, y and $t \geq 0$. Hence by (31.1.2) and (31.1.3) the claim follows.   ∎

We generate a random sample of size $n = 15$ from BVE$(\lambda_1, \lambda_2, \lambda_{12})$ with parameters

$$\lambda_1 = \lambda_2 = \lambda_{12} = 1$$

and use our test $V_{15}(\kappa_1, \kappa_2)$ to see if the data are, indeed, coming from a DBMRL distribution. For illustration purpose we have used $\kappa_1 = \kappa_2 = 0.5$ and $\kappa_1 = \kappa_2 = 0.000001$. The data along with the values of $V_{15}^{(i)}(0.5, 0.5)$ and $V_{15}^{(i)}(0.000001, 0.000001)$, $1 \leq i \leq 15$, are given in Table 31.1. For this data we computed the following:

$$V_{15}(0.5, 0.5) = 0.03825271 \ ,$$

$$\widehat{\text{var}}[\sqrt{15} V_{15}(0.5, 0.5)] = 0.000209525 \text{ and}$$

$$Z \simeq \frac{\sqrt{15} V_{15}(0.5, 0.5)}{\widehat{\text{var}}^{1/2}[\sqrt{15} V_{15}(0.5, 0.5)]} = 10.23504869 \ .$$

$$V_{15}(0.000001, 0.000001) = 0.078231309 \ ,$$

$$\widehat{\text{var}}[\sqrt{15} V_{15}(0.000001, 0.000001)] = 0.001388412 \text{ and}$$

$$Z \simeq \frac{\sqrt{15} V_{15}(0.000001, 0.000001)}{\widehat{\text{var}}^{1/2}[\sqrt{15} V_{15}(0.000001, 0.000001)]} = 8.13143224 \ .$$

To find Z we use Result 4. Since both Z's are significantly large, we reject H_0 of Gumbel's Bivariate Exponential distribution in favor of DBMRL distribution using both the statistics $V_{15}(0.5, 0.5)$ and $V_{15}(0.000001, 0.000001)$. Hence, we may conclude that the (X, Y) data in Table 31.1 are from a DBMRL distribution.

31.5   Monte Carlo Simulation and Small Sample Power Studies

A Monte Carlo study was carried out to estimate critical points for the Gumbel's bivariate exponential distribution [given in (31.2.1)] with $\lambda_1 = \lambda_2 = 1$ and $\delta = 0.5$ corresponding to significance levels close to $\alpha = 0.01$, 0.025, 0.05, 0.075 and 0.01 of the $V_n(0.000001, 0.000001)$ test. The choice for $\kappa_1 = \kappa_2 = 0.000001$ looks strange. The authors in [1] recently obtained the univariate version of the $V_n(\kappa_1, \kappa_2)$ test. In the univariate case, it was established that very small value of κ is optimum. Thus we are just

TABLE 31.1
Simulated data from BVE(1,1,1) distribution

i	X	Y	$V_{15}^{(i)}(0.5, 0.5)$ $\times^{14} P_3$	$V_{15}^{(i)}(0.000001, 0.000001)$ $\times^{14} P_3$
1	0.85852	0.85852	80.59427454	161.8363395
2	0.43298	0.94952	84.90275610	172.0766731
3	0.33060	0.49190	81.87476729	160.9910866
4	0.26532	0.26532	76.30241523	159.1414596
5	0.80902	0.44406	87.00980970	169.6721485
6	0.02275	0.02805	103.9329628	233.8508113
7	0.34428	0.46632	76.90321255	153.3542841
8	0.54898	0.89990	81.56440351	165.9546981
9	0.30006	0.30006	68.36474244	144.3404084
10	0.82375	0.82375	79.89188247	161.5428588
11	0.16056	0.11486	87.84843498	182.8036349
12	0.04258	0.02538	91.97060134	188.3816610
13	0.53273	0.76707	78.85154984	158.2564007
14	0.84560	0.47653	83.50552844	165.8039532
15	0.08040	0.28136	89.54074294	182.6832909

using the same value in the bivariate case. This study was done for $n = 3, 4, \ldots, 15$, each value being based on 5000 samples of the required size. Table 31.2 gives the upper tail critical values of $(^n P_3) \cdot V_n(0.000001, 0.000001)$. That is, Table 31.2 contains values of $C_{\alpha,n}$ such that $\Pr_{H_0}\{(^n P_3) \cdot V_n(0.000001, 0.000001) \geq C_{\alpha,n}\} = \alpha$.

Now, the following tables contain small sample powers of $V_n(0.000001, 0.000001)$ test for $\alpha = 0.05$ and 0.1 for BVE$(1, 1, \lambda_{12})$ alternatives. Tables for other α's are available from the authors. The probabilities which are estimates are $\Pr_{H_1}\{(^n P_3) \cdot V_n(0.000001, 0.000001) \geq C_{\alpha,n}\}$ and $C_{\alpha,n}$'s are the simulated critical points for $\alpha = 0.05$ and 0.1.

31.6 Summary

The general mutivariate extension of the $V_n(\kappa_1, \kappa_2)$ test would be straight forward. Attempts are now being made to develop a testing procedure based on the TTT-Transform and explore its relationship with the $V_n(\kappa)$ and $V_n(\kappa_1, \kappa_2)$ tests. Testing for censored data would also be of interest. Results of these studies will be reported in follow-up papers.

Acknowledgements

Research sponsored by the Air Force Office of Scientific Research, Air Force Systems command, USAF, under grant number AFOSR-F49620-92-J-0371. The US Government is authorized to reproduce and distribute reprints for Governmental purposes notwithstanding any copyright notation thereon.

TABLE 31.2
Simulated critical values of $V_n(0.000001, 0.000001)$ test

$n\backslash\alpha$	0.1	0.075	0.05	0.025	0.01
3	0.8879	0.9582	1.0521	1.2219	1.3865
4	2.3157	2.6176	2.9758	3.4495	3.9984
5	4.6321	5.2687	5.9646	7.1548	8.7370
6	7.8619	8.6357	10.0334	11.9432	14.2981
7	12.4794	13.9094	16.1014	19.3270	23.3420
8	17.1287	19.5125	22.6873	27.5668	33.5947
9	23.5958	27.0544	30.8230	37.1562	44.0314
10	31.6597	35.1651	40.5038	49.9352	58.8471
11	42.0092	46.5009	53.4880	64.2117	79.4868
12	52.7890	59.8441	68.6338	80.1233	93.9363
13	66.8896	74.3914	83.5436	99.1128	118.8902
14	82.3424	92.6905	105.3850	124.6485	145.7704
15	96.5865	107.2951	121.0001	146.7341	174.9272

TABLE 31.3
Simulated values of power of $V_n(0.000001, 0.000001)$ test

$n\backslash\lambda_{12}$	$\alpha = 0.05$					
	0.5	1.0	1.5	2.0	2.5	3.0
3	0.0638	0.0746	0.0912	0.1164	0.1218	0.1280
4	0.0702	0.0938	0.1220	0.1562	0.1746	0.1858
5	0.0728	0.1194	0.1554	0.1858	0.2362	0.2502
6	0.0842	0.1354	0.1928	0.2400	0.2786	0.3072
7	0.0796	0.1318	0.1912	0.2470	0.2898	0.3242
8	0.0838	0.1608	0.2234	0.2974	0.3496	0.3898
9	0.0934	0.1724	0.2622	0.3320	0.3910	0.4300
10	0.0946	0.1790	0.2770	0.3562	0.4324	0.4726
11	0.0978	0.2002	0.3004	0.3794	0.4350	0.5030
12	0.0914	0.1952	0.3024	0.3824	0.4738	0.5192
13	0.1036	0.2120	0.3302	0.4122	0.5108	0.5766
14	0.0924	0.2004	0.3284	0.4282	0.5054	0.5748
15	0.1122	0.2401	0.3702	0.4706	0.5670	0.6220

TABLE 31.4
Simulated values of power of $V_n(0.000001, 0.000001)$ test

$n\backslash\lambda_{12}$	$\alpha = 0.1$					
	0.5	1.0	1.5	2.0	2.5	3.0
3	0.1138	0.1266	0.1460	0.1726	0.1846	0.1908
4	0.1362	0.1736	0.2244	0.2568	0.2856	0.2944
5	0.1370	0.2040	0.2518	0.3040	0.3492	0.3736
6	0.1430	0.2122	0.2942	0.3352	0.3868	0.4298
7	0.1424	0.2246	0.2954	0.3758	0.4172	0.4532
8	0.1680	0.2668	0.3436	0.4268	0.4858	0.5290
9	0.1664	0.2764	0.3888	0.4612	0.5304	0.5726
10	0.1724	0.2802	0.3968	0.4782	0.5554	0.5954
11	0.1784	0.3064	0.4170	0.5010	0.5626	0.6220
12	0.1764	0.3080	0.4412	0.5210	0.6174	0.6586
13	0.1734	0.3144	0.4460	0.5360	0.6242	0.6806
14	0.1784	0.3188	0.4588	0.5668	0.6380	0.6966
15	0.1864	0.3498	0.4868	0.5960	0.6860	0.7302

References

1. Bandyopadhyay, D. and Basu, A.P. (1990), A class of tests for exponentiality against decreasing mean residual life alternatives, *Communications in Statistics - Theory and Methods*, **19**, 905-920.

2. Bryson, M.C. and Siddiqui, M.M. (1969), Some criteria for ageing, *Journal of the American Statistical Association*, **64**, 1472-1483.

3. Galambos, J. and Kotz, S. (1978), *Characterizations of Probability Distributions*, Lecture Notes in Mathematics, No. 675, Springer-Verlag, Berlin.

4. Hoeffding, W.A. (1948), A class of statistics with asynmptotically normal distribution, *Annals of Mathematicl Statistics*, **19**, 293-325.

5. Sen, P.K. (1977), Some invariance principles relating to jackknifing and their role in sequential analysis, *Annals of Statistics*, **5**, 316-239.

6. Zahedi, H. and Ebrahimi, N. (1988), Testing for bivariate new better than used in expectation distributions, *Technical Report*, Northern Illinois University, De Kalb, IL.

Part IV

Survival Analysis and Multivariate Models

32

Conditional Survival Models

Barry C. Arnold

University of California, Riverside

ABSTRACT Bivariate survival models can sometimes be characterized in terms of conditional survival functions of the form $P(X > x|Y > y)$ and $P(Y > y|X > x)$. Several well known bivariate models can be developed in this way. Examples are given, together with analogous multivariate extensions. Related characterizations involving $P(X > x|Y > y)$ and $E(Y|X > x)$ are also discussed.

Keywords: Bivariate exponential, Bivariate extreme, Bivariate generalized Pareto, Burr XII, Functional equations

CONTENTS

32.1 Introduction

How should one develop flexible parametric families of distributions to model multivariate phenomena? Historically, attention has been focussed on distributions with specified marginal features. This approach is so entrenched that when one speaks of, for example, a multivariate exponential distribution it is often tacitly assumed that the distribution will have exponential marginals. In [2], the authors call into focus the difficulties inherent in visualizing marginal features of multivariate densities and argue that conditional properties are more easily grasped. In that monograph, a compendium of conditionally specified models were introduced. Bivariate conditionally specified distributions, in their formulation, had joint densities $f_{X,Y}(x,y)$ with all conditionals of X given $Y = y$ belonging to a particular parametric family and all conditionals of Y given $X = x$ belonging to a second, possibly different, parametric family. In the context of bivariate survival models, it is more natural to condition on component survivals, i.e. on events such as $\{X > x\}$ and $\{Y > y\}$ rather than conditioning on particular values of X and Y. Models generated by such conditional specifications will be surveyed in the present paper. Initially the focus will be on bivariate distributions. Multivariate extensions will be sketched in Section 32.5.

32.2   Conditional Survival in the Bivariate Case

Consider bivariate random variables (X, Y) (discrete or absolutely continuous) with the set of possible values for X (respectively Y) denoted by $\mathcal{X}$ (respectively $\mathcal{Y}$). Suppose that for each $(x, y) \in \mathcal{X} \times \mathcal{Y}$ we are given $f_X(x|Y > y)$ and $f_Y(y|X > x)$. Reasonable questions to ask at this juncture include:

(i) Are the given families of conditional densities compatible?

(ii) If they are, do they determine a unique joint density for (X, Y)?

Since conditional survival functions are uniquely determined by conditional densities, an equivalent formulation of the problem is available. The survival function formulation will be seen to be more readily resolved than the conditional density formulation. Thus we ask about compatibility of putative families of conditional survival functions of the forms

$$P(X > x|Y > y),\ (x, y) \in \mathcal{X} \times \mathcal{Y} \text{ and } P(Y > y|X > x),\ (x, y) \in \mathcal{X} \times \mathcal{Y}.$$

The compatibility issue is readily resolved as follows.

THEOREM 32.2.1

Two families of conditional survival functions

$$P(X > x|Y > y) = a(x, y), \quad (x, y) \in \mathcal{X} \times \mathcal{Y}$$

and $\qquad\qquad P(Y > y|X > x) = b(x, y), \quad (x, y) \in \mathcal{X} \times \mathcal{Y} \qquad (32.2.1)$

are compatible if and only if their ratio factors, i.e. if and only if there exist functions $u(x), x \in \mathcal{X}$ *and* $v(y), y \in \mathcal{Y}$ *such that*

$$\frac{a(x, y)}{b(x, y)} = \frac{u(x)}{v(y)}, \quad (x, y) \in \mathcal{X} \times \mathcal{Y} \qquad\qquad (32.2.2)$$

where $u(x)$ *is a one dimensional survival function (non-increasing, right continuous and of total variation 1).*

PROOF   If $a(x, y)$ and $b(x, y)$ are to be compatible there must exist corresponding marginal survival functions $P(X > x) = u(x)$ and $P(Y > y) = v(y)$. Writing the event $P(X > x, Y > y) = a(x, y)v(y) = b(x, y)u(x)$ yields (32.2.2).   ∎

In, for example, [4] an analogous theorem was presented involving $f_{X|Y}(x|y)$ and $f_{Y|X}(y|x)$. In that setting, existence of a compatible joint distribution involved a factorizaton result parallel to (32.2.2). However after proving existence of a solution, additional assumptions were required to guarantee uniqueness. In the present setting, life is simpler. If there is any pair $u(x), v(y)$ for which (32.2.2) holds, it is readily verified that they are unique. Thus two families of survival functions (32.2.1) will *uniquely* determine a joint distribution (via a joint survival function) if their ratio factors as in (32.2.2).

32.3 Conditional Survival Functions in Given Parametric Families

Rather than specify the precise form of $P(X > x|Y > y)$ we might only require that for, each $y \in \mathcal{Y}$, it be a member of a specified parametric family of survival functions. An analogous requirement, that for each $x \in \mathcal{X}$, $P(Y > y|X > x)$ should be a member of a possibly different parametric family of survival functions would also be imposed. What kind of joint survival functions will be determined by such constraints? We begin with an example.

Suppose that for each $y > 0$, the conditional survival function for X given $Y > y$ is exponential with scale parameter $u(y)$. Analogously suppose that Y given $X > x$ is also always exponential. Thus for some functions $u(y)$ and $v(x)$ we have

$$P(X > x|Y > y) = \exp[-u(y)x]$$

and
$$P(Y > y|X > x) = \exp[-v(x)y] . \qquad (32.3.1)$$

The equations (32.3.1) are to hold for every $x > 0, y > 0$. If there is to be a joint survival function for (X, Y) consistent with (32.3.1) it must have associated marginal survival functions $\phi_1(x) = P(X > x)$ and $\phi_2(y) = P(Y > y)$ and we must have

$$\phi_2(y) \exp[-u(y)x] = P(X > x, Y > y)$$

$$= \phi_1(x) \exp[-v(x)y] \qquad (32.3.2)$$

where ϕ_1, ϕ_2, u and v are unknown functions. If we take logarithms of both sides in (32.3.2) and define $\tilde{\phi}_2(y) = \log \phi_2(y)$ and $\tilde{\phi}_1(x) = \log \phi_1(x)$ our functional equation becomes

$$\tilde{\phi}_2(y) - u(y)x = \tilde{\phi}_1(x) - v(x)y . \qquad (32.3.3)$$

This is a special case of the Stephanos-Levi-Civita-Suto functional equation. See [2, p. 15] for a detailed discussion. Equations of this form can be solved by differentiation or, if analyticity is not assumed, by differencing. In order for (32.3.3) to hold we must have

$$u(y) = \alpha + \gamma y$$

and

$$v(x) = \beta + \gamma x \qquad (32.3.4)$$

for some constants α, β, γ. Substituting this back in equations (32.3.2) and solving for $\phi_2(y)$ and $\phi_1(x)$ we eventually obtain the following expression for the joint survival function of (X, Y).

$$\bar{F}(x, y) = P(X > x, Y > x) = \exp[\delta + \alpha x + \beta y + \gamma xy], \quad x > 0, \ y > 0 . \qquad (32.3.5)$$

It is evident that δ must be 0 and in order for (32.3.5) to represent a valid joint survival function we must have $\frac{\partial}{\partial x}\frac{\partial}{\partial y}\bar{F}(x, y) \geq 0 \ \forall x, y > 0$. This means we must take $\alpha\beta \geq -\gamma$. In addition we need $\alpha, \beta < 0$ and $\gamma \leq 0$. Reparameterizing in terms of marginal scale parameters and an interaction parameter we have

$$\bar{F}(x, y) = \exp -(\frac{x}{\sigma_1} + \frac{y}{\sigma_2} + \theta\frac{xy}{\sigma_1\sigma_2}), \quad x > 0, \ y > 0 \qquad (32.3.6)$$

where $\sigma_1, \sigma_2 > 0$ and $0 \leq \theta \leq 1$. This is recognizable as Gumbel's Type I Bivariate Exponential distribution (see [8]). The distribution has exponential marginals (set x or y equal to zero in (32.3.6) to see this). Gumbel noted that the correlation is always non-positive (analogous non-positive correlation was encountered in the exponential conditionals distribution discussed in [5]). In the present case one finds

$$\rho(X, Y) = -1 + \int_0^\infty \frac{e^{-y}}{1 + \theta y} dy \tag{32.3.7}$$

and

$$-0.404 \le \rho(X, Y) \le 0 . \tag{32.3.8}$$

Gumbel provided the following expressions for conditional densities, means and variances. For $y > 0$,

$$f_{X|Y}(x|y) = \frac{1}{\sigma_1}[(1 + \theta\frac{x}{\sigma_1})(1 + \theta\frac{y}{\sigma_2}) - \theta]e^{-(1+\theta\frac{y}{\sigma_2})\frac{x}{\sigma_1}}, \quad x > 0 , \tag{32.3.9}$$

$$E(X|Y = y) = \sigma_1 \frac{[1 + \theta + \frac{\theta y}{\sigma_2}]}{(1 + \frac{\theta y}{\sigma_2})^2} \tag{32.3.10}$$

and

$$var(X|Y = y) = \sigma_1^2 \frac{(1 + \theta + \theta\frac{y}{\sigma_2})^2 - 2\theta^2}{(1 + \theta\frac{y}{\sigma_2})^4} . \tag{32.3.11}$$

In [9], the authors characterized the Gumbel Bivariate exponential distribution as the only one with the property that

$$E(X - x|X > x, Y > y) = E(X|Y > y) \tag{32.3.12}$$

and

$$E(Y - y|X > x, Y > y) = E(Y|X > x) . \tag{32.3.13}$$

The fact that the distribution has exponential conditional survival functions (i.e. that (32.3.1) holds) does not appear to have been remarked on in the literature.

The key feature of the above example was that the assumed conditional survival functions $P(X > x|Y > y)$ and $P(Y > y|X > x)$ admitted closed form expressions. Most of the examples in which we can successfully carry out the kind of conditional survival characterization which led to the Gumbel bivariate exponential model, may be viewed as special cases of the following characterization paradigm.

Let $\bar{\Phi}$ denote a specific survival function. The corresponding family of generalized ($\bar{\Phi}$) survival functions is of the form

$$\bar{F}(x; \mu, \sigma, \gamma, \delta) = [\bar{\Phi}((\frac{x - \mu}{\sigma})^\delta)]^\gamma \tag{32.3.14}$$

where $\mu \epsilon \mathbf{R}$, $\sigma > 0$, $\delta > 0$, $\gamma > 0$.

Now let $\bar{\Phi}_1$ and $\bar{\Phi}_2$ be two survival functions with associated families of generalized ($\bar{\Phi}_i$) $i = 1, 2$, survival functions as in (32.3.14). We seek to identify all possible bivariate survival functions for (X, Y) such that for each $y \epsilon \mathcal{Y}$

$$P(X > x|Y > y) = [\bar{\Phi}_1((\frac{x - \mu_1(y)}{\sigma_1(y)})^{\delta_1(y)})]^{\gamma_1(y)} \tag{32.3.15}$$

and for each $x \epsilon \mathcal{X}$

$$P(Y > y|X > x) = [\bar{\Phi}_2((\frac{y - \mu_2(x)}{\sigma_2(x)})^{\delta_2(x)})]^{\gamma_2(x)} \tag{32.3.16}$$

for some unknown functions $\mu_1(y), \sigma_1(y), \delta_1(y), \gamma_1(y), \mu_2(x), \sigma_2(x), \delta_2(x)$ and $\gamma_2(x)$. More modest goals involve seeking survival functions which satisfy (32.3.15), (32.3.16) when some of the functions $\mu_1(y), \ldots, \gamma_2(x)$ are assumed to be known constants, and perhaps in addition assuming that $\bar{\Phi}_1 = \bar{\Phi}_2$. Our exponential example corresponded to the choice $\bar{\Phi}_1(x) = \bar{\Phi}_2(x) = e^{-x}$, $x > 0$ and $\mu_1(y) = 0$, $\mu_2(x) = 0$, $\delta_1(y) = \delta_2(x) = \gamma_1(y) = \gamma_2(x) = 1$. The functional equation to be solved to determine models satisfying (32.3.15) and (32.3.16) is obtained by multiplying the right hand sides of (32.3.15) and (32.3.16) by the corresponding unknown marginal survival functions so that they may be equated (as was done in equation (32.3.2)). The tractability of the resulting functional equation depends on the nature of the survival functions $\bar{\Phi}_1$ and $\bar{\Phi}_2$ and depends on how many and which ones of the unknown functions $\mu_1(y), \ldots, \gamma_2(x)$ are assumed to be constants. No general theory appears to be available but an interesting list of cases in which the program can be successfully carried out are catalogued in the next section.

32.4 Specific Examples of Distributions Characterized by Conditional Survival

32.4.1 Weibull conditional survival functions

Suppose that (X, Y) has support $\mathbf{R}^+ \times \mathbf{R}^+$ and for each $y > 0$

$$P(X > x | Y > y) = exp - [x/\sigma_1(y)]^{\gamma_1(y)}, \ x > 0 \qquad (32.4.1)$$

and for each $x > 0$

$$P(Y > y | X > x) = exp - [y/\sigma_2(x)]^{\gamma_2(x)}, \ y > 0 . \qquad (32.4.2)$$

If these conditional survival functions are to be compatible there must exist corresponding marginal survival function $\bar{F}_1(x)$ and $\bar{F}_2(y)$. If we multiply (32.4.1) by $\bar{F}_2(y)$ and equate it to (32.4.2) multiplied by $\bar{F}_1(x)$ and take logarithms we encounter the following functional equation

$$\log \bar{F}_2(y) - [\frac{x}{\sigma_1(y)}]^{\gamma_1(y)} = \log \bar{F}_1(x) - [\frac{y}{\sigma_2(x)}]^{\gamma_2(x)} . \qquad (32.4.3)$$

It is conjectured that no solution to (32.4.3) exists unless $\gamma_1(y)$ and $\gamma_2(x)$ are constant functions. If $\gamma_1(y) = \gamma_1$, $\forall \ y$ and $\gamma_2(x) = \gamma_2 \ \forall \ x$ then (32.4.3) is a special case of the Stephanos-Levi-Civita-Suto equation (analogous to (32.3.3) earlier) and we readily conclude that

$$\sigma_1(y)^{\gamma_1} = (\alpha + \gamma \ y^{\gamma_2})^{-1}$$

$$\sigma_2(x)^{\gamma_2} = (\beta + \gamma \ x^{\gamma_1})^{-1}$$

and eventually arrive at a joint survival function of the form

$$\bar{F}(x, y) = exp - [(\frac{x}{\sigma_1})^{\gamma_1} + (\frac{y}{\sigma_2})^{\gamma_2} + \theta(\frac{x}{\sigma_1})^{\gamma_1}(\frac{y}{\sigma_2})^{\gamma_2}] \ x > 0, \ y > 0 \qquad (32.4.4)$$

where $\sigma_1, \sigma_2 > 0$ and $0 \leq \theta \leq 1$. The Gumbel bivariate exponential (32.3.6) corresponds to the choice $\gamma_1 = \gamma_2 = 1$ in (32.4.4).

32.4.2   Logistic conditional survival functions

Although many survival models are associated with non-negative random variables, there
is nothing in principle to stop us from considering random variables which can assume
negative values. For example, we might postulate logistic conditional survival functions.
Thus for each $y \epsilon \mathbf{R}$, we assume

$$P(X > x | Y > y) = [1 + e^{(x - \mu_1(y))/\sigma_1(y)}]^{-1}, \quad x \epsilon \mathbf{R} . \tag{32.4.5}$$

and for each $x \epsilon \mathbf{R}$,

$$P(Y > y | X > x) = [1 + e^{(y - \mu_2(x))/\sigma_2(x)}]^{-1}, \quad y \epsilon \mathbf{R} . \tag{32.4.6}$$

If we seek a general solution we are led to an equation analogous to (32.4.3). Only the case
$\sigma_1(y) = \sigma_1$ in (32.4.5) and $\sigma_2(x) = \sigma_2$ in (32.4.6) is tractable. Eventually we encounter
the class of bivariate survival functions with logistic conditional survival functions given
by

$$\bar{F}(x, y) = [1 + e^{(x - \mu_1)/\sigma_1} + e^{(y - \mu_2)/\sigma_2} + \theta e^{[(x - \mu_1)/\sigma_1 + (y - \mu_2)/\sigma_2]}]^{-1} \tag{32.4.7}$$

where $\mu_1, \mu_2 \epsilon \mathbf{R}^+$ and $\theta \epsilon [0, 2]$. The constraint $\theta \epsilon [0, 2]$ is needed to guarantee that
$\frac{\partial}{\partial x} \frac{\partial}{\partial y} \bar{F}(x, y) \geq 0 \; \forall \; x, y$.

32.4.3   Burr XII (or generalized Pareto) conditional survival functions

In this case we assume that for each $y > 0$,

$$P(X > x | Y > y) = [1 + [\frac{x}{\sigma_1(y)}]^{c_1(y)}]^{-k_1(y)}, \quad x > 0 \tag{32.4.8}$$

and for each $x > 0$

$$P(Y > y | X > x) = [1 + (\frac{y}{\sigma_2(x)})^{c_2(x)}]^{-k_2(x)}, \quad y > 0 \tag{32.4.9}$$

for positive functions $c_1(y), k_1(y), \sigma_1(y), c_2(x), k_2(x)$ and $\sigma_2(x)$. If (32.4.8) and (32.4.9)
are to both hold, marginal survival functions $\bar{F}_1(x)$ and $\bar{F}_2(y)$ must exist and we must
have

$$\bar{F}_2(y)[1 + (\frac{x}{\sigma_1(y)})^{c_1(y)}]^{-k_1(y)} = \bar{F}_1(x)[1 + (\frac{y}{\sigma_2(x)})^{c_2(x)}]^{-k_2(x)} . \tag{32.4.10}$$

If we define

$$b_1(y) = \sigma_1(y)^{-c_1(y)}$$

and

$$b_2(x) = \sigma_2(x)^{-c_2(x)}$$

then we may rewrite (32.4.10) in the form

$$\bar{F}_2(y)[1 + b_1(y)x^{c_1(y)}]^{-k_1(y)} = \bar{F}_1(x)[1 + b_2(x)y^{c_2(x)}]^{-k_2(x)} . \tag{32.4.11}$$

The class of all solutions to (32.4.11) is difficult to describe. Special cases amenable to
further analysis are (i) when $c_1(y) = c_1$ and $c_2(x) = c_2$ and (ii) when $k_1(y) = k_1$ and
$k_2(x) = k_2$.

In case (i), equation (32.4.11) is analogous to equation (4.1) of [2] . It can be transformed
to a form equivalent to equation (3.1) of [6]. Two families of solutions consequently exist.

Substituting the solutions back into the expressions for the joint survival function (32.4.11) we obtain the following. In Family I,

$$\bar{F}(x,y) = [1 + (\frac{x}{\sigma_1})^{c_1} + (\frac{y}{\sigma_2})^{c_2} + \theta(\frac{x}{\sigma_1})^{c_1}(\frac{y}{\sigma_2})^{c_2}]^{-k}, \quad x > 0, \ y > 0 \qquad (32.4.12)$$

for positive constants $c_1, \sigma_1, c_2, \sigma_2, k$ and $\theta \in [0,2]$. Again the condition $\theta \in [0,2]$ is needed to ensure a positive density. This family of bivariate Burr XII or generalized Pareto distributions was first described in [7] (see also [1] for a multivariate version).

The second family of solutions to (32.4.11) with $c_1(y) = c_1$ and $c_2(x) = c_2$ lead to joint survival functions of the form

$$\bar{F}(x,y) = \exp\{-\theta_1 \log(1 + (\frac{x}{\sigma_1})^{c_1}) - \theta_2 \log(1 + (\frac{y}{\sigma_2})^{c_2})$$

$$-\theta_3 \log(1 + (\frac{x}{\sigma_1})^{c_1}) \log(1 + (\frac{y}{\sigma_2})^{c_2})\}, \quad x > 0, \ y > 0 \ . \quad (32.4.13)$$

for $\theta_1 > 0, \theta_2 > 0, \theta_3 \geq 0, \sigma_1 > 0, \sigma_2 > 0, c_1 > 0, c_2 > 0$. The family (32.4.13) of bivariate Burr XII (or generalized Pareto) distributions has not previously been described in the literature.

If we impose condition (ii) in (32.4.11), i.e. that $k_1(y) = k_1$ and $k_2(x) = k_2$, it is not hard to verify that we must have $k_1 = k_2 = k$ and then the resulting functional equation to be solved may be transformed to one that is equivalent to (32.4.3). The only readily obtainable solutions will then have $c_1(y) = c_1$ and $c_2(x) = c_2$ and we will be led to solutions which are already included in the family (32.4.12). Thus the two parametric families (32.4.12) and (32.4.13) represent the totality of known bivariate distributions with Burr XII conditional survival functions.

32.4.4 Extreme conditional survival functions

A smallest extreme value distribution has a survival function of the form

$$\bar{F}(x) = \exp[-e^{(x-\mu)/\sigma}], \quad -\infty < x < \infty \qquad (32.4.14)$$

where $\mu \in \mathbf{R}$ and $\sigma > 0$. In this context we seek to identify bivariate distributions for which for each $y \in \mathbf{R}$,

$$P(X > x | Y > y) = \exp[-e^{(x-\mu_1(y))/\sigma_1(y)}], \quad x \in \mathbf{R} \qquad (32.4.15)$$

and for each $x \in \mathbf{R}$,

$$P(Y > y | X > x) = \exp[-e^{(y-\mu_2(x))/\sigma_2(x)}], \quad y \in \mathbf{R} \ . \qquad (32.4.16)$$

If (32.4.15) and (32.4.16) hold then marginal survival functions must exist and if we introduce the notation

$$\psi_1(y) = \log P(Y > y)$$

$$\psi_2(x) = \log P(X > x)$$

we conclude that the following functional equation must be satisfied

$$\psi_1(y) - e^{[x-\mu_1(y)]/\sigma_1(y)} = \psi_2(x) - e^{[y-\mu_2(x)]/\sigma_2(x)} \ . \qquad (32.4.17)$$

It is conjectured that no solutions to (32.4.17) exist with nonconstant functions $\sigma_1(y)$ or $\sigma_2(x)$. If $\sigma_1(y) = \sigma_1 \ \forall \ y$ and $\sigma_2(x) = \sigma_2 \ \forall \ x$ then (4.17) becomes an equation of the form

$$\psi_1(y) - \rho_1(x)\lambda_1(y) = \psi_2(x) - \rho_2(y)\lambda_2(x)$$

where $\rho_1(x) = e^{x/\sigma_1}$ and $\rho_2(y) = e^{y/\sigma_2}$ are known functions. It is thus a Stephanos-Levi-Civita-Suto equation and admits a solution of the form

$$\lambda_1(y) = \alpha + \gamma e^{y/\sigma_2}$$
$$\lambda_2(x) = \beta + \gamma e^{x/\sigma_1}$$
$$\psi_1(y) = -\beta e^{y/\sigma_2}$$
$$\psi_2(x) = -\alpha e^{x/\sigma_1}$$

Substituting these back into (32.4.17), which involves expressions for the logarithm of the joint survival function, we arrive after reparameterization at a model of the form

$$\bar{F}(x,y) = \exp[-e^{(x-\mu_1)/\sigma_1} - e^{(y-\mu_2)/\sigma_2} - \theta e^{(x-\mu_1)/\sigma_1 + (y-\mu_2)/\sigma_2}] \ , \quad x, y \ \epsilon \ \mathbf{R} \ . \quad (32.4.18)$$

Here $\mu_1, \mu_2 \ \epsilon \ \mathbf{R}, \sigma_1, \sigma_2 \ \epsilon \ \mathbf{R}$ and $0 \leq \theta \leq 1$. This bivariate extreme distribution is a simple transform of Gumbel's Type I bivariate exponential distribution (they share the same copula, i.e., have the same uniform marginals representation).

32.5 Multivariate Extensions

For any k-dimensional random variable $\underline{X} = (X_1, \ldots, X_k)$ and any $i = 1, 2, \ldots, k$ we use the symbol $\underline{X}_{(i)}$ to denote the $(k-1)$-dimensional random vector obtained from $\underline{X}$ by deleting the i'th coordinate. Analogously the real vector $\underline{x}_{(i)}$ is obtained from the real vector $\underline{x}$ by deleting the i'th coordinate. A natural analog to Theorem 32.2.1 would involve the question of compatibility of the following k conditional survival functions

$$P(X_i > x_i | \underline{X}_{(i)} > \underline{x}_{(i)}) = a_i(x_i, \underline{x}_{(i)}), \ i = 1, 2, \ldots, k \ . \quad (32.5.1)$$

Here and henceforth when we write $\underline{a} > \underline{b}$ for two vectors it is to be interpreted as holding coordinatewise, i.e. $a_i > b_i$ for each coordinate i. The condition for compatibility is clearly that for each $i \neq j$ the ratio $a_i(x_i; \underline{x}_{(i)})/a_j(x_j; \underline{x}_{(j)})$ should factor in the following manner:

$$\frac{a_i(x_i; \underline{x}_{(i)})}{a_j(x_j; \underline{x}_{(j)})} = \frac{u_j(\underline{x}_{(j)})}{u_i(\underline{x}_{(i)})} \quad (32.5.2)$$

where the $u_j(\underline{x}_{(j)})$'s are $(k-1)$-dimensional survival functions.

As in Sections 32.3 and 32.4, the next step is to consider joint survival functions specified by the requirement that for each i, $P(X_i > x_i | \underline{X}_{(i)} > \underline{x}_{(i)})$ should belong to some particular parametric family of one dimensional survival functions with parameters which might depend on $\underline{x}_{(i)}$. An exponential example is again a good starting point.

Thus, we might seek the most general class of k-dimensional survival functions with support $(0, \infty)^k$ such that for each i and for each $\underline{x}_{(i)} \ \epsilon \ \mathbf{R}_+^{k-1}$

$$P(X_i > x_i | \underline{X}_{(i)} > \underline{x}_{(i)}) = \exp[-\lambda_i(\underline{x}_{(i)})x_i], \quad x_i > 0 \quad (32.5.3)$$

for some positive functions $\lambda_i(\underline{x}_{(i)})$, $i = 1, 2, \ldots, k$. If we multiply the expressions in (32.5.3) by $\bar{F}_i(\underline{x}_{(i)})$, the $(k-1)$-dimensional survival functions which must exist for compatibility, and take logarithms, we are led to a system of Stephanos-Levi-Civita-Suto functional equations whose only solutions lead to k-dimensional versions of (32.3.6) of the following form:

$$\bar{F}(x_1,\ldots,x_k) = \exp - \sum_{\underline{s} \epsilon \xi_k} \theta_{\underline{s}}(\prod_{j=1}^k x_j^{s_j}), \quad \underline{x} > \underline{0} \tag{32.5.4}$$

where ξ_k is the set of vectors of 0's and 1's of dimension k with at least one coordinate being a 1. Certain constraints must be imposed in the $\theta_{\underline{s}}$'s which appear in (32.5.4) to guarantee that it represents a genuine survival function. For example we must have $\theta_{\underline{s}} \geq 0 \; \forall \; \underline{s}$ and $\theta_{\underline{s}} > 0 \; \forall \; \underline{s}$ which include only one coordinate equal to 1. In addition $\theta_{(1,1,00\ldots0)} \leq \theta_{(1,0\ldots0)}\theta_{(0,1,0\ldots0)}$ etc., since the bivariate marginals will necessarily be of the form (32.3.6) whose interaction parameter θ was constrained to be ≤ 1. Equation (32.5.4) is a Gumbel Type I multivariate exponential model. It was obtained not by marginal specification but by requiring appropriate (i.e., exponential) form for all the conditional survival functions.

Using analogous arguments it is not difficult to identify the form of the k-dimensional analogs of the bivariate survival functions displayed in equations (32.4.4), (32.4.7), (32.4.12), (32.4.13) and (32.4.18). For example, the analog to (32.4.12) is

$$\bar{F}(\underline{x}) = [1 + \sum_{\underline{s} \epsilon \xi_k} \theta_{\underline{s}}(\prod_{j=1}^k x_j^{c_j s_j})]^k, \quad \underline{x} > \underline{0} \tag{32.5.5}$$

where ξ_k is as defined following equation (32.5.4) and where the $\theta_{\underline{s}}$ are suitably constrained to guarantee that (32.5.5) is a valid k-dimensional survival function.

32.6   Conditional Distributions

An analogous collection of results can be developed using distribution functions instead of survival functions. We revert to the bivariate case realizing that straightforward multivariate extensions will exist. Thus we might ask whether two families of conditional distributions of the form

$$P(X \leq x | Y \leq y) = a(x, y)$$

and

$$P(Y \leq y | X \leq x) = b(x, y)$$

are compatible. As in Theorem 32.2.1 the answer is yes, provided that the ratio $a(x,y)/b(x,y)$ factors appropriately. Next we could ask about the nature of bivariate distributions which are constrained to have $P(X \leq x | Y \leq y)$ for each y and $P(Y \leq y | X \leq x)$ for each x, belong to specified parametric families of distributions. One example will suffice. Suppose we ask that the conditional distributions be power-function distributions. Thus we are considering (X, Y) with $0 \leq X \leq 1, 0 \leq Y \leq 1$ and for each $y \; \epsilon \; (0, 1)$

$$P(X \leq x | Y \leq y) = x^{\alpha_1(y)}, \quad 0 < x < 1 \tag{32.6.1}$$

while for each $x \; \epsilon \; (0, 1)$

$$P(Y \leq y | X \leq x) = y^{\alpha_2(x)}, \quad 0 < y < 1 . \tag{32.6.2}$$

If we (no surprise here) multiplying (32.6.1) and (32.6.2) by the corresponding marginal distribution functions which must exist for compatibility, equate them and take logarithms we get a familiar functional equation

$$\log P(Y \leq y) + \alpha_1(y) \log x = \log P(X \leq x) + \alpha_2(x) \log y . \qquad (32.6.3)$$

From this we readily conclude that

$$\alpha_1(y) = \alpha + \gamma \log y$$

and

$$\alpha_2(x) = \beta + \gamma \log x .$$

Substituting these in (32.6.3) we can determine the marginal distributions and exponentiating either side of (32.6.3) we obtain the following form of the joint distribution function of (X, Y)

$$F_{X,Y}(x, y) = P(X \leq x, Y \leq y) = x^\alpha y^\beta e^{\gamma(\log x)(\log y)}, \ 0 < x < 1, \ 0 < y < 1 . \qquad (32.6.4)$$

In order for (32.6.4) to represent a genuine joint distribution function we must require that $\alpha > 0$, $\beta > 0$ and γ must be negative and satisfy

$$-\gamma \leq \alpha\beta . \qquad (32.6.5)$$

Perhaps the easiest way to verify this is to recognize that if (X, Y) has the joint distribution (32.6.4) then defining $U = -\log X, V = -\log Y$ we find that (U, V) has a Gumbel Type I distribution with joint survival function (32.3.6). The constraint (32.6.5) is equivalent to the constraint $\theta \leq 1$ in (32.3.6).

32.7   An Alternative Specification Paradigm Involving Conditional Survival

In a sense, the specification of both families of conditional survival functions involves redundant information. Rather than specifying $P(X > x|Y > y)$ for every x, y and $P(Y > y|X > x)$ for every x, y it is enough to specify every function $P(X > x|Y > y)$ and $P(Y > y|X > x)$ for just one value of x. Alternatively some functional of the family of survival functions $P(Y < y|X > x)$, $x \in \mathcal{X}$ might be adequate, in conjunction with knowledge of $P(X > x|Y > y)$ for every x, y, to completely specify the joint distribution of (X, Y). In particular a conditional regression specification may suffice.

Thus we seek all bivariate distributions such that for given functions $a(x, y)$ and $\psi(x)$ we have

$$P(X > x|Y > y) = a(x, y), \quad (x, y) \in \mathcal{X} \times \mathcal{Y} \qquad (32.7.1)$$

and

$$E(Y|X > x) = \psi(x), \quad x \in \mathcal{X} . \qquad (32.7.2)$$

We will say that $a(x, y)$ and $\psi(x)$ are compatible if there exists a joint survival function $P(X > x, Y > y)$ satisfying (32.7.1) and (32.7.2). Natural questions that arise are: (i) Under what conditions are functions $a(x, y)$ and $\psi(x)$ compatible? (ii) If they are compatible, when do they determine a unique distribution? (iii) For a given $a(x, y)$ can we identify the class of all compatible choices for $\psi(x)$?

To simplify computations in the remainder of this section we will assume that the support of (X, Y) is the positive quadrant (i.e. $x > 0, y > 0$). This is not an un-natural restriction for survival models.

If $a(x, y)$ and $\psi(x)$ are to be compatible then there must exist a corresponding marginal survival function for Y which we will denote by $h(y)[= P(Y > y)]$. For each $x > 0, \psi(x)$ may be obtained by integrating the conditional survival function $P(Y > y|X > x)$ with respect to y over $[0, \infty)$. Thus for each $x > 0$,

$$\psi(x) = \int_0^\infty P(Y > y|X > x)dy$$

$$= \int_0^\infty P(X > x, Y > y)/P(X > x)dy$$

$$= \int_0^\infty \frac{a(x, y)h(y)}{a(x, 0)h(0)}dy$$

$$= \int_0^\infty \frac{a(x, y)}{a(x, 0)}h(y)dy \qquad (32.7.3)$$

since $h(0) = P(X > 0) = 1$. Thus $h(y)$ is obtainable by solving an integral equation with known kernel $a(x, y)$, i.e.

$$\psi(x)a(x, 0) = \int_0^\infty a(x, y)h(y)dy . \qquad (32.7.4)$$

For certain choices of the kernel $a(x, y)$ (which is the specified conditional survival function) this equation can be solved.

Suppose that we postulate an exponential conditional survival function of the form

$$a(x, y) = P(X > x|Y > y)$$

$$= \exp[-(\alpha + \beta y)x], \quad x > 0, y > 0 \qquad (32.7.5)$$

where $\alpha > 0$ and $\beta > 0$. Taking the limit as $y \to 0$, we find

$$a(x, 0) = e^{-\alpha x} \qquad (32.7.6)$$

For a given conditional mean survival function $\psi(x)$, we must obtain the corresponding survival function $h(y) = P(Y > y)$ by solving (32.7.4) which upon substituting (32.7.5) and (32.7.6) assumes the form:

$$\psi(x)e^{-\alpha x} = \int_0^\infty e^{-\alpha x - \beta x y}h(y)dy$$

i.e.,

$$\psi(x) = \int_0^\infty e^{-\beta x y}h(y)dy \qquad (32.7.7)$$

Next note that $\psi(0) = E(Y|X > 0) = E(Y) = \int_0^\infty h(y)dy$. Consequently if we define a density function on $(0, \infty)$ by

$$\tilde{h}(y) = h(y)/\psi(0) , \qquad (32.7.8)$$

we conclude that the Laplace transform of $\tilde{h}$ i.e.

$$M_{\tilde{h}}(t) = \int_0^\infty e^{-ty}\tilde{h}(y)dy \qquad (32.7.9)$$

satisfies (from (32.7.7))

$$M_{\tilde{h}}(t) = \psi(\frac{t}{\beta})/\psi(0) . \qquad (32.7.10)$$

Thus, provided that ψ is a completely monotone function (specifically a Laplace transform of a finite measure on $(0, \infty)$ with a decreasing density), then it uniquely determines $\tilde{h}$ by (32.7.10) which by normalization yields $h(y)[= P(Y > y)]$. In this case we can identify the form of every function ψ which is compatible with the family of conditional survival functions given by (32.7.5). We only need to check that the product $h(y)a(x, y)$ is a valid survival function.

For example, if we took

$$\psi(x) = (\gamma + \delta x)^{-1} ,$$
(32.7.11)

which is a completely monotone function, then we would have, from (32.7.10),

$$M_{\tilde{h}}(t) = (\gamma + \delta\frac{t}{\beta})^{-1}/\gamma^{-1}$$

$$= (1 + \frac{\delta}{\beta\gamma}t)^{-1} .$$
(32.7.12)

This is recognizable as the Laplace transform of an exponential density. So we can conclude that

$$\tilde{h}(y) = \frac{\beta\gamma}{\delta} \exp(-\frac{\beta\gamma}{\delta}y), \ y > 0 .$$
(32.7.13)

The survival function for Y, i.e. $h(y)$, is obtained by normalizing (32.7.13) to have the value 1 at $y = 0$. Thus

$$h(y) = P(Y > y) = \exp(-\frac{\beta\gamma}{\delta}y) .$$
(32.7.14)

Finally the unique joint survival function with $P(X > x|Y > y)$ given by $a(x, y)$ in (32.7.5) and $E(Y|X > x)$ given by $\psi(x)$ in (32.7.11) is of the form:

$$P(X > x, Y > y) = \exp[-\alpha x - \frac{\beta\gamma}{\delta}y - \beta xy], \ x > 0, y > 0 .$$
(32.7.15)

In order that (32.7.14) represent a valid joint survival function we need $0 < \frac{\delta}{\gamma} < \alpha$. In this case we arrive at the Gumbel (I) bivariate exponential distribution (displayed earlier with a different parameterization in (32.3.6)).

Acknowledgements

This report had its genesis during a discussion with Subhash Kochar. He asked whether there might not be results dealing with $f_X(x|Y > y)$ and $f_Y(y|X > x)$ analogous to the results for conditional densities $f_{X|Y}(x|y)$ and $f_{Y|X}(y|x)$ reported in [2] . Rephrasing his question to consider $P(X > x|Y > y)$ and $P(Y > y|X > x)$ led to the results in this report.

References

1. Arnold, B.C. (1990), A flexible family of multivariate Pareto distributions, *Journal of Statistical Planning and Inference*, **24**, 249-258.

2. Arnold, B.C., Castillo, E., and Sarabia, J.M. (1992), *Conditionally specified distributions*, Lecture Notes in Statistics No. 73, Springer-Verlag, Berlin.

3. Arnold, B.C., Castillo, E., and Sarabia, J.M. (1993), Multivariate distributions with generalized Pareto conditionals, *Statistics & Probability Letters*, **17**, 361-368.

4. Arnold, B.C and Press, S.J. (1989), Compatible conditional distributions, *Journal of the American Statistical Association*, **84**, 152-156.

5. Arnold, B.C. and Strauss, D. (1988), Bivariate distributions with exponential conditionals, *Journal of the American Statistical Association*, **83**, 522-527.

6. Castillo, E. and Galambos, J. (1987), Lifetime regression models based on a functional equation of physical nature, *Journal of Applied Probability*, **24**, 160-169.

7. Durling, F.C. (1975), The bivariate Burr distribution, In *Distributions in Scientific Work*, Vol. 1, (Eds., G.P. Patil, S. Kotz and J.K. Ord), 329-335, Reidel, Dordrecht.

8. Gumbel, E.J. (1960), Bivariate exponential distributions, *Journal of the American Statistical Association*, **55**, 698-707.

9. Nair, K.R.M. and Nair, N.V. (1988), On characterizing the bivariate exponential and geometric distributions, *Annals of the Institute of Statistical Mathematics*, **40**, 267-271.

33

Multivariate Survival Analysis Using Random Effect Models

Jon E. Anderson

University of Georgia

ABSTRACT This paper reviews contributions to multivariate survival analysis using random effect models. We focus on three modeling approaches: the proportional hazards frailty model, the scale change model, and the copula models. Within these models we examine the dependence structures, and estimation methods. We also review an interesting application, and give some directions for future research.

Keywords: Association, Frailty models, NPML estimate, Scale change model

CONTENTS

33.1 Introduction

The developments discussed in this article relate to situations where more than one time to event variable is measured on an experimental unit. For example, in bone marrow transplants, investigators are concerned about a potential connection between the transplant rejection time and the time a type of infection occurs. Here the experimental unit is a patient, and the time to event measurements are for two different events. Time till failure of components within the same machine provides another example. These examples, and many other applications, exhibit association among the survival times on the same experimental unit. The random effect approach to modeling assumes the observations within an experimental unit are independent conditional on an unobserved covariate. These models are well suited to this kind of data because they induce association when averaged over the distribution of the unobserved covariate.

We begin with a review of the multivariate survival literature in Section 33.2, and then procede by outlining several modeling approaches. We conclude this review by giving some future research directions in Section 33.6.

33.2   Literature Review

The earliest models for multivariate survival data extended the standard univariate exponential survival model to the bivariate case. The exponential distribution plays an important role in reliability and survival analysis; it is relatively simple and provides an adequate representation to many survival phenomena. In [16], [28], and [5], the authors tried to extend the exponential to the bivariate case. The extension was not straightforward. Marshall and Olkin found an absolutely continuous bivariate distribution cannot have both the bivariate loss of memory property and exponential marginals.

The bivariate exponential proved difficult and little progress was made on modeling bivariate survival data until the development of the proportional hazards model in [8]. This model served as the basis for much of the subsequent work in multivariate survival analysis. Cox suggested modeling the failure time t of individual i with covariates $x' = (x_1, \ldots, x_p)$ as,

$$h_i(t, x) = h(t)\exp(\beta' x), \qquad (33.2.1)$$

where β is a p-vector of parameters, and $h(t)$ is a baseline hazard function. The model in (33.2.1) applies only to univariate problems, but subsequent work extended this model to bivariate data.

33.2.1   Random Effect Extensions to the Cox Model

In [20], the authors first extended the proportional hazards model to bivariate data on twins or matched pairs. Their extension allows the baseline hazard, $h(t)$ in (33.2.1) to vary from pair to pair. This means a member of pair k has hazard function,

$$h_k(t; x) = \alpha_k h(t)\exp(\beta' x).$$

In particular they considered the exponential with covariates,

$$h_k(t; x) = \alpha_k \exp(\beta' x),$$

and the Weibull with covariates,

$$h_k(t; x) = \alpha_k t^{\eta-1}\exp(\beta' x).$$

By allowing $h(t)$ to vary stochastically from pair to pair, they introduced a random effect for each pair. In [38], the author extended the Holt and Prentice formulation by assuming the random effects α_k follow a gamma distribution. He showed this procedure yields estimates of β with greater precision than the Holt and Prentice formulation.

In [7], the author, presented the first association measure derived from Cox's model. In a study of the association between the lifespans of fathers and their sons, he modified Cox's model so that conditional on ω_k the lifespans are independent and the hazard functions for fathers and sons in pair k are of the form,

$$h_k(t) = h(t)e^{\omega_k},$$

where ω_k is a random variable representing an unknown factor common to both father and son. Using this model he constructed θ, an association measure defined by the ratio of age specific incidence rates for sons given their fathers status by,

$$\theta = \frac{h_s(t_1 \mid T_2 = t_2)}{h_s(t_1 \mid T_2 > t_2)}. \qquad (33.2.2)$$

The function, $h_s(t_1 \mid T_2 = t_2)$, is the hazard for sons at time t_1, given their father died at t_2. The denominator is the hazard for sons at t_1, given that the father lives past t_2. Notice that θ is constant for all (t_1, t_2). In Clayton's paper, emphasis shifted from covariate effects to the underlying dependence.

In [35], the authors defined the random variable $z = e^{w_k}$ as frailty. It represents all unobserved risk factors for an individual. Therefore individuals with large z values tend to have a larger risk of death than those with small z values. This construction allows modeling individual lifetimes with a common hazard function and a random frailty variable z that reflects the differing robustness of each individual.

33.3 Random Effect Models

This modeling approach assumes that variables within an experimental unit are independent conditional on another random variable. This random variable then varies over a distribution to produce quantities like the population survival or hazard function. This structure also induces dependence among the observations within an experimental unit. An useful analogy is the intraclass correlation coefficient where the correlation depends on between and within unit variability.

33.3.1 Proportional Hazards Frailty Model

As a univariate example of such a model, the authors in [36], [22], [31], as discussed above, incorporated a random effect into a baseline survival model to adjust for the heterogeneity in the population. This random effect, called frailty, enters the model as a multiplicative adjustment to a baseline hazard function. Thus, the survival time for an individual, T, has cumulative hazard function,

$$H(t \mid Z = z) = zH(t),\tag{33.3.1}$$

where $H(t)$ is an unobserved cumulative baseline hazard function, the same for all individuals. This individual also receives z, an unobserved, independent realization of the random variable Z from distribution $G(z)$. The random variable Z, also called frailty, represents genetic and environmental influences, risk factors, and susceptibilities for each individual. The distribution of these susceptibilities, $G(z)$, determines the heterogeneity of these risks in the population. The survival function conditional on z is thus, $S(t) = \exp\{-zH(t)\}$.

To extend this structure to the bivariate case, consider random variables T_1, T_2 so that the bivariate survival function, conditional on z is,

$$S(t_1, t_2) = \exp\{-z[H_1(t_1) + H_2(t_2)]\|\},$$

where $H_i(t_i)$ represents the baseline cumulative hazard functions for the ith event. Note that z is an unobserved frailty common to each member of the pair. The population survival function is,

$$S(t_1, t_2) = \int_0^\infty \exp\{-z[H_1(t_1) + H_2(t_2)]\|\}dG(z),$$

where $G(z)$ is the distribution of the random variable Z in the population. The previous expression is the Laplace transform of Z evaluated at $H_1(t_1) + H_2(t_2)$. Thus the population bivariate survival function is a mixture. The n-dimensional multivariate survival function induced by this model is then the Laplace transform of Z evaluated at $\sum_{i=1}^n H_i(t_i)$.

Association

In this section we discuss association in general, and time dependent measures in particular, for bivariate survival problems. In a typical bivariate problem we face a choice of analysis; we can perform either a regression or correlation analysis. The same choice applies to bivariate survival problems, except that survival studies usually are more suitable to association methods. For example, there is often no clear distinction between dependent and independent variables. Even if prediction or regression is ultimately used, an association analysis may be an appropriate exploratory technique.

In [29], section 3, the author shows that the proportional hazards frailty model produces positively associated random variables T_1, T_2. To examine this dependence further, we search for association measures to help describe the dependence structure. Time dependent association measures indexed by age or time have great potential usefulness in bivariate survival problems because the dependence between the variables may change over time. For example, the two variables could display strong dependence at early ages, but weaken gradually over time. This means time dependent association measures may capture the dependence well even in very general dependence structures.

In [31], the author provided a new advancement in association by extending Clayton's association measure in (33.2.2). Instead of a ratio of hazards, Oakes defined the association measure, $\theta^*(t_1, t_2)$, in terms of the bivariate survival function as,

$$\theta^*(t_1, t_2) = \frac{S S_{12}}{S_1 S_2} \, ,$$

where $S = S(t_1, t_2)$, $S_{12} = \partial^2 S(t_1, t_2)/\partial t_1 \partial t_2$, and $S_i = \partial S(t_1, t_2)/\partial t_i$ for $i = 1, 2$. This new measure differs from Clayton's θ in that this measure depends on time (t_1, t_2). This measure is an example of a local association measure. That is, a function that measures association in an instant past time (t_1, t_2).

In [4], the authors presented several time dependent association measures applied to twin data, to illustrate the usefulness of time dependent association measures. Here we describe one of the measures, the conditional expected residual life measure, in detail. Given a general bivariate distribution for (X,Y), statisticians use the function $E(Y \mid X = x)$ to describe the dependence between X and Y. In bivariate survival problems, conditional expectations also describe dependence between the random variables. As an example, for positively associated twin pairs, the expected life for twin 1 given twin 2 lives 80 years should be greater than the expected life for twin 1 without any information about twin 2. An association measure using this reasoning can be formed by dividing the conditional expectation for T_1 given information about T_1 and T_2 by the marginal conditional expectation for T_1 given only information about T_1,

$$\phi(t_1, t_2) = \frac{E(T_1 \mid T_1 > t_1, T_2 > t_2) - t_1}{E(T_1 \mid T_1 > t_1) - t_1}.$$

The numerator is the expected residual life of T_1 given $T_1 > t_1, T_2 > t_2$. That is, the life expectancy for T_1 beyond t_1 given $T_1 > t_1, T_2 > t_2$. The denominator is the expected residual life for T_1 given $T_1 > t_1$. This measure describes how much the knowledge that $T_2 > t_2$ influences the expectation of T_1. Values of $\phi(t_1, t_2)$ very different from 1 indicate strong influence for T_2, and therefore strong association between T_1 and T_2. If T_1, T_2 are positively associated, as t_2 increases, $\phi(t_1, t_2)$ should also increase. We can evaluate $\phi(t_1, t_2)$ for any (t_1, t_2), but some choices may be more informative than others. We set $t_1 = t_2 = t$ and evaluate $\phi(t, t)$ over many values of t. This measure provides a convenient interpretation for twin research because $\phi(t, t)$ measures the percentage of additional expected life for twin 1 due to their twin being alive at t.

The numerator of ϕ in terms of a proportional hazards frailty model is given by,

$$\frac{\int_{t_1}^{\infty} S(s,t_2)ds}{S(t_1,t_2)} = \frac{\int_{t_1}^{\infty}\int_0^{\infty} \exp\{-z[H_1(s)+H_2(t_2)]\}\, dG(z)\, ds}{S(t_1,t_2)}, \tag{33.3.2}$$

and the denominator from,

$$\frac{\int_{t_1}^{\infty} S(s)ds}{S(t_1)} = \frac{\int_{t_1}^{\infty}\int_0^{\infty} \exp\{-zH_1(s)\}\, dG(z)\, ds}{S(t_1)}. \tag{33.3.3}$$

For many frailty distributions, expressions like 33.3.2 and 33.3.3 require numerical integration because closed form expressions are not available. However, for gamma $G(z)$ and a exponential baseline hazard function, we can obtain a closed form expression for $\phi(t_1,t_2)$. For a gamma distribution with shape parameter α, and mean μ, the bivariate survival function S necessary to compute $\phi(t_1,t_2)$ is given by,

$$S(t_1,t_2) = \frac{(\alpha/\mu)^{\alpha}}{[\alpha/\mu + \sum_{i=1}^{2} H_i(t_i)]^{\alpha}}.$$

This gives,

$$\phi(t_1,t_2) = \frac{(\alpha/\mu + t_1 + t_2)}{(\alpha/\mu + t_1)}.$$

The previous expression holds for $\alpha > 1$. We give a graph of this function for $\alpha = 1.75$, $\mu = 1.0$, in the top half of Figure 33.1. The lower portion of this figure gives $\phi(t_1,t_2)$ along the line $t_1 = t_2 = t$. We note that $\lim_{t \to \infty} \phi(t,t) = 2$. Thus, the association has a finite limit at very old ages.

Identifiability

In this section we investigate if the parameters can be estimated from the observations on the random variables in this model. In this section we review the literature relating to identifiability in the proportional hazards frailty model. In [9], the authors show that the frailty distribution $G(z)$, and the baseline hazard $h(t)$ can be identified if $E(z) < \infty$, and there is a covariate that has two or more values. In [18], the authors remove the need for covariates by assuming a functional form for $h(t)$. They show that for univariate data and a functional form for $h(t)$, all the parameters in the model can be identified if the distribution $G(z)$ satifies certain conditions. The conditions on $G(z)$ typically depend on the choice of baseline hazard function. For example, the model with a Weibull baseline hazard function of the form, $h(t) = \gamma t^{\gamma-1}$ can be identified if the distribution $G(z)$ has a finite mean. They also show that a model with a Gompertz baseline hazard function can be identified if $E(z^2) < \infty$.

For bivariate data, the model is already identified from univariate data for $G(z)$ with finite mean. However, in [22], the author showed that for a model with a stable frailty distribution, the additional observation in an experimental unit enables identifiability.

Parametric Estimation

In this section we discuss estimation methods for the proportional hazards frailty model. Our task is to estimate $h_1(t)$, $h_2(t)$ and $G(z)$ from pairs of observed survival data, (t_{1k}, t_{2k}), and $(\delta_{1k}, \delta_{2k})$, for $k = 1, \ldots, K$ pairs. If t_{ik} is an event time, $\delta_{ik} = 1$. If t_{ik} is right censored, $\delta_{ik} = 0$. These data are K realizations of the survival variables T_1, T_2 and censoring variables C_1, C_2. The joint distribution of the survival variables is assumed independent of the censoring variables. For a model specification using parametric forms for $h_1(t), h_2(t)$ and $G(z)$, maximum likelihood is used to estimate parameters. The likelihood of pair k, conditional on z is given by,

$$L_{k|z} = \prod_{i=1}^{2} h_i(t_{ik} \mid z)^{\delta_{ik}} S(t_{ik} \mid z)$$

$$= \left\{ \prod_{i=1}^{2} h_i(t_{ik})^{\delta_{ik}} \right\} z^{\delta_{+k}} \exp\left\{ -z \sum_{i=1}^{2} H_i(t_{ik}) \right\}, \qquad (33.3.4)$$

where δ_{+k} is the number of events in pair k. Integrating equation (33.3.4) over the frailty distribution yields the unconditional likelihood,

$$L_k = \left\{ \prod_{i=1}^{2} h_i(t_{ik})^{\delta_{ik}} \right\} \int_0^\infty z^{\delta_{+k}} \exp\left\{ -z \sum_{i=1}^{2} H_i(t_{ik}) \right\} dG(z).$$

Therefore, the likelihood for the entire sample is the product over all pairs, $L = \prod L_k$. The log likelihood $l = \log\{L\}$, provides the basis for estimation. For integrals with closed form expresssions, partial derivatives are computed, and solved for the maximum likelihood estimates using Newton-Raphson techniques. If there is no closed form expression, numerical integration is used.

Semiparametric Estimation

If we are unable or unwilling to specify a parametric form for $G(z)$, a non-parametric estimate of $G(z)$ is available. In this section we discuss the non-parametric maximum likelihood(NPML) estimate of $G(z)$. In [25], the author showed that the NPML estimate of a mixing distribution like $G(z)$ is a finite mixture. This means it is a distribution with a finite number of jumps. Results in [27] show that for univariate problems the mixing distribution can have at most m points of increase, where m is the number of distinct observations. In the bivariate case we have K bivariate observations, therefore from Lindsay's results the NPML estimate of $G(z)$ can have at most K points. Previous applications of the NPML estimate in the proportional hazards frailty model have been given in [18], [6], and [4].

The distribution will be discrete with at most K mass points: $\mu_1, \ldots, \mu_K$, with probabilities $\pi_1, \ldots, \pi_K$. To start, h_1 and h_2 are assumed known; the EM algorithm is used to estimate the mixture. The algorithm is motivated by missing data theory so that under the proportional hazards frailty model, complete data would consist of the following vector: $(t_{ik}, \delta_{ik}, r_{jk})$ for $j = 1, \ldots, N$. The survival time for individual i in pair k is t_{ik}, δ_{ik} is the censoring indicator such that $\delta_{ik} = 1$ for an event, and $\delta_{ik} = 0$ for a censored observation. Under the proportional hazards frailty model, pair k receives a random draw from $G(z)$ yielding z_k. The mass point indicator function r_{jk}, is given by,

$$r_{jk} = \begin{cases} 1 & \text{if } z_k = \mu_j \\ 0 & \text{otherwise.} \end{cases}$$

Therefore r_{jk} keeps track of which mass point, μ_j was assigned to pair k. If we knew all the r_{jk}, we could compute the sufficient statistics, instead we find their expected value. This forms the E-step of the EM algorithm. It will suffice to compute, $E(r_{jk} \mid t_{1k}, t_{2k}, \delta_{1k}, \delta_{2k}, \mu, \pi)$, because the minimal sufficient statistics can be derived from these quantities.

$$E(r_{jk} \mid t_{1k}, t_{2k}, \delta_{1k}, \delta_{2k}, \mu, \pi) = \frac{\pi_j \, \mu_j^{\delta+k} \exp\{-\mu_j \sum_{i=1}^{2} H_i(t_{ik})\}}{\sum_{s=1}^{N} \pi_s \mu_s^{\delta+k} \exp\{-\mu_s \sum_{i=1}^{2} H_i(t_{ik})\}}$$

$$= w_{jk}$$

The M-step uses the expected values of the mass point indicators to compute maximum likelihood estimates of π and μ. The $\hat{\pi}_j$ are given by, $\hat{\pi}_j = w_{j+}/w_{++}$, for $j = 1, \ldots, N$. To obtain estimates for the mass points μ_j, consider the log likelihood for a given mass point μ_j,

$$l_j = \sum_{k=1}^{K} \sum_{i=1}^{2} r_{jk} \{-\mu_j H_i(t_{ik}) + \delta_{ik}[\log \mu_j + \log h_i(t_{ik})]\}.$$

This likelihood is maximized with respect to μ_j, and after substituting the expectations, w_{jk} gives,

$$\hat{\mu}_j = \frac{\sum_{k=1}^{K} \sum_{i=1}^{2} w_{jk} \delta_{ik}}{\sum_{k=1}^{K} \sum_{i=1}^{2} w_{jk} H_i(t_{ik})}.$$

The EM process begins with initial starting values for π and μ. The E-step produces the w_{jk}, and these are used in the M-step to produce new estimates of π and μ. The process is repeated until convergence is achieved. Estimates of the baseline hazard parameters are found using a function maximization routine that searches over the baseline hazard parameter space for the maximum likelihood estimates. At each step in this search the NPML estimate of $G(z)$ is produced.

33.3.2 Scale Change Model

Another random effect modeling approach given in [3] is the scale change model. This model differs from the proportional hazards frailty model in that the random effect z enters the baseline hazard function to change the time scale. For this model the cumulative hazard function for an individual conditional on $Z = z$ is given by,

$$H(t \mid Z = z) = H(zt), \tag{33.3.5}$$

where H is an unobserved cumulative baseline hazard function, the same for all individuals. As in the proportional hazards frailty model, each individual receives a z, an independent realization of Z from distribution $G(z)$. This means the survival function for an individual, conditional on $Z = z$, is given by,

$$S(t \mid Z = z) = S(zt) = \exp\{-H(zt)\}. \tag{33.3.6}$$

Thus, for this model the random effect enters the model through a scale change of time rather than through a multiplicative adjustment to the baseline hazard function.

We extend the univariate scale change model in equation (33.3.5) to bivariate data by assuming both members of pair share a common realization z from $G(z)$. This means the hazard function for a member of pair conditional on z is given by,

$$h(t \mid Z = z) = zh(zt),$$

where z is shared by both members of pair, and $h(t)$ is a baseline hazard function common for all individuals. We assume T_1, T_2 are conditionally independent given $Z = z$. Thus, bivariate survival function for pair, conditional on z is given by,

$$S(t_1, t_2 \mid Z = z) = \exp\{-[H_1(zt_1) + H_2(zt_2)]\}. \tag{33.3.7}$$

We obtain the population bivariate survival function, $S(t_1, t_2)$, by integrating equation (33.3.7) over the distribution $G(z)$. This gives,

$$S(t_1, t_2) = \int_0^\infty \exp\{-[H_1(zt_1) + H_2(zt_2)]\} dG(z). \tag{33.3.8}$$

Association

One goal of this research is to study the bivariate survival function for the scale change model, and investigate the implications on association. In [28], section 3, the authors showed that the proportional hazards frailty model induces non-negative association among members of a pair. The next lemma and theorem establish that the scale change model also produces non-negative associations.

LEMMA 33.3.1
The expression $P(T_1 \le t_1, T_2 \le t_2) \ge P(T_1 \le t_1)P(T_2 \le t_2)$ is equivalent to $S(t_1, t_2) \ge S_1(t_1)S_2(t_2)$, where S is the bivariate survival function, and S_1 and S_2 are univariate survival functions for the scale change model.

PROOF   If we expand, $P(T_1 \le t_1, T_2 \le t_2)$ in terms of survival functions we get, $1 + S(t_1, t_2) - S(0, t_2) - S(t_1, 0)$. Similarly, $P(T_1 \le t_1)P(T_2 \le t_2) = 1 - S_1(t_1) - S_2(t_2) + S_1(t_1)S_2(t_2)$. If we note that, $S(0, t_2) = S_2(t_2)$, and that, $S(t_1, 0) = S_1(t_1)$, then by substitution we have, $S(t_1, t_2) \ge S_1(t_1)S_2(t_2)$. ∎

THEOREM 33.3.2
Given baseline cumulative hazard functions, $H_1(t)$ and $H_2(t)$, such that the bivariate survival function for the variables T_1, T_2 is given in equation(33.3.8), then the variables T_1, T_2 are positively quadrant dependent. That is, $P(T_1 \le t_1, T_2 \le t_2) \ge P(T_1 \le t_1)P(T_2 \le t_2)$.

PROOF   From [26], if $P(T_1 \le t_1, T_2 \le t_2) \ge P(T_1 \le t_1)P(T_2 \le t_2)$ for all t_1, t_2, then T_1, T_2 are positively quadrant dependent. The family of distributions that satisfy the above condition is denoted by F_1. Lemma 33.3.1 shows that Lehmann's positively quadrant dependence condition is equivalent to satisfying $S(t_1, t_2) \geq S_1(t_1)S_2(t_2)$ for the scale change model.

For the scale change model, the condition, $S(t_1, t_2) \geq S_1(t_1)S_2(t_2)$, is equivalent to, $E_G[S_1(t_1 \mid z)S_2(t_2 \mid z)] \geq E_G[S_1(t_1 \mid z)]E_G[S_2(t_2 \mid z)]$. The pair of variables, $(z, z) \in F_1$ and by lemma 1 (iii) of [26], $(S_1(t_1 \mid z), S_2(t_2 \mid z)) \in F_1$ for decreasing functions of z, S_1 and S_2. Therefore, the variables $S_1(t_1 \mid z)$ and $S_2(t_2 \mid z)$ are positively quadrant dependent. By lemma 3 of [26], if the variables $(S_1(t_1 \mid z), S_2(t_2 \mid z)) \in F_1$, then $E[S_1(t_1 \mid z)S_2(t_2 \mid z)] \geq E[S_1(t_1 \mid z)]E[S_2(t_2 \mid z)]$. Thus, $S(t_1, t_2) \geq S_1(t_1)S_2(t_2)$, for all t_1, t_2, and thus T_1, T_2 are positively quadrant dependent for the scale change model. ∎

The local odds ratio as discussed in [31], and in a previous section is defined as, $\theta^*(t_1, t_2) = S\,S_{12}/S_1\,S_2$, where $S = S(t_1, t_2)$, $S_{12} = \partial^2 S(t_1, t_2)/\partial t_1 \partial t_2$, and $S_i = \partial S(t_1, t_2)/\partial t_i$ for $i = 1, 2$. When the conditions of the dominated derivative theorem hold, [11], we can bring the partial derivatives inside the integral giving:

$$S_\nu(t_1, t_2) = \frac{\partial S(t_1, t_2)}{\partial t_\nu}$$

$$= -\int_0^\infty z\,h_\nu(zt_\nu)\exp\{-\sum_{i=1}^2 H_i(zt_i)\}dG(z), \nu = 1, 2.$$

$$S_{12}(t_1, t_2) = \frac{\partial^2 S(t_1, t_2)}{\partial t_1 \partial t_2}$$

$$= \int_0^\infty z^2 h_1(zt_1)h_2(zt_2)\exp\{-\sum_{i=1}^2 H_i(zt_i)\}dG(z).$$

Thus, $\theta^*(t_1, t_2)$ can be estimated from data by estimating the baseline hazard functions h_1, h_2 and the function $G(z)$. We then compute the partial derivatives given above using these estimates.

Theorem 33.3.2 shows that the scale change model produces positively associated variables. Thus, we should expect $\theta^*(t_1, t_2) \geq 1$ for many (t_1, t_2). However, $\theta^*(t_1, t_2)$ can be less than one. To examine this, the partial derivatives are expanded to obtain,

$$\theta^*(t_1, t_2) = 1 + \frac{Cov[zh_1(zt_1)zh_2(zt_2) \mid T_1 > t_1, T_2 > t_2]}{E[zh_1(zt_1) \mid T_1 > t_1, T_2 > t_2]E[zh_2(zt_2) \mid T_1 > t_1, T_2 > t_2]}.$$

Thus, the covariance in the second term determines if $\theta^*(t_1, t_2) > 1$. In general, the behavior of $\theta^*(t_1, t_2)$ depends on the baseline hazard functions and $G(z)$, but the following example gives $\theta^*(t_1, t_2) < 1$ for some (t_1, t_2). Let $h_1 = t$ and $h_2(t) = 1/t^2$ for some range of t. This implies that the covariance becomes $Cov[z^2 h_1(t_1), (1/z)h_2(t_2) \mid T_1 > t_1, T_2 > t_2]$, and thus $\theta^*(t_1, t_2) < 1$ for some values of (t_1, t_2).

In Figure 33.2 we give $\theta^*(t_1, t_2)$ for the scale change model with a Gompertz $b = .3$ baseline hazard function and a gamma $G(z)$ with $\alpha = 2$ and $\mu = 1$, where the gamma $G(z)$ has parametrization,

$$g(z) = \frac{(\alpha/\mu)^\alpha}{\Gamma(\alpha)} z^{\alpha-1} \exp\{-(\alpha/\mu)z\}.$$

In [31], the author shows that $\theta^*(t_1, t_2) = 1 + 1/\alpha$ for the proportional hazards frailty model with a gamma $G(z)$ and any choice of baseline hazard function. The difference between $\theta^*(t_1, t_2)$ for the scale change and proportional hazards frailty models suggests that the scale change model can produce bivariate structures different from those for the proportional hazards frailty model.

Identifiability

In this section identifiability issues of the scale change model for uncensored data are described. In [17], the authors examined identifiability of the scale change model with covariates, but little is known about the identifiability of this model with no covariates. The scale change model with no covariates is not identifiable in general without further restrictions on h and G. The results in [3] show that for broad families of baseline hazard functions, conditions exist for $G(z)$ to obtain identifiability of the scale change model for univariate data.

Parametric Estimation

For the bivariate model, survival data (t_{1k}, t_{2k}), and $(\delta_{1k}, \delta_{2k})$, are observed for $k = 1, \ldots, K$ pairs. If t_{ik} is an event time, $\delta_{ik} = 1$. If t_{ik} is right censored, $\delta_{ik} = 0$. These data are K realizations of the survival variables T_1, T_2 and censoring variables C_1, C_2. As with univariate data, we assume the joint distribution of the survival variables is independent of the censoring variables.

For a model with parametric forms for $h(t)$ and $G(z)$, maximum likelihood is used to estimate parameters. Likelihood methods will typically require numerical integration to evaluate the likelihood function at the current parameter values. These likelihoods can be maximized by using FORTRAN routines, or packages like S-Plus or Mathematica.

The likelihood for bivariate data permits each member of a pair to have its own baseline hazard function denoted $h_i(t_i)$ for $i = 1, 2$, but assumes both members of a pair share the same Z from $G(z)$. The likelihood of pair k, conditional on $Z = z$ is given by,

$$L_{k|z} = \prod_{i=1}^{2} \{z h_i(z t_{ik})\}^{\delta_{ik}} \exp\{-H_i(z t_{ik})\} .$$

Integrating over $G(z)$ yields the unconditional likelihood for pair k,

$$L_k = \int_0^\infty \left\{ \prod_{i=1}^{2} h_i(z t_{ik})^{\delta_{ik}} \right\} z^{\delta_{+k}} \exp\{-\sum_{i=1}^{2} H_i(z t_{ik})\} dG(z),$$

and thus the likelihood L is given by, $L = \prod_{k=1}^{K} L_k$. If covariates are observed, the univariate time scale adjustment with $\psi(x_{ik}, \beta) = \exp\{\beta' x_{ik}\}$, is used for $i = 1, 2$.

Semiparametric Estimation

In this section we discuss the nonparametric maximum likelihood (NPML) estimate of $G(z)$ for the scale change model. As with the proportional hazards frailty model, baseline hazard functions $h_1(t_1), h_2(t_2)$, are assumed known, and the EM algorithm is used to estimate the mass points and the probabilities in the finite mixture. In the scale change model, complete bivariate data would consist of the following vector: $(t_{ik}, \delta_{ik}, r_{jk})$ for $j = 1, \ldots, N$, where t_{ik} and δ_{ik} are the survival time and censoring indicator respectively for individual i in pair k. For each observation t_{ik}, we know r_{jk} for $j = 1, \ldots, N$. The likelihood for these data is given by,

$$L = \prod_{j=1}^{N} \prod_{k=1}^{K} \prod_{i=1}^{2} \pi_j^{r_{jk}} \left\{ [\mu_j h_i(\mu_j t_{ik})]^{\delta_{ik}} \exp\{-H_i(\mu_j t_{ik})\} \right\}^{r_{jk}} .$$

Initial values for μ and π enable us to calculate the expected sufficient statistics in the E-step of the algorithm. The E-step gives,

$$E(r_{jk} \mid t_{1k}, t_{2k}, \pi, \mu, \delta_{1k}, \delta_{2k}) = P(z_k = \mu_j \mid t_{1k}, t_{2k}, \pi, \mu, \delta_{1k}, \delta_{2k})$$

$$= \frac{\pi_j \left\{ \prod_{i=1}^{2} [\mu_j h_i(\mu_j t_{ik})]^{\delta_{ik}} \right\} \exp\{-\sum_{i=1}^{2} H_i(\mu_j t_{ik})\}}{\sum_{s=1}^{N} \pi_s \left\{ \prod_{i=1}^{2} [\mu_s h_i(\mu_s t_{ik})]^{\delta_{ik}} \right\} \exp\{-\sum_{i=1}^{2} H_i(\mu_s t_{ik})\}}$$

$$= w_{jk}.$$

The M-step finds updated estimates of μ and π given the current values of the w_{jk}. The $\hat{\pi}_j = w_{j+}/w_{++}$, and the maximum likelihood estimates for μ_j are found by maximizing the portion of the log likelihood for mass point j denoted by,

$$l_j = \sum_{k=1}^{K} \sum_{i=1}^{2} \log \left\{ [\mu_j h_i(\mu_j t_{ik})]^{\delta_{ik}} \exp\{-H_i(\mu_j t_{ik})\} \right\}^{r_{jk}} ,$$

with respect to μ_j. Closed form solutions are generally not available, therefore we use a small number of Newton-Raphson iterations during each M-step, and repeat these E and M steps until achieving convergence. Maximum likelihood estimates for the baseline hazard parameters are found using a routine that searches over the baseline hazard parameter space. At each step in this search an NPML estimate of $G(z)$ is calculated.

33.4   Copula Models

A copula function is a distribution function on the unit square with uniform marginal distributions. Archimedean copula models are a special case of copula models with representation,

$$S(t_1, t_2) = \phi[\phi^{-1}(S_1(t_1)) + \phi^{-1}(S_2(t_2))]$$

where $0 \le \phi(u) \le 1$, $\phi(0) = 1$, $\phi(u)' < 0$, and $\phi(u)'' > 0$. If ϕ is a Laplace transform, the copula models reduce to the proportional hazards frailty model from the relation $\phi^{-1}(S_1(t_1)) + \phi^{-1}(S_2(t_2)) = H_1(t_1) + H_2(t_2)$, where S_1, S_2 are marginal survival functions. The dependence between the survival times is controlled by a parameter α in this formulation. In [32], the authors showed that for the copula representation to hold for $n > 2$ dimensional survival functions, ϕ must be proportional to a Laplace transform. Thus, in higher dimensions, copula models and the proportional hazards frailty model coincide. In [14], [15] and [31], the authors studied these models because the marginal distributions can be modeled separately from the dependence structure. This is a significant advancement over the frailty models that do not permit this separation of marginal distributions and dependence structure.

33.4.1   Association

Kendall's coefficient of concordance, τ, is defined by,

$$\tau = P[(T_1' - T_1)(T_2' - T_2) \geq 0] - P[(T_1' - T_1)(T_2' - T_2) < 0],$$

where $(T_1', T_2'), (T_1, T_2)$ are two independent draws from the bivariate distribution. Kendall's τ is bounded between -1 and 1 and has a similar interpretation as the correlation coefficient. In the [14] and [15], the authors showed that Kendall's τ is given by,

$$\tau = 4 \int_0^1 \rho(s)ds + 1,$$

where $\rho(s) = \phi^{-1}(s)/\phi^{-1}(s)'$, and ϕ is given in the copula model.

In [31], the author provided linkage to the work of Genest and MacKay because for the archimedean distributions discussed in Genest and MacKay, θ^* depends on time only through $S(t_1,t_2)$ so that,

$$\theta^*(t_1,t_2) = \theta(S(t_1,t_2)).$$

The function θ also has a local association interpretation. For two pairs $T^{(a)} = (T_1^{(a)}, T_2^{(a)})$ and $T^{(b)} = (T_1^{(b)}, T_2^{(b)})$; let $\tilde{T}^{(ab)}$ denote the component-wise minimum. Then θ is defined as,

$$\theta(S(t_1,t_2)) = \frac{P\left\{(T^{(a)}, T^{(b)}) \text{ Concordant} \mid \tilde{T}^{(ab)} = t_1,t_2\right\}}{P\left\{(T^{(a)}, T^{(b)}) \text{ Discordant} \mid \tilde{T}^{(ab)} = t_1,t_2\right\}}.$$

Another method for examining association in bivariate survival data is to plot the contours of the bivariate density function. The bivariate survival function produced from a gamma frailty distribution has the form,

$$S(t_1,t_2) = \{S(t_1)^{1-\alpha} + S(t_2)^{1-\alpha} - 1\}^{1/(1-\alpha)}, \quad \alpha > 1.$$

The first row of Figure 33.3 gives the bivariate density contours for $\log(T_1)$ and $\log(T_2)$ for this model when $\alpha = 1.22$, and 5.67. From this row we see that as α increases, the association between $\log(T_1)$ and $\log(T_2)$ also increases.

Another member of the copula model family was given by Frank(1979). The bivariate survival function has the form,

$$S(t_1,t_2) = \log_\alpha\{1 + (\alpha^{S_1(t_1)} - 1)(\alpha^{S_2(t_2)} - 1)/(\alpha - 1)\},$$

where $\alpha > 0$, and the logarithm is to the base α. Contours of the bivariate density for this model are given in the second row of Figure 33.3 for $\alpha = 0.39$ and 0.0001. The contours in this figure show that several choices for dependence structure are available within the copula model framework. We refer readers to [33] for details on these and other copula models.

33.4.2   Estimation Methods

One of the attractive features of copula models is that the marginal distributions and the dependence structure may be estimated separately. In [33], and [34], the authors developed three estimation procedures for these models: a fully parametric maximum likelihood method, a two-stage parametric maximum likelihood method, and a two-stage partially parametric maximum likelihood method. The two-stage methods estimate the marginal distributions ignoring dependence in the first stage, and then maximizing the profile likelihood with respect to α in the second stage, holding the margins fixed.

The two-stage parametric approach uses parametric forms for the marginal distributions, but the partially parametric approach uses Kaplan-Meier estimates for the margins. In [23], the author proposed a similar two-stage procedure using the nonparametric cumulative hazard function estimate of [30] for the margins instead of the Kaplan-Meier survival function estimates of Shih and Louis. However, Hougaard did not examine the properties of his two-stage method. In [33] and [34], the authors found that the estimates of α from the two-stage approaches are asymptotically efficient because the score functions of marginal parameters are orthogonal to the score function of α evaluated at independence.

33.5 An Application of Random Effect Models

In [35], the authors presented a detailed application of random effect models to survival times of Danish twins. The authors used these models to examine the gerontological question: Is longevity inherited? This example is motivated by the debate among demographers, gerontologists, and others, about the influence of genetic factors on aging. In the last one hundred years the average life expectancy for the United States population has increased steadily, and now exceeds 75 years. The debate now is how much more improvement can be achieved in the future. The debate is split into two main camps. A view promoted by the author in [12] and [13] suggests that there exists a genetically determined cap on the lifespan of each individual. They may expire before their maximum possible age from accidents, poor eating habits, smoking, or many other reasons, but if they can approach their genetically determined cap, they will expire very soon. This view is contrasted in work by [37] . This work suggests that if progress continues in reducing mortality rates through better medical and public health programs, average lifespans will continue to increase.

The data used to examine this question are from the Danish Twin Register of Odense Denmark. Specifically, twins born between 1870-1880, known to be alive at age 30 were used in this study. Proportional hazards frailty models were extended in several ways to more accurately model the twin data. Their Model 4 provides a good example of extending the proportional hazards frailty model structure to the question of interest. Here the authors denote the maximum lifespan as t^*, and include heterogeneity through the frailty variable Z. The model is specified by, $h(t_{ij} \mid t^*, Z = z) = z h_0(t_{ij})$ for $t_{ij} \leq t^*$ and $S(t_{ij} \mid t^*, Z = z) = 0$ for $t_{ij} > t^*$, for identical twin pair i, and members $j = 1, 2$. This model has proportional hazards frailty structure up to time t^*, with no chance of survival past t^*. The authors examined this and similar models for the Danish twin data and found little evidence to support a genetically-determined cap on lifespan. For other analyses of the Danish twin data, we refer readers to [24] and [4].

33.6   Future Directions

In this section we outline a few promising areas of research related to random effect
models.

33.6.1   Stochastic Process Random Effects

In this approach a random effect model similar to the proportional hazards frailty model
is used, except that the random effect is allowed to change over time as a stochastic
process. In [2], the author denotes this stochastic process as $z(t,\omega)$, a function of t for
a given path ω in the sample space Ω. This implies that the cumulative hazard function
given a sample path $z(t,\omega)$ is,

$$H(t \mid z(t,\omega)) = \int_0^t z(s,\omega)h(s)ds, \qquad (33.6.1)$$

where $h(s)$ is an unobservable baseline hazard function. As in the proportional hazards
frailty model, each individual receives a $z(t,\omega)$, an independent realization of the stochastic process $z(t)$. The survival function for an individual, conditional on $z(t,\omega)$, is given
by,

$$S(t \mid z(t,\omega)) = \exp\{-H(t \mid z(t,\omega))\}. \qquad (33.6.2)$$

Thus, for this model the random effect enters the model through a multiplicative adjustment to the baseline hazard function at every time t. We refer readers to [39] and [40] for
other approaches to randomly changing frailty models.

   This type of model may be reasonable for study of human longevity and aging because
individuals may face a different set of susceptibilities and risks at older ages than at
younger ages. The bivariate structures produced by this kind of model have yet to be
explored.

33.6.2   Multivariate Random Effects

The modeling approaches described in this review consider a univariate random effect
z with distribution $G(z)$. An alternative might be to consider a random vector, $\mathbf{z} =
(z_1,\ldots,z_p)$ with distribution $G(\mathbf{z})$. This approach has great potential for interesting
dependence structures and greater modeling flexibility.

33.6.3   Goodness of Fit Methods

Better goodness-of-fit methods need development throughout survival analysis, but especially in random effect models because closed-form expressions are often not available.
The best hope for these methods is to use the counting process, and martingale technology
described in [10] and [1]. Modifications to these existing methods should greatly improve
our ability to examine model adequacy for both marginal and joint distributions.

References

1. Andersen, P., Borgan, O., Gill, R., and Keiding, N. (1993), *Statistical Models Based on Counting Processes*, Springer-Verlag, New York.

2. Anderson, J. (1994), Survival analysis using a stochastic process random effect model, *Technical Report*, Department of Statistics, University of Georgia.

3. Anderson, J. and Louis, T.A. (1994), Survival analysis using a scale change random effects model, *Technical Report*, Department of Statistics, University of Georgia.

4. Anderson, J.E., Louis, T.A., Holm, N.V., and Harvald, B. (1992), Time dependent association measures for bivariate survival distributions, *Journal of the American Statistical Association*, **87**, 641-650.

5. Block, H.W. and Basu, A.P. (1974), A continuous bivariate exponential extension, *Journal of the American Statistical Association*, **69**, 1031-1037.

6. Butler, S. and Louis, T.A. (1992), Random effects models with non-parametric priors, *Statistics in Medicine*, **11**, 1981-2000.

7. Clayton, D.G. (1978), A model for association in bivariate life-tables and its application in epidemiological studies of familial tendency in chronic disease incidence, *Biometrika*, **65**, 141-151.

8. Cox, D.R. (1972), Regression models and life tables (with discussion), *Journal of the Royal Statistical Society, Series B*, **34**, 187-220.

9. Elbers, C. and Ridder, G. (1982), True and spurious duration dependence: The identifiability of the proportional hazards model, *Review of Economic Studies*, **49**, 403-409.

10. Fleming, T. and Harrington, D. (1991), *Counting Processes and Survival Analysis*, John Wiley & Sons, New York.

11. Fraser, D.A.S. (1976), *Probability and Statistics*, Duxbury Press, California.

12. Fries, J.F. (1980), Aging, natural health, and the compression of morbidity, *New England Journal of Medicine*, **303**, 130-135.

13. Fries, J.F. (1983), The compression of morbidity, *Milbank Memorial Fund Quarterly/Health and Society*, **61**, 397-419.

14. Genest, C. and MacKay, R.J. (1986a), Copules archimes et familles de lois bidimensionnelles dont les marges sont donnees, *Canadian Journal of Statistics*, **14**, 145-159.

15. Genest, C. and Mackay, R.J. (1986b), The joy of copulas, *The American Statistician*, **40**, 280-283.

16. Gumbel, E.J. (1960), Bivariate exponential distributions, *Journal of the American Statistical Association*, **55**, 698-707.

17. Heckman, J.J. and Honore, B.E. (1989), The identifiability of the competing risks model, *Biometrika*, **76**, 325-330.

18. Heckman, J.J. and Singer, B. (1982), The identification problem in econometric models for duration data, In *Advances in Econometrics*, (Ed., W. Hildebrand), 39-77, Cambridge University Press, England.

19. Heckman, J.J. and Singer, B. (1984), A method for minimizing the impact of distributional assumptions in econometric models for duration data, *Econometrica*, **52**, 271-319.

20. Holt, J.D. and Prentice, R.L. (1974), Survival analyses in twin studies and matched pair experiments, *Biometrika*, **61**, 17-30.

21. Hougaard, P. (1984), Life table methods for heterogeneous populations distributions describing the heterogeneity, *Biometrika*, **71**, 75-83.

22. Hougaard, P. (1986), Survival models for heterogeneous populations derived from stable distributions, *Biometrika*, **73**, 387-396.

23. Hougaard, P. (1989), Fitting a multivariate failure time distribution, *IEEE Transactions on Reliability*, **38**, 444-448.

24. Hougaard, P., Holm, N., and Harvald, B. (1992), Measuring the similarities between the life times of adult Danish twins born 1881-1930, *Journal of the American Statistical Association*, **87**, 17-24.

25. Laird, N.M. (1978), Non-parametric maximum likelihood estimation of a mixing distribution, *Journal of the American Statistical Association*, **73**, 805-811.

26. Lehmann, E.L. (1966), Some concepts of dependence, *Annals of Mathematical Statistics*, **37**, 1137-1153.

27. Lindsay, B.G. (1983), The geometry of mixing likelihoods: a general theory, *Annals of Statistics*, **11**, 86-94.

28. Marshall, A.W. and Olkin, I. (1967), A multivariate exponential distribution, *Journal of the American Statistical Association*, **62**, 30-44.

29. Marshall, A.W. and Olkin, I. (1988), Families of multivariate distributions, *Journal of the American Statistical Association*, **83**, 834-841.

30. Nelson, W.B. (1972), Theory and applications of hazard plotting for censored failure data, *Technometrics*, **14**, 945-965.

31. Oakes, D. (1989), Bivariate survival models induced by frailties, *Journal of the American Statistical Association*, **84**, 487-493.

32. Schweizer, B. and Sklar, A. (1983), *Probabilistic Metric Spaces*, North-Holland, Amsterdam.

33. Shih, J. and Louis, T.A. (1992), Inferences on the association parameter in copula models for bivariate survival data, *Technical Report*, Division of Biostatistics, University of Minnesota, Minneapolis, MN.

34. Shih, J.H. (1992), *Models and Analysis for Multivariate Failure-Time Data*, Ph.D. Dissertation, Division of Biostatistics, University of Minnesota, Minneapolis, MN.

35. Vaupel, J., Harvald, B., Holm, N., Yashin, A.I., and Xiu, L. (1992), Survival analysis in genetics: Danish twin data applied to a gerontological question, In *Survival Analysis: State of the Art* (Eds., J. Klein and P. Goel), NATO ASI Series, 121-137, Kluwer Academic Publishers, The Netherlands.

36. Vaupel, J.W., Manton, K.G., and Stallard, E. (1979), The impact of heterogeneity in individual frailty on the dynamics of mortality, *Demography*, **16**, 439-454.

37. Vaupel, J.W. and Owen, J.M. (1986), Anna's life expectancy, *Journal of Policy Analysis*, **5**, 383-389.

38. Wild, C.J. (1983), Failure time models with matched data, *Biometrika*, **70**, 633-641.

39. Woodbury, M. and Manton, K. (1977), A random walk model of human mortality and aging, *Theoretical Population Biology*, **11**, 37-48.

40. Yashin, A., Manton, K., and Vaupel, J. (1985), Mortality and aging in a heterogeneous population: A stochastic process model with observed and unobserved vari-

ables, *Theoretical Population Biology*, **27**, 154-175.

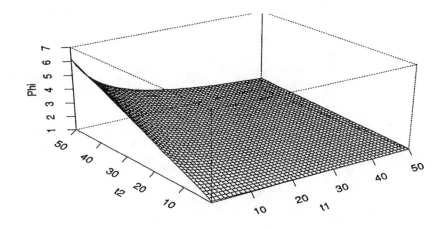

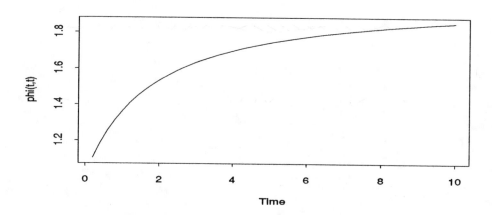

FIGURE 33.1
$\phi(t_1,t_2)$ for gamma $G(z)$ at $\mu = 1$ and $\alpha = 1.75$. The top portion gives the surface $\phi(t_1,t_2)$, and the lower portion shows $\phi(t_1,t_2)$ along $t_1 = t_2 = t$

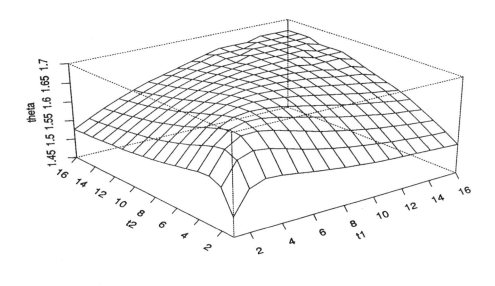

FIGURE 33.2
Odds ratio measure $\theta^*(t_1,t_2)$ for the scale change model with a Gompertz baseline
hazard function and a gamma $G(z)$

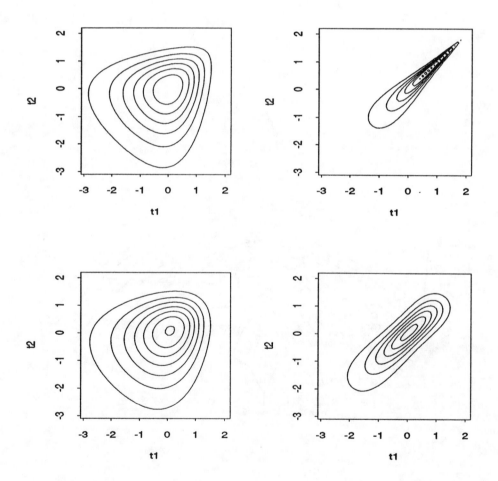

FIGURE 33.3
Bivariate density functions for copula models. The first row is for gamma frailty, and the second row is for Frank's family

34

On the Dependence of Structure of Multivariate Processes and Corresponding Hitting Times (A Survey)

Nader Ebrahimi

Northern Illinois University

ABSTRACT A direct approach to derive dependence properties among the hitting times of multivariate processes has been initiated in [2] and explored further in [5], [6], and [3]. The aim of this paper is to review all the concepts and results obtained for multivariate processes which help us to identify positive and negative dependent structures between components of a multivariate process and their hitting times.

Keywords: Bivariate Poisson process, Brownian motion, Diffusion process, Empirical process, Hitting time, Homogeneous Poisson process, Multivariate stationary Gaussian process, NBU process, Non-stationary process, Positive frequency of order 2, Positive orthant dependence, Positive regression dependence, Positively associated, Strong Markov process, Total positive of order 2

CONTENTS

34.1 Introduction

The reliability, $\bar{F}(t)$, of a system is the probability that the system will preserve its characteristics within specified limits during a specified time interval $[0, t]$. If a system failure is an event in which at least one characteristic of the system shifts outside certain permissible limits, and if T is the failure time then

$$\bar{F}(t) = P(T > t). \tag{34.1.1}$$

Suppose that the system reliability is determined by a finite number of characteristics. For i, $i = 1, 2, \ldots, k$, denote the value of the ith characteristic at time t by $X_i(t)$ and assume that it is within permissible limits if $X_i(t) < a_i$, where $a_1, \ldots, a_k$ are fixed and known values. One may, for example, look upon a_i as the breaking threshold of total damages, $X_i(t)$, by time t. Obviously, the random time, $T_i(a_i)$, at which the ith characteristic first crosses its limit is given by

$$T_i(a_i) = \begin{cases} Inf\{t \in \Lambda : X_i(t) \geq a_i\} \\ \infty \end{cases} \quad \text{if } X_i(t) < a_i \text{ for all } t \in \Lambda, \qquad (34.1.2)$$

where the index set Λ is a subset of $R_+ = [0, \infty]$. In this setting, the failure time of the system, T is given by

$$T = Min(T_1(a_1), \ldots, T_k(a_k)). \qquad (34.1.3)$$

In view of (34.1.2) and (34.1.3),

$$\bar{F}(t) = P(T_i(a_i) > t, i = 1, \ldots, k) \qquad (34.1.4)$$

Formulation of system reliability by means of Equations (34.1.2) to (34.1.4) is relevant to engineering disiplines relating to structural safety, mechanical vibration, etc.

In general it is possible to assess the system reliability provided that one can jointly model $X_1(t), \ldots, X_k(t)$, see [7]. However, such information may sometimes be unavailable or difficult to obtain. Thus, in many practical situations it is desirable or necessary to check whether or not the reliability of the system, $\bar{F}(t)$, meets a given specification when $P(T_i(a_i) > t), i = 1, \ldots, k$, are known. If a *lower bound* in the foregoing already meets or exceeds $\bar{F}(t)$, then one knows for sure that the system meets the specification. As we already mentioned such conclusions are gratifying particularly when the evaluation of (34.1.4) is not always feasible (For example, $X_1(t), \ldots, X_k(t)$ are hetergenous and (or) dependent.)

Concepts of dependence are very useful in connection to computing a bound for $\bar{F}(t)$ in terms of $P(T_1(a_1) > t), \ldots$, and $P(T_k(a_k) > t)$. For example if we know that $P(\cap_{i=1}^k (T_i(a_i) > t)) \geq \sqcap_{i=1}^k P(T_i(a_i) > t)$, then one can assess $P(T_i(a_i) > t), i = 1, \ldots, k$, and derive a bound for $\bar{F}(t)$. Besides bounds information about the dependent structure may bring forth new inequalities for multivariate processes. For more applications see also Section 34.4.

There are many concepts of dependence. In this manuscript we present a survey of some of these concepts. Moreover, we prove various theorems which not only clarify some properties of these dependent concepts but also help us to identify positive or negative dependence among two or more processes and their hitting times.

34.2   Concepts of Dependence for Stochastic Processes

Often the goal of works on multivariate processes is to study distributional aspects of two or more hitting time. (See for example Equation (34.1.4)) Conventionally, such random times have been studied for isolated processes or for families of processes exhibiting specific dependence structure such as the Markovian property. Whereas, this approach has been very useful, stochastic modeling of hitting times per se to account for positive or negative dependence among them is equally fruitful.

Ebrahimi has initiated one such direct approach to the study of the dependence structures of hitting times for bivariate processes. His approach has been explored further by [5,6] , and [3].

Suppose that we are given a k-dimensional ($k \geq 2$) stochactic vector process $\{X(t) = (X_1(t), \ldots, X_k(t)); t \in \Lambda\}$, where the index set Λ always be a subset of $R_+ = [0, \infty]$. The state space of $X(t)$ is the cartesian product $E = E_1 \times E_2 \times \ldots \times E_k$, which will be a subset of k-dimensional Euclidean space R^k.

We now present several concepts of positive (negative) dependence for any k-dimensional stochastic vector process $X(t)$.

DEFINITION 34.2.1 *The $k \geq 2$ different processes $\{X_1(t); t \in \Lambda\}, \cdots, \{X_k(t); t \in \Lambda\}$ are positively orthant dependent (POD) (negatively orthant dependent (NOD)) if*

$$P(\bigcap_{i=1}^{k}[X_i(t_i) > a_i]) \geq (\leq) \prod_{i=1}^{k} P[X_i(t_i) > a_i] , \qquad (34.2.1)$$

$$P(\bigcap_{i=1}^{k}[X_i(t_i) \leq a_i]) \geq (\leq) \prod_{i=1}^{k} P(X_i(t_i) \leq a_i) , \qquad (34.2.2)$$

for all $a_i \in E_i$ and $t_i \in \Lambda$, $i = 1, \cdots, k$, and $\{X_1(t); t \in \Lambda\}, \cdots, \{X_k(t); t \in \Lambda\}$ are each (univariate) POD (NOD). For $j = 1, \cdots, k$, we say that a one dimensional process $X_j(t)$ is POD (NOD) if for any $0 \leq s_1 < s_2 < \cdots < s_n, s_i \in \Lambda$ and $a_i \in E_j, i = 1, \cdots, n,$

$$P(\bigcap_{i=1}^{n}[X_j(s_i) > a_i]) \geq (\leq) \prod_{i=1}^{n} P(X_j(s_i) > a_i) ,$$

and

$$P(\bigcap_{i=1}^{n}(X_j(s_i) \leq a_i)) \geq (\leq) \prod_{i=1}^{n} P(X_j(s_i) \leq a_i) .$$

Also, their hitting times $T_1(a_1), \cdots, T_k(a_k)$ are POD (NOD) if

$$P(\bigcap_{i=1}^{k}(T_i(a_i) > t_i)) \geq (\leq) \prod_{i=1}^{k} P(T_i(a_i) > t_i) , \qquad (34.2.3)$$

and

$$P(\bigcap_{i=1}^{k}(T_i(a_i) \leq t_i)) \geq (\leq) \prod_{i=1}^{k} P(T_i(a_i) \leq t_i) , \qquad (34.2.4)$$

for every $a_i \in E_i$ and $t_i \in \Lambda, i = 1, \cdots, k.$

From Definition (34.2.1), it is clear that if $X_i(t_i) = X_i, i = 1, \cdots, k$, then $X_1, \cdots, X_k$ are POD (NOD) if $P(\bigcap_{i=1}^{k}(X_i > x_i)) \geq (\leq) \prod_{i=1}^{k} P(X_i > x_i)$ and $P(\bigcap_{i=1}^{k}(X_i \leq x_i)) \geq (\leq) \prod_{i=1}^{k} P(X_i \leq x_i)$ which coincides with the classical definitions of POD and NOD. (See [4].)

For $k = 2$, POD (NOD) will also be called positive quadrant dependent (PQD). (Negative quadrant dependent (NQD). See [8].)

DEFINITION 34.2.2 *The processes $\{X_1(t); t \in \Lambda\}, \cdots, \{X_k(t); t \in \Lambda\}$ are associated $(k \geq 2)$ if*

$$cov(f(X_i(t_i), i = 1, \ldots, k) , g(X_i(t_i), i = 1, \ldots, k)) \geq 0 , \qquad (34.2.5)$$

for all non-decreasing real valued functions f and g such that the covariance exists and all $t_i \in \Lambda, i = 1, \ldots, k$, and $\{X_1(t); t\varepsilon\Lambda\}, \ldots, \{X_k(t); t \in \Lambda\}$ are each univariate associated. For $j = 1, \ldots, k$, a one dimensional process $\{X_j(t); t \in \Lambda\}$ is said to be associated if for *any* $0 \leq s_1 < s_2 < \cdots s_n, cov(f(X_j(s_i), i = 1, \ldots, n), g(X_j(s_i), i = 1, \cdots, n)) \geq 0$. *Also, we say their hitting times* $T_1(a_1), \cdots, T_k(a_k)$ *are associated if*

$$cov(f(T_1(a_1), \ldots, T_k(a_k)) \, , \, g(T_1(a_1), \ldots, T_k(a_k))) \geq 0 \, , \qquad (34.2.6)$$

for all non-decreasing functions f *and* g *such that the covariance exists and all* $a_i \in E_i, i = 1, \ldots, k$.

From Definition (34.2.2) it is clear that if $X_i(t_i) = X_i$, the sequence of random variables $X_1, \cdots, X_k$ are associated if $cov(f(X_1, \cdots, X_k), g(X_1, \cdots, X_k)) \geq 0$ which coincides with Definition given by [1].

DEFINITION 34.2.3   *The processes* $\{X_1(t) : t \in \Lambda\}, \cdots, \{X_k(t); t \in \Lambda\}(k \geq 2)$ *are negatively associated (NA) if for every disjoint subsets* B_1 *and* B_2 *of* $\{1, 2, \cdots, k\}$,

$$cov(f(X_i(t_i), i \in B_1) \, , \, g(X_i(t_i), i \in B_2)) \leq 0 \, , \qquad (34.2.7)$$

for all real valued non-decreasing functions f *and* g. *Furthermore, each* $\{X_i(t); t \in \Lambda\}, i = 1, \cdots, k$, *are NA. For* $j = 1, \cdots, k$ *we say that a one dimensional process* $\{X_j(t); t \in \Lambda\}$ *is NA if for any* $0 \leq s_1 < s_2 < \cdots < s_n$ *and any two disjoint subsets* A_1 *and* A_2 *of* $\{1, 2, \ldots, n\}, cov(f(X_j(s_i), i \in A_1), g(X_j(s_i), i \in A_2))) \leq 0$. *Also, their hitting times* $T_1(a_1), \ldots, T_k(a_k)$ *are NA if for every pair of disjoint subsets* B_1 *and* B_2 *of* $\{1, 2, \ldots, k\}$,

$$cov(f(T_i(a_i), i \in B_1) \, , \, g(T_i(a_i), i \in B_2)) \leq 0 \, , \qquad (34.2.8)$$

for all non-decreasing real valued functions f *and* g *for which the covariance exists and all* $a_i \in E_i, i = 1, \ldots, k$.

In Definition (34.2.3) if $X_i(t_i) = X_i$, then the $X_1, \ldots, X_k$ are NA if $cov(f(X_i, i \in B_1), g(X_i, i \in B_2)) \leq 0$ for any two disjoint subsets B_1 and B_2 of $\{1, \ldots, k\}$.

It is clear that if the univariate process has stationary independent increments and if for every $t \in \Lambda$ the density function of $X(t)$, $f_{X(t)}$, is Positive frequency of order $2(PF_2)$, then $\{X(t); t \in \Lambda\}$ is associated. (A function $h(x)$ is said to be PF_2 if $\log h(x)$ is a concave function.) Also if the process $\{X(t); t \in \Lambda\}$ is a Markov process and $P[X(t) = y|X(s) = x]$ is totally positive or order 2 (TP_2) in x and y for all $s \leq t$, then $\{X(t); t \in \Lambda\}$ is associated. (A function $h(s, t)$ is said to be TP_2 if $\begin{vmatrix} h(s_1, t_1) & h(s_1, t_2) \\ h(s_2, t_2) & h(s_2, t_2) \end{vmatrix} \geq 0$ for all $s_1 < s_2, t_1 < t_2$.) For more properties of PF_2 and TP_2 see [1]. Also, if for every finite set of time points $t_1 < t_2 < \cdots < t_n, t_i \in \Lambda, i = 1, \cdots, n$, the random variables $X(t_1), X(t_2) - X(t_1), \cdots, X(t_n) - X(t_{n-1})$ are independent, then the process $\{X(t); t \in \Lambda\}$ is said to be a process with independent increments. Moreover, a process $\{X(t); t \in \Lambda\}$ is said to possess stationary increments if $X(t + s) - X(t)$ has the same distribution for all $t, s, t + s \in \Lambda$.

Having laid down some concepts of dependence, several comments regarding independence among the components of $X(t)$ are in order. First, we will say that $X_1(t), \cdots, X_k(t)$ are independent if for all $t_i \in \Lambda, i = 1, \cdots, k, X_1(t_1), \cdots, X_k(t_k)$ are independent in the usual sense of independence among k random variables. Second, we say that hitting times $T_i(a_i), i = 1, \cdots, k$, are independent if for all $a_i \in E_i, i = 1, \cdots k, T_1(a_1), \cdots, T_k(a_k)$ are independent.

First, we list several properties of POD(NOD) and associated (NA) of stochastic processes. Here Δ is either POD, NOD, associated or NA. For proofs of all these properties see [3].

(i) If $X_1(t), \cdots, X_k(t)$ are Δ, then any subset is also Δ;

(ii) If $X_1(t), \cdots, X_k(t)$ are Δ, then $g_1(X_1(t)), \cdots, g_k(X_k(t))$ are Δ for all non-decreasing (non-increasing) functions $g_1, \cdots, g_k$.

(iii) If $X(t) = (X_1(t), \cdots, X_k(t))$ and $Y(t) = (Y_1(t), \cdots, Y_k(t))$ are POD(NOD), $X(t)$ and $Y(t)$ are independent, and $X(t)$ has independent components, then $X(t) + Y(t)$ is POD(NOD).

(iv) If $X(t) = (X_1(t), \cdots, X_k(t))$ and $Y(t) = (Y_1(t), \cdots, Y_k(t))$ are associated (NA), $X(t)$ and $Y(t)$ are independent, then $X(t) + Y(t)$ is associated (NA).

(v) If $X(t) = (X_1(t), \cdots, X_k(t))$ and $Y(t) = (Y_1(t), \cdots, Y_k(t))$ are both non-negative and associated (NA) processes and if $X(t)$ and $Y(t)$ are independent, then $Z(t) = X(t)Y(t) = (X_1(t)Y_1(t), \cdots, X_k(t)Y_k(t))$ is associated (NA).

(vi) If $X_1(t), \cdots, X_k(t)$ are Δ and $X_i(t)$ has continuous sample path, $i = 1, \cdots, k$, then $Y_1(t), \cdots, Y_k(t)$ are Δ, where $Y_i(t) = \int_0^t g_i(u)X_i(u)du, i = 1, \cdots, k$, and $g_i(u)$'s are non-negative continuous functions.

Next we list several properties of POD(NOD) of hitting times. For proofs of all these properties see [5] and [6].

(i) If $T_1(a_1), \cdots, T_k(a_k)$ are POD(NOD) and g_i are increasing functions, then $W_1(b_1), \ldots, W_k(b_k)$ are POD (NOD), where

$$W_i(b_i) = Inf\{t : g_i(X_i(t)) > b_i\}, \qquad i = 1, \cdots, k ;$$

(ii) If $T_1(a_1), \cdots, T_k(a_k)$ are POD(NOD) and if $h_i : R_+ \to R_+$ are strictly increasing, $i = 1, \cdots, k$. Then, $V_1(b_1), \cdots, V_k(b_k)$ are POD (NOD), where

$$V_i(b_i) = Inf\{t : X_i(h_i(t)) \geq b_i\}, \qquad i = 1, \cdots, k .$$

(iii) Suppose that in the context of (b), the functions $h_1(t), \cdots, h_k(t)$ are arbitrary trajectories of k mutually independent increasing processes that are independent of the processes $X_i(t), i = 1, \cdots k$ as well. Then, the conclusion of part (b) holds.

34.3   Theoretical Results

In this section we summarize all the existing results without the proofs. For proofs of all the results we refer to papers by [2,3,5] and [6] and references therein.

THEOREM 34.3.1
If (a){$X_1(t); t \in \Lambda$} and {$X_2(t); t \in \Lambda$} are binary processes with state spaces {a, b}, {c, d} respectively, (b) max (a, b) and max (c, d) are absorbing states, and (c)$T_1(a_1)$ and $T_2(a_2)$ are PQD (NQD). Then {$X_1(t); t \in \Lambda$} and {$X_2(t); t \in \Lambda$} are PQD (NQD).

THEOREM 34.3.2
Consider a one dimensional process {$X_1(t); t \in \Lambda$} such that $X_1(t)$ is Δ, where Δ is either POD, NOD, associated or NA. Then $T(a_1), \cdots, T(a_k)$ are Δ, where $T(a_i) = Inf\{n : X_1(n) \geq a_i\}, i = 1, \cdots, k$. Here the index set $\Lambda = \{0, 1, 2, \cdots, \}$.

It should be mentioned that in Theorem 34.3.2 similar results can be obtained for a general process provided that the process has continuous sample paths.

THEOREM 34.3.3

Let $\{X_n(t) = (X_{n1}(t), \cdots, X_{nk}(t)); t \in \Lambda\}, n \geq 1$, be a sequence of k-dimensional processes such $T_{n1}(a_1), \cdots, T_{nk}(a_k)$ are POD (NOD), where $T_{ni}(a_i) = Inf\{t : X_{ni}(t) \geq a_i\}, i = 1, \cdots, k$. If $X_n(t)$ converges weakly to another k-dimensional process $\{Y(t); t \geq 0\}$ (with respect to any [11] topology) as $n \to \infty$ and if $X_n(t), n = 1, 2, \cdots$ and $Y(t)$ all have sample paths that are right continuous on R_+ with finite limits at all t, then $T_1(b_1), \cdots, T_k(b_k)$ are POD (NOD). Here $T_i(b_i) = Inf\{t : Y_i(t) \geq b_i\}, i = 1, \cdots, k$.

In the case of random variables, it is well-known that a random variable is TP_2 with itself and therefore it is PQD with itself. We now give a counter example to demonstrate the fact that a one-dimensional process $\{X_1(t); t \geq 0\}$ is not always PQD with itself.

Example 1

Consider a discrete time process $\{X_1(n); n \geq 0\}$ such that $X_1(0)$ and $X_1(1)$ have the following joint distribution:

		$X_1(1)$		
		0	1	2
	0	0	0	$\frac{1}{3}$
$X_1(0)$	1	0	$\frac{1}{3}$	0
	2	$\frac{1}{3}$	0	0

Using the table, it is easy to check that

$$0 = P(X_1(0) < 1, X_1(1) < 2) < P(X_1(0) < 1)P(X_1(1) < 2) .$$

That is $\{X_1(n); n \geq 0\}$ is not PQD with itself. Furthermore,

$$P(T_1(1) > 0, T_1(2) > 1) = P(X_1(0) < 1, X_1(1) < 2)$$
$$= 0 < P(T_1(1) > 0)P(T_1(2) > 1),$$

which means that the corresponding hitting times are not also PQD. This discussion motivates us to look for a process which is PQD with itself. Indeed, in the rest of this section, we study only the interesting special case of a one-dimensional process that is positively dependent with itself.

THEOREM 34.3.4

Consider a one dimensional process $\{X_1(t); t \in \Lambda\}$ such that $T_1(a) = Inf\{t : X_1(t) \geq a\}$. If the hitting time process $\{T_1(a); a \in E_1\}$ has the property that for any a and $b, a < b, T(b) - T(a)$ and $T(a)$ are independent, then $X_1(t)$ is positive regression dependent (PRD) with itself. ($X_1(t)$ is said to be PRD with itself if $P(T(d) > x_2|T(c) = x_1)$ increases (decreases) in x_1 for all $x_2 \in R_+$ and all $c, d \in E_1, c < d$.)

It should be mentioned that the concept of PRD is stronger than the concept of PQD or Associated. Thus if the assumptions of Theorem 34.3.4 holds we also get $T(a)$ and $T(b)$ are associated for all $a, b \in E_1$.

Before we state our next several results we need to know the property known as the strong Markov property. To describe this property we introduce a class of random times called Markov times. A Markov time τ for the process $\{X(t); t \geq 0\}$ is a random variable with values in $[0, \infty]$, with the property that for every fixed time s, the occurrence or non-occurrence of the event $\{\tau \leq s\}$ can be determined by a knowledge of $\{X(t); 0 \leq t \leq s\}$.

For example, if one can not decide whether or not the event $\{\tau \leq 5\}$ has happened by observing only $\{X(t); 0 \leq t \leq 5\}$, then τ is not a Markov time. On $\{\tau < \infty\}$, if the conditional distribution of $\{X(t+\tau); t \geq 0\}$ given the past up to the time τ is the same as the distribution of the process $\{X(t); t \geq 0\}$ starting at $X(\tau)$, then $\{X(t); t \geq 0\}$ possess the strong Markov property.

THEOREM 34.3.5
Let $\{X_1(t); t \geq 0\}$ be a strong Markov process with continuous sample paths, then $X_1(t)$ is PRD with itself.

THEOREM 34.3.6
Let $\{X_1(t); t \geq 0\}$ be a strong Markov process with state space $[0, \infty]$. If $(a)X_1(0) = 0, (b)$ with probability 1, the sample paths of $X_1(t)$ does not have positive jumps. Then, $\{X_1(t); t \geq 0\}$ is PRD with itself.

THEOREM 34.3.7
Let $\{X_1(t); t \geq 0\}$ be a strong Markov process with state space $\{0, 1, 2, \ldots\}$. If $(a)X_1(0) = 0, (b)\{X_1(t); t \geq 0\}$ is free of positive skips, that is the sample paths can not have positive jumps greater than one, then $\{X_1(t); t \geq 0\}$ is PRD with itself.

34.4 Examples and Applications

Using results from Sections 34.2 and 34.3, one can identify the dependence structure of the processes and also their hitting times.

Example 1
Any one dimensional diffusion process whose paths are continuous with probability one is PRD with itself. Moreover, if $L(t) = L(0, t)$ be the local time process (at 0) of one dimensional Brownian motion, then $L(t)$ is PRD with itself. ▯

Example 2
Let $\{X(t); t \geq 0\}$ be a diffusion process on the interval $[0, y]$ which has a reflecting boundary at zero. Let $\{N_k(t); t \geq 0\}$ be a sequence of birth-death processes on $\{0, 1, 2, \cdots\}$ with the assumption $\mu_0 = 0$ for its death rates. If $\alpha_k N_k(t)$ converges weakly to $X(t)$, then $X(t)$ is PRD with itself. ▯

Example 3
NBU processes have been discussed by [10]. Any NBU process which satisfies conditions of Theorems 34.3.5, 34.3.6, or 34.3.7 is PRD with itself. ▯

Example 4
Consider the one-dimensional empirical process $\{F_n(t); u \leq t \leq 1\}$ constructed in the usual way from i.i.d. copies of a uniform random variable on $[0,1]$. This process is PRD with itself. ▯

Example 5

Consider a system with two components which is subjected to shocks. Let $N(t)$ be the number of shocks received by time t and let $Z_1(t) = \sum\limits_{i=1}^{N(t)} X_i, Z_2(t) = \sum\limits_{i=1}^{N(t)} Y_i$ be total damages to components one and two by time t respectively. Note that X_i and Y_i are damages to components 1 and 2 by shock i, respectively. If $(X_1, Y_1), (X_2, Y_2), \cdots$ are independent, X_i and Y_i are PQD, $i = 1, 2, \cdots$, then $W_1(a_1)$ and $W_2(a_2)$ are PQD. Here $W_1(a_1) = Inf\{t : Z_1(t) \geq a_1\}$, $W_2(a_2) = Inf\{t : Z_2(t) \geq a_2\}$ and $N(t)$ is a Poisson process which is independent of X_i and $Y_i, i = 1, 2, \cdots, k$. ▯

Example 6

Let $\{(N_1(t), N_2(t)); t \geq 0\}$ be a bivariate Poisson process (cf. [1]) with joint probability function

$$P(N_1(t) = k_1, N_2(t) = k_2) = exp(-\lambda t(p_{11} + p_{10} + p_{01}))$$

$$\times \left[\sum_{m=0}^{Min(k_1, k_2)} \frac{(\lambda t p_{11})^m (\lambda t p_{10})^{k_1-m} (\lambda t p_{01})^{k_2-m}}{m!(k_1 - m)!(k_2 - m)!} \right].$$

Then $T_1(a_1)$ and $T_2(a_2)$ are PQD, where

$$T_i(a_i) = Inf\{t : N_i(t) \geq a_i\}, i = 1, 2.$$

▯

Example 7

Consider two different repair policies. The first policy is that we replace a failed unit with a new identical unit. We let $N(t)$ denote the number of replacements up to time t. The second policy consists of repairing the unit to its condition just prior to failure, that is, a minimal repair. We donote the number of minimal repairs up to time t by $W(t)$. Then, $\{N(t); t \geq 0\}$ and $\{W(t); t \geq 0\}$ are PQD. If $T_1(a) = Inf\{t; N(t) \geq a\}$ and $T_2(b) = Inf\{t : W(t) \geq b\}$, then $T_1(a)$ and $T_2(b)$ are PQD. ▯

Example 8

Consider the non-stationary process $X(t)$ given by

$$X(t) = \alpha(t)Y(t), t \geq 0$$

where $\alpha(t)$ is a deterministic continuous function much that $\alpha(t) \geq 0$ and $Y(t)$ is non-negative stationary process in the sense that $cov(Y(t), Y(t+h)) = K(h), h \geq 0$. If $Y(t)$ is associated, then $X(t)$ is also associated. Moreover, $T(a_1), \ldots, T(a_k)$ are also associated. Here $T(a_i) = Inf\{n : X(n) \geq a_i\}$. ▯

Example 9

Consider a k-dimensional multivariate stationary Gaussian process $\{X(t) = (X_1(t), \cdots, X_k(t)) : t \in \{0, 1, 2, \cdots\}\}$ such that

$$cov(X_i(t), X_i(s)) = R_i(t - s) \geq 0 ,$$

$$cov(X_i(t), X_j(s)) = h_{ij}(t - s) \geq 0$$

for all $t \geq s$. Then $X(t)$ is associated. Moreover, the corresponding hitting times are associated. ☐

Application 1

Consider a series system with k components. Assume that the component i fails if the total damages to the component i exceeds of threshold $a_i, i = 1, \cdots, k$. Let $X_i(t)$ be the total damages to the ith component at time t. Then, the life-time of the system is given by the random variable $T = \min_{1 \leq i \leq k} T_i(a_i)$, where $T_i(a_i) = Inf\{t : X_i(t) \geq a_i\}, i = 1, \ldots, k$. If $T_1(a_1), \ldots, T_k(a_k)$ are POD, then

$$P(T > t) \geq \sqcap_{i=1}^{k} P(T_i(a_i) > t) .$$

Similar bounds can be obtained for parallel systems.

Application 2

Consider the following stress-strength model for two systems. Let $Z_i(t), i = 1, 2$, be the strength of system i at time t. We will assume that the two systems receive shocks from common source. Using a cumulative damage shock model (see [1]), we let $N(t)$ denote the number of shocks occuring by time t and U_i are i.i.d. positive random variables denoting the damage to either system due to the i-th shock $(i = 1, \ldots)$. Hence the stress expreienced by either system at time t is given by the process $X(t) = \sum_{i=1}^{N(t)} U_i$. Now from Example 5, $X(t)$ is PQD with itself. Assuming that $Z_1(t)$ and $Z_2(t)$ are independent we get that the life-times of two systems $T_1(0)$ and $T_2(0)$ are PQD. Here,

$$T_i(0) = Inf\{t : X(t) - Z_i(t) \geq 0\} , \; i = 1, 2 .$$

That is,

$$P(T_i(0) > t_i; i = 1, 2) \geq \sqcap_{i=1}^{2} P(T_i(0) > t_i) .$$

Application 3

Consider a geometric Brownian motion,

$$X(t) = a \, exp(B(t) + (\theta - \frac{1}{2})t), \; a > 0 .$$

where $B(t)$ is Brownian motion. This process has been used as a model of the behavior of prices of assest. In the sequential estimation of θ, suppose that we stop observing the process $X(t)$ until it hits a barrier b. If $T(b) = Inf\{t : 0 < t \leq \infty, X(t) = b\}$, then the maximum likelihood estimate of θ is (see [12]) given by

$$\hat{\theta}(T(b)) = (log(b/a))T(b) + \frac{1}{2} .$$

In order to study the stability of $\hat{\theta}$ with regard to the stopping rule, we might look at the variance of $\hat{\theta}(T(b_1)) - \hat{\theta}(T(b_2))$, that is, the function

$$K(b_1, b_2) = Var[\hat{\theta}(T(b_1)) - \hat{\theta}(T(b_2))]$$

for any b_1 and b_2. Clearly the dependence structure of the process $\{T(b) : b \geq 0\}$ will provide some useful information about $K(b_1, b_2)$. In view of Theorem 34.3.5, $X(t)$ is

PRD with itself and it follows that $\hat{\theta}(T(b_1))$ and $\hat{\theta}(T(b_2))$ are PRD random variables. We therefore obtain

$$K(b_1, b_2) \leq Var(\hat{\theta}(T(b_11))) + Var(\hat{\theta}(T(b_2))) \ .$$

Application 4

In [9], the author has considered the waiting time until the occurrence of a "cluster of size k," $k \geq 2$, in a homogeneous Poission process. In [10], the authors argued that such a waiting time is the hitting time $T(k - \frac{1}{2})$ of a new better than used process (NBU) $Z(t)$ given in Example (2.4) of their paper. Now, one can verify that $Z(t)$ is PQD with itself. Consequently, given $k_1 < k_2$,

$$P(T(k_1 - \frac{1}{2}) > x_1, T(k_2 - \frac{1}{2}) > x_2) \geq \Pi_{i=1}^2 P(T(k_i - \frac{1}{2}) > x_i).$$

References

1. Barlow, R.E. and Proschan, F. (1981), *Statistical Theory of Reliability and Life Testing: Probability Models*. To begin with, Silver Spring, MD.

2. Ebrahimi, N. (1987), Bivariate processes with positive or negative dependent structures, *Journal of Applied Probability*, **24**, 115-122.

3. Ebrahimi, N. (1994), On the dependence structure of multivariate processes and corresponding hitting times, *Journal of Multivariate Analysis* (To appear).

4. Ebrahimi, N. and Ghosh, M. (1981), Multivariate negative dependence, *Communications in Statistics*, **A10**, 307-337.

5. Ebrahimi, N. and Ramalingam, T. (1988), On the dependence structure of hitting times of univariate processes, *Journal of Applied Probability*, **25**, 355-362.

6. Ebrahimi, N. and Ramalingam, T. (1989), On the dependence structure of hitting times of multivariate processes, *Journal of Applied Probability*, **26**, 287-295.

7. Ebrahimi, N. and Ramalingam, T. (1993), Estimation of system reliability in Brownian stress-strength models based on sample paths, *Annals of the Institute of Statistical Mathematics*, **45**, 9-19.

8. Lehmann, E.L. (1966), Some concepts of dependence, *Annals of the Institute of Statistical Mathematics*, **37**, 1136-1153.

9. Leslie, R.T. (1969), Recurrence times of clusters of Poisson points, *Journal of Applied Probability*, **6**, 372-388.

10. Marshall, A. and Shaked, M. (1983), New better than used processes, *Advances in Applied Probability*, **15**, 601-615.

11. Skorohod, A.V. (1956), Limit theorems for stochastic processes, *Theory of Probability and Its Applications*, **1**, 261-290.

12. Sorensen, M. (1983), On maximum likelihood estimation in randomly stopped diffusion-type processes, *International Statistical Review*, **51**, 93-110.